AF333065

HEAT TRANSFER - THEORETICAL ANALYSIS, EXPERIMENTAL INVESTIGATIONS AND INDUSTRIAL SYSTEMS

Edited by **Aziz Belmiloudi**

Heat Transfer - Theoretical Analysis, Experimental Investigations and Industrial Systems
Edited by Aziz Belmiloudi

CBS, Edition **2016**

Published by InTech

Janeza Trdine 9, 51000 Rijeka, Croatia

Publishing Process Manager Iva Lipovic

Technical Editor Teodora Smiljanic

Cover Designer InTech Design Team

Image Copyright Pakhnyushcha, 2010. Used under license from Shutterstock.com

Additional hard copies can be obtained from orders@intechopen.com

Heat Transfer - Theoretical Analysis, Experimental Investigations
and Industrial Systems, Edited by Aziz Belmiloudi
p. cm.
ISBN 978-953-307-226-5

Contents

Preface

These last years, spectacular progress has been made in all aspects of heat transfer. Heat transfer is a branch of engineering science and technology that deals with the analysis of the rate of transfer thermal energy. Its fundamental modes are the conduction, convection, radiation, convection vs. conduction and mass transfer. It has a broad application area to many different branches of science, technology and industry, ranging from biological, medical and chemical systems, to common practice of thermal engineering (e.g. residential and commercial buildings, common household appliances, etc), industrial and manufacturing processes, electronic devices, thermal energy storage, and agriculture and food process. In engineering practice, an understanding of the mechanisms of heat transfer is becoming increasingly important since heat transfer plays a crucial role in the solar collector, power plants, thermal informatics, cooling of electronic equipment, refrigeration and freezing of foods, technologies for producing textiles, buildings and bridges, among other things. Engineers and scientists must have a strong basic knowledge in mathematical modelling, theoretical analysis, experimental investigations, industrial systems and in- formation technology with the ability to quickly solve challenging problems by developing and using new more powerful computational tools, in conjunction with experiments, to investigate design, parametric study, performance and optimization of real-world thermal systems.

In this book entitled "Heat transfer: Theoretical analysis, Experimental investigations and Industrial systems", the authors provide a useful treatise on the principal concepts, new trends and advances in technologies and practical design engineering aspects of heat transfer, pertaining to theoretical and experimental investigations, calculations and industrial utilizations, in particular in the form of innovative experiments and systems, measurement system analysis, as well to complement or develop new theoretical models. The present book contains large number of studies in both fundamental and application approaches with various modern and emerging engineering applications.

These include "Heat Transfer in micro systems" (chapters 1 to 7), which concerns emerging areas of research such as micro- and nanoscale science and technology, with various applications, such as fluidized bed reactors, micro-fin tubes and refrigeration equipment, electronic components and microchips, micro- bubble and molecular crystals; "Boiling, freezing and condensation heat transfer" (chapters 8 to 12), which focuses on the transfer of heat through a phase transition (it is of great significance in industry), with various applications as pool boiling, nuclear reactor safety, food products and water-ethanol mixtures; "Heat transfer and its assessment" (chapters 13 to 20), which covers theoretical and experimental analysis, visualization, assessment and enhancement, with various applications, such as thermoacoustic refrigerator, nuclear reactors, electronic components, fouling process and oscillatory flows; "Heat transfer calculations" (chapters 21 to 25) which concerns exprimentations and numerical simulations, with various applications such as fluidized beds and turbine cascades.

The editor would like to express his sincere thanks to all the authors for their contributions in the different areas of their expertise. Their domain knowledge combined with their enthusiasm for scientific quality made the creation of this book possible. The editor sincerely hopes that readers will find the present book interesting, valuable and current.

Aziz Belmiloudi
European University of Brittany (UEB),
National Institut of Applied Sciences of Rennes (INSA),
Mathematical Research Institute of Rennes (IRMAR),
Rennes, France.

Part 1

Heat Transfer in Micro Systems

1

Integrated Approach for Heat Transfer in Fluidized Bed Reactors

Zeeshan Nawaz[1], Shahid Naveed[2], Naveed Ramzan[2] and Fei Wei
*[1]Beijing Key Laboratory of Green Chemical Resection Engineering and Technology
(FLOTU) Department of Chemical Engineering, Tsinghua University,
[2]Department of Chemical Engineering, University of Engineering and Technology, Lahore,
[1]P. R. China
[2]Pakistan*

1. Introduction

Fluidization of gas-solid system is long-standing subject of basic research. As a result of fluid-particle intensive contact, an isothermal system with superior heat and mass transfer abilities favors its use for chemical reactions, mixing, drying, and other applications. Inspire of huge benefits, number of problems are still associated with gas-solid fluidization system. Solid properties, in particular, particle size and size distribution significantly affect the interaction/contacting between particles, their movement and the fluid-particle mixing. Distinct macroscopic phenomena of plug formation, channeling and particle agglomeration was observed with increasing superficial gas velocity in conventional fluidized bed of fine/nano-particles (group C in the Geldart classification) due to strong interparticle forces.

The concept of powder–particle fluidized bed (PPFB) was first introduced by Kato et al. in early 1990s [1]. It is known to be a useful technique to fluidize group C particles without external aid like acoustic, centrifugal, magnetic, stirring and/or vibrating fields, etc. In the PPFB process, fine powders (group C) are fluidized with coarse particles (group B). The bi-modal fluidized bed system at steady state gives a certain stable hold-up of fine powders in the bed [2]. Many investigators, Sun and Grace [3], and Xue et al. [4] studied bi-modal fluidization in bubbling or turbulent regimes, while, Wei et al. [5] and Du [6, 7] concentrate on CFBs. The investigation of Xue et al. showed that by adding coarse particles fluidization quality of fine particles could improve [4]. Extensive studies of Wei et al. indicated that the addition of coarse particles to a fluid catalytic cracking (FCC) riser decreased the lateral solids mixing and had insignificant influence on axial solids mixing [5]. Du et al. studied the axial and lateral mixing by using tracer particles of different sizes in a FCC riser and found that the axial solids back mixing increased, while radial solids mixing decreased with the increase of particle size and density. Recently, Zeeshan et al. explained bi-modal/bi-particle fluidized bed system hydrodynamics, attrition, mixing behaviour, and its applications [8].

Stable and uniform heat transfer in Fluidized Bed Reactors (FBR) without providing provisions of external or internal source is a difficult task for designers. As continuous heat supply and deduction is a necessary part of FBR operation for controlling highly endothermic and exothermic reactions, respectively. The state-of-the-art idea of bi-modal particle (Gas-Solid-Solid) fluidization is given by FLOTU, in order to overcome above said reaction barriers in a fluidized bed technology. In this chapter, a comprehensive overview of

GSS-FRB research was discussed with practical example of direct propane dehydrogenation reaction. Direct propane dehydrogenation is complicated in engineering constraints due to equilibrium limitations and endothermic nature of reaction, therefore, continuous supply of heat is required. The superiority of GSS-FBR operation was discussed and compared from hydrodynamics to reaction results. The results of fixed bed micro-reactor and integrated bi-modal particle fluidized bed reactors were compared, and parametrically characterized. The significant features of this technique was also highlighted and proposed intensified design as a promising opportunity for highly endothermic and exothermic reactions through FBR, with both economic and operational benefits.

2. Integrated approach

In order to get fluidized bed technology in operation for high endothermic or exothermic reactions Zeeshan et al. proposed a unique design of co-fluidized bed reactor, where coarse particles sized catalyst was co-fluidized with fine FCC catalyst [9, 10]. Coarse particles serve as principal catalyst for desired reaction (endothermic/exothermic), while fine particles serve as a heat carrier/heat absorbent. For an endothermic reaction: the fine catalyst particles take heat from regenerator, transfer heat to principal catalyst (coarse particles) in a fluidized bed reactor and then may serve as secondary catalyst in other reactors like olefins inter-conversion, MTO, MTP, etc as shown in Figure 1.

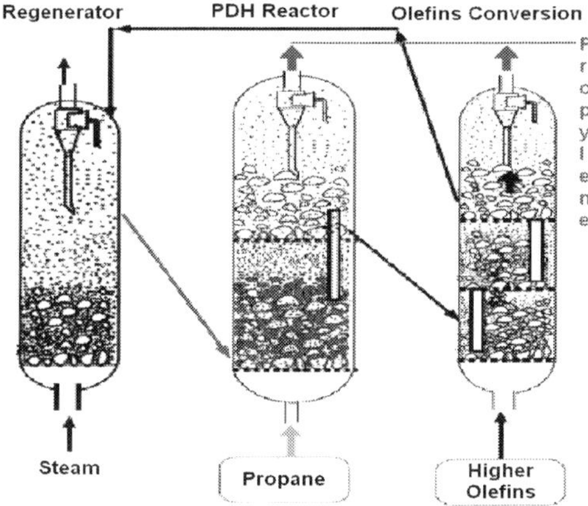

Fig. 1. Proposed process design and reactors sequence

3. Features of technology

Gas-solid-solid (GSS) fluidization system is a unique piece of equipment and its features related to the better mixing quality. The mixing and hydrodynamic behavior of FCC particles with coarse particles was investigated in 2D and 3D co-fluidized beds respectively. The reason of enhancing mixing and fluidization properties by adding coarse particles is particularly due to the movement of coarse particles, those breaks the strong interparticle forces between FCC particles and destroy bubble wake (bubble disassociation strategy), time to time. The information about the design of 2D and 3D co-fluidized bed reactors, and experimental specification can be find elsewhere [8].

To process the solid particulates in fluidized bed and slurry phase reactors, attrition is an inevitable consequence and is therefore one of the preliminary parameters for the catalyst design for a fluidized bed reactor. The mechanical degradation propensity of the zeolite catalysts (particles) were investigated in a bi-modal distribution environment using a Gas Jet Attrition - ASTM standard fluidized bed test (D-5757) [11]. The experimentation was conducted in order to explore parameters affecting attrition phenomena in a bimodal fluidization. In a bimodal fluidization system, two different types of particles were co-fluidized isothermally. Detailed information about experimentation design can be found elsewhere [12].

3.1 Mixing behaviour

The formation, breakage and growth of bubbles are definitive for fluidization, heat exchange and mixing efficiency. Fluidization occurs particularly as a result of a dynamic balance between gravitational forces and forces of a fluid through bed. For the systems where the fluidizing particles have significantly different densities and sizes, those create instable local pockets of very high void fraction termed bubbles. This system is known as self-excited nonlinear system. Previously, it is known that periodic perturbation of a system parameter may result in chaos suppression. Here we examine uniform mixing in co-fluidization (GSS), and noted a unique bubble braking phenomena that is principally responsible for uniform mixing. After the formation of bubble, the bubble flow pattern and pressure distribution are sensitive to the block arrangement.

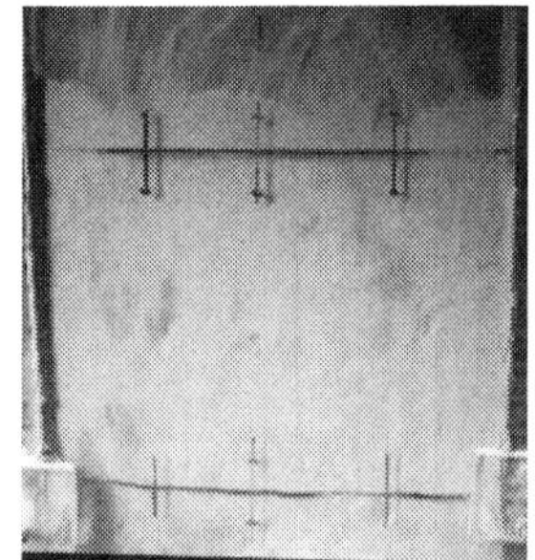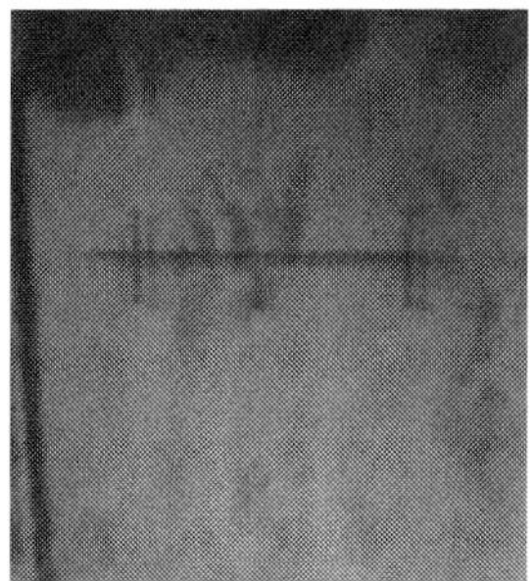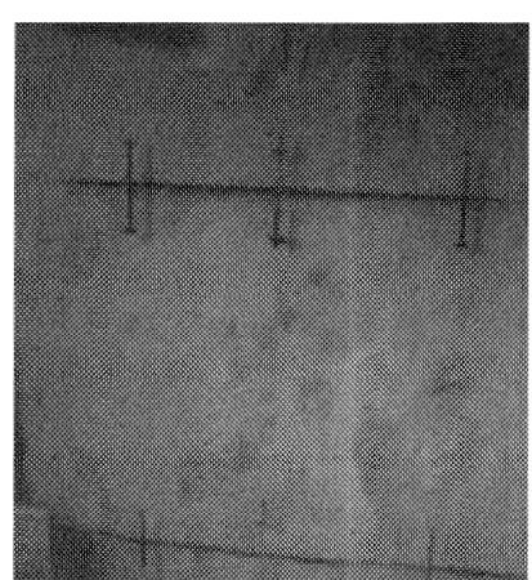

Fig. 2. Unique bubble braking phenomena of GSS fluidization

In bi-modal particle system (GSS FBR) the large bubbles were broken down into small bubbles. This occurs when coarse particles will break large bubble wake's, and generated bubbles give swirl. Momentum was transferred by the bubble to solids in both axial and lateral directions that enhance mixing. The coarse particles produce a high frequency of bubbles and low pressure amplitude around the orifice in comparison with additional distributer, and in ultimate promote smoother fluidization of FCC catalyst with high/uniform mixing. However, coarse particles throughout the bed are in idealistic symmetrical flow, and no jammed was observed. This phenomenon was captured from 2D co-fluidized bed and shown in Figure. 2 [8].

3.2 Hydrodynamics of co-fluidization

After fundamental description of the role of the bubbles in a bi-modal fluidized bed, hydrodynamic of FCC particles with coarse particles was investigated. On the whole,

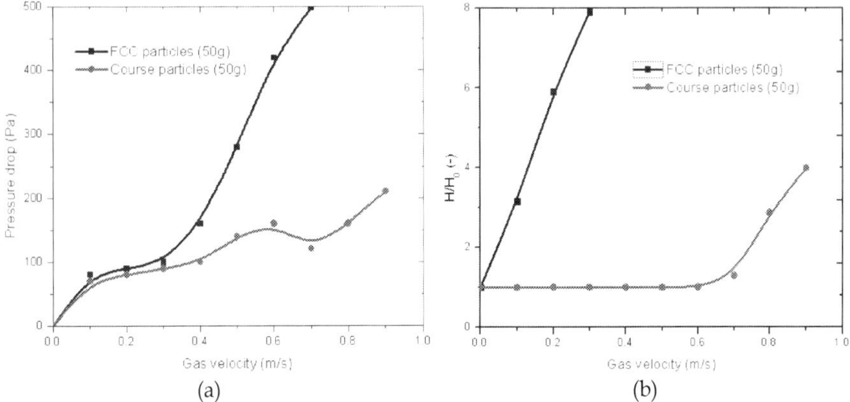

Fig. 3. (a) Bed pressure drop curve of FCC and coarse particles, (b) Bed expansion curve of FCC and coarse particles

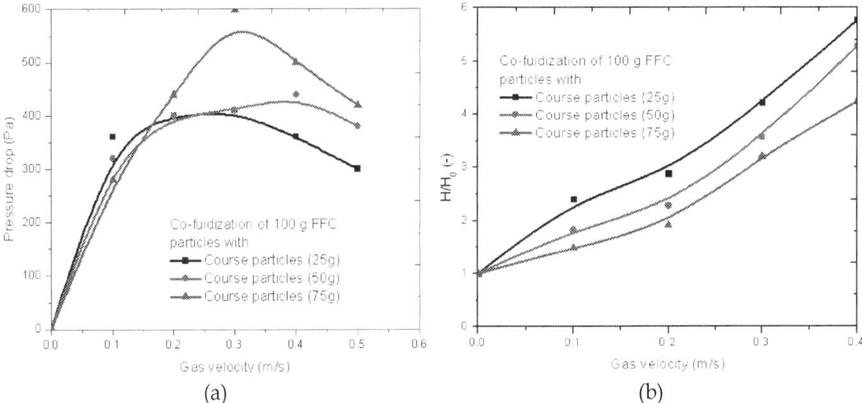

Fig. 4. (a) Bed pressure drop during co-fluidization of FCC and coarse particles, (b) Bed expansion during co-fluidization of FCC and coarse particles

fluidization properties and mixing was improved. Therefore, here we focus our attention on the overall behavior of the bed. The fluidization of FCC catalyst shows crack formation at low superficial gas velocities. With the increase of the gas velocity to around 0.1 m/s, the FCC particles become fluidizing and their fine counter parts air borne at the upper part of the bed. The experimental results of the pressure drop and bed height ratio, of FCC and coarse particles bed (independently) are shown in Figure 3 (a) and (b), respectively. This is the reason that the bed height ratio of FCC particle is become violent with small increase in velocity. The upper part of bed demonstrates very dilute bed, but in turbulent fluidization behavior. The bed surface was severely disturbed by large gas bubble eruptions and fine particle ejections, which lead to enormous elutriation. While, coarse particles also feel difficulty in fluidization at lower gas velocities, till 0.3 m/s. With the increase of gas velocity, the pressure drop curves do not show a plateau. Figure 3 (b) shows that the expansion of FCC particles bed is much higher than that of the coarse particles bed. This gap can be modified by decreasing the particle size difference. The physical mixture of FCC particles with coarse particles was made in following ratio: The FCC particles were fixed to 100 g, with coarse particles were added as 25, 50 and 75 g. Details about the gross bed behavior, pressure drop and expansion characteristics are shown in Figure 4 (a) and (b). The addition of coarse particles in different quantity to FCC particles has improved the gross fluidization behavior significantly. Furthermore, the minimum fluidization velocity of the mixture decreases. In co-fluidization system, it's difficult to measure ΔP across single type of particles (either for small or coarse) as they are well mixed.

3.3 Attrition study

Several experimental studies and empirical models were developed in order to characterize the extent of attrition in a fluidized environment. Gwyn et. al. [12] developed the following empirical relationship (see equation 1) by considering shearing time of attrition, using single particle system with high velocity air jets in a fluidized bed. The Gwyn constants K and m, were the function of material properties and size, and determined experimentally; further details can be find elsewhere [12].

$$W = Kt^{m} \tag{1}$$

However, to date, no study has focused on particles attrition phenomena (systematically explained in Figure 5) in bi-modal particles fluidized bed (co-fluidization environments) and multi-particle sized system, i.e. the most feasible design for handling extremely endothermic or exothermic reactions, while, disadvantage of attrition become serious. Therefore the study focused attention not only on attrition calculations, but explore its incremental phenomena and develop a more generalizes relationship for calculating attrition debris in bi-model particle system.

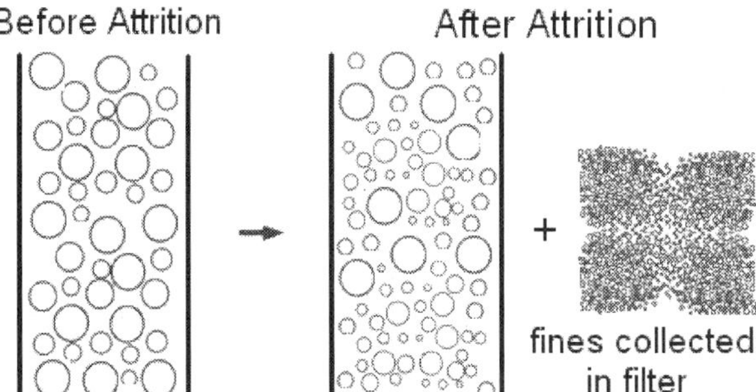

Fig. 5. Systematic image of attrition tube before and after attrition run

AJI in certain cases was as high as 0.28, means that 28 % of the small particles were lost due to attrition. The attrition evaluation was generally pronounced by the attrition rate which depends upon certain thresh hold for the size limit of fines collected. In practice, attrition mass was more important than the collected fine's size distribution, therefore ASTM standard D-5757 attrition test method was selected for the present study and operated under standard operating procedure, in the density range of 1-3 g/cm^3 [12]. The flow regime changes with density variations and fluidization velocities, but settling chamber provides

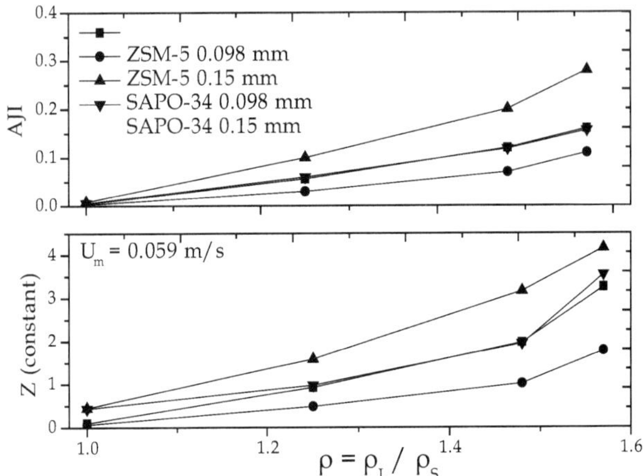

Fig. 6. Effect of relative density on newly defined constant Z and AJI after 5 hr

slipping area to keep the particles airborne in these circumstances, and only let the fines go out. Therefore ASTM standard fluidized bed was selected and operated at fixed superficial gas velocity i.e. 0.059 m/s according to the standard operating procedure. The exact values of Gwyn constants for each particle (small zeolite catalysts) at standard conditions were evaluated in accord with an exact protocol.

Effect of relative density was considered to define this new constant Z and its influence can be observed in Figure 6. There it was assumed that S was constant and its value is 1. This newly developed function (equation 2) has the ability to successfully explain small particles attrition of zeolites and other synthetic catalysts used in bimodal fluidized bed reactors. This empirical relationship will become more generalized and accurate because Gwyn constants of each particle (catalyst) were determined individually. Therefore the applicability of this function was not only limited to the measurement of zeolites attrition in a bimodal fluidized bed environment but can be used for other catalysts. The narrow span of deviation i.e. ± 0.50 g in attrition debris after 5.0 hours operation was observed. The model was generalized as changes related to material properties, and by considering these properties the relationship was made generalized. For each type of material first determine the values of their constants for single particle system as explained by Gwyn, and then using proposed model the attrition loss of that sample in a bi-model fluidization at designated conditions can be calculated.

$$x = (\rho)^5 \cdot (S)^{0.44} \cdot (K) \cdot (t)^{1.53m} \tag{2}$$

The attrition phenomena of small particles in a bimodal particles fluidized bed was found different from single particle fluidized bed phenomena's, because in this system fracture mechanism plays vital role in increasing attrition. Moreover, the power of impact was observed to be the function of large particle density and size ratios.

4. Case study

The proposed bi-modal fluidized bed has wide application in chemical industry in particular for the exothermic and endothermic reactions. At present a complete case study of direct propane dehydrogenation is discussed using bi-modal fluidized bed technology [10]. The propane dehydrogenation is the most economical route to propylene, but very complex due to endothermic reaction requirements, equilibrium limitations, stereo-chemistry and in engineering constraints. The state of the art idea of bi-model particle (Gas-Solid-Solid) fluidization was applied, in order to overcome alkane dehydrogenation reaction barriers in a fluidized bed technology. In this study, the propane dehydrogenation reaction was studied in an integrated fluidized bed reactor, using Pt-Sn/Al$_2$O$_3$-SAPO-34 novel catalyst (ZeeFLOTU) at 590 °C. The results of fixed bed micro-reactor and integrated bi-model particle fluidized bed reactors were compared, and parametrically characterized. The results showed that the propylene selectivity is over 95 %, with conversion between 31-24 %. This significant enhancement is by using novel bi-model particle fluidization system, owing to uniform heat transfer throughout the reactor and transfer coke from principal catalyst to secondary catalyst, which increases principal catalyst's stability. Experimental investigation reveals that the novel Pt-Sn/Al$_2$O$_3$-SAPO-34 catalyst and proposed intensified design of fluidized bed reactor is a promising opportunity for direct propane dehydrogenation to propylene, with both economic and operational benefit. The experimental setup is shown in Figure 7 while catalyst and experimental design information can found elsewhere [10, 13-18].

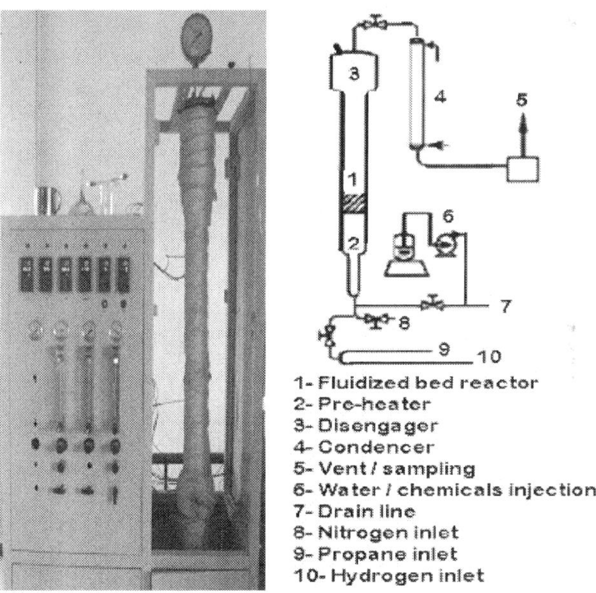

Fig. 7. Hot-model bi-model particle Fluidized Bed Reactor (FBR) apparatus.

The experimental results of propane dehydrogenation using novel catalyst Pt-Sn/Al$_2$O$_3$-SAPO-34 using bi-modal pilot scale fluidized bed reactor are shown in Figure 8. The comparison demonstrates that bi-model results are far better than fixed and single particle fluidized bed [10]. The influence of fluidization mode in an integrated bi-model particle fluidized bed was also investigated for propane dehydrogenation to propylene. It is observed that the propylene selectivity in a fluidized bed reactor was improved after 1 hr operation, when the reactor reaches steady state conditions. The steady and uniform conversion and yield is also achieved. Actually, in fixed bed reactor coke deposition is high as compared to fluidized bed reactor (see Table 1). In the two particles co-fluidized system it was observed that the coke deposited on SAPO-34 (fine catalyst particles) is higher than the metal incorporated SAPO-34 (principal catalyst). Therefore, it's easy for principal dehydrogenation catalyst (Pt-Sn/Al$_2$O$_3$-SAPO-34) to sustain its activity for longer duration.

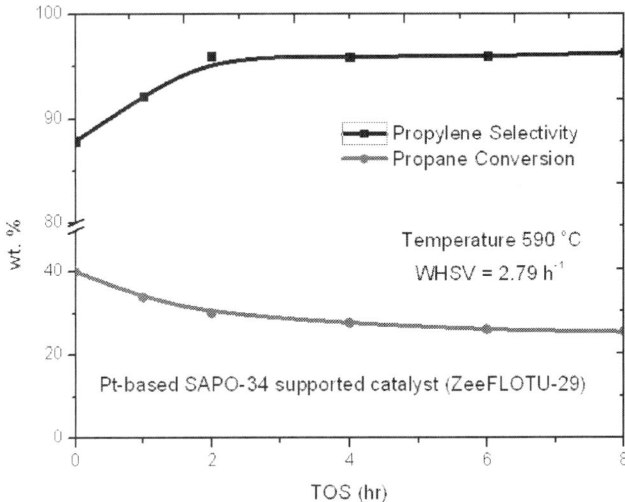

Fig. 8. Performance comparison of novel catalyst in fixed bed with proposed GSS-fluidized bed reactor

Reactor	[a]Coke (wt. %)	[b]Deactivation (%)
Fixed bed	0.41	54
Fluidized Bed	0.24	45

[a]O$_2$-pulse coke analysis
[b]Deactivation = [(X$_0$-X$_f$)/X$_0$ x 100];
where, X$_0$ is the initial conversion at 5 min. and X$_f$ is the final propane conversion
Table 1. Deactivation rate and amount of coke formed on principal catalyst (Pt-Sn/Al-SAPO-34)

Moreover, in the continuous processing the small catalysts (those serve as heat carrier) were continuously regenerated and the process efficiency was improved. Above 96 % propylene selectivity was obtained at 8 hr time-on-stream. Sustainable conversion with lower deactivation rate (see table 1) was also observed. The lower propylene yield initially was due to lower conversion and selectivity, which increase gradually with time. Therefore we can say that the impressive results were obtained using this integrated fluidized bed reactor. It was interesting to find that the reaction stability and activity of catalyst will become superior, but also superior in coke management. Nevertheless, deactivation and/or activity loss of the bi-metallic catalysts is due to coke deposition and Pt sintering [13-15]. Therefore, in coke analysis of bi-model particle fluidized bed catalyst it was noted that large amount of coke is deposited over non-metallic SAPO-34, that is in a continuous recirculation through the regenerator, in continuous setup. It's an effective way to protect catalyst activity for longer time with stable activity, and so called coke management.

The overall picture of selective propane dehydrogenation to propylene over above said catalyst at 590 °C in bi-modal fluidized bed reactor is shown in the OPE plot in Figure 9. The data was plotted with respect to yields and selectivity. The best propane conversion range to have high propylene yield and selectively is observed to be between 24-28% conversions. In designated operating range the propylene yield is above 25 % and selectivity is as high as 96 %. While at higher conversions both propylene yield and selectivity dropped sharply, with the increase in ethane formation. It is further noted that the higher conversion favours both cracking and hydride transfer reaction with the decrease in dehydrogenation rate. Moreover, the deactivation of catalyst may also lead to cracking.

The performance of the Pt-Sn/Al$_2$O$_3$-SAPO-34 is evaluated in a continuous mode of reaction-regeneration for three cycles. The results are shown in Table 2. The catalysts were regenerated with nitrogen mixed steam for 4 hr at 600 °C. After regeneration, the Pt was re-dispersed using C$_2$Cl$_2$H$_4$ solution, injected with nitrogen at 500 °C. The detailed chlorination method can be finding elsewhere [13, 15]. After the regeneration and re-dispersion of Pt, catalyst was reduced in hydrogen environment, and reused for next reaction cycle at identical conditions. The results clearly demonstrate hydrothermal stability of the catalyst. Therefore, the robustness of proposed design of bi-model particle (Gas-Solid-Solid) fluidized bed reactor and Pt-Sn/Al$_2$O$_3$-SAPO-34 is successfully proved.

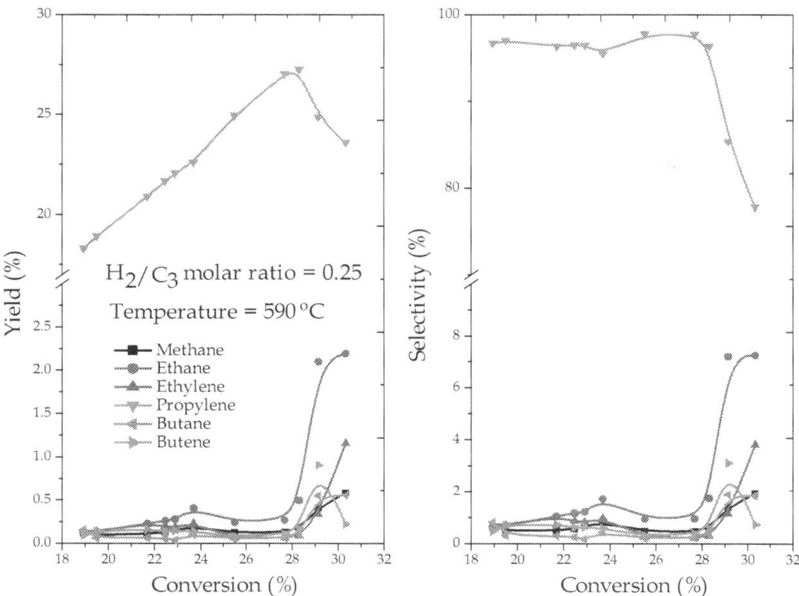

Fig. 9. OPE with respect to yield and selectivity

Al-SAPO-34	Cycle I				Cycle II				Cycle III			
supported	Conversion		Selectivity		Conversion		Selectivity		Conversion		Selectivity	
TOS	1 hr	8 hr	1 hr	8 hr	1 hr	8 hr	1 hr	8 hr	1 hr	8 hr	1 hr	8 hr
Pt-Sn-based	29.1	26.2	85.4	96.8	28.3	25.5	85.9	96.9	27.2	24.6	87.9	97.4

Table 2. Influence of hydrothermal treatment and catalysts performance in a continuous operation (Reaction conditions: T = 590 °C, WHSV = 5.6 h-1, H2/C3H8 molar ratio = 0.25)

5. Heat transfer analysis of bi-modal (G-S-S) fluidization system

In order to evaluate the heat transfer efficiency of the bi-modal fluidized bed reactor, the analysis is derived from an energy balance and a mass balance with the following simplifying assumptions: the powder particles are not considered individually but as a single phase, motion and temperature gradients are considered only in the direction of flow, the course particle bed is considered to be dilute packing as per total volume of the bed, then the heat transfer from the powder (fine particles) to packing (course particles) is considered as question. Under such assumptions, a plug flow model is established to study heat transfer. In the hot model experiments, since the fine particle mass flux could not be measured, the powder-to-gas mass flux ratio, $R_{sg}(G_s/G_g)$ is analyzed. The best value for R_{sg} has to be found by trial and error, by experiments. A suitable R_{sg} should satisfy the condition of higher conversion and selectivity, and is closed to axial distribution of temperature in experiments [19]. The experiments demonstrate that the temperature is almost uniform with +/- 10 °C in the co-fluidized bed to keep propane conversion in reasonable range. The Gs/Gg values calculated were also calculated for the same superficial gas velocity. It is worthwhile to note that the temperature of the pre-heated propane injection (550 °C) was taken to 590 °C for a highly endothermic reaction and maintained because of the continuous supply of heat through fine particles. The dimensionless R_h is defined as follows:

$$R_h = \sum (\Phi_i Cp_{g,i}) + W_s Cp_s / \Phi_{propane} Cp_{g,propane} \tag{3}$$

where i stands for propane, propylene and hydrogen. The higher R_h is found to be beneficial for supplying reaction heat and its increases with the increase in fine particle mass flux. While the fine particle replaces a portion of feed gases as heat carrier enhances hear transfer rate. The axial distribution of temperature of the GSS-FBR is far superior to that in the adiabatic packed-bed reactor for endothermic reaction.

6. Summary

An integrated design of Fluidized Bed Reactors (FBR) having superior heat supply scheme without providing provisions of external or internal source is discussed for endothermic reactions. The state-of-the-art idea of bi-modal particle (Gas-Solid-Solid) fluidization is presented and briefly discussed in all regards. The superior operational benefits, in particular heat transfer were enhancing using bi-modal fluidized bed technology. The technology was discussed with a case study of propane dehydrogenation using GSS-FRB. A generalized heat transfer information for the bi-modal fluidization system with respect to case study. The intensified design of GSS-FBR is found to be a promising opportunity for endothermic and exothermic reactions, as heat supply or removal is always a problem.

7. References

[1] K. Kato, T. Takarada, N. Matsuo, T. Suto, N. Nakagawa, Int. Chem. Engg. 34, 605 (1994).

[2] T. Nakazato, J. Kawashima, T. Masagaki and K. Kato, Adv. Powder Technol. 17, 433 (2006).

[3] G. Sun, J.R. Grace, Chem. Eng. Sci. 45, 2187 (1990).

[4] H. Xue, W. Zhang, X. Yang, Eng. Chem. Metall. (China), 17, 235 (1996).

[5] F. Wei, Y. Cheng, Y. Jin, Z.Q. Yu, Can. J. Chem. Eng. 76, 19 (1998).

[6] B. Du, Master Thesis of Tsinghua University, Beijing, China, 1999.

[7] B. Du, F. Wei, Chem. Eng. Process. 41, 329 (2002).

[8] Zeeshan Nawaz, Yujian Sun, Yue Chu and Fei Wei, "Mixing behavior and hydrodynamic study of GAS-SOLID-SOLID Fluidization system: Co-Fluidization of FCC and coarse particles", FLUIDIZATION XIII, May 16-21, 2010, Gyeong-ju, Korea.

[9] Zeeshan Nawaz, PhD Thesis of Tsinghua University, Beijing, China, 2010.

[10] Zeeshan Nawaz, Yue Chu, Wei Yang, Xiaoping Tang, Yao Wang and Fei Wei, Study of Propane Dehydrogenation to Propylene in an integrated Fluidized Bed Reactor using Pt-Sn/Al-SAPO-34 novel catalyst, Industrial & Engineering Chemistry Research, 49 (2010) 4614-4619.

[11] ASTM Standard D-5757-Revised, ASTM, Philadelphia PA, (2006).

[12] Zeeshan Nawaz*, Tang Xiaoping, Xiaobo Wei and Fei Wei, Attrition behavior of small particles in bimodal particles fluidized bed environment: Influence of density and size ratio, Korean J. Chem. Eng., 27(4), 1025-1031 (2010) (DOI: 10.1007/s11814-010-0240-5)

[13] Zeeshan Nawaz, Fei Wei and Shahid Naveed, Highly stable Pt-Sn-based, SAPO-34 supported, Al binded catalyst, and Integrated Fluidized Bed Reactor Design for Alkane Dehydrogenation, Pak. Patent Application No. 1127/2009; Chinese Patent Application No. 200910091226.6.

[14] Zeeshan Nawaz, Tang Xiaoping, Qiang Zhang, Wang Dezheng and Fei Wei, A highly selective Pt-Sn/SAPO-34 catalyst for propane dehydrogenation to propylene, Catalysis Communications, 10 (2009) 1925-1930.

[15] Zeeshan Nawaz and Wei Fei, Hydrothermal study of Pt-Sn-based SAPO-34 supported novel catalyst used for selective propane dehydrogenation to propylene, Journal of Industrial and Engineering Chemistry, 16 (2010) 774-784.

[16] Zeeshan Nawaz, Xiaoping Tang, Yao Wang and Fei Wei, Parametric characterization and influence of Tin on the performance of Pt-Sn/SAPO-34 catalyst for selective propane dehydrogenation to propylene, Industrial & Engineering Chemistry Research, 49 (2010) 1274–1280.

[17] Zeeshan Nawaz and Wei Fei, The Pt-Sn-based SAPO-34 supported novel catalyst for n-butane dehydrogenation, Industrial & Engineering Chemistry Research, 48 (2009) 7442-7447.

[18] Zeeshan Nawaz, Tang Xiaoping, Shahid Naveed, Qing Shu and Fei Wei, Experimental Investigation of Attrition Resistance of Zeolite Catalysts in Two Particle Gas-Solid-Solid Fluidization System, Pakistan Journal of Scientific & Industrial Research, 53 (2010) 169-174.

[19] Shuiyuan Huang, Zhanwen Wang, Yong Jin, Studies on gas-solid-solid circulating Fuidized-bed reactors, Chemical Engineering Science 54 (1999) 2067-2075.

Two-Phase Heat Transfer Coefficients of R134a Condensation in Vertical Downward Flow at High Mass Flux

A.S. Dalkilic[1] and S. Wongwises[2]
*[1]Heat and Thermodynamics Division, Department of Mechanical
Engineering, Yildiz Technical University, Yildiz, Istanbul 34349
[2]Fluid Mechanics, Thermal Engineering and
Multiphase Flow Research Lab. (FUTURE), Department of
Mechanical Engineering, King Mongkut's University of
Technology Thonburi, Bangmod, Bangkok 10140
[1]Turkey
[2]Thailand*

1. Introduction

The transfer process of heat between two or more fluids of different temperatures in a wide variety of applications is usually performed by means of heat exchangers e.g. refrigeration and air-conditioning systems, power engineering and other thermal processing plants. In refrigeration equipment, condensers' duty is to cool and condense the refrigerant vapor discharged from a compressor by means of a secondary heat transfer fluid such as air and fluid. Examinations and improvements on the effectiveness of condenser have importance in case there is a design error which can cause the heat transfer failure occurrence in the condenser.

Industries have been trying to improve the chemical features of alternative refrigerants of CFCs due to the depletion of the earth's ozone layer. Because of the similar thermo-physical properties of HFC-134a to those of CFC-12, an intensive support from the refrigerant and air-conditioning industry has given to the refrigerant HFC-134a as a potential replacement for CFC-12. On the other hand, even though there isn't much difference in properties between these two refrigerants, the overall system performance may be affected significantly by these differences. For that reason, the detailed investigation of the properties of HFC-134a should be performed before it is applied.

A large number of researchers focused on the heat transfer and pressure drop characteristics of refrigerants over the years, both experimentally and analytically, mostly in a horizontal straight tube. The CFCs' heat transfer and pressure drop studies inside small diameter vertical either smooth or micro-fin tubes for downward condensation has been concerned comparatively little in the literature. Briggs et al. (1998) have studied on in-tube condensation using large diameter tubes of approximately 20.8 mm with CFC113. Shah (1979) used some smooth tubes up to 40 mm i.d. at horizontal, inclined and vertical positions to have a well-known wide-range applicable correlation. Finally, it has been compared by researchers commonly for turbulent condensation conditions and it is considered to be the most predictive condensation model for the annular flow regime in a tube. Oh and Revankar (2005) developed the Nusselt theory (1916) to investigate the PCCS condenser using 47.5 mm i.d. and 1.8 m long vertical tube in their experimental study for the validation of their theoretical calculations. Maheshwari et al. (2004) simulated PCCS condensers used in water-cooled reactors using a 42.77 mm i.d. vertical tube during downward condensation presence of non-condensable gas. The prediction of heat transfer considering mass transfer along the tube length was performed by means of a computer code in their study.

Convective condensation in annular flow occurs for many applications inside tubes such as film heating and cooling processes, particularly in power generation and especially in nuclear reactors. The one of the most important flow regimes is annular flow which is characterized by a phase interface separating a thin liquid film from the gas flow in the core region. This flow regime is the most investigated one either analytically or experimentally, because of its practical significance and common usage.

Akers et al. (1959) focused on the similarity between two phase and single phase flows and developed a correlation using two-phase multiplier to predict frictional two phase pressure drop which is same rationale as the Lockhart-Martinelli (1949) two-phase multiplier. His model is known as "equivalent Reynolds number model" in the literature. According to this model, an equivalent all liquid flow, which produces the same wall shear stress as that of the two phase flow, is replaced instead of annular flow inside a tube. Many researchers benefitted from his model such as Moser et al. (1998) and Ma and Rose (2004). Moser et al. (1998) predicted heat transfer coefficient in horizontal conventional tubes developing a model. Ma and Rose (2004) studied heat transfer and pressure drop characteristics of R113 in a 20.8 i.d. vertical smooth and enhanced tubes.

Dobson et al. (1998) used zeotropic refrigerants in their condensation tests including the wide range of mass flux in horizontal tubes and benefitted from the two-phase multiplier approach for annular flow. In their study, heat transfer coefficient increased with increasing mass flux and vapor quality in annular flow due to increased shear and thinner liquid film than other flow regimes. Sweeney (1996) modified their model their model for R407C considering the effect of mass flux.

Cavallini et al. (1974) have some significant theoretical analysis on the in-tube condensation process regarding the investigation of heat transfer and pressure drop characteristics of refrigerants condensing inside various commercially manufactured tubes with enhanced surfaces using a number of correlations in the literature. Lately, Cavallini et al. (2003) prepared a review paper on the most recent condensation works in open literature including the condensation inside and outside smooth and enhanced tubes.

Valladares (2003) presented a review paper on in-tube condensation heat transfer correlations for smooth and micro-fin tubes including the comparison of experimental data belong to various experimental conditions from different researchers. Wang and Honda (2003) made a comparison of well-known heat transfer models with experimental data belong to various refrigerants from literature and proposed some models, valid for the modified annular and stratified flow conditions, for micro-fin tubes. Bassi and Bansal (2003) compared various empirical correlations and proposed two new empirical models for the determination of condensation heat transfer coefficients in a smooth tube using R134a with lubricant oil. Jung et al. (2003, 2004) did condensation tests of many refrigerants such as R12, R22, R32, R123, R125, R134a, and R142b inside a smooth tube. They paid attention not only comparison of their experimental data with various well-known correlations but also proposition on a new correlation to predict condensation heat transfer coefficients.

Generally, the condensation heat transfer coefficients and pressure drops in tubes have been computed by empirical methods. The modifications of the Dittus-Boelter single-phase forced convection correlation (1930) are used in the literature, as in Akers et al. (1959), Cavallini and Zecchin (1974), and Shah (1979). Dalkilic et al. (2010a) made a comparison of thirteen well-known two-phase pressure drop models with the experimental results of a condensation pressure drop of R600a and R134a in horizontal and vertical smooth copper tubes respectively and revealed the main parameters of related models and correlations. Dalkilic et al. (2008a) used the equivalent Reynolds number model (1998) to propose a new correlation for the two-phase friction factor of R134a and also discussed the effect of main parameters such as heat flux, mass flux and condensation temperature on the pressure drop. Dalkilic et al. (2009b) made a comparison of eleven well-known correlations for annular flow using a large amount of data obtained under various experimental conditions to have a new correlation based on Bellinghausen and Renz's method (1992) for the condensation heat transfer coefficient of high mass flux flow of R134a. The effects of heat flux, mass flux and condensation temperature on the heat transfer coefficients also exist in their paper. Dalkilic et al. (2009c) showed the significance of the interfacial shear effect for the laminar condensation heat transfer of R134a using Carey's analysis (1992), which is the improved version of Nusselt's theory (1916), and proposed a new correlation based on Bellinghausen and Renz's method (1992) for the condensation heat transfer coefficient during annular flow of R134a at low mass flux in a vertical tube. Dalkilic et al. (2008b) investigated thirty-three void fraction models and correlations from the available literature and compared them each other using relevant data. The friction factors, based on the analysis of Ma et al. (2004), are obtained from various void fraction models and correlations and a comparison was made with each other and also with those determined from graphical information provided by Bergelin et al. (1946). The presentation for the effect of void fraction alteration on the momentum pressure drop was also shown in their paper. Dalkilic et al. (2009a) compared some simple void fraction models of the annular flow pattern for the forced convection condensation of pure R134a taking into account the effect of the different saturation temperatures in high mass flux conditions. The calculated film thickness from void fraction models and correlations and those from Whalley's annular flow model (1987) were compared each other using their experimental database. Dalkilic and Wongwises (2010b) used Barnea et al. (1982)'s mathematical model, based on the momentum balance of liquid and vapour phases, in order to determine the condensation film thickness of R134a in their paper. The discussions for the effects of heat flux, mass flux and condensation temperature on the film thickness and condensation heat transfer coefficient were also made for laminar and turbulent flow conditions. Six well-known flow regime maps from the literature were found to be in good agreement for the annular flow conditions in the test tube in spite of their different operating conditions. Dalkilic et al. (2010e) used Kosky and Staub's model (1971) to predict flow pattern transitions and validate the results of void fraction models and correlations proposed in their previous publications and also show the identification of flow regimes in data corresponding to annular flow downward condensation of R134a in a vertical smooth copper tube. Furthermore, investigation of twelve number of well-known flow regime correlations from the literature is performed to identify the flow regime occurring in the test tube. Dalkilic et al. (2010d) calculated the average predicted heat transfer coefficient of the refrigerant using Kosky and Staub's model (1971) and the Von Karman universal velocity distribution correlations by means of different interfacial shear stress equations valid for annular flow in horizontal and vertical tubes in order to validate

Chen et al.'s annular flow theory (1987). The discussions for the effects of heat flux, mass flux and condensation temperature on the pressure drop were also made in their paper. A new correlation including dimensionless parameters such as the equivalent Reynolds number, Prandtl number, R number (ρ-μ ratio), Lockhart and Martinelli parameter, Bond number and Froude number was proposed in their paper using a large number of data points for the determination of turbulent condensation heat transfer coefficient. Dalkilic et al. (2010c) proposed a new experimental approach on the determination of condensation heat transfer coefficient in a vertical tube by means of von Karman's universal velocity distribution and Kosky and Staub's annular flow film thickness model (1971). They benefitted from thirteen numbers of frictional pressure drop models and thirty five numbers of void fraction models in their model. Dalkilic and Wongwises (2009d) reported a detailed review of research on in-tube condensation by reason of its significance in refrigeration, air conditioning and heat pump applications. The heat transfer and pressure drop investigations for the in-tube condensation were included and almost all relevant research subjects were summarized, such as condensation heat transfer and pressure drop studies according to tube orientation (horizontal, vertical, inclined tubes) and tube geometry (smooth and enhanced tubes), flow pattern studies of condensation, void fraction studies, and refrigerants with oil. Besides to the above, various other conference papers (2008, 2009, 2010) were used to support and validate their proposed models and correlations in their papers. It can be seen from above studies, Dalkilic and Wongwises studied in-tube condensation process comprehensively using their experimental facility whose test section is working as a double-tube heat exchanger.

In spite of the existence of some available information in the literature, there still remains room for further research. As a result, the major aim in the present chapter of the book is to investigate the appliance of well-known empirical annular flow correlations to the annular flow condensation at high mass flux in a vertical double tube heat exchanger. The independency of annular flow heat transfer empirical correlations from tube orientation (1987) and the general applicability for a vertical short tube is also proved in this chapter.

2. Data reduction

Fig. 1 shows the steady-state physical model of downward film condensation in a vertical tube. Nusselt-type analysis is valid under some assumptions such as: laminar film flow; saturated state for the vapour of R134a; condensed film of R134a along the tube surface; constant physical properties corresponding to inlet pressure and temperature conditions; no entrainment. It should be noted that there is an interfacial shear effect at the interface occurred due to the much greater vapour velocity than the film velocity. The solution of the problem can be started from the calculation of the force balance in Eq. (1) for the differential element in the control volume neglecting the inertia and downstream diffusion contributions as shown in Dalkilic et al.'s study (2009c).

The force balance for the differential element in the control volume can be expressed as follows:

$$\rho_l g dx dy dz + \tau_\delta(y + dy)dx dz + P(z)dx dy = \tau_\delta(y)dx dz + P(z + dz)dx dy \qquad (1)$$

It should be noted that the solution of the investigated case in this chapter will be different from Nusselt's solution (1916) due to the high mass flux condition of the condensate flow in the tube.

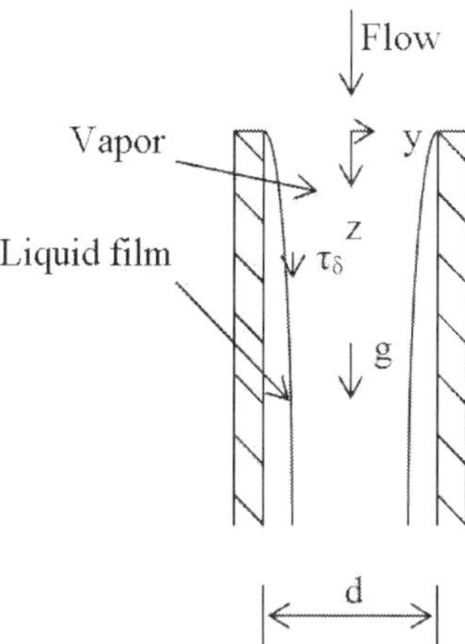

Fig. 1. System model for analysis of downward condensation

2.1 Experimental heat transfer coefficient

The details of determination of the experimental heat transfer coefficient by means of
experimental setups can be seen in many papers given in references.

$$h_{exp} = \frac{Q_{TS}}{A_i(T_{ref,sat} - T_{w,i})}$$
(2)

where h_{exp} is the experimental average heat transfer coefficient, Q_{TS} is the heat transfer rate
in the test section, T_{wi} is the average temperature of the inner wall, $T_{ref,sat}$ is the average
temperature of the refrigerant at the test section inlet and outlet, and A_i is the inside surface
area of the test section:

$$A_i = \pi dL$$
(3)

where d is the inside diameter of the test tube. L is the length of the test tube.

2.2 Uncertainties

The uncertainties of the Nusselt number and condensation heat transfer coefficient in the
test tube, which belongs to a refrigerant loop consisting of an evaporator, test section and
condenser, varied from ±7.64% to ±10.71%. The procedures of Kline and McClintock (1953)
were used for the calculation of all uncertainties. Various uncertainty values of the study
can be seen from Table 1 in detail.

Based on this usual method, suppose that a set of measurement is made and the uncertainty
in each may be expressed with same odds. These measurements are then used to calculate
some desired result of the experiments. The result P is a given function of the independent
variables x_1, x_2, x_3, ...x_n. Thus:

$$P = P(x_1,\ x_2,\ x_3,.....,x_n)$$
(4)

Parameters	Uncertainty
$T_{ref,\,sat}$ (°C)	0.19
x_i	±6.96-8.24%
ΔT (K)	±0.191
$(T_{w,out}-T_{w,in})_{TS}$ (K)	±0.045
$(T_{w,in}-T_{w,out})ph$ (K)	±0.13
m_{ref} (g s^{-1})	±0.023
$m_{w,TS}$ (g s^{-1})	±0.35
$m_{w,pre}$ (g s^{-1})	±0.38
q_{TS} (W m^{-2})	±6.55-8.93%
q_{pre} (W m^{-2})	±12-14.81%
h_{ref} (W m^{-2}K^{-1})	±7.64-10.71%.
ΔP (kPa)	±0.15

Table 1. Uncertainty of experimental parameters

Let w_p be the uncertainty in the result, and w_1, w_2, w_3, ...,w_n be the uncertainties in each independent variables. If the uncertainties in the independent variables are given with some odds, then the uncertainty in the result can be given as followed:

$$w_p = \pm \left[\left(\frac{\partial P}{\partial x_1} w_1 \right)^2 + \left(\frac{\partial P}{\partial x_2} w_2 \right)^2 + \left(\frac{\partial P}{\partial x_3} w_3 \right)^2 + \dots\dots + \left(\frac{\partial P}{\partial x_n} w_n \right)^2 \right]^{1/2} \tag{5}$$

2.2.1 Vapor quality (x_{in})

The vapor quality entering the test section is calculated from an energy balance, which gives the total heat transfer rate from hot water to liquid R134a in the evaporator as the sum of sensible and latent heat transfer rates, on the pre-heater:

$$x_{in} = \frac{1}{i_{ref,fg}} \left[\frac{Q}{m_{ref,T}} - C_{p_{ref}} \left(T_{ref,sat} - T_{ref,pre,in} \right) \right] \tag{6}$$

$$w_{x_{in}} = \pm \left[\left(\frac{\partial x_{in}}{\partial Q} w_Q \right)^2 + \left(\frac{\partial x_{in}}{\partial m_{ref,T}} w_m \right)^2 + \left(\frac{\partial x_{in}}{\partial T_{ref,sat}} w_{T_{ref,sat}} \right)^2 + \left(\frac{\partial x_{in}}{\partial T_{ref,pre,in}} w_{T_{ref,pre,in}} \right)^2 \right]^{1/2} \tag{7}$$

$$w_{x_{in}} = \pm \left[\left(\frac{1}{m_{ref,T} i_{ref,fg}} w_Q \right)^2 + \left(\frac{Q}{i_{ref,fg} m_{ref}^2} w_{m_{ref,T}} \right)^2 + \left(\frac{C_{p,ref}}{i_{ref,fg}} w_{T_{ref,sat}} \right)^2 + \left(\frac{Cp_{ref}}{i_{ref,fg}} w_{T_{ref,pre,in}} \right)^2 \right]^{1/2} \tag{8}$$

2.2.2 Heat flux (Q)

The outlet quality of R134a from the evaporator is equal to the quality of R134a at the test section inlet. The total heat transferred in the test section is determined from an energy balance on the cold water flow in the annulus:

$$Q = m_w C_{p\,w}(T_{w,out} - T_{w,in}) \tag{9}$$

$$w_Q = \pm\left[\left(\frac{\partial Q}{\partial m_w}w_{m_w}\right)^2 + \left(\frac{\partial Q}{\partial T_{w,in}}w_{T_{w,in}}\right)^2 + \left(\frac{\partial Q}{\partial T_{w,out}}w_{T_{w,out}}\right)^2\right]^{1/2} \tag{10}$$

2.2.3 Average heat transfer coefficient (h_i)

The refrigerant side heat transfer coefficient in Eq. (2) is determined from the total heat transferred in the test section which is a vertical counter-flow tube-in-tube heat exchanger with refrigerant flowing in the inner tube and cooling water flowing in the annulus:

$$w_{h_i} = \left[\left(\frac{\partial h_i}{\partial Q}w_Q + \frac{\partial h_i}{\partial\left(T_{ref,sat} - T_{ref,w,i}\right)}w_{(T_{ref,sat} - T_{ref,w,i})}\right)\right] \tag{11}$$

2.2.4 Nusselt number (Nu)

Nusselt number can be expressed as follows:

$$Nu = \frac{h_i L}{k_l} \tag{12}$$

$$w_{Nu} = \pm\left[\left(\frac{\partial Nu}{\partial h_i}w_{h_i}\right)^2\right]^{1/2} \tag{13}$$

2.2.5 Temperature difference

The temperature measurements for the determination of outer surface temperature of the test tube are evaluated as an average value for the uncertainty analysis and calculation procedure can be seen as follows:

$$T_0 = \frac{T_1 + T_2 + T_3 + \ldots\ldots + T_{10}}{10} \tag{14}$$

$$w_{T_0} = \pm\left[\left(\frac{\partial T_0}{\partial T_1}w_{T_1}\right)^2 + \left(\frac{\partial T_0}{\partial T_2}w_{T_2}\right)^2 + \ldots\ldots + \left(\frac{\partial T_0}{\partial T_{10}}w_{T_{10}}\right)^2\right]^{1/2} \tag{15}$$

$$w_{T_0} = \pm\left[\left(\frac{1}{10}w_{T_1}\right)^2 + \left(\frac{1}{10}w_{T_2}\right)^2 + \ldots\ldots + \left(\frac{1}{10}w_{T_{10}}\right)^2\right]^{1/2} \tag{16}$$

3. Heat transfer correlations in this chapter

The comparison of convective heat transfer coefficients with Shah correlation (1979) has been done by researchers commonly for turbulent condensation conditions especially in vertical tubes (2000) and it is found to be the most comparative condensation model during annular flow regime in a tube (2002). Shah correlation (1979) is based on the liquid heat transfer coefficient and is valid for $Re_1 \geq 350$. A two-phase multiplier is used for the annular flow regime of high pressure steam and refrigerants (2000, 2002) in the equation.

Dobson and Chato (1998) used a two-phase multiplier to develop a correlation for an annular flow regime. A correlation is also provided for a wavy flow regime. The researchers used commonly their correlations for zeotropic refrigerants. Its validated mass flux (G) covers the values more than 500 kg m^{-2} s^{-1} for all qualities in horizontal tubes.

Sweeney (1996) developed the Dobson and Chato model (1998) using zeotropic mixtures for annular flow.

Cavallini et al. (1974) used various organic refrigerants for the condensation inside tubes in both vertical and horizontal orientations and developed a semi empirical correlation as a result of their study.

Bivens and Yokozeki (1994) modified Shah correlation (1979) benefitted from various flow patterns of R22, R502, R32/R134a, R32/R125/R134a.

Tang et al. (2000) developed the Shah [(1979) equation for the annular flow condensation of R410A, R134a and R22 in i.d. 8.81 mm tube with $Fr_{so}>7$.

Fujii (1995) modified the correlation in Table 2 for shear-controlled regimes in smooth tubes. There is also another correlation belong to Fujii (1995) for gravity controlled regimes.

Chato (1961) used a two-phase multiplier to modifiy Dittus-Boelter's correlation (1930) for an annular flow regime.

Traviss et al. (1972) focused on the flow regime maps for condensation inside tubes. Their correlation was suggested for the condensation of R134a inside tubes specifically. Their model takes into account of the variation in the quality of the refrigerant using Lockhart-Martinelli parameter.

Akers and Rosson (1960) developed the Dittus–Boelter (1930)'s single-phase forced convection correlation. Their correlation's validity range covers the turbulent annular flow in small diameter circular tubes and rectangular channels.

Tandon et al. (1995) developed the Akers and Rosson (1960) correlation for shear controlled annular and semi-annular flows with $Re_g > 30000$. There is another correlation of Tandon et al. (1995) for gravity-controlled wavy flows with $Re_g < 30000$.

4. Results and discussion

It should be noted that it is possible for researcher to identify the experimental data of condensation by means of both flow regime maps and sight glass at the inlet and outlet of the test section. Dalkilic and Wongwises (2009b, 2010b, 2010e) checked the data shown in all figures and formulas that they were in an annular flow regime by Hewitt and Robertson's (1969) flow pattern map and also by sight glass in their experimental setup. The vapor quality range approximately between 0.7-0.95 in the 0.5 m long test tube was kept in order to obtain annular flow conditions at various high mass fluxes of R134a.

In Table 2, the list of correlations is evaluated to show the similarity of annular flow correlations which are independent of tube orientation (horizontal or vertical). Chen et al. (1987) also developed a general correlation to discuss this similarity in their article by relating the interfacial shear stress to flow conditions for annular film condensation inside tubes.

Researcher	Model/Correlation
Shah (1979)	$\mathrm{Re}_l \geq 350 \quad \mathrm{Re}_l = \dfrac{Gd(1-x)}{\mu_l} \quad h_{shah} = h_{sf}\left(\dfrac{1.8}{Co^{0.8}}\right) \quad P_{red} = \dfrac{P_{sat}}{P_{critic}}$ $h_{sf} = h_l(1-x)^{0.8} \qquad Co = \left(\dfrac{1}{x}-1\right)^{0.8}\left(\dfrac{\rho_g}{\rho_l}\right)^{0.5} \qquad h_l = \dfrac{k_l}{d}\left(0.023\left(\dfrac{\mathrm{Re}_l}{1-x}\right)^{0.8}\mathrm{Pr}_l^{0.4}\right)$
Dobson and Chato (1998)	For $G > 500$ kg m^{-2} s^{-1} $\quad Nu = 0.023\,\mathrm{Re}_l^{0.8}\,\mathrm{Pr}_l^{0.4}\left[1+\dfrac{2.22}{X^{0.89}}\right] \quad Ga = \dfrac{\rho_l(\rho_l-\rho_g)gd^3}{\mu_l^2}$ $X = \left(\dfrac{1-x}{x}\right)^{0.9}\left(\dfrac{\rho_g}{\rho_l}\right)^{0.5}\left(\dfrac{\mu_l}{\mu_g}\right)^{0.1} \qquad Fr_{so} = c_3\,\mathrm{Re}_l^{c_4}\left(\dfrac{1+1.09X^{0.039}}{X}\right)^{1.5}\dfrac{1}{Ga^{0.5}}$ $\mathrm{Re}_l = \dfrac{Gd(1-x)}{\mu_l}$ For $\mathrm{Re}_l > 1250$ and $Fr_{so} > 18 \quad Nu = 0.023\,\mathrm{Re}_l^{0.8}\,\mathrm{Pr}_l^{0.3}\dfrac{2.61}{X^{0.805}}$
Sweeney (1996)	$Nu = 0.7\left(\dfrac{G}{300}\right)^{0.3} Nu_{Dobson-Chato}$
Cavallini et al. (1974)	$Nu_l = 0.05\,\mathrm{Re}_{eq}^{0.8}\,\mathrm{Pr}^{0.33} \quad \mathrm{Re}_{eq} = \mathrm{Re}_g\left(\mu_g/\mu_l\right)\left(\rho_l/\rho_g\right)^{0.5}\mathrm{Re}_l \quad \mathrm{Re}_g = Gdx/\mu_g$
Bivens and Yokozeki (1994)	$Nu = Nu_{Shah}\left(0.78738+\dfrac{6187.89}{G^2}\right) \qquad P_{red} = P_{sat}/P_{critic}$ $Nu_{Shah} = 0.023\,\mathrm{Re}_l^{0.8}\,\mathrm{Pr}_l^{0.4}\left[1+\dfrac{3.8}{P_{red}^{0.38}}\left(\dfrac{x}{1-x}\right)^{0.76}\right]$
Tang et al. (2000)	$Nu = 0.023\,\mathrm{Re}_l^{0.8}\,\mathrm{Pr}_l^{0.4}\left[1+4.863\left((-\ln(P_{red})\dfrac{x}{1-x}\right)^{0.836}\right]$ $\mathrm{Re}_l \leq 1250\ c_3= 0.025\ c_4= 1.59 \quad Fr_{so} = c_3\,\mathrm{Re}_l^{c_4}\left(\dfrac{1+1.09X^{0.039}}{X}\right)^{1.5}\dfrac{1}{Ga^{0.5}}$ $\mathrm{Re}_l>1250\ c_3= 1.26\ c_4= 1.04$
Fujii (1995)	$Nu_l = 0.0125\left(\mathrm{Re}_l\sqrt{\rho_l/\rho_g}\right)^{0.9}\left(\dfrac{x}{1-x}\right)^{0.1x+0.8}\mathrm{Pr}_l^{0.63}$
Chato (1961)	$Nu = 0.023\,\mathrm{Re}_l^{0.8}\,\mathrm{Pr}_l^{0.4}\left[\dfrac{2.47}{X^{1.96}}\right]$
Traviss et al. (1972)	$Nu = \mathrm{Re}_l^{0.9}\,\mathrm{Pr}_l\dfrac{F_1(X)}{F_2(\mathrm{Re}_l,\mathrm{Pr}_l)} \qquad F_1(X) = 0.15\left[\dfrac{1}{X}+\dfrac{2.83}{X^{0.476}}\right]$ $\mathrm{Re}_l>1125 \qquad F_2 = 5\mathrm{Pr}_l+5\ln(1+5\mathrm{Pr}_l)+2.5\ln(0.00313\,\mathrm{Re}_l^{0.812})$
Akers and Rosson (1960)	$Nu = 0.0265\,\mathrm{Re}_{eq}^{0.8}\,\mathrm{Pr}_l^{1/3} \quad \mathrm{Re}_{eq} = \dfrac{G_{eq}d}{\mu_l} \quad G_{eq} = G\left[(1-x)+x\left(\rho_l/\rho_g\right)^{0.5}\right]$
Tandon et al. (1995)	$\mathrm{Re}_g > 30000 \qquad Nu = 0.084\,\mathrm{Re}_g^{0.67}\,\mathrm{Pr}_l^{1/3}\left[\dfrac{1}{Ja_l}\right]^{1/6} \qquad Ja_l = \dfrac{Cp_l\Delta T_{sat}}{i_{fg}}$

Table 2. Annular flow heat transfer correlations and models

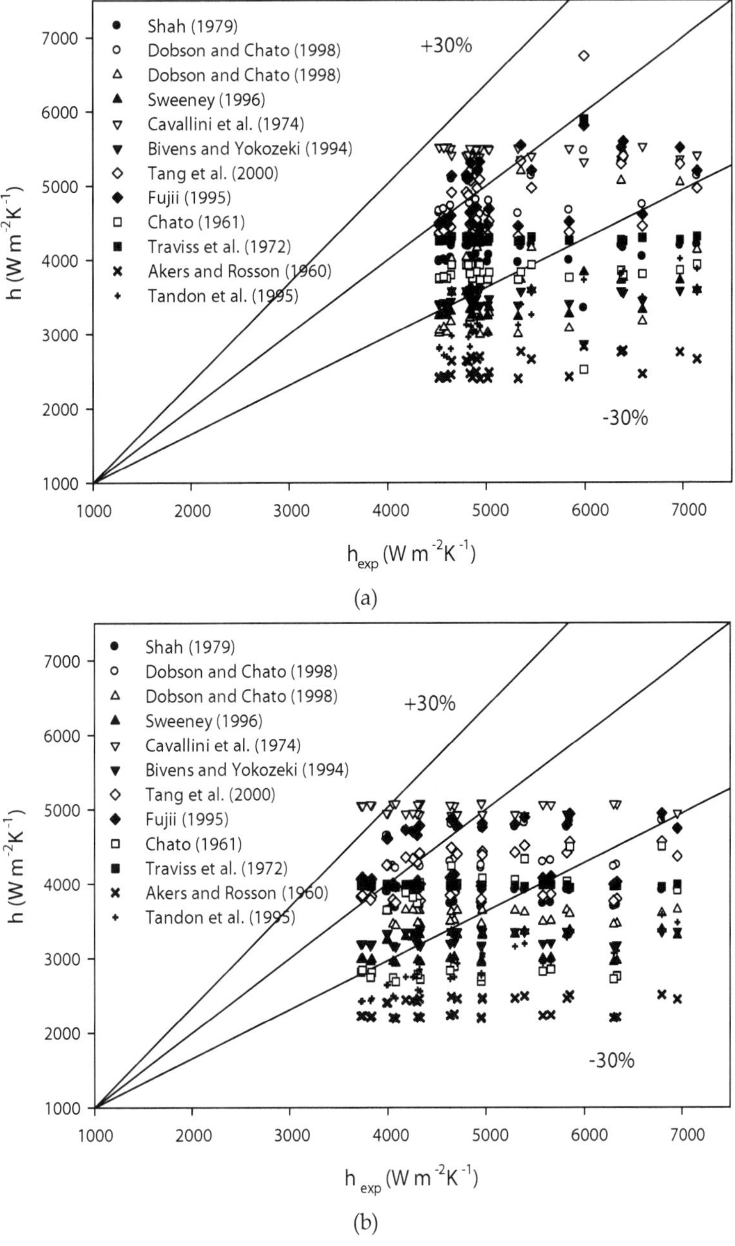

(a)

(b)

Fig. 2. Comparison of experimental condensation heat transfer coefficient vs. various correlations for G=300 kg m^{-2} s^{-1} (a) T$_{sat}$=40 °C and (b) T$_{sat}$=50 °C

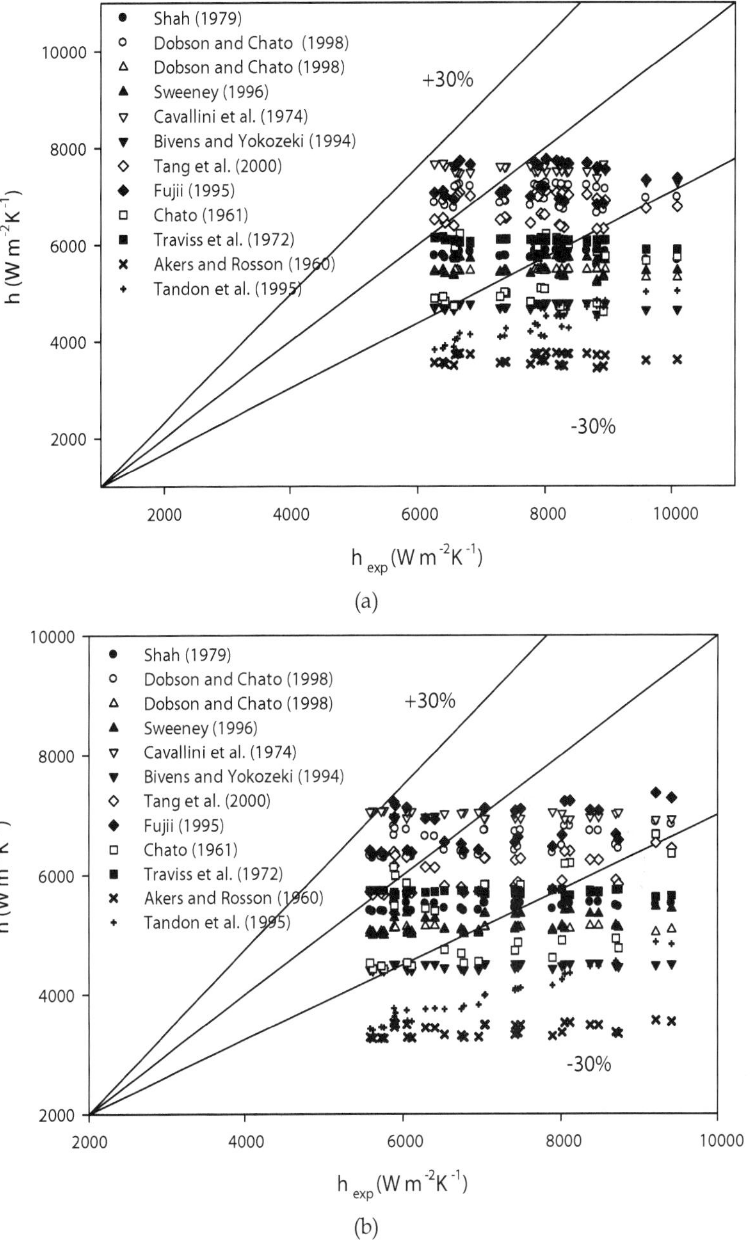

Fig. 3. Comparison of experimental condensation heat transfer coefficient vs. various correlations for G=456 kg m^{-2} s^{-1} (a) T$_{sat}$=40 °C and (b) T$_{sat}$=50 °C

The comparison of experimental heat transfer coefficients with various annular flow correlations are shown in Figs. 2-3 in a 30% deviation line for the condensation temperatures of 40 and 50 °C and mass fluxes of 300 and 456 kg m^{-2}s^{-1} respectively. It can be clearly seen from these figures that the Dobson and Chato (1998) correlation, Cavallini et al. (1974) correlation, Fujii (1995) correlation are in good agreement with the experimental data. In addition to this, the majority of the data calculated by Shah (1979) correlation, Dobson and Chato (1998) correlation, Tang et al. (2000) correlation, Traviss et al. (1972) correlation fall within ±30%, whereas, Tandon et al. (1995) correlation and Akers and Rosson (1960) correlation are found to be incompatible with the experimental data. Nonetheless, the Chato (1961) correlation, the Sweeny (1996) correlation, and the Bivens and Yokozeki (1994) correlation are found to have poor agreement with experimental data.

In the literature, some correlations are developed for gravity-controlled regimes such as Fujii (1995) and Tandon et al. (1995) correlations. In this chapter, the correlations, proposed for gravity-controlled regimes, are not found to be in good agreement with the data as expected for annular flow regime. These kinds of correlations do not exist in this chapter due to their large deviations and validity for wavy flow. Chen et al. (1987) reported that the vapor shear stress acting on the interface of vapor-liquid phases affects the forced convective condensation inside tubes especially at high vapor flow rates. On that account, Fujii (1995) and Tandon et al. (1995)'s shear-controlled correlations were used to predict condensation heat transfer coefficient of R134a. Furthermore, Valladares (2003) obtained similar results on these explanations in this chapter for the condensation of various refrigerants in horizontal tubes by Valladares (2003).

5. Conclusion

In this chapter of the book, the method to determine the average heat transfer coefficient of R134a during condensation in vertical downward flow at high mass flux in a smooth tube is proposed. The comparison between the various annular flow heat transfer correlations and experimental heat transfer coefficients is shown with ±30% deviation line. It can be noted that the Dobson and Chato (1998) correlation, the Cavallini et al. (1974) correlation, the Fujii (1995) correlation are found to have the most predictive results than others in an 8.1 mm i.d. copper tube for the mass fluxes of 300 and 456 kg m^{-2}s^{-1} and condensation temperatures of 40 and 50 °C. As a result of the analysis in this chapter, it is proven that annular flow models are independent of tube orientation provided that annular flow regime exists along the tube length and capable of predicting condensation heat transfer coefficients inside the vertical test tube although most of these correlations were developed for the annular flow condensation in horizontal tubes.

6. Acknowledgements

The authors would like to thank King Mongkut's University of Technology Thonburi (KMUTT), the Thailand Research Fund, the Office of Higher Education Commission and the National Research University Project for the financial support. Especially, the first author wishes to thank KMUTT for providing him with a Post-doctoral fellowship.

7. Nomenclature

A	surface area, m^2
Cp	specific heat, $J\ kg^{-1}K^{-1}$
d	internal tube diameter, m
Fr	Froude number
G	mass flux, $kg\ m^{-2}s^{-1}$
Ga	Galileo number
g	gravitational constant, $m\ s^{-2}$
h	convective heat transfer coefficient, $W\ m^{-2}K^{-1}$
i	entalphy, $J\ kg^{-1}$
i_{fg}	latent heat of condensation, $J\ kg^{-1}$
Ja	Jakob number
k	thermal conductivity, $W\ m^{-1}K^{-1}$
L	length of test tube, m
m	mass flow rate, $kg\ s^{-1}$
Nu	Nusselt number
P	pressure, MPa
Pr	Prandtl number
Re	Reynolds number
T	temperature
Q	heat transfer rate, W
q	mean heat flux, $kW\ m^{-2}$
y	wall coordinate
x	mean vapor quality
y	radial coordinate
z	axial coordinate
X	Lockhart-Martinelli parameter
w	uncertainty

Greek Symbols

ΔT	vapor side temperature difference, $T_{ref,sat}-T_{wi}$, °C
ρ	density, $kg\ m^{-3}$
μ	dynamic viscosity, $kg\ m^{-1}s^{-1}$
τ	shear stress, $N\ m^{-2}$

Subscripts

eq	equivalent
exp	experimental
g	gas/vapor
i	inside
in	inlet
l	liquid
out	outlet
pre	preheater
red	reduced
ref	refrigerant

sat   saturation
so   Soliman
T   total
TS   test section
w   water
w_i inner wall
δ film thickness

8. References

Akers, W.W., Rosson, H.F. (1960). Condensation inside a horizontal tube, *Chemical Engineering Progress Symposium Series*, Vol. 56, No. 30, 145-149.

Akers, W.W., Deans, A., Crosser, O.K. (1959). Condensing heat transfer within horizontal tubes. *Chemical Engineering Progress Symposium Series*, Vol. 55, No. 29, 171-176.

Barnea, D., Shoham, O., Taitel, Y. (1982). Flow pattern transition for vertical downward two phase flow. *Chemical Engineering Science*, Vol. 37, No. 5, 741-744.

Bassi, R., Bansal, P.K. (2003). In-tube condensation of mixture of R134a and ester oil: emprical correlations. *International Journal of Refrigeration*, Vol. 26, No. 4, 402-409.

Bellinghausen, R., Renz, U. (1992). Heat transfer and film thickness during condensation of steam flowing at high velocity in a vertical pipe. *International Journal of Heat and Mass Transfer*, Vol. 35, No. 3, 683-689.

Bivens, D.B., Yokozeki, A. (1994). Heat transfer coefficient and transport properties for alternative refrigerants. *Proc. 1994 Int. Refrigeration Conference*, 299-304, Indiana.

Bergelin, O.P., Kegel, P.K., Carpenter, F.G., Gazley, C. (1946). Co-current gas–liquid flow. II. Flow in vertical tubes. *ASME Heat Transfer and Fluid Mechanic Institute*, 19-28.

Briggs, A., Kelemenis, C., Rose, J.W. (2000). Heat transfer and pressure drop measurements for in-tube condensation of CFC-113 using microfin tubes and wire inserts. *Experimental Heat Transfer*, Vol. 13, No. 3, 163-181.

Briggs, A., Kelemenis, C., Rose, J.W. (1998). Condensation of CFC-113 with downflow in vertical, internally enhanced tubes. *Proceedings of 11th IHTC*, August 23-28, Kyongju.

Carey, V.P. (1992). Liquid-Vapor Phase Change Phenomena, *Hemisphere Publishing*, 1992.

Cavallini, A., Zecchin, R. (1974). A dimensionless correlation for heat transfer in forced convection condensation. *6th International Heat Transfer Conference*, Vol . 3, 309-313, Tokyo.

Cavallini, A., Censi, G., Del Col, D., Doretti, L., Longo, G.A., Rossetto, L., Zilio, C. (2003). Condensation inside and outside smooth and enhanced tubes - a review of recent research. *International Journal of Refrigeration*, Vol. 26, No. 4, 373-392.

Chato, J.C. (1961), Laminar Condensation inside horizontal and inclined tubes. *ASHRAE Journal*, Vol. 4, 52-60.

Chen, S.L., Gerner, F.M., Tien, C.L. (1987). General film condensation correlations. *Experimental Heat Transfer*, Vol. 1, No. 2, 93-107.

Dalkilic, A.S., Laohalertdecha, S., Wongwises, S. (2008). A comparison of the void fraction correlations of R134a during condensation in vertical downward laminar flow in a smooth and microfin tube. *Proceedings of the Micro/Nanoscale Heat Transfer International Conference*, Parts A-B, 1029-1040, Tainan.

Dalkilic, A.S., Laohalertdecha, S., Wongwises, S. (2008). Two-phase friction factor obtained from various void fraction models of R-134a during condensation in vertical downward flow at high mass flux. *Proceedings of the ASME Summer Heat Transfer Conference*, Vol. 2, 193-206, Jacksonville.

Dalkilic, A.S., Laohalertdecha, S., Wongwises, S. (2008). Two-phase friction factor in vertical downward flow in high mass flux region of refrigerant HFC-134a during condensation. *International Communications in Heat and Mass Transfer*, Vol. 35, No. 9, 1147-1152. (a)

Dalkilic, A.S., Laohalertdecha, S., Wongwises, S. (2008). Effect of void fraction models on the two-phase friction factor of R134a during condensation in vertical downward flow in a smooth tube. *International Communications in Heat and Mass Transfer*, Vol. 35, No. 8, 921-927. (b)

Dalkilic, A.S., Laohalertdecha, S., Wongwises, S. (2009). Experimental investigation on the condensation heat transfer and pressure drop characteristics of R134a at high mass flux conditions during annular flow regime inside a vertical smooth tube. *ASME Summer Heat Transfer Conference*, July 19-23, San Francisco.

Dalkilic, A.S., Agra, O. (2009). Experimental apparatus for the determination of condensation heat transfer coefficient for R134a and R600a flowing inside vertical and horizontal tubes. *ASME Summer Heat Transfer Conference*, July 19-23, San Francisco.

Dalkilic, A.S., Laohalertdecha, S., Wongwises, S. (2009). Experimental research on the similarity of annular flow models and correlations for the condensation of R134a at high mass flux inside vertical and horizontal tubes. *ASME International Mechanical Engineering Congress and Exposition*, November 13-19, Lake Buena Vista.

Dalkilic, A.S., Wongwises, S. (2009). A heat transfer model for co-current downward laminar film condensation of R134a in a vertical micro-fin tube during annular flow regime. *The Eleventh UK National Heat Transfer Conference, Queen Mary University of London*, September 6-8, London.

Dalkilic, A.S., Laohalertdecha, S., Wongwises, S. (2009). Effect of void fraction models on the film thickness of R134a during downward condensation in a vertical smooth tube. *International Communications in Heat and Mass Transfer*, Vol. 36, No. 2, 172-179. (a)

Dalkilic, A.S., Laohalertdecha, S., Wongwises, S. (2009). Experimental investigation of heat transfer coefficient of R134a during condensation in vertical downward flow at high mass flux in a smooth tube. *International Communications in Heat and Mass Transfer*, Vol. 36, No. 10, 1036-1043. (b)

Dalkilic, A.S., Yildiz, S., Wongwises, S. (2009). Experimental investigation of convective heat transfer coefficient during downward laminar flow condensation of R134a in a vertical smooth tube. *International Journal of Heat and Mass Transfer*, Vol. 52, No. 1-2, 142-150. (c)

Dalkilic, A.S., Wongwises, S. (2009). Intensive literature review of condensation inside smooth and enhanced tubes. *International Journal of Heat and Mass Transfer*, Vol. 52, No. 15-16, 3409-3426. (d)

Dalkilic, A.S., Laohalertdecha, S., Wongwises, S. (2010). Comparison of condensation frictional pressure drop models and correlations during annular flow of R134a inside a vertical tube. *ASME-ATI-UTI Thermal and Environmental Issues in Energy Systems*, May 16-19, Sorrento.

Dalkilic, A.S., Wongwises, S. (2010). Experimental study on the flow regime identification in the case of co-current condensation of R134a in a vertical smooth tube. *ASME International Heat Transfer Conference*, August 8-13, Washington.

Dalkilic, A.S., Agra, O., Teke, I., Wongwises, S. (2010). Comparison of frictional pressure drop models during annular flow condensation of R600a in a horizontal tube at low mass flux and of R134a in a vertical tube at high mass flux, *International Journal of Heat and Mass Transfer*, Vol. 53, No. 9-10, 2052-2064. (a)

Dalkilic, A.S., Wongwises, S. (2010). An investigation of a model of the flow pattern transition mechanism in relation to the identification of annular flow of R134a in a vertical tube using various void fraction models and flow regime maps. *Experimental Thermal and Fluid Science*, Vol. 34, No. 6, 692-705. (b)

Dalkilic, A.S., Laohalertdecha, S., Wongwises, S. (2010). New experimental approach on the determination of condensation heat transfer coefficient using frictional pressure drop and void fraction models in a vertical tube. *Energy Conversion and Management*, Vol. 51, No. 12, 2535-2547. (c)

Dalkilic, A.S., Wongwises, S. (2011). Experimental study on the modeling of condensation heat transfer coefficients in high mass flux region of refrigerant HFC-134a inside the vertical smooth tube during annular flow regime. *Heat Transfer Engineering*, Vol. 32, No. 1, 1-12. (d)

Dalkilic, A.S., Wongwises, S. (2010). Validation of void fraction models and correlations using a flow pattern transition mechanism model in relation to the identification of annular vertical downflow in-tube condensation of R134a. International Communications in Heat and Mass Transfer Vol. 37, No. 7, 827-834. (e)

Dittus, F.W., Boelter, L.M.K. (1930). Heat transfer in automobile radiators of the tubular type, *University of California Publications on Engineering, Berkeley, CA*, Vol. 2, No. 13, 443-461.

Dobson, M.K., Chato, J.C. (1998). Condensation in smooth horizontal tubes. *Journal of Heat Transfer-Transactions of ASME*, Vol. 120, No. 1, 193-213.

Fujii, T. (1995). Enhancement to condensing heat transfer-new developments, *Journal of Enhanced Heat Transfer*, Vol. 2, No. 1-2, 127-137.

Hewitt, G.F., Robertson, D.N. (1969). Studies of two-phase flow patterns by simultaneous x-ray and flash photography. Rept AERE-M2159, UKAEA, Harwell.

Jung, D., Song, K., Cho, Y., Kim, S. (2003). Flow condensation heat transfer coefficients of pure refrigerants. *International Journal of Refrigeration*, Vol. 26, No. 1, 4-11.

Jung, D., Cho, Y., Park, K. (2004). Flow condensation heat transfer coefficients of R22, R134a, R407C and R41A inside plain and micro-fin tubes. *International Journal of Refrigeration*, Vol. 27, No. 5, 25-32.

Kim, S.J., No, H.C. (2000). Turbulent film condensation of high pressure steam in a vertical tube. *International Journal of Heat and Mass Transfer*, Vol. 43, No. 21, 4031-4042.

Kline, S.J., McClintock, F.A. (1953). Describing uncertainties in single sample experiments. *Journal of the Japan Society of Mechanical Engineers*, Vol. 75, No. 1, 3-8.

Kosky, P.G., Staub, F.W. (1971). Local condensing heat transfer coefficients in the annular flow regime, *AIChE Journal*, Vol. 17, No. 5, 1037-1043.

Liebenberg, L., Bukasa, J.P., Holm, M.F.K., Meyer, J.P., Bergles, A.E. (2002). Towards a unified approach for modelling of refrigerant condensation in smooth tubes. *Proceedings of the International Symposium on Compact Heat Exchangers*, 457-462, Grenoble.

Lockhart, R.W., Martinelli, R.C. (1949). Proposed correlation of data for isothermal two-phase, two-component flow in pipes. *Chemical Engineering Progress*, Vol. 45, No. 1, 39-48.

Ma, X., Briggs, A., Rose, J.W. (2004). Heat transfer and pressure drop characteristics for condensation of R113 in a vertical micro-finned tube with wire insert. *International Communications in Heat and Mass Transfer*, Vol. 31, No. 5, 619-627.

Maheshwari, N.K., Sinha, R.K., Saha, D., Aritomi, M. (2004). M., Investigation on condensation in presence of a noncondensible gas for a wide range of Reynolds number. *Nuclear Engineering and Design*, Vol. 227, No. 2, 219-238.

Moser, K., Webb, R.L., Na, B. (1998). A new equivalent Reynolds number model for condensation in smooth tubes. *International Journal of Heat Transfer*, Vol. 120, No. 2, 410-417.

Nusselt, W. (1916). Die oberflachen-kondensation des wasserdampfer. *Zeitschrift des Vereines Deutscher Ingenieure*, Vol. 60, No. 27, 541-569.

Oh, S., Revankar, A. (2005). Analysis of the complete condensation in a vertical tube passive condenser. *International Communications in Heat and Mass Transfer*, Vol. 32, No. 6, 716-722.

Shah, M.M. (1979). A general correlation for heat transfer during film condensation inside pipes. *International Journal of Heat and Mass Transfer*, Vol. 22, No. 4, 547-556. (3)

Sweeney, K.A. (1996). The heat transfer and pressure drop behavior of a zeotropic refrigerant mixture in a micro-finned tube. *M.S. thesis*, Dept. of Mechanical and Industrial Engineering, University of Illinois at Urbana-Champaign.

Tandon, T.N., Varma, H.K., Gupta, C.P. (1995). Heat transfer during forced convection condensation inside horizontal tube. *International Journal of Refrigeration*, Vol. 18, No. 3, 210-214.

Tang, L., Ohadi, M.M., Johnson, A.T. (2000). Flow condensation in smooth and microfin tubes with HCFC-22, HFC-134a, and HFC-410 refrigerants. part II: Design equations. *Journal of Enhanced Heat Transfer*, Vol. 7, No. 5, 311-325.

Travis, D.P., Rohsenov, W.M., Baron, A.B. (1972). Forced convection inside tubes: a heat transfer equation for condenser design. *ASHRAE Transactions*, Vol. 79, No. 1, 157-165.

Valladeres, O.G. (2003). Review of in-tube condensation heat transfer coefficients for smooth and microfin tubes. *Heat Transfer Engineering*, Vol. 24, No. 4, 6-24.

Wang, H.S., Honda, H. (2003). Condensation of refrigerants in horizontal microfin tubes: comparison of prediction methods for heat transfer. *International Journal of Refrigeration*, Vol. 26, No. 4, 452-460.

Whalley, P.B. (1987). Boiling, condensation, and gas-liquid flow, *Oxford University Press*.

3

Enhanced Boiling Heat Transfer from Micro-Pin-Finned Silicon Chips

Jinjia Wei and Yanfang Xue
Xi'an Jiaotong University
China

1. Introduction

A computer is mainly composed of chips on which a large number of semiconductor switches are fabricated. The requirement for increasing signal speed in the computer has focused the efforts of electronics industry on designing miniaturized electronic circuits and highly integrated circuit densities in chips. The integration technologies, which have advanced to the very large scale integration (VLSI) level, lead to an increased power dissipation rate at the chip, module and system levels. Sophisticated electronic cooling technology is needed to maintain relatively constant component temperature below the junction temperature, approximately 85°C for most mainframe memory and logic chips. Investigations have demonstrated that a single component operating 10°C beyond this temperature can reduce the reliability of some systems by as much as 50% (Nelson, 1978).

Traditionally, convection heat transfer from electronic hardware to the surroundings has been achieved through the natural, forced, or mixed convection of air; however, even with advances in air-cooling techniques, the improvements will not suffice to sustain the expected higher heat fluxes. As an effective and increasingly-popular alternative to air cooling, directly immersing the component in inert, dielectric liquid can remove a large amount of heat dissipation, of which pool and forced boiling possess the attractive attribute of large heat transfer coefficient due to phase change compared with single-phase.

An ideal boiling performance should provide adequate heat removal within acceptable chip temperatures. Direct liquid cooling, involving boiling heat transfer, by use of dielectric liquids has been considered as one of the promising cooling schemes. Primary issues related to liquid cooling of microelectronics components are mitigation of the incipience temperature overshoot, enhancement of established nucleate boiling and elevation of critical heat flux (CHF). Treated surface has been found to have great potential in enhancement of boiling heat transfer from electronic, significantly reducing the chip surface temperature and increasing CHF. Treated surfaces are used for nucleate boiling enhancement by applying some micro-structures on the chip surface to make the surface capable of trapping vapour and keeping the nucleation sites active or increasing effective heat transfer area. Since the 1970s, a number of active studies have dealt with the enhancement of boiling heat transfer from electronic components by use of surface microstructures that were fabricated directly on a silicon chip or on a simulated chip. These include a sand-blasted and KOH treated surface (Oktay 1982), a "dendritic heat sink" (brush-like structure) (Oktay and

Schemekenbecher 1972), laser drilled cavities (3-15μm in mouth dia.) (Hwang and Moran 1981), re-entrant cavities (0.23-0.49 mm in mouth dia.) (Phadke et al. 1992).

Messina and Parks (1981) used flat plate copper surfaces sanded with 240 and 600 grit sandpaper to boil R-113 and found the sandpaper finished surfaces were very efficient in improving boiling heat transfer and elevating CHF as compared to a smooth surface, with 240 grit more efficient than 600 grit. Anderson and mudawar (1989) roughened a 12.7-mm square copper surface by longitudinal sanding with a 600 grit silicon wet/dry sand paper to examine the effect of roughness on boiling heat transfer of FC-72. The roughness was 0.6-1.0 μm and the roughened surface produced an earlier boiling incipience than smooth surface and shifted the boiling curve toward a reduced wall superheat. However, the CHF value was not affected by the roughness as compared with the results obtained by Messina and Parks (1981). Chowdhury and Winterton (1985) found nucleate boiling heat transfer improved steadily as the surface roughness level was increased. However, when they anodized a roughened surface covered with cavities of around 1 μm size, which had hardly any effect on the roughness, they observed that the nucleate boiling curves were virtually independent of roughness. They asserted that it was not roughness in itself but the number of active nucleation sites that influenced nucleate boiling heat transfer.

Oktay and Schmeckenbecher (1972) developed a brush-like structure called "dendritic heat sink" mounted on a silicon chip surface, and the thickness of the dendrite was 1 mm. The incipience boiling temperature in saturated FC-86 could be reduced to 60°C due to the high density of re-entrant and possibly doubly re-entrant cavities provided by the dendritic heat sinks, and an increase in CHF compared with a smooth surface was attributed, by the authors, to the deferred creation of Taylor instability on the dendritic surface.

Chu and Moran (1977) developed a laser-treated surface on a silicon chip, which consisted of drilling an array of cavities ranging in average mouth diameter from 3 to 15 μm staggered 0.25-mm centers. Boiling data in FC-86 revealed that the wall superheat at any particular heat flux decreased, and the critical heat flux was increased by 50%.

Phadke et al. (1992) used a re-entrant cavity surface enhancement for immersion cooling of silicon chip. The pool boiling heat transfer characteristics of the cavity enhanced surfaces were superior to those of a smooth surface, resulting in a substantial decrease in both the temperature overshoot and the incipient boiling heat flux.

Kubo et al. (1999) experimentally studied boiling heat transfer of FC-72 from micro-reentrant cavity surfaces of silicon chips. The effects of cavity mouth size (about 1.6μm and 3.1μm) and the cavity number density ($811/cm^2$ and $9600/cm^2$) were also investigated. The heat transfer performance of the treated surface was considerably higher than that of the smooth surface. The highest performance was obtained by a treated surface with larger cavity mouth diameter and cavity number density.

Nakayama et al. (1982) developed a tunnel structure, in which parallel rectangular cross-sectional grooves with the dimensions of 0.25×0.4 mm^2 (width×depth) were firstly gouged with a pitch of 0.55 mm on a copper surface (20×30 mm^2), then covered by a thin copper plate having rows of 50-to-150 μm diameter pores. R-11 was boiled and the wall superheat was reduced as compared to a smooth surface. They attributed the boiling enhancement to the liquid suction and evaporation inside the grooves. Later, Nakayama et al. (1989) used a 5-mm high porous copper stud with micro-channels to enhance boiling heat transfer of dielectric fluid FC-72. The porous stud could reduce the threshold superheat for the boiling incipience and increasing CHF. The boiling heat transfer levelled off with further increasing stud height.

Anderson and Mudawar (1989) also attached mechanically manufactured cavities, micro-fins and micro-pin-fins to vertical 12.7 mm square copper chips immersed in a stagnant pool of FC-72. They found that large artificial cavities with the mouth diameter of 0.3 mm were incapable of maintaining a stable vapour embryo and had only a small effect on boiling heat transfer compared with a smooth surface, while micro-finned and micro-pin-finned surfaces significantly enhanced the nucleate boiling mainly due to a heat transfer area increase. The micro-pin-finned surface with the fin dimensions of $0.305 \times 0.305 \times 0.508$ mm³ (width×thickness×height) provided CHF values in excess of 50 W/cm² and 70 W/cm² for the liquid subcoolings of 0 and 35K, respectively.

In 1990's, You and his co-researchers made a noticeable progress in nucleate boiling enhancement by use of a series of micro-porous surfaces. You et al. (1992) applied a 0.3-3.0 μm alumina particle treatment on a simulated electronic chip surface with spraying method and tested in FC-72. Compared with a smooth reference surface, a reduction of 50% in incipient and nucleate boiling superheats and an increase of 32% in the CHF were realized. O'Connor and You (1995) further used the spraying application to apply the alumina particles (0.3-5.0 μm) on a simulated electronic chip surface. The enhancement of nucleate boiling heat transfer showed excellent agreement with those observed by You et al. (1992) with an exception of a much higher CHF increase (47% increase) due to the increased heater thickness (1 mm aluminium nitride) which provided CHF data free from thermal conductance/capacitance effects.

In their subsequent studies, O'Connor and You (1995), O'Connor et al. (1996) painted 3-10μm silver flakes or 8-12 μm diamond particles on the copper surface. Chang and You (1996, 1997) used 1-50μm copper particles and 1-20 μm aluminium particles to form porous coatings. These micro-porous coating surfaces showed almost identical high boiling enhancement with a reduced incipient superheat, increased nucleate boiling heat transfer coefficient and CHF as compared to an unenhanced surface. These performance enhancements were due to the creation of micro-porous structures on the heater surfaces which significantly increased the number of active nucleation sites.

Bergles and Chyu (1982) reported a pool boiling from a commercial porous metallic matrix surface. Working fluids were R-113 and water. The excellent steady boiling characteristics of this type of surface were confirmed, however, high wall superheat were required in most cases to initiate the boiling. From the previously mentioned investigations, it is apparent that surface microstructure of the correct size plays an important role in the enhancement of boiling heat transfer. Most treated surfaces can reduce the boiling incipience temperature, improve the nucleate boiling heat transfer and increase CHF. However, the enhancement often deteriorated greatly in the high heat flux region, especially near CHF resulting in a too high wall temperature at the CHF point as compared to the maximum allowable temperature for the normal operation of LSI chips, making the enhancement not so sound in practical high-powered electronics cooling application.

Mudawar's group (Ujereh et al. 2007) studied the nucleate pool boiling enhancement by use of carbon nanotube (CNT) arrays, and found CNTs were quite effective in reducing incipience superheat and enhancing the boiling heat transfer coefficient.

Li et al. (2008) reported a well-ordered 3D nanostructured macroporous surfaces which was fabricated by elecrodeposition method for efficiently boiling heat transfer. Since the structure is built based on the dynamic bubbles, it is perfect for the bubble generation applications such as nucleate boiling. The result indicated that at heat flux of $1W/cm^2$, the heat transfer coefficient is enhanced over 17 times compared to a plain reference surface.

El-Genk and Ali (2010) experimentally studied the enhancement of saturation boiling of degassed PF-5060 dielectric liquid on microporous copper dendrite surfaces. These surface layers were deposited by electrochemical technique. The result showed that the thickest layer (145.6μm) of Cu nanodendrite surface is very promising for cooling electronic components, while keeping the junction temperature relatively low and no temperature excursion.

However, it is still a challenge for these treated surfaces to increase CHF by a large margin for the application of cooling with high-heat-flux chip.

The present work is to develop a surface treatment that can provide a nearly invariant high heat transfer rate throughout the whole nucleate boiling region and elevate CHF greatly within an acceptable chip temperature. For the previous micro structured surfaces, the reason for the severe deterioration of heat transfer performance at high heat fluxes is that a large amount of vapors accumulate in the structures which prevent the bulk of liquid from contacting the superheated wall for vaporization. The enhancement, due to the increased thermal resistance of the large amount of vapours trapped in the microstructures, tapers off noticeably as the heat flux approaches the CHF (See Fig. 1).

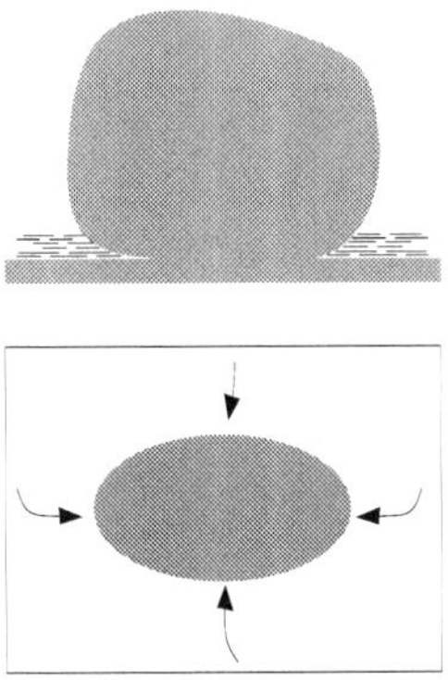

Fig. 1. Schematic of heat transfer phenomenon of smooth or previous porous structures

Therefore, the high-efficiency microstructures should provide a high driving force and a low-resistance path for the easy access of bulk liquid to the heater surface despite of large bubbles covering on the surface at high heat fluxes. Subsequently, Wei et al. (2003) developed a new model for the heat transfer and fluid flow in the vapour mushroom region of saturated nucleate pool boiling. The vapour mushroom region is characterized by the formation of a liquid layer interspersed with numerous, continuous columnar vapour stems underneath a growing mushroom-shaped bubble shown in Fig. 2. And, the liquid layer between the vapour mushroom and the heater surface has been termed as the macrolayer, whereas the thin liquid film formed underneath vapour stems is known as the microlayer. Thus, three highly efficient heat transfer mechanisms were proposed in Wei et al. (2003)'s model, which regards the conduction and evaporation in the microlayer region, the conduction and evaporation in the macrolayer region and Marangoni convection in the macrolayer region as the heat transfer mechanism. Furthermore, Wei et al. (2003)'s numerical results showed that the heat transfer can be efficiently transferred to the vapour-liquid interface by the Marangoni convection. At the same time, the evaporation at the triple–point (liquid-vapour-solid contact point) plays a very important role in the heat transfer with a weighting fraction of about 60% over the heat flux ranges investigated, and the relative evaporations at the bubble-liquid interface and the stem-liquid interface are about 30% and 10% respectively. However, the vapour stem will eventually collapse and result in shut off of the Marangoni convection and microlayer evaporation in the vapour mushroom region of saturated pool nucleate boiling heat transfer. On the above situation, further investigations were also carried out by Wei et al. (2003) for the cases in which Marangoni convection or/and microlayer evaporation were not considered. The result indicated that the highest wall temperature can be obtained in the cases of no Marangoni convection and microlayer evaporation. So, this indicates that both the Marangoni convection and microlayer evaporation play important roles in the mushroom region of saturated pool nucleate boiling heat transfer.

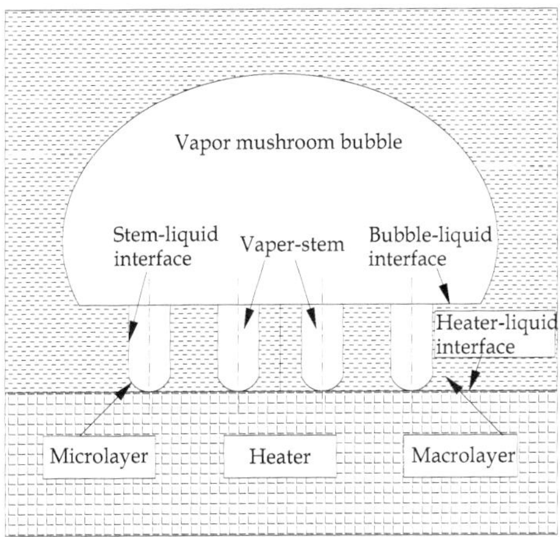

Fig. 2. Schematic of vapour mushroom structure near heated surface

Therefore, to overcome the above problems occur, we developed a micro-pin-finned surface with the fin thickness of 10-50 μm and the fin height of 60-120 μm. The fin gap was twice the fin thickness. The generated bubbles staying on the top of the micro-pin-fins can provide a capillary force to drive plenty of fresh liquid into contact with the superheated wall for vaporization through the regular interconnected structures formed by micro-pin-fins, as well as improve the microlayer evaporation and the Marangoni convection heat transfer by the motion of liquid around the micro-pin-fins (See Fig. 3).

So, the boiling heat transfer performance of FC-72 for the micro-pin-finned surfaces was firstly carried out in pool boiling test system. The maximum cooling capacity of this type of cooling module is determined by either the occurrence of CHF or complete vapor-space condensation (Kitching et al.1995). Then, some researchers such as Mudawar and Maddox (1989), Kutateladze and Burakov (1989), Samant and Simon (1989), and Rainey et al. (2001), have found that both of fluid velocity and subcooling had significant effects on the nucleate boiling curve and the critical heat flux of their thin film heater. Therefore, the combined effects of fluid velocity and subcooling on the flow boiling heat transfer of FC-72 over micro-pin-fined surfaces were investigated for further enhancement of boiling heat transfer to cool high-heat-flux devices.

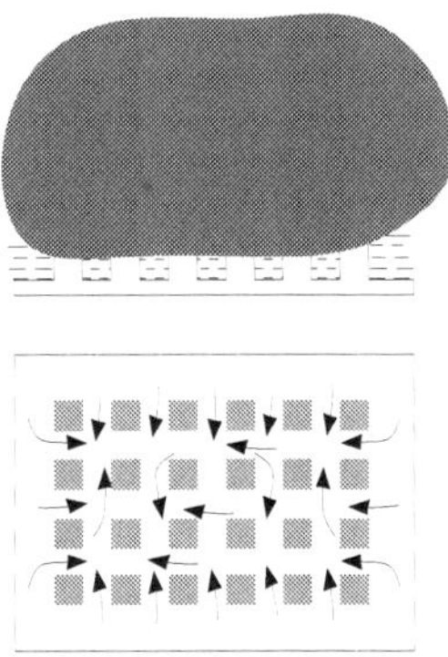

Fig. 3. Schematic of boiling heat transfer phenomena of micro-pin-fins at high heat flux near CHF

2. Experimental apparatus and procedure

2.1 Test facility of pool boiling

The first test facility for pool boiling heat transfer is shown schematically in Fig. 4. The test liquid FC-72 was contained within a rectangular stainless vessel with an internal length of 120 mm, width of 80mm, and height of 135 mm (1.3L), which was submerged in a thermostatic water bath (42L) with a temperature adjusted range of 5-80°C. The bulk temperature of FC-72 within the test chamber was maintained at a prescribed temperature by controlling the water temperature inside the water bath. Additional liquid temperature control was provided by an internal condenser, which was attached at the ceiling of the test vessel and through which water was circulated from cooling unit. The pressure inside the test vessel was measured by pressure gauge and a nearly atmospheric pressure was maintained by attaching a rubber bag to the test vessel. For the visual observation of boiling phenomena, the test chamber was fitted with glass windows in both the front and back. The test heater assembly consisting of a test chip bonded on a pyrex glass plate and a vacuum chuck made of brass was immersed horizontally in the test chamber with the test chip facing upward. The local temperatures of the test liquid at the chip level, and 40mm and 80mm above the chip level were measured by T-type thermocouples the hot junctions of which were located on a vertical line 25mm apart from the edge of test chip.

Details of the test section are shown in Fig. 5. The test chip was a P doped N-type silicon chip with the dimensions of 10×10×0.5mm³. The specific resistance of the test chip was 1-2Ωcm, and the thermal conductivity was about 156W/m.K at room temperature.

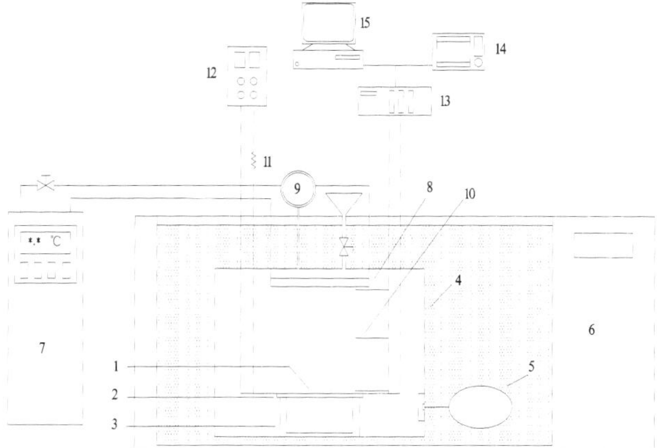

1, Test chip; 2,Glass plate; 3,Vacuum chuck; 4, Test vessel; 5,Rubber bag; 6,Water bath; 7,Cooling unit; 8, Condenser; 9,Pressure gauge; 10,Thermocouples; 11, Standard resistance; 12, DC power supply; 13, Digital multimeter; 14,Imange acquisition System; 15, Computer

Fig. 4. Schematic diagram of experimental apparatus for pool boiling

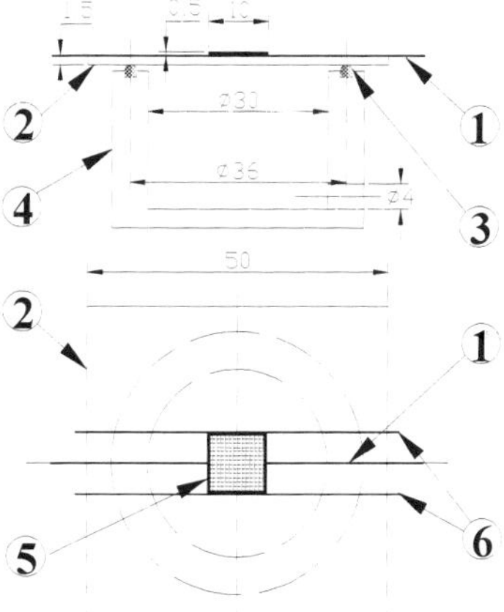

1, Thermocouple; 2, Pyrex glass plate; 3, O-ring; 4, Vacuum chuck; 5, Silicon chip;
6, Copper lead wire

Fig. 5. Test section for pool boiling

The chip was Joule heated by a direct current. Two 0.25-mm diameter copper wires for power supply and voltage drop measurement were soldered with a low temperature solder to the side surfaces at opposite ends. In order to secure the Ohmic contact between the test chip and the copper wire, a special solder with the melting point of 300°C was applied to the chip with ultrasonic bonding method before soldering the copper wires.

The local wall temperatures at the center and about 1.5mm from the edge on the centreline of the chip were measured by two 0.12-mm diameter T-type thermocouples bonded under the test chip. The test chip was bonded on the top of a 50×50×1.2 mm³ pyrex glass plate using epoxy. Then the glass plate on which the test chip was bonded was pressed firmly to the O-ring on a brass vacuum chuck when the inside of the chuck was evacuated by using a vacuum pump. This facilitated an easy exchange of the test chip and minimization of conduction heat loss due to conduction and convection from the rear surface of the chip. The side surfaces of the chip were covered with an adhesive to minimize heat loss. Therefore, only the upper surface of the chip was effective for heat transfer.

2.2 Test facility of forced flow boiling

The second test facility for the forced flow boiling heat transfer is shown schematically in Fig. 6. It is a closed-loop circuit consisting of a tank, a scroll pump, a test section, two heat exchangers and a turbine flowmeter. The tank served as a fluid reservoir and pressure regulator during testing.

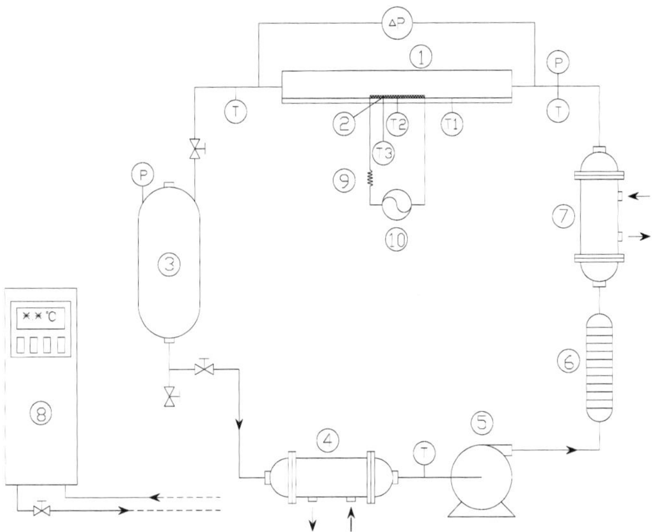

1, Test Section; 2, Test chip; 3, Tank; 4, Condenser; 5, Pump; 6, Flowmeter; 7, Pre-heater; 8, Cooling unit; 9, Standard resistance; 10, Direct current

Fig. 6. Forced flow boiling test loop

The condenser prior to the pump was used to cool the fluid and prevent cavitation in the pump. The pre-heater prior to the test section was used to control the test section inlet temperature. The pump was combined with a converter to control the mass flow rate. To ensure proper inlet pressure control, a pressure transducer was installed at the inlet of the test section. The pressure drop across the test section was also measured by a pressure difference transducer. The flowmeter and the sensors for pressure and pressure difference have the function of outputting 4~20mA current signals and were measured directly by a data acquisition system.

The test silicon chip is bonded to a substrate made of polycarbonate using epoxy adhesive and fixed in the horizontal, upward facing orientation on the bottom surface of a 5mm high and 30mm wide horizontal channel as shown in Fig. 7. The chip is located 300mm from the inlet of the test section so that the fluid flow on it is estimated to be fully developed turbulent flow for the present fluid velocity range.

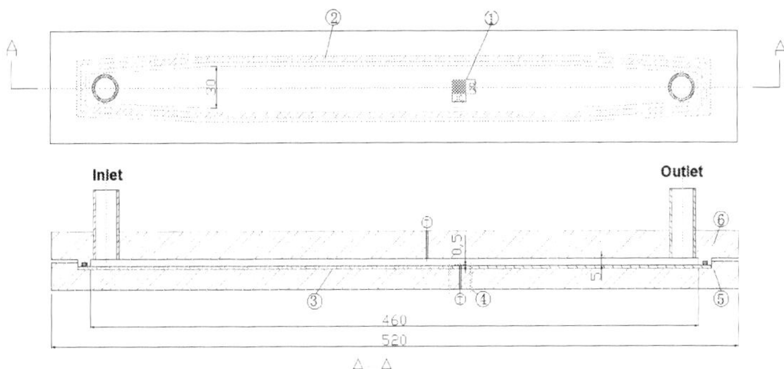

1, Test chip; 2, O-ring; 3, Polycarbonate plate; 4, Lead wires; 5, Lower cover; 6, Upper cover

Fig. 7. Test section for forced flow boiling

2.3 Experimental conditions and test procedure

Experiments were performed at three fluid velocities of 0.5, 1 and 2m/s for the second test system and the liquid subcoolings of 3, 15, 25, 35 and/or 45K for both of test systems. Six kinds of silicon chips, one with a smooth surface and five with square micro-pin-fins having fin dimensions of 30×60, 30×120, 30×200, 50×60 and 50×120 μm² (thickness×height) are tested. The fin pitch p was twice the fin thickness. The micro-pin-finned chip was fabricated by the dry etching technique. These chips were named chips S, PF30-60, PF30-120, PF30-200, PF50-60 and PF50-120, respectively. The scanning electron microscope (SEM) images of the five micro-pin-finned chips are shown in Fig. 8a-e, respectively. A smooth chip was also tested for comparison.

In the above two test systems, the test chips were Joule heated by using a programmable d.c. power supply. The power supply was connected in series to a standard resistance (1Ω) and the test chip. The standard resistance was used to measure the electric current in the circuit. Power input to the test chip was increased in small steps up to the high heat flux region of nucleate boiling. The heat flux q was obtained from the voltage drop of the test chip and the electric current. In order to prevent real heater burnout, an overheating protection system was incorporated in the power circuit. If the wall temperature sharply increases by more than 20K in a short time, the data acquisition algorithm assumed the occurrence of CHF condition and the power supply was immediately shut down. The CHF value was computed as the steady state heat flux value just prior to the shut down of the power supply. The uncertainties in the chip and bulk liquid temperature measurements by the thermocouple and the resistance thermometry for the pool boiling is estimated to be less than 0.1K and that to be less than 0.3K for the forced flow boiling. The uncertainty in the calculated value of q for the pool boiling is mostly due to the heat loss and estimated to be less than 15 and 5.0 percents for natural convection and nucleate boiling regions, respectively. For the forced flow boiling, the uncertainty in the calculated value of q is estimated to be less than 16 and 6.0 percents for the forced convection and nucleate boiling regions respectively. It is relevant to note here that q includes the heat transferred to the bulk liquid by conduction through the copper lead wires and the polycarbonate substrate.

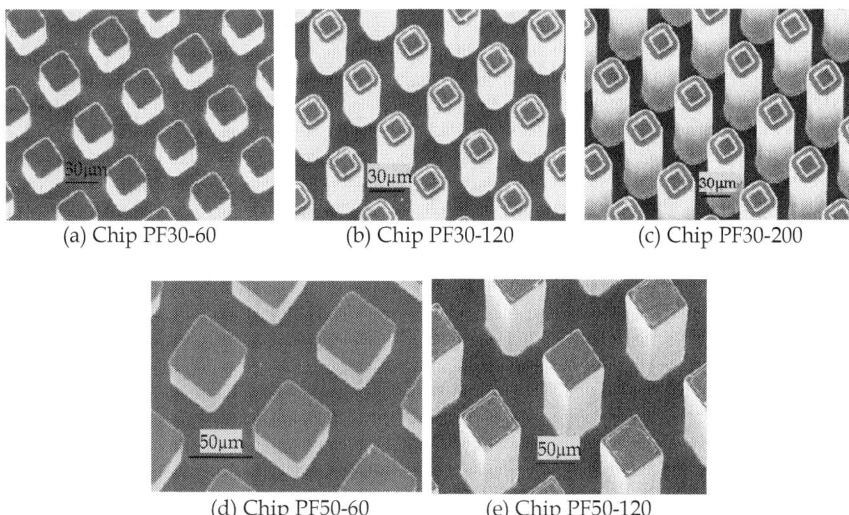

(a) Chip PF30-60   (b) Chip PF30-120   (c) Chip PF30-200

(d) Chip PF50-60   (e) Chip PF50-120

Fig. 8. SEM Images of micro-pin-fins

The experiment was repeated two or three times for all chips. The time interval between the subsequent runs was greater than 0.5 h. The boiling curves showed a good repeatability for all cases except for the boiling incipience point. Thus, only the results for the third runs are presented in the next section.

3. Results and discussion

3.1 Pool boiling performance of micro-pin-finned surfaces

Figure 9 shows the boiling curves of micro-pin-finned surfaces of PF30-60 (fin thickness of 30μm and height of 60μm) and PF30-200 (fin thickness of 30μm and height of 200μm) at $\Delta Tsub$=3K. The boiling curve of the smooth surface chip S is also shown for comparison. All chips follow almost the same q versus ΔT_{sat} relation in the non-boiling region despite that chip PF30-200 has a large fin height of 200 μm. This indicates that the fin height up to 200 μm is not effective in the enhancement of natural convection heat transfer. However, at the nucleate boiling region, the micro-pin-finned surfaces show considerable heat transfer enhancement compared to chip S. Furthermore, the boiling curves of chips PF30-60 and PF30-200 are very steep and the wall superheats show a very small change with increasing heat flux q up to the critical heat flux (CHF) point. It is supposed that the increased surface activated in the nucleate region to be much larger for chips PF30-60 and PF30-200, hence the boiling heat transfer is enhanced more greatly. The q_{CHF} increases in the order of chips S, PF30-60 and PF30-200. For the micro-pin-finned chips in the present study, the wall temperature at the CHF point is lower than 85°C.

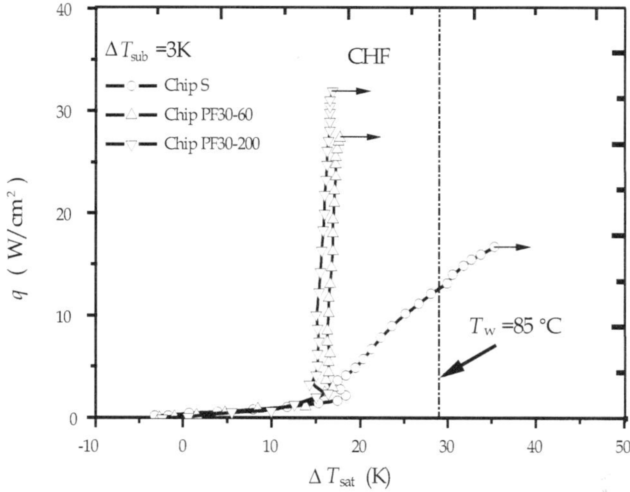

Fig. 9. Boiling curves of chips PF30-60, PF30-200 and S, ΔT_{sub}=3K

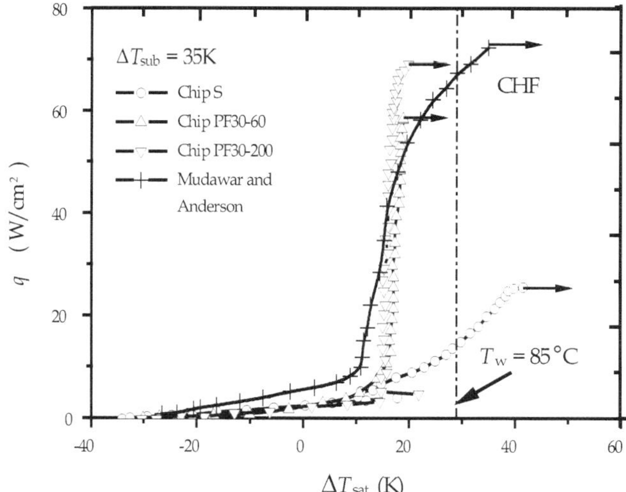

Fig. 10. Boiling curves of chips PF30-60, PF30-200 and S, ΔT_{sub}=35K

Figures 10 and 11, respectively, show the boiling curves of micro-pin-finned chip PF30-60 and PF30-200 at ΔT_{sub}=35K and ΔT_{sub}=45K. In Fig. 10, the boiling curves of the micro-pin-finned surface reported by Mudawar and Anderson (1989) and a smooth surface are also shown for comparison. The test surface of Mudawar and Anderson (1989) had square micro-pin-fins with the dimensions of 0.305×0.508 mm² (thickness×height). These fin dimensions are one order of magnitude greater than those of chip PF30-60 in the present experiments. The ΔT_{sat} values in the low-heat-flux region are smaller than that for chips PF30-60 and PF30-200, whereas the wall superheats at the CHF point $\Delta T_{sat,CHF}$ are greater than that for the latter. While the measured heat flux in the free convection region for chip S and the micro-pin-finned chips in this study is almost the same, the heat flux for the micro-pin-finned chip reported by Mudawar and Anderson (1989) is about 40% higher than that for chip S, indicating that the increased area of their micro-pin-finned surface over a smooth surface is effective in natural convection heat transfer.

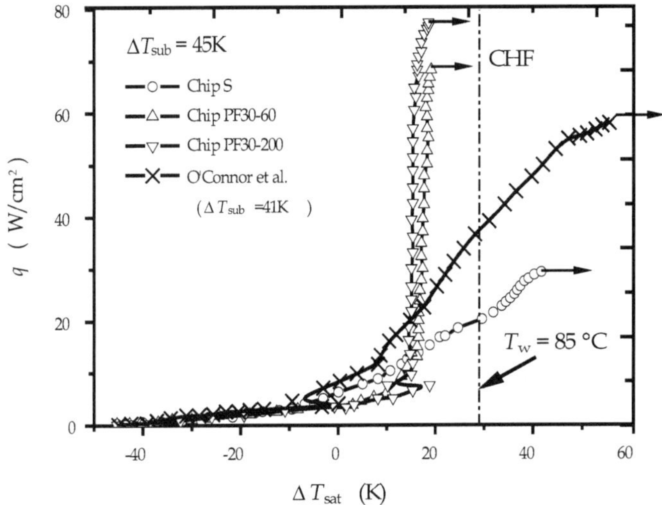

Fig. 11. Boiling curves of chips PF30-60, PF30-200 and S, ΔT_{sub}=45K

Although the q_{CHF} reported by Mudawar and Anderson (1989) for the micro-pin-finned surface is higher than that of chips PF30-60 and PF30-200, the boiling curve near the CHF point shows a much smaller slop than the micro-pin-finned chips in the present study.

In Fig. 11 boiling occurs at a much smaller value of ΔT_{sat} and the temperature overshoots at the boiling incipience are small, compared to the case of Fig. 10. The boiling curves of the microporous surface developed by O'Connor et al. (1996) and a smooth surface are also shown for comparison. The test surface of O'Connor et al. (1996) had a porous layer consisting of 8-12 μm diamond particles produced by painting technique. The porous surface shows a severe deterioration of boiling heat transfer performance at high heat flux region and the value of q_{CHF} is smaller than that of the micro-pin-finned chips. However, the micro-pin-finned surfaces in the present study show a sharp increase in the heat flux with increasing wall superheat from the boiling incipience to the critical heat flux. The increase of CHF for micro-pin-finned surfaces could reach more than twice that of a smooth chip. Comparison of the experimental results reveals that the wall temperature at the CHF point $T_{w,CHF}$ is higher than 85°C for chip S and the previous results reported by Mudawar and Anderson (1989) and O'Connor et al. (1996). On the other hand, the micro-pin-finned chips PF30-60 and PF30-200 show $T_{w,CHF}$ smaller than 85°C.

As stated previously, the upper limit of temperature for a reliable operation of electronic chips is given by 85°C. Thus the maximum allowable heat flux q_{max} is given by the CHF if $T_{w,CHF}$<85°C and by q at T_w=85°C if $T_{w,CHF}$>85°C. Figure 12 shows the variation of maximum heat flux of chips PF30-60 and PF30-200 with liquid subcooling ΔT_{sub}. The experimental data by Mudawar and Anderson (1989) and O'Connor et al (1996) are also shown for comparison. We can see that the maximum heat flux of the micro-pin-finned surfaces in the present study is much higher than that of the porous and other large scale micro-pin-finned and smooth surfaces, and increases greatly with the liquid subcooling. The micro-pin-finned surface of Mudawar and Anderson (1989) shows almost the same q_{max} with chip PF30-200. The porous surface of O'Connor et al (1996) shows the q_{max} value in between those for micro-pin-finned chips and smooth chip. The difference of q_{max} between the porous surface and micro-pin-finned surfaces increases greatly with subcooling, indicating that the subcooling effect is larger for the micro-pin-finned surfaces.

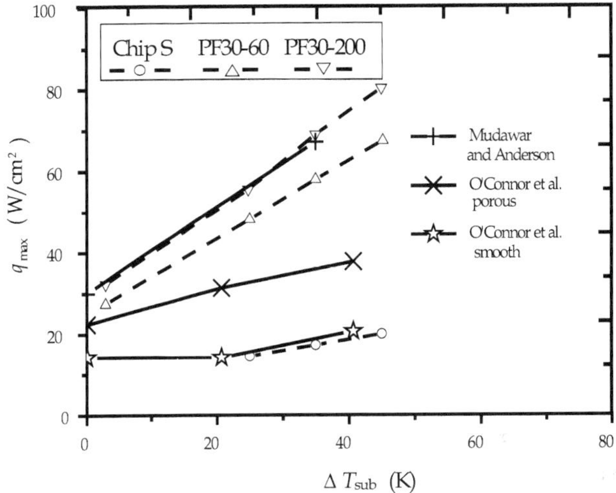

Fig. 12. Variation of q_{max} with ΔT_{sub}

3.2 Flow boiling performance of micro-pin-finned surfaces

Figure 13 shows the comparison of boiling curves for all surfaces with ΔT_{sub}=15K. For the fluid velocity V<2m/s, the wall superheat in the nucleate boiling region decreases in the order of chip S, PF50-60, PF30-60, PF30-120 for the same fluid velocity, again showing that the boiling heat transfer can be enhanced by increasing total surface area. However, the boiling curve of chip PF50-120 shows the smallest wall superheat despite of the surface area ratio of 3.4. Although the condition for the etching process is set to the same, the etching size is different for different micro-pin-finned chips, which may affect the etching results. From the scanning-electron-micrograph (SEM) image of micro-pin-fin (Fig. 8), we found that the roughness on the fin flank near the fin root usually increases with increasing etching depth. Chip PF50-120 with the largest fin thickness and height is observed to have a large roughness scale on the fin flank. It is considered that the roughness causes the earlier boiling incipience and thus smaller wall superheat in the nucleate boiling region. For the fluid velocity of 2m/s, the nucleate boiling curves of all micro-pin-finned chips are more or less affected by forced convection heat transfer since the slopes of boiling curves become smaller than those at lower velocities, but the nucleate boiling curves of 50-μm thick micro-pin-fins show a much larger slope than those of 30-μm thick micro-pin-fins. According to enhanced boiling heat transfer mechanism for micro-pin-fins developed by Wei et al. (2003), the microlayer evaporation and the Marangoni convection caused by thermal capillary force in the micro-pin-fin formed interconnect tunnel play an important role for boiling heat transfer. It is considered that bulk fluid flow may affect the Marangoni convection greatly and the smaller fin gap of 30 μm may generate a larger flow resistance for Marangoni convection around the fin sidewalls, resulting in a lower heat transfer performance. The larger slope shifts the boiling curve of chips PF50-60 and PF50-120 to a smaller wall superheat than that of chips PF30-60 and PF30-120, respectively, at high heat flux for V=2m/s. For comparison, Lie et al. (2007)'s saturated boiling curves for two pin-finned surfaces with the larger fin thicknesses of 100 and 200 μm are also shown in Fig. 13.

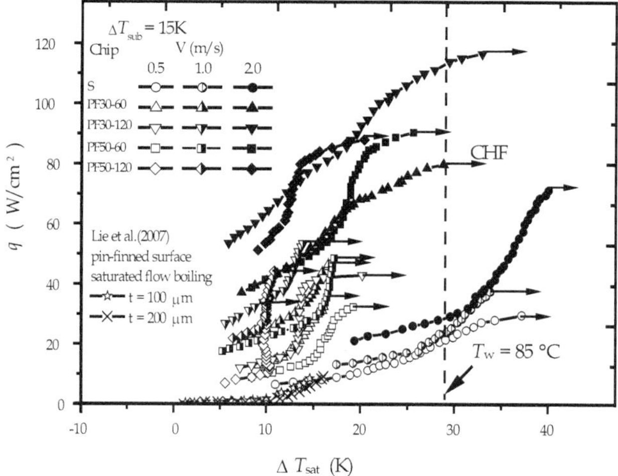

Fig. 13. Comparison of flow boiling curves for all chips, ΔT_{sub}=15K

The boiling curves of Lie et al. (2007) show a much smaller slope and a larger wall superheat compared with that of the present micro-pin-finned surfaces. Generally, the wall superheat deceases in the order of 200, 100, 50 and 30-μm micro-pin-fins for the fin height of about 70μm. We have found that the boiling heat transfer performance for the micro-pin-finned surface with the fin thickness of 10-50 μm is much better than the pin-finned surface with the larger fin thickness of about 300 μm used by Anderson and Mudawar (1989), and the fin thickness of 30-50μm is a suitable range of effectively enhancing boiling heat transfer. The optimum fin gap size is considered to be determined by the balance of the capillary force caused by evaporation of bubbles for driving the micro-flow in the gap of micro-pin-fins and the flow resistance. Large fin gap usually has small flow resistance but generates small microconvection heat transfer proportion, and *vice versa*. The present experimental study again shows the larger fin thickness above 100 μm is not so remarkably effective compared with the fin thickness of 30-50 μm.

Figures 14 and 15 show the comparison of boiling curves for all surfaces with ΔT_{sub}=25 and 35K, respectively. The trend of experimental data is basically the same as the case of ΔT_{sub}=15K shown in Fig. 13 except that the values of the CHF increase considerably in the order of 25 and 35 K.

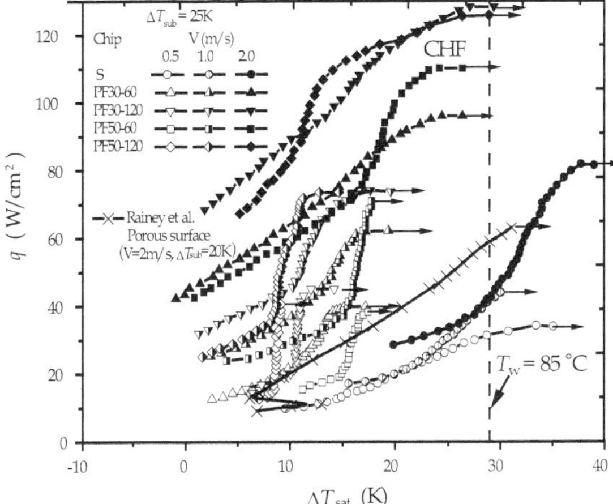

Fig. 14. Comparison of flow boiling curves for all chips, ΔT_{sub}=25K

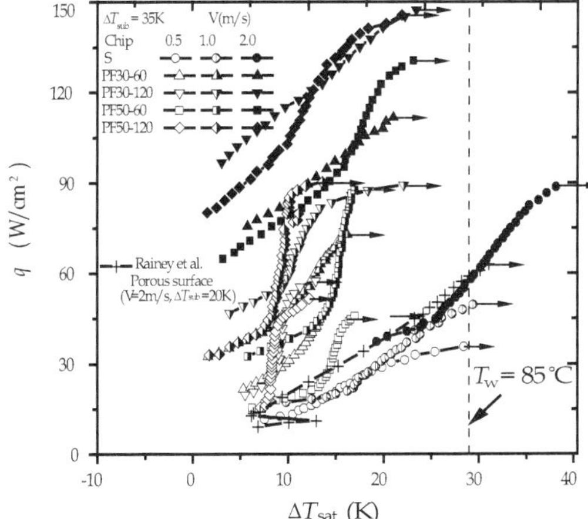

Fig. 15. Comparison of flow boiling curves for all chips, ΔT_{sub}=35K

The highest value of CHF (=148 W/cm²), about 1.5 times as large as that for the smooth surface, was obtained by chips PF30-120 (fin thickness of 30 μm and height of 120 μm) and PF50-120 (fin thickness of 50 μm and height of 120 μm) at liquid subcooling ΔT_{sub} = 35 K and flow velocity V = 2 m/s. For comparison, Rainey et al.(2001) 's boiling curve with the liquid subcooling of 20 K for the microporous surface at V = 2 m/s is also shown in Figs. 14

and 15. Although having an earlier boiling incipience, the microporous surface shows a larger wall superheat than the micro-pin-finned surfaces in the nucleate boiling region, and the heat flux at 85°C is less than half the CHF of chips PF30-120 and PF50-120.

We plot the relation ship of q_{max} with fluid velocity for all surfaces with fluid subcooling as a parameter in Fig. 16. The fluid velocity has a very large effect on q_{max}. For the low fluid subcooling of 15K and the velocity larger than 1m/s, the rate of q_{max} enhancement is increased remarkably, which was also supported by many researches such as Mudawar and Maddox (1989), Rainey et al. (2001), and etc., who had noted that the transition from low to high velocities was characterized by an increase in the rate of CHF enhancement with velocity. However, for the larger liquid subcoolings of 25 and 35K, the transition from low to high velocity is characterized by a decrease in the rate of q_{max} enhancement with velocity.

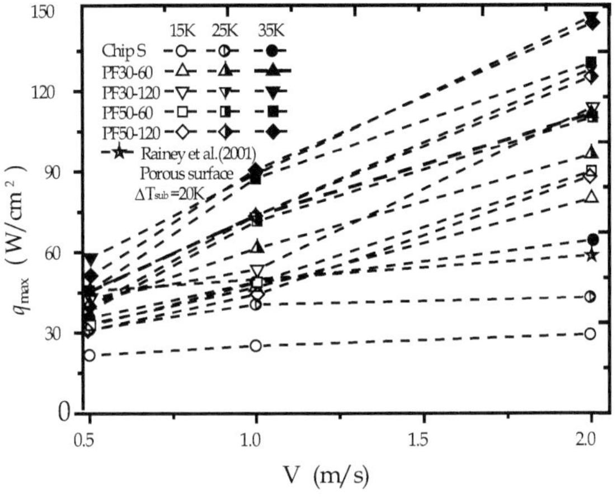

Fig. 16. Effects of fluid velocities and subcoolings on CHF

For a low fluid subcooling, as explained by Mudawar and Maddox (1989), the low velocity q_{max} was caused by dryout of the liquid sublayer beneath a large continuous vapor blanket near the downstream edge of the heater; however in the high velocity q_{max} regime, the thin vapor layer covering the surface was broken into continuous vapor blankets much smaller than the heater surface, decreasing the resistance of fluid flow to rewet the liquid sublayer and thus providing an additional enhancement to q_{max} and subsequent increase in slope. For a large fluid subcooling, the bubble size becomes small and the heater surface was not fully occupied with vapor layer for the fluid velocity range in this study. Therefore, there is no obvious slope change as seen in Fig. 16. Moreover, from the slope of boiling curves we can see that the effect of the fluid velocity on micro-pin-finned surfaces is more noticeably compared with the porous surface (Rainey et al. 2001) and the smooth surface, and the enhancement of q_{max} sharply increases with fluid velocity. For chips PF30-120 and PF50-120, the q_{max} reaches nearly 148W/cm² at V=2m/s and ΔT_{sub}=35K.

4. Conclusions and future research

All micro-pin-fined surfaces have considerable heat transfer enhancement compared with a smooth surface and other microstructured surfaces, and the maximum CHF can reach nearly 148W/cm^2 by chips PF30-120 and PF50-120 at V= 2 m/s and ΔT_{sub} = 35 K. The wall temperature for the micro-pin-finned surfaces is less than the upper temperature limit for the normal operation of LSI chip, 85°C. Therefore, micro-pin-finned surfaces are very prospective for high-efficiency electronic cooling.

Electronics cooling by using boiling heat transfer in space and on planetary neighbors has become an increasing significant subject. For the boiling heat transfer in microgravity, the buoyancy effect becomes weak, resulting in a longer stay time for the bubble departure. These may prevent the effective access of fresh bulk liquid to the heater surface in time, resulting in a lower boiling heat transfer performance at high heat flux (Wan and Zhao 2008). How to improve boiling heat transfer effectively in microgravity is an important issue. According to the excellent boiling heat transfer performance of the micro-pin-finned surfaces and the enhanced boiling heat transfer mechanism, it is supposed that although the bubbles staying on the top of the micro-pin-fins can not be detached soon in microgravity, the fresh bulk liquid may still access to the heater surface through interconnect tunnels formed by the micro-pin-fins due to the capillary forces, which is independent of the gravity level. Therefore, it is our great interest to study the boiling heat transfer performance of micro-pin-finned surfaces in microgravity in the future.

5. Acknowledgement

This work is supported by the program for new century excellent talents in university (NCET-07-0680).

6. References

Anderson, T.M. & Mudawar, I. (1989). Microelectronic cooling by enhanced pool boiling of a dielectric fluorocarbon liquid. ASME J. Heat Transfer, 111:752-759

Bergles, A.E. & Chyu, M.C. (1982). Characteristics of nucleate pool boiling from porous metallic coatings. ASME J. Heat Transfer, 104:279-285

Chang, J.Y. & You, S.M. (1996). Heat orientation effects on pool boiling of micro-porous-enhanced surfaces in saturated FC-72. ASME J. Heat Transfer, 118:937-943

Chang, J.Y. & You, S.M. (1997). Enhanced boiling heat transfer from micro-porous surfaces: effects of a coating composition and method. Int. J. Heat Mass Transfer, 40:4449-4460

Chowdhury, S.K.R. & Winterton, R.H.S. (1985). Surface effects in pool boiling. Int. J. Heat Mass Transfer, 28:1881-1889

Chu, R.C. & Moran, K.P. (1977). Method for customizing nucleate boiling heat transfer from electronic units immersed in dielectric coolant. US Patent 4,650,507

El-Genk, M.S. & Ali, A.F. (2010). Enhancement of saturation boiling of PF-5060 on microporous copper dendrite surfaces. ASME J. Heat Transfer, 132:071501(1-9)

Hwang, U.P. & Moran, K.F. (1981). Boiling heat transfer of silicon integrated circuits chip mounted on a substrate. In: ASME HTD, 20:53-59

Kitching, D.; Ogata, T. & Bar-cohen, A. (1995). Thermal performance of a passive immersion cooling multichip module. J. Enhanced Heat Transfer, 2:95-103

Kubo, H.; Takamatsu, H. & Honda, H. (1999). Effects of size and number density of micro-reentrant cavities on boiling heat transfer from a silicon chip immersed in degassed and gas-dissolved FC-72. Enhanced Heat Transfer, 6:151-160

Kutateladze,S.S. & Burakov, B.A. (1989).The critical heat flux for natural convection and forced flow of boiling and subcooled dowtherm. Problems of Heat Transfer and Hydraulics of Two-Phase Media, Pergamon,Oxford, pp. 63–70

Li, S.; Furberg, R.; Toprak, M.S.; Palm, B. & Muhammed, M. (2008). Nature-inspired boiling enhancement by novel nanostructure macroporous surfaces. Adv. Funct. Mater., 18:2215-2220

Lie, Y.M.; Ke, J.H.; Chang, W.R.; Cheng, T.C. & Lin, T.F. (2007). Saturated flow boiling heat transfer and associated bubble characteristics of FC-72 on a heated micro-pin-finned silicon chip. Int. J. Heat Mass Transfer, 50:3862-3876

Messina, A.D.; & Park, E.L. (1981). Effects of precise arrays of pits on nucleate boiling. Int. J. Heat Mass Transfer, 24:141-145

Mudawar, I. & Anderson, T.M. (1989). High flux electronic cooling by means of pool boiling – Part I; parametric investigation of the effects of coolant variation, pressurization, subcooling and surface sugmentation. Heat Transfer Electron ASME-HTD, 111:25-34

Mudawar,I. & Maddox,D.E. (1989). Critical heat flux in subcooled flow boiling of fluorocarbon liquid on a simulated electronic chip in a vertical rectangular channel. Int. J. Heat and Mass Transfer, 32: 379–394.

Nakayama, W.; Nakajima, T.; Ohashi, S. & Kuwahara, H. (1989). Modeling of temperature transient of micro-porous studs in boiling dielectric fluid after stepwise power application. ASME HTD, 111:17-23

Nelson, L.A.; Sekhon, K.S.; & Fritz, J.E. (1978). Direct heat pipe cooling of semiconductor devices. In: Proceedings of the 3rd international heat pipe conference, CA, USA, pp.373-376

Nakayama, W.; Daikoku, T. & Nakajima, T. (1982). Effects of pore diameters and system pressure on saturated pool nucleate boiling heat transfer from porous surfaces. ASME J. Heat Transfer, 104:286-291

O'Connor, J.P. & You, S.M. (1995). A painting technique to enhance pool boiling heat transfer in saturated FC-72. ASME J. Heat Transfer, 117:387-393

O'Connor, J.P.; You, S.M. & Chang, J.Y. (1996). Gas saturated pool boiling heat transfer from smooth and microporous surfaces in FC-72. ASME J. Heat Transfer, 118:662-667

Oktay, S. (1982). Departure from natural convection (DNC) in low-temperature boiling heat transfer encountered in cooling microelectronic LSI devices. In: Proc. 7th Int. Heat transfer Conf., 4:113-118

Oktay, S. & Schmeckenbecher, A. (1972). A method for forming heat sinks on semiconductor device chips. US Patent 3,706,127

Phadke, N.K.; Bhavnani, S.H.; Goyal, A.; Jaeger, R.C. & Goodling, J.S. (1992). Re-entrant caviy surface enhancements for immersion cooling of silicon multichip packages. IEEE Trans. Comp., Hybrids, Manuf. Technol.,15:815-822

Rainey, K.N.; Li, G. & You, S.M. (2001). Flow Boiling Heat Transfer From Plain and Microporous Coated Surfaces in Subcooled FC-72. ASME J. Heat Transfer, 123(5): 918-925

Samant, K.R. & Simon, T.W. (1989). Heat transfer from a small heatedregion to R-113 and FC-72. ASME J. Heat Transfer, 111:1053–1059

Ujereh, S.; Fisher, T.; & Mudawar, I. (2007). Effects of carbon nanotube arrays on nucleate pool boiling. Int. J. Heat Mass Transfer, 50:4023-4038

Wan, S.X. & Zhao, J.F. (2008). Pool boiling in microgravity: recent results and perspectives for the project DEPA-SJ10. Microgravity Sci. Technol., 20:219-224

Wei, J.J.; Yu, B. & Wang, H.S. (2003). Heat transfer mechanisms in vapor mushroom region of saturated nucleate pool boiling. Int. J. Heat and Fluid Flow, 24:210-222

You, S.M.; Simon, T.W. & Bar-Cohen, A. (1992). A technique for enhancing boiling heat transfer with application to cooling of electron equipment. IEEE Trans CHMT, 15:823-831

4

Heat Transfer in Minichannels and Microchannels CPU Cooling Systems

Ioan Mihai
"Stefan cel Mare" University of Suceava
Romania

1. Introduction

The 70-eth of previous century brought the miniaturization of electronic components and the development of systems towards micro and nano manufacturing and in time, more frequent and diverse applications occurred in other domains such as biomedical devices, MEMS and cooling nano technologies. Overheating of these micro components and micro devices led to the use of mini and microchannels in the above mentioned technologies. The aim is to eliminate as fast as possible the maximum heat quantity from these systems in order to ensure an increased reliability and functional stability (Kim & Kim, 2007). Using CPU's at high temperatures can lower cause system crashes in the short term and in the long term cause the life of your CPU to be greatly reduced. In extreme cases your CPU could burn out or melt onto the motherboard. Evacuation of a large heat flow through conduction and forced convection of the air cannot be adequate achieved by classical methods. Thus, we can conclude that the CPU cooling systems must ensure proper cooling of the CPU. It should be interesting to make a comparative study regarding the maximum values reached by the temperature that develops inside the CPU cores. In figures 1 and 2 we can see the CPU's rated maximum temperature, sometimes called critical temperature.

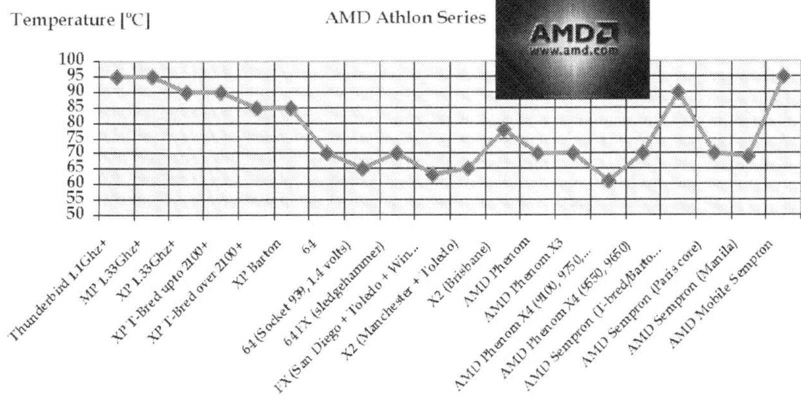

Fig. 1. Critical temperatures for AMD series CPUs

Due to practical reasons we ascertained that it is recommended to maintain the functional values (indicated by the computer's software) at a value measuring approximately 20 °C below the maximal values indicated in the two diagrams.

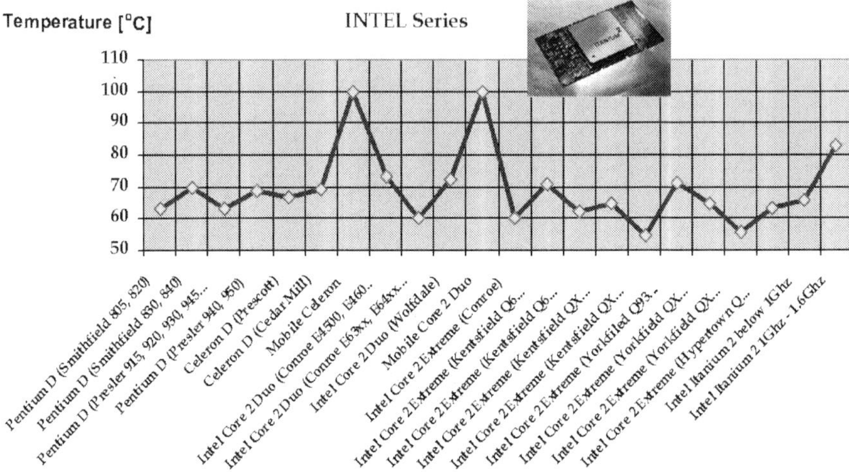

Fig. 2. Maximal values of the temperature for INTEL series CPUs

In both cases we notice that the maximal values of the temperature vary from a maximum of 100 ºC, for an Intel Mobile Celeron and Core 2 Duo, to a minimum of 54,8 °C for an Intel Core 2 Extreme (Kentsfield QX6800), whereas for an AMD Athlon CPU we have a maximum of 95 °C and a minimum of 61 °C, for the Phenom X4 models (9100, 9750, 9850). Together with CPU development, as it results from Figure 3, the heat flow increases significantly, depending on the frequency variation.

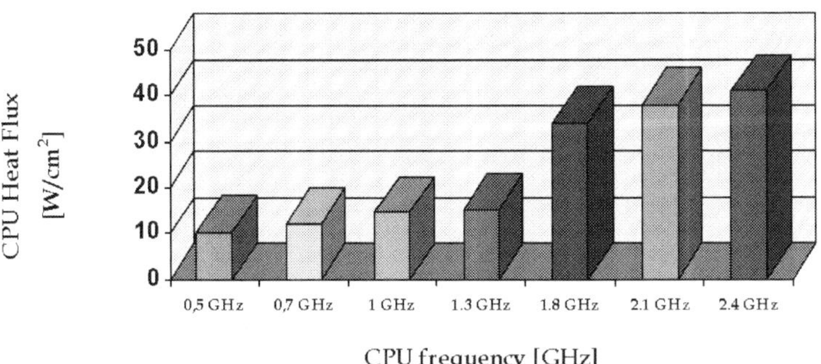

Fig. 3. Heat flow modification depending on the frequency variation

We can emphasize that this increase is exponential. In Figure 4 we show a graphical representation of the temperature flow and the CPU die size variation starting 1975 until the present day.

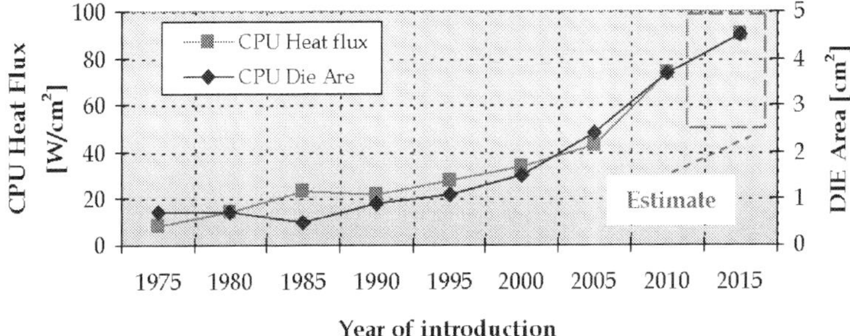

Fig. 4. Variation of the thermal flow and the CPU die size depending on the production year

By analyzing the information that was described above, one can observe that, no matter what is the manufactured CPU, an important density of thermal flow develops on the CPU die. High values of the thermal flow imply taking heat dissipation measures. If this requirement is not met then the risk of CPU thermal inducted damage might occur. Therefore, different cooling methods are needed which, under the circumstances of reaching extremely low temperatures, can lead to a significant increase in the CPU's processing speed.

Most recently, the attention was focused on the study of flow processes and heat transfer in microdevices. In these systems (Hadjiconstantinou & Simek, 2002), the flow and heat transfer processes are of nano and microscopic type and differ as basic mechanism from the macroscopic ones due to dimensional characteristics and molecular type phenomena.

2. CPU cooling methods and their limits

Until now (Viswanath et al., 2000, Meijer et al., 2009) indicate two CPU cooling systems, as they are described in Figure 5.

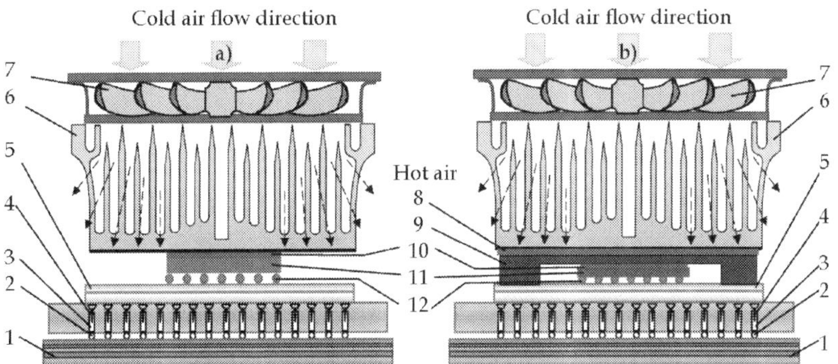

Fig. 5. Types of CPU cooling systems, where: (a) Package with Exposed Die and Heat Sink and (b) Lidded package and Heat Sink

Description of the annotations in Figure 5: 1 – Motherboard planes, 2 – solder ball, 3 – Receptacle, 4 – Pin, 5 – Substrate, 6 – Heat Sink, 7 – cooler, 8 – TIM (Thermal Interface Material) 2, 9 – Heat Spreader, 10 – TIM 1, 11 – CPU die, 12 – C4 Solder ball.

Strictly from a thermal transfer perspective, as (Meijer et al., 2009) also points out, the functions of the described elements are:

- Heat Sink – Dissipates heat to the environment;
- TIM 2 – Transmits heat to the Heat Sink;
- Heat Spreader:
  -Evens out heat spot;
  -Spreads heat horizontally;
  -Protects the die;
- TIM 1 – Transmit heat to the spreader;
- Die:
  -Heat generation;
  -Hot spot;

The Heat Sink place can be replaced by other cooling systems. A description of these systems (Pautsch, 2005) is represented in Table 1. Also, we mention each one's performance.

Cooling Technology	Cray Product	Power [W]	Power Density Removal [W/cm^3]
Air Cooling – Open loop	SV1	40	127,0
Air Cooling – Closed loop	HPAC	75	203,2
Conduction to Liquid – Cold plate	T3E	80	457,2
Conduction to Liquid – Top Hat	MTA	100	508,0
Single Phase Forced convection (immersion)	T90	75	381,0
Single Phase Impingement	SS1	120	762,0
Heat Pipe Assisted Heat Sink	Rainier	140	66,04
Spray Evaporative Cooling	X1/X1E	200	1016,0
Film Evaporative Cooling (Spray cooling with Phase Change)	Future	275	1397,0 Projected

Table 1. CPU cooling systems

In addition to the methods described in Table 1 (Banton & Blanchet, 2004) indicates other possibilities regarding major cooling technology trends, such as:

- Spray cooling with mercury inside – first circa 1999;
- Hollow core liquid cooled electronic modules – Mercury circa 1998;
- Microchannel cooling;
- Refrigeration, air chillers;
- Heat pipes;
- Thermo-electric coolers (TEC).

Recently, development boosted in the area concerning cooling methods that ensure dissipation of increasingly larger power flows. In this respect, (Meijer et al., 2009) indicate various possibilities to increase the cooling process efficiency by either operating on TIM or using one of the following: Ultra Thin High Efficiency Heat Sinks and Manifold Micro-channel Heat Sinks, Radially Oscillating Flow Hybrid Cooling Systems, Oscillating Flow Liquid Cooling and the Phonon Transport Engineering method.

A representation of the limits that can be reached by dissipating the heat flow to the maximum is described in Figure 6. In the figure one can observe the CPU cooling methods, the data belonging to (Pautsch, 2005) for the CRAY technologies.

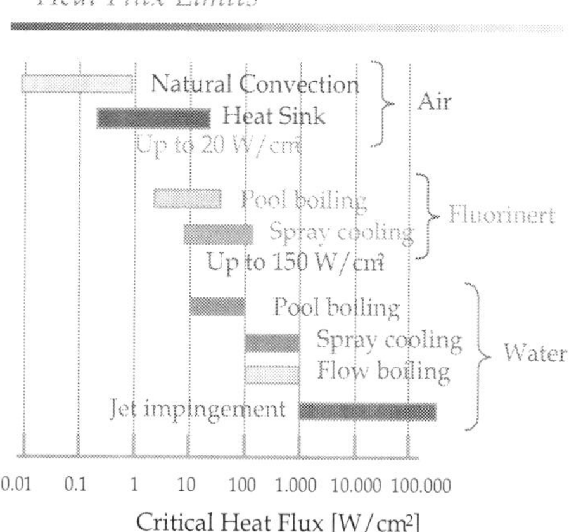

Fig. 6. Maximal values of the Critical Heat Flow (Pautsch) depending on the cooling methods that were used

With respect to the information presented above, we can ascertain that the functioning of a CPU system is extremely complex, and that a direct dependence between working speed and cooling degree exists and it was experimentally revealed in this regard.

3. Considerations regarding mini and micro channels

3.1 Presence of the mini and micro channels within CPU cooling systems

Heat and mass transfer is accomplished across the channel walls in many man-made systems, such as micro heat exchangers, desalination units, air separation units, CPU cooling etc.A channel serves (Kandlikar et al., 2005) to bring a fluid into intimate contact with the channel walls and to bring fresh fluid to the walls and remove fluid away from the walls as the transport process is accomplished. A classification of the channels conducted by (Kandlikar et al., 2005, Kandlikar & Grande, 2003) is described in Table 2.

Type of channel	Overall dimensions
Conventional channels	>3mm
Minichannels	3mm$\geq D$>200μm
Microchannels	200μm$\geq D$>10μm
Transitional Microchannels	10μm$\geq D$>1μm
Transitional Nanochannels	1μm$\geq D$>0.1μm
Nanochannels	0.1μm$\geq D$

Table 2. Channel classification scheme (Kandlikar et al., 2005); D value represents the smallest channel dimension.

Increasingly advanced technologies implemented in order to cool the CPU are using, under various types, mini or micro channels. Thus, (Escher et al., 2009) suggests using the Ultra-Thin Manifold Micro-Channel Heat sink; in Figure 7a one can observe the working criterion and, in Figure 7b, a SEM image of the micro channels.

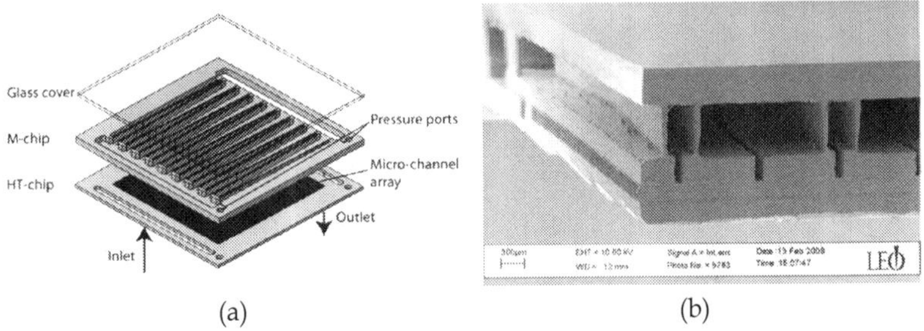

(a)   (b)

Fig. 7. Micro channels used within CPU systems (Escher et al., 2009): (a) basic schematics; (b) SEM image of the micro channels

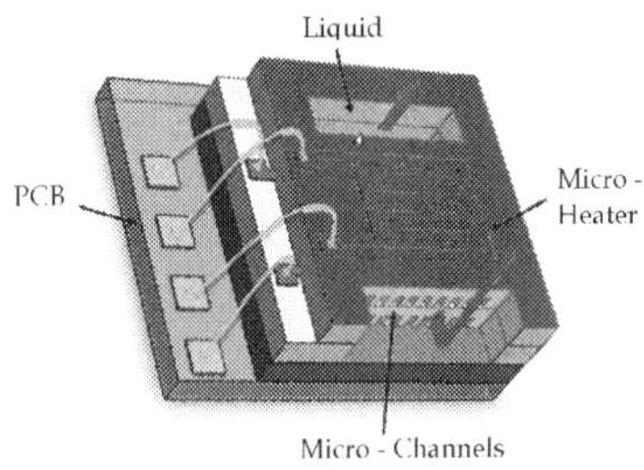

Fig. 8. Pyramid chip stack

The usage of the micro channels in the CPU systems is reminded by (Meijer et al., 2009) for the structures type Pyramid chip stack.
In Figure 8 we present such a structure, emphasizing on the micro channels that are present in the lower part.
IBM Zurich Research Laboratory in 2009 in its research report (Meijer et al., 2009) presents images of the micro channels that can be found in Figure 9.

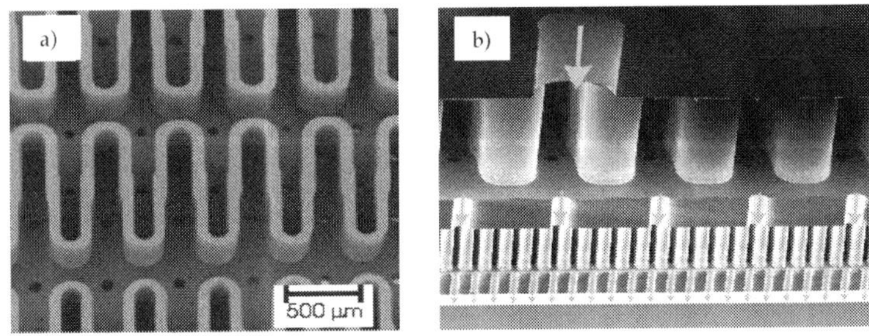

Fig. 9. Micro channels: (a) overview of the micro channels, (b) SEM cross-section of two-level jet plate with diameter of 35µm

Inside the *Heat transfer* laboratory of the "Stefan cel Mare" University of Suceava, Romania, we developed mini heat exchangers that integrate micro channels. Image 10a shows a plate from a mini heat exchanger and image 10b shows an AFM image (19.7 x 80.0 µm) of a micro channel. Also, image 10b shows, on the right size, the threshold corresponding to the copper material and, on the left side, the threshold corresponding to the composite material. In figure 10c the graph shows the values measured by the AFM.

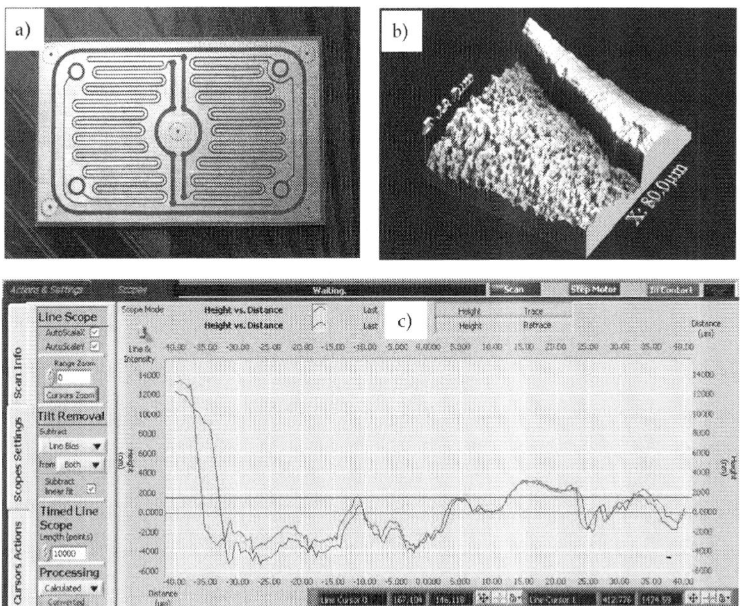

Fig. 10. Images specific to a mini heat exchanger with micro channels: (a) plate, (b) AFM image, (c) AFM graph with measurement data

The data in diagram 10.c show a difference between the lower part of the 1250nm channel and the maximum limit which corresponds to the 13600nm copper surface. This indicates a medium micro channel height of 14850 nm. In order to perform accurate surface topography measurements under various conditions, a Nanofocus μscan laser profilometer was employed. The images that were obtained for the micro channels, and a detail image, can be seen in Figure 11.

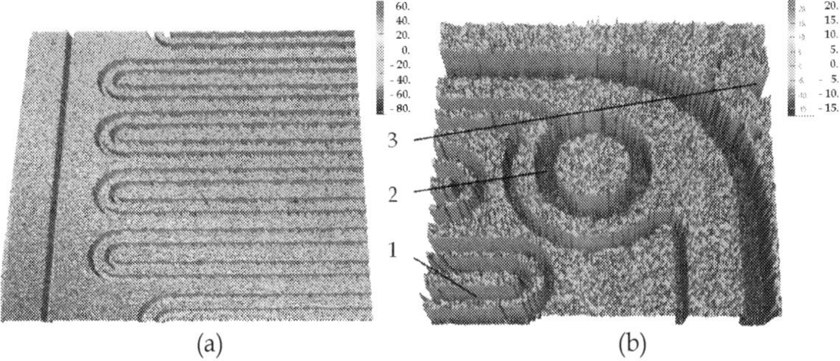

(a)   (b)

Fig. 11. Images obtained with the Nanofocus μscan profilometer: (a) micro channels, (b) micro channels detail image, were: 1 - microchannels, 2 - admission channelling, 3 - centre holes

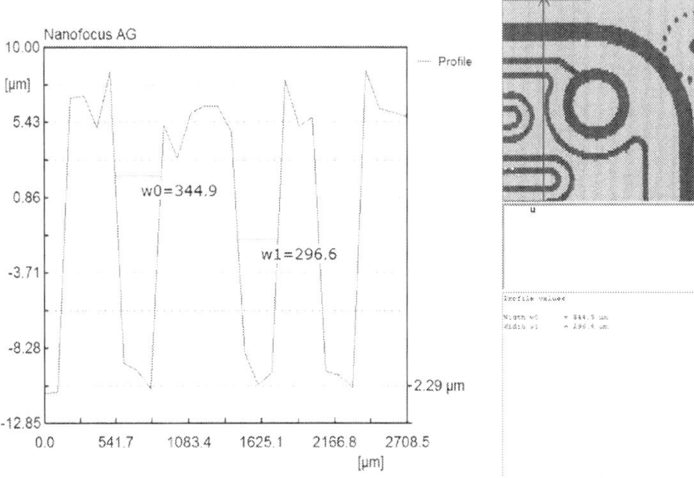

Fig. 12. Heat exchanger micro channels parameters that were obtained with the help of a Nanofocus µscan profilometer

The values obtained with the help of a profilometer, as they are described in figure 12, indicate a height approximately equal to the one indicated by the AFM, resulting in an average of 14,85 µm which is practically doubled by overlapping two plates. The width of a micro channel measured at the middle point of its height is 296.6 µm.

From what was described above we can ascertain that mini and micro channels are often used in various areas, one of them being CPU cooling.

3.2 Experimental evidence of defects occurred in interface layer of cooler-CPU

When the contact between two surfaces is imperfect, the specific thermal resistance of interface layer suddenly increases, so it became of frequent use to apply diverse materials between the CPU and radiator. Previous research showed that when applied in thin layers, CPU thermal grease may form micro and nanochannels on the surface. These can have a significant impact on the heat transfer between CPU and the cooling system radiator. These materials should both fill the gaps occurred due to surfaces roughness, material's fatigue, loading pressure etc. and transfer as much heat as possible during a short period of time (Mihai et al., 2010).

Inside the *Heat transfer* laboratory of the "Stefan cel Mare" University of Suceava, using an atomic force microscope AFM-universal SPM, Park Scientific instruments firm, processing unit module SPC 400, electronic control unit SFM 220A, we scanned the interface layer of different processors. The images obtained for an AMD processor are presented in Figures 13a (scale 1x1 µm) and 13b (scale 1.6x1.6 µm) respectively.

By analyzing the obtained images, with the help of the AFM, two aspects can be ascertained. Thus, from Figure 13a one can observe that flowing channels may occur in interface layer, having in this case, an estimated width of only 2µm for a height of 2000 Å. The second assertion can be deducted from Figure 13b, the appearance of the steady flow zones being clearly visible. Whatever the situation is, a change in the CPU-TIM-cooler contact appears, change which leads to the alteration of the heat exchange transfer. First, a special attention is paid to assess the effect of mechanical and thermal properties of the contacting bodies, applied contact pressures and surface roughness characteristics as well as the use of different thermal interface materials on the maximum temperature experienced by the CPU. Second, it can be appreciated that good wetting of the mating surfaces and the retention of asperity micro-contacts can become critical elements in effectively removing the heat generated by the CPU (Mihai et al., 2010). It is commonly recognized that roughness effect can have an impact on microchannel and microtube performance, both in terms of pressure drop and heat transfer. From those described above we can deduct that, if modifications in the TIM layer appear due to a faulty installation, inadequate push pressure values, material aging, faulty TIM appliance techniques, too high temperatures in its exploitation etc., the thermal conduction coefficient and, subsequently, the thermal resistivity changes, leading to CPU damage.

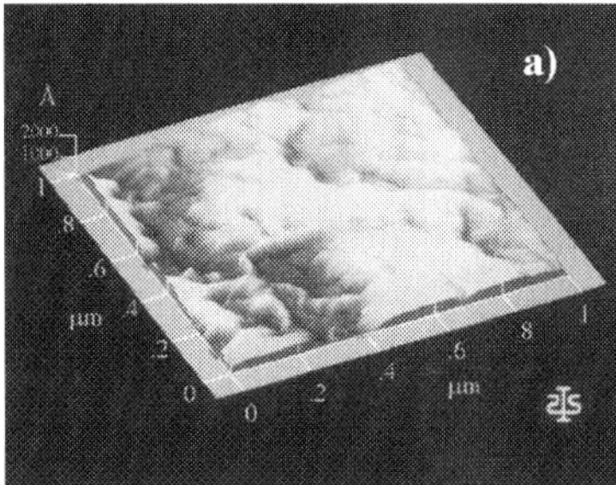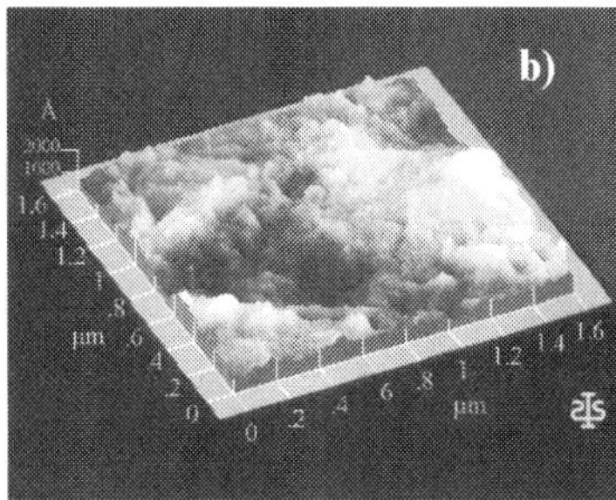

Fig. 13. (a) (b) AFM images of interface layer CPU – Heat Sink: (a) - with the micro channels highlighted, (b) – with the concavities and protuberances emphasized

3.3 Aspects regarding TIM dilatation
In the Heat Transfer Laboratory from the Suceava University, a test rig was conceived and built (Figure 14a) in order to study CPU thermal grease behaviour when subjected to high temperatures, close to those leading to CPU failure.

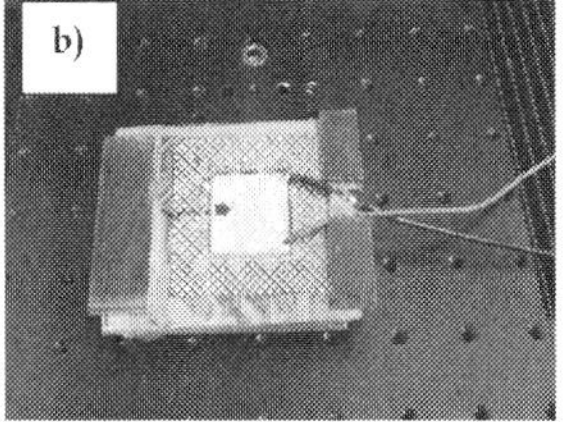

Fig. 14. (a) Test rig Image and (b) CPU and heating system

The description of the annotations in Image 14a is: 1 – laser head, 2 – CPU, 3 – work bench with coordinates, 4 – joystick, 5 - stand, 6 – DC Power supply for the resistors, 7 – nanofocus software. In order to simulate the CPU heating process, under the CPU were mounted 3 electric resistors. The assembly allows reaching a maximum temperature of 110 °C. A thin layer of thermal grease Keratherm Thermal Grease KP97 is applied on the CPU surface and then the temperature is gradually elevated. After the thermal condition is stabilized, the layer is subjected to profilometer scanning. The surface micro-topography is then analyzed for thermal grease volume and surface variations shown.

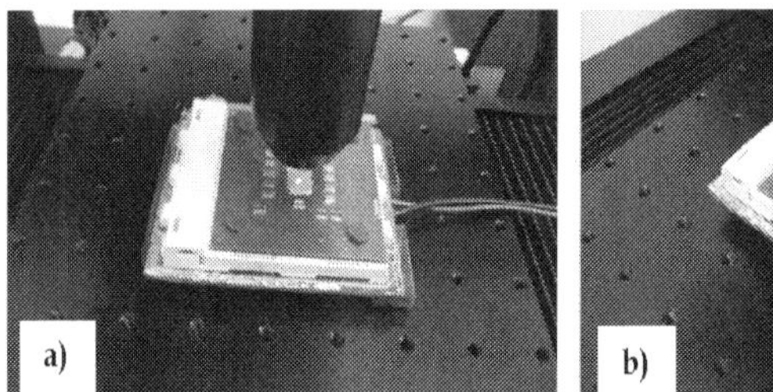

Fig. 15. Experimental investigations of thermal grease layer. (a) without crystal glass and (b) with crystal glass

The experimental determinations were conducted for three hypotheses:
- CPU without a thermal grease layer applied;
- CPU with a thermal grease layer applied (Figure 15a);
- Having a 2,7 mm crystal glass plate placed on the thermal grease covered CPU (Figure 15b), and a 12,5 N/cm² pressure applied in order to simulate cooler interface.

For the last two hypotheses, changing the voltage of the power supply led to the three resistors heating up to various predetermined values. The temperature was measured at the surface of the TIM material and CPU with the help of a laser thermometer.

During experimental measurements, several issues were investigated:
- In order to analyze TIM behaviour we simulated the CPU functioning with freshly deposited thermal grease on the surface;
- investigations on whether or not micro or nanometer channels appear in an incipient phase at the TIM surface during its heating stage;
- emphasizing the "pump-up" occurrence of the TIM caused by the expansion effect;
- detection of surface discontinuities appearing during heating (localized lack of material);
- monitoring thermal grease layer profile shape, roughness and waviness evolution.

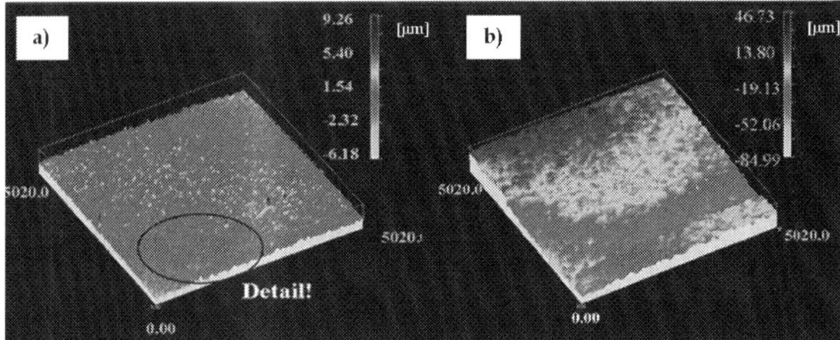

Fig. 16. (a) CPU without thermal grease; (b) Thermal grease layer at 20 °C

Next we will describe the images that were obtained by using the assembly specified in Figure 14a. On the left side image of figure 16a, a 3D image of the CPU surface obtained by laser profilometry is shown without the TIM being applied. In Figure 16b, on the CPU surface, a thermal grease layer was applied. It can be observed that, in contrast with the hypothesis described in Figure 14a, the asperity of the surface is modified.

The detail from Figure 16a allows observing the AMD sign. In Figure 17 the profile shape alteration is shown together with the heating of the CPU and, subsequently, the heating of the thermal grease surface.

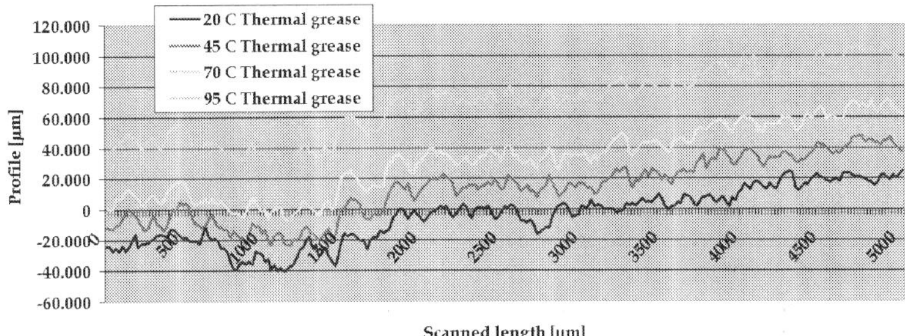

Fig. 17. Profile Shape variation depending on the temperature that was obtained by laser profilometry

The results obtained by profilometry of the applied thermal grease paste are shown in Figures 18a,b and for the roughness and waviness, for temperatures of 20, 45, 70 and 95 °C, in Figures 19a,b.

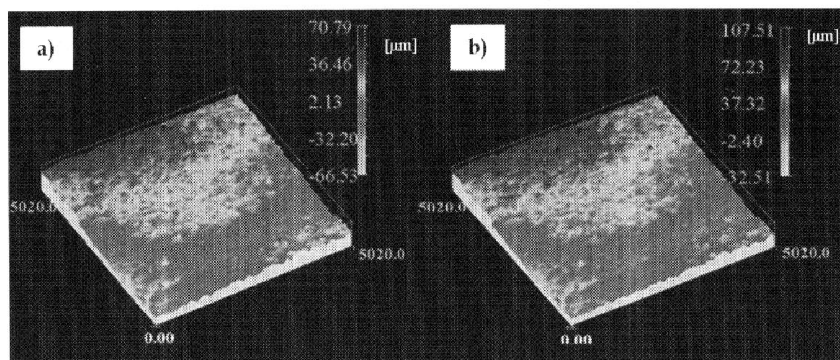

Fig. 18. Thermal grease layer at: (a) 45 °C and (b) 95 °C

For the second instance we applied a crystal glass over the TIM layer, the crystal glass being pushed downwards with a pressure measuring the same value that we previously mentioned.

The reason in using this crystal glass is brought about by our desire to simulate the TIM behaviour under heat, when flattening the heat sink irregularities in a real case scenario. In Figure 20 we show the images that were obtained by profilometry.

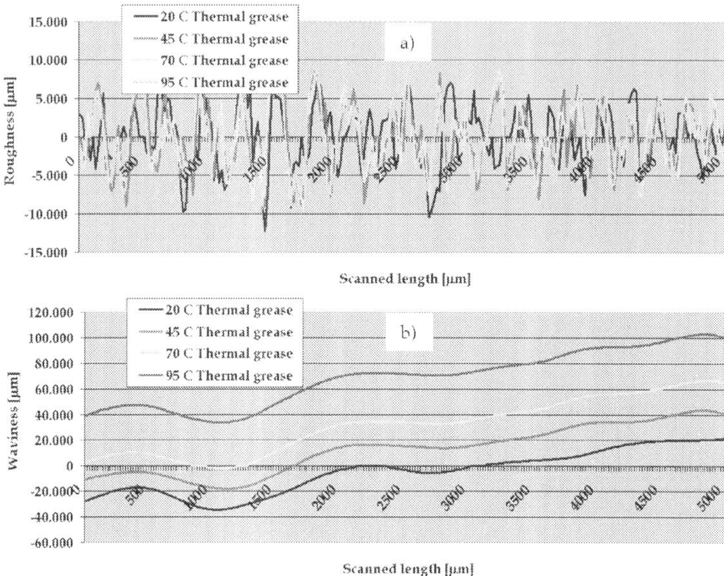

Fig. 19. The results obtained by profilometry for TIM. (a) roughness and (b) waviness

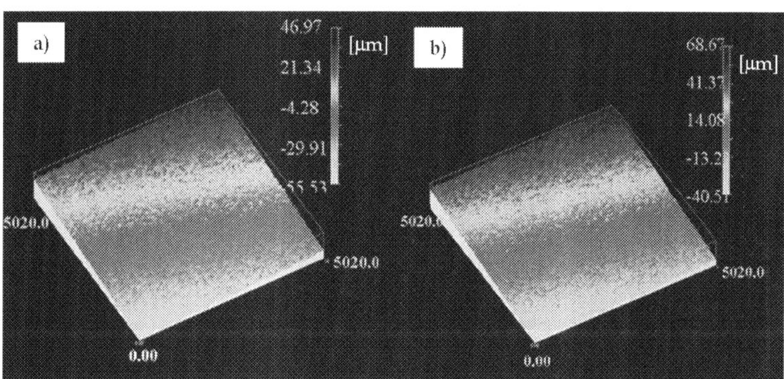

Fig. 20. Thermal grease under crystal glass: (a) at 20 °C and (b) at 95 °C

Images 20a and 20b show the variation in the roughness size when putting the TIM under pressure caused by an object with a low irregularity coefficient. The detailed results of the Profile Shape size are described in the diagram from Figure 21.

It can immediately be noticed that variations in thermal grease layer profile are considerably smaller after placing the crystal plate, by comparison with the free surface situation.

In order to compare how the same material reacts when is subjected to a heating process, having a much lower initial roughness, we drew graphs for roughness and waviness obtained with the help of the laser profilometer. These values can be observed in Figure 20.

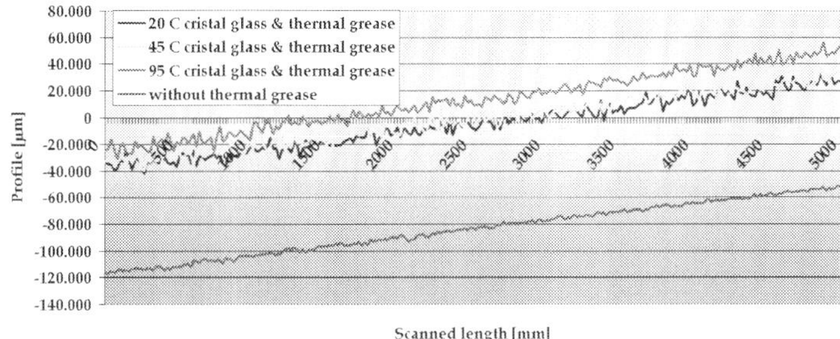

Fig. 21. The change in the Profile Shape size when the crystal glass was applied over the thermal grease

In order to compare the different behaviours of the same material, having an initial much lower roughness, when it is subjected to a heating process, we drew graphs for the roughness and waviness values that were obtained with the help of the laser profilometer. These values can be analyzed in figure 22.

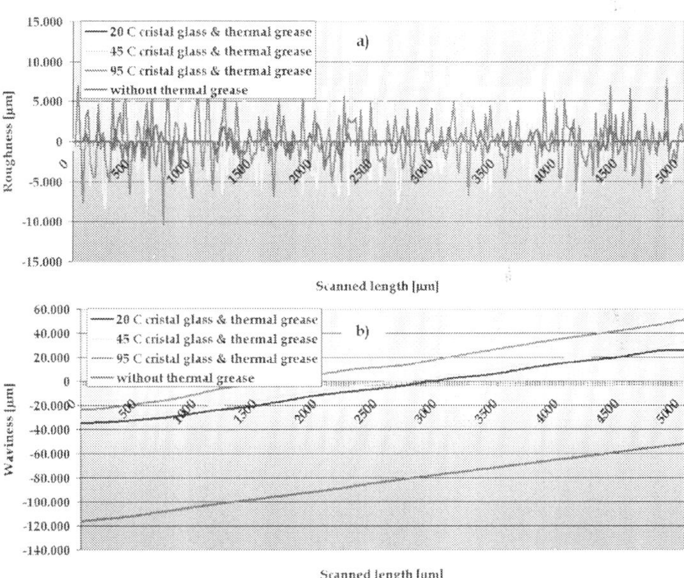

Fig. 22. Results obtained for TIM under a crystal glass. (a) Roughness and (b) Waviness

4. Heat exchange in the mini, micro and nano channels of the CPU cooling systems

4.1 Assessment of flowing regime

Most recently, the attention was focused on the study of flow processes and heat transfer in microdevices. In these systems, the flow and heat transfer processes are of nano and microscopic type and differ as basic mechanism from the macroscopic ones due to dimensional characteristics and molecular type phenomena. The early transition from laminar flow to turbulent flow, and the several times higher friction factor of a liquid flowing through microchannel than that in conventional theories. Heat transfer for a developing compressible flow was studied by (Kavehpour et al., 1997), using the same first-order slip flow and temperature jump boundary conditions as in heat transfer for a fully developed incompressible flow. The flow was developing both hydrodynamically and thermally and two cases were considered: uniform wall temperature or uniform wall heat flux. Viscous dissipation was neglected, but the compressibility of the working agent was taken into account. We can ascertain that Navier-Stokes and energy equations are inappropriate for micro or nano channels because of failure of the continuum assumption for micro flow. Different computational methods were developed; however, up to date, two main directions in the literature (Kandlikar et al., 2005) are known:

- the working fluid is considered as a collection of molecules;
- the fluid is considered an indefinitely divisible continuum.

The continuum model is usually used in modelling macrosystems. In order to perform the calculus relating to the micro channels of a heat sink it is necessary to take into account a series of parameters, such as: flow rate of the cooling fluid, temperature of the fluid and the channel wall, inlet and outlet pressure required for cooling fluid, hydraulic diameter of the channel, number of channels. Flowing regime in micro and nano channels is usually estimated by means of Knudsen number, given by the relation:

$$Kn = \frac{k}{H} \tag{1}$$

where k is the mean free path and H is the characteristic dimension of the flow. In time, depending on the values of Kn number, (Kandlikar et al., 2005, Niu et al., 2007, Hadjiconstantinou & Simek, 2002), several computing models presented in Table 3 were completed.

Knudsen number	Regime	Navier–Stokes equations viability
Kn<0.001	"Continuum"	Accurately modelled by the compressible Navier–Stokes equations; classical no-slip boundary conditions
0.001<Kn<0.1	"Slip-flow"	Navier–Stokes equations applicable; a velocity slip and a temperature jump considered at the walls, as rarefaction effects become sensitive at the wall first.
0.1<Kn<10	"Transition"	The continuum approach is no longer valid; the intermolecular collisions are not yet negligible and should be taken into account.
Kn>10	"Free-molecular"	Intermolecular collisions are negligible compared to collisions between the gas molecules and the walls.

Table 3. Flow regimes for different Kn Numbers

From calculus it resulted that the heat transfer coefficient takes extremely great values. The justification is based on extremely small valued of hydraulic flowing diameter.
Obviously, a decrease of hydraulic diameter leads to:

- diminished flow and consequently reduced fluid flowing capacity;

- diminished heat transfer phenomenon, leading to increase of temperature values.

As Knudsen number increases, fluid modelling is moving from continuum models to molecular models.

Until now (Kandlikar et al., 2005), a series of collision models were developed for the *inverse power law* (IPL) condition:

- **HS** - The simplest collision model is the *hard sphere* (HS) model, which assumes that the total collision cross-section is constant and for which the viscosity is proportional to the square root of the temperature. Actually, the HS model may be described with the exponent $\eta = \infty \rightarrow$ and $\omega = 1/2$.
- **VHS** - Bird, who proposed the variable hard sphere (VHS) model for applications to the Monte Carlo method (DSMC), has improved the HS model. The VHS model may be considered as a HS model with a diameter d that is a function of the relative velocity between the two colliding molecules.
- **MM** - Another classic model is the Maxwell molecules model with $\eta = 5$ and $\omega = 1$. Actually, the HS and the MM models may be considered as the limits of the more realistic VHS model, since real molecules generally have a behaviour which corresponds to an intermediate value $1/2 \le \omega < 1$.
- **VSS** - Koura and Matsumoto (1991; 1992) introduced the variable soft sphere (VSS) model, which differs from the VHS model by a different expression of the deflection angle taken by the molecule after a collision. The VSS model leads to a correction of the mean free path and the collision rate values (less than 3%). For $a = 1$, the VSS model reduces to the VHS model.

4.2 Applications of heat transfer through rectangular micro and nano channels in the case of compressible fluids

As initial statements, on consider that the fluid properties do not modify in time, the fluid is compressible and the flow is of hydrodynamic type. The mathematical model chosen for calculus considers the walls with steady thermal flow. Up to date, two main assumptions are used for computing:

- walls with steady thermal flow;
- constant temperature of the walls.

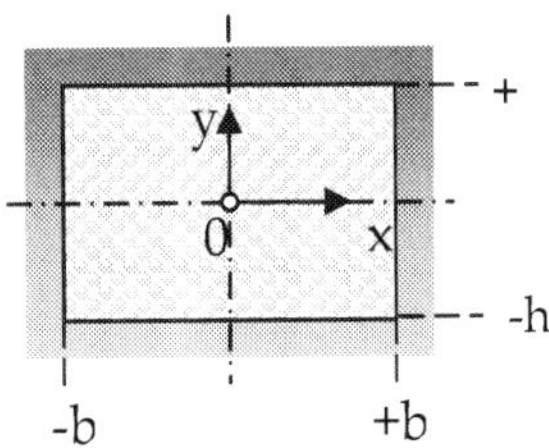

Fig. 23. Rectangular Channel

MathCad application was used in computing. The computing model is VSS and HS as proposed by (Kandlikar et al., 2005). A rectangular channel is considered, as seen in Figure 23, having the dimensions: 2b = 0,4 ·10⁻⁶ m, 2h = 0,2 ·10⁻⁶ m, l = 4,7 ·10⁻³ m, through which air flows. The inlet and outlet pressures taken into account for computing are presented in Table 4.

Pressure [N/m²]								
p_{outlet}	$p_{inlet\ 1}$	$p_{inlet\ 2}$	$p_{inlet\ 3}$	$p_{inlet\ 4}$	$p_{inlet\ 5}$	$p_{inlet\ 6}$	$p_{inlet\ 7}$	$p_{inlet\ 8}$
1,001 ·10⁴	1,006 · 10⁴	1,225 · 10⁴	1,450 · 10⁵	1,675 · 10⁵	1,900 · 10⁵	2,125 · 10⁵	2,350 · 10⁵	2,575 · 10⁵

Table 4. Inlet and outlet pressure

The viscosity coefficient for air was consider $\omega=0.77$, the exponent for VSS model $\alpha=1,36$, the dynamic viscosity $\mu_0=171,910 \times 10^{-7}$ Pa·s, the molecular mass $M=28,9710 \times 10^{-3}$ kg/kmol, and the specific constant $R=9,907 \times 10^3$ J/kg/K. An even temperature of $T=365$ K was considered for the wall. The viscosity was computed for T, $\mu(T)= 20,702 \times 10^{-6}$ Pa·s, coefficient $k_{2VSS}=1,034$, hydraulic diameter $D_H=266,667$ μm and subsequently, relation (2), the inlet/outlet mean free path is found:

$$\lambda_{i/o,VSS} = \frac{k_2 \cdot \mu(T) \cdot \sqrt{RT}}{p_{i/o}} \tag{2}$$

Consequently, based on *hydraulic diameter, microchannel depth, VSS and HS* model, the Knudsen number is computed with relations:

$$Kn_{i/o} = \frac{\lambda_{i/o\_VSS}}{D_H}; \quad Kn_{i/o\_P} = \frac{\lambda_{i/o\_VSS}}{2h}; \quad Kn_{i/oHS} = \frac{\lambda_{i/o\_HS}}{D_H}; \quad Kn_{i/o\_PHS} = \frac{\lambda_{i/o\_HS}}{2h}; \tag{3}$$

It is regarded that the ratio $a^*=h/b=55 \times 10^{-3}$ cannot be neglected. From the specialized literature we assumed, for the case of diffusive reflection, the following coefficients: $a_1=0,32283$, $a_2=1,9925$ and $a_3=4,2853$. Using the inlet over outlet pressure ratio $\Pi=p_i/p_o$, the mass flow (Kandlikar et al., 2005) we obtained a second-order model with help from the expression:

$$m_{NS2} = \frac{4 \cdot h^4 \cdot p_o^2 \cdot}{a^* \cdot \mu \cdot R \cdot T \cdot l}\left[\frac{a_1}{2} \cdot \left(\Pi^2 - 1\right) + a_2 \cdot Kn_{i/o/^{pr}/HS,} \cdot \left(\Pi - 1\right) + a_3 \cdot Kn_{i/o/^{pr}/HS,}^2 \cdot \ln\Pi \right] \tag{4}$$

The Nusselt number was found using the relation:

$$Nu^{-1} = \frac{\zeta^*}{4} + \frac{a_1 + a_2 \cdot \zeta^* + a_3 \cdot \zeta^{*2}}{a_4 \cdot \left(1 + 3\zeta^*\right)^2} \tag{5}$$

where $a_1 \div a_4$ are constant values.

In the expression above ζ^* represents a dimensionless coefficient of slip that results from:

$$\zeta^* = 4 \cdot Kn \frac{2-\sigma}{\sigma} \tag{6}$$

were σ represent the tangential momentum accommodation coefficient, dimensionless.

We will now describe the calculus results that were obtained, for the above referenced data, by using Mathcad. The variation of the parameters Kn and λ_{VSS}, λ_{HS} from relation 3 with respect to inlet pressure, is presented in Figure 24.

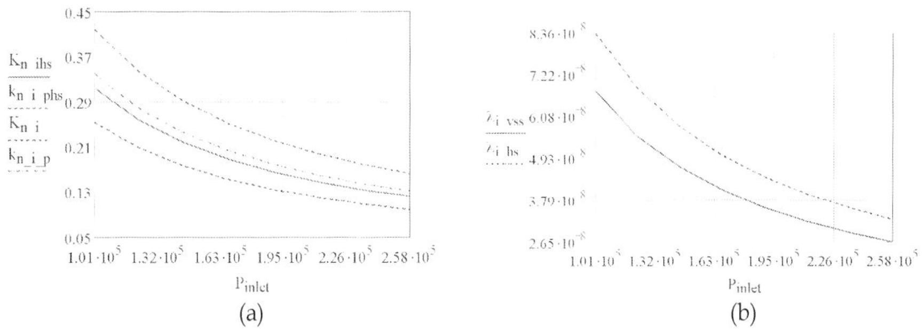

Fig. 24. Variation of parameters: (a) $Kn=f(p_{inlet})$ and (b) $\lambda_i=f(p_{inlet})$

Figure 25a presents the computed mass flow m [kg/s] with respect to Π, and figure 25b the results obtained for the gas: He, $0.05 \leq Kn_o \leq 0.09$, $P_o = 1.9 \cdot 105$ Pa.

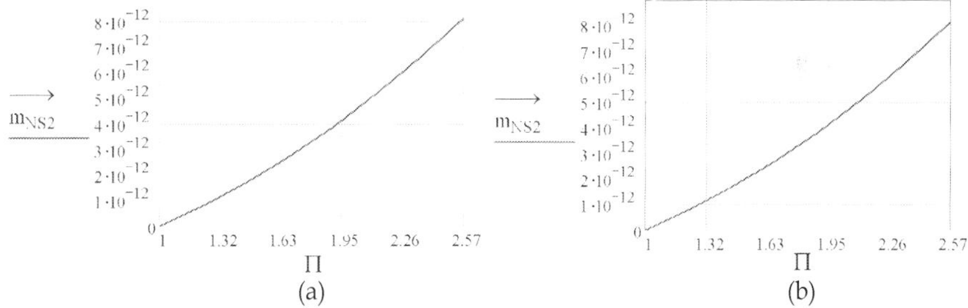

Fig. 25. Variation of the mass flow depending on Π: (a) values calculated for air and (b) values ascertained for He, by (Colin et al., 2004)

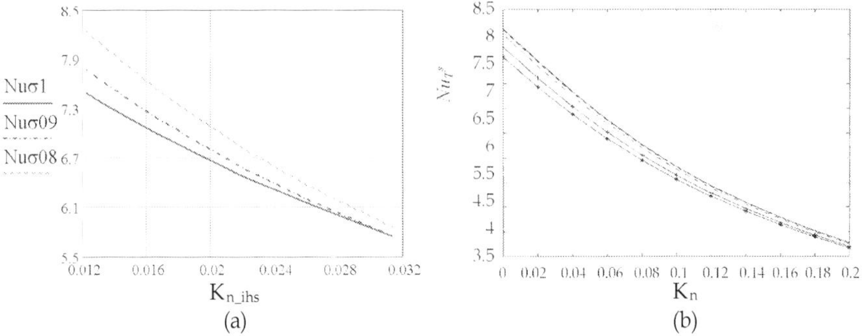

Fig. 26. Nusselt number variation depending on Knudsen number: (a) for air and (b) for air determinates by (Hadjiconstantinou & Simek, 2002).

By analyzing the diagrams one can notice that there is, in comparison to the inlet over outlet pressure ratio, the same order of magnitude for the mass flow values and almost the same curve shape in both cases, indicating the same variation pattern.

In order to study the Nusselt number dependency with regard to the inlet pressure but, at the same time, modify the tangential momentum accommodation coefficient, we drew, according to the previously resulted calculus, the diagrams described in figure 26a that have $\sigma=1$, $\sigma=0,9$ and, respectively, $\sigma=0,8$. The obtained values can be compared to the ones already existing in figure 26b, with the slip-flow calculation (constant wall), in the range $0<Kn<0.2$ and, for air, $Pr=0.7$.

We can make a representation of the Nusselt number variation depending on the inlet pressure. Figure 27 shows an increase in the Nusselt number together with the pressure value, fact which indicates the intensification of the convective heat exchange together with flow increase. It can also be seen a splay of the curves once the pressure increases. A higher increase in the tangential momentum accommodation leads to a lower coefficient in the Nusselt number values. Currently, several mathematical models are completed, and the VSS and HS models were adopted. The results obtained for rectangular channels with air have the same magnitude order as the ones obtained by (Kandlikar et al., 2005, Niu et al., 2007, Hadjiconstantinou & Simek, 2002) and the shape of the graphs from Figure 23 and Figure 24 is identical with the one obtained by authors (Kandlikar et al., 2005, Niu et al., 2007, Hadjiconstantinou & Simek, 2002). The validation of the mathematical model adopted in the present work is therefore completed.

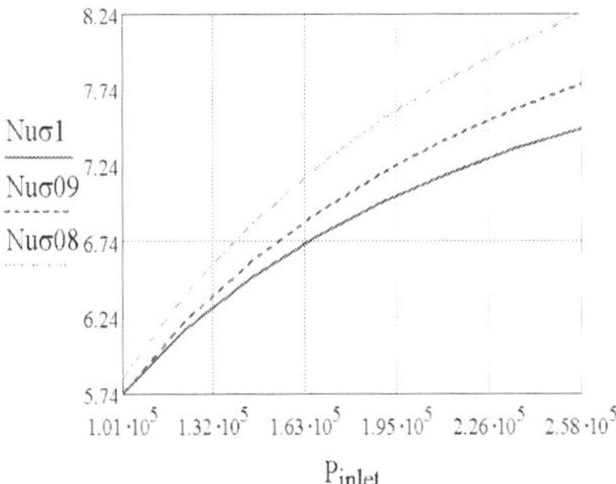

Fig. 27. Nu variation depending on P_{inlet}

4.3. Heat transfer through the material interface of CPU cooling systems

Next we will describe the heat exchange calculation at the cooler-CPU interface in the hypothesis of the adequate or inadequate thermal transfer surfaces. Heat exchange at the CPU-cooler interface can be interpreted as conductive, unidirectional with linear heat sources (Lienhard & Lienhard, 2003, Grujicic, 2004).

It is obvious that the heat source is the core of CPU, a case in which the conducted heat flow depends on the processor type (Mihai et al.), the emanation power and the efficiency of the cooling system.

In order to maintain the CPU temperature within a stable working environment, it is desirable that we absorb a major quantity of heat emission, task which is assigned to the thermal compound paste.

The junction-to-coolant thermal resistance is (Yovanovich et al., 1997, Lasance & Simons, 2005, Holman, 1997, Lee, 1998), in fact, composed of an internal, largely conductive resistance R_{jc} and an external, primarily convective resistance R_{ex} as it can be noticed in figure 28.

Total thermal spreading resistance for single-chip packages:

$$R_T = R_{jc} + R_{sp} + R_{ex} + R_{fl} \quad \left[\frac{K}{W}\right] \tag{7}$$

were:

- **The internal resistance Rjc** - is encountered in the flow of dissipated heat from the active chip surface through the material used to support and bond the chip and on to the case of the integrated-circuit package;
- **Spreading resistance Rsp** - arising from the three-dimensional nature of heat flow in the heat spreader and heat sink base;
- **External Resistance Rex** - for convective cooling;
- **Flow Resistance Rfl** – the transfer of heat to a flowing gas or liquid that does not undergo phase change is an increase in the coolant temperature from the inlet temperature to an outlet temperature.

In order to calculate the heat exchange within the CPU cooling system we imply using an aluminium heat sink with a number of fins N_f = 28, Height of fins h_f = 0.0355 m, Thickness of fins t_f = 0.0013 m, Shroud spacing h_b = 0.003 m, Base-Plate dimensions LxW=0,06 x 0,06 m, Power dissipation P = 22.5W, Heat sink width w = 0.06 m, Heat sink height h_{hs} = 0.037 m. We will now show the constants that were taken into account for the application that was developed in Mathcad: Ambient temperature (inside the calculator, through measurement) Ta = 316 K, Thermal conductivity of aluminium and air k_{Al} = 170.00 W/mK respectively k_{air}=0.026 W/mK, air density ρ=1.2 kg/m³, heat capacity of air C_p = 1.0 kJ/kgK, kinematics viscosity nu=1.5·10⁻⁵ m²/sec, Prandtl No. for air Pr = 0.7, Loss coefficient K = 0.9. In Figure 29 we show the graphical representation of the temperature variation depending on the CPU system structure when internal heat sources are present.

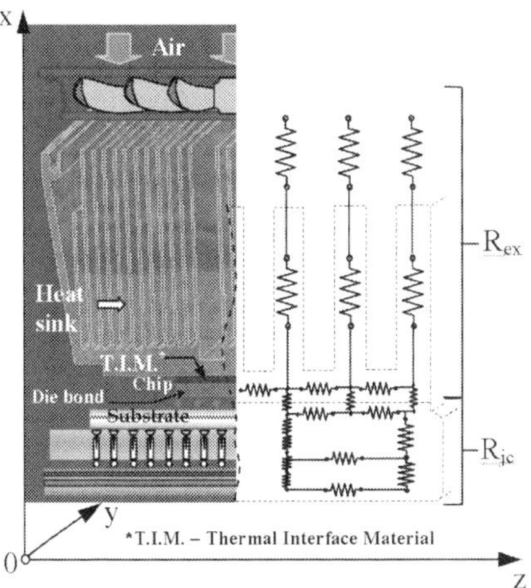

Fig. 28. Primary thermal resistances in a single-chip package

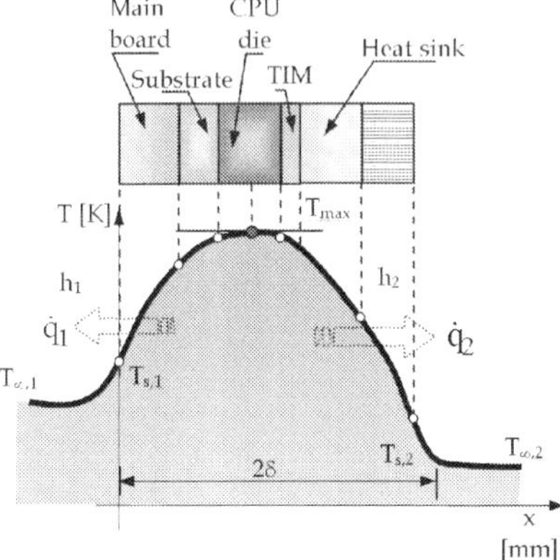

Fig. 29. Temperature variation in the CPU cooling system

In order to conduct the calculations we used the mathematical model proposed by J.P. Holman (Simons, 2004, Guenin, 2003). The area and the perimeter of the radiator are:

$$Area = w \cdot \left(h_b + h_f \right) - N_f \cdot t_f \cdot h_f \qquad \left[m^2 \right] \tag{8}$$

$$Perimeter = 2 \cdot \left(w + h_b + h_f \right) + N_f \cdot \left(t_f + 2 \cdot h_f \right) \quad [m] \tag{9}$$

Calculate hydraulic diameter of heat sink/shroud passage area:

$$D_{hyd} = \frac{4 \cdot Area}{Perimeter} \quad [m] \tag{10}$$

Holman indicates the initial use of a certain "guess value", marked as Vel. We gave this Vel an initial value Vel=0.2. The same author indicates the use of the relation below:

$$V = root \left[\left[Vel - \left[2 \cdot \frac{g \cdot P \cdot h_{hs}}{T_a \cdot Area \cdot C_p \cdot \rho \left[\left[\frac{64 \cdot nu}{Vel \cdot D_h} \right] \cdot \frac{h_{hs}}{D_h} + K \right]} \right]^{\frac{1}{3}} \right] \cdot Vel \right] \tag{11}$$

Calculate Reynolds' number:

$$Re = V \cdot \frac{D_{hyd}}{nu} \tag{12}$$

The frictional heating factor between two reciprocating parts in contact is:

$$f = \frac{64}{Re} \tag{13}$$

Reynolds' value on the direction of air flow in the radiator has the following expression:

$$\mathrm{Re}_x = \frac{W \cdot h_{hs}}{nu} \tag{14}$$

Wall heat flux:

$$Q_w = \frac{P}{\left(N_f \cdot 2 \cdot h_f + w \right) \cdot h_{hs}} \quad \left[\frac{W}{m^2} \right] \tag{15}$$

Calculate heat sink temperature rise:

$$\Delta T = \frac{Q_w \cdot \dfrac{h_{hs}}{k_{air}}}{\left(0.6795 \cdot \mathrm{Re}_x \right)^{\frac{1}{3}} \cdot \mathrm{Pr}^{\frac{1}{3}}} \quad [K] \tag{16}$$

Due to the fact that in the heat exchange process the convective effect steps in, Holman suggests for Nusselt number:

$$Nu_{(x)} = 0.453 \cdot \mathrm{Re}_x^{\frac{1}{2}} \cdot \mathrm{Pr}^{\frac{1}{3}} \tag{17}$$

The heat transfer coefficient:

$$h = \frac{\displaystyle\int_0^{h_{hs}} \left[Nu \cdot \frac{k_{air}}{x} \right] \cdot dx}{h_{hs}} \quad \left[\frac{W}{m^2} \right] \tag{18}$$

Calculate fin efficiency:

$$\eta_{fin} = \left(h_f + \frac{t_f}{2} \right)^{\frac{3}{2}} \cdot \left(\frac{h}{k_{al} \cdot t_f \cdot h_f} \right)^{\frac{1}{2}} \tag{19}$$

In order to determine the temperature field we will (Bejan, A. & Kraus A.D., 2003) use the Fourier equation:

$$\frac{\partial T}{\partial \tau} = \alpha \nabla_T^2 + \frac{q_v}{\rho c_p} \tag{20}$$

where $q_v = q_v(x,y,z,\tau)$ represents the CPU generated source density, measured in [W/m³]. By integrating the Fourier equation for the unidirectional, stationary regime, we obtain the expression of the temperature distribution in the wall:

$$T_{(x)} = -q_v \frac{x^2}{2k} + \left(\frac{T_{s,2} - T_{s,1}}{2\delta} + \frac{q_v \delta}{k} \right) \cdot x + T_{s,1} \quad [K] \tag{21}$$

Were $T_{s,1}$ $T_{s,2}$ being the temperatures of the exterior parts of the wall. The maximum temperature Tm in wall is achieved through x = x_m, resulting from condition: $\dfrac{dT}{dx} = 0$, that is:

$$x_m = \delta + \frac{k}{q_v} \cdot \frac{T_{s,2} - T_{s,1}}{2\delta} \qquad [m] \tag{22}$$

The maximum temperature zone [12] is found within plate $(0 \leq x_m \leq 2\delta)$, providing the following condition is observed:

$$-1 \leq \frac{k}{2 \cdot q_v \cdot \delta^2} \cdot \left(T_{s,2} - T_{s,1}\right) \leq 1 \tag{23}$$

If we replace x = x_m in equation (6) the maximum wall temperature is obtained:

$$T_{max} = \frac{q_v \cdot \delta^2}{2k} + \frac{k}{8 \cdot q_v \cdot \delta^2}\left(T_{s,2} - T_{s,1}\right)^2 + \frac{T_{s,1} + T_{s,2}}{2} \; [K] \tag{24}$$

If $T_{\infty,1,2}$ being the coolant temperatures (see figure 7), the limit conditions of third type are:

$$\text{If } x=0, \; ; -k\frac{dT}{dx}\bigg|_{x=0} = -h_1\left(T_{s,1} - T_{\infty,1}\right) \tag{25}$$

$$\text{If } x=2\delta, \; -k\frac{dT}{dx}\bigg|_{x=2\delta} = h_2\left(T_{s,2} - T_{\infty,2}\right) \tag{26}$$

We can determine the wall surfaces temperature [12]:

$$T_{s,1} = T_{\infty,1} + \frac{T_{\infty,2} - T_{\infty,1} + 2 \cdot \delta \cdot q_v\left(\dfrac{1}{h_2} + \dfrac{\delta}{k}\right)}{1 + \dfrac{h_1}{h_2} + 2\dfrac{h_1}{k}\delta} \qquad [K] \tag{27}$$

$$T_{s,2} = T_{\infty,2} + \frac{T_{\infty,1} - T_{\infty,2} + 2\delta q_v\left(\dfrac{1}{h_1} + \dfrac{\delta}{k}\right)}{1 + \dfrac{h_2}{h_1} + 2\dfrac{h_2}{k}\delta} \qquad [K] \tag{28}$$

We deem that the law of heat spreading throughout the entire volume is observed.
By first using the 1-19 expressions we calculate all the parameters that were previously mentioned. Taking into account the previously calculated measures, we determine, with relations 27 and 28, measures $T_{s,1}$ and $T_{s,2}$. With the help of relation 22, distance x_m which refers to the CPU core, where the temperature is highest, is calculated. The next step allows establishing the maximal temperature value T_{max} with relation 24 for verification of ulterior relations. Using relation 21 the maximum temperature field is determined, in plane z-y of CPU, through insertion of two matrices, which give the distance as well as the square distance in each knot. We thus moved away from a one-dimensional transfer to a bi-dimensional transfer. Knowing the maximum temperature field for each point of the matrix, the same law of heat transfer applies, on direction "x". The mathematic model proposed takes into account the thermal conduction coefficient "k" which is dependent of the type of material, inserting the corresponding values for each knot in the matrix. Sometimes it is common to use the transition from Cartesian coordinates to cylindrical coordinates. In order

to validate the suggested model we shall make a comparison between the obtained results and similar cases.

4.4 Results obtained through calculation

Following the calculation steps, performed with the help of Mathcad, as they were described above, if we regard the internal source of heat as being directly proportional to the energy generated by each kind of processor, then we can obtain the temperature variation corresponding to the CPU die area. The calculation results, as they are described in Figure 30a, are subsequent to the situations when TIM is unchanged. With regard to TIM imperfections taking the shape of nano or micro channels, such as those described in figure 13a, we ascertain by means of figure 30b that a temperature increment occurs, in amount of approximately 10 ⁰C, which might lead to CPU damage.

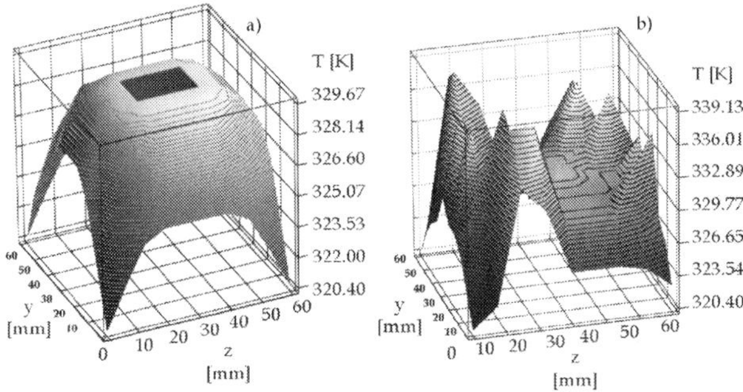

Fig. 30. The field of isotherms that corresponds to interface CPU: (a) for the same thermal conductivity coefficient and (b) in which case the coefficient of thermal conduction is altered.

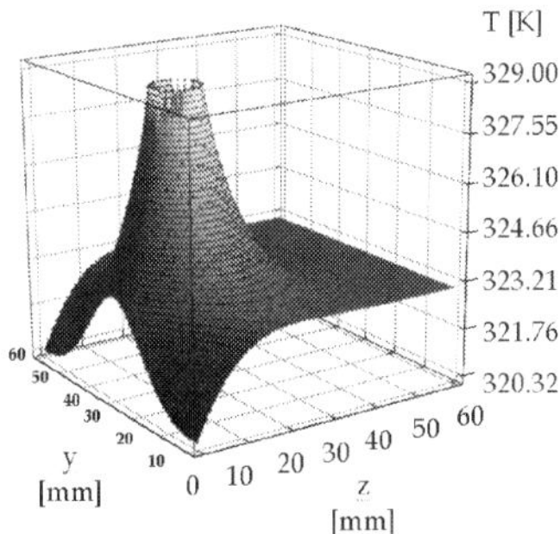

Fig. 31. Calculus in cylindrical coordinates for the field of isotherms that corresponds to for the same thermal conductivity coefficient interface TIM-CPU

Using a different calculation method, when TIM is unchanged, we obtain figure 31, thus noticing the preservation of the parabolic aspect below 323,21 K. However, the CPU area shows a conical shape that is specific to temperature increase. The values that were calculated in Mathcad are significantly close to the Cartesian model, as it can be noticed when comparing the obtained values to those comprised in figure 30a.

In order to study the way in which the temperature changes in the CPU cooling assembly, we conducted simulations using the ANSYS environment. The obtained results (Figure 32) were compared to other similar data. An overview of the CPU die – heat sink that was obtained by (Meijer, 2009) is referred to in figure 33. We can see that there is a uniform temperature field distribution and that the maximal value obviously relates to the CPU die area.

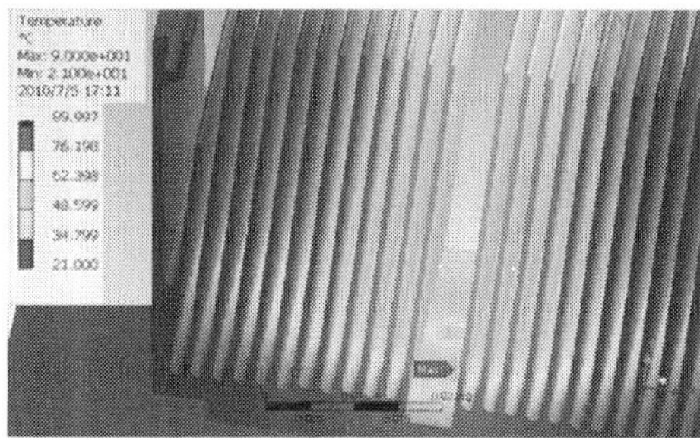

Fig. 32. The temperature field in the cooling assembly – view towards the heat sink obtained by Mihai

Fig. 33. Thermal modelling of the heat exchange for the CPU die – Heat Sink assembly (Meijer, 2009)

5. Conclusions

Considering the information that we described, we can conclude that there is a large variety of mini, macro and even nano channels inside the CPU cooling systems. In most cases they have a functional role in order to ensure the evacuation of the maximum amount of heat possible, using various criterions and effects such as Joule-Thompson or Peltier. We proved that the thermal interface material (TIM) plays an important role with regard to ensuring that the heat exchange is taking place. The AFM images of the CPU-cooler interface, showing that channels with complex geometry or stagnant regions can occur, disturbing the thermal transfer. Experimental investigations showed (figure 13) that even in an incipient phase, microchannels having $0,05 \div 0,01 \mu m$ in width, form in the TIM, at depths of at most 1000 Å, phenomenon explained as being a result of plastic characteristics upon deposition on CPU surface. Although the proportions of the channels that appear accidentally due to various reasons have nanometrical sizes, they can lead to anomalies in the CPU functioning, anomalies which are caused by overheating. The purpose of the measurements conducted by laser profilometry was to verify whether profile, waviness and roughness parameters show different variations under load and in addition to evaluate dilatation for increasing temperature.

These kind of experimental determinations allow us to make the following assessments:

i. Unwanted dilatation phenomena were experimentally outlined. This leads to a "pump up" effect for the material trapped at CPU – cooler interface, phenomenon also illustrated in (Viswanath et al., 2000);

ii. No surface discontinuities (localized lack of material) were observed during or after heating;

iii. It was clearly showed that shape deviations can appear when the material is freely applied on CPU surface, before cooler positioning (figure 17), but most of these variations are flattened after cooler placement as shown in figures 21.

iv. Thermal grease surface roughness evolution was monitored and it was illustrated that its mean values show no major changes after temperature increase, which indicates a good thermal stability of the used material .

Currently, several mathematical models are completed, and the VSS and HS models were adopted, indicating the role of thermal contact resistance. The conducted calculations are relevant in this respect in order to study what happens when the TIM is deteriorated. The mathematical results clearly indicate that any strain in the interface material leads to a change in thermal contact resistance, with an effect on CPU overheating. The results obtained for rectangular channels with air have the same magnitude order as the ones obtained by (Colin, 2006) and the shape of the graphs identical with the one obtained by authors (Niu et al., 2007). The validation of the mathematical model adopted is therefore completed. In the future additional research is required with regard to TIM stability, in order to counter the development of nano or micro channels.

6. References

Banton, R. & Blanchet D. (2004). Utilizing Advanced Thermal Management for the Optimization of System Compute and Bandwidth Density, *Proceeding of CoolCon MEECC Conference, pp. 1-62*, PRINT ISSN #1098-7622 online ISSN #1550-0381, Scottsdale, Arizona, (May 2004), Publisher ACM New York, NY, USA

Bejan, A. & Kraus A.D. (2003). *Heat transfer handbook*, Publisher John Wiley & Sons Inc. Hoboken, ISBN 0-471-39015-1, New Jersey, USA

Colin, S.; Lalonde, P. & Caen, R. (2004). Validation of a Second-Order Slip Flow Model in Rectangular Microchannels, *Heat Transfer Engineering*, Volume 25, No. 3., (mars 2004) 23 – 30, ISSN 0145-7632 print / 1521-0537 online

Colin S. (2006). Single-phase gas flow in microchannels, In: *Heat transfer and fluid flow in minichannels and microchannels*, Elsevier Ltd, 9-86, ISBN: 0-0804-4527-6, Great Britain

Escher, W.; Brunschwiler, T., Michel, B. & Poulikakos, D. (2009). Experimental Investigation of an Ultra-thin Manifold Micro-channel Heat Sink for Liquid-Cooled Chips, *ASME Journal of Heat Transfer*, Volume 132, Issue 8, (August 2010) 10 pages, ISSN 0022-1481

Escher, W.; Michel, B. & Poulikakos, D. (2009). A novel high performance, ultra thin heat sink for electronics, *International Journal of Heat and Fluid Flow*, Volume 31, Issue 4, (August 2010), 586-598, ISSN 0142-727X

Grujicic, M.; Zhao, C.L. & Dusel, E.C. (2004). The effect of thermal contact resistance on heat management in the electronic packaging, *Applied Surface Science*, Vol. 246 (December 2004), 290–302, ISSN 0169-4332

Guenin, B. (2003). Calculations for Thermal Interface Materials, *Electronics Cooling*, Vol. 9, No. 3, (August 2003), 8-9, Electronic Journal

Hadjiconstantinou, N. & Simek, O. (2002). Constant-Wall-Temperature Nusselt Number in Micro and Nano-Channels, *Journal of Heat Transfer*, Vol. 124, No. 2, (April 2002) 356-364, ISSN 0022-1481

Holman, J.P. (1997). *Heat transfer*, 8th ed., published by McGraw Hill, pp. 42-44, New York:, 1997. ISBN 0-07-029666-9

Kandlikar, S. & Grande, W. (2003). Evolution of Microchannel Flow Passages— Thermohydraulic Performance and Fabrication Technology, *Heat Transfer Engineering*, Vol. 24, No. 1, (Mars 2003), 3-17, ISSN 1521-0537

Kandlikar, S.; Garimella, S., Li D., Colin, S., King, M. (2005). *Heat transfer and fluid flow in minichannels and microchannels*, Elsevier Publications, ISBN: 0-08-044527-6, Great Britain

Kavehpour, H. P.; Faghri, M., & Asako, Y. (1997). Effects of compressibility and rarefaction on gaseous flows in microchannels, *Numerical Heat Transfer* part A, Volume 32, Issue 7, November 1997, 677–696, ISSN 1040-7782, Online ISSN: 1521-0634

Kim, D-K. & Kim, S. J. (2007). Closed-form correlations for thermal optimization of microchannels, *International Journal of Heat and Mass Transfer*, Vol. 50, No. 25-26. (December 2007) 5318–5322, ISSN 0017-9310

Lasance, C., & Simons, R. (2005). Advances in High-Performance Cooling For Electronics, *Electronics Cooling*, Vol.11, No. 4, (November 2005), 22-39, Electronic Journal

Lee, S. (1998). Calculating spreading resistance in heat sinks, *Electronics Cooling*, Vol. 4, No. 1., (January 1998), 30-33, Electronic Journal

Lienhard, J.H.IV. & Lienhard, J.H.V. (2003). *A heat transfer textbook*, 3 rd ed., published by Phlogiston Press, ISBN/ASIN: 0971383529, Cambridge-Massachusetts, USA

Meijer, I.; Brunschwiler T., Paredes S. & Michel B. (2009). Advanced Thermal Packaging, *IBM Research GmbH Presentation*, (nov.2009), pp.1-52, *Zurich Research Laboratory*

Mihai, I.; Pirghie, C. & Zegrean, V. (2010). Research Regarding Heat Exchange Through Nanometric Polysynthetic Thermal Compound to Cooler–CPU Interface, *Heat Transfer Engineering*, Volume 31, No. 1. (January 2010) 90 – 97, ISSN 1521-0537

Niu X.D.; Shu C. & Chew Y.T. (2007). A thermal lattice Boltzmann model with diffuse scattering boundary condition for micro thermal flows, *Computers & Fluids*, No. 36, (March 2006) 273-281, ISSN 0045-7930

Pautsch G. (2005). Thermal Challenges in the Next Generation of Supercomputers, *Proceeding of CoolCon MEECC Conference, pp. 1-83*, PRINT ISSN #1098-7622 online ISSN #1550-0381, Scottsdale, Arizona, (May 2005), Publisher ACM New York, NY, USA

Simons, R.E. (2004). Simple Formulas for Estimating Thermal Spreading Resistance, *Electronics Cooling*, Vol. 10, No. 2, (May 2004), 8-10, Electronic Journal

Viswanath, R.; Wakharkar, V., Watwe, A., & Lebonheur, V. (2000). Thermal Performance Challenges from Silicon to Systems, *Intel Technology Journal, Vol. Q3*, (Mars 2000), pp. 1-16, ISSN 1535-864X

Yovanovich, M.M.; Culham, J.R., & Teertstra, P. (1997). Calculating Interface Resistance, *Electronics Cooling*, Vol. 3, No. 2, (May 1997), 24-29, Electronic Journal

5

Microchannel Heat Transfer

C. W. Liu[1], H. S. Ko[2] and Chie Gau[2]
*[1]Department of Mechanical Engineering, National Yunlin University of Science and
Technology, Yunlin 64002
[2]Institute of Aeronautics and Astronautics, National Cheng Kung University, Tainan
70101,
Taiwan*

1. Introduction

Microchannel Heat transfer has the very potential of wide applications in cooling high power density microchips in the CPU system, the micropower systems and even many other large scale thermal systems requiring effective cooling capacity. This is a result of the micro-size of the cooling system which not only significantly reduces the weight load, but also enhances the capability to remove much greater amount of heat than any of large scale cooling systems. It has been recognized that for flow in a large scale channel, the heat transfer Nusselt number, which is defined as hD/k, is a constant in the thermally developed region where h is the convective heat transfer coefficient, k is thermal conductivity of the fluid and D is the diameter of the channel. One can expect that as the size of the channel decrease, the value of convective heat transfer coefficient, h, becomes increasing in order to maintain a constant value of the Nusselt number. As the size of the channel reduces to micron or nano size, the heat transfer coefficient can increase thousand or million times the original value. This can drastically increase the heat transfer and has generated much of the interest to study microchannel heat transfer both experimentally and theoretically.

On the other hand, the lab-on-chip system has seen the rapid development of new methods of fabrication, and of the components — the microchannels that serve as pipes, and other structures that form valves, mixers and pumps — that are essential elements of microchemical 'factories' on a chip. Therefore, many of the microchannels are used to transport fluids for chemical or biological processing. Specially designed channel is used for mixing of different fluids or separating different species. It appears that mass or momentum transport process inside the channel is very important. In fact, the transfer process of the mass is very similar to the transfer process of the heat due to similarity of the governing equations for the mass and the heat (Incropera et al., 2007). It can be readily derived that the Nusselt number divided by the Prandtl number to the nth power is equal to the Sherdwood number (defined as the convective mass transfer coefficient times the characteristic length and divided by the diffusivity of the mass) divided by the Schmidt number (defined as the kinematic viscosity divided by the diffusivity of the mass) to the nth power. Understanding of the heat transfer can help to understand the mass transfer or even the momentum transfer inside the microchannel (Incropera et al., 2007).

However, the conventional theories, such as the constitutive equations describing the stress and the rate of deformation in the flow, or the Fourier conduction law, are all established based on the observation of macroscopic view of the flow and the heat transfer process, but do not consider many of the micro phenomena occurred in a micro-scale system, such as the rarefaction or the compressibility in the gas flow, and the electric double layer phenomenon in the liquid flow, which can significantly affect both the flow and the heat transfer in a microchannel. Therefore, both the flow and the heat transfer process in a microchannel are significantly different from that in a large scale channel. A thorough discussion and analysis for both the flow and the heat transfer process in the microchannels are required. In addition, experimental study to confirm and validate the analysis is essential. However, accurate measurements of flow and heat transfer information in a microchannel rely very much on the exquisite fabrication of both the microchannel and the microsensors by the MEMS techniques. Successful fabrication of these complicated microchannel system requires a good knowledge on the MEMS techniques. Especially, accurate measurement of the heat transfer inside a microchannel heavily relies on the successful fabrication of the microchannel integrated with arrays of miniaturized temperature and pressure sensors in addition to the fabrication of micro heaters to heat up the flow.

It appears that microfluidics has become an emerging science and technology of systems that process or manipulate small (10^{-9} to 10^{-18} liters) amounts of fluids, using channels with dimensions of tens to hundreds of micrometres (George, 2006; Vilkner et al., 2004; Craighead, 2006). Various long or short micro or nanochannels have used in the system to transport fluids for chemical or biological processing. The basic flow behavior in the microchannel has been studied in certain depth (Bayraktar & Pidugu, 2006; Arkilic & Schmidt., 1997; Takuto et al., 2000; Wu & Cheng, 2003). The major problem in the past is the difficulty to install micro pressure sensors inside the channel to obtain accurate pressure information along the channel. Therefore, almost all of the pressure information is based on the pressures measured at the inlet and the outlet outside of the channel, which is used to reduce to the shear stress on the wall. The measurements have either neglected or subtracted an estimated entrance or exit pressure loss. These lead to serious measurement error and conflicting results between different groups (Koo & Kleinstreuer, 2003). The friction factor or skin friction coefficient measured in microchannel may be either much greater, less than or equal to the one in large scale channel. Different conclusions have been drawn from their measurement results and discrepancies are attributed to such factors as, an early onset of laminar-to turbulent flow transition, surface roughness (Kleinstreuer & Koo 2004; Guo & Li 2003), electrokinetic forces, temperature effects and microcirculation near the wall, and overlooking the entrance effect. In addition, when the size or the height of the microchannel is much smaller than the mean free path of the molecules or the ratio of the mean free path of the molecules versus the height of the microchannel, i.e. Kn number, is greater than 0.01, one has to consider the slip flow condition on the wall (Zohar et al. 2002; Li et al. 2000; Lee et al., 2002). It appears that more accurate measurements on the pressure distribution inside the microchannel and more accurate control on the wall surface condition are necessary to clarify discrepancies amount different work.

The lack of technologies to integrate sensors into the microchannel also occurs for measurements of the heat transfer data. All the heat transfer data reported is based on an average of the heat transfer over the entire microchannel. That is, by measuring the bulk flow temperature at the inlet and the outlet of the channel, the average heat transfer for this channel can be obtained. No temperature sensors can be inserted into the channel to acquire the local heat transfer data. Therefore, detailed information on the local heat transfer distribution inside the channel is not reported. In addition, the entry length information and the heat transfer process in the thermal fully developed region is lacking. Besides, the wall roughness inside the channel could not be controlled or measured directly in the tube. Therefore, its effect on the heat transfer is not very clear. This was attributed to cause large deviation in heat transfer among different work (Morini 2004; Rostami et al., 2002; Guo & Li, 2003; Obot, 2002). It appears that accurate measurements of the local heat transfer are required to clarify the discrepancies among different work.

Therefore, in this chapter, a comprehensive review of microchannel flow and heat transfer in the past and most recent results will be provided. A thorough discussion on how the surface forces mentioned above affect the microchannel flow and heat transfer will also be presented. A brief introduction on the MEMS fabrication techniques will be presented. We have developed MEMS techniques to fabricate a microchannel system that can integrate arrays of the miniaturized both pressure and temperature sensor. The miniaturized sensors developed will be tested to ensure the reliability, and calibrated for accurate measurements. In fact, fabrication of this microchannel system requires very complicated fabrication steps as mention by Chen et al. 2003a and 2003b. Successful fabrication of this channel which is suitable for measurements of both the local pressure drop and heat transfer data is a formidable task. However, fabrication of this complicated system can be greatly simplified by using polymer material (Ko et al., 2007). This requires fabrication of pressure sensor using polymer materials (Ko et al., 2008). The polymer materials that have a very low thermal conductivity can be fabricated as channel wall to provide very good thermal insulation for the channel and significantly reduce streamwise conduction of heat along the wall. This allows measurements of very accurate local heat transfer inside the channel. In addition, the height of the channel can be controlled at desired thickness by spin coating the polymer at desired thickness. The shape of the channel can be readily made by photolithography. All the design and fabrication techniques for both the channel and the sensor arrays will be discussed in this chapter. Measurements of both the local pressure drop and heat transfer inside the channel will be presented and analyzed. Therefore, the contents of the chapter are briefly described as follows:

1. Gas flow and the associated heat transfer characteristics in microchannels.
2. Liquid flow and heat transfer characteristics in microchannels including (a) the single phase and (b) the two phase flows.
3. MEMS fabrication techniques
4. Discussion on recent developments and challenges faced for MEMS fabrication of the microchannel system.
5. Working principle and fabrication of the miniaturized pressure and temperature sensors.
6. Fabrication of the complicated microchannel system integrated with arrays of either or both the miniaturized pressure and temperature sensors.
7. Local heat transfer and pressure drop inside the microchannels.

2. Gas flow characteristics in microchannels

Recent development of micromachining process which has been used to miniaturize the fluidic devices has become a focus of interest to industry, e.g. micro cooling devices, micro heat exchangers, micro valves and pumps, and lab-on-chips, more studies have been dedicated to this field. The fluid flows in micro scale capillary tube can be traced back to Knudsen at 1909. However, it has been very difficult to perform an experiment for micro scale flow and make detailed observation in a micro-channel due to the lack of techniques to fabricate a microchannel and make arrays of small sensors on the channel surface. Up to the present, most of the important information on micro scale thermal and flow characteristics inside the microchannel can not be obtained and measured. Instead, the flow and heat transfer experiments performed for micro scale flow in the past are mostly based on the measurements of pressures or temperatures at inlet and outlet of the channel and the mass flow rate, or the measurements on the surface of a relatively large scale channel. Therefore, some of peculiar transport processes which are not important in a large scale channel may play a dominant role to affect the flow and heat transfer process in the micro scale channel, e.g. the rarefaction effect of the gas flow. Therefore, the rarefaction of a gas flow in the microchannel should be taken into account in the analysis.

2.1 Theoretical analysis

In order to describe the rarefaction of gaseous flow, a ratio of the mean free path to the characteristic length of the flow called Knudsen number (Kn) has been used as a dimensionless parameter. The Knudsen number is defined as λ/D_c, where "λ" denotes the mean free path of gas molecules and "D_c" denotes the characteristic dimension of the channel. For convenience, it has been suggested (Tsien, 1948) that the rarefaction in gases can be typically classified into three flow regions by the magnitude of the Knudsen number, which are "the continuum flow regime", "the free-molecular flow regime" and "the near-continuum flow regime", as described as follows.

1. Continuum flow regime: This regime is defined for flow with Kn < 0.001. In this regime, the theories of the gas flow and fluid properties completely conform to the continuum assumption, and the Knudsen numbers approach to zero. In addition, the modified classical theories of the liquid flow are also suitable in this regime.

2. Near-continuum flow regime: this flow regime is defined in the range with $0.001 \leq Kn < 10$. The Knudsen number in this flow regime is still large enough that the flow is subject to a slight effect of rarefaction. The flow can be considered as a continuum in the core region except in the region adjacent to the wall where a small departure from the continuum such as velocity-slip or temperature jump is assumed. For convenience, one can further subdivide the flow into two regimes, i.e. the slip-flow regime and the transition-flow regime. In the slip-flow regime, the macroscopic continuum theory, therefore, is still valid due to small departures from the continuum. However, in order to conform to the real-gas behavior, it is necessary to adopt some appropriate corrections for the slip of fluid at the boundary. The slip-flow regime is defined in the range of $0.001 \leq Kn < 0.1$ while the transition-flow regime is defined in the range of $0.1 \leq Kn < 10$. In the transition-flow regime, the intermolecular collisions and the collisions between the gaseous molecules and the wall are of more or less equal importance. The flow configuration can be regarded as neither a continuum, nor a free-molecular flow. There is no simplified approach to attack this problem. Some conventional methods, such as, directly solving the complete sets of Boltzmann equations or using the empirical correlations from the experimental data, have been adopted.

3. Free-molecular flow regime: This flow regime is defined in the region with $10 \leq Kn$. The rarefaction effect dominates the entire flow field. The gas is so rarefied that

intermolecular collisions can be negligible. Hence, the flow characteristic is described by the kinetic theory of gas. Only interaction between gas molecules and boundary surface is considered.

Meanwhile, it has also been suggested (Tsien, 1946) that one can employ the kinetics theory of gases or the conventional heat transfer theory to study the gas flow in the continuum flow regime. When the gaseous rarefaction is within the range of the free-molecular flow regime, the kinetics theory of gases is suitable for use. However, in the range of the near-continuum flow regime, there has been no well-established method. In the slip-flow regime the gas flow can be considered as continuum. Hence, we can employ the macroscopic continuum theory to study the heat transfer in gases by taking account the velocity-slip and temperature-jump conditions at the wall. In the transition-flow regime the transport mechanisms in the rarefied gas are between the continuum and the free molecule flow regime, it is incorrect to consider the gas as a continuum or free molecule medium. Therefore, the theoretical study in the transition regime is very difficult. Many of the works (Ko et al., 2008, 2009, 2010; Bird et al., 1976a; Eckert and Drake, 1972; Yen, 1971; Ziering, 1961; Takao, 1961; Kennard, 1938) intend to develop some convenient methods to solve this problem, such as enlarging the validation of macroscopic continuum theory by using some corrections in boundary conditions or developing mathematical schemes to directly solve the highly nonlinear Boltzmann equation. However, these approaches are still not successful.

For theoretical study of the rarefied-gas flow, Kundt and Warburg (1875) have been the first to propose an important inference by experimental observation. They found an interesting phenomenon that the gaseous flow exhibits a velocity-slip on solid wall when the pressure in the system is sufficiently low. This phenomenon later has been confirmed by the analytical results from kinetics theory of gas by Maxwell (1890). In addition, Maxwell also defined a parameter "f_S" called tangential momentum accommodation coefficient to modify the departures from the theoretical assumptions and real-gas behavior in molecular collision processes. The value of f_S will presumably depend upon the character of the interaction between the gaseous molecules and the wall, such as the surface roughness or the temperature etc. In the observations of wall slip, Timiriazeff (1913) made the first direct measurements of wall slip. However, the most accurate measurements of velocity slip are undoubtedly made by Stacy and Van Dyke, respectively. Hence, a sound theory used to describe the rarefied gas behaviors has been established successfully. In the heat transfer studies, Smoluchowski (1910) has performed the first experiments for a heated rarefied gas flow and found the temperature-jump occurring on the solid wall.

Kennard (1938) has suggested that it could be analogous to the phenomenon of velocity slip and thus developed an approximate expression to describe this temperature discontinuity. In a flow field with a temperature of the gas flow different from the neighboring solid wall, there exists a temperature difference in a small distance "g", which is called temperature jump distance, between the gas and the solid wall. The jump distance "g" is inversely proportional to the pressure but directly proportional to the mean-free-path of the gas. Due to the very small jump distance, it looks as having a discontinuity in the temperature distribution between the gas flow and the neighboring solid wall. By using the thermal accommodation coefficient proposed by Knudsen (1934) and the concepts of heat transfer mechanism between gas molecules defined by Maxwell, a theory for the microscopic heat transfer occurred in the rarefied gas flows has been successfully established.

In addition, the gas flow in a micro-channel also involves other problems, such as compressibility and surface roughness effects. Therefore, other dimensionless parameters, such as the Mach number, Ma, and the Reynolds number, Re, should also be adopted. The relationship among these parameters has been derived and can be expressed as follows.

$$\text{Re} = \sqrt{\frac{k\pi}{2}} \frac{Ma}{Kn} \tag{2-1}$$

where k is the specific heat ratio (c_p/c_v) of the gas. Since both Ma and Kn vary with compressibility of gas in the channel, the value of Re should vary according to the above equation. The full set of governing equations for two dimensional, steady and compressible gas flows can be written as follows (Khantuleva et al., 1982):

$$\frac{\partial(\rho u)}{\partial x} + \frac{\partial(\rho v)}{\partial y} = 0 \tag{2-2}$$

$$\rho u \frac{\partial u}{\partial x} + \rho v \frac{\partial u}{\partial y} = -\frac{\partial p}{\partial x} + \mu[\frac{\partial^2 u}{\partial x^2} + \frac{\partial^2 u}{\partial y^2} + \frac{1}{3}(\frac{\partial^2 u}{\partial x^2} + \frac{\partial^2 v}{\partial x \partial y})] \tag{2-3}$$

$$\rho u \frac{\partial v}{\partial x} + \rho v \frac{\partial v}{\partial y} = -\frac{\partial p}{\partial y} + \mu[\frac{\partial^2 v}{\partial x^2} + \frac{\partial^2 v}{\partial y^2} + \frac{1}{3}(\frac{\partial^2 v}{\partial x^2} + \frac{\partial^2 u}{\partial x \partial y})] \tag{2-4}$$

$$\rho u C_p \frac{\partial T}{\partial x} + \rho v C_p \frac{\partial T}{\partial y} = u \frac{\partial p}{\partial x} + v \frac{\partial p}{\partial y} + k(\frac{\partial^2 T}{\partial x^2} + \frac{\partial^2 T}{\partial y^2}) + \mu[2(\frac{\partial u}{\partial x})^2 + 2(\frac{\partial v}{\partial y})^2 + (\frac{\partial v}{\partial x} + \frac{\partial u}{\partial y})^2 - \frac{2}{3}(\frac{\partial u}{\partial x} + \frac{\partial v}{\partial y})^2] \tag{2-5}$$

$$p = \rho R T = n k T \tag{2-6}$$

The boundary conditions for the velocity slip and temperature jump on the top and bottom walls are shown as follows (Wadsworth et al., 1993):

$$u - u_w = \frac{2 - \sigma_u}{\sigma_u} \lambda (\frac{\partial u}{\partial y})_w + \frac{3}{4} \frac{\mu}{\rho T} (\frac{\partial T}{\partial x})_w; \quad y = \pm h/2 \tag{2-7}$$

$$T - T_w = \frac{2 - \sigma_T}{\sigma_T} \frac{2\gamma}{(\gamma + 1)} \frac{\lambda}{\text{Pr}} (\frac{\partial T}{\partial y})_w; \quad y = \pm h/2 \tag{2-8}$$

where σ_u and σ_T are the momentum and the energy accommodation coefficient, respectively. λ, γ and h are the mean free path, the specific heat ratio and the height of the microchannel, respectively. Review of the recent literature indicates that compressible gas flow problems have been studied from the slip to the continuum flow regimes, however, different results are obtained in the micro-channels as described in the following paragraphs.

To analyze the rarefied gas characteristics in the near-continuum flow regime, the methods used (Takao, 1961; Kennard, 1938) in the classical kinetics theory of gas include (1) the small-perturbation approach, (2) the moment methods and (3) the model equation. The mathematical procedures of the small-perturbation approach are to use the perturbation technique to linearize the Boltzmann equation. Since this method can be used in both the near-continuum regime and the near free-molecules regime, therefore, it is suitable for practical applications. The moment methods are first to make adequate assumptions in the velocity distribution f such as to express f in terms of a power series, i.e. $f = f_0(1 + a_1(Kn) + a_2(Kn)^2 +...)$ as proposed by Chapman and Enskog. Then, substitute the assumed velocity distribution into the Boltzmann equation. The methods of the model equation are to construct a physics model, such as the B-G-K model proposed by Bhatnagar, Gross and Krook (1954), to simplify the expression of Boltzmann equation. Since the governing equation of the system is greatly simplified by the appropriate assumptions in the previous two methods, these approaches can be used for limited ranges of flows. In the numerical simulation (Bird, 1976a; Yen, 1971; Ziering, 1961), a very efficient computational scheme, i.e. DSMC (Direct Simulation Monte Carlo) method, has been developed. However, this method still suffers from the highly nonlinear behavior in the Boltzmann equation. Meanwhile, the use of different approach to solve even the same physical problem will encounter different difficulties due to the different advantages and limitations faced by each method. In addition, the predictions from the analysis should be confirmed by the experiments.

In the studies of numerical calculation, Beskok and Karniadkis (1994) have developed a scheme called "spectral element technique" to simulate the momentum and heat transfer processes of a rarefied gas subjected to either a channel-flow or an external-flow condition. The results have indicated that when the gas passes through a micro-channel at velocity-slip condition, it can cause a significant reduction in drag coefficient C_D on the walls. This is mainly caused by the thermal-creep effect when the Knudsen number increases significantly. Meanwhile, they have also addressed that the thermal-creep effect of the gas flow in a uniformly heated micro-channel can increase the mass flow rate, and the increase can be greatly enhanced by raising the inlet velocity. In addition, other effects, i.e. the compressibility and the viscous heating effects that may be occurred in the rarefied gas flow should also be considered. Chu et al. (1994) has used numerical analysis to evaluate the efficiency of heat removal when gas flows through an array of micro-channel under continuum or the velocity-slip condition. This numerical simulation is intended to study the cooling performance inside a micro-channel array that fabricated in a silicon chip. The numerical approaches have adopted the finite-difference methods incorporated with SOR (Successive over-relaxation) techniques to solve the problem with Neumann boundary conditions. The assumptions used include fully developed hydrodynamic condition, fully developed thermal condition and uniform heating on the bottom wall with the top wall well insulated. From the numerical results they have found that even though the temperature-jump causes decrease in Nusselt number that is contrary to continuum flow, the entire heat transfer performance were still higher than the case of continuum flow; this peculiar phenomenon is mainly due to the velocity-slip effects that induce greater mass flow per unit time into the channel. Therefore, the design of gas flow through a micro-channel array at the slip-flow regime as cooling is suggested. Fan and Xue (1998) have used the numerical method of the "DSMC" to simulate the gas flow in micro-channels at the slip-flow regime. They have assumed that the gas flow is simultaneously subjected to the effects of the velocity-slip and the compressibility. In addition, the effects of pressure ratio "P_0" between two ends of the micro-channel on the flow are also studied. Simulation analysis was carried out under different ratios of P_0, and the results indicated that the velocity-profiles of the flow near both ends of the channel are deviated from the parabolic profile. The mean flow velocity near the channel outlet increases greatly by increasing the ratio of P_0. The deviation from the parabolic profile is caused mainly by both the entrance and the exit effect of the microchannel, only the flow field far from the end of the micro-channel can conform to the fully developed flow conditions. The second account of flow acceleration is not only significantly affected by the velocity-slip, but also induced by the compressibility of gas. Since the compressibility effect causes decrease in both the density and the pressure near the exit of channel, and the greater decrease in the exit pressure can accelerate the flow again to make up the density drop. Therefore, acceleration of the flow in a microchannel can be increased by increasing the pressure ratio P_o. Meanwhile the slip flow characteristics in the channel can be observed from the simulation results for the shear stress and velocity distributions near the wall region. The results further exhibit that the compressibility induced by the increase of P_o can greatly affect the gas flow behavior when the flow in the microchannel is at the slip-flow regime.

2.2 Experimental measurements

For experiments of gas flow in micro-channels, Wu and Little (1983) have measured the friction factors for both laminar and turbulent gas flows in trapezoidal channels. The widths of the channels are from 130 to 200 μm and the depths are from 30 to 60 μm, respectively. The working fluids used include nitrogen, helium and argon gases. The friction factors, f, obtained in his experiment are larger than the theoretical prediction for the critical Reynolds number less than 400. The deviations of the data form the prediction are attributed to the very high degree of surface roughness and measurement uncertainty. For a nitrogen gas flow in micro-tubes, the effects of wall surface roughness on the pressure drop or the friction factors are studied by Choi et al. (1991) for both laminar and turbulent flow. The micro-tube diameters are from 3 to 81 μm and the wall roughness is from 0.00017 to 0.0116. It is found that the Poiseuille number, Po, which is defined as f × Re, is 53 in the laminar region when the diameter of the tube is less than 10 μm. The Po of 53 in his experiment is lower than the theoretical value of 64 for fully developed laminar flow in the large scale tube, where the Po is kept as a constant. In the experiments of turbulent flow region, the results indicate that the Colburn analogy is not valid when the diameter of micro-tubes is less than 80 μm.

Some of pressure drop measurements have a good agreement with the predictions of the conventional theory. Acosta et al. (1985) has measured the friction factors in rectangular micro-channels, and the results are very close to the friction factor predicted by the conventional theory in small aspect ratios channels. Lalonde et al. (2001) has studied the friction factor of air flow in a micro-tube with a diameter of 52.8 μm. The experimental data has a good agreement with the predictions from the conventional theory. Turner et al. (2001) has performed an experiment to measure the friction factor with different working fluids, such as nitrogen, helium and air in microchannels with hydraulic diameters varying from 4 to 100 μm. The walls of the rectangular channels consider both the rough and the smooth wall conditions. The results indicate that the friction factors in laminar region for both the rough and the smooth wall conditions have good agreement with the conventional theory.

In contrast to the results that agree with the conventional theory, Pfahler et al. (1990a, 1990b) and Pfahler et al. (1991) have performed experiments to obtain the friction factor for working fluids of helium and nitrogen in micro-channels with the heights varying from 0.5 to 40 μm. The results indicate a significant reduction of C_f (Po_{exp}/Po_{theo}) which is a function of channel depth. The C_f decreases with decreasing Re in the smallest channel. Yu et al. (1995) has performed the experiments of gas flow in a micro-channel with either a trapezoidal or a rectangular cross section. The hydraulic diameter varies between 1.01 and

35.91 μm. They have observed a friction factor smaller than the prediction of the conventional theory, and conclude that the deviation may be caused by both effects of compressibility and rarefaction of the gas. Harley et al. (1995) has performed the experiments for subsonic, compressible flow in a long micro-channel. The working fluids used are nitrogen, helium and argon gases. The channels are fabricated by silicon wafer, and the dimensions of the channels are 100 μm wide, 10 mm long with depths varied from 0.5 to 20 μm. The experimental data have been presented in terms of the Po with hydraulic diameter from 1 to 36 μm. The measured friction factors agree with the theoretical prediction, but become smaller when the depth of channel decreases to 0.5 μm. The reduction in the friction factor is attributed to the occurrence of slip flow. The compressibility effects are also found by Li et al. (2000) who have performed an experiment of nitrogen gas flow in five different micro-tubes with diameters from 80 to 166 μm. The pressure drop along the tube became nonlinear when the Much number is higher than 0.3.

In order to understand more detailed pressure information inside a micro-channel, arrays of the pressure sensors should be integrated in the micro-channel for measurement of pressure distribution. Pong et al. (1994) are the first to present that a rectangular micro-channel can be fabricated with integrated arrays of pressure sensors for pressure distribution measurements. Both the helium and the nitrogen gas are used as the working fluid in his study. The channels are from 5 to 40 μm wide, 1.2 μm deep and 3000 μm long. The experimental results indicate that the pressure distribution is not linear and is lower than the prediction based on the continuum flow analysis in the micro-channel. The non-linear effects are caused by both effects of rarefaction and compressibility of the gas due to the high pressure loss. Liu et al. (1995) have used the similar channel as in Pong et al. (1994) but having different shapes to perform the experiments. The channel has a uniform cross section and has the dimensions of 40 μm wide, 1.2 μm deep and 4.5 mm long. The pressure drop distribution found is also nonlinear. For the channel with non-uniform cross section, sudden pressure changes are found at locations where variations of the cross section occur. In the mean time, analysis of the channel flow has also been performed with the assumptions of a steady, isothermal, and continuum flow with wall slip condition. However, the analysis can not explain the small pressure gradients measured near the inlet and the outlet of the channel.

Shih et al. (1996) has repeated the experiments of Pong by using a similar micro-channel with dimensions of 40 μm wide, 1.2 μm deep and 4000 μm long to measure the pressure distribution and mass flow rate for helium or nitrogen gas flow. The results of helium have a good agreement with the analysis based on the Navier-Stokes equations with slip boundary condition. The boundary condition of a slip flow on the wall is given by

$$u_w = \psi Kn(\partial u / \partial y) \tag{2-9}$$

where ψ is momentum accommodation coefficient. In general, $\psi = 1$ has been used for engineering calculation. All the experimental data indicate a non-linear dependence of the pressure drop with the mass flow rate. Li et al. (2000) and Lee et al. (2002) have performed experiments for channels with orifice and venture elements. The dimensions of channels are 40 μm wide, 1 μm deep and 4000 μm long. The working fluid used is nitrogen which has an inlet pressure up to 50 Psig. The mass flow rates are measured as a function of the pressure drop. The results indicate that the pressure distribution is non-linear and the pressure drop is a function of mass flow rate. The experimental data are used to compare with the prediction from the Navier-Sotkes equation with a slip boundary condition. The friction factors for both channels with either the orifice or the venture are all lower than theoretical prediction.

It appears that contradictory results have been found in the previous studies. More accurate measurements of the pressure drop and heat transfer inside a microchannel are required. This requires fabrication of a micro-channel system, integrated with arrays of micro pressure sensors or temperature sensors, fabricated by surface micromachining process. However, the microchannel fabricated previously with arrays of pressure sensor is limited to a channel height of 1.2 μm due to the use of oxide sacrificial layer which is deposited by chemical vapor deposition (CVD) process. Much thicker deposition of the oxide layer is not possible with the current technology. In addition, the channel structure is very weak due to fabrication of the channel wall with a very thin film, only gas flow is allowed for the experiment. Therefore, in order to provide a channel which has a much greater height and is suitable for liquid flow conditions with a strong wall, an entirely new fabrication process for the channel should be considered and designed.

3. Liquid flow characteristics in microchannels

The liquid flow can be regarded as a continuum even in a very small channel. However, liquid flow can become boiling when the wall temperature is higher than the vaporization temperature of the liquid. Therefore, the liquid flow regimes can be divided into the single phase flow and the two phase flow regime. The real behaviors of heat transfer in the laminar or the transition flow (before turbulent) regime are deviated significantly from the prediction using the continuum theory due to the nonlinear terms of the surface forces in the Navier-Stokes equations. The surface forces play a major role in the micro-scale liquid flow, which can be significantly affected by the geometry, the electro-kinetic transport process, the hydrophilic or hydrophobic of the surface condition etc. inside the microchannel.

3.1 Experimental results

Single-phase liquid flow is considered incompressible in a micro-channel. However, the geometric configurations, such as the aspect ratio, the geometric cross-section of the channel or the surface roughness etc., can significantly affect the characteristics of the flow and the heat transfer process in a microchannel. Harms et al. (1997, 1999) have observed a friction factor well predicted by the conventional theory in the laminar region. Webb et al. (1998) have observed that the conventional theory is able to predict the single phase heat transfer and the friction factor for a rectangular channel. Pfund et al. (1998) have studied the water flow in rectangular micro-channels at Reynolds numbers between 40 and 4000. The friction factor has a good agreement with the conventional theory in the laminar flow region, but increase by the surface roughness in the turbulence region. Xu et al. (1999, 2000) have fabricated the rectangular micro-channels by bonding an aluminum plate or a silicon wafer with a Plexi glass. The channels were etched on a silicon or aluminum substrate. The hydraulic diameters of the aluminum channels are from 46.8 to 344.3 μm and for silicon channels are from 29.59 to 79.08 μm, respectively. The experimental results for liquid flow in micro-channels have very good agreement with the prediction from the Navier-Stokes equation for a Newtonian flow in laminar region. Qu et al. (2000, 2002) has performed experiments for water in silicon micro-channels with trapezoidal cross section having hydraulic diameter from 51 to 169 μm. The pressure drop measured has a good agreement with the prediction based on conventional theory. More experiments have indicated that the deviation from the prediction is attributed to the roughness of the channel wall and viscosity of the fluid. The friction factors obtained from these experiments are higher than the predictions from the conventional theory. Li et al. (2000, 2003) have fabricated different micro-tubes made by glass, silicon or stainless steel with diameters ranging from 79.9 to 166.3 µm, 100.25 to 205.3 µm and from 128.6 to179.8 µm, respectively. The results of the friction factor measured for DI water, in glass and silicon micro-tubes where tube wall can be considered smooth, has good agreement with the conventional theory. The deviation of the data in the stainless steel tube is attributed to the surface roughness. They have concluded that the relative roughness of the wall can not be neglected for micro-tube in the laminar flow region. Sharp et al. (2000) have considered laminar flow of water in micro-tubes with hydraulic diameters ranging from 75 to 242 µm. Their data agree with the conventional theory. Wu et al. (2003) have provided the experimental data of friction factor for DI water in smooth silicon micro-channels with trapezoidal cross section having hydraulic diameter from 25.9 µm to 291 µm. The results of their data have a good agreement with the prediction from the conventional theory. They conclude that the Navier-Stokes equations are still valid for laminar flow of DI water in microchannel with smooth wall and hydraulic diameters as small as 26 µm.

Some work reported the friction factors that are very different from the theoretical prediction. Yu et al. (1995) has performed experiments of water flow in silica micro-tubes with diameters ranging from 19 to 102 µm and the Reynolds numbers between 250 and 20000. The friction factors are lower than the theoretical predictions. Jiang et al. (1995, 1997) have studied water flow through rectangular or trapezoidal channels. The dimensions of the channels are 35 to 120 µm wide and 13.4 to 46 µm deep. The friction factor data are greater than the theoretical prediction, but become lower when the Reynolds numbers are between 1 and 30. It appears that the deviations of the friction factor measured from the prediction may be attributed to the surface behaviors of the liquid flow, especially the surface roughness of the channel wall, the surface potential and the electro-kinetic effect induced by the electrical double layer (EDL) etc. as discussed in the following section.

3.2 Analysis of electric double layer effect

If the liquid contains a very few amount of ions (ex. impurities), the electrostatic charges on the non-conducting solid surface will attract the counter-ions in the liquid flow. The rearrangement of the charges on the solid surface and the balancing charges in the liquid is called the electrical double layer. The thickness of the EDL is significantly affected by the ion concentration, the liquid flow polarity, the surface roughness and the surface potential. A thicker EDL possibly induced by a lower ion concentration, a polar liquid, a poor surface roughness or a higher surface potential could cause a larger friction factor and pressure gradient. This can significantly reduce the flow velocity, and the heat transfer of a liquid flow in the microchannel. This is true for infinitely diluted solution such as the millipore water, the thickness of the EDL is considerably large (about 1 µm). However, for solution with high ionic concentration, the thickness of the EDL becomes very small, normally a few nanometer. In this case, therefore, the EDL effects on the flow in microchannels can be negligible.

To account for the EDL effect for polar liquid flow in the microchannel, most of the work performed in the past is the theoretical simulation where the physical models can be formulated based on (1) the Poisson-Boltzmann equations for the EDL potential, (2) the

Laplace equations with the applied electrostatic field, and (3) the Navier-Stokes equations modified to include effects of the body force due to the interaction between electrical and zeta potential. However, the numerical results are always lower than the empirical data due to the unusual and complex surface behaviors described above. In addition, the aspect ratio and the geometric cross-section of the channels can also affect the thickness of the EDL. In general, the friction factor increases with decreasing the aspect ratio of the channels. A microchannel with a cross section of circular shape usually has the lowest friction factor. The friction factor in a silicon channel is larger than in a glass channel due to the different surface potential of the channel walls with millipore water.

The Poisson-Boltzmann equations for the EDL potential in a rectangular microchannel are described as follows (Beskok & Karniadakis, 1994):

$$\frac{\partial^2 \psi}{\partial y^2} + \frac{\partial^2 \psi}{\partial z^2} = -\frac{\rho_e}{\varepsilon_0 \varepsilon} = \frac{2n_\infty ze}{\varepsilon_0 \varepsilon} \sinh(\frac{ze\psi}{k_b T}) \tag{3-1}$$

$$n_i = n_{i\infty} \exp(-\frac{z_i e \psi}{k_b T}) \tag{3-2}$$

$$\rho_e = ze(n_+ - n_-) = -2zen_\infty \sinh(\frac{ze\psi}{k_b T}) \tag{3-3}$$

where ψ and ρ_e are the electrical potential and the net charge density per unit volume. ε is the dielectric constant of the solution. ε_0 is the permittivity in vacuum. $n_{i\infty}$ and z_i are the bulk ionic concentration and the valence of type i ions, respectively; e is the charge of the proton; k_b is the Boltzmann constant; T is the absolute temperature.

To account for the electric field effect, the Navier-Stokes equation describing the flow motion can be rewritten as following:

$$\frac{\partial^2 u}{\partial y^2} + \frac{\partial^2 u}{\partial z^2} = \frac{1}{\mu}\frac{dp}{dx} - \frac{1}{\mu}E_x \rho_e \tag{3-4}$$

where E_x is an induced electric field (or called electrokinetic potential) and p is the hydraulic pressure in the rectangular microchannel.

At a steady state, the net electrical current is zero, which means:

$$I = I_s + I_c = 0 \tag{3-5}$$

$$I_s = 4 \int_{h-\frac{1}{k}}^{h} \int_{w-\frac{1}{k}}^{w} u(y,z)\rho_e(y,z)dy\,dz \tag{3-6}$$

where I_s and I_c are the streaming and the conduction currents, respectively. In addition, the net charge density is non-zero essentially only in the EDL region whose characteristic thickness is given by $1/k$ (k is the Debye-Huckel parameter).

The conduction current, that is the transport of the excess charge in the EDL region of a rectangular microchannel, driven by the electrokinetic potential is given by:

$$I_c = 4\,\lambda_o\,E_x\,(h+w)\frac{1}{k} \tag{3-7}$$

$$k = [2z^2e^2n_\infty / (\varepsilon\,\varepsilon_o k_b\,T)]^{1\!/\!2} \tag{3-8}$$

where λ_o is the bulk electrical conductivity ($1/\Omega$ m). h and w are the height and the width of the microchannel, respectively. Substituting Eq.(3-6) and Eq.(3-7) into Eq.(3-5), the electrokinetic potential (E_x) can be written as follows:

$$E_x = -\frac{\int_{h-1/k}^{h}\int_{w-1/k}^{w} u(y,z)\,\rho_e(y,z)\,dy\,dz}{\lambda_o\,(w+h)(1/k)} \tag{3-9}$$

Both the Poisson-Boltzmann equation, Eq.(3-1) and Navier-Stokes equation, Eq.(3-4), can be solved numerically in order that both the EDL and the velocity fields in the rectangular microchannel can be determined.

3.3 Comparison with the data

Despite the theoretical prediction, some work presents occurrence of the electrical double layer of water flow in a micro-channel. Ren et al. (2001) have performed experiments to measure the interfacial electrokinetic effects of a liquid flow through rectangular silicon micro-channels with diameters of 28.1, 56.1 and 80.3 µm. Both the DI water and the KCl solutions with two different concentrations of 10-4 and 10-2 M are used as working fluid. The measured pressure drops for the pure DI water and the lower KCl concentration solution are significantly higher than that for higher concentration solution and the theoretical prediction. The authors have concluded that a significant increase in the friction factor is attributed to occurrence of the electrical double layer (EDL) which increases the pressure drop in the small micro-channels. Similar results have also been obtained by Li et al. (2001).

To compare with the experimental results, the analytical predictions for both the flow and the heat transfer developed from continuum assumption indicate large discrepancy when the characteristic length of the micro-channel becomes small enough. In the studies of liquid flow, many investigators (Ren et al., 2001; Fan et al., 1998; Chen, 1996; Chu et al., 1994; Choi et al., 1991; White et al., 1991; Pfahler et al., 1990, 1991) have concluded that even though the liquids can be regarded as a continuum in a very small system, the real behaviors of heat transfer at the laminar or the transition (before turbulent) condition are deviated from the predictions based on the conventional theory. Usually, for the data published, the uncertainties of flow rate measured and friction factor estimated are 2-5 % and 10-15 %, respectively. For most heat transfer studies, the uncertainties are under ± 20 %. In summary, the geometric effects, such as the aspect ratio, the cross-section shape or the surface roughness etc., can significantly affect the characteristics of both the flow and the heat transport in a microchannel. The onset of transition to turbulent flow in smooth microchannels does not occur if the Reynolds number is less than 1000. For a laminar flow, the Nusselt number varies as the square root of the Reynolds number. In turbulent flow, however, the numerical studies are not applicable and thus many empirical correlations have been proposed, but were not verified. However, satisfactory estimates of the heat transfer coefficients can be obtained with sufficient accuracy by using either experimental results in smooth channels with large hydraulic diameter or conventional correlations.

Tso and Mahulikar (1998) have obtained the heat transfer for laminar liquid flow through a microchannel in both the thermal-developing region and the thermal-developed region. It is found that the Nusselt number decreases with increasing the Reynolds number not only in the thermal-developed region, but also in the thermal entry region. The results also indicate that the pressure distribution along the microchannel exhibits a non-linear profile. Despite much of the studies has addressed that the liquid flow appears a greatly complicated relation between Nusselt number and Reynolds number, however, all the results are very based on the assumption of continuum flow. Therefore, more detailed analysis combined with experiments is still required to clarify the role of the EDL and different results among different works.

3.4 Two-phase flow phenomenon in the microchannel

The two-phase flow or flow-boiling phenomenon in the microchannel exhibits some unusual characteristics. It is found that the bubbles are not rapidly generated even at a very high heat flux from the heated microchannel (Qu et al., 2000). Therefore, further experimental investigations on the flow boiling in microchannels were made by others (Ren et al., 2001; Peng & Wang, 1993; Lin and Pisano, 1991, 1994). In addition, the effect of microchannel scale, geometric configuration, liquid velocity, liquid sub-cooling and liquid concentration on the flow boiling were investigated. It is found that the heat transfer enhanced by a large more volatile component concentration is greater than the pure more volatile liquid. The heat transfer coefficient at the onset of flow boiling and in the partial nucleate boiling region was greatly influenced by the liquid concentration, the geometric configuration, the size of microchannel, and the flow velocity and sub-cooling, but not in the fully nucleate boiling region. Peng and Wang (2001), and Hu (1998) found the so-called "bubble extinction" behavior due to an induced vigorous nucleate boiling mode on a normal-sized heater or abnormal-sized channels. The normal bubbles could not successfully grow and form, if the channel height is less than a critical liquid space required. In order to interpret the unusual behavior observed in microchannel boiling, Peng and Wang (1994) proposed the concepts of "evaporating space" and "fictitious boiling". In fact, the small bubbles that can form initially in microchannel will eventually collapse since the size of the bubble could not grow up exceed the critical radius of bubble (r_c) formulated by conventional nucleation theory. The fictitious boiling occurred was attributed to the crowded tiny bubbles that grow and then collapse rapidly in a cyclic manner, and thereby mimicking a boiling state that can transfer large amount of heat. The observations suggest that close to bubble nucleation temperature the liquid will vigorously oscillate in the microchannel due to the emergence of tiny bubble embryos. More detailed explanations are given in (Jiang et al., 2001; Peng et al., 1998).

The experiments by Peng and Wang (1993) for flow boiling of water have been carried out in a stainless steel microchannel with rectangular cross-section of 600 μm×700 μm. In a much smaller channel array, with hydrodynamic diameter of 40 and 80 μm, made on a silicon substrate by wet etch, three stable phase-change modes, i.e. local nucleation boiling, large bubble formation and annular flow, were observed depending on the input power level (Qu & Mudawar, 2003). However, bubbly flow, commonly observed in macrochannels, could not be developed in the microchannels. A stable annual flow was also observed in a microchannel heat sink contained 21 parallel channels having a $231\,\mu m \times 713\,\mu m$ cross-section (Lee et al., 2003).

Lee et al. (2003) proposed that a nearly rectangular microchannel heat sink with $14\,\mu m$ in depth integrated with a local heater and array of temperature sensors on silicon substrate was made to investigate the size and shape effects on the two-phase patterns in microchannel forced convection boiling. It is found that when the heat input power increases, the downstream movement of the transition region increases the void fraction and causes a lower devices temperature. However, at the high flow rate, the transition region almost occupies the entire channel, the increase in the heat input power results in a higher devices temperature. An annular pattern induced by flow boiling appears stably in triangular microchannels, but not in rectangular microchannels. Two-phase boiling or superheated flow has numerous promising applications such as in cooling of electronic components. The principle advantage of two-phase flow lies in the utilization of latent heat absorbed by the working fluid due to phase change from liquid to vapor without increasing the flow fluid temperature. In fact, two-phase flow heat transfer in microcahnnel is a very important and interesting problem indeed.

However, much of the attention at later time has been given to the study of dynamic flow boiling instability in microchannels (Cheng et al., 2009; Wang et al., 2008; Wang et al., 2007; Kandlikar, 2006; Wu & Cheng, 2003, 2004; Brutin et al., 2003; Hetsroni et al., 2002; Hetsroni et al., 2001). A periodic annular flow and the periodic dry steam flow were observed for boiling of water in 21 silicon triangular microchannels having a diameter of $129\,\mu m$ in (Hetsroni et al., 2001, 2002). However, two types of two-phase hydrodynamic instabilities, i.e. severe pressure drop oscillation and mild parallel channel instability were identified (Qu & Mudarwar, 2003) in the similar microchannels as in other work (Hetsroni et al., 2001). A simultaneous flow visualization and measurement was made on flow boiling of water in two parallel silicon microchannels of trapezoidal cross-section having hydraulic diameters of $158.8\,\mu m$ and $82.8\,\mu m$, respectively (Wu & Cheng, 2003). The results shows that two-phase flow and single-phase liquid flow appear alternatively in microchannels, which leads to large amplitude/long-period fluctuations with time in temperatures, pressures and mass flux. The flow pattern map in terms of heat flux versus mass flux showing stable and unstable flow boiling regimes in a single microchannel has been identified (Wu & Cheng, 2004). It is found that stable and unstable flow-boiling modes existed in microchannels, depending on four parameters, namely, heat/mass flux ratio, inlet water subcooling, channel geometry, and physical properties of the working medium (Wang et al., 2007). In addition, the magnitudes of temperature and pressure fluctuations in the unstable flow-boiling mode depend greatly on the configurations of the inlet/outlet connections with the microchannels (Wang et al., 2008). By fabricating an inlet restriction on each microchannel or the installation of a throttling valve upstream of the test section, reversed flow of vapor bubbles can be suppressed resulting in a stable flow-boiling mode. Based on the exit quality of the flow from a microchannel, more detailed flow regimes are identified (Cheng et al., 2009).

In the past, however, a very important issue, i.e. the surface wettability effect, has been overlooked in the study of boiling flow heat transfer in a microchannel. The boiling flow phenomenon found in the microchannel is only for certain surface wettability. By changing the material of the microchannel or surface wetting property, the boiling flow phenomenon may be completely different. This may cause discrepancy of flow patterns observed in different channels made by different materials. Phan et al. (2009) have found that the wettability of a surface has a profound effect on the nucleation, growth and detachment of bubbles from the bottom wall in a tank. For hydrophilic (wetted) surfaces, it has been found that a greater surface wettability increases the vapor bubble departure radius and reduces the bubble emission frequency. Moreover, lower superheat is required for the initial growth of bubbles on hydrophobic (un-wetted) surfaces. However, the bubble in contact with the hydrophobic surface cannot detach from the wall and have a curvature radius increasing with time. At higher heat flux, the bubble spreads over the surface and coalesces with bubbles formed at other sites, causing a large area of the surface to become vapour blanketed.

The wettability of channel surface has been studied by Liu et al. (2011) who have fabricated three different microchannels with identical sizes at 105 x 1000 x 30000 μm but at different wettability. The microchannels were made by plasma etching a trench on a silicon wafer. The surface made by the plasma etch process is hydrophilic and has a contact angle of 36° when measured by dipping a water droplet on the surface. The surface can be made hydrophobic by coating a thin layer of low surface energy material and has a contact angle of 103° after the coating. In addition, a vapor-liquid-solid growth process was adopted to grow nanowire arrays on the wafer so that the surface becomes super-hydrophilic with a contact angle close to 0°. Different boiling flow patterns on a surface with different wettability were found, which leads to large difference in temperature oscillations. Periodic oscillation in temperatures was not found in both the hydrophobic and the super-hydrophilic surface. During the experiments, the heat flux imposed on the wall varies from 230 to 354.9 kW/m² and the flow of mass flux into the channel from 50 to 583 kg/m²s. Detailed flow regimes in terms of heat flux versus mass flux are also obtained.

4. Basic MEMS fabrication techniques

4.1 Chemical vapor deposition

Chemical vapor deposition (CVD) is a typical technique to fabricate a thin film on a substrate. In a CVD process, gaseous reactants are introduced into a heated reaction chamber. The chemical reactive gases diffuse onto and absorbed by the substrate. Then thermal dissolution reaction of the reactive gases occurs which lead to deposition of a thin solid film on the heated substrate surfaces. Depending upon the relative pressure and the temperature used the CVD processes are categorized as: (1) the atmospheric pressure chemical vapor deposition (APCVD), (2) the low pressure chemical vapor deposition (LPCVD), and (3) the plasma-enhanced chemical vapor deposition (PECVD). The process temperatures of APCVD and LPCVD are ranged from 500°C to 850°C. In PECVD processes, a part of thermal energy is shared from the plasma source. Therefore, the process temperatures of the PECVD are lower on the order of 100°C to 350°C. The silicon based thin films such as poly-silicon, amorphous silicon, silicon dioxide, tetraethoxysilane (TEOS, $Si(C_2H_5O)_4$) or silicon nitride film can be fabricated by using the CVD process. The chemicals used and the reaction occurred in the CVD process for different kinds of films are listed in Table 1. The poly-silicon film can be used for fabrication of pressure or temperature sensors or micro-heaters. The TEOS oxide layer is fabricated as insulator between each sensor layer. In addition, deposition of the silicon nitride film can be used to prevent penetration of moisture into the sensors during liquid flow experiments which may cause damage of the micro-sensors or micro electronics integrated in the micro-channel.

Films	Chemical reactions
Poly-silicon	$SiH_4 \rightarrow Si + 2\,H_2$
Silicon dioxide	$SiH_4 + O_2 \rightarrow SiO_2 + 2\,H_2$ $SiCl_2H_2 + 2\,N_2O \rightarrow SiO_2 + 2\,N_2 + 2\,HCl$
TEOS (tetraethoxysilane)	$Si(OC_2H_5)_4 \rightarrow SiO_2 + $ by-products
Silicon nitride	$3\,SiH_4 + 4\,NH_3 \rightarrow Si_3N_4 + 12\,H_2$ $3\,SiCl_2H_2 + 4\,NH_3 \rightarrow Si_3N_4 + 6\,HCl + 6\,H_2$

Table 1. Chemical reactions used in the CVD process for different kinds of films.

4.2 Evaporation and sputtering deposition

Both evaporation and sputtering deposition are classified as physical vapor deposition (PVD) process which can form different kinds of films on a substrate directly from a source material. PVD is typically used for deposition of electrically conducting layers such a metal or silicide. Evaporation deposition of a thin film on a substrate is done by sublimation of a heated source material in a vacuum chamber. The vapor flux from the source can be condensed and coated on the substrate surface. The evaporation methods can be further categorized as the vacuum thermal evaporation (VTA), the electron beam evaporation (EBE), and the molecular beam epitaxy (MBE).

The simplest evaporator consists of a vacuum chamber with a crucible which can be heated to a high temperature, as shown in Figure 1(a) and 1(b) by a filament. The filament is used as a heater, which is made of Tungsten (W), a refractory (high temperature) metal. Evaporation is accomplished by gradually increasing the temperature of the filament until the source material melts. Filament temperature is then further raised to evaporate the source material from the crucible. The substrates are mounted on top of the crucible and are deposited with a thin film of evaporated material.

In the electron beam (E-beam) evaporation system, the high-temperature filament is replaced with an electron beam, as shown in Figure 1(c). A high-intensity beam of electrons, with energy up to 15 keV, is focused on the source material to be evaporated in a crucible. The energy from the electron beam only melts a portion of the source material, which eventually evaporates and condenses on the substrate to form a thin layer.

Sputtering deposition requires generation of plasma gas between high voltage electrodes, as shown in Figure 2, where positively ions can be accelerated and bombards on a target material (a cathode) so that flux of atoms can be sputtered and collected on the substrate. Usually, a physically inert gas, such as argon gas, is made into plasma by knocking out electrons of the molecules with high speed electrons emitted from the cathode. The sputtering deposition has the advantages of depositing various materials include not only for pure materials or metals, but also for compounds, alloys, refractory materials, or piezoelectric ceramics. In addition, puttering deposition has no shadowing effect as that occurred in evaporation deposition, which causes non-uniform deposition of a film. Therefore, sputtering deposition has been widely used for deposition of different kinds of films.

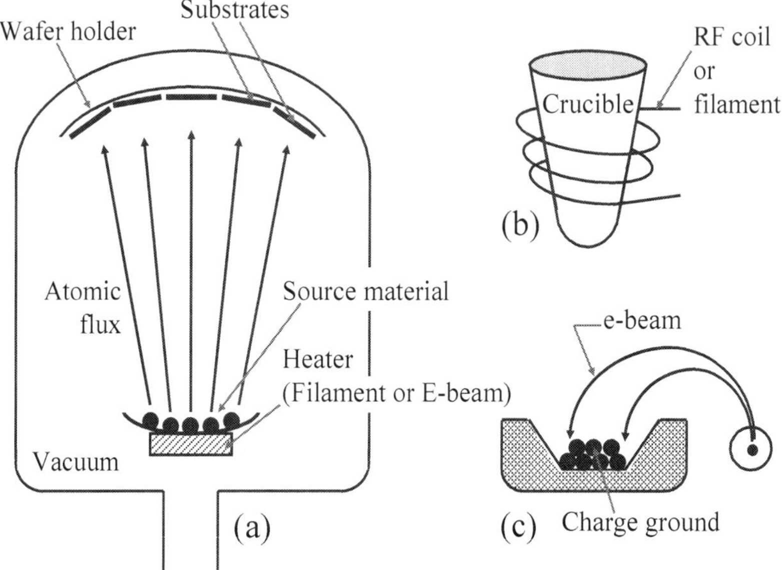

Fig. 1. (a) Schematic of the thermal evaporation system, (b) the use of filament or RF coil as heating source and (c) the use of electron beam as the heating source.

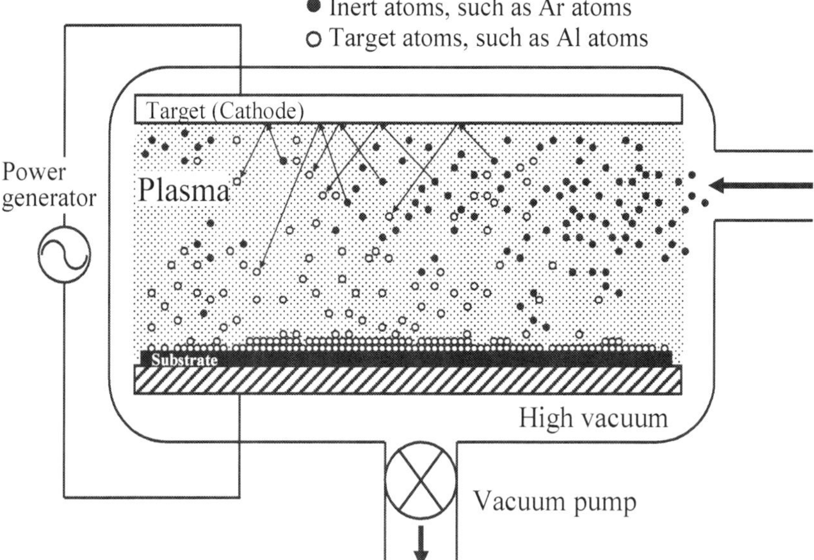

Fig. 2. Schematic of a sputtering system.

4.3 Photolithography

Lithography is the most important technique for transferring micro-patterns onto substrate. Depending upon resolution required, the light sources used for lithography process can be a mercury lamp, a laser light, an electron beam, X-ray or an ion beam. However, photolithography using mercury lamp or laser light is the most popular method for low-cost and fast prototyping of micro-fluidic fabrications. This technique uses a photosensitive polymer layer, the so called photo-resist (PR), to transfer a desired pattern from a photomask to the substrate. The mask is a transparent glass plate or a plastic sheet with metal (chromium, Cr) or ink patterns. The photolithography process is shown in Figure 3. First, the photo-resist is spin coated onto the substrate. After light exposure and a developing process of the PR a desired pattern can be transferred from the photomask to the PR, and then transferred to the underneath layer by a wet or a dry etch process. The resolution of a proximity photolithography (R) process which depends on the wavelength (λ) of the light source can be written as follows:

$$R = K_1 \frac{\lambda}{NA}$$

where K_1 is the optical system constant and NA is the numerical aperture of image lens system. The reduction of the resolution can be made by reducing the wavelength or increasing the NA. However, increasing the NA can lead to large reduction in the depth of the focus, which is very detrimental to the images. It appears that the resolution of photolithography is primarily determined by the wavelength of the light source.

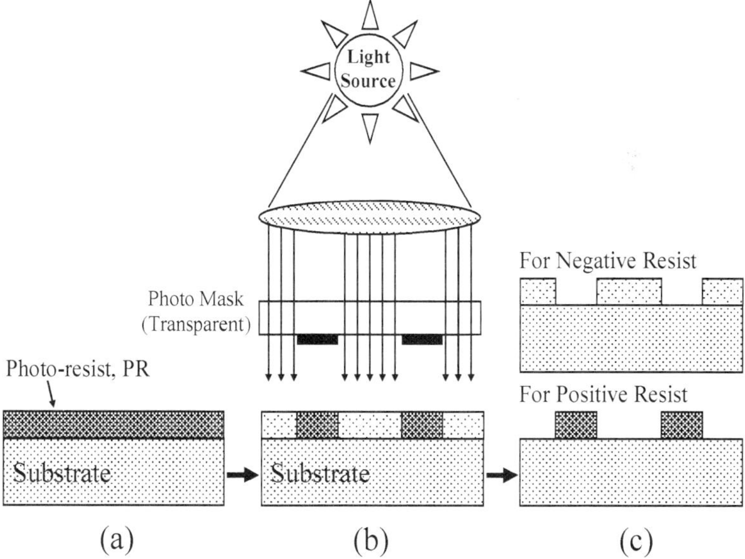

Fig. 3. Process of photolithography: (a) spin coat the PR, (b) light exposure and (c) developing the PR.

4.4 Anisotropic wet etching

Single crystal silicon can be anisotropically etched. In general, the etching rate is highly dependent on the crystal's orientation in the single-crystal silicon as shown in Figure 4. Most of the chemical solutions have a distinct slow etching rate on the crystal face of (111), which practically causes a etch stop on this surface. This allows formation of microstructures with sharp edges and corners that can be defined in a single crystal substrate. For the different orientation silicon substrates, the angle between other crystal faces with respect to the (111) crystal face are shown in Figure 5. Different etchants, such as KOH, NaOH, TMAH (Tetramethyl ammonium hydroxide) and EDP (Ethylenediamine pyrochatechol), may be used for anisotropic wet etch for the silicon. The angles between different facets are 54.7° and 90° for {100} to {111} and {110} to {111}, respectively. However, using KOH solution the etching selectivities for different facets are around 400 and 600 for {100}/{111} and {110}/{111}, respectively. Therefore, V-shaped or rectangular cavities can be readily fabricated and obtained by this method.

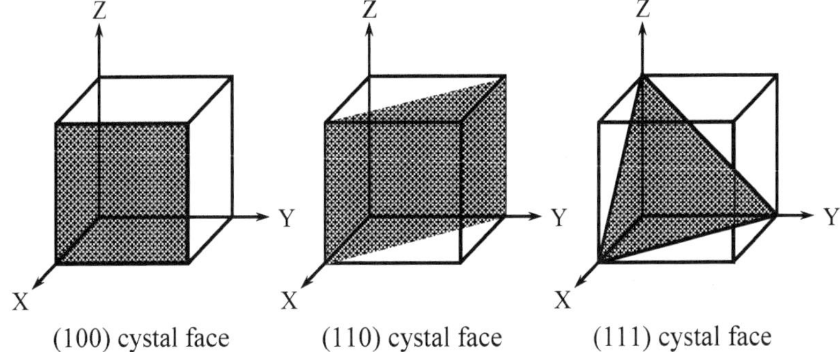

Fig. 4. The crystal planes in silicon lattice structure.

Anisotropic etching solution

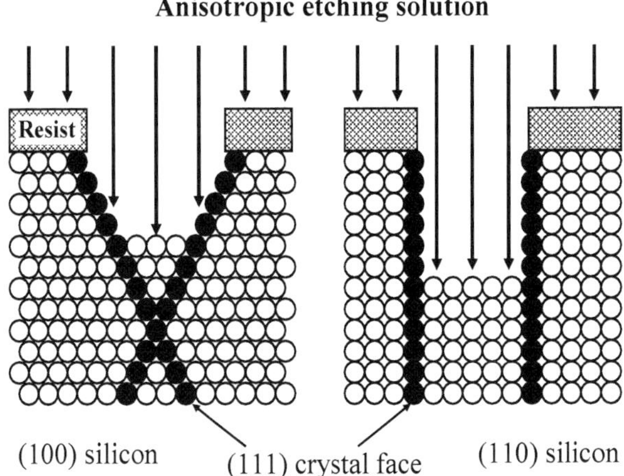

Fig. 5. Anisotropic etching in the single crystal silicon. [Madou, 2002]

4.5 Deep reactive ion etching

The process which can remove the materials from the surface of a substrate or a bulk substrate is called "ETCHING". Anisotropic dry etching process does not depend on crystal orientation of silicon wafer. This micromachining process involves the exposure of the substrate to an ionized gas. Etching occurs through chemical or physical interaction between the ions in the gas and the atoms of substrate. The most often applied techniques can be divided into three groups: (1) physical sputter etching or ion beam etching, (2) chemical plasma etching and (3) combined physical/chemical etching. The physical method as so called "sputtering etching or ion beam etching" can achieve an anisotropic profile structure. As opposed to sputtering etching, chemical plasma etching is completely isotropic etching profile and has excellent selective properties. The etching rate of plasma etching is much faster than sputtering etching and is uncontrollable. However, the reactive ion etching (RIE) method combines the physical and chemical etchings to achieve excellent high selectivity and anisotropy for micromachining of silicon substrate. The etching rate ranges from 20 to 200 nm/min. However, this low etching rate can not satisfy the micromachining of microchannel with a high aspect ratio structure.

The problems of reactive ion etching (RIE) used in the micromachining process are that the sidewalls of the trenches are not vertical and the etching speeds are too slow. Therefore, two modified RIE techniques, i.e. the Bosch process and the cryogenic process, are developed to solve this problem. In the first approach, the Bosch process consists of cyclic etching and deposition process, as shown in Figure 6. In the etching step, silicon is etched by SF_6 plasma. In the deposition step, the supply gas is switched to C_4F_8 so that a fluorocarbon polymer thin film with a thickness of 10 nm can be deposited on the side wall of the trench. In the next cycle, the polymer film at the bottom surface of the trench is removed by Ar ion bombardment, while the film at the sidewalls is intact, which can protect the sidewalls from the attack of the SF_6. Usually, the etching rate of Bosch process ranges from 1 to 12 µm/min. In the second approach, i.e. the cryogenic process, the substrate is cooled down to -110 to -195 ºC with liquid Helium. The cryogenic temperatures allow reactant gas such as SF_6 or O_2 to condense on the trench surface. The condensed film can protect the side-wall, while the condensed film on the bottom can be removed by the ion bombardment. The exposed bottom wall can be further etched into the substrate, as shown in Figure 7.

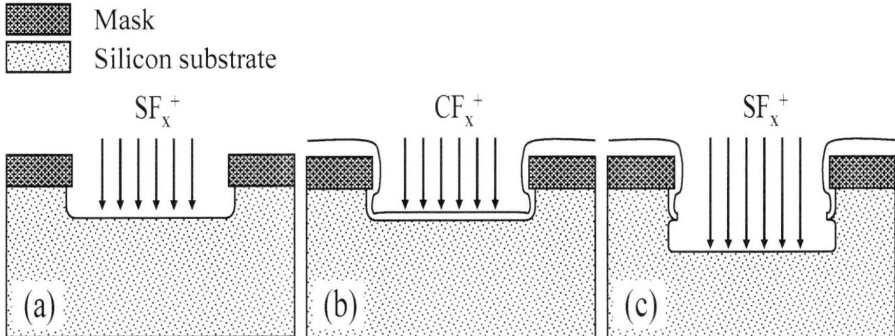

Fig. 6. Anisotropic dry etching: Bosch process. (a) SF_x^+ etching, (b) CF_x^+ deposition and (c) SF_x^+ etching.

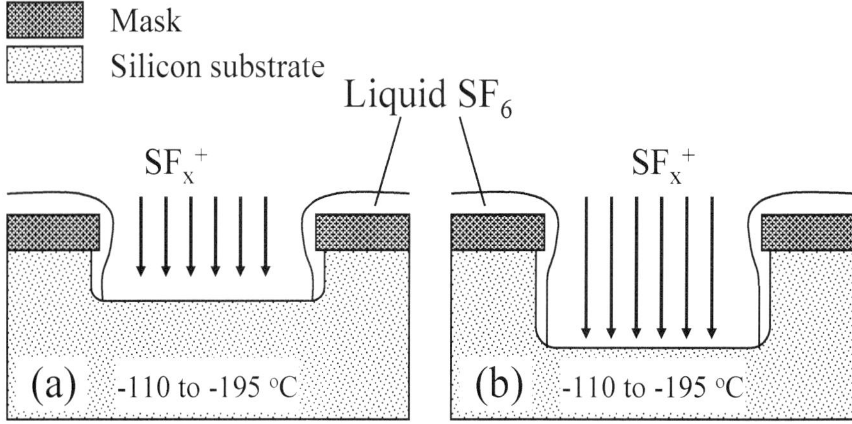

Fig. 7. Anisotropic dry etching: cryogenic process.

4.6 Doping

Doping is probably the best known semiconductor technique. Here, doping atoms are introduced into a silicon substrate in a defined way so that either n-type or p-type semi-conductor layer can be formed. The techniques play an important role in fabrication of semiconductor devices. In addition to varying the electrical properties, doping can also improve wear and corrosion of a semiconductor material. The doping atoms, like boron or phosphorous, can create an etching stop layer in a silicon substrate, allowing fabrication of a thin film or microstructure at desired locations by wet etch when doping atoms can be placed in the desired location.

4.6.1 Ion implantation

Ion implantation is one of the most expensive technology processes, after photolithography, due to the complex systems involved. Ion implantation involves shooting charged ions, which are externally accelerated in a vacuum, into the silicon wafer. The ions can penetrate up to a few micrometers below the surface. The silicon disc is irradiated uniformly with scanning focused ion beam. By measuring the beam current, the amount of dopant implanted can be precisely controlled. In addition, the doping concentration obtained has improved homogeneity, the doping parameters can be easily adjusted, and the doping profile under the wafer's surface can be accurately controlled.

An ion implantation process usually includes the initial ion implantation and the following thermal annealing process. For example, after dopants are implanted into the central portion of a LPCVD polysilicon layer, the implanted layer requires annealing at 950 °C for 30 minutes in a standard anneal furnace to re-crystallize the polysilicon layer and to diffuse the dopants uniformly across the layer (Ko et al., 2007; Liu, 2004; Qu et al., 2003). After annealing process, the doped polysilicon layer can be patterned. The variation of the resistivity of the doped LPCVD polysilicon with the boron concentration is shown in Figure 8. This empirical data can be used to design arrays of polysilicon sensors or heaters integrated with the microchannel system.

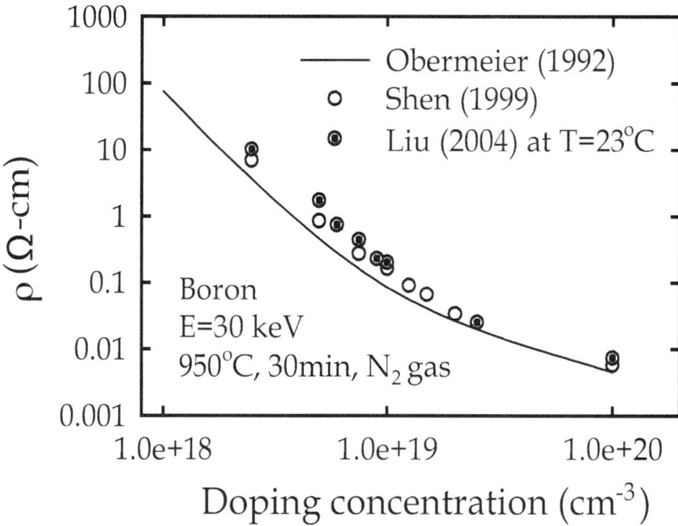

Fig. 8. The resistivity of boron-doped LPCVD polysilicon at different boron concentrations.

4.6.2 Dopant diffusion

In the diffusion method, silicon wafers are put into a furnace under a high temperature range from 850 to 1050 ºC, where dopant can be thermally diffused into the substrate. The sources of dopants can be either gases, liquids or solids. The solid dopant source can be BN, As2O3, P2O5, the liquid dopant source can BBr3, AsCl3, POCl3 while the gas dopant source can be B2H6, AsH3, PH3. However, the liquid dopant sources are the most frequently used. More detailed description can be found in the ref. (Sze, 2002). However, either of them has to be vaporized first and then the vapor can diffuse into the silicon substrate and form covalent bonding with the silicon lattice atoms. The main difficulty is the determination of the amount of dopant diffused into the substrate and the concentration of the doping atoms bonded in the silicon. Despite of the low cost, the method can only make a doping profile on the surface of a wafer, which restrains the wide applicability of this technique.

5. Fabrication of microchannel by MEMS techniques

5.1 Silicon based microchannel fabrication

5.1.1 Bulk micromachining process

Either isotropic or anisotropic etch process can be used to fabricate microchannels in a bulk material, such as a silicon or a glass substrate. A channel with a variety of cross-sectional shapes can be made by different etch processes, as shown in Figure 9. Different microchannels can be formed by either anisotropic etch of the {100} or {110} silicon wafers or by isotropic etch of silicon or glass substrate, respectively. Finally, either anodic or fusion bonding can be employed to bond and enclose the channels with a glass plate or a silicon wafer.

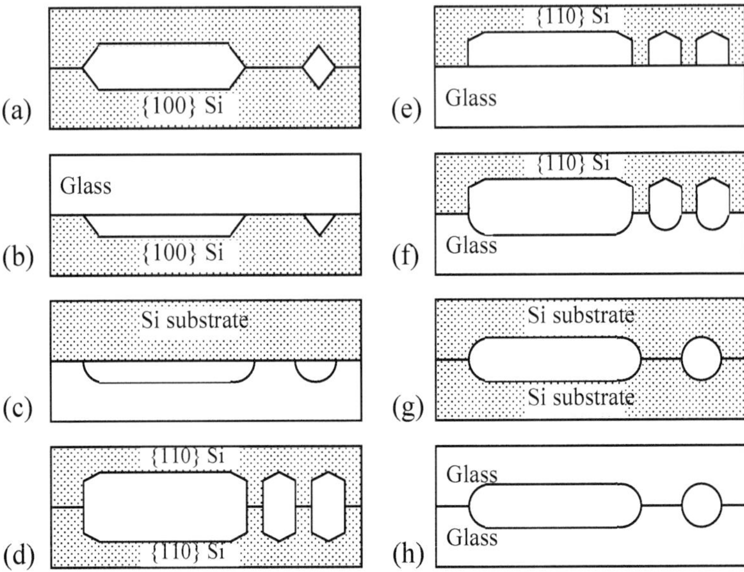

Fig. 9. Different microchannels made by bonding of bulk micromachined substrates in: (a) silicon-silicon, (b) glass-silicon, (c) silicon-glass, (d) silicon-silicon, (e) silicon-glass, (f) silicon-glass, (g) silicon-silicon, and (h) glass-glass.

5.1.2 Surface micromachining process

The surface micromachining is originated for deposition and patterning of thin layers into different structures, which may possibly use sacrificial layers to form microstructures. In this section, surface micromachining processes combined with deposition and etch of a polysilicon layer and the use of a sacrificial layer can be used to fabricate a microchannel. The fabrication process is described as follows:

1. Deposit and pattern the sacrificial layer, as shown in Figure 10(a).
2. Deposit and pattern the polysilicon layer as a microstructure, as shown in Figure 10(b).
3. Remove the sacrificial layer, followed by rinsing and drying the microstructure, as shown in Figure 10(c).
4. Seal the microstructure, as shown in Figure 10(d).

However, by using the surface micromachining the maximum height of the channel allowed for fabrication is only within a few micrometers (1 to 2 μm) due to the limitation of CVD deposition for the sacrificial layer. In comparison with the bulk micromachining process, the channel fabricated by the bulk micromachining process can be as large as a few hundred micrometers in height.

5.2 Polymer based microchannel fabrication
5.2.1 SU-8 resist

Photolithography of a thick resist is very beneficial for fabrication of a complicated microchannel system. Microchannel system can employ thick resists directly as a channel

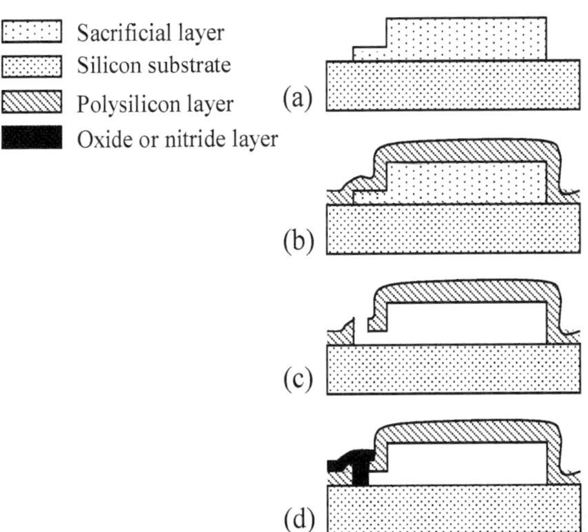

Fig. 10. The use of surface micromachining process to fabricate a channel with the use of a sacrificial layer.

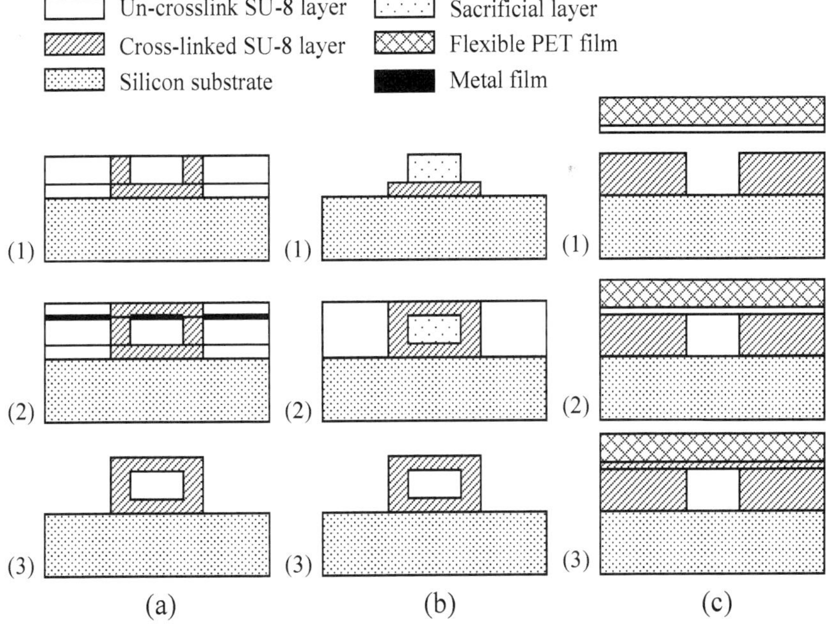

Fig. 11. Fabrication of SU-8 based microchannels with: (a) embedded metal mask, (b) removal of sacrificial layer and (c) direct use of the fusion bonding by SU-8.

structure or a template for channel molding or electroplating of metal to form a channel with high aspect ratios. The most popular thick photoresist used is EPON SU-8 epoxy resin which is a negative and transparent PR and can form a layer with a thickness from less than 1μm to few mm. The SU-8 appears to be the most suitable one since it can be readily spin coated on the substrate at desired thickness, and patterned into required shape of channel by photolithography. Therefore, different sizes and shapes of channel can be readily made. However, the use of SU-8 layer as part of the substrate has precluded any high temperature fabrication process since the material cannot withstand processing temperature higher than 200oC. The formation of a SU-8 channel structure should be at a low temperature process.

In many microfluidics applications, a single SU-8 layer can be used to form desired number of microchannels that may be integrated with arrays of microfluidic components or sensors. The top and bottom of the channels can be sealed with a SU-8 coated glass or plastic plate, using subsequent blanket exposure as shown in Figure 11.

5.2.2 PDMS molding

Poly-dimethylsiloxane (PDMS) is an excellent bio-compatible and an elastomeric polymer material, which is frequently used for applications in μTAS (micro total analysis system) or lab-on-a-chip systems. This is attributed to a number of useful properties of this material, such as low cost, low toxic, transparent from visible into near ultraviolet in wavelengths and chemical inertness.

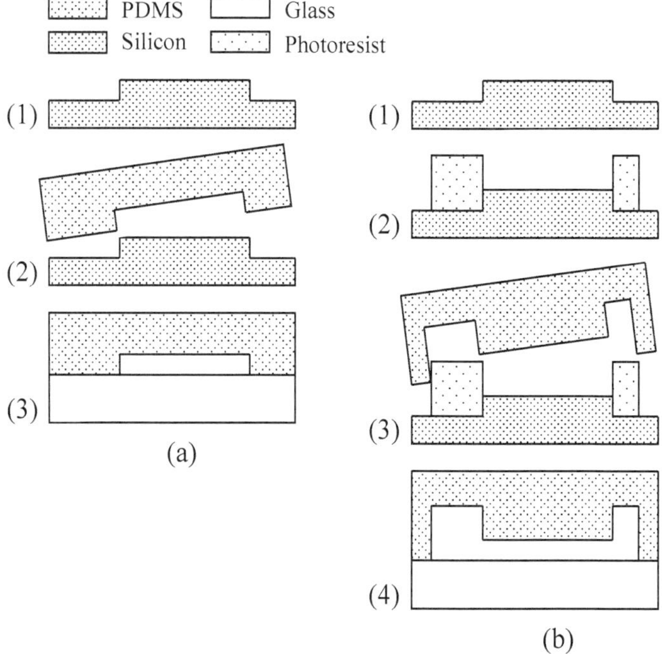

Fig. 12. Fabrication process for a (a) one-staged or (b) two-staged PDMS microchannels.

PDMS is made by mixing different prepolymers. The weight ratio of the base and the curing agent can be 10:1 or 5:1. Then, the PDMS mixture is poured into a master and stays for a few minutes to self-level. The whole set is then cured at a low temperature in the range from 60 to 80 °C for several hours. After peeling off, the structured PDMS membrane is bonded with the other substrates such as silicon, glass, plastic plate or PDMS plate etc. to complete the channels. Due to its simplicity in fabrication and rapid prototyping, PDMS molding is also called microcasting, which is a direct transfer of patterns. In many applications, the elastomeric PDMS can be used directly as a microfluidic device with microchannel connecting different components. Fabrication of PDMS into a microcannel is described as follows:

1. Etch the silicon or glass master, as shown in Figure 12(a-1).
2. Molding the PDMS microstructures, as shown in Figure 12(a-2).
3. Seal and laminate the PDMS microstructures, as shown in Figure 12(a-3).

5.2.3 Parylene

Parylene is a polymer that can be deposited by a CVD process at a low or room temperature. The CVD process allows coating a conformal film with a thickness ranging from several

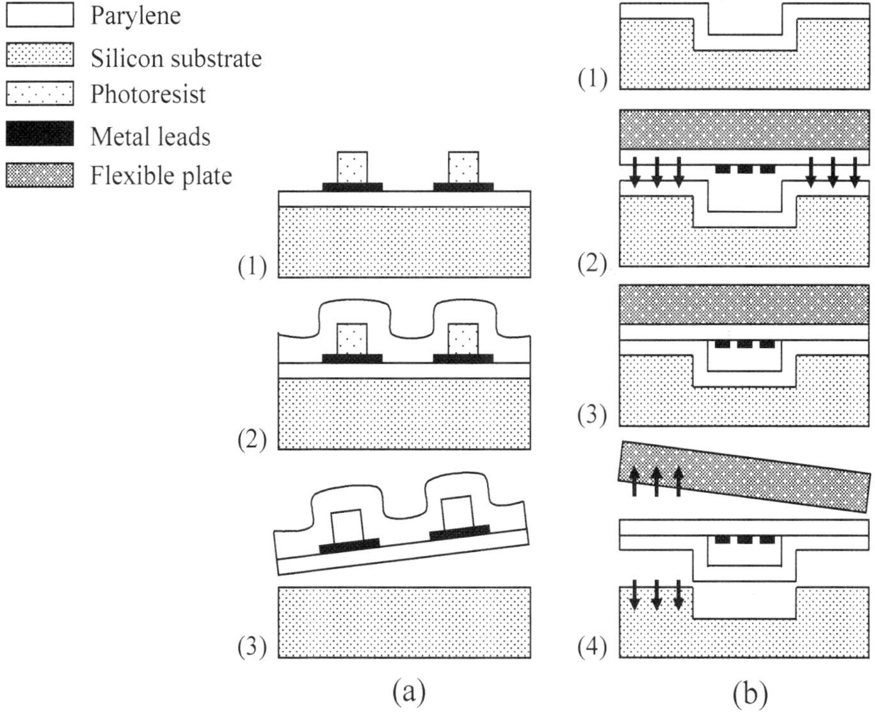

Fig. 13. Fabrication of Parylene based microchannels with: (a) removal of the sacrificial layer and (b) direct fusion bonding using Parylene.

micros to several millimeters. The basic types of Parylenes are Parylene N, C and D, all of which have good dielectric, electrical, physical and hydrophobic properties. However, Parylene D is modified from the same monomer by the substitution of the chlorine atom with two of the aromatic hydrogens. Parylene D is similar in properties to Parylens C, but with an additional property to withstand higher temperature. Deposition rates are fast, especially for Parylene C, which is normally deposited at a rate of about 10 µm/min. The deposition rates of Parylene N and Parylene D are slower. Parylene can be used in microfluidic devices as a structural material, which has a very low Young's modulus. Fabrication of microchannel using Parylene is described as follows:

1.   Deposit a Parylene base on a provisional substrate, as shown in Figure 13(a-1).
2.   Deposit and pattern a sacrificial layer on the Parylene base, as shown in Figure 13(a-2).
3.   After Secondary deposition of a Parylene film on the top, then remove the sacrificial layer, as shown in Figure 13(a-3).

6. Microsensors for pressure and temperature measurements

6.1 Miniaturized temperature sensors and heaters

6.1.1 Working principles

When the polysilicon is doped heavily with boron, a relatively low resistivity that is independent of temperature can be obtained. This heavily doped polysilicon layer can be used as a heater. However, when the polysilicon is doped with a less amount of boron, a linear relationship between the resistivity and the temperature of the layer can be obtained and this doped polysilicon layer can be used as a temperature sensor. The temperature variation of the resistivity for the doped polysilicon layer at different concentrations has been demonstrated (Ko et al., 2007; Liu, 2004; Qu et al., 2003). Shen proposed that the resistance change with temperature is slightly nonlinear at low boron concentration. The temperature coefficient of the resistance (TCR) of the doped polysilicon layer is negative over the range covered and the temperature dependence increases with decreasing doping concentration, as shown in Figure 14 made by Shen (2003). It appears that for the boron concentration of 10^{20} atoms/cm^3, the resistivity of the doped polysilicon layer is almost independent of temperature with a near zero TCR. The polysilicon doped at this concentration can be used as a heater which can provide a constant heating power for a long period of operation. Otherwise, for the boron concentration at 10^{19} atoms/cm^3, the resistivity of the doped polysilicon layer has a linear relationship with temperatures. Therefore, the polysilicon doped at this concentration can be used as a temperature sensor. All the temperature sensors and heaters required and made in a microchannel system should be based on different doping conditions of the concentrations described above.

6.1.2 Design consideration

The total resistance of a device is proportional not only to the resistivity (ρ) and thickness (t) of the doped polysilicon layer, but also to the ratio of l_R/w_R, i.e. shape effect. The definition of resistance is written as follows:

$$R = (\rho/t) \cdot (l_R/w_R) = R_S \cdot (l_R/w_R) \tag{6-1}$$

where $R_S = (\rho/t)$ is called the sheet resistance of the material. l_R and w_R are the length and the width of the resistor, respectively.

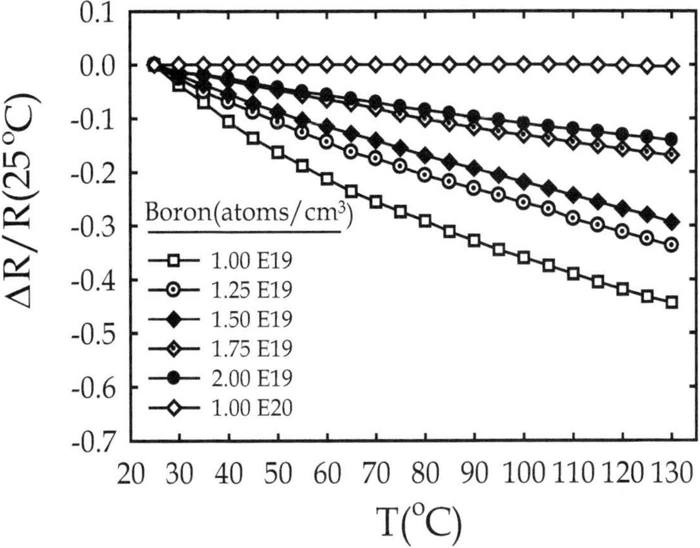

Fig. 14. The relative resistance variation with temperature at different concentrations of boron doped in polysilicon (Shen, 2003).

The TCR, α, is defined as follows:

$$\alpha = (\Delta R / R_o) / \Delta T = \Delta R / (R_o \cdot \Delta T) \tag{6-2}$$

$$\Delta R = R - R_o$$

$$\Delta T = T - T_o$$

where R_o is the resistance of the material at a reference temperature. R is the resistance of the material measured at certain temperature. The zero TCR means that the ΔR is near zero and the resistance of the material is constant within the range of the temperatures measured. The resolution ($R_{esoul.}$) of temperature sensor is defined as follows:

$$R_{esoul.} = \Delta R / (R_o \cdot \Delta T) \times R \tag{6-3}$$

For example, the size of temperature sensor designed is 80 µm×20 µm×0.4 µm in length, width and thickness, respectively. The material used as the sensor is polysilicon. The concentration of boron used for doping is 10^{19} atoms/cm³. The ρ is found to be 0.2034 Ω-cm, as shown in Figure 8. After a calculation, the sheet resistance (R_S), total resistance (R) and TCR of temperature sensor are 5.085×10³ Ω/squ., 20.34 kΩ at 23 oC and -4.286×10⁻³ /oC. Therefore, the resolution of the polysilicon sensor can be calculated and is found to be -87.18 Ω/oC. It appears that the resolution of the temperature sensor made by the doped polysilicon is much better than any types of the thermocouples.

6.1.3 Fabrication processes (Ko et al., 2007, 2009; Liu,2004)

1. A 0.3 μm thick LPCVD tetraethoxysilane (TEOS) oxide is deposited on the (100) wafer as a insulation layer between sensors to silicon substrate.
2. Then, a 0.3 μm thick LPCVD polysilicon film is deposited and then implanted heavily with boron with a dose of 3×10^{15} atoms/cm². This amount of dosage corresponds to a concentration of 10^{20} atoms/cm³ in the layer. After annealing at 950 ºC for 30 minutes, the doped polysilicon is patterned as the heaters. In this doping concentration, the TCR of resistors are near zero, as shown in Figure 15(a).
3. Next, after a deposition of a 0.3 μm thick LPCVD TEOS layer as insulation, the temperature sensors were made by depositing a 0.3 μm thick LPCVD polysilicon layer and then implanting with boron ions with a dose of 3×10^{14} atoms/cm². This amount of dosage corresponds to a concentration of 10^{19} atoms/cm³ in the layer. After annealing at 950 ºC for 30 minutes to re-crystallize the polysilicon layer and to uniformly diffuse the dopant ions across the layer, the doped polysilicon was patterned. The concentration of 10^{19} atoms/cm² will give a linear relationship between the resistance and the temperature of the doped polysicon layer. Thus, after proper calibration, the doped polysilicon layer can be used as temperature sensors, as shown in Figure 15(b).
4. Before metallization, a 0.3 μm thick LPCVD TEOS oxide is deposited as insulation. Then, contact holes were opened in this layer for metallization. The metallization was made by sputtering standard IC four layers of metals, i.e. Ti/TiN/Al-Si-Cu/TiN with a thickness of 0.04 μm/0.1 μm/0.9 μm/0.04 μm, respectively, onto the substrate surface. The metal layers were then patterned into circuits, as shown in Figure 15(c).

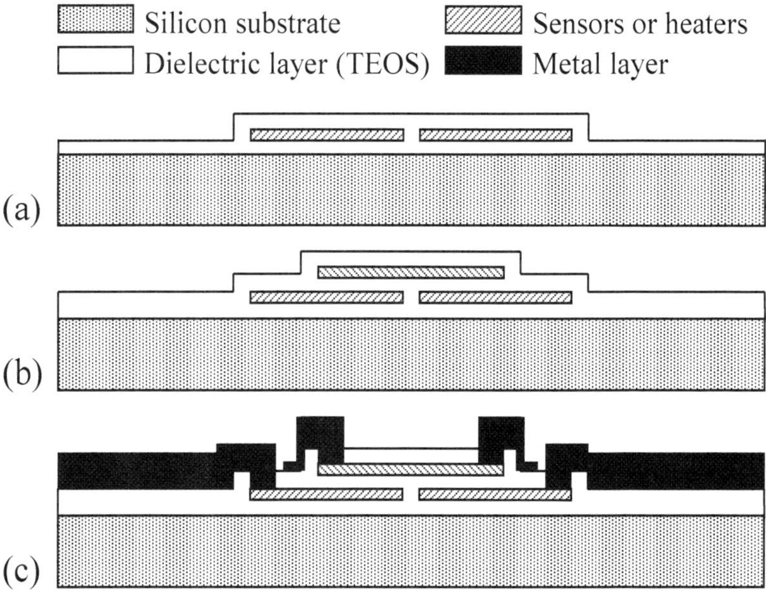

Fig. 15. Fabrication process of temperature sensor and heater (Ko et al., 2007; Liu, 2004).

6.1.4 Characteristics of temperature sensors and heaters

All the temperature sensors and the heater fabricated on the silicon substrate have to be calibrated in a constant temperature oven to ensure a measurement accuracy of ± 0.1°C. In fact, all the temperature sensors and the heaters can be moved onto a glass substrate for other purpose. The procedure to move all the sensors and heaters onto a glass substrate will be described in section 6.2 or 7.2. In that case, the substrate now is a very low thermal conductivity material, it is difficult for the heat in the oven to conduct into the substrate, especially in the region for the channel formation, and to provide a uniform and constant temperature condition required for the calibration. Therefore, the calibration procedure will take a much longer time than the case with silicon substrate. The electric resistance measured in the doped polysilicon layer at various boron concentrations and temperatures is compared with the published data (Ko et al., 2007; Liu, 2004), as shown in Figure 16. It is found that the resistance in the current work for the concentration at 10^{20} atoms/cm^3 agrees well with the published results. However, the data for the current case with the concentration at 10^{19} atoms/cm^3 is slightly different from other work, as indicated by the red circle as shown in Figure 16. This difference is attributed to the use of different implantation equipments from different companies, and the batch deposition of LPCVD polysilicon layer with a 5-10 % uniformity can occur in a single wafer or from wafer to wafer. The characteristic curves of the polysilicon temperature sensor and the heater used in the present channel system are also shown in Figure 17 (Liu, 2004). It is found that the output signal of the temperature sensor has a linear relation with the temperatures and a high resolution of 1.26 KΩ/°C. Therefore, this can provide a very accurate measurement in the temperature. In addition, the resistance of the heater does not vary with the temperature. This can readily provide a uniform and constant heating power required in the experiments.

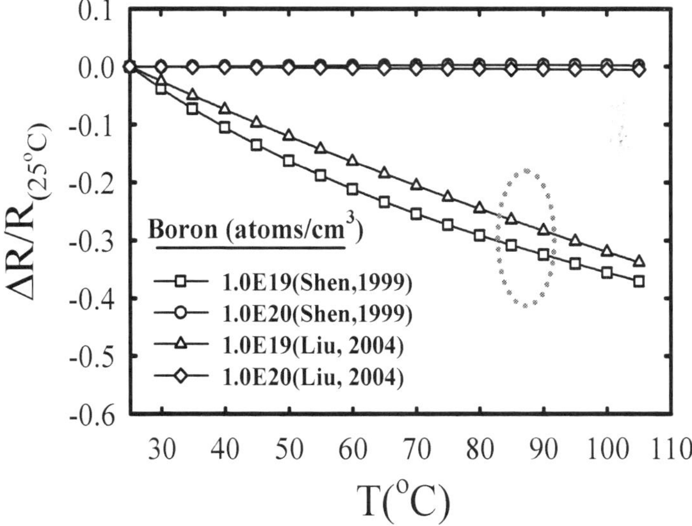

Fig. 16. Comparison of the resistance variation with temperatures between works for the polysilicon doped at different concentrations (Ko et al., 2007; Liu, 2004).

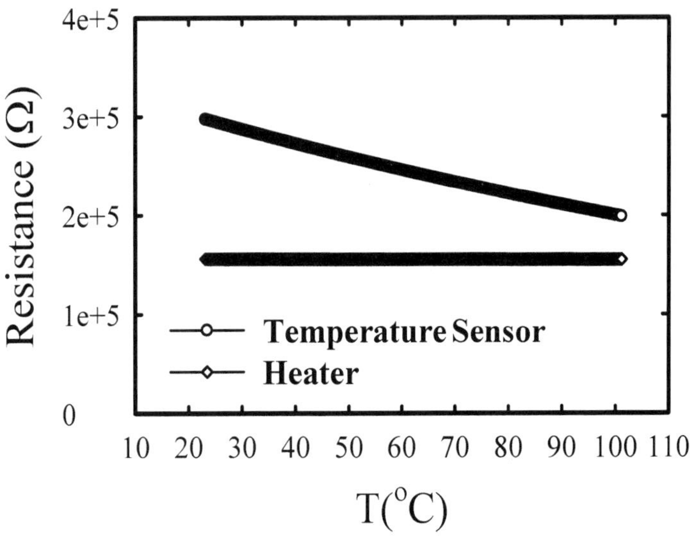

Fig. 17. Characteristic curves of the polysilicon temperature sensor and the heater that can be used in the present channel system (Liu, 2004).

6.2 Miniaturized pressure sensors
6.2.1 Working principles

Fabrication process of the micro pressure sensor has attracted much attention due to its wide industrial applications in various areas (Sze, 1994). In general, fabrication of pressure sensor adopts inorganic materials, e.g. piezoresistive sensor which is fabricated on a silicon substrate. Fabrication of these kinds of piezoresistive pressure sensors can be classified into bulk (Sze, 1994; Peake et al., 1969; Tufte et al., 1962) and surface (Guckel, 1991; Sugiyama et al., 1991; Clark et al., 1979) micromachining process. In the conventional designs of pressure sensor, both the bulk and the surface micromachining can be used. For the bulk micromachining technique, KOH solution is usually used to etch into the silicon substrate to form cavities, allowing for diaphragm deformation, in trapezoid shape with a 54.7° inclined wall. Since the silicon substrate is relatively thick and the bottom region consumed is much larger than the top region where pressure diaphragm defines. The number of sensors per wafer fabricated with bulk micromachining is much less than that fabricated with surface micromachining. Fortunately, this sensor is readily made and the yield of sensor is very high.

For the surface micromachining process, however, the number of sensors per wafer fabricated is much higher due to the fact that much smaller size of cavities for the pressure sensors can be made. However, the height of the cavities which allows deformation of diaphragm is made much smaller due to the use of sacrificial layer for later formation of the diaphragm. The sacrificial layer is deposited by chemical vapor deposition (CVD) process, which is relatively thin and is usually less than 2-3 µm. This has limited the height of the cavity defined by the sacrificial layer. Therefore, the pressure measurement range for the sensor made by the surface micromachining is much narrow. In addition, stiction of diaphragm can readily occur in the drying of DI water rinse process after removal of the sacrificial tetraethyl orthosilicate (TEOS) or phosphorus silicate glass (PSG) oxide. Special attention should be made to avoid the problem of stiction (Kim et al., 1998; Komvopoulos, 1996; Core et al., 1993; Legtenberg et al., 1993). In addition, the height of the cavity cannot be made large enough which will not give enough space for diaphragm deformation and allow the pressure measurement in a wider range. This is attributed to the use of sacrificial material, such as silicon oxide, that cannot be deposited thick enough by the CVD process.

In view of the disadvantages of the inorganic pressure sensor fabricated by bulk or surface micromachining of silicon material, the current work proposes fabrication of pressure sensor with polymer material, such as SU-8. In a review of literature, there are few other polymer materials that were found for fabrication of sensor system (Madou, 2002; Martin et al., 1998; Shirinov et al., 1996). A schematic diagram for a cross-section view of the sensor is shown in Figure 18 where the cavity can be filled with pressurized gas that causes deflection of the diaphragm on the top. All the wall material used, including the diaphragm for the sensor, is SU-8 except for the sensing material and circuit system on the top. The fabrication process presented in this paper can make arrays of pressure sensors and has the advantages of both the bulk and the surface micromachining process. The process can fabricate a greater number of sensors per wafer, and the sensors fabricated allow wider measurement range of pressure. The substrate can be any material. Here we select Pyrex glass. The fabrication process is very simple and the sensor can be readily made by spin coat SU-8 layer and patterning with lithography. In addition, the height of the cavity which allows deformation of diaphragm can be varied readily by spin coat different thickness of the SU-8 layer. Especially, the SU-8 diaphragm is formed without use of sacrificial layer and subsequent DI water rinse processes after removal. Therefore, fabrication process has completely absence of diaphragm stiction and has a much higher yield.

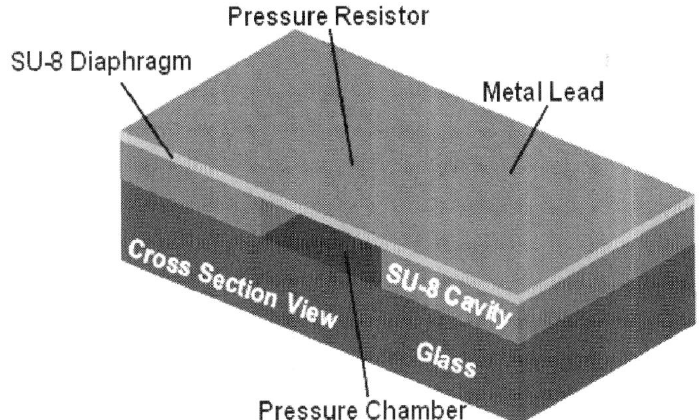

Fig. 18. A schematic diagram for a cross-section view of the pressure sensor designed.

The SU-8 material has been widely used for formation of microfluidic system (Arshak et al., 2006; Ribeiro et al., 2005; Lee et al., 2003), and relatively few in sensor system (Giordani et al., 2007). Therefore, current pressure sensor can also be readily applied and integrated into a microfluidic system or lab-on-chip (Pelletier et al., 2007; Trung et al., 2005; Vilkner et al.,

2004) where pressure information is required. Arrays of current pressure sensors can be readily fabricated along the channel or in a micromixer for more detailed flow information. They can also be fabricated at inlet, outlet or inside of a micropump for evaluation of its performance.

Theoretical modeling for the stress and deformation of the diaphragm is derived. Numerical calculation is provided for diaphragm design consideration. Finally, the current pressure sensors made can provide a much better thermal insulation than the ones made by the previous surface micromachining process of silicon because of the use of the extremely low thermal conductivity materials, such as the SU-8 and the Pyrex glass. The SU-8 material has a thermal conductivity of 0.2 W/mK (Monat et al., 2007) where the Pyrex glass has a value of 1.4 W/mK. This is highly important in the fabrication of a micro thermal system where thermal insulation should be seriously considered. More detailed fabrication techniques and performance evaluation of this sensor are also provided and discussed.

6.2.2 Design consideration

Before fabrication process, the size of the SU-8 diaphragm used in the pressure sensor should be determined from a proper analysis and design. The size of the SU-8 diaphragm is actually determined by the pressure range to be measured, the maximum strain that the diaphragm can sustain and the sensitivity required. The deformation of SU-8 diaphragm can be modeled as a square shell or plate with four edge clamped under a uniform normal pressure force as shown in Figure 19.

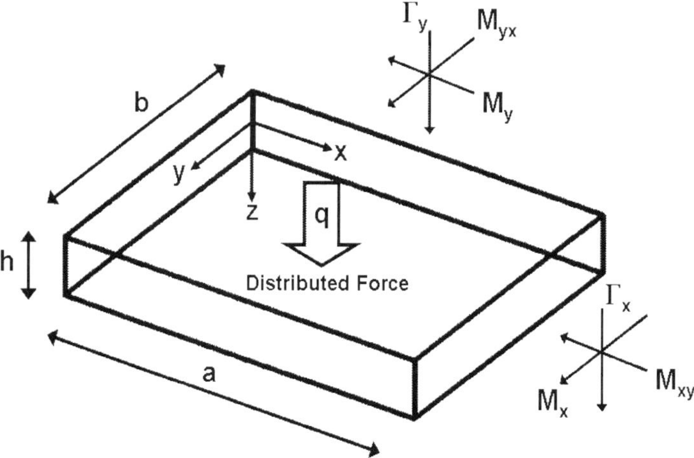

Fig. 19. Schematic diagram for the square plate deformed under pressure force.

From the solving of the governing equation for the deflection of the plate, the deflection of plate can obtain:

$$u = \frac{q_n}{D}\left(\frac{b}{\pi n}\right)^4 \sum_{n=1,3,5,\ldots}^{\infty} X_n(x)\sin\frac{n\pi y}{b} \tag{6-4}$$

By letting n= 1, 3 and 5 only, the maximum deflection can be found as follows:

$$u_{max} = C\,(1 - v^2)\,\frac{pb^4}{Eh^3}$$

(6-5)

where C is $\dfrac{0.032}{1 + \alpha^4}$, with the assumption that all edges of the plate are clamped, and α is ratio of the width to the length of the plate. However, C obtained by Guckel (1990) is 0.0152. The comparison is shown in Figure 20, which shows that for the diaphragm size of 200μm by 200μm and the thickness of the diaphragm varying from three to thirteen micron, the numerical result from ANSYS calculation agrees very well with the approximate solutions of Westergaard and Guckel, respectively. Therefore, the ANSYS is adopted for calculation in the design for proper size of diaphragm.

For a pressure range from 1 to 4 atms, the prediction for the maximum strain occurred in the plate versus the applied pressure force is presented in Figure 21 for different thickness of diaphragm. The maximum strain the SU-8 layer can sustain is found 0.77% (Guckel, 1991). Therefore, the rectangle presented in Figure 21 represents the safety region where the maximum strain is less than 0.77% for the diaphragm size designed. For a better sensitivity, i.e. more deflection, the size of 150μm×150μm wide at a thickness of approximately 9 μm is selected.

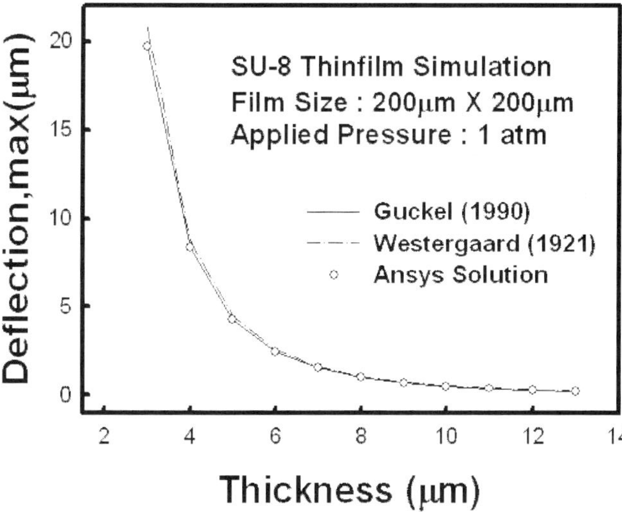

Fig. 20. Comparison of the numerical result from the ANSYS with the analytical results and other published data.

Polymer pressure sensor does not have the problems of narrow depth of cavity and stiction of diaphragm made by surface micromachining method. The most promising candidate for pressure sensor fabrication is the use of the photoresists, i.e. SU-8. The SU-8 can be readily spin coated on the substrate from a few microns to a hundred microns thick. This can give us a much deeper cavity depth if diaphragm layer can be properly deposited on the top to enclose the cavity, as shown in Figure 18. The diaphragm designed uses the same material as the cavity side wall in order to reduce difference of the thermal expansion coefficient for the cavity wall and the diaphragm which may cause deflection of diaphragm and significantly affect measurement accuracy. In this way, one would face another problem of selecting suitable sacrificial material to fill into the cavity to create a flat surface in order to spin coat the SU-8 diaphragm on the above. In addition, the sensor material currently available, which is to be deposited on the top of diaphragm, is piezoresistive film that is a doped polysilicon layer deposited by LPCVD at a high temperature process. The use of high temperature process of doped polysilicon for sensor formation has precluded the possible use of SU-8 as both diaphragm and side wall of cavity.

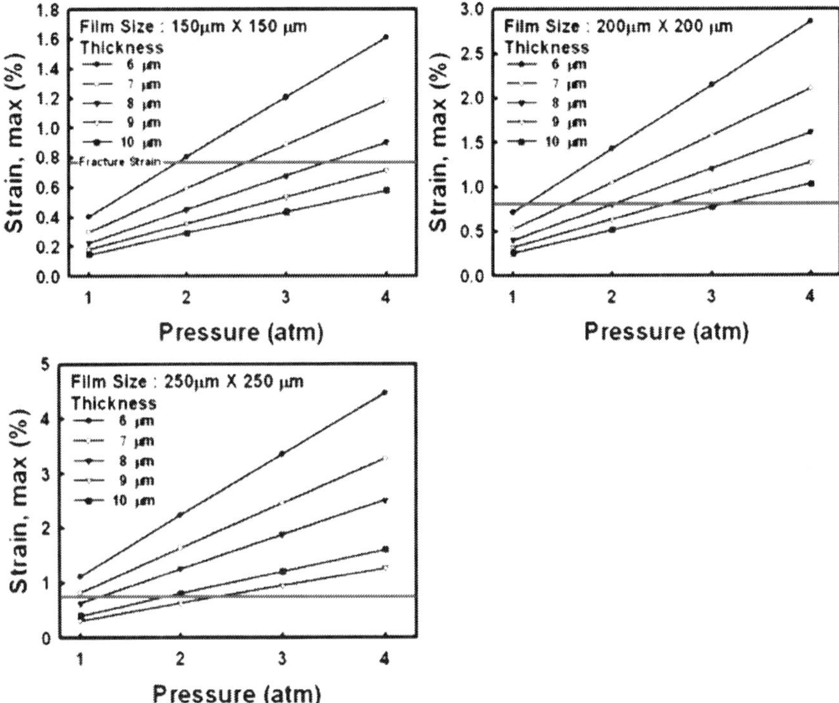

Fig. 21. The maximum strain variation with pressure for different sizes and thicknesses of SU-8 diaphragm.

However, this difficulty can be readily overcome by reversing the fabrication process of the pressure sensor. That is, one can first deposit the the piezoresistive layer and the metal lines on the silicon substrate which is a high temperature process, and then the diaphragm and the cavity wall which is a low temperature process. The piezoresistive layer can be made with the polysilicon layer implanted with very high concentration of boron or phosphorous.

In the current process, one selects very high concentration of boron such that the resistivity variation with the temperature in the polysilicon layer can be minimized. From the experimental data plotted in (Shen, 2004; Kanda, 1982; Mason, 1969), for boron concentration greater than 10^{20} atms/cm^3 the resistivity variation with temperature can be negligible small. Thus, this concentration of boron is adopted in the implantation process for the polysilicon layer. After implantation, the polysilicon is annealed for 30 mins at 950 °C. The next step is to spin coat SU-8 diaphragm, which has the flexibility to readily control the thickness. It is then followed by spin coat another layer of SU-8 for cavity wall at desired thickness. Thus, the formation of SU-8 layer will not go through a high temperature process. Finally, a Pyrex glass can be bonded with the patterned SU-8 layer on the top to enclose the cavities. Once the silicon substrate is completely removed by wet etch, a successful pressure sensors can be readily achieved.

6.2.3 Fabrication processes (Ko, 2009)

1. A 0.3 µm thick LPCVD TEOS oxide is deposited on the (100) wafer and used as protection mask for the upper layer devices during the later long period of TMAH wet etch to completely remove the Si wafer.
2. Next, a 0.3 µm thick LPCVD polysilicon film is deposited and then implanted heavily with boron with a dose of 3×10^{15} atoms/cm^2. This amount of dosage corresponds to a concentration of 10^{20} atoms/cm^3 in the layer. After annealing at 950 °C for 30 minutes, the doped polysilicon is patterned as the pressure sensors.
3. Before metallization, a 0.3 µm thick LPCVD TEOS oxide is deposited as insulation. Then, contact holes were opened in this layer for metallization. The metallization was made by sputtering standard IC four layers of metals, i.e. Ti/TiN/Al-Si-Cu/TiN with a thickness of 0.04µm/0.1µm/0.9µm/0.04µm, respectively, onto the substrate surface. The metal layers were then patterned into circuits, as shown in Figure 22(a).
4. A 9 µm thick SU-8 layer is spin coated on the substrate as the pressure sensing diaphragm. Then, a 50 µm thick SU-8 layer is spin coated again on the substrate and patterned into the active cavity to allow for movement of the pressure diaphragm. It is noted that there is a soft bake before layer exposure to evaporate the solvent and a post exposure bake to make the edge between the exposed and unexposed region more sharp and clear. Finally, instead of using the hard bake, a high intensity of light is used to illuminate to complete the cross-linking of the resin since the hard bake will need a temperature at 200 °C that will damage the underneath thick epoxy layer. For the soft bake, the SU-8 layer is first maintained at 65 °C for 7 minutes with a 5 °C/minute ramping rate starting from room temperature, and then baked at 95 °C for 15 minutes with a 5 °C/minute ramping rate starting from 65 °C to release the internal residual stress of SU-8 thick layer. In fact, the success of the SU-8 channel strongly depends on the baking process after light exposure.
5. It is now ready to move the devices made on the Si wafer onto a low thermal conductivity Pyrex glass. This is done first by bonding the silicon wafer with the Pyrex glass, as shown in Figure 22(b).
6. After the bonding process, the silicon substrate is ready for removal and cleared off. This is done by wet etch the silicon with TMAH solution at 90 °C for 5-6 hours. Instead of using KOH, the selection of TMAH is attributed to its relatively high selectivity for silicon versus oxide. This can avoid the sensors attacked by the etchant during the long period of wet etch process since the protection layer of the sensors are made of TEOS

oxide. A successful movement of the pressure sensors onto the Pyrex glass substrate is shown in Figure 23(a). The cavities of the pressure sensors are successfully made, as shown in Figure 23(b). No distortion was found.

7. Next, a 80 μm thick SU-8 layer is spin coated on the substrate and patterned by lithography to form a test section for the pressurized gas.

8. A PMMA plate is then bonded, using epoxy resin, to enclose the test section, as shown in Figure 22(c). The pressure sensor is completed.

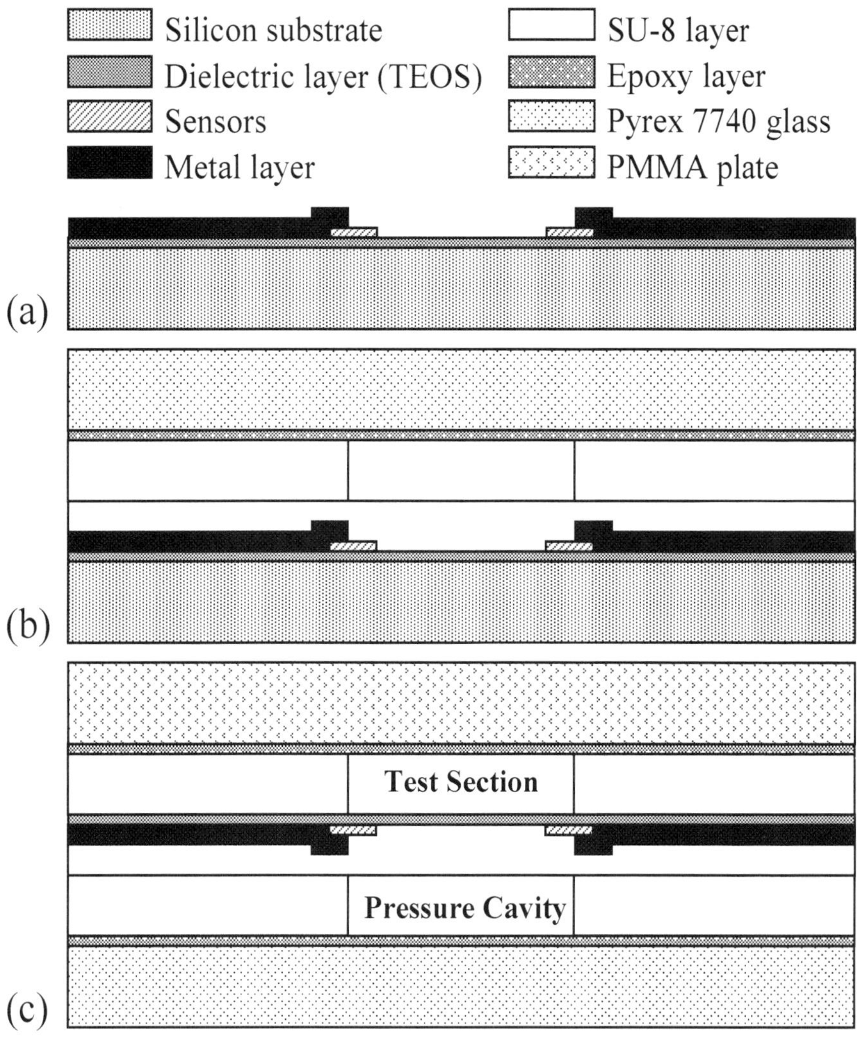

Fig. 22. Fabrication process of pressure sensor (Ko, 2009).

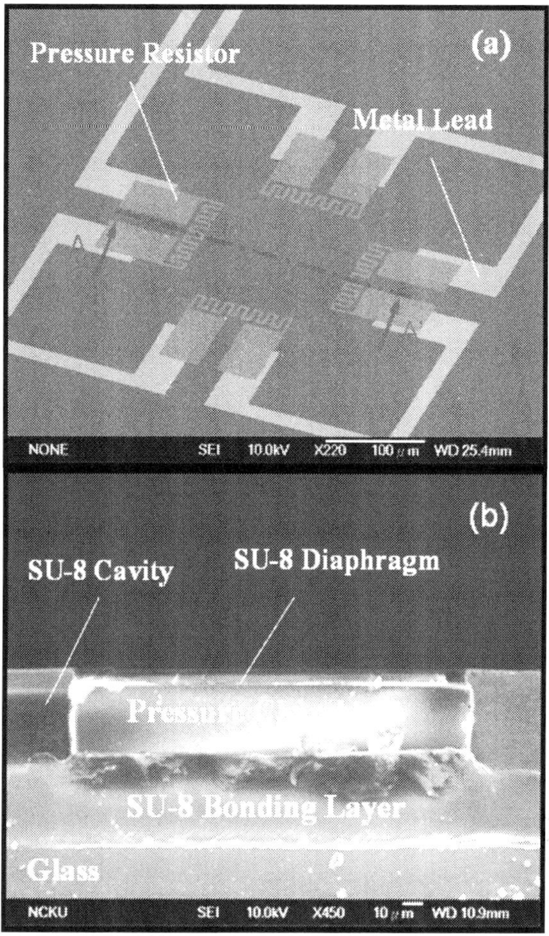

Fig. 23. Side view of SEM photographs. (a) Pressure sensor resistor embedded in SU-8 diaphragm and (b) pressure sensor diaphragm and cavity made by SU-8 lithography (Ko, 2009).

6.2.4 Characteristics of pressure sensors (Ko, 2009)

Since the arrays of pressure sensors were made of polysilicon layer doped with a designated concentration of 10^{20} atoms/cm^3 of boron, the temperature coefficient of resistance (TCR) of the sensors is almost zero which can eliminate the temperature effect for the sensors. In addition, the resistance of sensors designed is very large and is about 57-58 KΩ. The pressure effect on the resistivity of the sensor is expected very large. Therefore, the sensors have a very high signal resolution. The complicated temperature compensation currents and the electrical signal amplifier are not required and used in this sensor system. The characteristic curves for the sensors were obtained, as shown in Figure 24. The results show a very linear variation between the pressure and the resistance of the sensors. Since the diaphragm is made of SU-8 materials, which has a smaller Young's modulus than the polysilicon film, this leads to a higher strain in the SU-8 film. Therefore, the resolution for the present SU-8 diaphragm sensors is much higher than the polysilicon diaphragm sensors and is found to be 20.88 ohm/psig.

In order to test the reliability, i.e. the response, recovery and life time or fatigue of the current pressure sensors, a pressurized air is supplied to the sensor cavity for a period of 30 seconds and then released. The repeated sharp rise and descend of the pressure, as shown in Figure 25, suggests that the present SU-8 pressure sensor has very good reliability. Higher frequency of pressure oscillation still indicates that response and recovery test in the current sensor is still reliable except a slight drop of the oscillation signal is observed, as shown in Figure 26. This slight drop in oscillation signal is attributed to the frictional heating of the polymer diaphragm due to high frequency of vibrations. The frication heating increases the temperature and leads to reduction of the resistance of the polysilicon sensor. After cooling the sensor back to the ambient temperature, oscillation signal recovers back to the original value. The frictional heating may not be readily seen in polysilicon sensor due to the very high thermal conductivity of the material that can rapidly dissipate the frictional heating. The test for membrane under high frequency oscillation of pressure has proceeded for more than three thousand cycles which is more than 24 hours and does not indicates creeping or fatigue of the SU-8 membrane.

In addition, the sensor is put into oven at controlled temperatures for calibration at different temperatures. The results indicate that resistance variation at different temperatures is small and is less than 1% of the measured value. As one plots pressure variation versus temperature, as shown in Figure 27, the pressure variation is very small and less than 1 psig. The small pressure variation versus temperature may be attributed to the difference in the thermal expansion coefficient between the SU-8 and the Pyrex glass which causes small stress in the SU-8 side wall and the diaphragm. It is expected that replacing the Pyrex glass with other high temperature polymer plate, such as Poly(ether-ether-ketone) (PEEK), that can endure the high temperature annealing process of the SU-8 and have a closer thermal expansion coefficient to the SU-8, can further reduce the temperature effect on the pressure measurements.

Finally, in order to ensure that the SU-8 side wall or the diaphragm of the sensor is not permeable to the air, permeability test for the SU-8 to the air is performed. This is done by sending pressurized air into the test section (chamber), as shown in Figure 22(c), of the sensor at desired pressure level and closing both the inlet and outlet valves. The pressure variation in the test section is monitored by the sensor for more than 7 days, and the signals are sent into computer for plotting. It appears that the pressure inside the test section is almost kept constant except for a slight variation due to noise from the instrument. The variation of the pressure due to noise is less than 1.76% of the total pressure during a long period of 7 day observation. This result indicates that the pressurized air inside the test section will not leak slowly through the SU-8 diaphragm into the pressure chamber, nor the pressurized air (in either the test section or the pressure chamber) will leak slowly through the SU-8 side wall into the ambient. This permeability test also indicates that the SU-8 diaphragm will not creep under a constant pressure load of 350 kPa for 7 days. Therefore, one may expect that the SU-8 diaphragm will not creep as long as the pressure load is within the elastic deformation of the material unless when the material becomes aged.

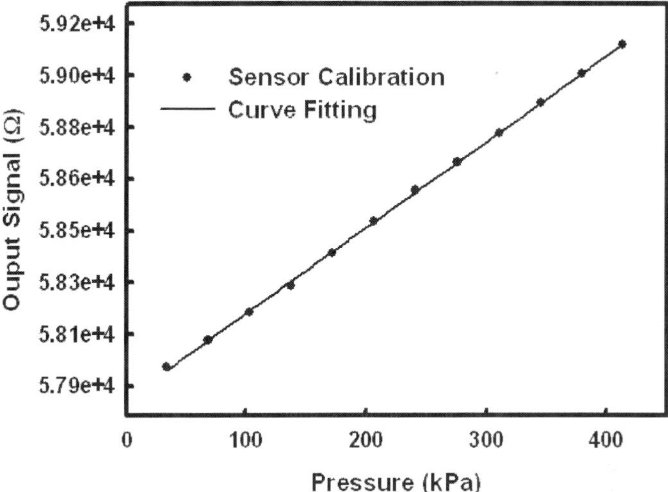

Fig. 24. A very linear variation between the pressure and the resistance of the sensors (Ko, 2009).

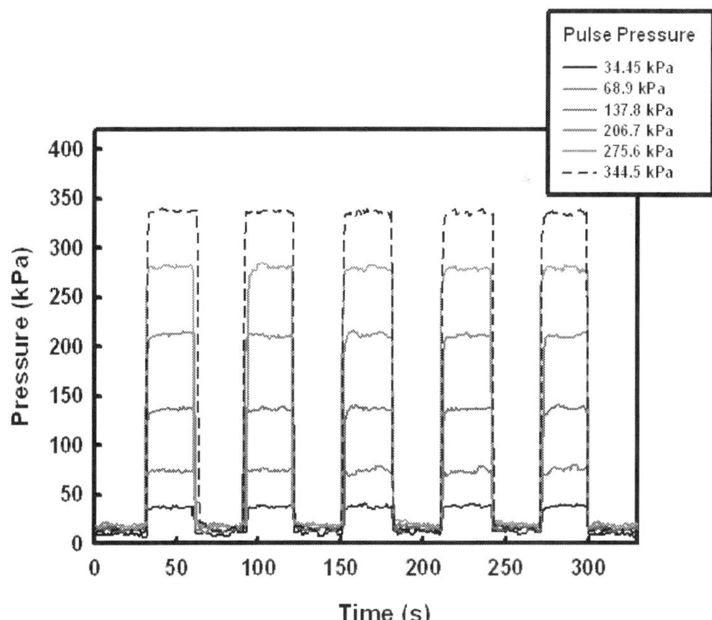

Fig. 25. Response and recovery test for pressure sensor at different pulsed pressures (Ko, 2009).

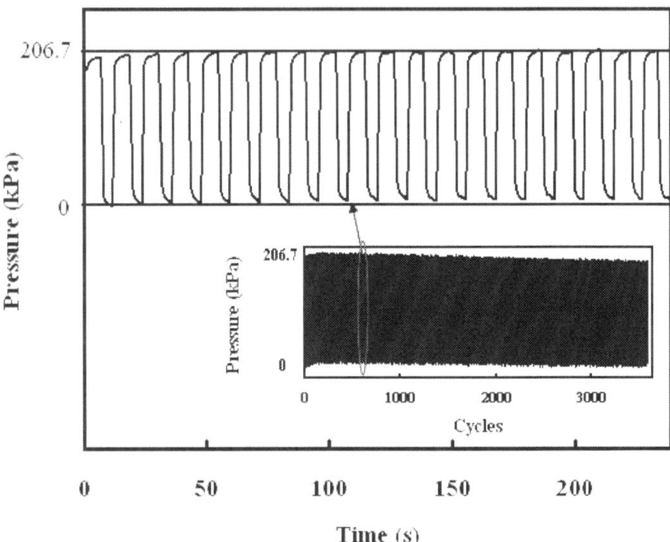

Fig. 26. Response and recovery test for pulsed pressure at high frequency (Ko, 2009).

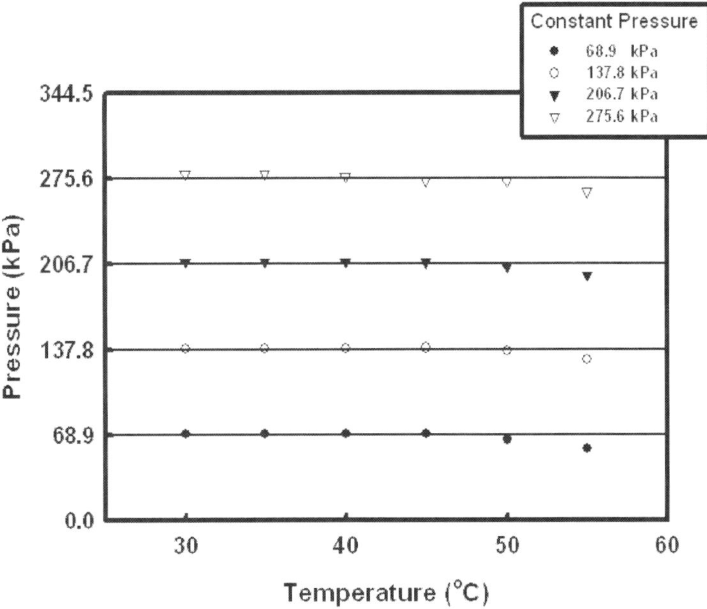

Fig. 27. Thermal stability test shows that the pressure signal is almost independent of temperature variation in the system (Ko, 2009).

7. Fabrication of microchannel integrated with arrays of miniaturized pressure sensors and temperature sensors

The development and fabrication of a micro-flow heated channel for studies of local flow and heat transfer process is very important not only for the basic research but also for practical application in MEMS or NEMS thermal system. Unfortunately, most of the current micro-channel used (Choondal and Suresk, 2000; Yu et al., 1995) or fabricated (Jiang et al., 1999; Tao and Mahulikar, 1998; Chu et al., 1994) could not provide a local or detailed flow and heat transfer information, but a global information (only for flow information at inlet and outlet). How to prevent or control heat loss are the most important considerations for heat transfer study. Most of the MEMS or NEMS techniques developed up to the present are to use silicon wafer, a very good thermal conductor that has a very high thermal conductivity (148 W/mK), as substrate to fabricate a microchannel (Pfahler et al., 1990; Pfahler et al., 1991; Arkilic and Breuer, 1994; Arkilic and Schmidt, 1997; Shih et al., 1996; Qu et al., 2000a and 2000b; Ren and Li, 2001; Ren et al., 2001a and 2001b). This will lead a large amount of heat loss from the inside channel to the outside ambient, a uniform heat flux boundary condition on the heated wall could not be maintained. Therefore, local heat transfer information in the channel cannot be obtained and understood.

The idea to fabricate a suspended channel over an air layer (Wu et al., 1998) to reduce the heat loss may be adopted to fabricate a heat transfer channel. However, this kind of channel that has a very thin wall can be readily deformed by the high-pressure flow. This deformation is expected to significantly affect the detailed flow and heat transfer process inside the channel. In addition, the techniques involved in the fabrication are suitable for fabrication with a very small height (in a few micrometers) channel.

The current work is to develop a novel fabrication procedure for a micro-channel system that can be used to study and obtain the flow and the local heat transfer coefficient inside the channel. One side of the channel wall can be heated uniformly and all the other sidewalls can be well insulated. In the meantime, the heated wall is integrated with an array of micro-temperature sensors and pressure sensors that can provide measurements of local temperature and local pressure of the heated wall and study the local flow and heat transfer process along the channel.

Initially, both the heater and the array of temperature sensors and pressure sensors that are made of polysilicon layer doped with different concentration of boron are deposited on a (100) silicon wafer by LPCVD. Then, an ultra thick SU-8 layer is spin coated on the substrate and patterned into the active cavity of the pressure diaphragm. After SU-8 lithography process, both the heater and the array of the sensor layers are moved to the surface of a low thermal conductivity Pyrex glass. This was made by bonding the deposited silicon wafer with the Pyrex glass to use a thick layer of epoxy resin. Then, the TMAH (25 wt%) solution at 90 ºC was used to etch and completely remove the silicon wafer. The next step is to spin a thick layer of SU-8 on the epoxy-glass substrate and form a micro-channel structure by lithography. The last step is to bond the epoxy-glass substrate with a thick PMMA plate to form a well-insulated and compact micro-channel. It is noted that both the SU-8 layer and PMMA plate are the very low thermal conductivity materials, as shown in Table 2. Design consideration and fabrication techniques involved in this processes will be discussed in the following sections.

7.1 Design consideration

In the design of a micro-channel for the current flow and heat transfer study, the most important consideration is to reduce or control the amount of the heat loss from the channel to the ambient and to provide a uniform heat flux input on the wall that can be readily measured. To provide a uniform heat flux on the wall is not difficult. This can be achieved by selecting a resistor material that is independent of the temperature and distributing the resistor material uniformly on the channel wall. To pass a steady current through the resistor, a uniform heat generation along the wall can be obtained. However, the heater may have an uncontrollable heat loss at different locations such that the some of heat generated in the heater may lose to the substrate and the real heat input into the channel flow is not uniform. Thus, a uniform heat flux boundary condition on the wall cannot be obtained. In addition, the thermal energy inside the channel flow may lose to the wall and subsequently lose to the substrate if the wall is very conductive. Unfortunately, most of the MEMS techniques developed up to the present use silicon wafer, a very good thermal conductor that has a very high thermal conductivity, as a substrate to fabricate a micro-channel. These difficulties make the measurements and analysis of the heat transfer inside a formidable task. Chen et al. (2001) etched an air layer underneath the channel system to reduce the heat loss. This air layer is thick enough to effectively reduce the heat loss. However, it makes the heated wall very thin that the channel may distort itself during high-pressure gas or liquid flowing through the channel. In addition, the channel fabricated is very complicated. Although the design of using the air layer underneath the channel as insulation is a good idea, the fabrication process is very complicated and difficult, and the yield of the fabrication is very low.

Material	Thermal Conductivity (W/mK) at 300K	Thermal Expansion Coefficient (ppm/K) at 300K
Silicon	148	2.6
Al-Cu-Si alloy	161	19.2
Polysilicon	34	2.4
Silicon nitride	16	0.8
Silicon oxide	1.38	10.3
Pyrex glass	1.4	2.8
Epoxy resin	0.15	62.5
PMMA plate	0.16	62
SU-8	0.2*	52*
Air	0.026	

Data can be found in data sheet for NATO™ SU-8 negative tone photoresists, formulations 50 and 100, released by Micro Chem Corp.

Table 2. Thermal physical properties of the fabrication materials (Touloukian, 1970-1979).

In general, the heat input into the channel flow system can be moved away from the bottom heated wall of the channel by the convective flow, i.e. air or water flow through the channel. However, some portion of the heat input, i.e. the heat loss, can be moved away to the ambient by the conduction of the substrate or the channel structure. In fact, most of heat loss occurs by the substrate conduction, especially for a substrate with a higher thermal conductivity. Thus, in order to effectively reduce the heat loss and maintain a uniform heat flux boundary condition on the heated wall, several alternatives can be selected. One can use the thermal insulation materials on the back of the silicon substrate to reduce the heat loss. In addition, to reduce the spanwise heat conduction along the substrate, the thickness of the substrate can be reduced by mechanical polishing or particular wet etching techniques. In addition, instead of using the air layer underneath the channel, one was thinking of using other low thermal conductivity materials, such as Pyrex glass, as substrate. However, by a careful calculation even the Pyrex glass could not provide enough thermal insulation for both the channel and the heater. Other materials such as PMMA, polyimide or epoxy resin that have a much lower thermal conductivity, as shown in Table 2, can be used to avoid the heat loss from the heater to the ambient outside. An alternative insulation is to move the devices above the silicon wafer onto the Pyrex glass. This requires bonding the device with the Pyrex glass using epoxy resin, which is a low thermal conductivity material. Therefore, the epoxy resin can be served as both the bonding agent and a part of the substrate. This can significantly reduce both the normal and the spanwise conduction along the substrate. A schematic diagram designed for the microchannel integrated with arrays of micro pressure and temperature sensors is shown in Figure 29.

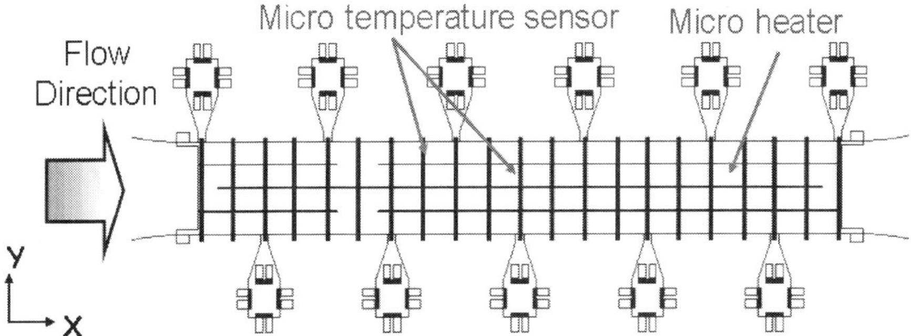

Fig. 28. Micro-channel system with arrangements of arrays of micro pressure sensors, micro temperature sensors and two sets of heaters used for study of micro scale heat transfer and pressure drop of the flow.

7.2 Fabrication process flow

The entire fabrication process, as shown in Figure 29, is described in the following:

1. A 0.3 µm thick LPCVD tetraethoxysilane (TEOS) oxide is deposited on the (100) wafer and used as protection mask for the upper layer devices during the later long period of TMAH wet etch to completely remove the Si wafer.
2. The temperature sensors were made by depositing a 0.3 µm thick LPCVD polysilicon layer and then implanting with boron with a dose of 3×10^{14} atoms/cm^2. This amount of dosage corresponds to a concentration of 10^{19} atoms/cm^3 in the layer. After annealing at 950°C for 30 minutes to re-crystallize the polysilicon layer and to uniformly diffuse the dopant ions across the layer, the doped polysilicon was patterned. The concentration of 10^{19} atoms/cm^2 will give a linear relationship between the resistivity and the temperature of the doped polysicon layer. Thus, after proper calibration, the doped polysilicon layer can be used as temperature sensors.

3. Next, after a deposition of a 0.3 µm thick LPCVD TEOS layer as insulation, a 0.3 µm thick LPCVD polysilicon film is deposited and then implanted heavily with boron with a dose of 3×10^{15} atoms/cm^2. This amount of dosage corresponds to a concentration of 10^{20} atoms/cm^3 in the layer. After annealing at 950°C for 30 minutes, the doped polysilicon is patterned as the pressure sensors.

4. Before metallization, a 0.3 µm thick LPCVD TEOS oxide is deposited as insulation. Then, contact holes were opened in this layer for metallization. The metallization was made by sputtering standard IC four layers of metals, i.e. Ti/TiN/Al-Si-Cu/TiN with a thickness of 0.04 µm/0.1 µm/0.9 µm/0.04 µm, respectively, onto the substrate surface. The metal layers were then patterned into circuits, as shown in Figure 29 (a).

5. A 9 µm thick SU-8 layer is spin coated on the substrate as the pressure sensing diaphragm, as shown in Figure 29 (b). Then, a 50 µm thick SU-8 layer is coated again on the substrate and patterned into the active cavity to allow for movement of the pressure diaphragm. It is noted that there is a soft bake before layer exposure to evaporate the solvent and a post exposure bake to make the edge between the exposed and unexposed region more sharp and clear. Finally, instead of using the hard bake, a high intensity of light is used to illuminate to complete the cross-linking of the resin since the hard bake will need a temperature at 200 °C that will damage the underneath thick epoxy layer (refers to the epoxy layer mentioned in item 5 below). For the soft bake, the SU-8 layer is first maintained at 65 °C for 7 minutes with a 5 °C/minute ramping rate starting from room temperature, and then baked at 95 °C for 15 minutes with a 5 °C/minute ramping rate starting from 65 °C to release the internal residual stress of SU-8 thick layer. In fact, the success of the SU-8 channel strongly depends on the baking process after light exposure.

6. It is now ready to move the devices made on the Si wafer onto a low thermal conductivity Pyrex glass. This is done first by bonding the silicon wafer with the Pyrex glass, as shown in Figure 29 (c), using epoxy resin as the bonding agent. Since the epoxy resin has a significantly lower thermal conductivity than the Pyrex glass, this can better reduce the heat loss from the channel to the ambient.

7. After the bonding process, the silicon substrate is ready for removal and cleared off. This is done by a wet etch process. The silicon is etched with TMAH solution at 90 °C for 5-6 hours. Instead of using KOH, the selection of TMAH is attributed to its relatively high selectivity for silicon versus oxide. This can avoid the sensors attacked by the etchant during the long period of wet etch process since the protection layer of the sensors are made of TEOS oxide. After successful movement of the pressure sensors onto the Pyrex glass substrate, no distortion was found, as shown in Figure 29(d).

8. Next, another SU-8 layer is spin coated on the epoxy-glass substrate at desired thickness and patterned by lithography to form a rectangular microchannel and cavities of the pressure sensors that connect with channel through narrow size tunnels. The side walls of the channel, the narrow tunnels, or the cavities of the pressure sensors are successfully made, as shown in Figures 30 (a), (b) and (c). It is noted that a similar post exposure bake should be made in order to obtain a nice shape of channel and cavities of the pressure sensor.

9. A PMMA plate is then bonded, using epoxy resin, with the epoxy-glass substrate to enclose the channel and the cavities, as shown in Figure 29 (e).

A picture for the completed channel is shown in Figure 30 (d).

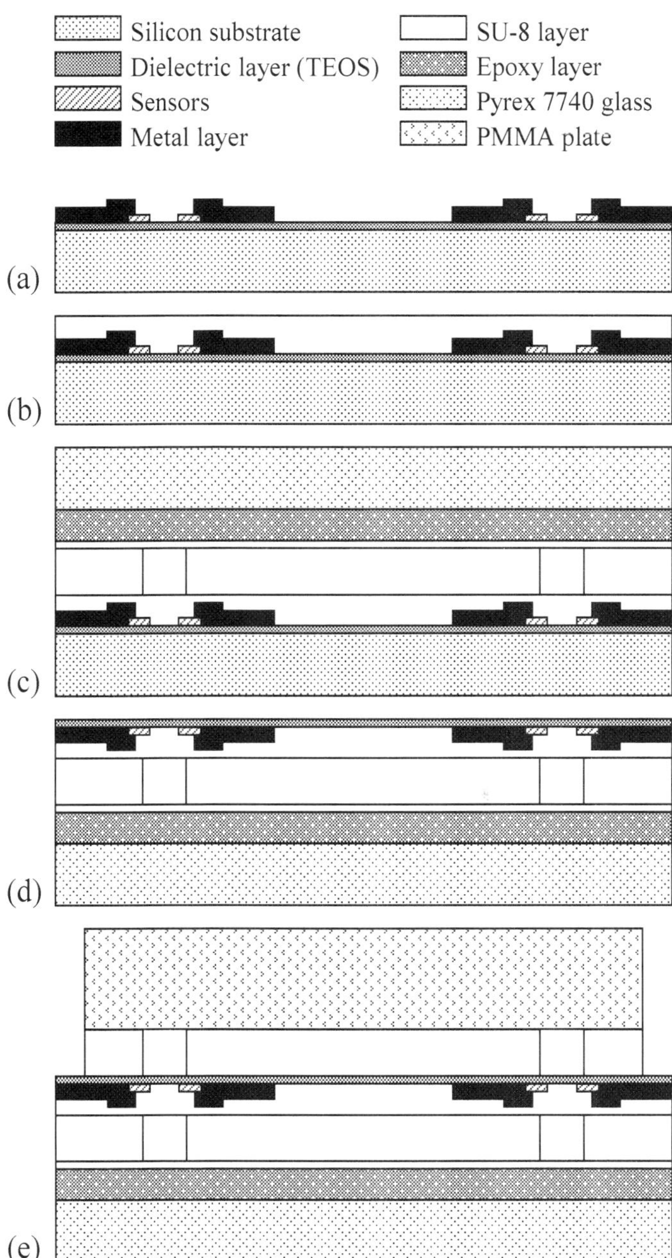

Fig. 29. Fabrication process for the entire microchannel system.

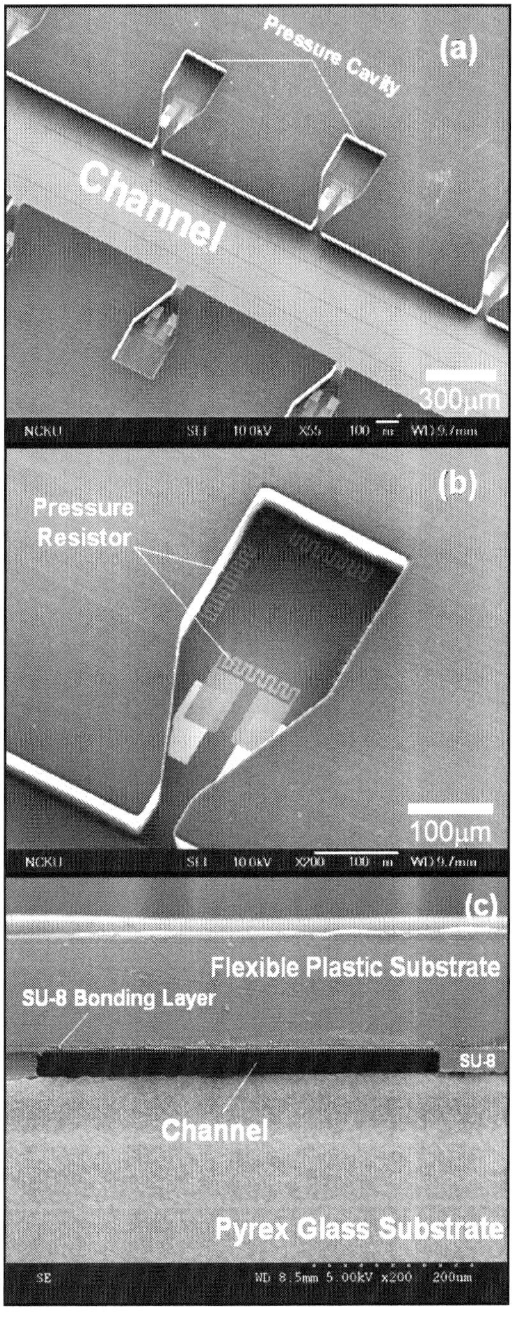
Pressure Cavity
(a)
Channel
300µm
NCKU   SEI   10.0kV   X55   100 m   WD 9.7mm
(b)
Pressure
Resistor
100µm
NCKU   SEI   10.0kV   X200   100 m   WD 9.7mm
(c)
Flexible Plastic Substrate
SU-8 Bonding Layer
SU-8
Channel
Pyrex Glass Substrate
SE   WD 8.5mm 5.00kV x200   200um

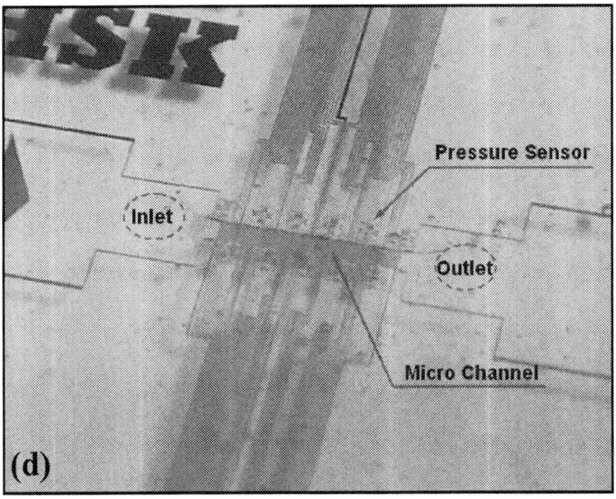

Pressure Sensor

Inlet

Outlet

Micro Channel

(d)

Fig. 30. SEM photographs of SU-8 channel integrated with arrays of micro pressure sensors: (a) global view, (b) close view and (c) cross section view and (d) the completed microchannel (Ko, 2009).

8. Local pressure drop and heat transfer characteristic inside a microchannel

8.1 Liquid flow characteristics

In order to compare with the experimental data, theoretical analysis of the flow characteristic inside the microchannel is performed. Since the microchannel has very wide in size as compared with the height, the microchannel flow can be assumed as parallel plate channel flow. Since the microchannel flow can rapidly become fully developed, as indicated in the later data measurement, the governing equation for fully developed flow in the microchannel can be greatly simplified. The solution can be readily found as follows (Incropera et al., 2007):

$$u(y) = \frac{y^2}{2\mu}\frac{dp}{dx} - \frac{H}{2\mu}\frac{dp}{dx}y = \frac{1}{2\mu}\frac{dp}{dx}(y^2 - Hy) \tag{8-1}$$

To integrate the equation transversely, the mean velocity of the flow in the channel can be obtained as follows:

$$U = -\frac{H^2}{12\mu}\frac{dp}{dx} \tag{8-2}$$

where U is the mean velocity of the flow. From Eq. (6-2), the longitudinal pressure distribution can be found as follows:

$$p_x = p_i - \frac{12\mu U}{H^2}x \tag{8-3}$$

or

$$\Delta p = \frac{12\mu Ul}{H^2}$$

The Reynolds number here is defined as the liquid density times the mean velocity of the flow and the hydraulic diameter (D_h) of the rectangular channel divided by the dynamic viscosity of the liquid as follows:

$$\mathrm{Re} = \frac{\rho U D_h}{\mu} \tag{8-4}$$

where is ρ the density of the water. The friction factor of flow can be calculated with a given pressure drop.

For the liquid flow, Qu et al. (2000, 2002) has found that the pressure drop in microchannel is significantly higher than the theoretical analysis due to the roughness effect. To eliminate the wall roughness effect, therefore, the roughness on the wall surface of the current micro-channel is minimized, and is less than ± 5 nm to 15 nm. This roughness is negligibly small in comparison with the height of the channel, which is either 68.2 µm or 23.7 µm. During the experiment, the pressure distributions measured at different Reynolds numbers are compared with the analysis, as shown in Figure 31(a) for the channel height at 34.7 µm. The pressure drop of the water is found significantly higher than the prediction by Eq. (8-3). A similar result was also found by Mala and Li (1998) who showed that the pressure drop of water in micro-tubes is significantly higher than the analysis. Under the same flow rate and tube diameter, the pressure drop in metal tubes is lower than in silica glass tubes. This can be attributed to the electric surface potential in the channel which attracts counterions in the liquid and causes formation of electric double layer on the wall. The thickness of the EDL in the micro channel for these three different solutions can be estimated by the inverse of the square root of the Debye-Huckel parameter (Mala et al. 1997) as follows:

$$\tau = \left(\frac{\varepsilon \varepsilon_o \kappa_b T}{2 z_i^2 e^2 n_\infty}\right)^{1/2} \tag{8-5}$$

where ε is the dielectric constant of the solution and ε_o is the permittivity of vacuum, κ_b is the Boltzmann constant, T is the absolute temperature, n_∞ and z_i are the bulk concentration and the valence of type i ions. Therefore, the EDL thickness is approximately 1 µm for water. In order to minimize the EDL effect, the working fluid used is changed to KCl solution at concentration of 10^{-4} M or 10^{-2} M, which can effectively remove the surface charges by the metal ions in the solution and significantly reduce the thickness of EDL. The EDL thickness is approximately 100 nm for the KCl solution with concentration of 10^{-4}M and 1 nm for the KCl solution with concentration of 10^{-2}M. The result indicates, as shown in Figure 31(a), that the pressure distribution for a higher KCl concentration (10^{-2} M) is significantly lower than for the pure DI water and is close to the analysis. It appears that the deviations for the experimental data of water from the prediction are attributed to surface charges and the EDL on the channel wall. The thicker the EDL, the greater is the deviation of the pressure drop from the analysis. The current finding for the reduction of EDL thickness which makes the pressure drop in the microchannel approach to the analysis agrees with the results in Ren et al. (2001). When the channel height increases to 68.2 µm, the pressure drops for the case of the pure water and the cases of different concentrations of KCl solutions collapse into a single line, as shown in Figure 31(b), and are closer but higher than the analysis. It appears that the EDL of different fluids does not have a significant effect on the pressure drop in a greater height of channel even though the EDL of different fluids on the wall has different thickness.

For a large scale channel, the entrance region can be clearly identified by the non-linear distribution of pressure distribution. However, in the entry region, no clear indication of non-linear distribution of pressure can be found, as shown in Figure 31(a) or 31(b). For large scale channel, the entry length can be estimated by the equation as follows Ren et al. (2001):

$$L_{in} = 0.02 \, D_h \, Re \tag{8-6}$$

For the channel height at 68.2 µm and Reynolds number at 128, the hydrodynamic entry length calculated from Eq. (8-6) is approximately 300 µm. However, pressure drop within 300 µm does not indicated greater or nonlinear variation, as shown in Figure 31(b). It shows that the hydrodynamic entry region in microchannel is very short and could not be identified from the current pressure distribution measurements. It appears that the boundary layer in the entrance region of a microchannel develops very fast and becomes fully developed very close to the entrance.

The same kinds of fluids are also performed in channels at the height of 23.7 µm. In the channel at the height of 23.7 µm, more deviations of the pressure drop data for different fluids occur, as shown in Figure 32(a) for pressure drop variation with the Reynolds numbers in the fully developed region. The pressure drop for water flow has the largest slop, and decreases when the concentration of KCl solution increases. These results clearly indicate the effect on the thickness of EDL relative to the height of channel. For water flow with a thicker EDL on the channel wall, the appearance of the EDL can increase the pressure drops by 13.4 % in the smaller height of channel (23.7 µm), but increase only 8.6 % in the greater height of channel (34.7 µm). For the channel height at 68.2 µm, all the pressure drop variations with the Reynolds number for different fluids collapse into a single line, as shown in Figure 32(b), and are closer but higher than the analysis. The higher pressure drop data for different fluids than the prediction by Eq. (8-3) can be attributed to the aspect ratio effect of the channel. The prediction by Eq. (8-3) is based on an analysis in a parallel plate channel, while the current channel with channel height of 68.2 µm or 23.7 µm has an aspect ratio of 1:7.3 or 1:21, respectively.

From the definition of the friction factor,

$$f = \frac{\tau_w}{\dfrac{1}{2}\rho U^2} = \frac{D_h}{2\rho U^2}\frac{dp}{dx} \tag{8-7}$$

where D_h is hydrodynamic diameter of the channel cross section and is equal to $2Hw/(H+w)$.

The product of friction factor, f, and the Reynolds number is defined as the Poiseuille number (Po) (Baviere, 2004) and can be written as follows for parallel plate channel:

$$Po = f\,Re = 24 \tag{8-8}$$

The parallel plate channel has an infinite width in z direction, the velocity is a two dimensional parabolic profile. However, when the channel width in the z direction is finite the velocity close to the side wall becomes three dimensional. The side wall effect can produce greater pressure drop. The smaller the channel width in the z direction, the greater the aspect ratio, the greater the pressure drop becomes. For large scale channel, the greater Po value for channel with greater aspect ratios has been calculated and correlated in terms of aspect ratio, q as follows Baviere et al. (2004):

$$Po(a) = 24\,(1 - 1.3553\,a + 1.9467\,\alpha^2 - 1.7012\,a^3 + 0.9564\,a^4 - 0.2537\,a^5) \tag{8-9}$$

From Eqs. (8-7), (8-8) and (8-9) the pressure gradient can be re-calculated for the aspect ratio less than 1:21, as shown in Figure 33(a) and Figure 33(b) indicated by the red dash line. The pressure gradient predicted by Eq. (8-9) in the fully developed region agrees very well with the data for working fluid at the much higher concentration of KCl or when channel height is very large (68.2 µm) and EDL effect is negligible small. For clarity, the Po number for both the data and the prediction by Eq. (8-8) has been plotted as shown in Figure 33. The agreement between data and the prediction at channel height of 68.2 µm indicate that wall roughness in the current channel has a negligible effect on the pressure drop. Even at a smaller height of channel, 23.7 µm, the agreement between data at high concentration of KCl with negligible effect of EDL and the prediction further indicates the negligible effect of the wall roughness in the current channel.

In order to compare with the pressure drop data published in the literature, the normalized friction constant which is most frequently used is adopted and written as follows.

$$C_f^* = \frac{Po_{exp}}{Po_{theory}} \tag{8-10}$$

where the friction constant Po_{exp} is found from the product of Re with the measured friction factor and Po_{theory} is directly evaluated from Eq. (8-8). Large discrepancy among different works can be found as shown in Figure 34(a) and Figure 34(b). The data selected here for comparison are only limited to rectangular channels which have a channel height close to the current channel. More detailed sizes of the channels and the materials the channels used are also listed in the table underneath the figure. In general, the normalized friction constants for all the work in the past are not only much greater than unity which is a theoretical prediction for large scale channel, but also greater than the data of the current work. This is attributed to the use of the inlet and outlet pressure to calculate the pressure drop in the channel which is used to evaluate the friction constant. Most of the work in the past can not obtain the pressure inside the channel, but measure the pressure outside the channel at the inlet and outlet (Pfund et al., 2000; Xu et al., 2000; Papautsky et al.,1999; Peng et al., 1995). Since there is large pressure loss for flow entering or exiting the channel, the pressure drop measured outside the channel at the inlet and the outlet is much greater than the actual pressure drop in the channel. Therefore, much greater friction constants obtained from others are expected.

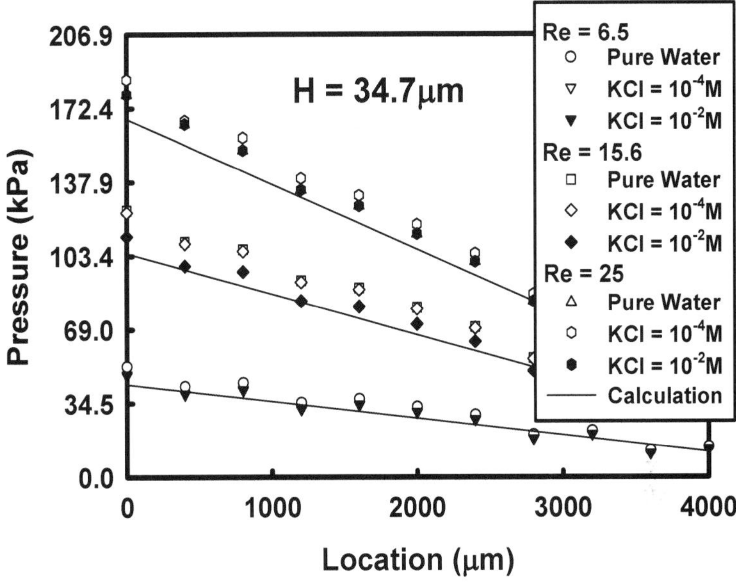

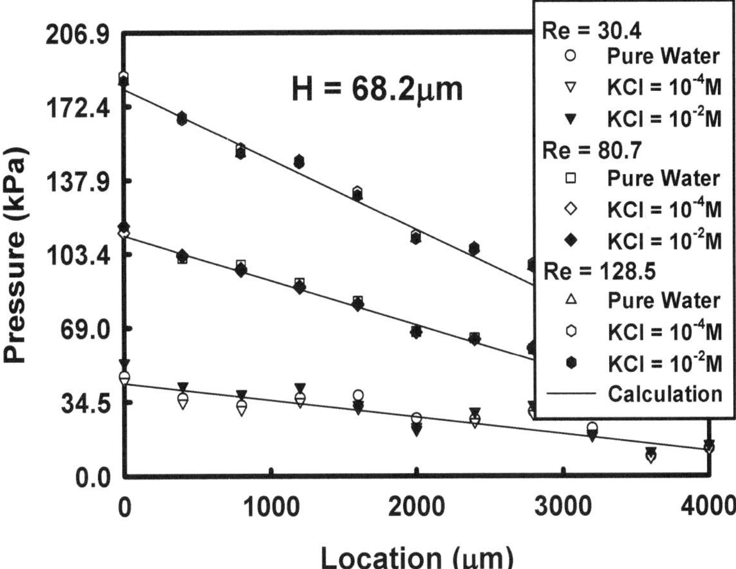

Fig. 31. Local pressure distributions at different Reynolds numbers in a micro-channel with channel height of (a) 34.7 μm and (b) 68.2 μm (Ko, 2009).

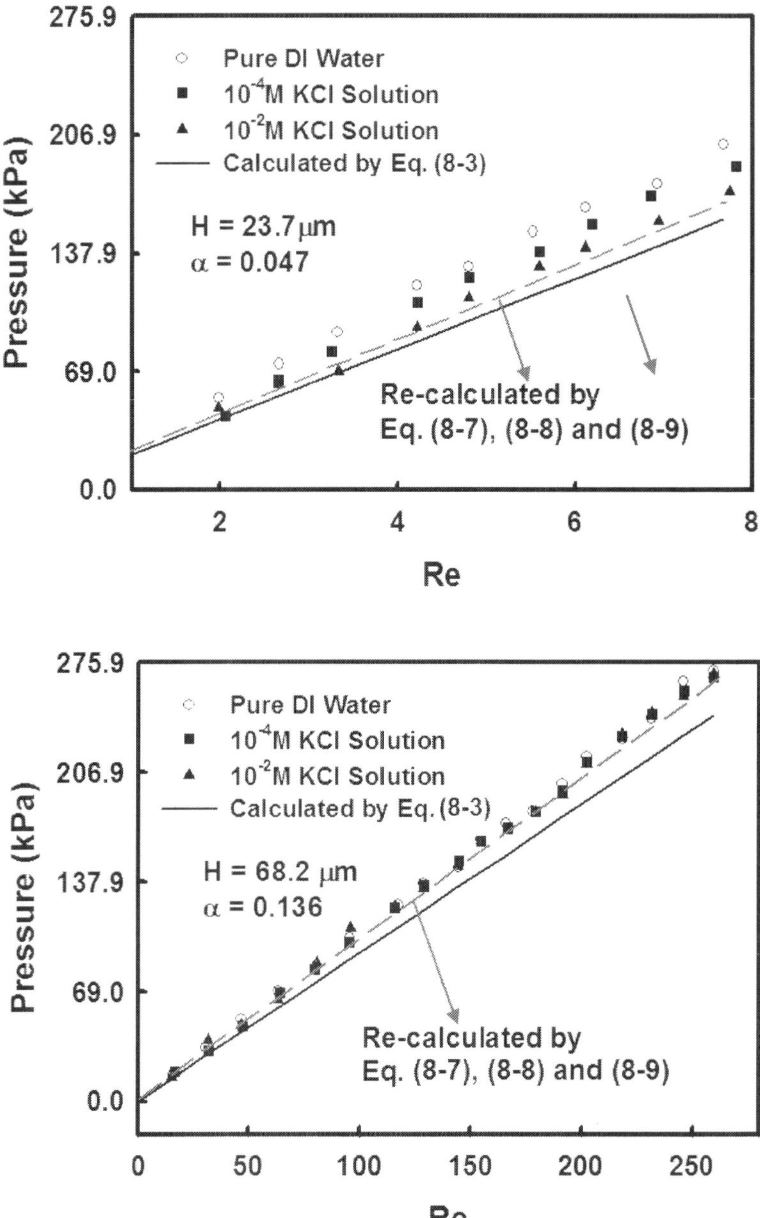

Fig. 32. Pressure drops at different Reynolds numbers in fully developed region a microchannel with (a) channel height of 23.7 μm and (b) channel height of 68.2 μm (Ko, 2009).

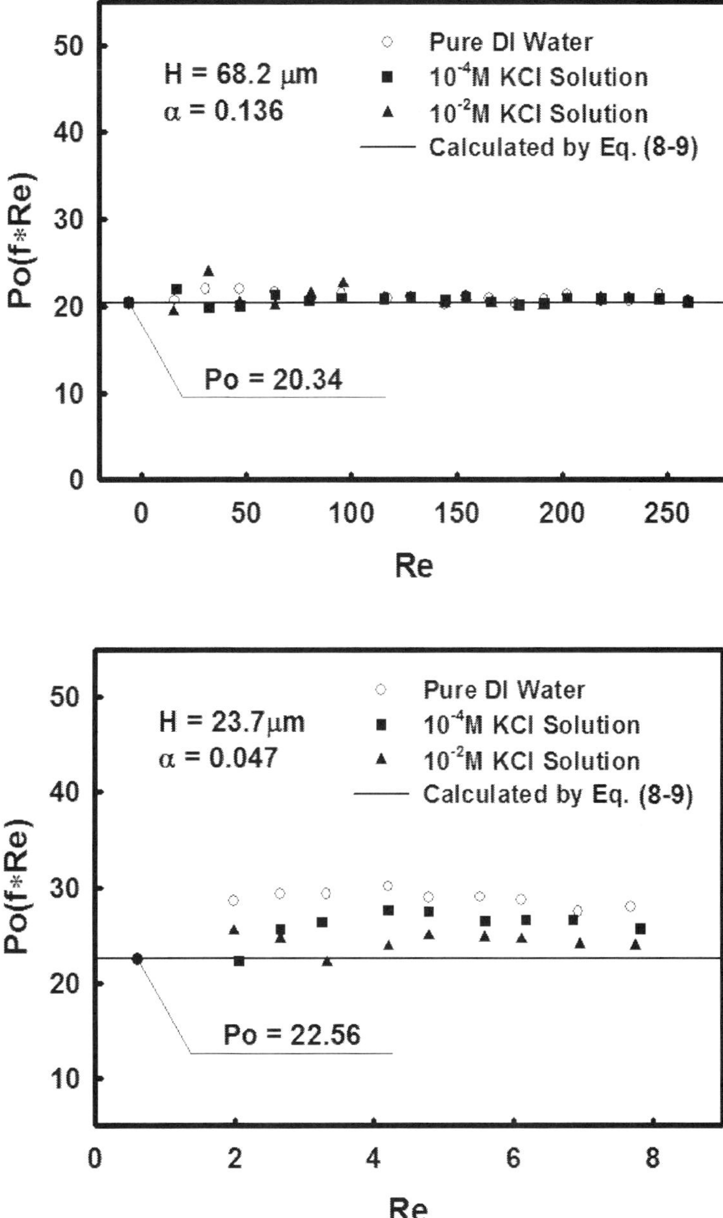

Fig. 33. The Poiseuille numbers at different Reynolds numbers in a micro-channel with (a) channel height of 68.2 µm and (b) channel height of 23.7 µm (Ko, 2009).

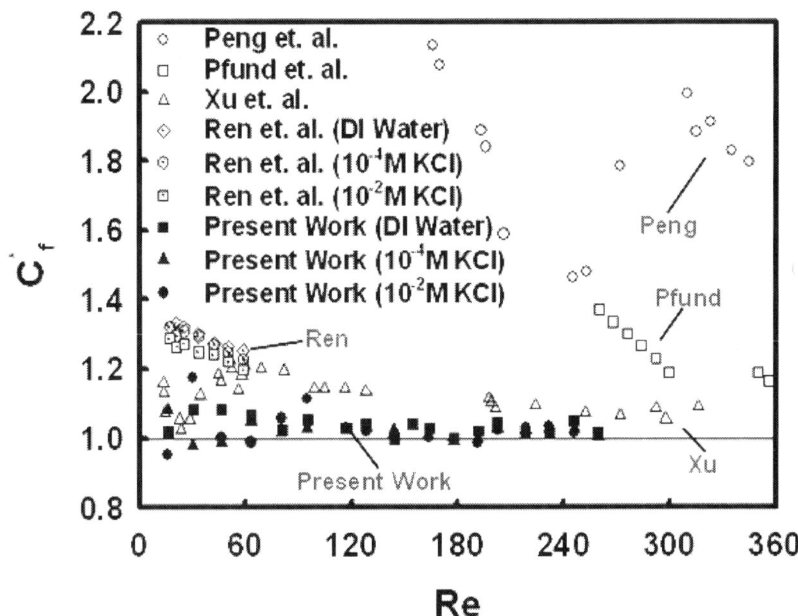

References	Channel Material	Working Fluid	Height (m)	Width (m)	Aspect Ratio	D_h (m)
Peng et al. (1995)	stainless steel	Water	100	200	0.5	133
Pfund et al.(2000)	Polycarbonate-Gasket-Polyimide	Water	263	10000	0.263	416.4
Xu et al. (2000)	Glass-Silicon	Water	43.7	415.3	0.106	79.08
Ren et al. (2001)	Silicon	Water/KCl	40.5	5000	0.0081	80.34
Present work (Ko, 2009)	Polymer	Water/KCl	68.2	500	0.1364	120.02

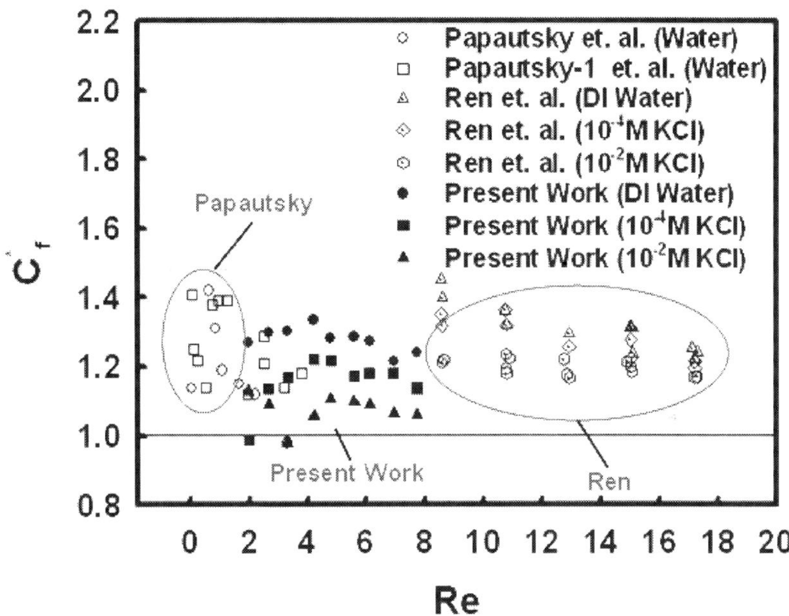

Reference	Channel Material	Working Fluid	Height (m)	Width (m)	Aspect Ratio	D_h (m)
Papautsky (1999)	Nickel	Water	24.65	600	0.0411	47.35
Papautsky-1(1999)	Nickel	Water	26.35	150	0.1757	44.82
Ren et al. (2001)	Silicon	Water / KCl	28.2	5000	0.0056	56.08
Present work (Ko, 2009)	Polymer	Water / KCl	23.7	500	0.0474	45.24

Fig. 34. (a) Comparison of the normalized friction constants between the current work and the published results for current channel with (a) 68.2 μm in height and (b) 23.7 μm in height.

In addition, data obtained from previous work scatter very much. The scatter in data among different works should not be attributed to the geometry of the channel which has a different aspect ratio since the comparison has already accounted for the aspect ratio effect. The scatter in data among different works is mostly attributed again to the use of the pressures at the inlet and the outlet outside the channel to evaluate the pressure drop inside the channel. This may lead to erroneous result. For example, the data from Xu et al., as shown in Figure 34(a) appears to be very good and is very close to the theoretical prediction. However, in a much smaller channel height, i.e. 15.4 µm, where the EDL effect can become significant and is expected to increase the pressure drop, the data from Xu et al. is identical to the theoretical prediction, as shown in Figure 34(a) of their paper (Xu et al., 2000). Most of the data did not discuss the effect of EDL except the work of Ren et al. (2001). Ren et al. has used a higher concentration of KCl solution to significantly reduce the thickness of the EDL and its effect on the pressure drop so that the friction constant in a smaller size of channel, as shown in Figure 34(b), with a higher concentration of KCl solution is smaller than the one with a lower concentration of KCl solution. However, all the data from Ren et al. are much greater than the theoretical prediction, as show in Figure 34(a) for a greater channel height and Figure 34(b) for a smaller channel height. The higher normalized friction constant obtained is attributed to the erroneous estimation of the entry length, which is used to estimate the pressure drop in the entry region, and the pressure loss at the exit. The normalized friction constant in Ren et al. is estimated from the pressure drop which is measured outside the channel at the inlet and the outlet and subtracted by the pressure drop in the entry region and the pressure loss at the outlet. Despite of this drawback, the data of Ren et al. consistently show that EDL can have a significant effect on the pressure drop when the channel height is small. The EDL effect can become small when a high concentration of KCl solution or a greater height of channel is used. These findings are also confirmed in our measurements. The agreement of the current pressure drop with the theoretical prediction suggests that much of the conflicting effects of the microchannel size on the pressure drop reported from previous work are clarified. From the comparison, one can further confirm the techniques and the fabrication of the micro-channel used in this study are suitable for precision pressure drop measurement.

8.2 Flow characteristics for air
8.2.1 Theoretical analysis
Theoretical analysis of micro flow inside the channel is performed firstly, which is then used to compare with the data of pressure measured and analyze the characteristics of micro flow inside the channel. Since the channel has a large value in its width (500 µm) in comparison with its height (from 23 µm to 68 µm), the channel flow in the present study can be assumed as laminar flow between two parallel plates at steady state. The configuration with the coordinate system and flow velocity of micro-channel flow are defined in Figure 35. Therefore, the governing equations of the flow in this micro-channel system can be written in the following:

Continuity equation:

$$\frac{\partial \rho u}{\partial x} + \frac{\partial \rho v}{\partial y} = 0 \qquad (8\text{-}11)$$

Navier-Stokes Equation:

$$\rho(u\frac{\partial u}{\partial x} + v\frac{\partial u}{\partial y}) = -\frac{\partial p}{\partial x} + \mu(\frac{\partial^2 u}{\partial x^2} + \frac{\partial^2 u}{\partial y^2}) \qquad (8\text{-}12)$$

$$\rho(u\frac{\partial v}{\partial x} + v\frac{\partial v}{\partial y}) = -\frac{\partial p}{\partial x} + \mu(\frac{\partial^2 v}{\partial x^2} + \frac{\partial^2 v}{\partial y^2}) \qquad (8\text{-}13)$$

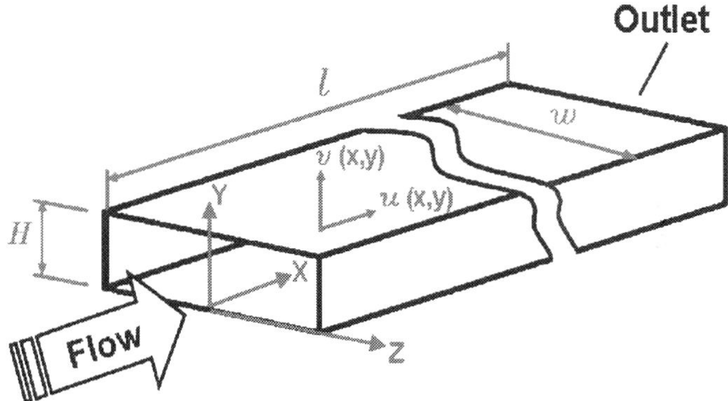

Fig. 35. Schematic diagram and the coordinate system used for the micro-channel flow system.

The micro-channel is actually fabricated by very low thermal conductivity materials which can be considered as well insulated channel. Therefore, the gas flow inside the channel can be assumed to be undergoing isentropic process as follows:

$$\frac{p}{\rho^k} = cons. \qquad (8\text{-}14)$$

If non-slip boundary condition on the wall is used, then the boundary condition can be written as:

$$U(y)_w = 0 \qquad (8\text{-}15)$$

In Eq. (8-14), k is the ratio of specific heat C_p versus C_v. For micro flow in parallel plate channel, the height is much smaller than either the length or the width of the channel. One can assume that the gradient terms, normal to the surface of channel, in Navier-Stokes Equation are much larger than those along the channel. As a result, one can neglect terms that represent x-direction diffusion of momentum (along the channel) relative to those of y-direction (normal to the channel surface). That is

$$u \gg v, \quad \frac{\partial}{\partial y} \gg \frac{\partial}{\partial x} \ and \ \frac{\partial^2 u}{\partial y^2} \gg \frac{\partial^2 u}{\partial x^2} \qquad (8\text{-}16)$$

Therefore, the Navier-Stokes Equation for the micro-channel flow can be simplified as

$$\frac{\partial p}{\partial x} = \mu \frac{\partial^2 u}{\partial y^2} \tag{8-17}$$

$$\frac{\partial p}{\partial y} = 0 \tag{8-18}$$

Eq. (8-17) and Eq. (8-18) means the pressure varies only in the streamwise direction. Therefore, Eq. (8-17) can be readily solved analytically. This equation is the same as channel flow in the fully developed region. The pressure distribution measurements presented in the later section indicate that the entry length of flow in micro-channel could hardly be observed. It means that the flow in most part of the micro-channel is fully developed except in the very short entry region where x is not large as compared with the width of the channel. The solution of Eq. (8-17) is a parabolic velocity profile as follows:

$$u(y) = \frac{y^2}{2\mu}\frac{dp}{dx} - \frac{H}{2\mu}\frac{dp}{dx}y = \frac{1}{2\mu}\frac{dp}{dx}(y^2 - Hy) \tag{8-19}$$

where H is the height of the channel. To integrate the above equation along y direction of the channel, the average velocity can be obtained. Therefore, the mass flow rate (Qm) of in the channel can be shown as follows:

$$Q_m = -\frac{H^3 w \rho}{12\mu}\frac{dp}{dx} = -\frac{H^3 w p^{\frac{1}{k}}}{12\mu p_o^{\frac{1}{k}} \rho_o^{-1}}\frac{dp}{dx} \tag{8-20}$$

where w is width of the channel and the subscript o refers to the condition at the outlet location of the last pressure sensor in the micro-channel. The above equation can be re-arranged as follows:

$$Q_m dx = -\frac{H^3 w p^{\frac{1}{k}}}{12\mu p_o^{\frac{1}{k}} \rho_o^{-1}} dp \tag{8-21}$$

One can integrate above equation from the entrance to any location, x, in the channel and obtain the pressures distribution inside the channel as follows:

$$p_x = \left[p_i^{\frac{k+1}{k}} - \frac{12 Q_m x (\frac{k}{k+1}) \mu p_o^{\frac{1}{k}} \rho_o^{-1}}{H^3 w} \right]^{\frac{k}{k+1}} \tag{8-22}$$

Where the subscript i refers to the condition at the inlet location of the first pressure sensor in the micro-channel. To integrate Eq. (8-21) from the entrance to the end of the channel, the mass flow rate can be obtained explicitly as follows:

$$Q_m = \frac{H^3 w p_0^{\frac{k+1}{k}}}{12(\frac{k}{k+1})\mu p_o^{\frac{1}{k}} \rho_o^{-1} l}\left[\left(\frac{p_i}{p_o}\right)^{\frac{k+1}{k}} - 1\right] \qquad (8\text{-}23)$$

where l is the distance from the first pressure sensor to the last pressure sensor in the micro-channel and is equal to 4000 μm. During the calculation of pressure distribution, the dynamic viscosity (μ) in the equation is based on the entrance temperature of the channel flow.

For the case when the heat transfer occurs (with a given heat flux boundary condition on the wall), the wall is no longer adiabatic. The above equation can be modified with k replaced by n and is written as follows:

$$p_x = \left(p_i^{\frac{n+1}{n}} - \frac{12 Q_m x (\frac{n}{n+1})\mu p_o^{\frac{1}{n}} \rho_o^{-1}}{H^3 w}\right)^{\frac{n}{n+1}} \qquad (8\text{-}24)$$

where n can be obtained by fitting the above equation to the pressure distribution data. In this way, the pressure variation with either the density or the temperature can be found by a polytropic process as follows:

$$\frac{p}{\rho^n} = const. \qquad (8\text{-}25)$$

8.2.2 Experimental results of the pressure drop for air flow

The wall roughness has been shown to play very important role on the pressure gradient, which contributed the scattering of friction factor data reported in others (Guo & Li, 2003; Kleinstreuer & Koo, 2004), in micro-channel flow. However, the wall information in most of the micro-channels reported in the literature is absent. Therefore, the wall roughness in the current channel is measured and studied. During etching and removing of the silicon substrate, the wall roughness of the channel has been minimized by proper etching process as described previously. The roughness of the current channel wall is measured by the profile meter using a small probe, scanning along axial location of the channel wall. The values of roughness are less than ± 5 nm to 15 nm. The relative roughness is much less than 1 % as compared with the height of the channel (the smallest height of the channel is 23 μm). Therefore, accurate results of pressure drop were obtained based on the small roughness of the wall surface in the micro-channel in this study. Typical results of the pressure distributions, both by theoretical prediction and experiment, for height of channel at 23 μm are shown with good agreement in Figure 36. Both experimental data and the theoretical prediction indicate that the pressure distribution inside is not linear, due to the compressibility effect of the air flow in the micro-channel, except when the flow speed in the channel is lower than 0.3 Mach, which can be assumed as incompressible condition. The agreement between experimental data and theoretical prediction is very good even in the upstream region close to the entrance. Therefore, this means that the pressure distribution in the region close to the entrance has reached a fully developed flow. This clearly indicates that the hydrodynamic entrance length in the micro-channel flow is very short as mentioned previously in the derivation of the pressure distribution equation. Therefore, the assumptions used in the analysis are thus confirmed. In addition, all the pressure data drop smoothly from the entrance to the exit. There is no any sudden change in the pressure gradient where flow may undergo an earlier onset of transition from laminar to turbulent flow as mentioned by others (Peng et al., 1994, 1996), which causes greater friction factor in micro-channel.

The variation of the pressure difference between the inlet and the outlet with the volume flow rate is also measured and compared with theoretical prediction, as shown in Figure 37 for the channel height varied from 23 μm to 55 μm. The agreement is still very good for any of the micro-channels. This result further indicates that there is not any reduction or enhancement in the friction on the wall either might be due to reduction in viscosity or surface roughness effect. There is again no any sudden change in the pressure drop which indicates that the flow may undergo an earlier onset of transition from laminar to turbulent flow as mentioned by others. The Reynolds number defined in Figure 37 is equal to the mean velocity of flow in the hydraulic diameter of the channel and divided by the kinematic viscosity of the air flow.

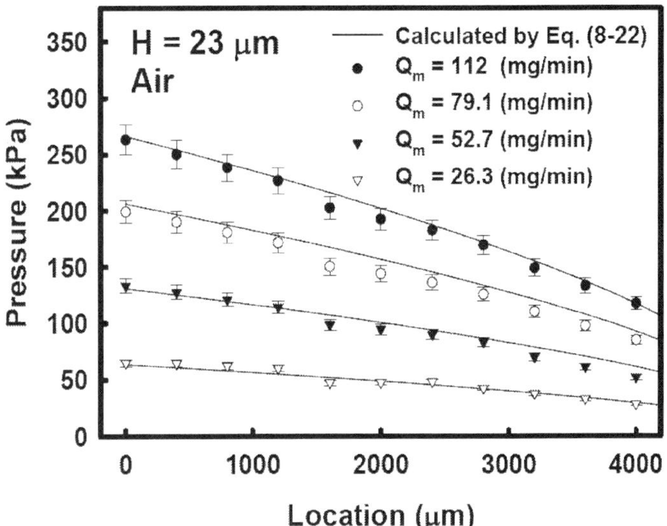

Fig. 36. Comparisons of the pressure distributions between the current data and the analytical results for the airflow along the micro-channel.

It is noted that the friction factor in micro-channel of air flow is not a constant if one converts the pressure distribution in Figure 36 into the friction factor. This result is contrary to many of other experimental work which obtains a friction factor at a constant value (Takuto et al., 2000; Liu et al., 2004; Kohl et al., 2005; Bayraktar & Pidugu, 2006). That means that air flow in their micro-channels has been unconsciously assumed as incompressible. To calculate Reynolds number for the amount of mass flow rate in the current channel, the largest Reynolds number is very close to 2000, as shown in Figure 5.4, which is in the laminar regime. In the large scale channel, this is an extremely low Reynolds number flow condition and the flow usually can be assumed as incompressible. However, in the micro-channel flow, the flow speed for Re = 2000 is 280 m/s for channel height of 55 µm. In this case, the March number for Re = 2000 is equal to 0.82, and the flow is highly compressible and can not be assumed as incompressible. To assume incompressible flow will lead to an erroneous skin friction coefficient. In fact, to assume incompressible flow, the friction factor obtained will be much smaller than in the large scale channel as presented in some of the other works.

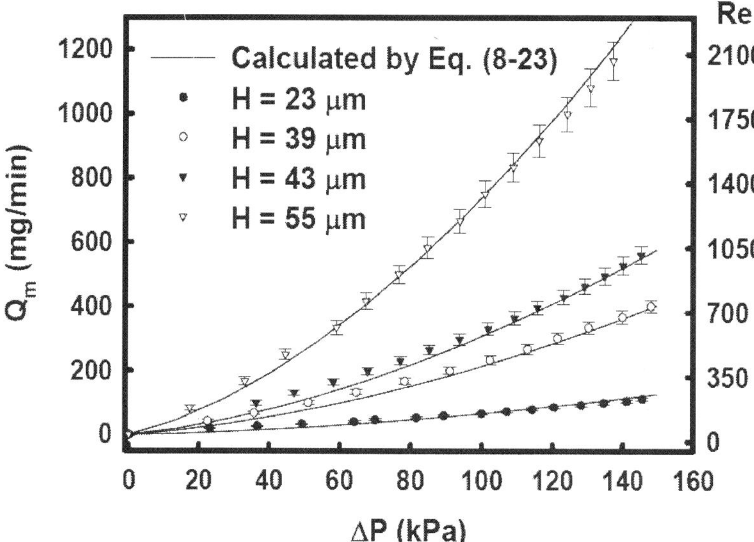

Fig. 37. Variation of the pressure difference between the inlet and the outlet with the mass flow rate or the Reynolds number.

With the constant heat flux imposed on the wall, the pressure distribution is also measured as shown in Figure 38. The pressure variation along the channel is very similar to the case when the channel wall is insulated except that the magnitude of the pressure is higher. Eq. (8-24) is used to fit the pressure distribution data by varying the value of n in this equation. The value of n determined from the fit can be used to find the relationship between the pressure and the density in the microchannel under the constant heat flux condition. This pressure versus density relationship is required for determination of the heat transfer coefficient as described in the next section.

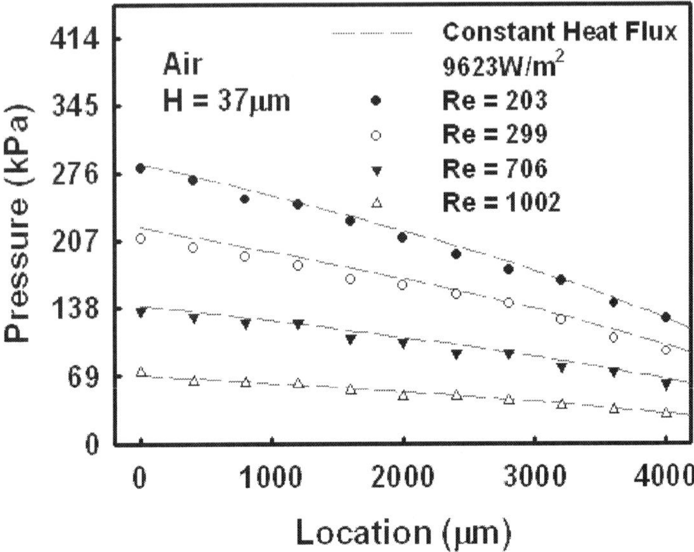

Fig. 38. Comparisons of the pressure distributions between the current data and the analytical results for the airflow along the microchannel for the case under constant wall heat flux, Eq. (8-24) is used to fit the data by varying the value of n.

8.3 Local heat transfer for air flow

In order to further study the heat transfer characteristics in the micro-channel, the current micro-channel is integrated with array of temperature sensors, pressure sensors and a set of heater. In the current experiment, the heat transfer coefficient, h_b, is defined by the equation as follows:

$$h_b = \frac{\dot{q}}{(T_w - T_b)} \tag{8-26}$$

where $\dot{q}$ is the heat flux imposed along the wall by the electric heater and is equal to the product of the voltage and the current passing through the heater divided by the total area of the heater. T_w is the wall temperatures measured by the temperature sensors along the channel at particular locations x. T_b is the bulk temperature of the flow which can be estimated from the amount of heat flux imposed on the wall which heats up the air flow and rises the temperature of the bulk flow. By applying a control volume and conservation of energy for the flow in the channel as discussed by Incropera et al. (2007), one can obtain a conservation equation of energy as follows:

$$\dot{q}\pi D_h dx = \dot{m}c_p dT_b \tag{8-27}$$

where $\dot{m}$ is the mass flow rate of the air in the channel. Rearranging and integrating the above equation from the inlet of the channel, one can obtain:

$$\int_{T_{in}}^{T_b} dT_b \int_0^x \frac{\dot{q}\pi D_h}{\dot{m}c_p} dx \tag{8-28}$$

Therefore,

$$T_b = T_i + \frac{\dot{q}\pi D_h}{\dot{m}c_p} x \tag{8-29}$$

The local Nusselt number, Nu_b, can be defined as:

$$Nu_b = \frac{h_b \times D_h}{k} \tag{8-30}$$

where k is thermal conductivity of the gas flow. Once heat transfer coefficient is obtained, the local Nusselt number can be found from the above equations.

8.3.1 Direct measurements of the heat transfer for air flow

The heat transfer experiments have been performed with air flow in a channel with channel height of 37 μm. The velocity of air flow is controlled by mass flow controller and is assumed in the laminar regime where the Re from 203 to 1006. The thermal conductivity of the air used in the Nusselt number is evaluated at the mean bulk temperature, i.e. an average of the bulk temperature of the flow at the inlet and the outlet of the channel. As shown in Figure 39 (a), the heat flux is varied from 5413 to 21652 W/m² and the Reynolds number is kept at 500. The Nusselt numbers in the micro-channel with channel height at 37 μm is significantly higher than that of theoretical prediction by the parallel plates channel (Rohsenow et al., 1985), and become closer as the heat flux on the wall decreases. For the case of heat flux equal or lower than 9623 W/m², all the Nusselt number results collapse into a single line, as shown in Figure 39 (b), in the downstream of the channel where flow has become fully developed. The Nusselt number results collapsing into a single curve in the fully developed region agrees with theoretical results for prediction of large scale channel except that the Nusselt number is higher. This indicates that other effects, such as property variation with temperature, compressibility of flow, should be taken into account in order to obtain a good correlation of the heat transfer data.

8.3.2 Correction for the bulk temperature

The reason for this deviation may be partially attributed to property variation, such as compressible gas flow and leads to increase in the heat transfer. In order to account for the compressibility of the gas which can expand and reduce the kinetic energy of the flow, and change the bulk temperature of the flow, the theoretical calculation equation for the bulk temperature should be re-derived from concept of energy balance, as follows:

$$d\dot{q} = \dot{m}(e_2 - e_1) + (\frac{U_2^2}{2} - \frac{U_1^2}{2}) \tag{8-31}$$

where e is enthalpy and the subscript indicates properties at different locations. For a differential control volume, the above equation can be written as:

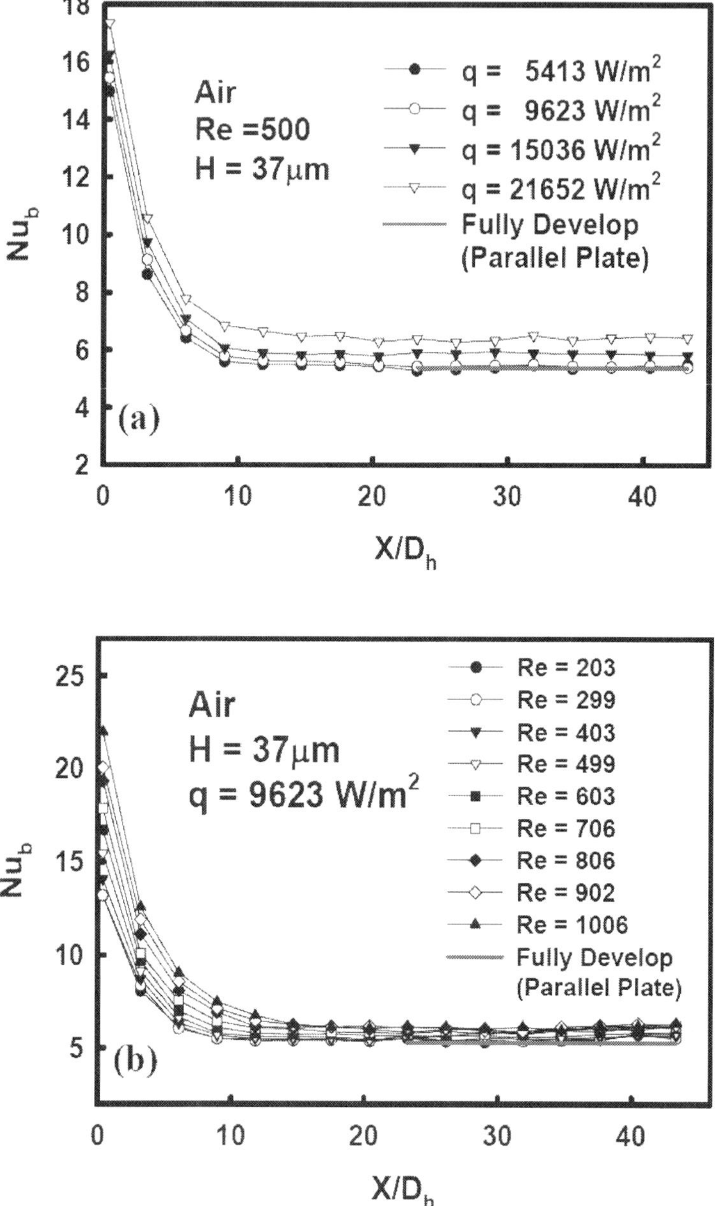

Fig. 39. Experimental results for local Nusselt number distributions under (a) different heating flux conditions and (b) different Reynolds number.

$$dq = \dot{m}\,de + d(\frac{U^2}{2}) \tag{8-32}$$

where $dq = \dot{q}_w \pi D_h$, and $de = c_p\,dT$, re-arranging the above equation, one obtains

$$\frac{dT_b}{dx} = \frac{\dot{q}_w \pi D_h}{\dot{m}c_p} - \frac{d(\frac{U^2}{2})}{c_p dx} \tag{8-33}$$

To integrate above equation from the entrance to the desired location, x, in the channel, one can obtain the bulk temperature of the flow as follows:

$$T_b = T_i + \frac{\dot{q}_w \pi D_h}{\dot{m}c_p} x - (\frac{U_x^2}{2} - \frac{U_i^2}{2})\frac{1}{c_p} \tag{8-34}$$

From the continuity equation,

$$\rho_1 A_1 U_1 = \rho_2 A_2 U_2 \tag{8-35}$$

The bulk temperature can be re-arranged as follows:

$$T_b = T_i + \frac{\dot{q}_w \pi D_h}{\dot{m}c_p} x - \frac{U_i^2}{2c_p}\left[(\frac{\rho_i}{\rho_x})^2 - 1\right] \tag{8-36}$$

Since the current micro-channel is actually fabricated by very low thermal conductivity materials, the channel wall can be considered as well insulated. Without heat transfer into the channel, the gas flow inside the channel can be assumed to be undergoing isentropic process. For the case when a desired heat flux is imposed on the bottom wall, the gas is assumed under polytropic process and the pressure density relationship can be written as follows:

$$\frac{p}{\rho^n} = cons. \tag{8-37}$$

where n is a constant and is determined from the heating condition on the wall of the channel. Therefore, the density in Eq. (8-36) can be replaced by the local pressure as follows:

$$T_m = T_i + \frac{\dot{q}_w \pi D_h}{\dot{m}c_p} x - \frac{U_i^2}{2c_p}\left[(\frac{p_i}{p_x})^{\frac{2}{n}} - 1\right] \tag{8-38}$$

8.3.3 Heat transfer data
From the local pressures measured by current pressure sensors along the micro-channel, the bulk temperature can be re-calculated from the above equation. Therefore, the local Nusselt

numbers along the channel can be obtained and are presented in Figure 40 for different Reynolds number conditions. All the Nusselt number results collapse into a single curve in the fully developed region, as shown in Figure 40 (a) for different heat flux conditions. Again, for the case of heat flux at 9623 W/m^2 the Nusselt numbers at different Reynolds numbers also collapse into a single curve in the fully developed region, as shown in Figure 40 (b), but the curve is slightly lower than the prediction in large scale channel. Theoretical calculation for the heat transfer in large scale rectangular channel at different aspect ratios in the fully developed region has been performed and reported (Incropera et al., 2007). It is found that in large scale channel the Nusselt number is 4.55 for rectangular channel with aspect ratio of 1:10, but is 5.38 for ideal parallel plate channel. In fact, the current micro-channel is not an ideal parallel plates channel and is a channel with aspect ratio of 1:13. Therefore, the lower Nusselt numbers of experimental results in the current channel than the prediction for parallel plates channel is expected and is attributed to the effect of aspect ratio.

In order to compare the heat transfer in the entrance region, the flow is assumed to be hydrodynamically fully developed before entering into the channel and the problem becomes a thermal entry length problem because the hydrodynamic entry length is very short as defined as Eq. (8-6). In the thermal entry length region, the heat transfer process can be correlated in terms of the inverse Graetz number, where

$$G_z = \frac{\mathrm{Re}\,\mathrm{Pr}}{x/D_h} \tag{8-39}$$

The thermal entry length in the large scale tube can be usually expressed as (Incropera et al., 2007),

$$Gz^{-1} \sim 0.05 \tag{8-40}$$

This expression or the entry length is found at the point when the Nusselt number results approach a constant value. From our heat transfer data, as shown in Figure 41, the entry length in the current micro-channel can be found and expressed as,

$$Gz^{-1} \sim 0.023 \tag{8-41}$$

It appears that the current heat transfer data does not have the same trend as in the large scale tube. This is attributed to the fact that the current micro-channel is not an ideal parallel plate channel. The correlation equation for parallel plates channel in the entry region (Naito, 1984) can be expressed as follows:

$$Nu_x = 0.461043Gz^{1/2}(1 + 6.35257Gz^{-1/2} - 33.4079Gz^{-1} + 419.158Gz^{-3/2} - 1537.84Gz^{-2}) \tag{8-42}$$

Comparison of the correlation equation for the local Nusselt numbers between the prediction for large scale channel and the experimental data for micro-channel with a channel height of 37 μm is made in Figure 41. The agreement in entrance region is very good except in the fully develop region the current data is slightly low due to the aspect ratio effect.

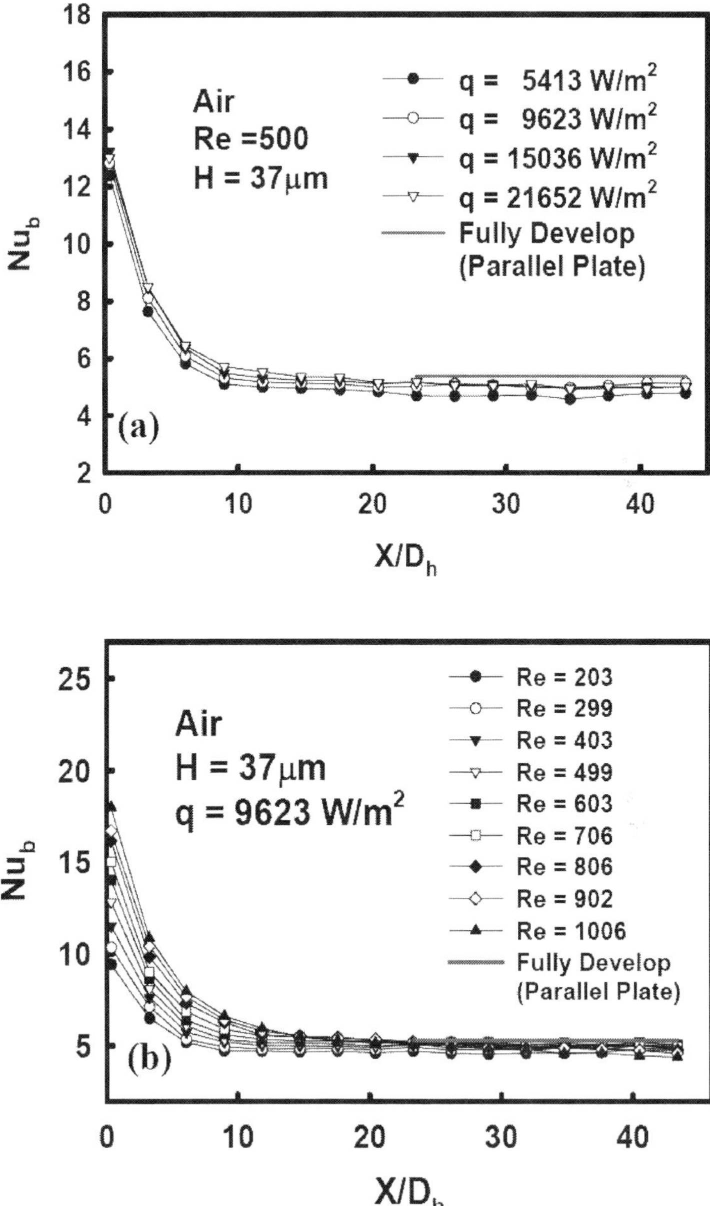

Fig. 40. Re-calculated results for local Nusselt number distributions under (a) different heating flux conditions and (b) different Reynolds number.

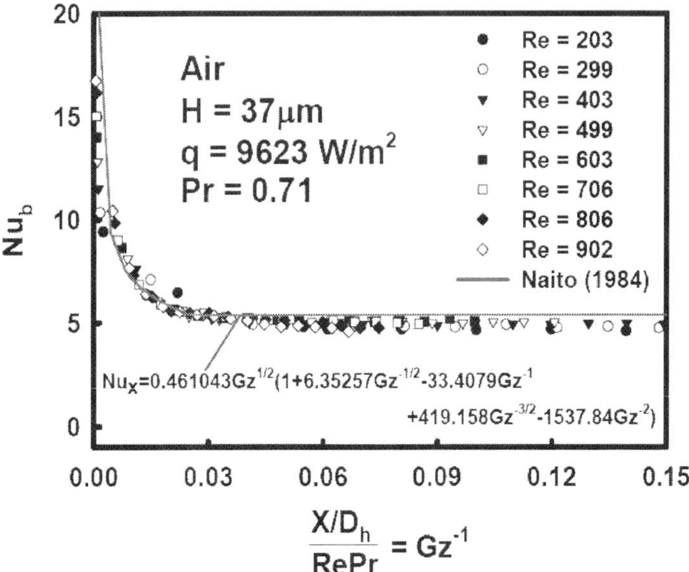

Fig. 41. Comparison of the local Nusselt numbers between the prediction for large scale channel and the experimental data.

Since some of the heat transfer data reported in the literature is the average heat transfer coefficient or the average Nusselt number, comparison with these data can be made when the average Nusselt number is defined as follows:

$$Nu_{ave} = \frac{1}{x} \int_0^x Nu_b \, dx \qquad (8\text{-}43)$$

The average Nusselt numbers in the current micro-channel are obtained in the range of laminar flow and compared, as shown in Figure 42, with the correlation predicted by Naito for air flow in large scale parallel plates channel. The average Nusselt number is very close, but smaller than the correlation obtained by Naito. This is attributed to the effect of aspect ratio. The result is of interest to compare current heat transfer data with the results of others in micro or large scale channel. To avoid influence of EDL in liquid flow, comparison is limited to air flow in channels. In general, when Re > 2300, the flow inside channel becomes turbulent. The average Nusselt number for Re > 2300 is much larger than for Re < 2300 (Morini, 2004; Incropera et al., 2007). For large scale channel, the Nusselt number in the two parallel plates is, in general, greater than that in circular tube (Incropera et al., 2007). This is indicated in Figure 43 where the average Nusselt number of the numerical results obtained by Naito and experimental results obtained by this study for two parallel plates, and is higher than the correlation obtained by Hausen for circular tube. The heat transfer correlations obtained by Choi (1991) is for air flow in mcirotubes. Their data is much smaller than the correlation of Hausen. The much lower value of Nusselt number and its large increase with Reynolds number in Choi's data was attributed to the streamwise conduction of heat along the micro-tube. The micro-tubes used in the past were made of stainless steel which is convenient for electric heating, and is expected to cause large streamwise conduction of heat in the finite thickness of the tube since the thermal conductivity of stainless steel is relatively high and is in the range from 13 to 15 W/mK depending upon the types of the steel used. However, the streamwise conduction of heat along the current micro-channel wall has been minimized due to the use of a thick layer of epoxy which has a very low thermal conductivity. In addition, wall roughness in the micro-tubes of Choi is expected large enough which may result the discrepancy. However, the roughness of their micro-tubes is not reported. When one goes into turbulent flow region, the heat transfer data from different work in micro-channel scatters very much, but are all higher than those in large scale channel (Wu & Little, 1983; Yu et al., 1995; Morini, 2004; Kays et al., 2005). It is noted that data from both the work of Dittus-Boelter and Gnielinski (Rohsenow et al., 1985) is for air flow in large scale tubes. The higher heat transfer data in micro-channel can be attributed to wall roughness effect in the micro-tubes. It has been realized (Kays et al., 2005; Incropera et al., 2007) that wall roughness can significantly enhance turbulent heat transfer in the large scale tube. Thus, wall roughness is expected to significantly enhance turbulent heat transfer in micro-channel. Unfortunately, wall roughness on each of the mcirotube wall was not reported and was expected to be very high in the micro stainless steel tubes if no special treatment on the surface is provided.

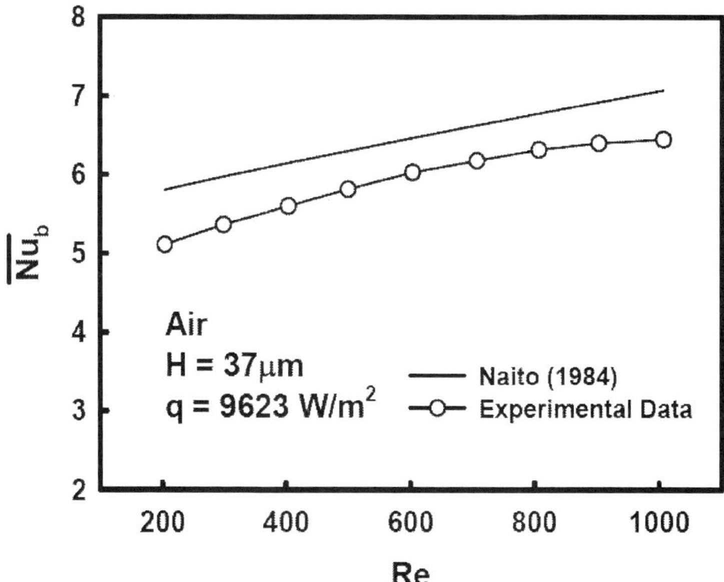

Fig. 42. Comparison of the average Nusselt numbers in the current micro-channel with the correlation predicted by Naito for air flow in large scale parallel plates channel.

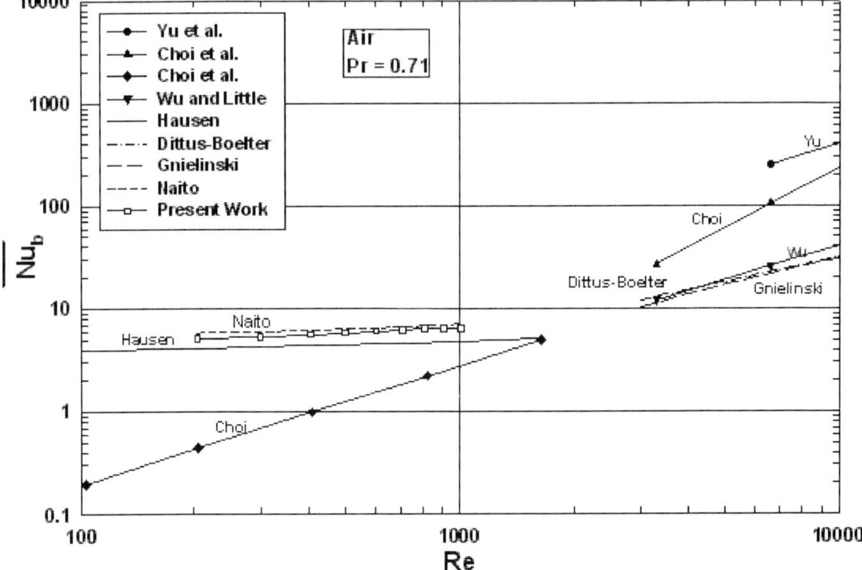

Fig. 43. Comparison of average Nusselt numbers between the current work and published data.

9. References

Acosta, R. E., Muller, R.H., and Tobias, W. C. (1985). Transport processes in narrow (Capillary) channels, *AIChE Journal*, Vol. 31, pp.473–482.

Arkilic, E. B. and Schmidt, M. A. (1997). Gaseous slip flow in long microchannels, *Journal of Microelectromechanical System*, Vol. 6(2), pp.167-178.

Arshak, K., Morris, D., Arshak, A., Korostysnska, O., Jafer, E., Waldron, D., and Harris, J. (2006). Development of polymer-based sensors for integration into a wireless data aquisition system suitable for monitoring environmental and physiological processes, *Biomolecular Engineering*, Vol. 23, pp.253-257.

Arulanandam, S., and Li, D. (2000). Liquid transport in rectangular microchannels by electroosmotic pumping, *Colloids and Surfaces A: Physicochemical and Engineering Aspects*, Vol. 161, pp.89-102.

Bayraktar, T. and Pidugu, S. B. (2006). Review: Characterization of liquid flows in microfluidic systems, *Int. J. Heat Mass Transfer*, Vol. 49, pp.815-824.

Beskok, A., and Karniadakis, G. E. (1994) Simulation of heat and momentum transfer in complex microgeometries, *Journal of Thermophysics and Heat Transfer*, Vol. 8, No. 4, pp. 246-259.

Bhatnagar, P. L., Gross, E. P. and Krook, M. (1954) A Model for Collision Processes in Gases. I. Small Amplitude Processes in Charged and Neutral One-Component Systems, *Phys. Rev.*, Vol. 94, pp.511–525.

Bird, G. A. (1976a). *Molecular gas dynamics and the direct simulation of gas flows*, Clarendon Press, Oxford, pp. 183-207.

Bird, G. A. (1976b). *Molecular gas dynamics*, Clarendon Press, Oxford, pp. 110-117.

Brutin, D., Topin, F., and Tadrist, L. (2003) Experimental study of unsteady convective boiling in heated minichannels, *International Journal of Heat and Mass Transfer*, Vol. 46, pp. 2957-2965.

Chen, H. R. (2001). *Design and fabrication of the microchannel system for thermal fluid study*, Ph D. Thesis, National Cheng Kung University, Tainan, Taiwan.

Chen, H. R., Gau, C., Dai, B. T. and Tsai, M. S. (2003a). A Novel Planarization Process for Sacrificial Polysilicon Layers in a Micro-Thermal System, *Sensors and Actuators A: Physical*, Vol. 108/1-3, Nov. pp.86-90.

Chen, H. R., Gau, C., Dai, B. T. and Tsai, M. S. (2003b). A Monolithic Fabrication Process for a Micro-Flow Heat Transfer Channel Suspended over an Air Layer with Arrays of Micro Sensors and Heaters, *Sensors and Actuators A: Physical*, Vol. 108/1-3, Nov. pp.81-85.

Chen, X. (1996). *Gas kinetics and its applications in heat transfer and flow*, TsingHua University Press, Beijing.

Cheng, P., Wang, G., and Quan, X. (2009). Recent work on boiling and condensation in microchannels, *ASME Journal of Heat Transfer*, Vol. 131, pp. 043211-043225.

Choi, S. B., Barron, R. F., and Warrington, R. O. (1991). Fluid flow and heat transfer in microtubes, Micromechanical Sensors, Actuators, and Systems, *ASME, DSC*, Vol. 32, pp. 123-134.

Choondal, B. S. and Suresh, V. G. (2000). A comparative analysis of studies on heat transfer and fluid flow in microchannels, *Proc. of the Micro-Scale Heat Transfer Conf.* pp.80-92.

Chu, W. K. H., Hsu, C. T., Wong, M., and Zohar, Y. (1994). Heat transfer in a microchannel flow, *Proceedings of the 1994 IEEE Electron Devices Meeting*, Hong Kong, pp. 38-43.

Clark, S. K., and Wise, K. D. (1979). Pressure sensitivity in anisotropically etched thin-diaphragm pressure sensors, *IEEE Trans. Electron Devices*, ED-26, pp. 1887-1896.

Core, T. A., Tsang, W. K., and Sherman, S. J. (1993). Fabrication technology for an integrated surface micromachined sensor, *Solid State Technolog*, pp. 39-47.

Craighead, H. (2006). Future lab-on-chip technologies for interrogating individual molecules, *Nature*, Vol. 442 (27), pp.387-393.

Eckert, E. R. G., and Drake, R. M. (1972). *Analysis of heat and mass transfer*, McGraw-Hill, New York, pp.467-495.

Fan, Q., and Xue, H. (1998). Compressible effects in microchannel flows, *Proceedings of the IEEE/CPMT Electronics Packaging Technology Conference*, pp.224-228.

George, M. (2006). The origins and the future of microfluidics, *Nature* Vol. 442, pp.368-373.

Giordani. N., Camberlein, L., Gaviot, E., Polet, F., Pelletier, N., and Bêche, B. (2007). Fast psychrometers as new SU-8 based microsystems, *IEEE Transactions on Instrumentation and Measurement*, Vol. 56, No. 1, pp.102-106, 2007.

Guckel, H. (1991). Surface micromachined pressure transducer, *Sens. Actuators A*, Vol. 28, pp. 133-146.

Guckel, H. et al. (1990). Microstructure sensors, *International Electron Devices Meeting*, pp. 613-616.

Guo, Z. and Li, Z. (2003). Size effect on microscale single-phase flow and heat transfer, *Int. J. Heat Mass Transfer* Vol. 46, pp.149–59.

Harley, J. C., Huang, Y. H., Bau, H., and Zemel, J. N. (1995). Gas flow in micro channels, *Journal of Fluid Mechanics*, Vol. 284, pp257-274.

Harms, T. M., Kazmierczak, M. J. and Gerner, F. M. (1999). Developing convective heat transfer in deep rectangular microchannels, *International Journal of Heat Fluid Flow*, Vol. 20, pp. 149–157.

Harms, T.M., Kazmierczak, M.J., Gerner, F.M., Holke, A., Henderson, H.T., Pilchowski, J. and Baker, K. (1997). Experimental investigation of heat transfer and pressure drop through deep microchannels in a (110) silicon substrate, *Proceedings of ASME Heat Transfer Division*, ASME HTD, Vol. 351-1, pp. 347–357.

Hetsroni, G., Mosyak, A., Segal, Z., and Ziskind, G. (2002). A uniform temperature heat sink for cooling of electronic devices, *International Journal of Heat and Mass Transfer*, Vol. 45, pp.3275-3286.

Hetsroni, G., Mosyak, A. and Segal, Z. (2001). Nonuniform temperature distribution in electronic devices cooled by flow in parallel microchannels, *IEEE Transactions on components and packaging technologies*, Vol. 24, pp.17-23.

Hu, H. Y., Peterson, G. P., Peng, X. F. and Wang, B. X. (1998). Interface fluctuation propagation and superposition model for boiling nucleation, *International Journal of Heat and Mass Transfer*, Vol. 41, pp.3483-3490.

Incropera F. P., DeWitt D. P., Bergman T. L. and Lavine A. S. (2007). *Fundamentals of Heat and Mass Transfer* 6th Ed., New York, John Wiley & Sons Inc.

Jiang, P. X., Fan, M. H., Si, G. S. and Ren, Z. P. (1997). Thermal-hydraulic performance of small scale micro-channel and porous-media heat exchangers, *International Journal of Heat and Mass Transfer*, Vol. 44, pp.1039–1051.

Jiang, L., Wong, M. and Zohar, Y. (1999). Phase change in microchannel heat sinks with integrated temperature sensors, *J. MEMS*, Vol., 8, pp.358-365.

Jiang, L., Wong, M., and Zohar, Y. (2001). Forced convection boiling in a microchannel heat sink, *Journal of Microelectromechanical System*, Vol. 10, pp.80-87.

Jiang, X. N., Zhou, Z. Y., Yao, J., Li, Y. and Ye, X. Y. (1995). Micro-fluid flow in micro channel, *Proceedings of the 8th International Conference on Solid-State Sensors and Actuators, and Eurosensors IX (Transducers '95)*, Stockholm, Sweden, pp.317-320, June 25-29.

Kanda, Y. (1982). A graphical representation of the piezoresistive coefficients in silicon, *IEEE Transactions on electron devices*, Vol. ED-29, No. 1, pp. 64-70.

Kandlikar, S. G. (2006). Nucleation characteristics and stability considerations during flow boiling in microchannels, *Experimental Thermal and Fluid Science*, Vol. 30, pp.441-447.

Kays, W. M., Crawford, M. E. and Weigand, B. (2005). *Convective Heat and Mass Transfer*, 4th ed., McGraw-Hill, New York.

Kennard, E. H. (1938) Kinetic theory of gases-with an introduction to statistical mechanics, McGraw- Hill, Inc., New York, pp.291-337.

Khantuleva, T. A. (1982). Nonlocal hydrodynamical models of gas flows in the transition regime. In: *Rarefied gas dynamics 1*, Belotserkovskii, O. M., Kogan, M. N., Kutateladze, S. S. and Rebrov, A. K. Ed., Plenum Press, New York, pp.229-235.

Khrustalev, D., and Faghri, A. (1993). Thermal analysis of a micro heat pipe, *Heat Pipes and Capillary Pumped Loops*, ASME, HTD, Vol. 236, pp.19-30.

Kim, C. J., Kim, J. Y., and Sridharan, B. (1998). Comparative evaluation of drying techniques for surface micromachining, *Sens. Actuators A*, Vol. 64, pp.17-26.

Kleinstreuer, C. and Koo, J. (2004). Computational Analysis of Wall Roughness Effects for Liquid Flow in Micro-Conduits, *J. Fluid Engineering*, Vol. 126, pp.1-9.

Knudsen, M. (1934). *The kinetic theory of gases*, Methuen, London.

Ko, H. S. and Gau, C. (2011). Micro scale thermal fluid transport process in a microchannel integrated with arrays of temperature and pressure sensors, *Microfluidics and Nanofluidics*, (in press).

Ko, H. S. and Gau, C. (2009) Bonding of a complicated polymer microchannel system for study of pressurized liquid flow characteristics with electric double layer effect, *Journal of Micromechanics and Microengineering*, Vol. 19, art no. 115024.

Ko, H. S. (2009). *Fabrication and development of micro fluidic system with embedded micro pressure and temperature sensors for study of thermal and fluid flow properties*, Ph D. thesis, National Cheng Kung University, Tainan, Taiwan.

Ko, H. S., Liu, C. W., Gau, C. and Jeng, D. Z. (2008). Flow characteristics in a microchannel system integrated with arrays of micro pressure sensors using polymer material, *Journal of Micromechanics and Microengineering*, Vol. 18, art no. 075016.

Ko, H. S., Liu, C. W., and Gau, C. (2007). Micro pressure sensor fabrication without problem of stiction for a wider range of measurement, *Sensors and Actuators A: Physical*, Vol. 138, pp.261–267.

Ko, H. S., Liu, C. W., and Gau, C. (2007). A novel fabrication for pressure sensor with polymer material and evaluation of Its performance, *Journal of Micromechanics and Microengineering*, Vol. 17, pp.1640-1648.

Kohl, M. J., Abdel-Khalik, S. I., Jeter, S. M. And Sadowski, D. L. (2005). An experimental investigation of microchannel flow with internal pressure measurements, *International Journal of Heat and Mass Transfer*, Vol. 48, pp.1518-1533.

Komvopoulos, K. (1996). Surface engineering and microtribology for micro-electromechanical systems, *Wear*, Vol. 200(1-2), pp.305-327.

Koo, J. and Kleinstreuer, C. (2003). Liquid flow in microchannels: experimental observations and computational analyses of microfluidics effects, *J. Micromech. Microeng.* Vol. 13 pp.568-579.

Kundt, A. and Warburg, E. (1875). On the friction and thermal conductivity in rerefied gases, *Phil. Mag.*, Vol. 50, pp.53.

Lalonde, P., Colin, S., and Caen, R. (2001). Mesure de debit de gaz dans les microsystemes, *Mechanical Industry*, Vol. 2, pp.355–362.

Lee, G. B., Lin, C. H. and Chang, G. L. (2003). Micro flow cytometers with buried SU-8/SOG optical waveguides, *Sensors and Actuators: Physical A*, Vol. 103, pp.165-170.

Lee, M., Wong, Y. Y., Wong, M. and Zohar, Y. (2003). Size and shape effects on two-phase flow patterns in microchannel forced convection boiling, *Journal of Micromechanics and Microengineering*, Vol. 13, pp.155-164.

Lee, W. Y., Wong, M. and Zohar, Y. (2002). Pressure loss in constriction microchannels, *Journal of MicroElectroMechanicalSystem*, Vol. 11, No. 3, pp.236-244.

Legtenberg, R., Elders, J. and Elwenspoek, M. (1993). Stiction of surface microstructure after rinsing and drying: model and investigation of adhesion mechanisms, *Proc. 7th Int. Conf. Solid-State Sensors and Actuators (Transducer'93)*, Yogohama, Japan, pp.198-201.

Li, D. (2001). Electro-viscous effects on pressure-driven liquid flow in microchannels, *Colloids and Surface A*, Vol. 195, pp.35-57.

Li, X., Lee, W. Y., Wong, M. and Zohar, Y. (2000) Gas flow in constriction microdevices, *Sensor and Actrators A*, Vol. 83, pp.277-283.

Li, Z. X., Du, D. X. and Guo, Z. Y. (2003). Experimental study on flow characteristics of liquid in circular microtubes, *Microscale Thermophysicals Engineering*, Vol. 7, pp.253–265.

Li, Z. X., Du, D. X. and Guo, Z. Y. (2000). Experimental study on flow characteristics of liquid in circular microtubes, *Proceedings of International Conference on Heat Transfer and Transport Phenomena in Microscale*, Begell House, New York, USA, pp.162–168.

Lin, L. and Pisano, A. P. (1991). Bubble forming on a micro line heater, *Micromechanical Sensors, Actuators, and Systems*, ASME, DSC, Vol. 32, pp.147-164.

Lin, L., Udell, K. S. and Pisano, A. P. (1994). Liquid-vapor phase transition and bubble formation in micro structures, *Thermal Science Engineering*, Vol. 2, pp.52-59.

Liu, C. W. (2004). *Fabrication development for micro channel system by MEMS technology with measurements of the inside thermal transport process*, Ph D. thesis, National Cheng Kung University, Tainan, Taiwan.

Liu, C. W., Gau, C. and Dai, B. T. (2004). Design and Fabrication Development of a Micro Flow Heated Channel with Measurements of the inside Micro-Scale Flow and Heat Transfer Process, *Biosensors and Bioelectronics*, Vol. 20, No. 1, pp.91-101.

Liu, J. and Tai, Y. (1995). MEMS for pressure distribution studies of gaseous flows in microchannels, *An Investigation of Micro Structures, Sensors, Actuators, Machines, and Systems*, pp. 209-215.

Liu, T. Y., Li, P.L., Liu, C.W., Gau, C. (2011). Boiling flow characteristics in microchannels with very hydrophobic surface to super-hydrophilic surface, *International Journal of Heat and Mass Transfer*, Vol. 54, pp.126-134.

Madou, M. J. (2002). *Fundamentals of microfabrication*, 2nd ed., CRC Press, pp.276-277.

Mala, G. M. and Li, D. (1999). Flow characteristics of water in microtubes, *International Journal of Heat and Fluid Flow*, Vol. 20, pp.142-148.

Mala, G. M., Yang, C. and Li, D. (1998). Electrical double layer potential distribution in a rectangular microchannel, *Colloids and Surfaces A: Physicochemical and Engineering Aspects*, Vol. 135, pp.109-116.

Mala, G. M., Li, D. and Dale, J. D. (1997). Heat transfer and fluid flow in microchannels, *International Journal of Heat and Mass Transfer*, Vol. 40, No. 13, pp.3079-3088.

Mala, G. M., Li, D., Werner, C., Jacobasch, H. J. and Ning, Y. B. (1997). Flow characteristics of water through a microchannel between two parallel plates with electrokinetic effects, *International Journal of Heat and Fluid Flow*, Vol. 18, No. 5, pp.489-496.

Mala, G. M., Li, D. and Dale, J. D. (1996). Heat transfer and fluid flow in microchannels, *Microelectromechanical Systems (MEMS)*, ASME, DSC, Vol. 59, pp.127-136.

Martin, B. W., Hagena, O. F., and Schomburg, W. K. (1998). Strain gauge pressure and volume-flow transducers made by thermoplastic molding and membrane transfer, *Proc. MEMS '98*, Jan. 25-29, Heidelberg, Germany (1998), pp. 361-366.

Mason, W. P. (1969). Use of solid-state transducers in mechanics and acoustics, *J. Aud. Eng. Soc.*, Vol. 17, pp.506-511.

Maxwell, J. C. (1890). On the condition to be satisfied by a gas at the surface of a solid body, *The Scientific Papers of James Clerk Maxwell*, Vol. 2, pp.704, Cambridge University Press, London.

Morini, G. L. (2004). Single-phase convective heat transfer in microchannels: a review of experimental results, *International Journal of Thermal Science*, Vol. 43, pp.631-651.

Monat, C., Domachuk, P., and Eggleton, B. J. (2007). Integrated optofluidics: A new river of light, *Nature*, Vol. 1, pp. 106-114.

Naito, E. (1984). Laminar heat transfer in the entrance region between parallel plates - the case of uniform heat flux, *Heat transfer, Japanese research*, Vol. 13(3), pp.92-106.

Obermeier, E. and Kopystynski, P. (1992). Polysilicon as a material for microsensor applications, *Sensors and Actuators A*, Vol. 30, pp.149-155.

Obot, N. T. (2002). Toward a better understanding of friction and heat/mass transfer in microchannel – A literatures review, *Microscale Thermophysical Engineering*, Vol. 6, pp.155-173.

Papautsky, I., Gale, B. K., Mohanty, S., Ameel, T. A. and Frazier, A. B. (1999). Effect of rectangular microchannel aspect ratio on laminar friction constant, SPIE, Vol. 3877, pp.147-158.

Peake, E. R., Zias, A. R. and Egan, J. V. (1969). Solid-state digital pressure transducer, *IEEE Trans. on Electron Devices*, Vol. 16, pp.870-876.

Pelletier, N., Bêche, B., Tahani, N., Zyss, J., Camberlein, L. and Gaviot, E. (2007). SU-8 waveguiding interferometric micro-sensor for gage pressure measurement, *Sensors and Actuators: Physical A*, Vol. 135, pp.179-184.

Peng, X. F., Hu, H. Y. and Wang, B. X. (1998a). Boiling nucleation during liquid flow in microchannels, *International Journal of Heat and Mass Transfer*, Vol. 41, pp.101-106.

Peng, X. F., Hu, H. Y. and Wang, B. X. (1998b), Flow boiling through V-shape microchannels, *Experimental Heat Transfer*, Vol. 11, pp.87-90.

Peng, X. F., Liu, D., Lee, D. J. and Yan, Y. (2000). Cluster dynamics and fictitious boiling in microchannels, *International Journal of Heat and Mass Transfer*, Vol. 43, pp.4259-4265.

Peng, X. F., and Peterson, G. P. (1996). Convective heat transfer and flow friction for water flow in microchannel structures, *International Journal of Heat and Mass Transfer*, Vol. 39, No. 12, pp.2599-2608.

Peng, X. F., Peterson, G. P. and Wang, B. X. (1994). Heat transfer characteristic of water flowing through microchannels, *Experimental Heat Transfer*, Vol. 7, pp.265-283.

Peng, X. F., Peterson, G. P. and Wang, B. X. (1996). Flow boiling of binary mixtures in microchanneled plates, *International Journal of Heat and Mass Transfer*, Vol. 39, No. 6, pp.1257-1264.

Peng, X. F., Tien, Y. and Lee, D. J. (2001). Bubble nucleation in microchannels: statistical mechanics approach, *International Journal of Heat and Mass Transfer*, Vol. 44, pp. 2957-2964.

Peng, X. F. and Wang, B. X. (1993). Forced convection and flow boiling heat transfer for liquid flowing through micro channel, *International Journal of Heat and Mass Transfer*, Vol. 36, No. 14, pp.3421-3427.

Peng, X. F. and Wang, B. X. (1994). Evaporation space and fictitious boiling for internal evaporation of liquid, *Science Foundation in China*, Vol. 2, pp.55-5.

Peng, X. F. and Wang, B. X. (1998a), Forced-convection and boiling characteristics in microchannels, *Proceedings of the 11th International Heat Transfer*, Vol. 1, pp.371-390.

Peng, X. F. and Wang, B. X. (1998b). Boiling characteristics in microchannels/micro structures, *The 11th Int. Symposium on Transport Phenomena*, ISTP-11, Hsinchu, Taiwan, pp 485-491.

Peng, X. F., Wang, B. X., Peterson, G. P. and Ma, H. B. (1995). Experimental investigation of heat transfer in flat plates with rectangular microchannels, *International Journal of Heat and Mass Transfer*, Vol. 38, No. 1, pp.127-137.

Pong, K. C., Ho, C. M., Liu, J. and Tai, Y. C. (1994). Nonlinear pressure distribution in uniform microchannel, *Application of Microfabrication to Fluid Mechanics*, ASME FED, Vol. 197, pp.51-56.

Pfahler, J., Harley, J. and Bau, H. (1991). Gas and liquid flow in small channels, *Micromechanical Sensors Actuators, and Systems*, ASME, DSC, Vol. 32, pp.49-59.

Pfahler, J., Harley, J., Bau, H. and Zemel, J. (1990a). Liquid transport in micron and submicron channels, *Sensors and Actuators A*, Vol. 21/23, pp.431-434.

Pfahler, J., Harley, J., Bau, H. and Zemel, J. (1990b). Liquid and gas transport in small channels, *Microstructures, Sensors, and Actuators*, ASME, DSC, Vol. 19, pp.149-157.

Pfund, D., Rector, D., Shekarriz, A., Popescu, A. and Welty, J. (2000). Pressure drop measurements in a microchannel, *Fluid Mechanics and Transport Phenomena*, Vol. 46, pp.1496-1507.

Pfund, D. A., Shekarriz, A., Popescu, A. and Welty J. R. (1998). Pressure drops measurements in microchannels, *Proceedings of MEMS*, ASME DSC, Vol. 66, pp.193–198.

Phan, H. T., Caney, N., Marty, P., Colasson, S. and Gavillet, J. (2009). Surface wettability control by nanocoating: The effects on pool boiling heat transfer and nucleation mechanism, *International Journal of Heat and Mass Transfer*, Vol. 52, pp.5459–5471.

Qu, W., Mala, G. M. and Li, D. (2000a). Pressure-driven water flows in trapezoidal silicon microchannels, *International Journal of Heat and Mass Transfer*, Vol. 43, pp.353-364.

Qu, W., Mala, G. M. and Li, D. (2000b). Heat transfer for water flow in trapezoidal silicon microchannels, *International Journal of Heat and Mass Transfer*, Vol. 43, pp.3925-3936.

Qu, W. and Mudawar, I. (2002). Experimental and numerical study of pressure drop and heat transfer in a single-phase micro-channel heat sink, *International Journal of Heat and Mass Transfer*, Vol. 45, pp.2549-2565.

Qu, W. and Mudawar, I. (2003a). Flow boiling heat transfer in two-phase micro-channel heat sinks-I. Experimental investigation and assessment of correlation methods, *International Journal of Heat and Mass Transfer*, Vol. 46, pp.2755-2771.

Qu, W. L. and Mudarwar, I. (2003b). Measurement and prediction of pressure drop in two-phase micro-channel heat sinks, *International Journal of Heat and Mass Transfer*, Vol. 46, pp. 2737-2753.

Ren, L. and Li, D. (2001). Electroosmotic flow in heterogeneous microchannels, *Journal of Colloid and Interface Science*, Vol. 243, pp.255-261.

Ren, L., Li, D. and Qu, W. (2001a). Electro-viscous effects on liquid flow in microchannels, *Journal of Colloid and Interface Science*, Vol. 233, pp.12-22.

Ren, L., Qu, W. and Li, D. (2001b). Interfacial electrokinetic effects on liquid flow in microchannels, *International Journal of Heat and Mass Transfer*, Vol. 44, pp.3125-3134.

Ribeiro, J. C., Minas, G., Turmezei, P., Wolffenbuttel, R. F. and Correia, J. H. (2005). A SU-8 fluidic microsystem for biological fluids analysis, *Sensors and Actuators: Physical A*, Vol. 123-124, pp.77-81.

Rohsenow, W. M., Hartnett, J. P. and Ganic, E. N. (1985). *Handbook of Heat Transfer Fundamentals* 2nd ed., McGraw-Hill, New York, pp.7-49.

Rostami, A. A., Mujumdar, A. S. and Saniei, N. (2002). Flow and heat transfer for gas flowing in microchannel: a review, *Heat and Mass Transfer,* Vol. 38, pp.359-367.

Sharp, K. V., Adrian, R. J. and Beebe, D. J. (2000). Anomalous transition to turbulence in microtubes, *Proceedings of International Mechanical Engineering Cong. Expo.*, 5th Micro-Fluidic Symp., Orlando, FL, pp. 150–158.

Shen, C. H. and Gau, C. (2004). Thermal chip fabrication with arrays of sensors and heaters for micro-scale impingement cooling heat transfer analysis and measurements, *Biosensors and Bioelectronics*, Vol. 20, No. 1, pp.103-114.

Shen, C. H. (2003). *Thermal chip fabrication with arrays of sensors and heaters with analysis and measurements of micro scale impingement cooling flow and heat transfer*, Ph D. Thesis, National Cheng Kung University, Tainan, Taiwan.

Shih, J. C., Ho, C. M., Liu, J. and Tai, Y. C. (1996). Monatomic and polyatomic gas flow through uniform microchannels, *ASME DSC*, Vol.59, pp.197–203.

Shirinov, A. V. and Schomburg, W. K. (1996). Polymer pressure sensor from PVDF, *Proc. Eurosensor*, Göteborg, Schweden Sep. 17th-20th, pp.84-85.

Smoluchowski, M. Von (1910). Uber den temperaturesprung bei warmeleitung in gasen, *Anz. Wiss. Krakau ser. A*, Vol. 5, pp.129.

Sugiyama, S., Shimaoka, K. and Tabata, O. (1991). Surface micromachined micro-diaphragm pressure sensors, *Proc. 6th Int. Conf. Solid-State Sensors and Actuators (Transducer'91)*, pp.188-191.

Sze, M. (1994). *Semiconductor Sensors*, John Wiley and Sons, Inc., New York, pp.17-80.

Sze, M. (2002). *Semiconductor Devices, Physics and Tecnology*, 2nd ed., John Wiley and Sons, Inc., New York, pp.171-200.

Takuto, A., Soo, K. M., Hiroshi, I. and Kenjiro, S. (2000). An experimental investigation of gaseous flow characteristics in microchannels, *Proceedings of the International Conference on Heat Transfer and Transport Phenomena in Microscale*, Banff, Canada, pp.155-161.

Takao, K. (1961). Rarefied gas flow between two parallel plates, in: *Rarefied gas dynamics*, Talbot, L., Ed., Academic Press, New York, Section 1, pp.465-473.

Tao, C. P. and Mahulikar, S. P. (1998). Laminar convection behavior in microchannels in conventional thermal entry length and beyond, *Proceedings of the IEEE/CPMT Electronics Packaging Technology Conference*, pp.126-132.

Touloukina, Y. S. (1970-1979). *Thermophysical properties of matter: the TPRC data series; a compressive compilation of data*, IFI/Plenum, New-York.

Trung, N., Nguyen, N. and Wu, Z. (2005). Micromixer-a review, *J. Micromech. Microeng.*, Vol. 15, pp.R1-R16.

Tsien, H. S. (1948). Superaerodynamics, the mechanics of rarefied gases, *Journal of Aeronautic Society*, Vol. 13, pp.653-664.

Tufte, O. N., Chapman, P. W. and Long, D. (1962). Silicon diffused-element piezoresistive diaphragms, *Journal of Applied Physics*, Vol. 33, pp.3322-3327.

Turner, S. E., Sun, H., Faghri, M. and Gregory, O. J. (2001). Compressible gas flow through smooth and rough microchannels, *Proceedings of IMECE*, New York, USA, HTD-24145.

Vilkner, T., Janasek, D. and Manz, A. (2004). Micro total analysis systems: recent developments, *Anal. Chem.*, Vol. 76, pp.3373-3386.

Wadsworth, D. C., Erwin, D. A. and Muntz, E. P. (1993). Transient motion of a confined rarefied gas due to wall heating or cooling, Journal of Fluid Mechanics, Vol. 248, pp.219-235.

Wang, G. D., Cheng, P. and Bergles, A. E. (2008). Effects of inlet/outlet configurations on flow boiling instability in parallel microchannels, *International Journal of Heat and Mass Transfer*, Vol. 51, pp.2267-2281.

Wang, G. D., Cheng, P. and Wu, H. Y. (2007). Unstable and stable flow boiling in parallel microchannels and in a single microchannel, *International Journal of Heat and Mass Transfer*, Vol. 50, pp.4297-4310.

Webb, R. L. and Zhang, M. (1998). Heat transfer and friction in small diameters channels, *Microscale Thermo physics Engineering*, Vol. 2, pp. 189–202.

White, F. M. (1991). Viscous fluid flow, McGraw-Hill, New York.

Wu, H. Y. and Cheng, P. (2003a). Friction factors in smooth trapezoidal silicon microchannels with different aspect ratios, *International Journal of Heat and Mass Transfer*, Vol. 46, pp. 2519–2525.

Wu, H. Y. and Cheng, P. (2003b). Visualization and measurements of periodic boiling in silicon microchannels, *International Journal of Heat and Mass Transfer*, Vol. 46, pp.2603-2614.

Wu, H. Y. and Cheng, P. (2003c). Liquid/two-phase/vapor alternating flow during boiling in microchannels at high heat flux, *International Communications in Heat and Mass Transfer*, Vol. 30, pp.295-302.

Wu, H. Y. and Cheng, P. (2004). Boiling instability in parallel silicon microchannels at different heat flux, *International Journal of Heat and Mass Transfer*, Vol. 47, pp.3631-3641.

Wu, P. and Little, W. H. (1983). Measurement of friction factor for the flow of gases in very fine channels used for micro-miniature Joule-Thompson refrigerators, *Cryogenics*, pp 272-277.

Xu, B., Ooi, K. T., Wong, N. T. and Choi, W. K. (2000). Experimental investigation of flow friction of liquid flow in microchannels, *Int Comm. Heat Mass Transfer*, Vol. 27, pp.1165-1176.

Xu, B., Ooi, K. T., Wong, N. T., Liu, C. Y. and Choi, W. K. (1999). Liquid flow in microchannels, *Proceedings of the 5th ASME/JSME Joint Thermal Engineering Conference*, San Diego, CA, pp.150–158.

Yang, C. and Li, D. (1997). Electrokinetic effects on pressure-driven liquid flows in rectangular microchannels, *Journal of Colloid and Interface Science*, Vol. 194, pp.95-107.

Yang, C., and Li, D. (1998). Analysis of electrokinetic effects on the liquid flow in rectangular microchannels, *Colloids and Surface A: Physicochemical and Engineering Aspects*, Vol. 143, pp.339-353.

Yang, C., Li, D. and Masliyah, J. H. (1998). Modeling forced liquid convection in rectangular microchannels with electrokinetic effects, *International Journal of Heat and Mass Transfer*, Vol. 41, pp.4229-4249.

Yen, S. M. (1971). Monte carlo solutions of nonlinear Boltzmann equation for problems of heat transfer in rarefied gases, *International Journal of Heat and Mass Transfer*, Vol. 14, pp. 1865-1869.

Yu, D., Warrington, R.O., Barron, R. and Ameel, T. (1995). An experimental and theoretical investigation of fluid flow and heat transfer in microtubes, *Proceedings of ASME/JSME Thermal Engineering Joint Conference*, Maui, pp.523–530.

Ziering, S. (1961). Plane Poiseuille Flow, in: *Rarefied gas dynamics,"* Talbot, L., Ed., Academic Press, New York, Section 1, pp.451-463.

Zohar, Y., Lee, Y. K., Lee, W. Y., Jiang, L. and Tong, P. (2002). Subsonic gas flow in a straight and uniform microchannel, *J. Fluid Mech.* Vol. 472, pp.125-151.

6

Heat Transfer in Molecular Crystals

V.A. Konstantinov
*B. Verkin Institute for Low Temperature Physics and Engineering
of the National Academy of Sciences of Ukraine, Kharkov
Ukraine*

1. Introduction

This short review does not pretend to comprehend all available information concerning the thermal conductivity of molecular crystals, in particular, at low temperatures (below $20K$). For such kind of information see, for example, Batchelder, 1977; Gorodilov et al., 2000; Jezowski et al., 1997; Ross et al., 1974; Stachowiak et al., 1994. The goal of this paper consists in presentation and generalization of the new experimental results and theoretical models, accumulated over the past 2-3 decades, to a certain extent changing existing view about the heat transfer in crystals. Quite recently, it was not doubted that high-temperature thermal conductivity of molecular crystals is proportional to the inverse temperature, $\Lambda \propto 1/T$. It was based on both the experimental data and assumptions being evident at first sight from which this dependence followed. In simple kinetic model, the phonon thermal conductivity can be represented as $\Lambda = 1/3Cvl$, where C (the heat capacity) and v (the sound velocity) can be considered to be constant at $T \geq \Theta_D$, and averaged phonon mean-free path l is inversely proportional to the temperature. More precise expression (see, for example, Berman, 1976; Slack, 1979) can be written in the form:

$$\Lambda = K \frac{ma\Theta_D^3}{\gamma^2 T} ,$$ (1)

where m is the average atomic (molecular) mass; a^3 is the volume per atom (molecule); $\gamma = -(\partial \ln \Theta_D / \partial \ln V)_T$ is the Grüneisen parameter, and K is a structure factor. In time, data on the deviation from $1/T$ dependence has accumulated, and in a number of cases some ideas qualitatively explaining the observed behaviour of thermal conductivity have been proposed. The problem has been, however, that the theory predicts the $1/T$ law at the constant volume of the sample, whereas the measurements were carried out at constant pressure. In this case, thermal expansion, been usually rather essential at high temperatures (the molar volume of molecular crystals may change up to 10-20% in the temperature interval from zero and up to the melting temperature) leads, as a rule, to additional decrease of Λ with rise of temperature. Moreover, in many cases, the phonons are not the only excitations determining the heat transfer and scattering process. The dependence of the thermal conductivity on the molar volume can be described using Bridgman's coefficient:

$$g = -(\partial \ln \Lambda / \partial \ln V)_T ,$$ (2)

It follows from Eqs. (1) and (2) that for crystals:

$$g = 3\gamma + 2q - 1/3,\tag{3}$$

were $q=(\partial\ ln\gamma/\partial\ lnV)_T$. Ordinarily, it is assumed that $\gamma\propto V$ and the second Grüneisen coefficient $q\cong1$ (Slack, 1979; Ross et al., 1984). Taking into account that $\gamma\cong2\div3$ for a number of simple molecular crystals (Manzhelii et al., 1997) it is expected that $g\cong8\div11$ and $\Lambda\propto V^{8\div11}$. It means that 1% change in volume may result in 8-11% change in thermal conductivity. Data measured at saturated vapour and atmosphere pressures can be considered as equivalent because the difference between them is much smaller than accuracy of experiment and they will be further denoted as isobaric ($P\cong0,\ MPa$) data.

Constant-volume investigations are possible for molecular solids having a comparatively large compressibility coefficient. Using a high-pressure cell, it is possible to grow a solid sample of sufficient density. In subsequent experiments it can be cooled with practically unchanged volume, while the pressure in the cell decreases. In samples of moderate densities the pressure drops to zero at a certain characteristic temperature T_0 and the isochoric condition is then broken; on further cooling, the sample can separate from the walls of the cell. In the case of a fixed volume, melting of the sample occurs in a certain temperature interval and its onset shifts towards higher temperatures as density of samples increases (For more experimental details see Konstantinov et al., 1999).

As the temperature increases, phonon scattering processes intensify, the mean-free path length l decreases and it may approach to the lattice parameter. The question of what occurs when the phonon mean-free path becomes comparable to the lattice parameter or its own wavelength is one of the most intriguing problems in the thermal conductivity of solids (see, for example, Auerbach & Allen, 1984; Feldman et al., 1993; Sheng et al., 1994). According to preferably accepted standpoint, in this case the vibrational modes assume a "diffusive" character, but the basic features of the kinetic approach retain their validity. Some progress in the description of the heat transport in strongly disordered materials has come about through the concept of the minimum thermal conductivity Λ_{min} (Slack, 1979; Cahill et al., 1992), which is based on the picture where the lower limit of the thermal conductivity is reached when the heat is being transported through a random walk of the thermal energy between the neighboring atoms or molecules vibrating with random phases. In this case Λ_{min} can be written as the following sum of three Debye integrals:

$$\Lambda_{min} = \left(\frac{\pi}{6}\right)^{1/3} k_B\, n^{2/3} \sum_i \mathrm{v}_i \left\{ \left(\frac{T}{\Theta_i}\right)^2 \int_0^{\Theta_i/T} \frac{x^3 e^x}{\left(e^x - 1\right)^2}\,dx \right\},\tag{4}$$

The summation is taken over three (two transverse and one longitudinal) sound modes with the sound speeds v_i; Θ_i is the Debye cutoff frequency for each polarization expressed in degrees K; $\Theta_i = v_i\left(\hbar/k_B\right)\left(6\pi^2 n\right)^{1/3}$; n is the number density of atoms or molecules. Although no theoretical justification exists as yet for this picture of the heat transport, the evidence for its validity has been obtained on a number of amorphous solids in which the high temperature thermal conductivity has been found to agree with the value predicted by this model. Indirect evidence has also been obtained in measurements of the thermal conductivity of highly disordered crystalline solids, in which no thermal conductivity smaller than that predicted by this model seems to have ever been observed (Cahill et al., 1992). It is evident, that thermal conductivity approaches its lower limit Λ_{min} in amorphous solids and strongly disordered crystals (Auerbach & Allen, 1984; Cahill et al., 1992; Sheng et al., 1994). This raises the question whether or not the three-phonon scattering processes in themselves may result in Λ_{min} in perfect crystals with rise of temperature.

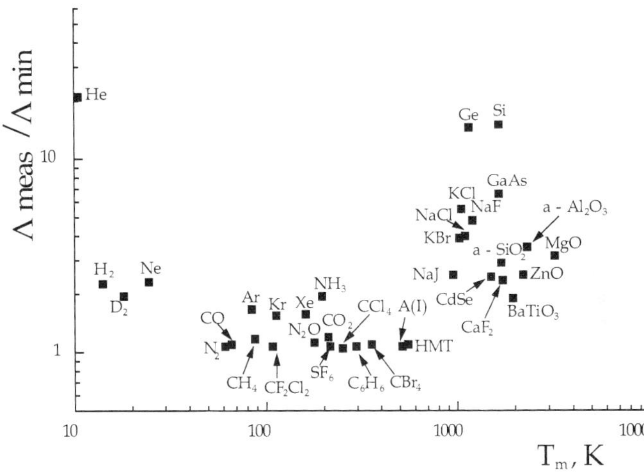

Fig. 1. The ratio $\Lambda_{meas}/\Lambda_{min}$ immediately below the corresponding melting temperatures (T_m) versus T_m for crystals with different types of chemical bonds.

To find an answer let us compare the measured thermal conductivity Λ_{meas} of a number of crystals with different types of chemical bonds and the lower limit to thermal conductivity Λ_{min} at the corresponding melting temperatures T_m (see Fig. 1). It is evident that the ratio $\Lambda_{meas}/\Lambda_{min}$ increases as the crystal bond becomes stronger. In van-der-Waals-type crystals $\Lambda_{meas}/\Lambda_{min} \cong 1.5 \div 2$, while in the crystals with diamond-type structure it is of the order of $10 \div 20$, i.e. van-der-Waals-type crystals are the most suitable objects for observing the thermal conductivity "minimum" due to umklapp processes only and this will be demonstrated further.

Crystals containing molecules or molecular ions are more complicated than crystals containing only atoms and ions, since the former possess translational, orientational and intramolecular degrees of freedom. An important common feature of simple molecular crystals is that in the condensed phases the intermolecular forces are much weaker than the intramolecular ones, so that the molecular parameters remain close to those in the gas. As a rule, the intramolecular forces and associated intramolecular vibration frequencies ($\sim 1000 cm^{-1}$) exceed by an order of magnitude the intermolecular ones (the corresponding lattice-mode frequencies are below $\sim 1000 cm^{-1}$). Such a large difference between the two types of frequencies makes it possible to safely regard the respective types of motion as independent. So as far as the lattice vibrations are concerned, the molecules can be treated as rigid bodies. In such an approximation each molecule participates in two types of motion: translational, when the molecular center of mass shifts, and rotational, when the center of mass rests. Many features in the dynamics of the simple molecular solids are related to the rotational motion of the molecules (Parsonage & Stavaley, 1972). At very low temperature, the structure of a crystal is perfectly ordered and the molecules can perform only small amplitude translational vibrations at the lattice sites and oscillations around selected axes (libration) in a manner so that the motion of neighboring molecules is correlated and collective translational and orientational excitations (phonons and librons) propagate throw the crystal. Calculation of anharmonic effects show that translational vibrations are characterized by relatively small amplitudes while the amplitude of librational vibrations in molecular crystals is sufficiently large even at $T\cong0$, so that the harmonic theory can hardly claim to give more than qualitative picture of the librational motion (Briels et al., 1985; Manzhelii et al., 1997). As the temperature increases, the rotational motion may, in principle, pass through the following stages depending on the relationship between the central and anisotropic forces: enlargement of the libration amplitudes, the appearance of jump-like reorientations of the molecules, increase of the frequency of reorientations, hindered rotation of the molecules, and, finally, nearly free rotation of the molecules. In the last two cases a phase transition takes place, as a rule, before the crystal melts, giving rise to a structure in which translational long-range order is preserved while the orientational order is lost. It is a characteristic property of crystals consisting of high-symmetry "globular" molecules like CH_4, N_2, adamantane ($C_{10}H_{16}$) or, in some degree, of "cylindrical" molecules like benzene, C_2F_6 or long-chain n-alkanes. They form high-temperature "plastic" or orientationally-disordered (commonly called ODIC: Orientational Disorder In Crystals) phases in which the rotational motion of molecules resembles their motion in the liquid state (Parsonage & Stavaley, 1972). In crystals consisting of molecules of a lower symmetry the long-range orientational order is preserved, as a rule, up to the melting points.

In harmonic approximation, phonons and librons (rotational excitations) are treated as independent entities. Real phonons are, however, coupled together and with rotational excitations by anharmonic terms of the crystal Hamiltonian (Manzhelii et al., 1997; Lynden-Bell & Michel, 1994). Therefore, translational and orientational types of motions in molecular crystals are not independent of one another, but rather they occur as coupled translational-orientational vibrations. That involves considerable difficulties to describe this case with analytical expressions. As a consequence, a simplified model where the translational and orientational subsystems are described independently is usually used (Manzhelii et al., 1997; Kokshenev et al., 1997). The coupling produces a shift of phonons frequency with respect to the harmonic value as well as a broadening of bands, associated to the finite phonon lifetime. In such approximation the TO coupling results in an additional contribution to the thermal resistance of the crystal $W=1/\Lambda$. This additional thermal resistance may decrease if the frequency of reorientations becomes sufficiently large, so that the TO coupling reduces. The relative simplicity of the investigated molecular crystals made possible an appropriate theoretical interpretation and provided establishing of the general relationships in heat transfer that result from the presence of rotational degrees of freedom of the molecules. In the experimental part of the paper the results of study of isochoric thermal conductivity of solidified inert gases, simple molecular crystals and their solutions at $T\geq\Theta_D$ will be considered then the models intended to explain temperature and volume dependences of thermal conductivity will be discussed.

2. Experimental results.

2.1 The solidified inert gases

The solidified inert gases *Ar*, *Kr* and *Xe* are convenient object for comparison of experimental data with theoretical calculations of thermal conductivity of a lattice since they are simplest solids closely conformable to theoretical models. This fact stimulated a considerable number of experimental and theoretical works (see, for example, the review of Batchelder, 1977). At $T \geq \Theta_D$ the phonon-phonon interaction is the only mechanism, which determines the magnitude and temperature dependence of the thermal conductivity Λ in perfect crystals. If the scattering is not too strong and the picture of elastic waves can be used, theory predicts the thermal conductivity $\Lambda \propto 1/T$ at fixed volume of the sample (Berman, 1976). The more rapid decrease of the thermal conductivity as $\Lambda \propto 1/T^2$ observed in these inert gases under saturated vapour pressure was originally ascribed to scattering process with participation of four or more phonons (Krupski et al., 1968). Later, Slack, 1972 suggested that the change of crystal volume with temperature by itself may lead to considerable deviations from the dependence $\Lambda \propto 1/T$ because of a "quasi-harmonic" change of the spectrum of vibrational modes, and Ecsedy & Klemens, 1977 showed theoretically that four-phonon processes are expected to be weak even at premelting temperatures. Subsequent thermal conductivity studies of solid *Ar*, *Kr* and *Xe* at fixed density (Clayton & Batchelder, 1973; Bondarenko et al., 1982; Konstantinov et al., 1988) confirmed that roughly the dependence $\Lambda \propto 1/T$ is valid when $T \geq \Theta_D$. Fig. 2 shows both isochoric for samples of different densities (Clayton & Batchelder, 1973) and measured under saturated vapour pressure $P \cong 0, MPa$, (Krupski et al., 1968) experimental data for solid argon in the W versus T

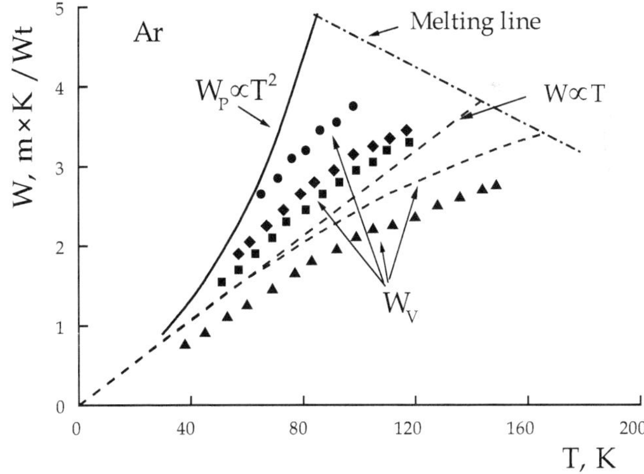

Fig. 2. Isobaric W_P (Krupski et al., 1968) and isochoric W_V (Clayton et al., 1973) thermal resistance $W=1/\Lambda$ of crystalline argon for samples of different densities.

coordinates, where $W=1/\Lambda$ is the thermal resistance of crystal. It is seen that appreciable deviations are observed at the highest temperatures, with the isochoric thermal conductivity varying markedly more slowly than $\Lambda \propto 1/T$ or $W \propto T$ dependence. A similar behaviour was also observed for krypton (Bondarenko et al., 1982) and xenon (Konstantinov et al., 1988). It was found that Λ can be described by the expression

$$\Lambda = \left(A/T + B \right)\left(V_0/V \right)^g , \tag{5}$$

where A and B are constants independent on the temperature; V_0 is molar volume at $T{=}0K$, and V is actual molar volume $V{=}V(T)$. The Bridgman's coefficients g for Ar, Kr and Xe were found in good agreement with calculated by Eq. 3 and they are equal to 9.7, 9.4 and 9.2 correspondingly.

The deviations observed were primary attributed to anharmonic renormalisation of the law of phonon dispersion at a fixed volume (Konstantinov et al., 1988). The quantitative calculation has not yet been carried out because of the complexity of model proposed. Later on the thermal conductivity of solid Ar, Kr and Xe was calculated within framework of the Debye model which allows for the fact that the mean-free path of phonons cannot become smaller than half the phonon wavelength (Konstantinov, 2001a).

2.2 Nitrogen-type crystals and oxygen

The N_2–type crystals (N_2, CO, N_2O and CO_2) consisting of linear molecules have rather simple and largely similar physical properties. In these crystals the anisotropic part of the molecular interaction is determined mostly by the electric-quadrupole forces. At low temperatures and pressures, these crystals have a cubic lattice with four molecules per unit cell. The axes of the molecules are along the body diagonals of the cube. In N_2 and CO_2 having equivalent diagonal directions the crystal symmetry is $Pa3$, for the noncentrosymmetrical molecules CO and N_2O the crystal symmetry is $P2_13$. In CO_2 and N_2O the anisotropic interaction is so strong that the crystals melt before the complete orientational disorder occurs. In N_2 and CO the barriers impeding the rotation of the molecules are an order of magnitude lower; as a result, the orientational disordering phase transitions occur at 35.7 and 68.13K, respectively. In the high-temperature phases, N_2 and CO molecules occupy the sites of the HCP lattice of the spatial group $P6_3/mmc$.

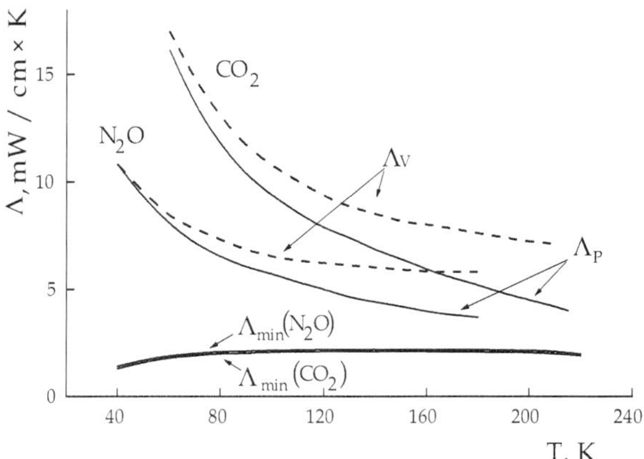

Fig. 3. Isochoric (dashed lines) and isobaric (solid lines) thermal conductivities Λ of CO_2 and N_2O.

The thermal conductivity of solid CO_2, N_2O, N_2 and CO was studied at saturated vapor pressure by Manzhelii et al., 1975. Isochoric thermal conductivity of CO_2 and N_2O was studied by Konstantinov et al., 1988b, and N_2 and CO by Konstantinov et al., 2005b; 2006a. The data for CO_2 and N_2O is shown in Fig. 3, solid and dashed lines depict isobaric and isochoric thermal conductivity respectively. The isochoric thermal conductivity is recalculated to the molar volumes which CO_2 and N_2O have at zero temperature and pressure. The lower limits to thermal conductivity, calculated accordingly to Eq. 4 are shown at the bottom. The Bridgman's coefficients g for nitrogen-type crystals were found in poor agreement with calculated by Eq. 3. The reason for it will be discussed later.

Whereas isobaric thermal conductivity of CO_2 roughly follows $1/T$ dependence, isochoric one deviates rather more strongly from the above dependence than in solidified inert gases. In N_2O both isochoric and isobaric thermal conductivities deviates strongly from $1/T$ dependence. To reveal the features to be associated with the anisotropic component of the molecular interaction it is necessary to compare molecular crystals with rare-gas solids in the reduced coordinates (de Bour, 1948). Such a comparison is of interest for the following reasons: the thermal resistance $W_{ph\text{-}ph}$ of an ideal crystal of an inert gas is due solely to phonon-phonon scattering. In CO_2 an additional phonon thermal resistance $W_{ph\text{-}lib}$ (or $W_{ph\text{-}rot}$) appears due to interaction phonons with librons (rotational excitations). In the case of N_2O the scattering resulting from dipole disordering is added to these phonon scattering mechanisms (W_{dip}). To a first approximation the total thermal resistance W is:

$$W = W_{ph-ph} + W_{ph-rot} + W_{dip} \tag{6}$$

It is convenient to make a comparison between the crystals mentioned because all of them have a FCC lattice. A modified version of the method of reduced coordinates was used. It is important to note that in this case there is no need to resort to some approximate model or other. As a rule, the reducing parameters used are the values of $T_{mol}=\varepsilon\sqrt{k}$, $W_{mol}=\sigma^2/k\sqrt{(m/\varepsilon)}$, and $V_{mol}=N\sigma^3$, were ε and σ are the parameters of the Lennard-Jones potential, m is molecular weight and k is the Boltzmann constant. It is reasonable to use as an alternative to this the values of the temperature and molar volume of abovementioned substances at the critical points T_{cr} and V_{cr}. The choice of the given parameters is explained as follows. For simple molecular substances T_{cr} and V_{cr} are proportional to ε and σ^3, respectively. However, the accuracy of determination is much higher for the critical parameters than for the parameters of the binomial potential. It should be mentioned that the quantities ε and σ depend substantially on the choice of binomial potential and the method used to determine it. Temperature dependence of isochoric thermal resistances of Xe, CO_2 and N_2O in the reduced coordinates is shown in Fig. 4. It is seen that all the contributions to the total thermal resistance are of the same order of magnitude and the deviations from the $W \propto T$ dependence increase from Xe to N_2O. The reason has to do with increasing of the phonon scattering and approaching of Λ to its lower limit Λ_{min} as it is seen in Fig. 3.

The behavior of thermal conductivity in the orientationally-ordered phases of N_2 and CO is very similar to CO_2 and N_2O. In the orientationally-disordered β-phases isochoric thermal conductivity of all samples of different density increases with rise of temperature, whereas isobaric one is nearly temperature independent (see in Fig. 5 experimental data for N_2; data for CO is very similar). In the framework of simple kinetic model, an increase of thermal conductivity with rise of temperature may be explained by an increase of the phonon mean-free path because of the weakening of the effect of some scattering mechanism. It is logically to assume that the interaction of phonons with rotational excitations provides such a mechanism. At the $\alpha{\rightarrow}\beta$ transition and over the orientationally-disordered β-phase a gradual transition from librations to hindered rotation takes place. In contrast to libration, free molecule rotation does not lead to phonon scattering. From the above it follows that there is a temperature interval where phonon scattering by the rotational excitations weakens with rise of temperature or, in other words, TO coupling decreases. The isobaric thermal conductivity is determined by partial compensation of this effect as a result of decreasing of thermal conductivity with rise of temperature due to thermal expansion.

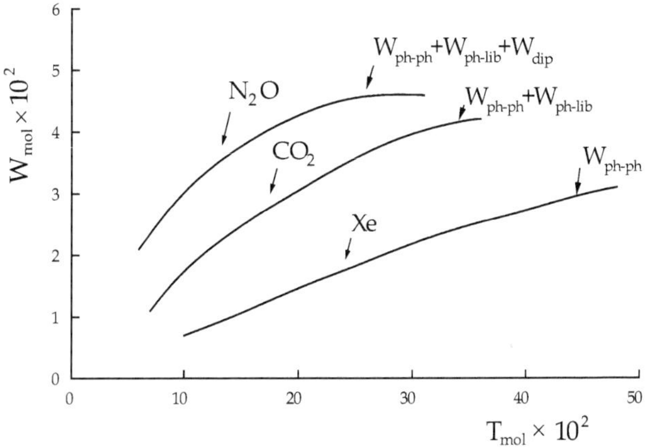

Fig. 4. Temperature dependences of isochoric thermal resistance of *Xe*, *CO₂* and *N₂O* in the reduced coordinates.

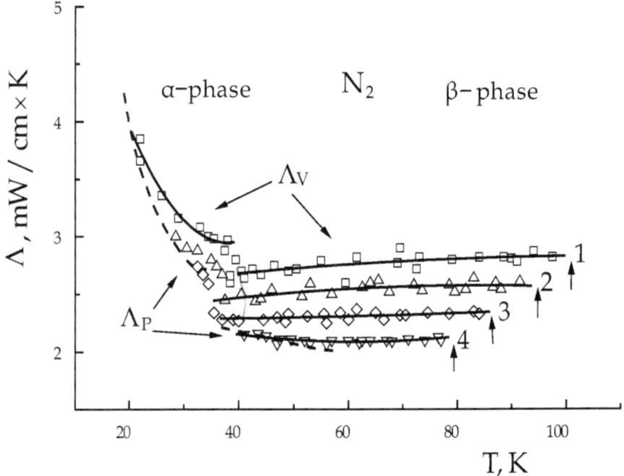

Fig. 5. Thermal conductivity of solid nitrogen. Solid lines are isochoric data for samples of different densities numbered from 1 to 4. Dashed lines are isobaric data.

Solid oxygen belongs to a small group of molecular crystals consisting of linear molecules. In contrast to the N_2-class crystals having the orientationally ordered $Pa3$-type structure, solid oxygen, much like halogens, has a collinear orientational packing because the valence rather than quadrupole forces predominate in the anisotropic interaction. Besides, in the ground state the O_2 molecule has the electron spin $S=1$ which determines the magnetic properties of oxygen. Another specific feature of solid O_2 is the fact that the energy of the magnetic interaction makes up a considerable portion of the total binding energy. This unique combination of molecular parameters has stimulated much interest in the physical properties of O_2, in particular its thermal conductivity. The thermal conductivity of solid O_2 was investigated under saturated vapor pressure in α, β and γ-phases over a temperature interval 1-52K (Ježowski, et al., 1993). The low-temperature α-O_2 phase is orientationally and magnetically ordered. On heating to 23.9K, the structure changes into the rhombohedral magnetically-ordered β-phase of the symmetry $R3m$. This is the simplest orientational structure, similar to α-O_2. On a further heating, orientational disordering (cubic cell, $Pm3m$ symmetry with $Z=8$) occurs at $T=43.8K$. Under atmospheric pressure oxygen melts at 54.4K. The thermal conductivity has a maximum in the α-phase at $T\approx6K$, drops sharply on a change to the β-phase, where it is practically constant, jumps again at the $\beta\rightarrow\gamma$ transition and increases in γ-O_2. The experimental results were interpreted as follows. In the magnetically ordered α-phase the heat is transferred by both phonons and magnons, and their contributions are close in magnitude: $\Lambda_{ph}\approx\Lambda_m$. On the $\alpha\rightarrow\beta$ transition the thermal conductivity decreases sharply (~60%) because the magnon component disappears during magnetic disordering. The weak temperature dependence of the thermal conductivity in the β-phase was attributed to the anomalous temperature dependence of the sound velocity in β-O_2 which is practically constant for the longitudinal modes and increases for the transverse ones. The growth of the thermal conductivity in γ-O_2 was attributed to the decay of the phonon scattering at the rotational excitations of the molecules and at the short-range magnetic order fluctuations at rising temperature. The isochoric thermal conductivity of γ-O_2 has been studied on samples of different density in the temperature interval from 44K to the onset of melting (Konstantinov et al., 1998b). More sharp increase of isochoric thermal conductivity was observed in γ-O_2 than in the isobaric case.

2.3 Methane and halogenated methanes

The solid halogenated methanes consisting of tetrahedral molecules are convenient objects to investigate the correlation between the rotational motion of molecules and the behavior of thermal conductivity. Methane (CH_4), and carbon tetrahalogenides (CF_4, CCl_4, CBr_4 and CJ_4) form high-temperature "plastic" or orientationally-disordered phases in which the rotational motion of molecules is similar to their motion in the liquid state. In crystals consisting of low-symmetry molecules such as chloroform ($CHCl_3$), methylene chloride (CH_2Cl_2) or dichlorodifluoromethane (CCl_2F_2) the anisotropic forces are much stronger and the long-range order persists in them up to the melting points. A special case is trifluoromethane CHF_3, where the second NMR momentum decreases sharply above $T=80K$ from 11.5G^2 to 3.0G^2 immediately prior to melting at $T_m=118K$, which suggests enhancement of the molecule rotation about the three-fold axes.

The molecule of methane can be presented as a regular tetrahedron with hydrogen atoms at the vertex positions and carbon atom in the center. The symmetry causes the molecule to exhibit permanent octupole electrostatic moment. At the equilibrium vapor pressure CH_4

solidifies at 90.7K and displays unchanged crystallographic structure called phase I down to 20.4K which is the temperature of phase transition to phase II. In both phases, the carbon atoms at the center of the tetrahedral molecule occupy sites of the face-centered cubic lattice. In the low-temperature phase II the orientation dependent octupole-octupole interaction leads to a partial orientational ordering. The crystal structure with six orientationally-ordered and two disordered sublattices belongs to the space group $Fm3c$. The orientationally-ordered molecules at D_{2d} site symmetry positions perform collective librations, while those at O_h positions rotate almost freely down to the lowest temperatures. In phase I all the tetrahedral molecules are orientationally-disordered, performing rotations which do not show any long-range correlation. $CH_4(I)$ is unique between ODIC molecular crystals since its molecular rotation is virtually free at premelting temperatures. The isobaric thermal conductivity of solid methane was measured within the temperature range of 21-90K in phase I (Manzhelii & Krupski, 1968) and within the temperature interval of 1.2–25K in phase II (Jeżowski et al., 1997). The results obtained revealed an existence of the strong phonon scattering mechanisms connected with rotational excitation of the methane molecules. The isochoric thermal conductivity was studied by (Konstantinov et al., 1999) on samples with molar volumes 30.5 and 31.1 $cm^3/mole$. The experimental data for the orientationally-disordered phase of $CH_4(I)$ is shown in Fig. 6.

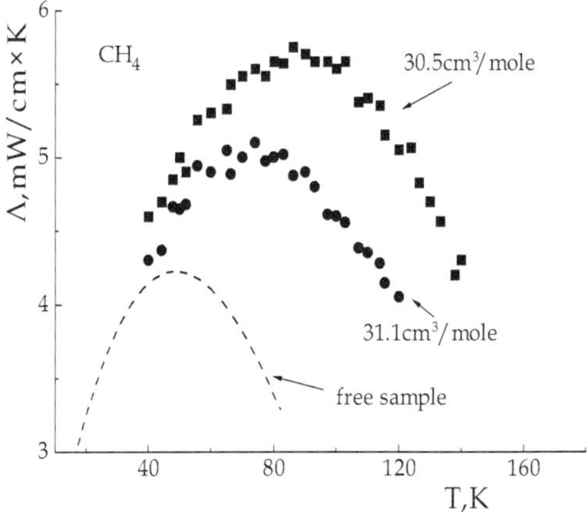

Fig. 6. The isochoric thermal conductivity of solid methane for samples having molar volumes 30.5 (■) and 31.1 (●) $cm^3/mole$ together with the isobaric data (dashed line).

It is seen that both isobaric and isochoric thermal conductivities first increase with rise of temperature, pass through a maximum and then decrease up to melting. Note, than regular „kinetic" maximum of thermal conductivity is observed at considerably lower temperature in phase II (Jeżowski et al., 1997). Maximum shifts towards higher temperatures as the density of the sample increases. The Bridgman coefficient is equal to 8.8±0.4.

The origin of such behavior of thermal conductivity is the same that for orientationally-disordered phases of other molecular crystals, it is decrease of phonon scattering on rotational excitations of molecules. However, in contrast to "plastic" phases of other molecular crystals where rotation is hindered, methane molecules rotate almost freely at premelting temperatures. Above the maximum, phonon-rotation contribution $W_{ph\text{-}rot}$ to the total thermal resistance W of methane tend to zero, and behavior of thermal conductivity is determined solely by increase of phonon-phonon scattering. It is clearly seen in Fig. 7, where the appropriate contributions were calculated using the method of reduced coordinates. The theoretical models proposed to describe thermal conductivity of solid methane will be discussed later.

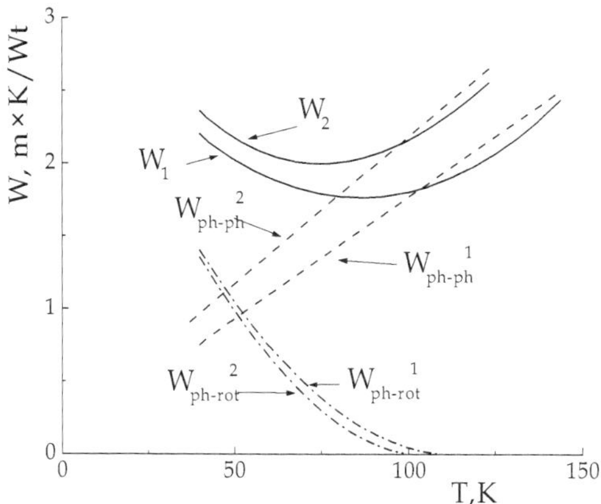

Fig. 7. Contributions of phonon-phonon scattering W_{ph-ph} and phonon scattering by rotational molecule excitations W_{ph-rot} to the total thermal resistance W of solid methane samples having molar volumes 30.5 (1) and 31.1 (2) $cm^3/mole$.

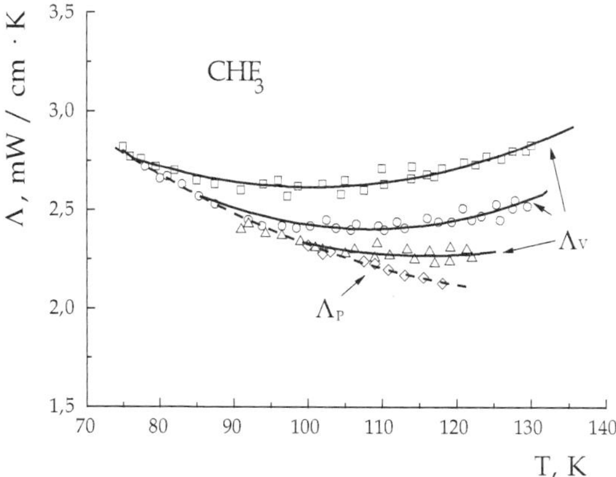

Fig. 8. The isochoric thermal conductivity of three solid CHF_3 samples of different densities. Solid lines show smoothed values of isochoric thermal conductivity. Dashed line and rhombus are for the thermal conductivity of a free sample.

Slow increase of isochoric thermal conductivity was also observed in the orientationally-disordered phases of CCl_4 (Konstantinov et al., 1991a) and CBr_4 (Ross et al., 1984) at recalculation the last experimental data to constant volume. The isochoric thermal conductivity in orientationally-ordered phases of halogenated methanes ($CHCl_3$, CH_2Cl_2 and CCl_2F_2) decreases with rise of temperature deviating markedly from $1/T$ dependence like case of CO_2 and N_2O (Konstantinov et al., 1991b; 1994; 1995). An interesting behavior of the isochoric thermal conductivity was found in trifluoromethane CHF_3 (Konstantinov et al., 2009a). CHF_3 melts at T_m=118K, the melting entropy being $\Delta S_f/R$ = 4.14. Neutron scattering investigations of the crystallographic structure of CHF_3 revealed only one crystalline phase of the spatial symmetry $P2_1/c$ with four differently oriented molecules in the monoclinic cell. The Debye temperature of CHF_3 is Θ_D=88±5K. Fig. 8 shows isochoric thermal conductivity of CHF_3 for three samples of different densities in the interval from 75K to the onset of melting. The isochoric thermal conductivity first decreases with increasing temperature, passes through a minimum at T~100K, and then starts to increase slowly. The weak growth of isochoric thermal conductivity with temperature in solid CHF_3 suggests that the translational-orientational coupling becomes weaker in this crystal at premelting temperatures owing to the intensive molecule reorientations about the three-fold axes.

Some parameters of the halogenated methanes discussed are presented in Table 1.

Substance		T_m, $T_{I\text{-}II}$	Structure	z	$\Delta S_f/R$	Θ_D, K	g	μ, D
CH_4	(I)	90.6	F_m3_m	4	1.24	96	8.8	0
	(II)	20.5	$P43m$	32		141		
CCl_4	(I)	250.3	F_m3_m	4	1.21		6.0	0
	(II)	225.5	$C2/c$	32		92	6.5	
CBr_4	(I)	363	F_m3_m	4	1.3	62	3.8	0
	(II)	320	$C2/c$	32			3.4	
CHF_2Cl	(I)	115.7	$P4_2/n$	8	4.25		4.5	1.41
	(II)	59	$P112/n$	8		70		
$CHCl_3$		210	$Pnma$	4	5.4	86*	3.9	1.01
CH_2Cl_2		176	$Pbcn$	4	3.13	115*	4.6	1.6
CF_2Cl_2		115	$Fdd2$	8	4.2	80*	5.0	0.51
CHF_3		118	$P2_1/c$	4	4.14	88*	4.6	1.6

* - Estimates obtained from IR and Raman spectra.

Table 1. Melting temperature T_m; phase transition temperature $T_{I\text{-}II}$; structure and the number of molecules per unit cell z; melting entropy $\Delta S_f/R$; Debye temperature Θ_D; Bridgman coefficient $g = -(\partial \ln \Lambda / \partial \ln V)_T$; dipole momentum of molecule μ.

2.4 Some special cases: SF_6 and C_6H_6

Sulphur haxafluoride SF_6 is often assigned to substances that have a plastic crystalline phase. Indeed, the relative entropy of melting $\Delta S_f/R$ of SF_6 is 2.61, which is close to the Timmermanns criterion. However, the nature of orientational disorder in the high-temperature phase of SF_6 is somewhat different from that of plastic phases in other molecular crystals, where the symmetries of the molecule and its surroundings do not coincide. The interaction between the nearest neighbors in the bcc phase is favorable for molecule ordering caused by the S-F bonds along the {100} direction, and the interaction with the next nearest neighbors is dominated by repulsion between the F atoms. According to X-ray and neutron diffraction data a strict order is observed in SF_6 (I) just above the phase transition point. The structural dynamical factor $\mathcal{R}$ characterizing the degree of the orientational order is close to unity in the interval 95-130K. This feature sets off SF_6 from other plastic crystals, such as methane, carbon tetrachloride, adamantane and so on, where the long-range orientational order becomes disturbed immediately after the phase transition. Orientational disordering in SF_6 starts to intensify only above 140K. As follows from the analysis of the terms of the Debye-Waller factor derived from neutron-diffractometric data for the high-temperature phase of SF_6, the F atoms have large effective libration amplitudes. As the temperature rises, the amplitudes increase to 20° and higher, but the F localization is still appreciable near {100} direction. This implies that the orientational structure of SF_6 (I) does not become completely disordered even at rather high temperatures. The disordering itself is dynamic by nature. The increasing amplitudes of librations are not the only factor responsible for the increasing orientational disordering at rising temperature. It is rather connected with dynamic reorientations, which become more intensive due to frustrations of the molecular interactions. Owing to these features, SF_6 offers a considerable possibility for investigating the influence of wide-range rotational states of the molecules on the thermal conductivity in a monophase one-component system, where such states can vary from nearly complete orientational ordering to frozen rotation.

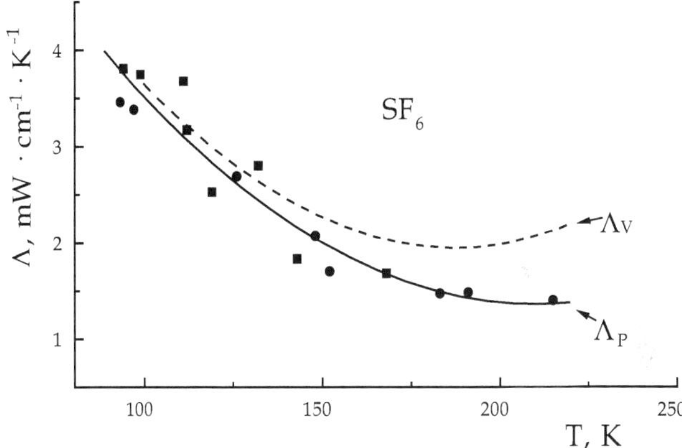

Fig. 9. Isochoric (V_m=58.25 $cm^3/mole$) and isobaric thermal conductivity of solid SF$_6$.

Isochoric thermal conductivity of SF$_6$ was studied by Konstantinov et al., 1992b, while isobaric one by Purski et al., 2003. The data are shown in Fig. 9. The isobaric thermal conductivity first decreases with increasing temperature and flattens out at premelting temperatures. The isochoric thermal conductivity first decreases with increasing temperature, passes through a minimum at $T\sim180K$, and then starts to increase. Such behavior was attributed to decreasing of phonon scattering on rotational excitations of molecules with rise of temperature.

Solid benzene has only one crystallographic modification under the saturated vapor pressure: it has the orthorhombic spatial symmetry $Pbca$ (D_{2h}^{15}) with four molecules per unit cell. Benzene melts at 278.5K with entropy of melting $\Delta S_f/R$ = 4.22, which is much higher that the Timmermanns criterion for $ODIC$ phases. The Debye temperature of C_6H_6 is 120K. In the interval 90-120K the second NMR moment of C_6H_6 drops considerably as a result of the molecule reorientations in the plane of the ring around the six-fold axis. The activation energy of the reorientational motion estimated from the spin-lattice relaxation time is 0.88 $kJ/mole$. The frequency of molecule reorientations at $85K$ is 10^4 sec^{-1}. On a further heating it increases considerably reaching 10^{11} sec^{-1} near T_m. The basic frequency of the benzene molecule oscillations about the six-fold axis at $273K$ is 1.05×10^{12} sec^{-1}. Isochoric thermal conductivity of C_6H_6 was studied by Konstantinov et al., 1992a, while isobaric one by Purski et al., 2003. The data are shown in Fig. 10.

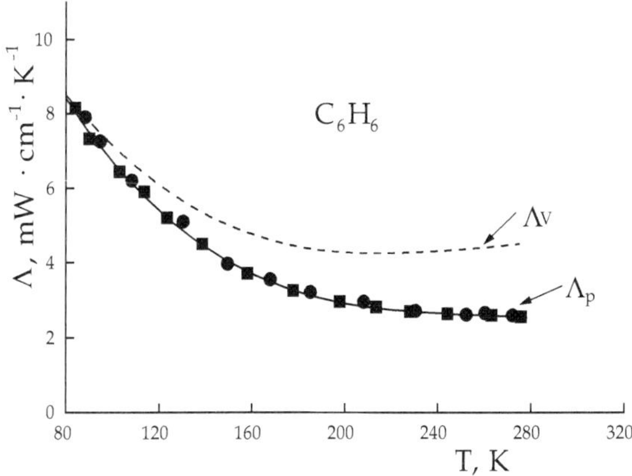

Fig. 10. Isochoric (V_m=70.5 $cm^3/mole$) and isobaric thermal conductivity of solid C_6H_6.

Like solid SF_6 the isobaric thermal conductivity of benzene first decreases with increasing temperature and flattens out at premelting temperatures. The isochoric thermal conductivity first decreases with increasing temperature, passes through a minimum at $T\sim210K$, and then starts to increase slowly. In contrast to SF_6 where the rotation is multi-axial, molecule of benzene rotates about released six-fold axis. The increase of thermal conductivity with rise of temperature was attributed like to previous cases to weakening of phonon scattering on rotational excitations of molecules.

2.5 Solid n-alkanes

Normal alkanes (n-paraffins) of the C_nH_{2n+2}-type form a class of substances that are intermediate in changing–over to long–chain polymers. N–alkanes have a comparatively simple structure and a molecular packing: in the solid state the axes of all molecules are always parallel to one another irrespective of a particular crystalline modification. Owing to their relative simplicity, normal alkanes are naturally considered as the starting point for understanding the structural and thermophysical properties of more complex long–chain compounds. N–alkanes exhibit an extremely diverse dynamic behavior both in the solid and liquid states. The melting temperature increases in this series of compounds with the length of the chain and its behavior is nonmonotonic: the n–alkanes with an odd number of carbon atoms (odd n–alkanes) melt at relatively lower temperatures than those with an even number of C atoms (even n–alkanes). An interesting effect is observed when the orthorhombic, monoclinic and triclinic structures alternate with the even and odd members of the series. The even n-alkanes with n=6-24 (n=the number of C atoms) crystallize at low temperatures forming a triclinic cell. Heptane (n=7) and nonane (n=9) have an orthorhombic structure at low temperatures. Glass-like cylindrical "rotational" phases with a hexagonal symmetry were found in rather narrow temperature interval below the melting points of odd n-alkanes starting with n=9. The region of existence of the "rotational" phase increases with the length of the chain. Hexagonal modifications also occur in even n-alkanes starting with n=22. Some parameters of n-alkanes discussed above are submitted in Table 2.

Substance	structure	$T_{\alpha\text{-}\beta}$,K	$\Delta S_{\alpha\text{-}\beta}/R$	T_m,K	$\Delta S_m/R$	$\Delta S_{\alpha\text{-L}}/R$	Λ_α/Λ_L
C_2H_6	$P2_1/n$, z=2	89.8	2.74	90.3	0.77	3.6	1.3
C_3H_8	$P2_1/n$, z=4	-	-	85.5	4.95	4.95	2.2
C_6H_{14}	P ī, z=1	-	-	177.8	8.85	8.85	1.9
C_9H_{20}	P ī, z=1	217.2	3.48	219.7	8.47	12.0	2.4
$C_{11}H_{24}$	P_{bcn}, z=4	236.6	2.9	247.6	10.8	13.7	2.4
$C_{13}H_{28}$	P_{bcn}, z=4	255.0	3.6	267.8	12.8	16.4	2.3
$C_{15}H_{32}$	P_{bcn}, z=4	270.9	4.1	283.1	14.7	18.8	2.1
$C_{17}H_{36}$	P_{bcn}, z=4	284.3	4.8	295.1	16.4	21.2	2.0
$C_{19}H_{40}$	P_{bcn}, z=4	296.0	5.6	304.0	18.8	24.3	2.0

Table 2. The structure of n-alkanes; the temperature $T_{\alpha\text{-}\beta}$ and the entropy $\Delta S_{\alpha\text{-}\beta}/R$ of the transition to the "rotational" phase; the temperature T_m and the melting entropy $\Delta S_m/R$; a complete change of the entropy and variations of thermal conductivity Λ_α/Λ_L during the ordered-phase – liquid transition.

The isochoric thermal conductivity of methane, ethane, propane and hexane was studied by Konstantinov et. al., 1999a; 2006c; 2009b and 2010 correspondingly. Isobaric thermal conductivity of "odd" numbered n-alkanes with n=9-19 was investigated by Forshman & Andersson, 1984. The isobaric thermal conductivity of n-alkanes discussed above is shown in Fig. 11 along with the thermal conductivities of the liquid phases of these compounds measured immediately after melting. It was noted (Forshman & Andersson, 1984) that the thermal conductivity of long–chain odd n-alkanes has some features in common. As the "rotational" phase melts, the thermal conductivity changes by about 35% and is independent of the chain length. The jump of the thermal conductivity on changing to the low temperature ordered phase decreases with the increasing length of the chain and makes ~85% for n-undecane and ~40% for n-nonadecane. The absolute value of thermal conductivity increases in the "rotational" phase with increasing of the chain length.

According to our studies the isobaric thermal conductivity exhibits closely similar behavior in short and long–chain n-alkanes. On the transition from the ordered phase to a liquid the thermal conductivity of the n-alkanes starting with propane changes nearly twice and is independent of the total transitions entropy and the chain length. This change is much smaller in the case of spherical and elliptic molecules: for example, $\Delta\Lambda/\Lambda_L$ is only 20-30% in methane and ethane. This can be related to the higher degree of orientational order in solid long–chain n-alkanes as compared to spherical molecules. The isochoric thermal conductivity of solid n-alkanes decreases with rise of temperature following a dependences weaker than $\Lambda \infty 1/T$. The deviation of the isochoric thermal conductivity from the dependence $\Lambda \infty 1/T$ in solid n-alkanes was explained proceeding from the concept of the lower limit to thermal conductivity (Konstantinov et. al., 2009b).

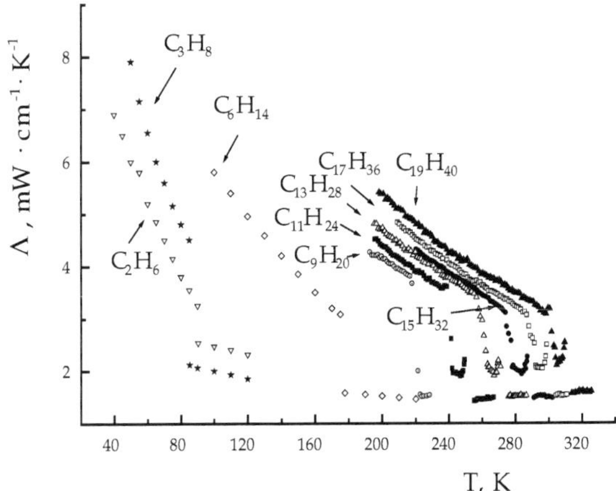

Fig. 11. Isobaric thermal conductivity of some n-alkanes.

2.6 Mixed molecular crystals

Mixed molecular crystals are convenient object for testing of the concept of the lower limit to thermal conductivity. In heavy solidified inert gases Ar, Kr and Xe the thermal conductivity approaches its lower limit at premelting temperatures. In the case when the thermal conductivity approaches Λ_{min} with rise of temperature, the effect of impurities should manifest itself in a specific manner. Impurities cannot considerably decrease the thermal conductivity at premelting temperatures at which Λ has already been close to the minimum value. As temperature decreases the contribution of impurities to the thermal resistance of crystal $W=1/\Lambda$ should increase. The isochoric thermal conductivity of solid $(CH_4)_{1-\xi}Kr_\xi$, $Kr_{1-\xi}Xe_\xi$ and $(CO_2)_{1-\xi}Kr_\xi$, $(CO_2)_{1-\xi}Xe_\xi$ solutions ($\xi=0\div1$) has been studied by Konstantinov et al., 2000, 2001a; 2002b and 2006b, respectively. Fig. 12 shows the temperature dependence of isochoric thermal conductivity for pure Kr and $Kr_{1-\xi}Xe_\xi$ solid solution reduced to samples for which condition of constant volume starts from $80K$.

It can be seen that the thermal conductivity of $Kr_{1-\xi}Xe_\xi$ solid solution decreases and its temperature dependence becomes weaker with an increase in Xe concentration. At $\xi=0.14$, the thermal conductivity virtually coincides with the lower limit to thermal conductivity calculated by Eq. 4.

Kr and CH_4 form a homogeneous solid solution with fcc structure above $30K$ at all ξ. They have similar molecule/atom radii and parameters of the pair potential. The Debye temperatures Θ_D of Kr and CH_4 are 72 and $143K$ respectively. At the same time the masses of the Kr atom and the CH_4 molecule are very different, 83.8 and 16 atomic units, respectively. The isochoric thermal conductivity of $(CH_4)_{1-\xi}Kr_\xi$ solid solutions has been studied between $40K$ and $\sim 150K$ over the wide range of concentrations, ($\xi=0.013, 0.032, 0.07, 0.115, 0.34, 0.71,$

0.855, 0.937, and 0.97). A gradual transition from the thermal conductivity of a highly perfect crystal to the minimum thermal conductivity was observed as the crystal becomes increasingly more disordered (see Fig. 13).

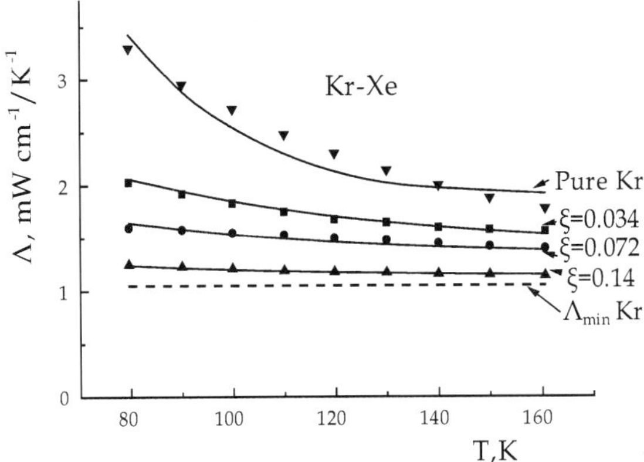

Fig. 12. Smoothed values of isochoric thermal conductivity of pure Kr and $Kr_{1-\xi}Xe_{\xi}$ solid solution for samples, whose volume is constant starting from $80K$. The dashed line is Λ_{min} of pure Kr, calculated according to Eq. 4.

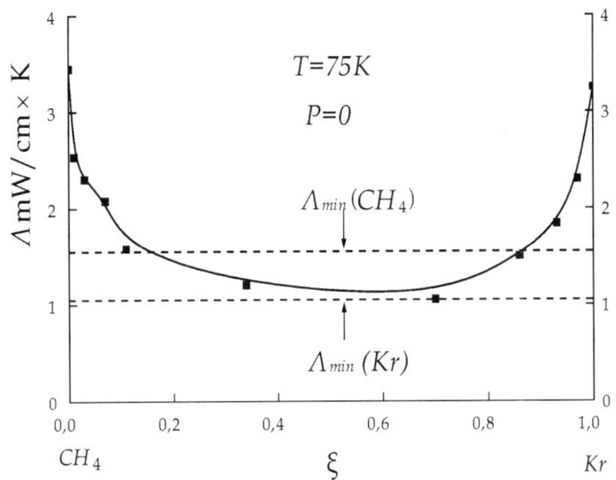

Fig. 13. The concentration dependence of the thermal conductivity of solid solution $(CH_4)_{1-\xi}Kr_{\xi}$ at $T=75K$ and $P=0$. The horizontal lines are Λ_{min} of pure Kr and CH_4 under the same conditions calculated by Eq. 4.

A strong decrease of the Bridgman coefficient was observed, from a value $g\cong9$ characteristic for pure Kr and CH_4 to $g\cong4$ at only a tiny impurity concentration.
An unusual effect of point defects on the thermal conductivity has been detected in $(CO_2)_{1-\xi}Xe_{\xi}$ and $(CO_2)_{1-\xi}Kr_{\xi}$ solid solutions (Konstantinov et al., 2006b). In pure CO_2 at $T >150K$ the

isochoric thermal conductivity decreases smoothly with increasing temperature. In contrast, the thermal conductivity of CO_2 /Kr and CO_2 /Xe solid solutions first decreases, passing through a minimum at 200–210K, and then starts to increase with temperature up to the onset of melting. Such behavior of the isochoric thermal conductivity was attributed to the rotation of the CO_2 molecules, which gain more freedom as the spherically symmetrical inert gas atoms are dissolved in the CO_2.

2.7 Short conclusion to the experimental part

It has been demonstrated that the $\Lambda \propto 1/T$ law is not obeyed in molecular crystals at all. Because of the strong translational-orientational (TO) coupling in the orientationally-ordered phases molecular librations contribute considerably to the thermal resistance $W=1/\Lambda$ of the crystal. As a result, the isochoric thermal conductivity approaches its lower limit Λ_{min} at premelting temperatures and shows significantly more slow dependence than $1/T$. In the case of hindered or almost free rotation of molecules in crystal the isochoric thermal conductivity increases with rise of temperature. It is originated due to weakening of TO coupling as the rotational motion of molecules attains more freedom. The isobaric thermal conductivity is determined by the partial compensation of these effects as a result of decreasing of thermal conductivity with rise of temperature due to thermal expansion.

3. Theoretical models.

3.1 The Debye model of thermal conductivity

In simple Debye model (see, for example, Berman, 1976) thermal conductivity Λ can be expressed as

$$\Lambda = \frac{\hbar^2}{2\pi^2 v^2 k_B T^2} \int_0^{\omega_D} l_\Sigma(\omega)\omega^4 \frac{e^{\frac{\hbar\omega}{k_B T}}}{\left(e^{\frac{\hbar\omega}{k_B T}} - 1\right)^2} d\omega, \tag{7}$$

where v is the sound velocity; ω_D is the Debye frequency $\omega_D = v(6\pi^2 n)^{1/3}$; $l_\Sigma(\omega)$ is the combined phonon mean-free path determined by the package of all scattering mechanisms

$$l_\Sigma(\omega) = \left(\sum_i l_i(\omega)^{-1}\right)^{-1}, \tag{8}$$

If the scattering governed by U-processes only the mean-free path of phonons is

$$l_u(\omega) = v / A\omega^2 T , \quad A = \frac{18\pi^3}{\sqrt{2}} \frac{k_B \gamma^2}{ma^2 \omega_D^3} ; \tag{9}$$

were m is the average atomic (molecular) mass; a^3 is the volume per atom (molecule); $\gamma = -(\partial \ln \Theta_D / \partial \ln V)_T$ is the Grüneisen parameter. It is easy to check that in high-temperature limit $T \geq \Theta_D$ Eqs. 7 and 9 reduce to Eq. 1. In the case of point defect scattering the mean-free path of phonons is

$$l_i(\omega) = v/B\omega^4 \;\; ; \;\; B = \frac{3\pi\Gamma}{2\omega_D^3} \;\; ; \;\; \Gamma = \xi(1-\xi)\left(\frac{\Delta M}{\overline{M}} + 6\gamma\frac{\Delta a}{a}\right)^2 \tag{10}$$

where ΔM is the mass difference of atoms (molecules) of the impurity and host; Δa is the change of the lattice parameter upon introduction of impurity.

3.2 Phonon-libron scattering

The contribution of phonon-libron scattering was considered by (Manzhelii et al., 1975), and (Kokshenev et al., 1997). In harmonic approximation decomposition of the interaction energy with respect to translational and angular displacements of molecules near equilibrium positions can be written as

$$H_{harm} = \sum_{\vec{q},\lambda}\hbar\omega_{\vec{q},\lambda}\left(b_{\vec{q},\lambda}^+ b_{\vec{q},\lambda} + \frac{1}{2}\right) + \sum_{\vec{k},j}\varepsilon_j\left(a_{\vec{k},j}^+ a_{\vec{k},j} + \frac{1}{2}\right), \tag{11}$$

where $b_{\vec{q},\lambda}$, $a_{\vec{k},j}$ are the operators of deletion of phonons and librons; λ and j are the numbers of phonon and libron branches. So long as the anharmonicity of molecular librations is substantial even at low temperatures a self-consistent theory was used for description of the libron subsystem (see, for example, Manzhelii et al., 1997). Only the interaction processes leading to the linear dependence of thermal resistance in high temperature $T \geq \Theta_D$ limit were taken into consideration

$$V = \frac{\gamma_1 aG}{\sqrt{N}}\sum_{q,q',k}\sqrt{qq'}\,a_k^+ b_q b_{q'}\Delta(k-q-q') + \gamma_2 G\sqrt{a/N}\sum_{q,k,k'}\sqrt{q}\,a_k a_{k'}^+ b_a\Delta(k-k'+q), \tag{12}$$

where $k\equiv(\vec{k},j)$, $q\equiv(\vec{q},\lambda)$; N is number of lattice points; $G=U\eta$ (U is a constant of molecular field, η is a parameter of the long-range orientational order). In Eq. 12 only the members were taken into consideration which allow for conservation law of energy and quasi-momentum. Three types of interactions were considered:

$$Ph_A + Ph_A \leftrightarrow Lib, \tag{13a}$$

$$Ph_A + \left(Ph_O \; or \; Lib\right) \leftrightarrow \left(Lib \; or \; Ph_O\right), \tag{13b}$$

$$Ph_A + Lib \leftrightarrow Lib', \tag{13c}$$

where Ph_A and Ph_O designate acoustic and optical phonons. The constants of interactions γ_1 and γ_2 can be expressed via characteristic temperatures Θ and Θ_l of translational and librational spectrum of crystal:

$$\gamma_1 = C_1\,\hbar^3 / Ma^2\Theta(J\Theta_l)^{1/2}\,k_B^{3/2}, \tag{14a}$$

$$\gamma_2 = C_2\,\hbar^3 / J\Theta_l\,(Ma^2\Theta)^{1/2}\,k_B^{3/2}, \tag{14b}$$

where J is a momentum of inertia of molecule. In quasi-harmonic approach the characteristic temperature is associated with a value of barrier G hindering of molecular rotation by the relation

$$\Theta_1 = (3\hbar^2 G/J\, k_B^2)^{1/2},\tag{15}$$

An additional thermal resistance of crystal determined by scattering processes of the type (13a) and (13b) can be presented as

$$W_1(T) = C_1 T \left(\frac{h}{k_B}\right)^3 \frac{\gamma_{lib}^2}{ma\Theta^3}\frac{ma^2}{J}S_1\,;\quad S_1 = \sum_v g_v\varphi_v\,,\tag{16a}$$

$$W_2(T) = C_2 T \left(\frac{h}{k_B}\right)^3 \frac{\gamma_{lib}^2}{ma\Theta^3}\frac{ma^2}{J}S_2\,;\quad S_2 = \sum_{v,v'} g_v g_{v'} f_{v,v'}\,,\tag{16b}$$

where γ_{lib} is Grüneisen parameter; g_v and $g_{v'}$ are degrees of degeneration of libron and optical branches; m and J are the mass and momentum of inertia of molecule; a is interatomic distance; φ_v and $f_{v,v'}$ are the surfaces of energy conservations in scattering processes like (13a) and (13b). In the simple Ziman model they can be written as:

$$\varphi_v = (\alpha_v^2 - 1)(6\alpha_v - 4 - \alpha_v^3)\,;\quad \alpha_v = \theta_v/\theta,\tag{17a}$$

$$f_{v,v'} = \sigma_{v,v'}^3(12 - \sigma_{v,v'}^2)\,;\quad \sigma_{v,v'} = \left|\Theta_v - \Theta_{v'}\right|/\Theta,\tag{17b}$$

where θ_v and $\Theta_{v'}$ are the frequencies of optical and phonon modes.
The total thermal resistance of crystal at presence of phonon-phonon and two phonon-libron mechanisms of scattering in the reduced coordinates is:

$$W^* = B^* T^* = B_0^* T^* \left[1 + (ma^2/J)(\alpha S_1 + \beta S_2)\right],\tag{18}$$

where coefficients B_0^*, α and β are equal for the group of one-type crystals. By this means that an additional phonon-libron scattering leads in the first approximation only to renormalization of coefficient A in Eq. 9.

3.3 Phonon scattering on rotational excitations of molecules

In the orientationally-disordered phases of molecular crystals there is no long-range order, what suggests that the distinct pure libration modes cannot propagate in the crystal. Nevertheless, the correlation effects are still strong immediately after the phase transition and the short-range orientational order persists. In this region there is an additional phonon scattering at the short-range orientational order fluctuations and it becomes weaker on a further temperature rise. To explain the behaviour of the thermal conductivity in the orientationally-disordered phases of solid methane and deuteromethane Manzhelii & Krupski, 1968 used the analogy between molecular and spin systems. In a number of magnetic crystals the thermal conductivity was observed to increase above the magnetic phase transition. Reason for these anomalies is the scattering of phonons by critical fluctuations of the short-range magnetic order above the Neel point (Kawasaki, 1963). In the orientationally-disordered phases of the molecular crystals an increase of the isochoric thermal conductivity with increasing temperature is due to weakening of phonon scattering by fluctuations of the short-range orientational order. By existing analogy, using the equations for one and two-phonon relaxation times of Kawasaki, 1963, the phonon mean-free path of each of the examined scattering mechanisms can be expressed as

$$l_I(\omega) = \rho \upsilon^5 / B^2 \Lambda_{rot} T \omega^2, \tag{19a}$$

$$l_{II}(\omega) = \pi \rho^2 \upsilon^8 / C^2 k_B C_{rot} T^2 \omega^4, \tag{19b}$$

where B and C are anisotropic molecular interaction constants; Λ_{rot} and C_{rot} are the thermal conductivity and the heat capacity of the rotational subsystem, respectively; ρ is the density. It is assumed that $B=C^2$ in the first approximation. The coefficient B can be roughly estimated from the dependence of the phase transition temperature T_f upon pressure $B = -\chi_T^{-1} \partial \left(\ln T_f \right) / \partial P$, where χ_T is isothermal compressibility. The thermal conductivity of the rotational subsystem can be calculated from the known gas-kinetic expression $\Lambda_{rot} = \frac{1}{3} C_{rot} a^2 \tau^{-1}$, where τ is the characteristic time of the site-to-site transfer of the rotational energy and can be estimated as a mean period of the librations. Taking into account Eqs. (8, 9, 19a, 19b), the phonon mean-free path in the orientationally-disordered phase can be written as

$$l_\Sigma(\omega) = \left(\frac{AT\omega^2}{\upsilon} + \frac{B^2 \Lambda_{rot} T \omega^2}{\rho \upsilon^5} + \frac{C^2 k_B C_{rot} T^2 \omega^4}{\pi \rho^2 \upsilon^8} \right)^{-1}, \tag{20}$$

The validity of such description was corroborated subsequently by the first principles calculations of the thermal conductivity of methane by Yasuda, 1978.

3.4 Phonons and "diffusive" modes

Eq. 8 is inapplicable if $l(\omega)$ becomes of the order of or less than half the phonon wavelength $\alpha\lambda / 2 = \alpha\pi v / \omega$, where α is a numerical factor of the order of unity (Roufosse & Klemens, 1974).

$$l(\omega) = \begin{cases} l_\Sigma(\omega), & 0 \le \omega \le \omega_0, \\ \alpha\pi\upsilon/\omega = \alpha\lambda/2, & \omega_0 < \omega \le \omega_D, \end{cases} \tag{21}$$

Taking into account Eqs. 9, 18, and 21 the phonon mobility edge ω_0 is equal for orientationally-ordered molecular crystals to

$$\omega_0 = 1/\alpha\pi AT. \tag{22}$$

Eq. 21 is the well-known Ioffe-Regel' criterion, which presumes localization, so that we shall assume excitations whose frequencies lie above the phonon mobility edge ω_0 to be "localized". Since completely localized states do not contribute to the thermal conductivity, localization is assumed to be weak and excitations can hop from site to site in a diffusion manner, as Cahill et al., 1992 supposed. This point of view is, on the hole, consistent with the results of Feldman et al., 1993, where the theory for intermediate case (where disorder is sufficient for oscillations not to propagate but insufficient for localization) was formulated for amorphous silicon. The idea was that the dominant scattering was correctly described by a harmonic Hamiltonian and reduced to a single-particle problem of decoupled oscillators. On this basis the thermal conductivity could be calculated exactly using a formula similar to the Kubo-Greenwood formula for the electric conductivity of disordered metals (Auerbach et al., 1984). The calculation performed by authors showed that in this case it is incorrect to indentify the Ioffe-Regel' edge with localization. Although the oscillations dominating in high-temperature heat transfer lie near the Anderson localization edge, with the exception of narrow band of states they are not completely localized. The Boltzmann theory assigns to each vibrational mode with wave vector $\vec{k}$ and propagation velocity $\vec{v} = \partial\omega / \partial\vec{k}$ a diffuseness $D_k = vl/3$, where l is the mean-free path length. The authors found that even though vibrations are not localized, a definite wave vector or velocity cannot be assigned to them. Nonetheless, they transfer heat and contribute to the thermal conductivity an amount of the order of $C_i(T)D_i/V$ for the ith mode, where the specific heat $C_i(T) = k_B$ for $T \geq \Theta_D$, D_i is the temperature-independent "diffuseness" of the mode. The numerical calculation is in many respects similar to the calculation performed according to Boltzmann's theory and it confirms the concept of a "minimum" thermal conductivity in the form it is discussed by Roufosse & Klemens, 1974; Slack, 1979 and Cahill et al., 1992.

If $\omega_0 \geq \omega_D$, then the mean-free path length of all modes is greater than $\lambda/2$ and Eqs. 7 and 22 come to Eq. 1 at $T \geq \Theta_D$. If $\omega_0 < \omega_D$, then the integral of thermal conductivity (7) splits into two parts which describe the contributions of low-frequency phonons and high-frequency "diffusive" modes to heat transfer:

$$\Lambda = \Lambda_{ph} + \Lambda_{dif} \tag{23}$$

In the high-temperature limit $T \geq \Theta_D$ these contributions are

$$\Lambda_{ph} = \frac{k_B\omega_0}{2\pi^2 v \, AT} \; ; \; \Lambda_{dif} = \frac{\alpha k_B}{4\pi v}(\omega_D^2 - \omega_0^2) , \tag{24}$$

The experimental results for a number of crystals were computer-fitted by the least-square method (see for details Konstantinov et al., 2003b). Figs. 14 and 15 show the contributions of phonons and "diffusive" modes to the isochoric thermal conductivity of Kr and CO_2.

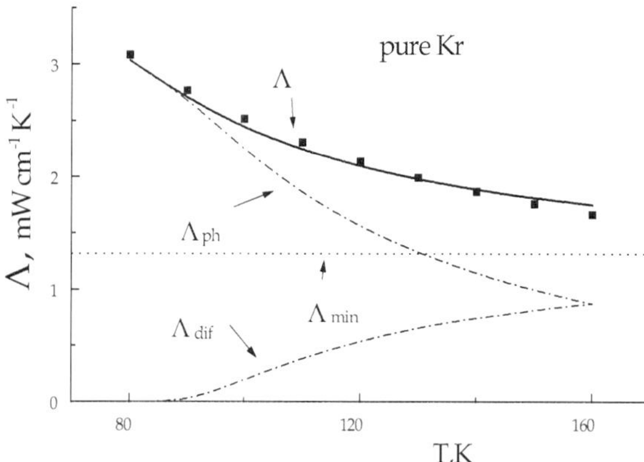

Fig. 14. Contributions Λ_{ph} and Λ_{dif} for Kr sample with molar volume 28.5 $cm^3/mole$.

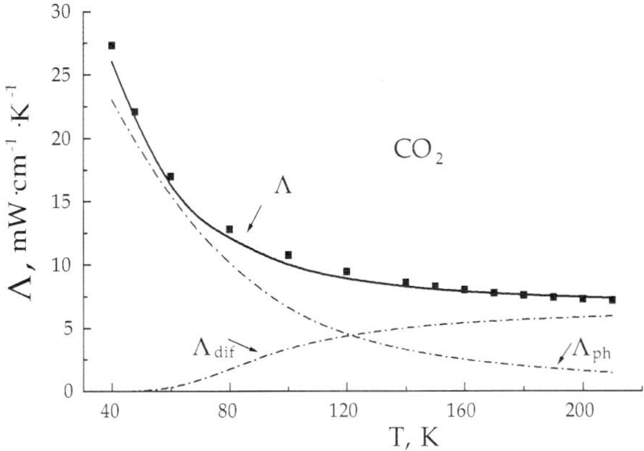

Fig. 15. Contributions Λ_{ph} and Λ_{dif} for CO_2 sample with molar volume 25.8 $cm^3/mole$.

It is seen that in Kr thermal conductivity possesses pure phonon character up to 80K. "Diffusive" modes arise above 80K and $\Lambda_{ph} \cong \Lambda_{dif}$ at 160K. In CO_2 where the strong additional phonon scattering arises due to TO coupling, "diffusive" modes arise at 50K and in the high temperature region most part of the heat is transferred by "diffusive" modes.

The applicability of such description is supported by the straightforward calculations of the thermal conductivity by the method of molecular dynamics using the Kubo-Greenwood formula. Recently, the thermal conductivity of solid argon with the Lennard-Jones potential has been described using two contributions made by low-frequency phonons with mean-free paths exceeding half the wavelength and high-frequency phonons with mean-free path of about half the wavelength (McGaughey & Kaviany, 2004).

In the orientationally-disordered phases ω_0 follows from Eqs. (20) and (21):

$$\omega_0 = -\frac{u}{\left(-\eta + \sqrt{u^3 + \eta^2}\right)^{1/3}} + \left(-\eta + \sqrt{u^3 + \eta^2}\right)^{1/3} \tag{25}$$

where the parameters u and η are

$$u = \frac{\pi\rho^2\upsilon^7}{3C^2 k_B C_{rot} T}\left(A + \frac{B^2 \Lambda_{rot}}{\rho\upsilon^4}\right), \quad \eta = -\frac{\rho^2\upsilon^7}{2\alpha C^2 k_B C_{rot} T^2} \tag{26}$$

The appropriate contributions to the thermal conductivity from the low-frequency phonons Λ_{ph} and the high-frequency "diffusive" modes Λ_{dif} are

$$\Lambda_{ph} = \frac{\hbar^2}{2\pi^2 v^2 k_B T^2} \int_0^{\omega_0} l_\Sigma(\omega)\omega^4 \frac{e^{\frac{\hbar\omega}{k_B T}}}{\left(e^{\frac{\hbar\omega}{k_B T}} - 1\right)^2} d\omega \tag{27a}$$

$$\Lambda_{\mathrm{dif}} = \frac{\alpha \hbar^2}{2\pi v k_B T^2} \int\limits_{\omega_0}^{\omega_D} \omega^3 \frac{e^{\frac{\hbar\omega}{k_B T}}}{\left(e^{\frac{\hbar\omega}{k_B T}} - 1\right)^2} \, d\omega \tag{27b}$$

Figs. 16 and 17 show the ratio ω_0/ω_D upon temperature and contributions Λ_{ph} and Λ_{dif} to the thermal conductivity of solid CO. The results were computer-fitted by the least-square method to the smoothed values of the thermal conductivity of the sample with V_m=27.93 $cm^3/mole$ individually in the a- and β-phases. The polarization–averaged sound velocity corresponding to this density was 1280 m/s. It was assumed that C_{rot} varies linearly from the value $2R$ to R during β-phase (Manzhelii et al. 1997). The varied parameters were a and A in the a–phase and a, A and B in the β–phase. The best agreement with the experimental results was achieved with a=1.55 and A= 4.3 $\times 10^{-16}$ s/K in the a–phase and with a=1.25, A= 1.0$\times 10^{-17}$ s/K, B=5.0 and C=2.24 in the β–phase. As the temperature rises, the ratio ω_0/ω_D decreases in the α-phase and increases in the β-phase. This increase can be attributed to decreasing of rotational correlations between the neighboring molecules become weaker.

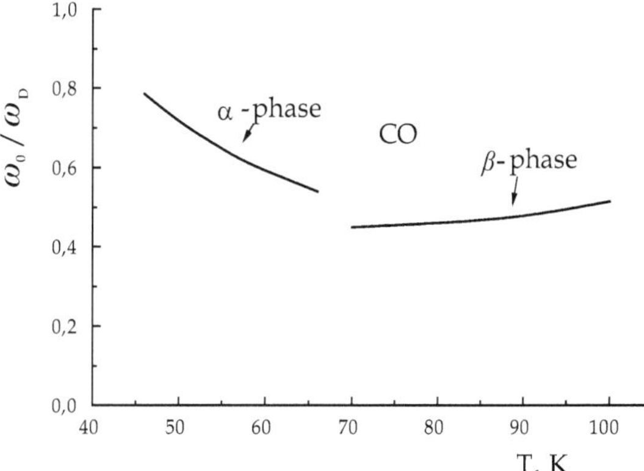

Fig. 16. The ω_0/ω_D ratio upon temperature.

It is seen that near T=45K most of the heat is transported by the phonons (the contribution of the "diffusive" modes is no more than 10%). However, immediately before the $a \rightarrow \beta$ transition over half of the heat is transported by the "diffusive" modes. In the orientationally-disordered phase the contribution of the "diffusive" modes immediately after phase transition is about two times larger than that of the phonons. As the temperature rises, the contribution of the "diffusive" modes decreases and that of the phonons increases because the scattering of the phonons by the short-range orientational order fluctuations becomes weaker due to their attenuation damping. The estimates show that both three-phonon scattering and one-phonon scattering became the dominant mechanisms. The lower limit of the thermal conductivity is reached when the mean-free paths of all the modes are $a\lambda/2$. It absolutely agrees with Eq. (4) if the polarization-averaged sound velocity v=$(v_l$+2$v_t)/3$ and a=1 are used.

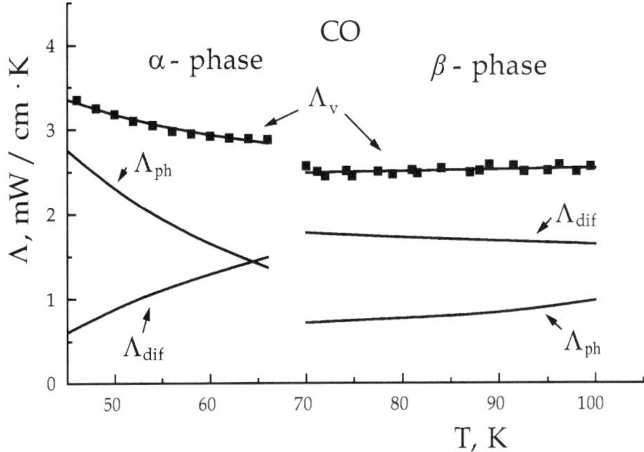

Fig. 17. Results of fitting Λ_V and the contributions to the thermal conductivity of CO sample with V_m=27.93 cm^3/mole from low-frequency phonons Λ_{ph} and "diffusive" modes Λ_{dif}.

3.5 Mixed crystals

In the case if thermal conductivity is governed by combined U-processes and point defect scattering it follows from Eqs. 8,10 and 21 that

$$\omega_0 = \frac{1}{(2\alpha\pi B)^{\frac{1}{3}}}\left[\sqrt[3]{1+\sqrt{1+u}} + \sqrt[3]{1-\sqrt{1+u}}\right] \tag{28}$$

where

$$u = \frac{4\alpha^2\pi^2 A^3 T^3}{27B} \tag{29}$$

The appropriate contributions to the thermal conductivity from the low-frequency phonons Λ_{ph} and the high-frequency "diffusive" modes Λ_{dif} are

$$\Lambda_{ph} = \frac{k_B}{2\pi^2 v}\frac{1}{\sqrt{ATB}} arctg\sqrt{\frac{B}{AT}}\omega_0 \tag{30a}$$

$$\Lambda_{dif} = \frac{\alpha k_B}{4\pi v}(\omega_D^2 - \omega_0^2) \tag{30b}$$

Fig. 18 presents the relative low-frequency phonon contribution Λ_{ph}/Λ to the thermal conductivity of pure Kr and the $(CH_4)_{1-\xi}Kr_\xi$ solid solution. A gradual transition from the thermal conductivity of a highly perfect crystal to the "lower limit" to thermal conductivity Λ_{min} was observed at $T\geq\Theta_D$ in the $(CH_4)_{1-\xi}Kr_\xi$ solid solution as the crystal becomes increasingly more disordered. It is seen that in Kr the thermal conductivity is of pure phonon character up to about 90K. As the CH_4 concentration increases, progressively more heat is transferred by the "diffusive" modes, but even at the highest concentration (29% of

CH_4 in Kr) and the highest temperatures ($150K$) an appreciable part of the heat (about 10%) is transferred by the low-frequency phonons. The values of α vary from 1 to 1.4. This supports the view about the vibrations localized in the $\lambda/2$ regions as the limiting case of the phonon picture (Cahill, 1992) and validity of Eq. (4) for prediction of the lower limit to the thermal conductivity of crystalline lattice. The ω_0/ω_D ratio decreases with rise of temperature and concentration of impurity CH_4 in Kr.

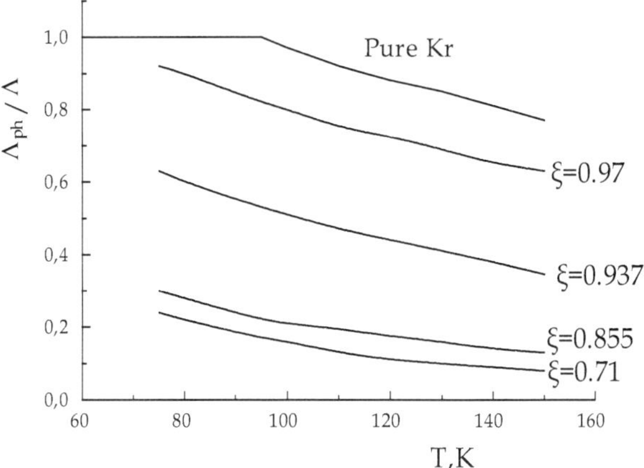

Fig. 18. The relative low-frequency phonon contribution to the thermal conductivity of pure Kr and $(CH_4)_{1-\xi}Kr_\xi$ solid solution.

3.6 The Bridgman coefficient g and molar volume dependence of thermal conductivity.

Subsequent studies of the thermal conductivity as a function of pressure (Ross et al., 1984) and as a function of density (Konstantinov et.al., 1988-2010) for a wide range of substances have shown that the values of the Bridgman coefficient vary, as a rule, in a range from 3–4 to 10–15. The general tendency is for g to decrease with increasing structural disorder; the weakest volume dependence of the thermal conductivity is found in glasses and polymers. In three cases — the Ih phase of ice, the phase NH_4F (I), and $CuCl$ — the Bridgman coefficient is negative. This is explained by anomalous behavior of the transverse modes, the velocity of which decreases with increasing pressure.

Values of the Bridgman coefficient g, measured at temperature T for some molecular substances are shown in Tab. 3. The problem of determining the Bridgman coefficient g in the present model reduces to finding the volume derivative of the expressions for Λ_{ph} and Λ_{dif}. Since $(\partial ln\Lambda/\partial lnV)_T = 3\gamma+2q-2/3$ we obtain

$$g = \frac{\Lambda_{ph}}{\Lambda}g_{ph} + \frac{\Lambda_{dif}}{\Lambda}g_{dif}\,, \tag{31}$$

where

$$g_{ph} = 5\gamma + 4q - 1, \tag{32a}$$

$$g_{dif} = -\left(\frac{\partial \ln \Lambda_{dif}}{\partial \ln V}\right)_T = -\gamma + \frac{1}{3} + \frac{2}{\omega_D^2 - \omega_0^2}(\omega_D^2 \gamma - \omega_0^2 \gamma_0), \qquad (32b)$$

$$\gamma_0 = 3\gamma + 2g - 1/3, \qquad (33)$$

Substance	T, K	g	Substance	T, K	g
Ar	90	9.7	HMT	300	8.9
Kr	120	9.4	Adamantane	320	6.4
Xe	160	9.2	Adamantane	300	9.8
CO_2	220	6.2	Cyclohexane	273	5.5
N_2O	180	5.7	Naphthalene	300	8.5
α-N_2	35	6.0	Anthracene	300	8.9
β-N_2	60	4.3	Sulphure	300	6.0
α-CO	60	5.2	NH_4Cl (II)	298	-6.2
β-CO	70	4.0	NH_4Cl (III)	160	8.6
γ-O_2	55	3.8	NH_4F (I)	298	-6.2
CH_4 (I)	90	8.8	NH_4F (II)	380	7.5
CCl_4 (Ib)	250	5.8	NH_4F (III)	386	18
CCl_4 (II)	225	6.5	H_2O (Ih)	120	-3.9
CBr_4 (I)	360	3.4	H_2O (VII)	286	4.8
CBr_4 (II)	300	3.8	H_2O (VIII)	246	4.8
CHF_3	118	4.6	C_2H_6	88	5.5
$CHCl_3$	210	3.9	C_2F_6	170	4.5
CH_2Cl_2	175	4.6	C_3H_8	85	7.5
CF_2Cl_2	115	5.0	C_6H_{14}	178	7.6
CHF_2Cl (I)	115	4.5			
C_6H_6	273	7.5			
SF_6 (I)	220	5.2			

Table 3. Values of the Bridgman coefficient g, measured at temperature T.

In our opinion, Eq. 31 has a general character. The main idea pursued in this paper is that the contributions to the molar volume dependence of the thermal conductivity from acoustic phonons and "diffusive" modes are sharply different. If the heat is transferred mainly by acoustic phonons (perfect crystals), then the Bridgman coefficient is described by Eq. 3. In the opposite case, when the thermal conductivity has reached its lower limit Λ_{min} and all the heat is transferred by "diffusive" modes (amorphous solids and strongly disordered crystals), then for $T>\Theta_D$ the lower limit to the thermal conductivity is described by $\Lambda_{min} \propto v/a^{2/3}$, and

$$g = \gamma + 1/3 \qquad (34)$$

This also follows from Eq. 32b in the limit $\omega_0 \to 0$. In the general case the Bridgman coefficient g is a weighted average over the acoustic and "diffusive" modes. Fig. 18 shows the Bridgman coefficients calculated for CO_2 and N_2O according to Eqs. 31-33. The temperature dependence of g was not investigated experimentally. For CO_2 and N_2O the Bridgman coefficients were determined only at the triple-point temperature. It is seen in Fig. 19 that the agreement between the experimental and computed values of g is completely satisfactory. It was also found for a number of molecular crystals and their solutions (Konstantinov et al., 2002b, 2003b).

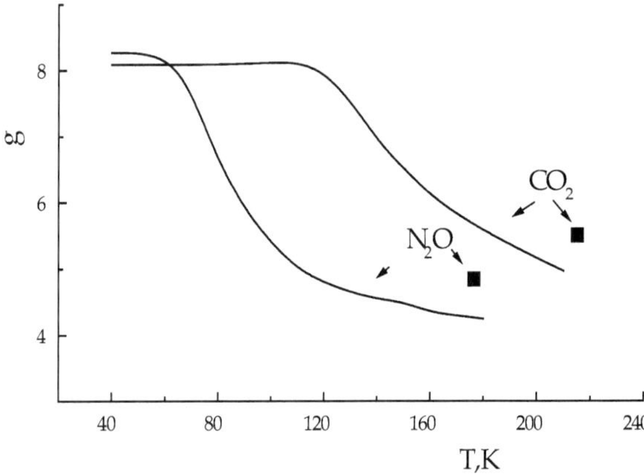

Fig. 19. Temperature dependence of the Bridgman coefficient g for solid CO_2 and N_2O calculated according to Eqs. 31-33. Dark squares are experimental values (Konstantinov et al., 1988b).

3.7 Lower limit to thermal conductivity and rotational energy transfer

It was found that the values of coefficient α for molecular crystals, which express the ratio of the lower limit to thermal conductivity obtained by fitting procedure to $\Lambda_{\min}$ calculated from Eq. 4 vary from 2 to 4 (Konstantinov et al., 2003). These values are much larger than for solidified inert gases, where α lies in the range 1.2-1.4. The most obvious reason for this difference is that the site-to-site rotational energy transfer must be taken into account. Eq. 4 for the lower limit of the thermal conductivity is valid for substances consisting of atoms other than molecules with rotational degrees of freedom. Slack, 1979 has taken into account the possibility of thermal energy transfer by optical phonons in crystals consisting of atoms of different kinds. In molecular crystals heat is transferred by mixed translational-orientational modes, whose specific heat for $T \geq \Theta_D$ saturates in proportion to the number of degrees of freedom. On this basis the following expression can be suggested for the lower limit of the thermal conductivity of molecular crystal Λ^*_{min} whose molecules have z rotational degrees of freedom.

$$\Lambda^*_{\min} = \frac{1}{2}\left(\frac{\pi}{6}\right)^{1/3}\left(1+\frac{z}{3}\right)k_B n^{2/3}(v_l + 2v_t),\tag{35}$$

It should be noted that although the Eq. 4 for Λ_{min} describes well, over all, the thermal conductivity of amorphous substances and strongly disordered crystals, it is nonetheless semi-empirical. The assumption that the minimum phonon mean-free path length is equal to half the wavelength is only one of many possibilities. Thus, Slack, 1979 assumed that it is equal to the phonon wavelength. In addition, the expression was obtained on the basis of a very simple Debye model which neglects phonon dispersion and the real density of states. The coefficient α is in such an event an integral factor that effectively takes account of the imperfection of the model.

3.8 Further problems

Up to now there is no consistent theory, describing thermal conductivity of molecular crystals from first principles proceeding from parameters of crystalline potential. Such description is only available for solid methane and deuteromethane (Yasuda, 1978). There is no absolute confidence that semi-empirical approach of Manzhelii et al., 1968 is completely adequate for describing of the thermal conductivity increase with rise of temperature in the case of hindered rotation of molecules, particularly in the case of one-axis rotation. Is there a clearly defined relationship between the frequency of molecular reorientations and the behavior of thermal conductivity?

What is the value of the lower limit to thermal conductivity in molecular crystals? Is there a need to allow for the site-to-site rotational energy transfer? There are not enough arguments for benefit of adequacy of Eq. 35 for describing the lower limit to thermal conductivity in molecular crystals.

There is a lack of data about the temperature dependence of Bridgman coefficient g in the wide temperature range to compare it with calculated values in the framework of the proposed model.

4. References

Auerbach, A. & Allen P.B., (1984). Universal high-temperature saturation in phonon and electron transport. *Physical Review B.*, Vol. 29, No. 6, pp. 2884-2890.

Batchelder, D.N. (1977). In *Rare Gas Solids*. Eds. Klein, M.L.; Venables, J.A., London, New York, San Francisco: Academic Press, Vol. II, 1977, pp. 883-920.

Berman, R. (1976). *Thermal Conduction in Solids*, Clarendon Press, Oxford (1976).

Bondarenko, A.I.; Manzhelii, V.G.; Popov, V.A.; Strzhemechnyi, M.A. & Gavrilko V.G. (1982). Isochoric thermal conductivity of crystalline *Kr* and *Xe*. Heat transfer by vacancies. *Soviet Journal Fizika Nizkikh Temperatur*, Vol. 8, No. 11, pp. 1215-1224.

de Bour, J. (1948). Quantum theory of condensed permanent gases I. Quantum-mechanical low of corresponding states. *Physica*, Vol. 14, No 1, pp. 139–148.

Briels, W.J.; Jansen, A.P.J. & van-der Avoird A. (1985). Translational-rotational coupling in strongly anharmonic molecular crystals with orientational disorder, *Chemie Physics et de Physico-Chemical Biologique*, Vol. 82, No. 2-3, pp. 125-136.

Cahill, D.G.; Watson, S.K. & Pohl R.O., (1992). Lower limit to thermal conductivity of disordered crystals. *Physical Review B.*, Vol. 46, No. 10, pp. 6131-6140.

Clayton, F. & Batchelder, D.N. (1973). Temperature and volume dependence of the thermal conductivity of solid argon. *Journal of Physics. C.*, Vol. 6, No. 7, pp. 1213-1228.

Ecsedy, D.J. & Klemens, P.G. (1977).Thermal resistivity of dielectric crystals due to four-phonon processes and optical modes. *Physical Review B.*, Vol. 15, No. 12, pp. 5957–5962.

Feldman, J.L.; Kluge, M.D.; Allen, P.B. & Wooten F. (1993). Thermal conductivity and localization in glasses: Numerical study of a model of amorphous silicon. *Physical Review B.*, Vol. 48, No. 17, pp. 12589-12602.

Forsman, H. & Andersson, P. (1984). Thermal conductivity at high pressure of solid odd numbered n-alkanes ranging from C_9H_{20} to $C_{19}H_{40}$. *Journal of Chemical Physics*, Vol. 80, No. 6, pp. 2804-2807.

Gorodilov, B.Ya.; Korolyuk O.A.; Krivchikov, A.I. & Manzhelii V.G. (2000). Heat Transfer in Solid Solutions Hydrogen-Deuterium. *Journal of Low Temperature Physics*, Vol. 119, No. 3/4, pp. 497-505.

Jeżowski, A.; Stachowiak, P.; Sumarokov V.V., Mucha J. & Freiman, Yu.A. (1993). Thermal conductivity of solid oxygen. *Physical Review Letters*, Vol. 71, pp. 97–100.

Jeżowski A.; Misiorek H.; Sumarokov V.V. & Gorodilov B.Ya. (1997). Thermal conductivity of solid methane. *Physical Review B.*, Vol. 55, No. 9, pp. 5578-5580.

Kawasaki K. (1963). On the behavior of the thermal conductivity near the magnetic transition point. *Progress of Theoretical Physics,* Vol. 29, No 6, pp. 801- 816.

Konstantinov, V.A.; Manzhelii, V.G.; Strzhemechnyi, M.A. & Smirnov, S.A. (1988a). The $\Lambda\propto 1/T$ law and isochoric thermal conductivity of rare gas crystals. *Soviet JournalFizika Nizkikh Temperatur*, Vol. 14, No. 1, pp. 48-54.

Konstantinov, V.A.; Manzhelii, V.G.; Tolkachev, A.M. & Smirnov, S.A. (1988b). Heat transfer in solid CO_2 and N_2O: dependence on the temperature and volume. *Soviet Journal Fizika Nizkikh Temperatur* Vol. 14, No. 2, pp. 412-417.

Konstantinov, V.A.; Manzhelii,V.G. & Smirnov, S.A. (1988c). Isochoric thermal conductivity of solid CO_2 with N_2O and Xe impurities. *Soviet Journal Fizika Nizkikh Temperatur*, Vol. 14, No. 7, pp. 412-413.

Konstantinov, V.A.; Manzhelii, V.G. & Smirnov, S.A. (1991a). Isochoric thermal conductivity and thermal pressure of solid CCl_4. *Physica Status Solidi (b)*. Vol. 163, pp. 368-374.

Konstantinov, V.A.; Manzhelii,V.G. & Smirnov, S.A. (1991b). Isochoric thermal conductivity of solid $CHCl_3$ and CH_2Cl_2. The role of rotational motion of molecules. *Low Temperature Physics*, Vol. 17, No. 7, pp. 462-464.

Konstantinov, V.A.; Manzhelii,V.G. & Smirnov, S.A. (1992a). Temperature dependence of isochoric thermal conductivity of crystalline benzene. *Ukrainian Physical Journal*, Vol. 37, No. 5, pp. 757-760.

Konstantinov, V.A.; Manzhelii,V.G. & Smirnov, S.A. (1992b). Influence of rotational motion of molecules on heat transfer in solid SF_6. *Low Temperature Physics*, Vol. 18, No.11 pp. 902-904.

Konstantinov, V.A. & Manzhelii, V.G. (1994). Phonon scattering and heat transfer in simple molecular crystals. In *"Die Kunst of Phonons"* ed. by T. Paskiewizc, T. Rapsewizc., New York, London: Plenum Press. 1994. pp. 321-332.

Konstantinov, V.A.; Manzhelii, V.G. & Smirnov, S.A. (1995). Isochoric thermal conductivity of solid freons of methane series: CF_2Cl_2 and CHF_2Cl. *Low Temperature Physics*, Vol. 21, No. 1. pp. 78-81.

Konstantinov, V.A.; Manzhelii, V.G.; Revyakin, V.P. & Smirnov, S.A. (1999a). Heat transfer in the orientationally disordered phase of solid methane. *Physica B.* Vol. 262. pp.421-425.

Konstantinov, V.G.; Revyakin, V.P. & Smirnov, S.A. (1999b). A setup for measuring of isochoric thermal conductivity of solidified gases and liquids. *Instruments and Experimental Techniques.* Vol. 42, No. 1, pp. 132-135.

Konstantinov V.A. (1999c). Heat transfer in simple molecular crystals at temperatures of the order of Debye ones and above them. *Functional Materials*,Vol. 6, No. 2, pp. 335–339.

Konstantinov, V.A.; Manzhelii, V.G.; Revyakin, V.P. & Smirnov, S.A. (2000). Manifestation of lower limit to thermal conductivity of solid Kr with CH_4 impurities. *Physica B.*, Vol. 291, pp. 59-65.

Konstantinov, V.A. Manifestation of the lower limit to thermal conductivity in the solidified inert gases. (2001a). *Journal of Low Temperature Physics.* Vol. 122, No. 3/4, pp. 459-465.

Konstantinov, V.A.; Manzhelii, V.G.; Pohl, R.O. & Revyakin V.P. (2001b). Search for the minimum thermal conductivity in mixed cryocrystals $(CH)_{1-\xi}Kr_\xi$. *Low Temperature Physics*, Vol. 27, No. 1. pp. 1159-1169.

Konstantinov, V.A.; Pohl, R.O. & Revyakin V.P. (2002a). Heat transfer in solid $Kr_{1-\xi}Xe_\xi$ solution. *Soviet Physics of Solid State*, Vol. 44, No. 5, pp. 824-829.

Konstantinov, V.A.; Revyakin, V.P. & Orel, E.S. (2002b). Molar volume dependence of thermal conductivity in mixed cryocrystals. *Low Temperature Physics*, Vol. 28, No. 2. pp. 136-139.

Konstantinov, V.A.; Revyakin, V.P. & Orel, E.S. (2003a). Heat transfer by low-frequency phonons and „diffusive" modes in cryocrystal solutions: the Kr-Xe system. *Low Temperature Physics*, Vol. 29, No. 9-10. pp. 759-762.

Konstantinov, V.A. (2003b). Heat transfer by low-frequency phonons and „diffusive" modes in molecular crystals. *Low Temperature Physics*, Vol. 29, No. 5. pp. 422-428.

Konstantinov, V.A.; Manzhelii, V.G.; Revyakin, V.P. & Sagan V.V. (2005a). Heat transfer in γ-phase of oxygen. *Journal of Low Temperature Physics*, Vol. 139, No. 5-6. pp. 703-709.

Konstantinov, V.A.; Revyakin, V.P.; Manzhelii V.G. & Sagan V.V. (2005b). Isochoric thermal conductivity of solid nitrogen. *Low Temperature Physics*, Vol. 31, No. 5. pp. 419-422.

Konstantinov, V.A.; Manzhelii, V.G.; Revyakin, V.P. & Sagan V.V. (2006a). Isochoric thermal conductivity of solid carbon oxide: the role of phonons and "diffusive" modes. *Journal of Physics: Condensed Matter*, Vol. 18, pp. 9901-9909.

Konstantinov, V.A.; Manzhelii, V.G.; Revyakin, V.P. & Sagan V.V. (2006b). Extraordinary temperature dependence of isochoric thermal conductivity of crystalline CO_2 doped with inert gases. *Low Temperature Physics*, Vol. 32, No. 11, pp. 1076-1077.

Konstantinov, V.A.; Revyakin, V.P. & Sagan V.V. (2006c). Rotation of the methyl groups and thermal conductivity of molecular crystals: ethane. *Low Temperature Physics*, Vol. 32, No. 7. pp. 689-694.

Konstantinov, V.A.; Revyakin, V.P. & Sagan V.V. (2009a). Heat transfer in solid halogenated methanes: trifluoromethane. *Low Temperature Physics*, Vol. 35, No. 4. pp. 286-289.

Konstantinov, V.A.; Revyakin, V.P. & Sagan V.V. (2009b).The isochoric thermal conductivity of solid n-alkanes: propane C_3H_8. *Low Temperature Physics*, Vol. 35, No. 7. pp. 577-579.

Konstantinov, V.A.; Revyakin, V.P. & Sagan V.V. (2010). Isochoric thermal conductivity of solid n-alkanes: hexane C_6H_{14}. *Low Temperature Physics*, Vol. 36, (in press).

Kokshenev, V.B.; Krupskii I.N. & Kravchenko Yu.G. (1997). High-temperature thermoconductivity of nitrogen-type crystals. *Brazilian Journal of Physics*, Vol. 27, No. 4, pp. 510-514.

Krupski, I.N. & Manzhelii V.G. (1968). Multi phonon interaction and thermal conductivity of crystalline argon, krypton and xenon. *Soviet Journal for Experimental and Theoretical Physics.* Vol. 55, No. 6, pp. 2075-2082.

Krupskii, I.N.; Koloskova L.A. & Manzhelii V.G. (1974). Thermal conductivity of deuteromethane. *Journal of Low Temperature Physics.* Vol. 14, No. 3 4, pp. 103 110.

Lynden-Bell, R.M. & Michel K.H. (1994). Translation-rotation coupling, phase transitions, and elastic phenomena in orientationally disordered crystals. *Review of Modern Physics*, Vol. 66, No. 3, pp. 721-762.

McGaughey, A.J.H. & Kaviany, M. (2004). Thermal conductivity decomposition and analysis using molecular dynamics simulations. Part I. Lennard-Jones argon. *International Journal of Heat and Mass Transfer*, Vol. 47, pp. 1783–1798.

Manzhelii, V.G. & Krupski, I.N. (1968). Thermal conductivity of solid methane. *Soviet Journal Physics of Solids*. Vol. 10, No 1, pp. 284-286.

Manzhelii V.G.; Kokshenev V.B.; Koloskova L.A. & Krupski I.N. (1975). Phonon-libron interaction and thermal conductivity of simplest molecular crystals. *Soviet Journal Fizika Nizkich Temperetur*, Vol. 1, No. 10, pp. 1302-1310.

Manzhelii V.G. & Freiman Yu.A. (1997). In: *Physics of cryocrystals*. Eds. Manzhelii, V.G. & Freiman Yu. A., Woodbury, New York: AIP PRESS. pp. 1-514.

Parsonage, N. & Stavaley, L. (1972). *Disorder in crystals*, Clarendon Press, Oxford.

Pursky, O.I.; Zholonko N.N. & Konstantinov, V.A. (2003). Influence of the rotational motion of molecules on the thermal conductivity of solid SF_6, $CHCl_3$, C_6H_6, and CCl_4. *Low Temperature Physics*, Vol. 29, No. 9/10, pp. 771-775.

Pursky, O.I. & Konstantinov V.A. (2008). Thermal conductivity of solid CHF_2Cl. *Physica B.*, Vol. 403, pp. 190-194.

Purski, O.I.; Zholonko, N.N. & Konstantinov V.A. (2000). Heat transfer in the orientationally disordered phase of SF_6. *Low Temperature Physics*, Vol. 21, No. 1. pp. 278- 281.

Ross, R.G.; Andersson, P.; Sundqvist, B. & Bäckström G. (1984). Thermal conductivity of solids and liquids under pressure. *Report of Progress in Physics*, Vol. 47, pp. 1347–1355.

Roufosse, M.C. & Klemens P.G. (1974). Lattice thermal conductivity of minerals at high temperatures. *Journal of Geophysical Researches*, Vol. 79, No. 5, pp. 703-705.

Sheng, P.; Zhou M. & Zhang Z. (1994). Phonon transport in strong-scattering media. *Physical Review Letters*, Vol. 72, No. 2, pp. 234-237.

Slack, G.A. (1972). In: *Proceedings of International Conference on phonon scattering in solids*, Eds. Albany H.J., Service de documentation du CEN, Saclay, pp. 24-32.

Slack, G.A. (1979). The thermal conductivity of nonmetallic crystals. *Solid State Physics*. Vol. 34, pp. 1-71.

Stachowiak, P.; Sumarokov V.V.; Mucha J. & Jezowski A. (1994). Thermal conductivity of solid nitrogen. *Physical Review B.*, Vol. 50, No. 1, pp. 543-546.

Yasuda H. (1978). Thermal conductivity of solid CH_4 and CD_4. *Journal of Low Temperature Physics*, Vol. 31, No. 1/2, pp. 223-256.

Nonlinear Bubble Behavior due to Heat Transfer

Ho-Young Kwak
Mechanical Engineering Department,
Chung-Ang University
Korea

1. Introduction

Previous studies of the forced oscillation of a spherical bubble in solution have been investigated by using the Rayleigh equation to obtain the time dependent bubble radius and a polytropic relation to obtain the gas pressure inside the bubble depending bubble volume (Lauterborn, 1976). In fact, the polytropic approximation with proper index values has been widely used for the gas undergoing quasi-equilibrium process in which there is heat transfer. However, the polytropic pressure-volume relationship fails to account the thermal damping due to heat transfer through the bubble wall because $P_b \, dV_b$ is a perfect differential and consequently its integral over a cycle vanishes (Prosperiretti et al., 1988) where P_b is the gas pressure inside the bubble and V_b is the bubble volume. Furthermore, the polytropic approximation assumes the uniform temperature for the gas intrinsically, which is valid only for a particular case and it is hard to tell whether the gas inside the bubble oscillating under ultrasound behaves isothermally or adiabatically (Loefstedt et al., 1993).

In this study, we have formulated a general bubble dynamics model, which is as follows. The density, velocity and pressure distributions for the gas inside a spherical bubble were obtained by solving the continuity and momentum equations analytically (Kwak et al., 1995, Kwak and Yang, 1995). With the set of analytical solutions for the conservation equations, the temperature distribution for the gas inside the bubble was also obtained by solving the energy equation for the gas. The heat transfer through the bubble wall was considered to obtain the instantaneous thermal boundary layer thickness from the mass and energy conservations for the liquid layer adjacent to the bubble wall by the integral method. The mass and momentum equations for the liquid outside the bubble wall provided the well known equation of motion for the bubble wall, the Rayleigh-Plesset equation in an incompressible limit or the Keller-Miksis equation in a compressible limit. The bubble dynamics model was applied to an evolving bubble formed form the fully evaporated droplet at the superheat limit (Kwak et al., 1995) and phenomena of sonoluminescence which is light emission associated with the catastrophic collapse of a micro-bubble oscillation under ultrasound (Young, 2005).

With uniform density, temperature and pressure approximations which are valid for the characteristic time scale of ms, the calculated values of the far field pressure signal from the evolving the bubble formed form the fully evaporated droplet at its superheat limit (Kwak et al., 1995) are in good agreement with the experimental results (Shepherd and Sturetevant, 1982). With uniform pressure approximation which is valid for the characteristic time scale of μs, the calculated values of the minimum velocity of the bubble wall, the peak temperature and pressure are excellent agreement with the observed ones for the sonoluminescing xenon bubble in sulfuric acid solutions (Kim et al., 2006). Furthermore, the calculated bubble radius-time curve displays alternating pattern of bubble motion which is apparently due to the heat transfer for the sonoluminescing xenon bubble, as observed in experiment (Hopkins et al., 2005). The bubble dynamics model presented in this study has also revealed that the sonoluminescence for an air bubble in water solution occurs due to the increase and subsequent decrease in the bubble wall acceleration which induces pressure non-uniformity for the gas inside the bubble during ns range near the collapse point (Kwak and Na, 1996). The calculated sonoluminescence pulse width from the instantaneous gas temperature for air bubble is in good agreement with the observed value of 150 ps (Byun et al., 2005). Due to enormous heat transfer the gas temperature inside the sonoluminescing air bubble at the collapse point is about 20000~40000 K instead of 10^7 K (Moss et al., 1994) which is estimated to be in the adiabatic case. Molecular dynamics (MD) simulation results for the sonoluminescing xenon bubble were compared to the theoretical predictions and observed results (Kim et al., 2007, Kim et al., 2008).

2. Temperature profile in thermal boundary layer

A sketch of the bubble model employed is given in Fig.1, which shows a spherical bubble in liquid temperature T_∞ and liquid pressure P_∞. Heat transfer is assumed to occur through the thermal boundary layer of thickness $\delta(t)$. The temperature profile in this layer is assumed to be quadratic (Theofanous et al., 1969).

$$\frac{T - T_\infty}{T_{bl} - T_\infty} = (1 - \xi)^2 \tag{1}$$

where T_{bl} is the temperature at the bubble wall and T_∞ is the ambient temperature in Eq.(1). The parameter ξ in Eq.(1) is given as $\xi = (r - R_b)/\delta$ and $R_b(t)$ is the instantaneous bubble radius. Such a second order curve satisfies the following boundary conditions:

$$T(R_b, t) = T_{bl}, \quad T(R_b + \delta, t) = T_\infty \quad \text{and} \quad \left(\frac{\partial T}{\partial r} \right)_{r = R_b + \delta} = 0 \tag{2}$$

The heat transfer conducted through this thermal boundary layer whose thickness is $\delta(t)$ can be obtained by applying the Fourier law at the bubble wall, or

$$\dot{Q}_b = k_l A_b \left(\frac{\partial T}{\partial r} \right)_{r = R_b} = - \frac{8 \pi R_b^2 k_l (T_{bl} - T_\infty)}{\delta} \tag{3}$$

where A_b is the surface area of bubble and k_l is the conductivity of liquid. The bubble model including such liquid phase zone has been verified experimentally (Suslick et al., 1986).

3. Conservation equations for the gas inside bubble

The hydrodynamics related to studying the bubble behavior in liquid involves solving the Navier-Stokes equations for the gas inside the bubble and the liquid adjacent the bubble

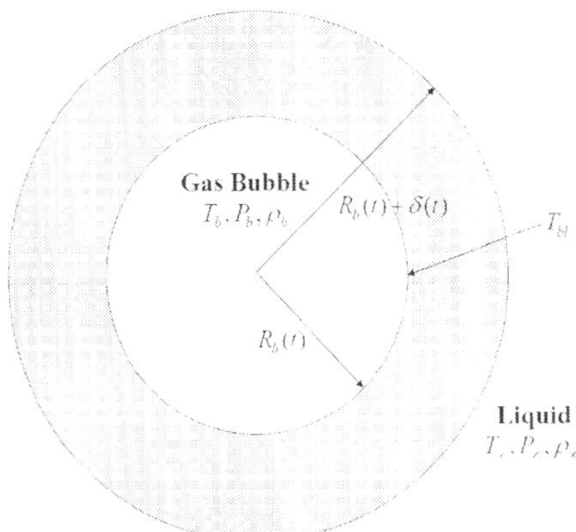

Fig. 1. A physical model with thermal boundary layer for a spherical bubble in liquid.

wall. Especially the knowledge of the behavior of gas or vapor inside evolving bubble is a key element to understand the bubble dynamics. Firstly, various conservation laws for the gas are considered to obtain the density, pressure and temperature distributions for the gas inside the bubble.

3.1 Mass conservation

The mass conservation equation for the gas inside a bubble is given by

$$\frac{D\rho_g}{Dt} + \rho_g \nabla \cdot \vec{u}_g = 0 \tag{4}$$

where ρ_g and u_g are gas density and velocity, respectively. With decomposition of the gas density in spherical symmetry as $\rho_0(t) + \rho_r(r,t)$, the continuity equation becomes

$$\left[\frac{d\rho_0}{dt} + \rho_0 \nabla \cdot \vec{u}_g\right] + \frac{D\rho_r}{Dt} + \rho_r \nabla \cdot \vec{u}_g = 0 \tag{5}$$

where ρ_0 is the gas density at the bubble center and ρ_r is the radial dependent gas density inside the bubble and the notation of the total derivative used here is $D/Dt = \partial/t + u(\partial/\partial r)$. The rate of change of the density of a material particle can be represented by the rate of volume expansion of that particle in the limit $V \to 0$ (Panton, 1996). Or

$$u_g = \frac{\dot{R}_b}{R_b} r \tag{6}$$

With this velocity profile, the density profile can be obtained as

$$\rho_g = \rho_0 + \rho_r \tag{7-1}$$

$$\rho_0 R_b^3 = \text{const.} \tag{7-2}$$

$$\rho_r = ar^2 / R_b^5 . \tag{7-3}$$

The constant, a is related to the gas mass inside a bubble (Kwak and Yang, 1995).

3.2 Momentum conservation

The momentum equation for the gas when neglecting viscous forces may be written as

$$\rho_g \frac{D\bar{u}_g}{Dt} = -\nabla P_b \tag{8}$$

The gas pressure P_b inside bubble can be obtained from this equation by using the velocity and density profiles given in Eq. (6) and (7), respectively. Or

$$P_b = P_{b0} - \frac{1}{2}\left(\rho_0 + \frac{1}{2}\rho_r \right) \frac{\ddot{R}_b}{R_b} r^2 \tag{9}$$

Note that the linear velocity profile showing the spatial inhomogeneities inside the bubble is a crucial ansatze for the homologous motion of a spherical object, which is interestingly encountered in another energy focusing mechanism of gravitational collapse (Jun and Kwak, 2000). The quadratic pressure profile given in Eq. (9), was verified recently by comparison with direct numerical simulations (Lin et al., 2002).

3.3 Energy conservation

Assuming that the internal energy for the gas inside a bubble is a function of gas temperature only as de = $C_{v,b}dT_b$, the energy equation for the gas inside the bubble may be written as

$$\rho_g \frac{De}{Dt} = \rho_g C_{v,b} \frac{DT_b}{Dt} = -P_b \nabla \cdot \bar{u}_g - \nabla \cdot \bar{q} \tag{10-1}$$

where Cv,b is the constant-volume specific heat and q is heat flux. The viscous dissipation term in the internal energy equation also vanishes because of the linear velocity profile. Since the solutions given in Eqs. (6), (7) and (9) also satisfy the kinetic energy equation, only the internal energy equation given in Eq. (10-1) needs to be solved. On the other hand, Prosperetti et al.(1988) solved the internal energy equation combined with the mass and momentum equation numerically to consider heat transport inside the bubble using a simple assumption. However, heat transfer through the liquid layer, which is very important in obtaining the temperature at the bubble wall, was not considered in their study.

Using the definition of enthalpy, the internal energy equation for the gas can also be written as

$$\rho_g \frac{Dh}{Dt} = \rho_g C_{p,b} \frac{DT_b}{Dt} = \frac{DP_b}{Dt} - \nabla \cdot \bar{q} \tag{10-2}$$

where $C_{p,b}$ is the constant-pressure specific heat. Eliminating DT_b/Dt from Eqs. (10-1) and (10-2), one can obtain the following heat flow rate equation for the gas pressure inside the bubble (Kwak et al., 1995, Kwak and Yang, 1995)

$$\frac{DP_b}{Dt} = -\gamma P_b \nabla \cdot \vec{u}_g - (\gamma - 1)\nabla \cdot \vec{q} \tag{11}$$

Rewriting Eq. (11), we have

$$(\gamma - 1)\nabla \cdot \vec{q} = -\frac{1}{R_b^{3\gamma}}\frac{D}{Dt}(P_b R_b^{3\gamma}) \tag{12}$$

which implies that the relation $P_b V^{3\gamma} =$ const. holds if $\nabla \cdot q = 0$ inside the bubble. Substituting Eq. (12) into Eq. (10), and rearranging the equation, we have

$$\frac{D}{Dt}\left\{\ln\left(\frac{P_b R_b^3}{T_b}\right)\right\} = \left(\frac{\rho_g R_g}{P_b} - \frac{1}{T_b}\right)\frac{DT_b}{Dt} \tag{13}$$

where R_g is gas constant. If the equation of state for ideal gas, $P_{bo} = \rho_g R_g T_{bo}$ holds at the bubble center, LHS in Eq. (13) vanishes. The result can be written as

$$P_{bo} R_b^3 / T_{bo} = const., \tag{14}$$

which is consistent with Eq.(7-2).
Note that the time rate change of the pressure at the bubble center can be written as with help of Eqs. (3) and (11).

$$\frac{dP_{bo}}{dt} = -\frac{3\gamma P_{bo}}{R_b}\frac{dR_b}{dt} - \frac{6(\gamma - 1)k_l(T_{bl} - T_\infty)}{\delta R_b} \tag{15}$$

The time rate change of the temperature at the bubble center can be obtained with help of Eq. (14). That is

$$\frac{dT_{bo}}{dt} = -\frac{3(\gamma - 1)T_{bo}}{R_b}\frac{dR_b}{dt} - \frac{6(\gamma - 1)k_l(T_{bl} - T_\infty)}{\delta R_b P_{bo}} \tag{16}$$

4. Temperature profiles inside the bubble

4.1 Uniform pressure profile inside the bubble

A temperature profile can be obtained by solving Eq. (12) with the Fourier law by assuming that the conductivity of gas inside the bubble is constant and the gas pressure inside the bubble is uniform (Kwak et al., 1995). That is

$$T_b = \frac{r^2}{6(\gamma - 1)k_g}\left[\frac{dP_b}{dt} + 3\gamma P_b \frac{\dot{R}_b}{R_b}\right] + T_{bo}(t) \tag{17}$$

where k_g heat conductivity of gas inside the bubble. The above equation can be written, with help of Eq.(12), as follows:

$$T_b = (T_{bo} - T_\infty)\left[1 - \left(\frac{r}{R_b}\right)^2 / (1 + \eta)\right] + T_\infty \tag{18}$$

where $\eta = \dfrac{k_g}{k_l}\dfrac{\delta}{R_b}$.

The temperature at the bubble wall can be obtained easily from the above equation. That is

$$T_{bl} = \left(T_\infty + \eta T_{bo}\right)/\left(1+\eta\right) \tag{19}$$

The above relation shows how the bubble wall temperature is related to the temperature at the bubble center and the ambient temperature. Assigning an arbitrary value on T_{bl} is not permitted as a boundary condition.

An uniform temperature distribution also occurs when there is no heat flux inside the bubble. This can be achieved when the bubble oscillating period is much longer than the characteristic time of heat diffusion so that the gas distribution function depends only on thermal velocity (thermal equilibrium case). In this limit, we may obtain the gas temperature inside the bubble by taking the value of the gas conductivity as infinity in Eq. (19). That is $T_b=T_{bl}=T_{bo}$, which validates the bubble dynamics formulation with an assumption of uniform vapor temperature inside the bubble (Kwak et al., 1995). The heat transfer through the thermal boundary layer adjacent to the bubble wall determines the heat exchange between the bubble and medium in this case.

However, the temperature gradient inside the bubble should exist, provided that characteristic time of bubble evolution is much shorter than the relaxation time of the vibration motion of the gases inside the bubble, which is of the order on 10^{-6}s for high gas temperature. If the temperature gradient inside the bubble exists inside the bubble, the heat transfer through the bubble wall depends on both the properties of the gas inside the bubble and the liquid in the thermal boundary layer. In this case one may rewrite Eq.(3), with help of Eq.(19). That is

$$\dot{Q}_b = -\dfrac{2A_b k_g (T_{bo} - T_\infty)}{R_b}/\left(1+\eta\right) \tag{20}$$

As long as the value of η is finite, there exists a temperature distribution inside the bubble. For a very small value of η, the heat flow rate from the bubble is solely determined from the temperature gradient of the gas inside the bubble (Prosperetti et al., 1988).

Assume the thermal conductivity for the gas inside the bubble is linearly dependent on the gas temperature such as

$$k_g = AT_b + B \tag{21}$$

For air $A=5.528\times10^{-5}$ W/mK^2 and $B=1.165\times10^{-2}$W/mK (Prosperetti et al., 1988) and for xenon, $A=1.031\times10^{-5}$ W/mK^2 and $B=3.916\times10^{-3}$ W/mK were used. With this approximation and Fourier law, one can obtain the following temperature profile by solving Eq. (12) with uniform pressure approximation, which is quite good until the acceleration and deceleration of the bubble wall is not significant. Thus

$$T_b(r) = \dfrac{B}{A}\cdot\left[-1 + \sqrt{(1+\dfrac{A}{B}T_{b0})^2 - 2\dfrac{A}{B}(T_{bl} - T_\infty)(\dfrac{r}{R_b})^2 / \eta^{'}}\,\right] \tag{22}$$

where $\dfrac{1}{\eta} = \dfrac{B}{k_l}\dfrac{\delta}{R_b}$.

The temperature distribution is given in Eq. (22) is valid until the characteristic time of bubble evolution is of the order of the relaxation time for vibrational motion of the molecules (Vincenti and Kruger, 1965) and/or is much less than the relaxation time of the translational motion of the molecules (Batchelor, 1967). The temperature at the bubble wall T_{bl} can also be obtained with the thermal boundary conditions given in Eq.(2). That is

$$T_{bl} = -\frac{B}{A}(1+\frac{1}{\eta}) + \frac{B}{A}\sqrt{\left(1+\frac{1}{\eta}\right)^2 + 2\frac{A}{B}\left(T_{bo} + \frac{A}{2B}T_{bo}^2 + \frac{T_\infty}{\eta}\right)} \tag{23}$$

For constant gas conductivity limit, or A→0 and B→ k_g the temperature distribution inside the bubble, Eq. (22), reduces to Eq. (18).

4.2 Non-uniform pressure profile

If the bubble wall acceleration has significant value, for example, the value exceed 10^{12} m/s^2, the second term is comparable to the first term in RHS of Eq.(9). This occurs for the sonoluminescing air bubble during few nanoseconds of collapse phase. Taking into account the bubble wall acceleration, the heat flow rate equation given in Eq. (12) may be rewritten as with help of Eqs. (6), (7) and (9)

$$(\gamma - 1)\nabla \cdot \vec{q} = -\left[\frac{dP_b}{dt} + 3\gamma P_{bo}\frac{\dot{R}}{R_b}\right] + \frac{1}{2}(\rho_o + \rho_r)\left[(3\gamma - 1)\frac{\dot{R}\ddot{R}}{R_b} + \frac{\dddot{R}}{R_b}\right]r^2 \tag{24}$$

Since the temperature rise due to the bubble wall acceleration is transient phenomenon occurred during few nanoseconds around the collapse point of the bubble, the above equation may be decomposed into

$$(\gamma - 1)\nabla \cdot \vec{q}_o = -\left[\frac{dP_b}{dt} + 3\gamma P_{bo}\frac{\dot{R}}{R_b}\right] \tag{25-1}$$

and

$$(\gamma - 1)\nabla \cdot (\vec{q} - \vec{q}_o) = \frac{1}{2}(\rho_o + \rho_r)\left[(3\gamma - 1)\frac{\dot{R}\ddot{R}}{R_b} + \frac{\dddot{R}}{R_b}\right]r^2 \tag{25-2}$$

Abrupt temperature rise and subsequent rapid quenching due to the bubble wall acceleration and the increase and decrease in the acceleration may be treated in another time scale (Davidson, 1972), different from the bubble wall motion. A solution of Eq.(25-2) with no temperature gradient at the bubble center is given as

$$T_b'(r) = -\frac{1}{40(\gamma - 1)k_g'}\left(\rho_o + \frac{5}{21}\rho_r\right)\left[(3\gamma - 2)\frac{\dot{R}_b\ddot{R}_b}{R_b} + \frac{\dddot{R}_b}{R_b}\right]r^4 + C(t) \tag{26}$$

The coefficient C may be determined from a boundary condition $k_g dT_b/dr = k_l dT_l/dr$ at the wall where T_l is the temperature distribution in the thermal boundary layer with different thickness δ'. Note that the boundary conditions employed for solving Eq. (25-1) and (25-2) are the same at the bubble center. However different properties of the gas was employed at the bubble wall so that the coefficient C(t) given in Eq. (26) is given as

$$C(t) = -\frac{1}{20(\gamma-1)}\left[(3\gamma-2)\dot{R}_b\ddot{R}_bR_b + \ddot{R}_bR_b^2\right]\left[\frac{\delta'}{k_l}\left(\rho_0 + \frac{5}{14}\rho_{r=R_b}\right) + \frac{R_b}{2k_g'}\left(\rho_0 + \frac{5}{21}\rho_{r=R_b}\right)\right] \quad (27)$$

The temperature distribution from Eq. (22) with low thermal conductivity k_g can be regarded as background one because the duration of the thermal spike represented by Eq. (26) is so short less than 500 ps. The gas conductivity at ultra high temperature k_g' may be obtained from collision integrals (Boulos et al., 1994). The value of δ' may be chosen so that proper bouncing motion results after the collapse and is about 0.1 µm. The final solution of the heat transport equation can be represented by the superposition of the temperature distributions caused by the uniform pressure and by the radial pressure variation induced by the rapid change of the bubble wall acceleration, as can be seen in equation (28); that is,

$$T(r) = T_b(r) + T_b'(r) \quad (28)$$

5. Navier-stokes equation for the liquid adjacent to the bubble wall

5.1 Bubble wall motion

The mass and momentum equation for the liquid adjacent bubble wall provides the well-known equation of motion for the bubble wall (Keller and Miksis, 1980), which is valid until the bubble wall velocity reaches the speed of sound of the liquid. That is

$$R_b\left(1 - \frac{U_b}{C_b}\right)\frac{dU_b}{dt} + \frac{3}{2}U_b^2\left(1 - \frac{U_b}{3C_b}\right) = \frac{1}{\rho_\infty}\left(1 + \frac{U_b}{C_b} + \frac{R_b}{C_b}\frac{d}{dt}\right)\left[P_B - P_s\left(t + \frac{R_b}{C_b}\right) - P_\infty\right] \quad (29)$$

The liquid pressure on the external side of the bubble wall P_B is related to the pressure inside the bubble wall P_b according to:

$$P_B = P_b - \frac{2\sigma}{R_b} - 4\mu\frac{U_b}{R_b} \quad (30)$$

The driving pressure of the sound field P_s may be represented by a sinusoidal function such as

$$P_s = -P_A \sin \omega t \quad (31)$$

where $\omega = 2\pi f_d$.

The Keller-Miksis equation reduces to the well known Rayleigh equation which is valid at the incompressible limit without forcing field (Batchelor, 1967). That is

$$R_b\frac{dU_b}{dt} + \frac{3}{2}U_b^2 = \frac{1}{\rho_\infty}\left(P_B - P_\infty\right) \quad (32)$$

5.2 Thermal boundary layer thickness

The mass and energy equation for the liquid layer adjacent to the bubble wall with the temperature distribution given in Eq. (1) provides a time dependent first order equation for the thermal boundary layer thickness (Kwak et al., 1995, Kwak and Yang, 1995). It is given by

$$\left[1+\frac{\delta}{R_b}+\frac{3}{10}\left(\frac{\delta}{R_b}\right)^2\right]\frac{d\delta}{dt}=\frac{6\alpha}{\delta}-\left[2\frac{\delta}{R_b}+\frac{1}{2}\left(\frac{\delta}{R_b}\right)^2\right]\frac{dR_b}{dt}$$
$$-\delta\left[1+\frac{1}{2}\frac{\delta}{R_b}+\frac{1}{10}\left(\frac{\delta}{R_b}\right)^2\right]\frac{1}{T_{bl}-T_\infty}\frac{dT_{bl}}{dt}$$

(33)

The above equation which was discussed in detail by Kwak and Yang (1995) determines the heat flow rate through the bubble wall. The instantaneous bubble radius, bubble wall velocity and acceleration, and thermal boundary thickness obtained from Eqs. (29) and (33) provide the density, velocity, pressure and temperature profiles for the gas inside the bubble with no further assumptions. The gas temperature and pressure at the bubble center can be obtained from Eqs. (15) and (16), respectively.

The entropy generation rate in this kind of oscillating bubble-liquid system, which induces lost work for bubble motion needs to be calculated by allowing for the rate change of entropy for the gas inside the bubble and the net entropy flow out of the bubble as results of heat exchange (Bejan, 1988). That is

$$\dot{S}_g=\frac{DS_b}{Dt}-\frac{\dot{Q}_b}{T_\infty}$$

(34)

6. Calculated examples

6.1 An evolving bubble formed from the fully evaporated droplet at its superheat limit - Uniform temperature and pressure distributions for the vapor inside the bubble

It is well known that one may heat a liquid held at 1 atm to a temperature far above its boiling point without occurrence of boiling. The maximum temperature limit at which the liquid boils explosively is called the superheat limit of liquid (Blander and Katz, 1975). It has been verified experimentally that, when the temperature of a liquid droplet in an immiscible medium reaches its superheat limit at 1 atm, the droplet vaporizes explosively without volume expansion and the fully evaporated droplet becomes a bubble (Shepherd and Sturtevant, 1982). Since the internal pressure of the fully evaporated droplet is very large (Kwak and Panton, 1985, Kwak and Lee, 1991), the droplet expands spontaneously. At the initial stage of this process, the fully evaporated droplet expands linearly with time. However, its linear growing fashion slows down near the point where the nonlinear growing starts. The pressure inside the bubble may be taken as the vapor pressure given temperature with saturated vapor volume at the start of the nonlinear growing. Since the vapor pressure inside the bubble is still much greater than the ambient pressure, the bubble expands rapidly so that it overshoots the mechanical equilibrium condition and its size oscillates. In this case, since the temperature of the vapor inside the bubble is so low that vibrational motion of the vapor is not excited and the characteristic time of bubble evolution of ms range is much longer than the relaxation time of the translational motion of vapor molecules, uniform temperature and pressure distribution for the vapor molecules inside the bubble are achieved (Kwak et al., 1995).

The calculated pressure wave signal from the evolving butane bubble in ethylene glycol at the ambient pressure of 1 atm and at a temperature of 378 K is shown in Fig. 2, together with the observed data (Shepherd and Sturtevant, 1982). In this case heat transfer occurs through the thermal boundary layer. Thermal damping due to finite heat transfer (Moody, 1984) is barely seen in this Figure. In Fig. 3(a), the time rate change of the vapor temperature during the bubble evolution is shown. As can be seen in this Figure,the bubble evolution is neither isothermal nor adiabatic. In Fig. 3(b), the entropy generation rate due to finite heat transfer for the evolving butane bubble is shown. As expected, the entropy generation during the bubble oscillation is always positive. More clear thermal damping can be observed from the far-field pressure signal of the evaporating droplet and evolving bubble formed from a cyclohexane droplet at its droplet, 492.0 K (Park et al., 2005) as shown in Fig.4. After first two volume oscillations, the original bubble has begun to disintegrate into a cloud of bubbles so that the far-field pressure signal becomes considerably smaller compared to the calculation results.

The far field pressure signal from the evolving bubble at a distance r_d from the bubble center can be written in terms of the volume acceleration of the bubble (Ross, 1976). Or

$$p(t) = \frac{\rho_\infty \ddot{V}_b}{4\pi r_d} = \frac{\rho_\infty}{r_d}(2R_b\dot{R}_b^2 + R_b^2\ddot{R}_b) \tag{35}$$

For the uniform temperature and pressure distribution, the bubble behavior can be calculated from Eqs. (15), (16), (19), (32) and (33) with appropriate initial conditions.

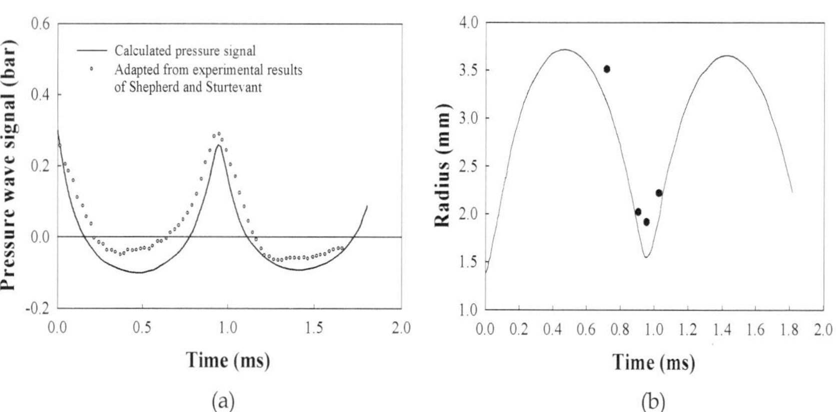

Fig. 2. Pressure wave signal from a oscillating butane bubble in ethylene glycol at 1.013 bar (a) and radius-time curve for the butane bubble (b) with the observed results (full circles).

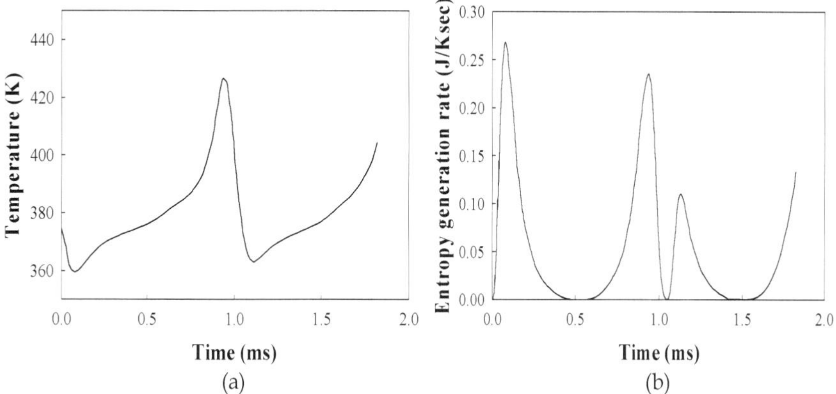

Fig. 3. Time dependence of the temperature inside the bubble (a) and time dependent entropy generation rate for the butane bubble (b) shown in Fig. 2.

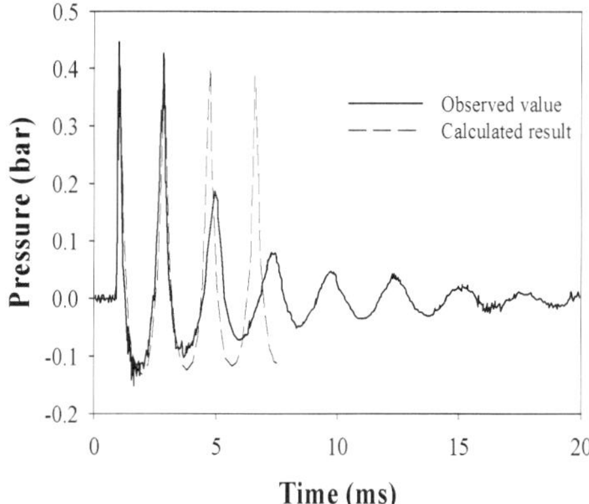

Fig. 4. Far-field pressure signal from an evolving bubble formed from a fully evaporated cyclohexane droplet at its superheat limit, 492.0 K in glycerine.

6.2 An air bubble oscillation under ultrasonic field---Non-uniform temperature and uniform pressure distribution for the gas inside bubble

If the gas temperature is above 2000 K, the vibrational modes of polyatomic molecules are expected to be excited. Since the relaxation time of the vibrational motion is rather long (10^{-6} s) compared with that of the translational motion, the perfect thermal equilibrium cannot be achieved for the duration in which mechanical equilibrium prevails. Certainly the temperature gradient for the gas inside bubble exists in this situation which is the case of an air bubble of micro size oscillates under ultrasonic field of amplitude below 1.2 atm and frequency of 26.5 kHz (Kwak and Yang, 1995).

The calculated radius–time curves for the bubble with an equilibrium radius of 8.5 μm , driven by the ultrasonic field with a frequency of 26.5 kHz and an amplitude of 1.075 atm which is certainly below the sonoluminescence threshold is shown in Fig.5. As shown in Table 1, the calculated values of the maximum radius and the period for each bouncing motion are in good agreement with the observed one (Loefstedt et al., 1993) can be seen. However, the bubble radius-time curve obtained by the Rayleigh equation with a polytropic relation of $P_b V_b^{1.4}$=const. shows 10 number of bouncing motions rather than 7.

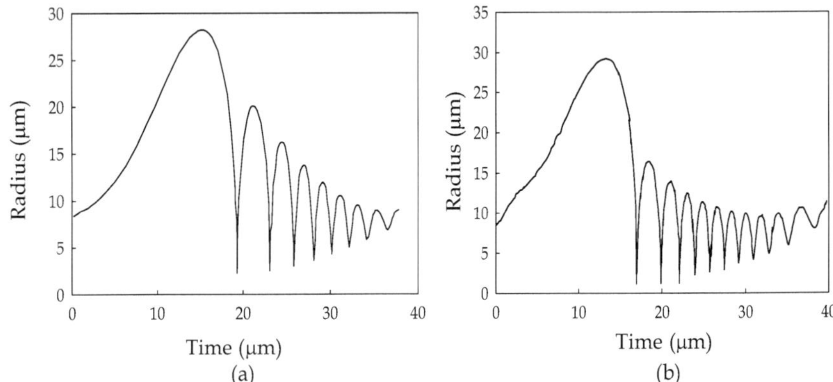

Fig. 5. Theoretical radius-time curves for an air bubble of R_0=8.5 μm at P_A=1.075 atm and f=26.5 kHz in water by our model (a) and by the Rayleigh-Plesset equation with polytropic relation (b).

Also the magnitude of the maximum bubble radius at the first bounce is significantly less than the observed one. The radius, the center temperature and pressure of the bubble at the collapse point are 2.23 μm, 2354 K and 518 atm, respectively (Kwak and Yang, 1995). The minimum bubble wall velocity at the bubble collapse point is about -111.5 m/s. In this case it is better to use Eq. (23) to obtain the bubble wall temperature and Eq. (29) to obtain the instantaneous bubble radius and velocity.

Bouncing number	Maximum bubble size (μm)		Corresponding period (μs)	
	Measured value	Calculated value	Measured value	Calculated value
1	19.9	20.0	3.56	3.62
2	15.6	16.2	2.43	2.76
3	13.0	13.6	2.24	2.24
4	11.1	12.1	2.06	2.07
5	10.4	10.5	1.87	1.90
6	9.7	9.5	1.87	1.82

Table. 1. Calculated and measured maximum bubble size and the corresponding period of bouncing motion after the first bubble collapse for air bubble of R_0 = 8.5 μm at P_A = 1.075 atm and f = 26.5 kHz.

6.3 Sonoluminescing xenon bubble in sulfuric acid solutions---Non-uniform temperature and almost uniform pressure distribution for the gas inside the bubble

Sonoluminecence (SL) phenomena associated with the catastrophic collapse of a gas bubble oscillation under ultrasonic field (Gaitan et al., 1992) have been studied extensively during last 20 years or so for their exotic energy focusing mechanism (Putterman and Weinger, 2000, Young, 2005). The SL from gas bubble in water is characterized by ten to hundred picoseconds flash (Gompf et al., 1997, Hiller et al., 1998), the bubble wall acceleration exceeding 10^{12} m/s² (Kwak and Na, 1996, Wininger et al., 1997), and submicron bubble radius at the collapse point (Weninger et al., 1997). On the other hand, the SL in sulfuric acid solution revealed rather longer flash widths of ns (Jeon et al., 2008), mild wall acceleration of 10^{10} m/s² (Hopkins et al., 2005) and micron bubble radius at the collapse point under similar conditions of ultrasonic field for the case of sonoluminescence in water. The calculated radius-time curve along with observed results for a xenon bubble with $R_0=$ 15 μm , driven by the ultrasonic field with a frequency of 37.8 kHz and an amplitude of 1.5 atm in aqueous solution of sulfuric acid is shown in Fig. 6. With air data for the thermal conductivity, the calculated radius-time curve which exactly mimics the alternating pattern of for the observed result shows two different states of bubble motion. With xenon data for the thermal conductivity, however, slight different pattern for the bubble motion was obtained. These calculation results imply that the bubble behavior, consequently the sonoluminescence phenomena depends crucially on the heat transfer in the gas medium as

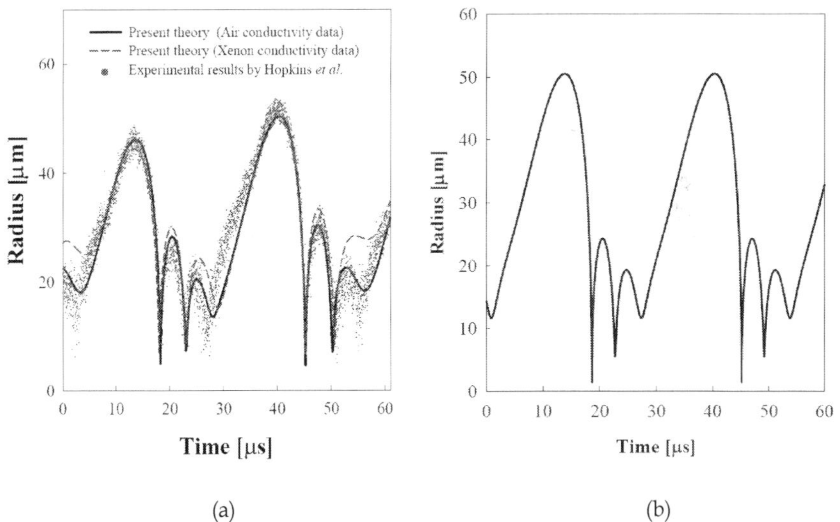

Fig. 6. Theoretical radius-time curve (a) along with observed one by Hopkins *et al.*(2005) for xenon bubble of R_0 =15.0 μm at P_A =1.50 atm and f_d =37.8 kHz in sulfuric acid solutions and the curve calculated by polytropic relation (b).

well as in the liquid layer and that the xenon bubble may contain a lot of air molecules. On the other hand the Rayleigh-Plesset equation with a polytropic relation, a conventional method used to predict the sonoluminescence phenomena cannot predict the two states of bubble motion as shown in Fig. 6(b). The alternating pattern for the bubble motion may happen due to the entropy generation by the finite heat transfer through the bubble wall, which produces lost work: less entropy generation in one cycle having lower maximum bubble radius provides more expansion work to the bubble next cycle, while larger amplitude motion experiencing more entropy generation provides less expansion work to the subsequent motion (Kim et al., 2006).

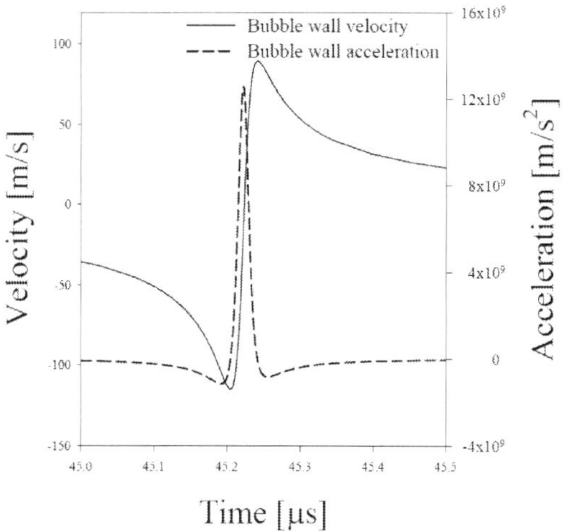

Fig. 7. Calculated bubble wall velocity (line) and acceleration (dash) near the collapse point for the bubble shown in Fig. 6.

Figure 7 shows the time-dependent bubble wall velocity and the variation of the bubble wall acceleration around the collapse point for the bubble shown in Fig. 6. The calculated magnitude of the minimum velocity at the collapse point for the light emitting cycles is about 115 m/s which is close to the observed velocity of 80 m/s. Whereas the maximum bubble wall velocity for non-light-emitting cycle is about 88 m/s, which is also close to the observed results of 60 m/s (Hopkins et al., 2005). However, the magnitude of the minimum velocity calculated by the Rayleigh-Plesset equation with the polytropic relation, which is about 900 m/s, is much higher than the observed value.

Figure 8 shows the calculated time-dependent bubble center temperature and the temporal emissive power with the average temperature for the light-emitting cycle of the bubble shown in Fig.6. The peak temperature calculated at the bubble center is about 8200 K, which is excellent agreement with the observed value of 6000~7000 K. In fact, the average temperature at the collapse point is about 6000 K because considerable temperature drop occurs at the bubble wall as shown in the insert. However, the pressure inside the bubble is almost uniform as expected as shown in Fig. 8 (b).

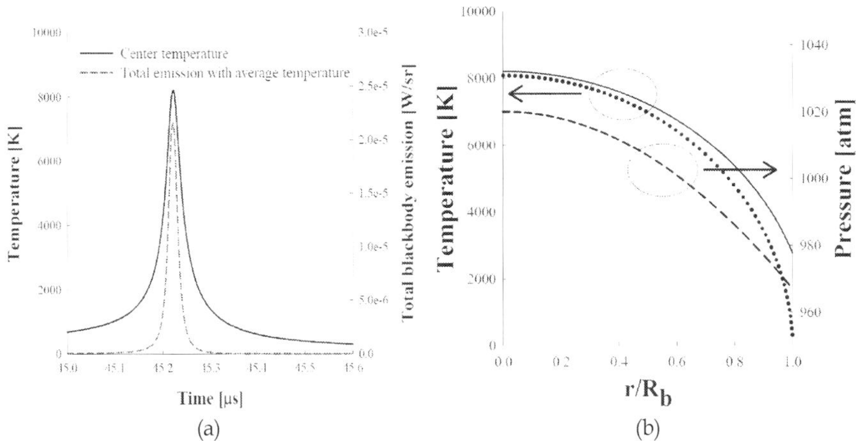

Fig. 8. Time dependent gas temperature at the bubble center and the corresponding total blackbody emission (a) with the average temperature and the temperature and pressure distributions at the collapse point for the bubble shown in Fig.6. The dotted curve in (b) indicates the temperature distribution obtained by Prosperetti eat al.'s (1988) boundary condition at the bubble wall.

6.4 Sonoluminescing air bubbles in water---Non-uniform temperature and non-uniform pressure distribution for the gas inside the bubble

When the bubble wall acceleration exceeds 10^{12} m/s², the pressure distribution for the gas inside the bubble is no longer uniform, as clearly confirmed by Eq. (9). In fact, the SL was observed for an air bubble in water or in sulfuric acid solution when the bubble wall acceleration exceeds 10^{12} m/s² ((Kim et al., 2006).

Figure 9 shows the density, pressure and temperature distributions inside the bubble at 400 ps before the collapse. Certainly, uniform pressure approximation is no longer valid during the collapsing phase for a sonoluminescing air bubble in water solution. Time dependent bubble wall acceleration and the gas temperature at the bubble center are shown in Fig. 10. As can be seen in this Figure, sudden increase and subsequent decrease in acceleration of the bubble wall results in rapid quenching of the gas followed by the substantial temperature rise up to 25000 K, which can be regarded as a thermal spike. Considerable increase in the gas temperature due to the bubble wall acceleration can be seen clearly in this Figure. The maximum bubble wall acceleration achieved near the bubble collapse point is over 10^{11} g (Kwak and Na, 1996), which is consistent with the observed value (Weininger et al., 1997). With uniform pressure approximation, the maximum temperature achieved is only 5800 K. On the other hand, the temperature of the gas inside the bubble goes to infinity without heat loss to the environment, that is k_g goes to zero. One may obtain the maximum temperature up to 10^7 K if one uses k_g= 0.01 W/mK, which indicates that heat transfer is very important also in this case. The heat flux at the collapse point is as much as 10 GW/m², however, the heat flow rate is about 2 mW.

The intense local heating and high pressure inside bubble and liquid adjacent to the bubble wall from such collapse that it can give arise to unusual effects in chemical reactions (Suslick, 1990), and the sonochemical process has been proven to be useful technique in making specialty nanomaterials (Kim et al., 2009). Note that the estimated temperature and pressure in the liquid zone around the collapsing bubble is about 1300 K and 1000 atm, respectively (Kwak and Yang, 1995).

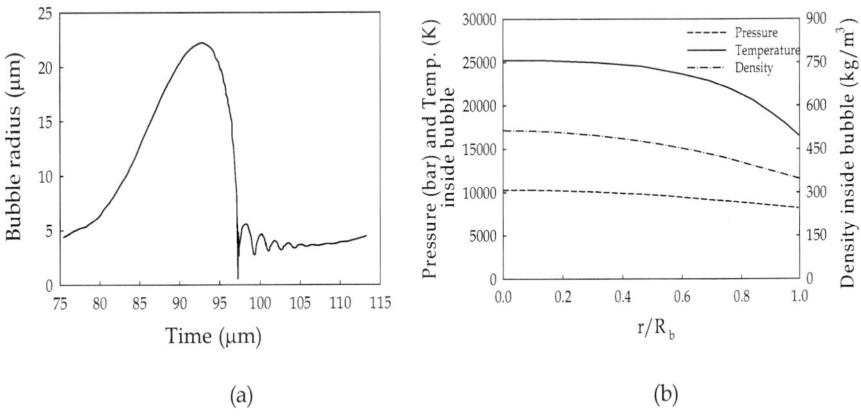

(a)   (b)

Fig. 9. Theoretical radius-time curve for an air bubble of R_0 =4.5 μm at P_A =1.30 atm and f_d =26.5 kHz in water (a) and density , pressure and temperature distributions inside this bubble at 400 ps prior to the collapse point.

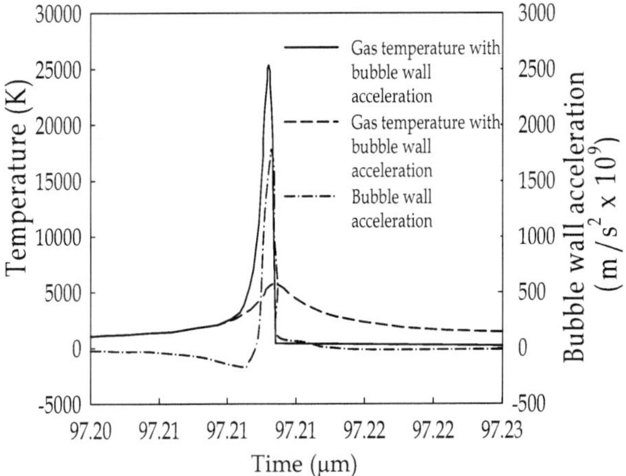

Fig. 10. Time dependent bubble wall acceleration and gas temperature around the collapse point for the case shown in Fig. 9.

7. Molecular dynamics simulation for sonoluminescing xenon bubble

In previous sections, the gas temperature inside the bubble depends crucially on the thermal conductivity of gas inside the bubble as well as the thermal conductivity of liquids in the thermal boundary layer. However, transport property values for the gas and liquid at the extreme condition are not available. As a consequence of this fact, validation of theoretical results was attempted through the MD simulation. The bubble radius was determined by the pressure obtained from MD simulation using Keller-Miksis equation. The mass transfer through the bubble wall, which does not affect the thermal properties at the collapse point very much (Kim and Kwak, 2007) was not considered in this simulation.

7.1 Molecular dynamics simulation

The principles and procedures for molecular dynamics simulation are well documented in standard text books (Haile, 1992, Rapaport, 1995). In this study, numerical integration of one million hard spheres was done in order to count the impulsive collisions between molecules moving by Newton's second law. As is well known, the kinetic events driven by Newton's law can be described by the Vlasov equation or the collisionless Boltzmann equation (Krall and Trivelpiece, 1973), which also yields the conservation equations for mass, momentum, and internal energy. This fact implies that MD simulation of hard sphere molecules might capture the physics related to SL phenomena. Furthermore, the soft parts of the potential are not significant for high energy collisions which occur near the collapse point (Ruuth et al., 2002).

7.2 Collision between molecule and bubble wall

The molecules inside the bubble may hit the molecules at the layer of the gas-liquid interface. In this simulation, the interface was assumed to be a hard wall, so that when a molecule hits the bubble wall, it reflects from the wall, as shown in Fig. 11. When considering the velocity of the bubble wall and assigning the thermal energy of the temperature at the bubble wall T_{bL} the velocity of the reflected molecules $\vec{v}_f$ may be obtained by the following equation for the heat bath boundary condition. In this case, the direction of the reflected particle may be assigned randomly or specularly (Kim et al., 2007).

$$\left| \vec{v}_f - U_b \hat{r}_i \right|^2 / 2 = (3/2)(k_B T_{bl} / m) \tag{36}$$

where U_b is the instantaneous bubble wall velocity, m is the mass of a molecule and k_B is the Boltzmann constant.

For the adiabatic boundary condition, the particles were assumed to be reflected from the wall with speed equal to the pre-collision speed in the local rest frame of the wall. The direction of reflected particles was determined according to the reflection law at the planar interface in this case. Explicitly, the velocity of the reflected particles is given by

$$(\vec{v}_f - U_b \hat{r}_i) = (\vec{v}_i - U_b \hat{r}_i) - 2\left[(\vec{v}_i - U_b \hat{r}_i) \cdot \hat{r}_i \right] \hat{r}_i \tag{37}$$

However, it has been found that neither the adiabatic nor the heat bath boundary condition is appropriate for treating the collapsing process of sonoluminescing gas bubbles. The heat bath boundary condition means that heat flow exists at the bubble wall while adiabatic means no heat flow at the boundary. The degree of the adiabatic change during the collapsing process can be described by the following effective accomodation coefficient (Yamamoto et al., 2006):

$$\alpha = (T_{in} - T_{out}) / (T_{in} - T_{bl}).$$ (38)

where T_{in} is the temperature of the gas particle moving toward the bubble wall and T_{out} is the temperature of particle leaving the wall. Two limiting cases are $\alpha = 0$ for adiabatic boundary condition and $\alpha = 1$ for the heat bath boundary condition.

For the adiabatic boundary case, a molecule does not lost its kinetic energy after a collision made with a molecule at the bubble wall. On the other hand, for the heat bath boundary case, a molecule having kinetic energy of T_i lost its kinetic energy to become a molecule having kinetic energy of T_{bl} after collision. The value of α defined in Eq. (38) determines the instantaneous gas temperature and pressure as well as the bubble radius and wall velocity via the heat transfer through the bubble wall.

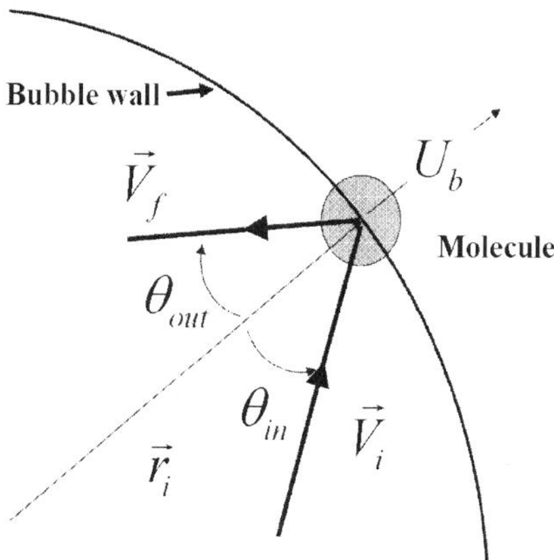

Fig. 11. A collision model between molecules and the bubble wall where v_i and v_f denote the velocity if incident and reflected molecule, respectively.

7.3 Calculation procedures

In this study, a sonoluminescing gas bubble in sulfuric acid solution was considered. The number of molecules in the sonoluminescing bubble may be obtained from the equation of state for an ideal gas.

$$P_\infty \left(\frac{4\pi}{3} R_0^3 \right) = N k_B T_\infty$$ (39)

where N is the number of molecules inside the bubble and P_∞ and T_∞ are the ambient pressure and temperature, respectively. For a sonoluminescing xenon bubble with an equilibrium radius of 0.7 μm in water at T_∞=300 K and P_∞=1 atm, the number of molecules occupying the bubble is about 3.5×10^7, which is difficult to handle with today's computing power.

One may use a scaling transformation for the molecular volume such as $V_gN = const.$ where V_g is the volume of gas molecule and N is the number of molecules occupying the bubble and that reduces the number of molecules that are handled in the MD simulation (Metten and Lauterborn, 2000). With this scaling transformation, of course, the mass and the energy of the system should be conserved so that $mN = const.$ and $TN = const.$

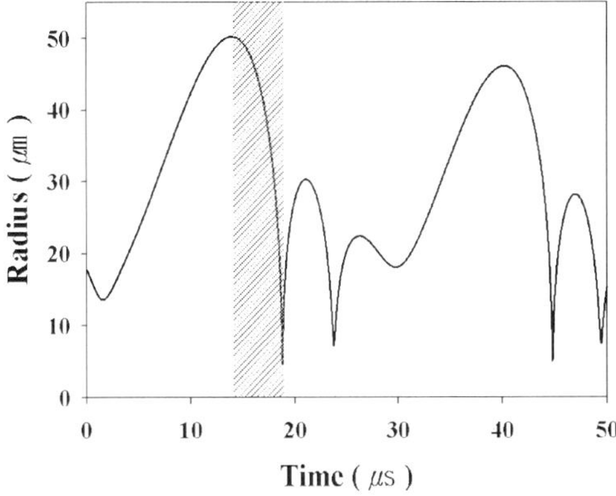

Fig. 12. Theoretical radius-time curves for a xenon bubble of R_o = 15 µm at P_A = 1.5 atm and f_d = 37.8 kHz in a sulfuric acid solution. Shaded area indicates the time span for MD simulations.

In this study, we chose number of simulated molecules to be 10^6, so the hard sphere diameter of a molecule increases to 71.79 σ for the xenon bubble of 15 µm in radius, where σ is the hard sphere diameter of the xenon molecule, taken to be 0.492 nm . The number 71.79 can be obtained from a relation $\sigma^3 N = \sigma_s^3 N_s$ with the number of simulated molecules of N_s =10^6 and the total number of xenon molecules inside the bubble of equilibrium radius of 15 µm, N= 3.46×10^{11}. The conversion factors for the bubble radius, pressure and temperature, which were used in this MD simulation are σ, ε/ σ³ and ε/ k_B, respectively.

7.4 Calculation results and discussion

MD simulation for a collapsing xenon bubble with equilibrium radius of 15.0 µm, driven by an ultrasound frequency of 37.8 kHz and amplitude of 1.5 atm in sulfuric acid solution, was started at its maximum radius of 50.3 µm. The xenon bubble was chosen in this study because observed data for the bubble wall velocity and the peak pressure and temperature at the collapse point have been reported (Hopkins et al., 2005), which is same as the case shown in Fig. 6. One to three million particles were used for a scaled-down MD simulation of the bubble that actually had 35 million molecules. The initial gas state with temperature of 297.8 K and pressure of 3.64×10^{-2} atm was obtained from theoretical results obtained using a set of solutions of the Navier-Stokes equations for the gas inside the bubble with consideration of heat transfer through the bubble wall (Kwak and Yang, 1995, Kwak and Na, 1996). The cross-hatched region in Fig.12 indicates the time period of MD simulation for the xenon bubble which shows an on/off luminescence pattern in sulfuric acid solution (Hopkins et al., 2005).

In Fig.13, the bubble radius-time curve(a) and the time-dependent bubble wall velocity (b) near the collapse point, calculated by MD simulations with α =0.15 are plotted along with the theoretical results. The time-dependent gas temperature at the bubble wall, used in the MD simulation for employing the heat bath boundary condition, was calculated by theory. A very similar trend between MD simulation and theoretical calculation in the bubble radius-time curve and the time-dependent bubble wall velocity were obtained. It is noted that the theoretical minimum velocity at the collapse point, which is about -115 m/s, is close to the observed value of -80 m/s (Hopkins et al., 2005). The calculated value of the minimum velocity by MD is about -130 m/s. The best agreement between the MD simulation and the theoretical results was obtained with a value of α =0.15 among the trial of α values, 0.1, 0.15, 0.3 and 0.5. The minimum radius of the at the collapse point for the xenon bubble is about 4.56 μm so that the packing fraction at the point is about 0.058.

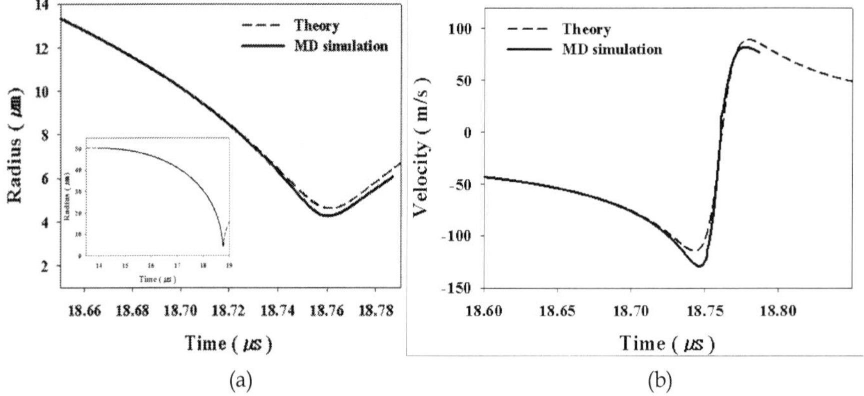

Fig. 13. Calculated radius–time curves (a) and time-dependent bubble wall velocity (b) obtained by MD simulation with α =0.15 and theory for the bubble shown in Fig. 12.

The time-dependent gas temperature and pressure inside the bubble around the collapse point, which were obtained by MD simulation with α =0.15 are shown in Fig.14. The peak temperature and pressure estimated by MD simulation are about 7,500 K and 1,260 atm, respectively, and are very close to the theoretically predicted values of 8,200 K and 1,010 atm, respectively. The MD simulation results shown in Figs.13 and 14 indicate that the collapsing process of a sonoluminescing micron bubble undergoes an almost adiabatic change although a large amount of heat transfer through the bubble wall occurs. At the moment of collapse, the heat flux at the bubble wall is much as 0.6 GW/m². A quite different collapsing process was obtained from the MD simulation with the heat bath boundary condition (α =1.0) (Kim et al., 2007): the compression was started very slowly at first and the final stage occurs very rapidly so that the estimated full width at half maximum (FWHM) of the luminescence pulse is about 5 ns, which is quite less than the value of the theoretical estimate of 20 ns (Kim et al., 2006).

Once the time dependent temperature for the gas inside the bubble around the collapse point, the FWHM of the light pulse can be estimated from the temperature profile with assumption of mechanism of the light emission (Kwak and Na, 1996, 1997). In Fig.15, the

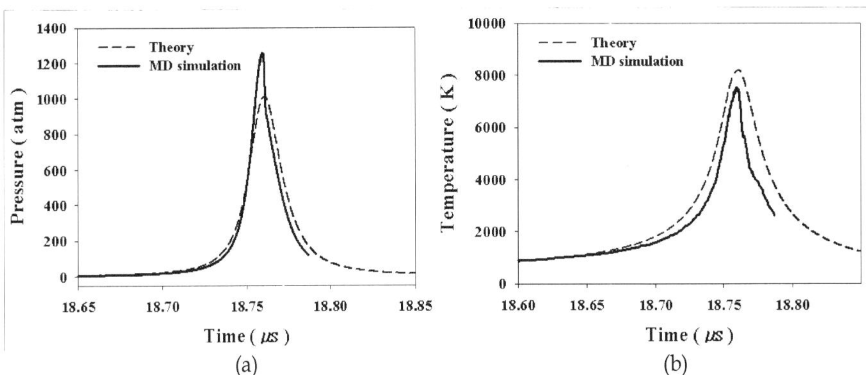

Fig. 14. Time-dependent center pressure (a) and temperature (b) obtained by MD simulation with α =0.15 and theory near the collapse point for the bubble shown in Fig. 12.

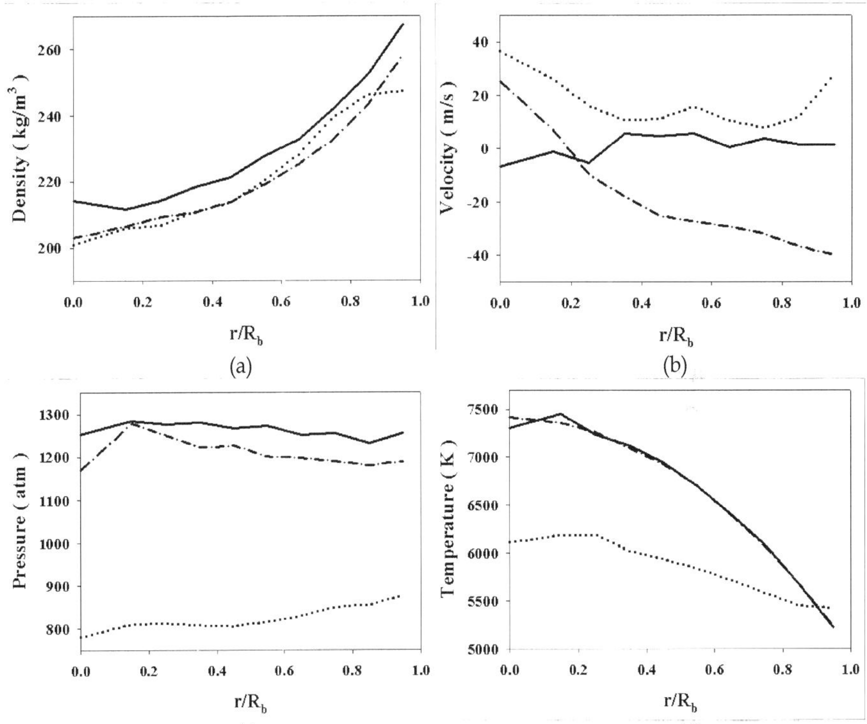

Fig. 15. Gas density(a), velocity(b), pressure(c), and temperature(d) distributions near the bubble collapse point obtained by MD simulation for the bubble shown in Fig.2 before 2.5 ns $(- \bullet - \bullet)$, at $(-)$, and after 2.5 ns $(\bullet \bullet \bullet \bullet)$ the collapse.

MD simulation results with α =0.15 for distribution of gas density(a), velocity(b), pressure(c), and temperature(d) are plotted before and at the collapse point for the bubble shown in Fig.12. The gas density increases radially in both compression and expansion phases, as predicted by theory. The magnitude of the gas velocity increases radially and has its maximum value at the bubble wall and a uniform null value at the collapse point, which can be predicted by the linear velocity profile obtained theoretically. The pressure has an almost uniform distribution before collapse. The temperature shows a quadratic profile with its minimum value near the bubble wall because of heat transfer, which can also be predicted by theory (Kwak and Yang, 1995, Kwak and Na, 1996, Kim et al., 2006).

8. Conclusions

A bubble dynamics model with consideration of heat transfer inside the bubble as well as in the thermal boundary layer of liquid adjacent to the bubble wall was discussed. Thermal damping due to finite heat transfer turns out to be very important factor for the evolving bubble formed from a fully evaporated droplet at its superheat limit and for the sonoluminescing xenon bubble showing two states of bubble motion in sulfuric acid solutions. Molecular dynamics simulation has been done to validate our theoretical model of bubble dynamics. In conclusion, the nonlinear behavior of an ultrasonically driven bubble and the sonoluminescence characteristics from the bubble in sulfuric acid solutions have been found to be correctly predicted by a set of solutions of the Navier-Stokes equations for the gas inside the bubble with considering heat transfer through the bubble wall.

9. References

Batchelor, G.K. (1967). *An Introduction to Fluid Dynamics*, Cambridge University Press, London.

Barber, B.P. & Putterman, S.J., Light scattering measurements of the repetitive supersonic implosion of a sonoluminescing bubble, *Phys. Rev. Lett.* Vol. 69, 3839, (1992).

Bejan, A. (1988). *Advanced Engineering Thermodynamics*, John Wiley & Sons, New York.

Blander , M. & Katz, J. L. (1975). Bubble nucleation in liquids, *AIChE J.* Vol. 21, No. 5, pp.833-848.

Boulos, M.I., Fauchais, P. & Pfender, E. (1994). *Thermal Plasma Vol.1* , Plenum Press, New York.

Byun,K., Kim, K.Y. & Kwak, H. (2005). Sonoluminescence characteristics from micron and submicron bubble, *J. Korean Phys. Soc.* Vol. 47, No. 6, pp.1010-1022.

Davidson, R.C. (1972). *Methods in Nonlinear Plasma Theory*, Academic Press, New York.

Gaitan, K.F., Crum, L.A., Church, C.C. & Roy, R.A.(1992). Sonoluminescence and bubble dynamics for a single, stable, cavitation bubble, *J. Acoust. Soc. Am.*, Vol.91, pp.3166-3183.

Gompf, B., Guenter, R., Nick, G., Pecha, R. & Eisenmenger, W. (1997). Resolving sonoluminescence pulse width with time-correlated single photon counting, *Phys. Rev. Lett.*, Vol. 79, pp. 1405-1407.

Haile , M. (1992). *Molecular Dynamics Simulation* ,John Wiley & Sons.

Hiller, R., Putterman, S.J. & Wininger, K.R. (1998). Time resolving spectra of sonoluminescence, *Phys. Rev. Lett.* Vol.80, pp. 1090-1093.

Hopkins, S.D., Putterman, S.J., Kappus, B.A., Suslick, K.S. & Camara, C.G. (2005). Dynamics of a sonoluminescing bubble in sulfuric acid, *Phys. Rev. Lett.* Vol.95, paper # 254301.

Jeon, J., Lim, C. & Kwak, H. (2008). Measurement of pulse width from sonoluminescing gas bubble in sulfuric acid solution, *J. Phys. Soc. Jpn*, Vol. 77, paper # 033703.

Jun, J. & Kwak, H. (2000). Gravitational collapse of Newtonian stars, *Int. J. Mod. Phys. D*, Vol.9, pp.35-42.

Keller, J.B. & Miksis, M. (1980). Bubble oscillations of large amplitude, *J. Acoust. Soc. Am.*, Vol.68, pp.628-633.

Kim, H., Kang K. & Kwak, H. (2009). Preparation of supported Ni catalysts with a core/shell structure and their catalytic tests of partical oxidation of methane, *Int. J. Hydrogen Energy*, Vol. 34, pp. 3351-3359.

Kim, K., Byun, B. & Kwak, H. (2006). Characteristics of sonoluminescing bubbles in aqueous solutions of sulfuric acid, *J. Phys. Soc. Jpn.*, Vol. 75, paper # 114705.

Kim, K. & Kwak, H. (2007). Predictions of bubble behavior in sulfuric acid solutions by a set of solutions of Navier-Stokes equations, Chem Eng. Sci., Vol. 62, pp. 2880-2889.

Kim, K., Kwak, H. & Kim, J.H. (2007). Molecular dynamics simulation of collapsing phase for a sonoluminescing gas bubble in sulfuric acid solutions : A comparative study with theoretical results, *J. Phys. Soc. Jpn.* Vol.76, 024301.

Kim, K.Y., Lim, C., Kwak, H. & Kim J.H. (2008). Validation of molecular dynamic simulation for a collapsing process of sonoluminescing gas bubble, *Mol. Phys.*, Vol. 106, No. 8, pp. 967-975.

Krall, N.A. &Trivelpiece, A.W.(1973). *Principles of Plasma Physics* ,McGraw-Hill, New York.

Kwak, H. & Lee, S. (1991). Homogeneous bubble nucleation predicted by molecular interaction model, *ASME J. Heat Trans.* Vol. 113, pp. 714-721.

Kwak, H. & Na, J. (1996). Hydrodynamic solutions for a sonoluminescing gas bubble, *Phys. Rev. Lett.*, Vol.77, pp.4454-4457.

Kwak, H. & Na, J. (1997). Physical processes for single bubble sonoluminescence, *J. Phys. Soc, Jap.* Vol.66, pp.3074-3083.

Kwak, H., Oh, S. & Park, C. (1995). Bubble dynamics on the evolving bubble formed from the droplet at the superheat limit, *Int. J. Heat Mass Transfer*, Vol.38, pp.1707-1718.

Kwak, H. & Panton, R.L. (1985). Tensile strength of simple liquids predicted by a model of molecular interactions, *J. Phys. D: Appl. Phys.*, Vol. 18, pp.647-659.

Kwak, H. & Yang,H. (1995). An aspect of sonoluminescence from hydrodynamic theory, J. Phys. Soc. Jpn. Vol. 64, 1980-1992.

Lauterborn, W. (1976). Numerical investigation of nonlinear oscillation of gas bubbles in liquids, *J. Acoust. Soc. Am.*, Vol. 99, pp. 283-293.

Lin, H., Storey, B.D. & Szeri, A.J.(2002). Inertially driven inhomogenieties in violently collapsing bubbles: the validity of the Rayleigh-Pessset equation, *J. Fluid Mech.*, Vol.452, pp.145-162.

Loefstedt, R., Barber, B.P. & Putterman, S.J. (1993). Toward a hydrodynamic theory of sonoluminescence, *Phys. Fluids*, Vol. A5, pp. 2911-2928.

Metten, B. & Lauterborn, W. (2000). In Nonlinear Acoustics at the Turn of the Millennium: ISNA 15, 15th International Symposium on Nonlinear Acoustics edited by W. Lauterborn and T. Kurz, AIP Conf. Proc. No. 524, pp. 429-432.

Moody, F.J. (1984). Second law thinking-example application in reactor and containment technology, In *Second Law Aspects of Thermal Design* (Edited by A. Bejan and R.C. Reid) Vol. HTD-33, pp. 1-9. ASME, New York.

Moss,W.C., Clarke, D.B., White, J.W. & Young, D.A.(1994). Hydrodynamic simulations of bubble collapse and picosecond sonoluminescence, *Phys. Fluids*, Vol.6, No. 9, pp. 2979-2985.

Na, J., Byun, G. & Kwak, H. (2003) Diffusive stability for sonoluminescing gas bubble, *J. Kor. Phys. Soc.* Vol. 42,pp.143-152.

Panton, R.L. (1996). *Incompressible Flow*, John Wiley & Sons, 2nd Edition, New York.

Park, H., Byun, K. & Kwak, H. (2005). Explosive boiling of liquid droplets at their superheat limits, *Chem. Eng. Sci.*, Vol. 60, pp. 1809-1821.

Prosperetti, A., Crum, L. A. & Commander, K. W. (1988). Nonlinear bubble dynamics, *J. Acoust. Soc. Am.*, Vol.88, pp.1061-1077.

Putterman, S.J. & Weininger, K.R. (2000). Sonoluminescence: How bubbles turn sound into light. Annual Review of Fluid Mech., Vol. 32, pp. 445-476.

Rapaport,D.C.(1995). *The Art of Molecular Dynamics Simulation*, Cambridge University Press, London.

Ross, D. (1976). *Mechanics of Underwater Noise*, Pergamon Press, Oxford.

Ruuth, S.J., Putterman, S. & Merriman, B. (2002). Molecular dynamics simulation of the response of a gas to a spherical piston: Implications for sonoluminescence, *Phys. Rev. E*, Vol. 66, 036310.

Shepherd, J.E. & Sturtevant, B. (1982). Rapid evaporation at the superheat limit, *J. Fluid Mech.*, Vol. 121, pp. 379-402.

Suslick, K.S. (1990). Sonochemistry, Science, Vol. 247, pp. 1439-1445.

Suslick, K.S., Hammerton, D.A. & Cline, R.E. (1986). The sonochemical hot spot, *Am. Chem. Soc.*, Vol. 108, pp. 5641-5642.

Theofanous, T., Biasi, L. & Isbin, H.S.(1969). A theoretical study on bubble growth in constant and time-dependant pressure fields, *Chem. Eng. Sci.*, Vol.24, pp.885-897.

Vincenti, W.G. & Kruger, C.H. (1965). *Introduction to Physical Gas Dynamics*, New York.

Weininger, K.R., Barber, B.P. & Putterman, S.J. (1997). Pulsed Mie scattering measurements of the collapse of a sonoluminescing bubble, Phys. Rev. Lett., Vol. 78, pp. 1799-1802.

Yamamoto,K., Takeuchi,H. & Hyakutake, T. (2006). Characteristics of reflected gas molecules at a solid surface, *Phys. of Fuids*, Vol.18,046103.

Young, F.R.(2005). *Sonoluminescence* , CRC Press, Boca Raton.

Part 2

Boiling, Freezing and Condensation Heat Transfer

8

Nucleate Pool Boiling in Microgravity

Jian-Fu ZHAO

Key Laboratory of Microgravity (National Microgravity Laboratory)/CAS;
Institute of Mechanics, Chinese Academy of Sciences (CAS)
China

1. Introduction

Nucleate pool boiling is a daily phenomenon transferring effectively high heat flux. It is, however, a very complex and illusive process because of the interrelation of numerous factors and effects as the nucleate process, the growth of the bubbles, the interaction between the heater's surface with liquid and vapor, the evaporation process at the liquid-vapor interface, and the transport process of vapour and hot liquid away from the heater's surface. Among many sub-processes in boiling phenomenon, gravity can be involved and play much important roles, even enshroud the real mechanism underlying the phenomenon. Our present knowledge on nucleate pool boiling phenomenon has been built with the aid of numerous meticulous experiments in normal gravity environment on the ground where gravity is a dominant factor. Gravity strongly affects boiling phenomenon by creating forces in the systems that drive motions, shape boundaries, and compress fluids. Furthermore, the presence of gravity can mask effects that ever present but comparatively small. Advances in the understanding of boiling phenomenon have been greatly hindered by masking effect of gravity. Microgravity experiments offer a unique opportunity to study the complex interactions without external forces, such as buoyancy, which can affect the bubble dynamics and the related heat transfer. Furthermore, they can also provide a means to study the actual influence of gravity on the boiling. On the other hand, since many potential applications exist in space and in planetary neighbours due to its high efficiency in heat transfer, pool boiling in microgravity has become an increasing significant subject for investigation. Therefore, the microgravity researches will be conductive to revealing of the mechanism underlying the phenomenon, and then developing of more mechanistic models for the related applications both on Earth and in space.

Research on boiling heat transfer in microgravity has a history of more than 50 years with a short pause in the 1970s and has been advanced with the development of various microgravity facilities and with increased experimental opportunities, especially in the last two decades. On the progress in this field, many comprehensive reviews and monographs are available now. Among many others, Straub (2001), Di Marco (2003), Kim (2003), and Ohta (2003a, b) summarized the experimental and theoretical works all over the world, which provided the status of this field at the beginning of our research.

In the past decade, two research projects on nucleate pool boiling in microgravity have been conducted aboard the Chinese recoverable satellites by our group in the National

Microgravity Laboratory/CAS. Ground-based experiments both in normal gravity and in short-term microgravity in the drop tower Beijing have also been performed. The major findings are summarized in the present chapter, while a brief review on the results of the space experiments has also been provided by Zhao (2010) recently.

2. Pool boiling on wire in microgravity

A TCPB (Temperature-Controlled Pool Boiling) device was developed to study heat transfer of pool boiling on thin wires both on the ground and aboard the 22nd Chinese recoverable satellite (RS-22) (Wan et al., 2003). A platinum wire of 60 μm in diameter and 30 mm in length was simultaneously used as a resistance heater and a resistance thermometer to measure the temperature of the heater surface. The heater resistance, and thus the heater temperature, was kept constant by a feedback circuit, which was similar to that used in constant-temperature hot-wire anemometry. Each step of the heater temperature lasted about 30 seconds in order to obtain steady pool boiling according to Straub (2001). The boiling chamber was filled with degassed R113 and was pressurized in an airproof container. A bellows connected with the chamber and the surrounding housing allowed the pressure in the chamber to be practically constant.

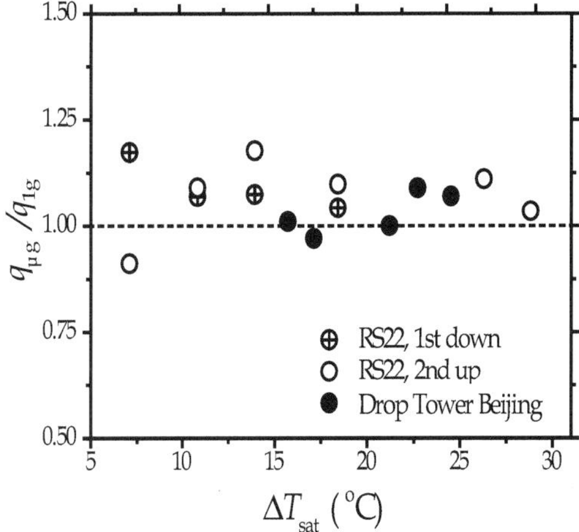

Fig. 1. Microgravity efficiency on heat transfer of nucleate boiling in microgravity (Zhao et al., 2009d).

Several preliminary experimental runs at subcooling condition were conducted in short-term microgravity utilizing the drop tower Beijing, which provides a course of about 3.6 s for microgravity experiments (Zhao et al., 2004). The space experiment was carried out aboard the 22nd Chinese recoverable satellite (RS-22) in September 2005 (Liu, 2006). The level of residual gravity was estimated in the range of $10^{-3} \sim 10^{-5}\ g_0$. Before and after the space flight, ground control experiments using the same facility were also conducted. Comparing with those in normal gravity, the heat transfer of nucleate boiling was slightly enhanced in short- and long-term microgravity (Fig. 1), while about 20% and 40% decrease of heat flux was observed for two-mode transition boiling in short- and long-term microgravity, respectively.

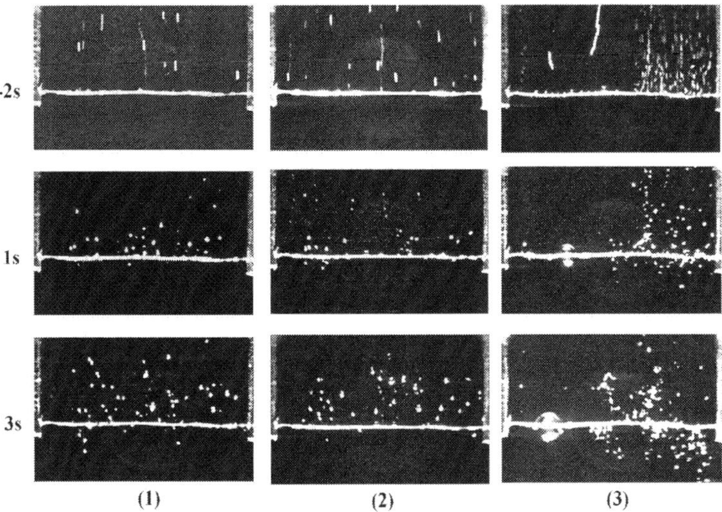

Fig. 2. Bubble behaviors on thin wire in different gravity conditions (Zhao et al., 2004).

In the drop tower tests, bubble behaviors were dramatically altered by the variation of the acceleration (Fig. 2). It was difficult to observe the lateral oscillation of bubbles along the wire in nucleate boiling regime in normal gravity, but this kind of motion was always able to observe in both short- and long-term microgravity. It could lead to the lateral coalescence between adjacent bubbles, and then detached the coalesced bubble from the wire. Sometimes, the coalesced bubble could enclose the wire and a bright spot appeared there. It couldn't, however, last long period and the boiling continued as nucleate boiling. In the two-mode transition boiling regime, the Taylor instability disappeared in microgravity, and then the surface tension reformed the shape of the wavy film appeared in normal gravity to a large spheroid bubble encircling the wire. Then the film part receded after releasing the drop capsule, while the part of nucleate boiling expanded along the wire. The centre of the large spheroid bubble wiggled along the wire and its size increased slowly. Sometimes, the wire near the centre of the large spheroid bubble brightened up, but no real burn-out was observed in the short-term microgravity experiments.

In the space experiment in long-term microgravity, special bubble behaviors were observed firstly (Zhao et al., 2007). There existed three critical bubble diameters in the discrete vapor bubble regime in microgravity, which divided the observed vapor bubbles into four regions (Fig. 3): Tiny bubbles were continually forming and growing on the surface before departing slowly from the wire when their sizes exceeded the first critical value. The bigger bubbles, however, were found staying on the surface again when their diameters were larger than the second critical value. If they grew further larger than the third critical value, departure would be observed once again. Furthermore, the first critical value exhibited no obvious difference between in normal gravity and in microgravity.

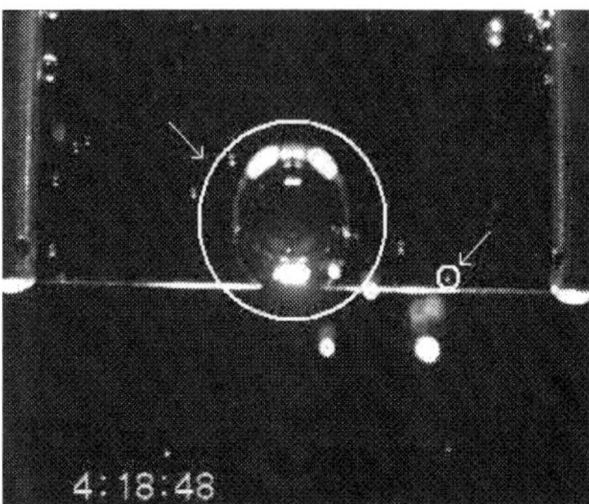

Fig. 3. Special bubble behaviors on thin wire in long-term microgravity (Zhao et al., 2007).

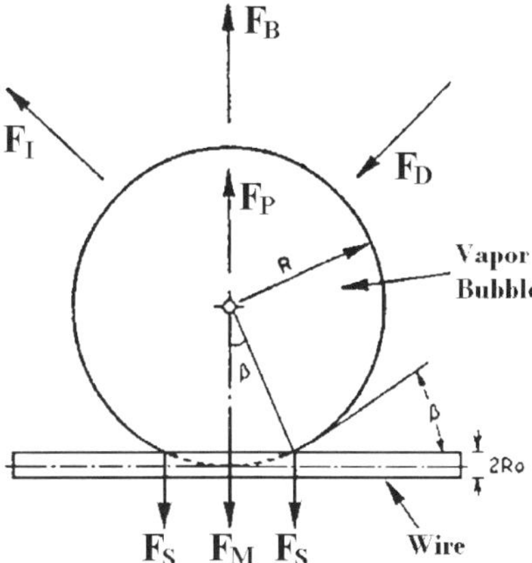

Fig. 4. Forces acted upon a vapour bubble growing on thin wire (Zhao et al., 2008).

Among the commonly used models for bubble departure, no one can predict the whole observation. A qualitative model was proposed by Zhao et al. (2008), in which the Marangoni effect was taken into account (Fig. 4)

$$f(y) = C_4 y^4 + C_3 y^3 + C_1 y + C_0 \tag{1}$$

where,

$$y = \tau^{1/2} \tag{2}$$

$$C_4 = \frac{4}{3}\pi E^3 (\rho_L - \rho_V) g \tag{3}$$

$$C_3 = -2K\pi |\sigma_T| E^2 \nabla T \tag{4}$$

$$C_1 = 4\sigma R_0 \sin^2 \beta + \frac{\pi}{3}\rho_L E^4 \qquad (5)$$

$$C_0 = R_0 E^3 \rho_L \sin^2 \beta \left(\frac{1}{3} - \frac{3}{8}C_d \right) \qquad (6)$$

$$E = \frac{1}{2\sqrt{\pi}} Ja\sqrt{\alpha_L} \qquad (7)$$

where τ, σ_T, σ, ρ, β, α, R_0, C_d and Ja denote the growing time of bubble, surface tension and its temperature coefficient, density, contact angle, heat diffusivity coefficient, wire radius, drag coefficient and the Jacob number, respectively. K is an empirical parameter to count the departure from the linear theory for the case of finite Reynolds and Marangoni numbers. The subscripts L and V denote liquid and vapour phases, respectively.

According to Eq. (1), the following conclusion can be obtained: If $f(y)<0$, the departure force is larger than the resistant force, so the bubble will stay on the heater's surface; if $f(y)>0$, the departure force is smaller than the resistant force, so the bubble will depart from the heater's surface. Fig. 5 also shows the predictions of Eq. (1) in microgravity. In normal gravity, the function for the total forces acting on the growing bubble, $f(y)$, has only one zero-value point, indicting only one critical diameter for bubble departure. When the residual gravity decreases to no more than $1.36\times10^{-4}\ g_0$, the second and third zero-value points will be predicted by the new model. Comparing the prediction at $g=10^{-4}\ g_0$ with the observation, the agreement is quite evident.

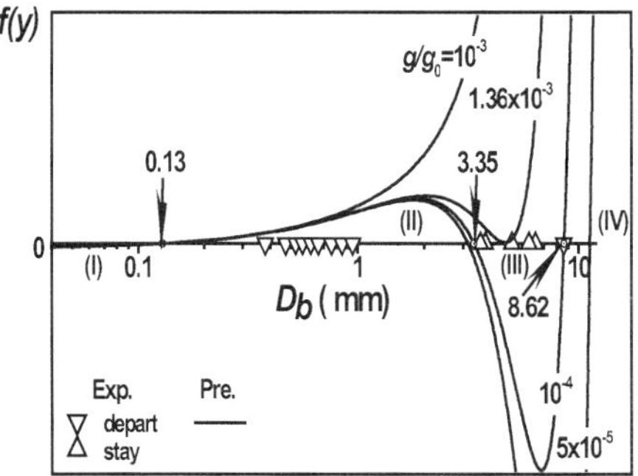

Fig. 5. Bubble departure in the discrete vapor bubble regime in microgravity. (Zhao et al., 2008).

The scaling of CHF with the gravity based on the data obtained both in the present study and in other researches reported in the literature was shown in Fig. 6. It was found that the Lienhard-Dhir-Zuber model (Lienhard & Dhir, 1973), established on the mechanism of hydrodynamic instability, can provide a relative good prediction on the trend of CHF in different gravity conditions, though the value of dimensionless radius $R'=R\sqrt{(\rho_L - \rho_G)g/\sigma}$ was far beyond the initial application range of the model. This observation was consistent with Straub (2001).

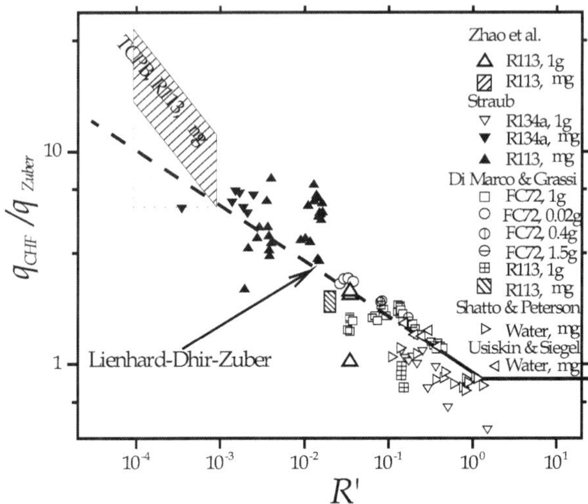

Fig. 6. Scaling of CHF with gravity (Zhao et al., 2009d).

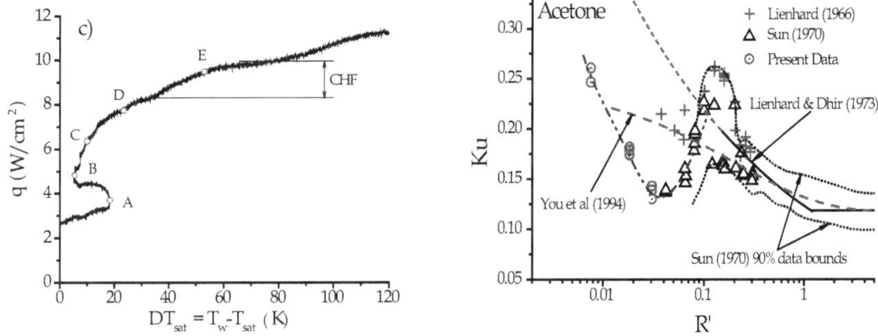

Fig. 7. Scaling behaviours of CHF on wires at saturated condition in normal gravity (Zhao et al., 2009b, c).

However, comparing the trend of CHF in Fig. 6 with the common viewpoint on the scaling of CHF, which was built upon a large amount of experimental data with variable heater diameter on the ground, it was inferred, as pointed out by Di Marco & Grassi (1999), that the dimensionless radius R', or equivalently the Bond number, may not be able to scale adequately the effects and to separate groups containing gravity due to the competition of different mechanisms for small cylinder heaters. Furthermore, Zhao et al. (2009b, c) revisited the scaling behaviours of CHF with respect to R' at small value of the Bond number in normal gravity conditions. It has been found that interactions between the influences of the subcooling and size on CHF will be important for the small Bond number, and that there may exist some other parameters, which may be material-dependant, in addition to the Bond number that play important roles in the CHF phenomenon with small Bond number (Fig. 7)

A parameter, named as the limited nucleate size d_{LN}, and a non-dimensional coefficient $\Gamma = d_{LN}/d_{wire}$ were introduced to interpret this phenomenon (Zhao et al., 2009b). It was assumed that the limited nucleate size is not dependent with gravity but with the other parameters of the boiling system, such as the material parameters of the working fluid and the heater, the heater surface condition, an so on. If Γ is small enough, the initial vapour bubbles will be much smaller than the heater surface and then the occurrence of the CHF will be caused by the mechanism of hydrodynamic instability. On the contrary, it will be caused by the mechanism of local dryout if Γ is so large that the initial bubble larger than the wire diameter d_{wire} may easily encircle the heater. Further researches, however, are needed for the delimitation of the two mechanisms.

3. Pool boiling on plate in microgravity

A QSPB (quasi-steady pool boiling) device was developed to study heat transfer of pool boiling on plane plate both in normal and in microgravity, which was flown aboard the Chinese recoverable satellite SJ-8 in September 2006 (Yan, 2007).

To avoid large scatterance of data points measured in steady state boiling experiments and to obtain continuous boiling curves in the limited microgravity duration, a transient heating method was adopted, in which the heating voltage was controlled as an exponential function with time, namely

$$U = U_0 \exp\left(\tau/\tau_0\right) \tag{8}$$

where τ denotes the heating time, and the period τ_0 determines the heating rate. In the space experiment aboard SJ-8 and the ground control experiments before the space flight, the period was set for $\tau_0 = 80$ s in order to make the heating process as a quasi-steady state, which was verified in the preliminary experiments on the ground. Furthermore, the period used in the present study was about 3~4 order of magnitude larger than those in Johnson (1971), which guaranteed the fulfillment of quasi-steady condition, though different structure of the heater and working fluid employed here.

The heater used in the study had an Al_2O_3 ceramic substrate with a size of $28 \times 20 \times 1$ mm^3 embedded in a PTFE base with a thickness of 25 mm. An epoxy-bonded composite layer of mica sheets and asbestos was set between the ceramic substrate and the PTFE base to reduce the heat loss. The effective heating area with an area of 15×15 mm^2 was covered by a serpentine strip of multi-layer alloy film with a width of 300 μm and a thickness about 10 μm. The space between the adjacent parallel strips is about 70 μm. In addition, the multi-layer alloy film also served simultaneously as a resistance thermometer. The averaged temperature of the heater surface in the experiments was calculated using the correlation between the temperature and the resistance of the multi-layer alloy film, which was calibrated prior to the space flight. In the data reduction, the data of the averaged temperature of the heater surface were filtered to remove noise effects. The total heat flux was transported into both the liquid and the Al_2O_3 ceramic substrate, while the heat loss to the PTFE base and the surrounding was neglected. The filtered temperature data was used to compute the increase of the inner energy of the Al_2O_3 ceramic substrate using appropriate numerical computations. Subtracting the increase of the inner energy of the Al_2O_3 ceramic substrate from the total heat flux input provided the heat flux to the liquid and the transient mean heat transfer coefficient.

Degassed FC-72 was used as the working fluid. The pressure was controlled by a passive control method similar with that used in the TCPB device. Venting air from the container to the module of the satellite decreased the pressure inside the boiling chamber from its initial value of about 100 kPa to the same as that in the module of the satellite, i.e. 40 ~ 60 kPa. An auxiliary heater was used for adjusting the temperature of the bulk liquid from the ambient temperature to about the middle between the ambient and saturation temperature at the corresponding pressure. Except the first run without pre-heating phase, each of the following runs consists of pre-heating, stabilizing and boiling phases, and lasts about one hour. The corresponding experimental conditions are listed in Table 1, in which the estimated values of the critical heat flux (CHF) and the corresponding superheats are also listed. Figs. 8 and 9 show some typical processes of bubble growth, heating history, and the corresponding boiling curves in the space experiments.

Run#	pressure p (kPa)	subcooling ΔT_{sub} (K)	CHF q_{CHF} (W/cm²)	superheat ΔT_{sat} (K)
I-1	90.8	36.9	8.3 ~ 10.0	28 ~ 66
I-2	97.3	25.8	6.6 ~ 9.1	34 ~ 76
I-3	102.3	21.8	7.0 ~ 7.6	40 ~ 56
I-4	105.7	19.5	7.7 ~ 8.2	20 ~ 29
I-5	111.7	18.4	8.6 ~ 8.9	11 ~ 17
II-1	57.2	24.5	5.7 ~ 6.9	24 ~ 42
II-2	91.1	18.8	7.4 ~ 9.5	26 ~ 55
III-1	65.5	27.5	6.3 ~ 6.6	30 ~ 35

Table 1. Space experimental conditions and the estimated CHF values (Zhao et al., 2009a).

Because of the residual gravity which was estimated in the range of $10^{-3} \sim 10^{-5}$ g_0, there could exist a week single-phase natural convection before the incipience of boiling. Due to the experimental schedule, different behaviours of the incipience of boiling were observed in the space experiment. In the first run I-1, a great amount of vapor appeared abruptly and explosively at the incipience of boiling. Surface tension then compelled the vapor to form several segregate bubbles. An obvious drop of the heater temperature was observed in the curve of the heating history, correspondingly. This drop caused an additional heat flux from the ceramic substrate to the liquid, and may result in a local maximum of the heat flux to the liquid in the transition region from the incipience to quasi-steady nucleate boiling despite of the monotonous increasing of the heating rate. On the contrary, a gradual growth of the first bubble was observed in the following runs. The process of bubble growth even appeared an obvious standstill after its first appearance. Correspondingly, no over-shooting or drop of the heater temperature can be observed in the curves of the heating history in the following runs. The first appearance of bubbles in the first five runs was observed at 21.89 s, 8.68 s, 8.12 s, 4.54 s, and 4.84 s, respectively. Comparing with the first run, the nucleate boiling occurred significantly earlier in the following runs. Considering the experimental procedure, it may indicate that there could be residual micro-bubbles in cavities after the preceding runs. These micro-bubbles would make the cavities easier to be activated, and the boiling would thus be initiated at a lower wall superheat. Furthermore, bubbles attached on the surface seemed to be able to suppress the activation of the cavities in the neighborhoods according to the detailed analyses of the video images.

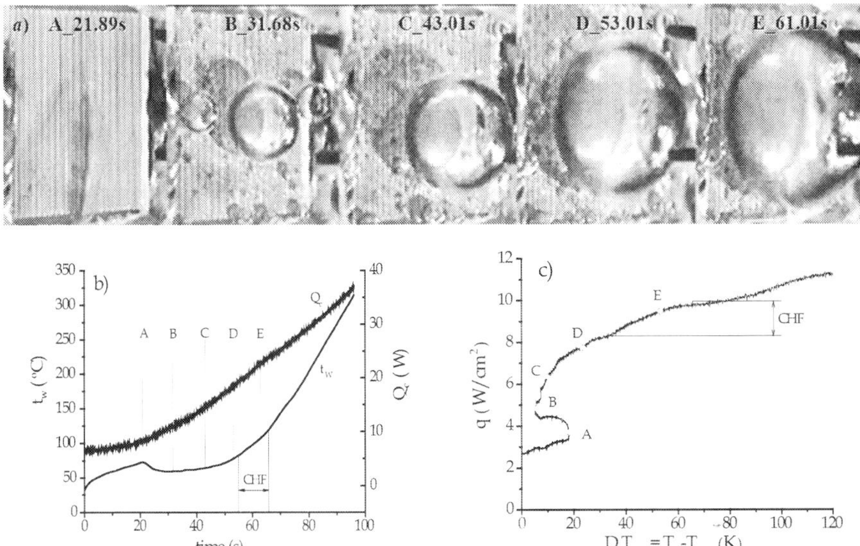

Fig. 8. Bubble dynamics (a), heating history (b), and boiling curve (c) in the run I-1 aboard SJ-8 (Zhao et al., 2009a).

It was observed that primary bubbles generated consistently, slid on the surface, and coalesced with each other to form a larger coalesced bubble. Although the video images were taken only from the sole direction of 45° with respect to the heater surface, it was able to be observed that some primary bubbles generated under the coalesced bubble. The coalesced bubble also engulfed small bubbles around it. It can be inferred that, as pointed out by Ohta et al. (1999), a macro-layer may exist underneath the coalesced bubble, where primary bubbles are forming.

For the cases of higher subcooling, the coalesced bubble with a relative smooth surface was observed oscillating near the center of the heater surface. Higher was the subcooling, smaller and smoother at the same heating time. The coalesced bubble shrank to an elliptical sphere under the action of surface tension. Its size increased with the increase of the surface temperature, but it was very difficult to cover the whole surface. Thus, the bottom of the coalesced bubble may dry out partly at high heat flux, while the other places, particularly in the corners of the heater surface were still in the region of nucleate boiling. Unfortunately, dry spot was not able to be observed directly in the present study. The fact, however, that there existed a much smooth increase of the averaged temperature of the heater surface and no turning point corresponding to CHF in boiling curves indicated a gradual transition to film boiling along with the developing of the area of local dry area, as described by Oka et al. (1995). In this case, it was difficult to determine the accurate value of CHF. However, the trend of the increasing heater temperature with the heating time provided some information of CHF. Supposing the rapid increase of heater temperature corresponds to the beginning of the transitional boiling while a constant slope of the temperature curve to the complete transition to film boiling, the range of CHF and the corresponding superheat were estimated, which were also marked in Fig. 8.

The bubble behaviours and the characteristics of the boiling curves at lower subcooling were different from those at higher subcooling. In these runs, *e.g.* the run I-4 shown in Fig. 9, the size of the coalesced bubble increased quickly, and a strong oscillation appeared on its surface. Higher was the pressure, stronger the surface oscillation. Furthermore, before the abrupt transition to film boiling, the heat flux remained increasing though the surface temperature rose slowly or even fell down along with the heating time. The above observations can be interpreted as follows. Because of the decrease of surface tension with the increase of the saturation temperature and the corresponding pressure, local dry spots underneath the coalesced bubble with a strong surface oscillation can not develop steadily. They may be re-wetted by the surrounding liquid, and nucleate boiling will remain on the heater surface. Furthermore, even more nucleate sites could be activated under the action of the strong oscillation of the coalesced bubble. Thus, heat transfer was enhanced.

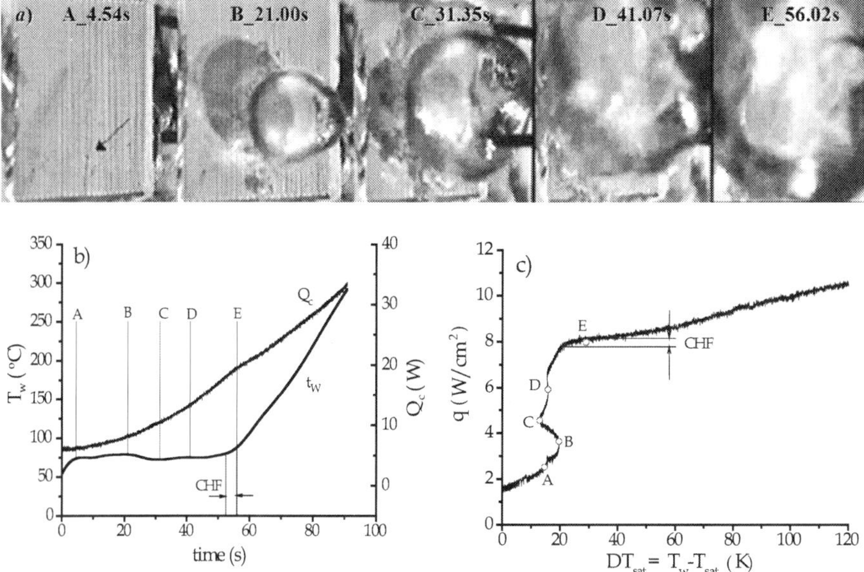

Fig. 9. Bubble dynamics (a), heating history (b), and boiling curve (c) in the run I-4 aboard SJ-8 (Zhao et al., 2009a).

Comparisons of boiling curves in microgravity showed that heat transfer was deteriorated with the decrease of subcooling at the same pressure but enhanced with the increase of pressure at the same subcooling. The estimated values of CHF in microgravity increased with the subcooling at the same pressure, and also increased with pressure at the same subcooling. These trends are similar with those observed in normal gravity. The value of CHF in microgravity, however, was only about one third of that at the similar pressure and subcooling in terrestrial condition. Unfortunately, the pressure and temperature of the liquid cannot be isolated completely because of the passive control of the pressure inside the boiling chamber used here. Thus, there existed some cross-influences of pressure and subcooling on CHF.

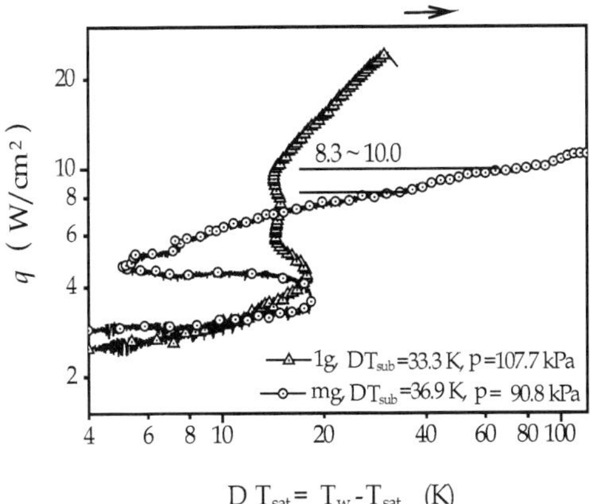

$$D\,T_{sat} = T_w - T_{sat} \quad (K)$$

Fig. 10. Comparison of boiling curves in different gravity (Zhao et al., 2009a).

In Fig. 10, boiling curves in different gravity were compared with each other at the similar pressure and subcooling conditions. Generally, boiling heat transfer in microgravity was deteriorated comparing with that in normal gravity, particularly at high superheats or heat fluxes. Much obvious enhancement, however, could be observed just beyond the incipience, which was consistent with those in steady state pool boiling experiments, such as reported by Lee et al. (1997). It was also observed that the incipience of boiling occurred in microgravity at the same superheat as that in normal gravity, which was in agreement with Straub (2001).

Recently, a new serial of experiments of pool boiling of FC-72 with non-condensable gas on smooth surface (denoted as chip S) in short-term microgravity have been conducted utilizing the drop tower Beijing (Xue et al., 2010). The boiling vessel was filled with about 3 L of FC-72 as the working liquid. The test chip was a P doped N-type silicon chip with the dimensions of 10×10×0.5 mm³, which was set horizontally upward. The chip was Joule heated by a direct current. Two 0.25-mm diameter copper wires were soldered by a low temperature solder to the chip side surfaces at the opposite end for power supply. A programmable DC power supply was used to provide constant heating electric current for the chip. A nearly atmospheric pressure is maintained by attaching a rubber bag to the test vessel. A K-type thermocouple was used for measuring the local temperature of the test liquid at the chip level, and directly connected to a temperature display monitor for visual observation through a CCD camera. Besides, the local wall temperature at the center of the chip and the local temperature of the test liquid at about 40 mm from the edge of the chip were measured by two 0.13 mm-diameter T-type thermocouples which were connect with a data acquisition system (DI710-UHS). A high speed video camera (VITcam CTC) imaging 250 frames per second at a resolution of 1024×640 pixels with a shutter speed of 1/2000s was used along with a computar lens (MLM-3XMP) to obtain images of the boiling process. The high speed camera was installed in front of the test vessel at a direction angle of 30° with respect to the heater surface. Due to the short duration of microgravity (nearly 3.6 s), boiling was initiated before the release of the drop capsule and keep at a steady state for enough duration. Then the drop capsule released, and the experiment was run in microgravity. After the recovery of the drop capsule, this experimental run was finished.

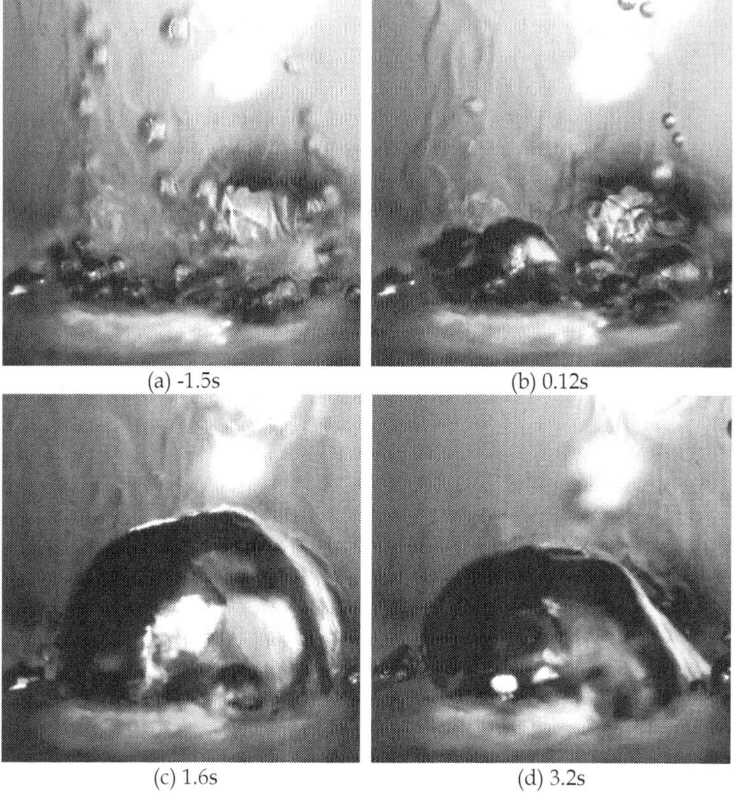

(a) -1.5s   (b) 0.12s

(c) 1.6s   (d) 3.2s

Fig. 11. Bubble behaviors on chip S (Wei et al., 2010).

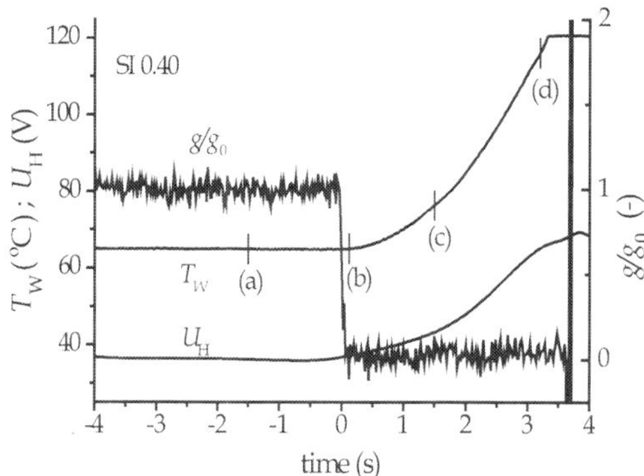

Fig. 12. Variations of surface temperature, heating voltage, and gravity for chip S (Wei et al., 2010).

Steady- or quasi-steady nucleate pool boiling was observed in the experiments for low and intermediate heat flux in the short-term microgravity conditions. At low heat fluxes in microgravity condition, the vapor bubbles increase in size but little coalescence occurs among bubbles due to large space between adjacent bubbles on the heater surface, thus the steady nucleate pool boiling can be obtained. As the heat flux increases, the vapor bubbles number as well as their size significantly increase in microgravity. Coalescence occurs continuously among adjacent bubbles. Departure of the coalesced bubbles from the heater surface caused by the surface oscillation of the coalesced bubble in lateral direction results in a constant heater temperature and heat flux in microgravity compared to that in normal gravity. The steady-state pool boiling still can be maintained. At high heat fluxes, a large coalesced bubble forms quickly and covers the heater surface completely in microgravity, followed by shrinking to an oblate in shape and smooth in contour due to the highly subcooled condensation (Fig. 11). An obvious increase of the heater temperature (Fig. 12), which indicates deterioration of boiling heat transfer, is then observed. Furthermore, the wall temperature exceeded the upper limit cutoff of the thermocouple instrument near the end of the short-term microgravity. It is possible for the occurrence of local dry-out or transition to film boiling at the bottom of the large coalesced bubble.

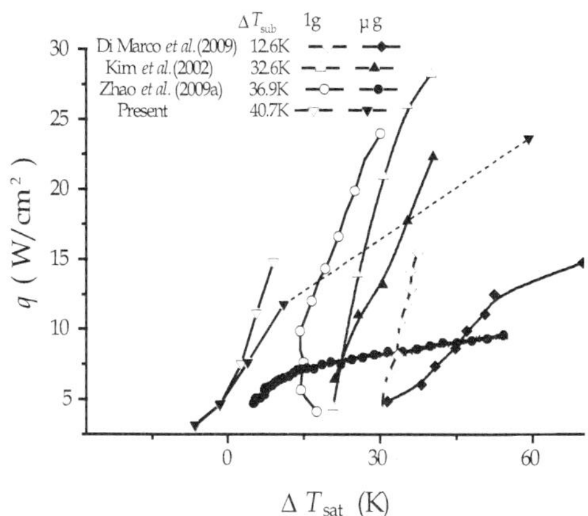

Fig. 13. Comparisons of pool boiling curves of FC-72 on plate of the present results with other data reported in the literature (Xue et al., 2010).

Figure 13 plots the heat transfer data of FC-72 on plate for the present results and other data reported in the literature under microgravity together with the pool boiling curves obtained under normal gravity. The last data point with the highest heat flux is not the value of the steady state due to the short duration of microgravity, and then a dashed line was used to connect it with the other data points. It can be clearly see that the influence of subcooling on nucleate pool boiling heat transfer in microgravity appears similar to that in earth gravity, i.e. the heat transfer increase distinctly as the subcooling increases. This agrees with the results of studies made by Lee et al. (1997). But, since the gravity level also greatly influences the average heater surface temperature, especially in the high heat flux region, the tendency of the influence of subcooling on nucleate pool boiling in microgravity is still required to further investigate. In particular, the results of Zhao et al. (2009a) in microgravity changes greatly from that in normal gravity and the value of CHF is only about one third of that at the similar pressure and subcooling in terrestrial condition. Besides, the trend of the present data result is consistent with that the Di Marco & Grass (2009), and the slopes of the heat transfer curves decrease in the high heat flux region in microgravity. However, the result of Kim et al. (2002) in microgravity changes slightly from that in normal gravity, which is different from the present and other studies. Thus, it can be inferred that the heat transfer of nucleate pool boiling in microgravity is related with the liquid subcooling, the size of the heater, the heating method, content of non-condensable gas, and so on.

4. Pool boiling on micro-pin-finned surface in microgravity

Very recently, a serial of experiments on boiling enhancement in microgravity by use of micro-pin-fins which were fabricated by dry etching have been performed in the drop tower Beijing (Wei et al., 2010). This project was motivated by the following observations in space experiments of boiling. Vapour bubbles cannot depart easily from the heater surface in microgravity, and then can grow attaching to the surface and coalesced with each other. As the increase of their sizes, the coalesced bubbles can cover the heater surface and prevent the fresh liquid from moving to the heater surface, thus local dryout may occur, resulting in deterioration of heat transfer. On the contrary, if plenty of fresh liquid can be supplied to the superheated wall for vaporization, the efficient nucleate pool boiling can be maintained and then no deterioration of heat transfer can occur. Following the enhanced boiling heat transfer mechanisms for the micro-pin-finned surfaces (Wei et al., 2009), it is supposed that although the bubbles staying on the top of the micro-pin-fins can not be detached soon in microgravity, the fresh bulk liquid may still access to the heater surface through interconnect tunnels formed by the micro-pin-fins due to the capillary forces, which is independent of the gravity level.

The experimental facility and schedual are similar with those used in Xue et al. (2010). The test chip was a P doped N-type silicon chip with the dimensions of 10×10×0.5 mm³. Micro-pin-fins were fabricated on the chip surface for enhancing boiling heat transfer. The fin thickness is 50 µm and fin height is 60 µm (denoted as chip PF50-60). The test chip was heated by setting a constant electric current for the desired heat flux to initiate boiling on the heater surface. After the heat transfer reached a steady state in about two minute, the free falling of drop capsule started which could provide approximately 3.6 s effective microgravity environment. The high-speed video camera could work for a duration time of 8 s, which was divided into two half sections by an external trigger signal. The bubble behaviours in normal gravity before the release of the drop capsule was recorded in the first half section, while those in microgravity after the release was recorded in the other one. Moreover, the data measurement and the video recording were operated simultaneously.

The transition of vapor bubble behaviors and the mean surface temperature of the micro-pin-finned chip responding to the variation of gravity level for the heating current of 0.42 A (corresponding to a heat flux of 19.4 W/cm²), similar to that shown in Figs. 11 and 12, are shown in Figs. 14 and 15, respectively. The positions of Fig. 14 a-d are marked on the curves of the mean temperature of Chip PF50-60 shown in Fig. 15. The liquid subcooling keeps at about 41K, also as the same as that shown in Figs. 11 and 12.

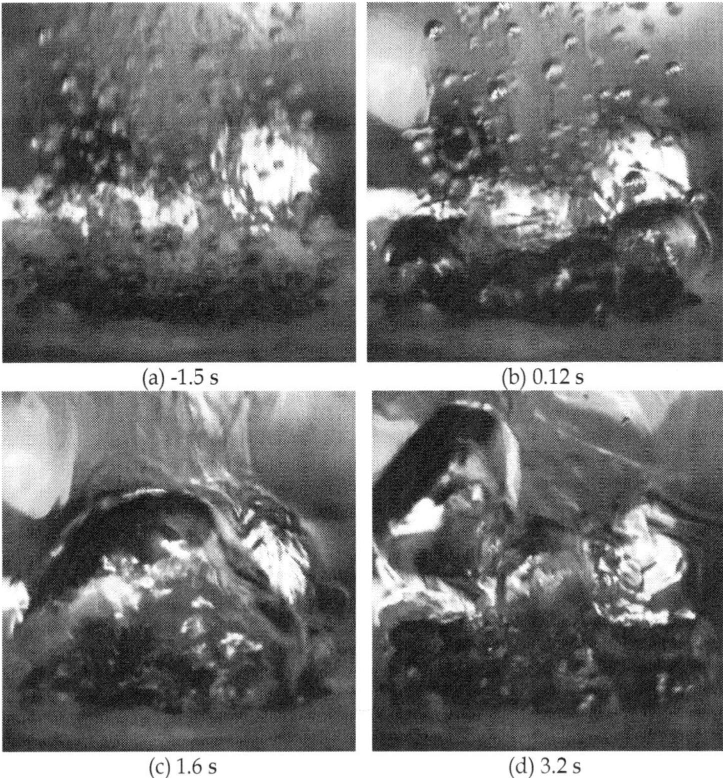

(a) -1.5 s   (b) 0.12 s

(c) 1.6 s   (d) 3.2 s

Fig. 14. Bubble behaviors on chip PF50-60 (Wei et al., 2010).

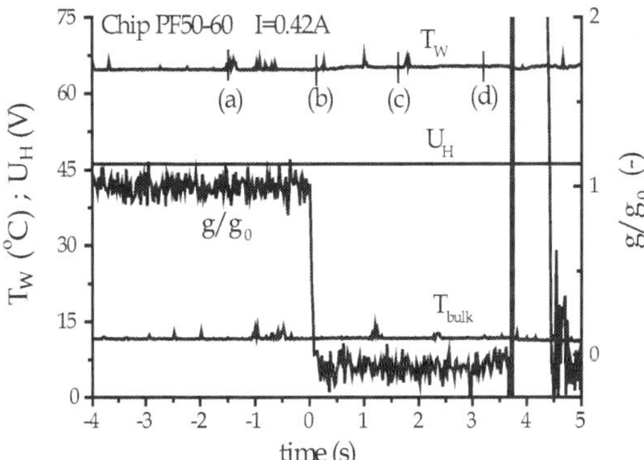

Fig. 15. Variations of surface temperature, heating voltage, and gravity for chip PF50-60 (Wei et al., 2010).

Just as the case of smooth chip, the bubbles generate and departure continuously from the heating surface caused by buoyancy forces in normal gravity before the release of the drop capsule (Fig. 14a). However, the bubble number are much larger than that for the smooth chip, indicating that the micro-pin-finned surface can provide larger number of nucleation sites for enhancing boiling heat transfer performance. At about 0.12 s after entering the microgravity condition, the vapour bubbles begin to coalesce with each other to form several large bubbles attaching on the chip surface (Fig. 14b). Some small bubbles are in the departure state when entering the microgravity condition, so we can still see them departing from the heater surface at this time. With increasing time, the bubbles coalesce to form a large spherical bubble (Fig. 14c). However, the large bubble covering on the heater surface does not cause obvious increase of wall temperature (Fig. 15).

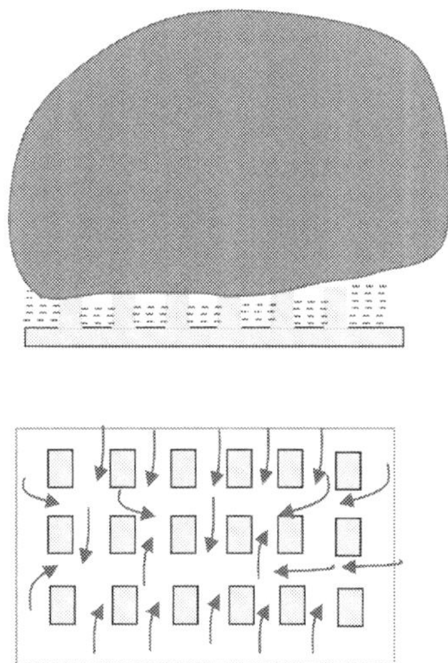

Fig. 16. Bulk liquid supply and micro-convection caused by capillary force (Wei et al., 2009).

The capillary force generated by the interface between the large bubble and the liquid of the micro-layer beneath the bubble drives plenty of fresh liquid to contact with the superheated wall for vaporization through the regular interconnected structures formed by the micro-pin-fins, as well as improves the micro-convection heat transfer by the motion of liquid around the micro-pin-fins, as shown schematically in Fig. 16. The sufficient supply of bulk liquid to the heater surface guarantees the continuous growth of the large bubble. Therefore, contrary to boiling on chip S, there is no deterioration of boiling heat transfer performance for the micro-pin-finned surface in microgravity, and the heater surface temperature can keep almost constant in both gravity and microgravity conditions.

In summary, the micro-pin-fined surface structure can provide large capillary force and small flow resistance, driving a plenty of bulk liquid to access the heater surface for evaporation in high heat flux region, which results in large boiling heat transfer enhancement. Since the capillary force is no relevant to the gravity level, the micro-pin-fined surface appears to be one promising enhanced surface for efficient electronic components cooling schemes not only in normal gravity but also in microgravity conditions, which is very helpful to reduce the cooling system weight in space and in planetary neighbors.

5. Future researches on boiling in microgravity in china

A new project DEPA-SJ10 has been planned to be flown aboard the Chinese recoverable satellite SJ-10 in the near future (Wan & Zhao, 2008). In the project, boiling at a single artifical cavity will be used as a model for studying subsystems in nucleate pool boiling of pure substances. Transient processes of bubble formation, growth and detachment will be observed, while the temperature distribution near the active nucleation site will be measured at subcooling and saturated conditions. The main aim is to describe bubble behavior and convection around the growing vapor bubble in microgravity, to understand small scale heat transfer mechanisms, and to reveal the physical phenomena governing nucleate boiling.

Numerical simulation on single bubble boiling has also been proposed, in which the single bubble boiling is set as a physical model for studying the thermo-dynamical behaviors of bubbles, the heat transfer and the corresponding gravity effect in the phenomenon of nucleate pool boiling (Zhao et al., 2010). According to some preliminary results, it was indicated that the growing bubble diameter is approximately proportional to the 0.4-th power of the growing time. The detach diameter of bubble is proportional to the -1/3-th power of the gravity, while the growing period to the -4/5-th power of the gravity. The heat flux is approximately proportional to the 1.5-th power of wall superheat with a fixed number density of active nucleation sites in all the studied gravity levels. The heat transfer through the micro-wedge region has a very important contribution to the whole performance of boiling.

Further experimental investigation on the performance of micro-pin-finned surface has also planned to be conducted in the drop tower Beijing, which aims to study the behaviour at very high heat flux around the critical heat flux phenomenon, as well as to determine the optimal structure of the micro-pin-fins.

These projects will be helpful for the improvement of understanding of such phenomena themselves, as well as for the development of space systems involving boiling phenomenon.

6. Conclusion

Nucleate pool boiling is a daily phenomenon transferring effectively high heat flux. It is, however, a very complex and illusive process. Among many sub-processes in boiling phenomenon, gravity can be involved and play much important roles, even enshroud the real mechanism underlying the phenomenon. Microgravity experiments offer a unique opportunity to study the complex interactions without external forces, such as buoyancy, which can affect the bubble dynamics and the related heat transfer. Furthermore, they can also provide a means to study the actual influence of gravity on the boiling. On the other hand, since many potential applications exist in space and in planetary neighbors due to its high efficiency in heat transfer, pool boiling in microgravity has become an increasing significant subject for investigation.

In the past decade, two research projects on nucleate pool boiling in microgravity have been conducted aboard the Chinese recoverable satellites. Ground-based experiments both in normal gravity and in short-term microgravity in the drop tower Beijing and numerical simulations have also been performed. The major findings are summarized in the present chapter.

Steady boiling of R113 on thin platinum wires was studied with a temperature-controlled heating method, while quasi-steady boiling of FC-72 on a plane plate was investigated with an exponentially increasing heating voltage. It was found that the bubble dynamics in microgravity has a distinct difference from that in normal gravity, and that the heat transfer characteristic is depended upon the bubble dynamics. Lateral motions of bubbles on the heaters were observed before their departure in microgravity. The surface oscillation of the merged bubbles due to lateral coalescence between adjacent bubbles drove it to detach from the heaters. Considering the influence of the Marangoni effects, the different characteristics of bubble behaviors in microgravity have been explained. A new bubble departure model has also been proposed, which can predict the whole observation both in microgravity and in normal gravity.

Slight enhancement of heat transfer on wires is observed in microgravity, while diminution is evident for high heat flux in the plate case. These different characteristics may be caused by the difference of liquid supply underneath the growing bubbles in the above two different cases. It is then suggested that a high performance of heat transfer will be obtained in nucleate pool boiling in microgravity if effective supply of liquid is provided to the bottom of growing bubbles. A series of experiments of pool boiling on a micro-pin-finned surface have been carried out utilizing the drop tower Beijing. Although bubbles cannot detach in microgravity but stay on the top of the micro-pin-fins, the fresh liquid may still access to the heater surface through interconnect tunnels formed between micro-pin-fins due to the capillary forces, which is independent of the gravity level. Therefore, no deterioration of heat transfer in microgravity is observed even at much high heat flux close to CHF observed in normal gravity.

The value of CHF on wires in microgravity is lower than that in normal gravity, but it can still be predicted well by the correlation of Lienhard & Dhir (1973), although the dimensionless radius in the present case is far beyond its initial application range. The scaling of CHF with gravity is thus much different from the traditional viewpoint, and a possible mechanism is suggested based on the experimental observations.

7. Acknowledgement

The studies presented here were supported financially by the National Natural Science Foundation of China (10972225, 50806057, 10432060), the Chinese Academy of Sciences (KJCX2-SW-L05, KACX2-SW-02-03), the Chinese National Space Agency, and the support from the Key Laboratory of Microgravity/CAS for experiments utilizing the drop tower Beijing. The author really appreciates Prof. W. R. Hu, Mr. S. X. Wan, Mr. M. G. Wei, and all research fellows who have contributed to the success of these studies. The author also wishes to acknowledge the fruitful discussion and collaboration with Prof. H. Ohta (Kyushu University, Japan), Prof. J. J. Wei (Xi'an Jiaotong University, China).

8. References

Di Marco, P., 2003. Review of reduced gravity boiling heat transfer: European research. J. Jpn. Soc. Microgravity Appl., 20(4), 252–263.

Di Marco, P., Grassi, W., 1999. About the scaling of critical heat flux with gravity acceleration in pool boiling. In: Proc. XVII UIT Nat. Heat Transfer Conf., Ferrara, pp.139-149.

Di Marco, P., Grassi, W., 2009. Effect of force fields on pool boiling flow patterns in normal and reduced gravity. Heat Mass Transfer, 45: 959-966.

Johnson, H.A., 1971. Transient boiling heat transfer to water. Int. J. Heat Mass Transfer, 14, 67–82.

Kim, J., 2003. Review of reduced gravity boiling heat transfer: US research. J. Jpn. Soc. Microgravity Appl., 20(4), 264–271.

Kim, J., Benton, J., Wisniewski, D., 2002. Pool boiling heat transfer on small heaters: effect of gravity and subcooling. Int. J. Heat and Mass Transfer, 45: 3919-3932.

Lee, H.S., Merte, H., Jr., Chiaramonte, F., 1997. Pool boiling curve in microgravity. J. Thermophy. Heat Transfer, 11(2), 216–222.

Lienhard, J.H., Dhir, V.K., 1973. Hydrodynamic prediction of peak pool boiling heat fluxes from finite bodies. J. Heat Transfer, 95, 152–158.

Liu, G., 2006. Study of subcooled pool boiling heat transfer on thin platinum wires in different gravity conditions. M. Sc. Thesis, Institute of Mechanics, Chinese Academy of Sciences, Beijing, China.

Oka, T., Abe, Y., Mori, Y.H., Nagashima, A., 1995. Pool boiling of n-pentane, CFC-113 and water under reduced gravity: parabolic flight experiments with a transparent heater. J. Heat Transfer Trans. ASME, 117: 408-417.

Ohta, H., Kawasaki, K., Azuma, H., Yoda, S., and Nakamura, T., 1999. On the heat transfer mechanisms in microgravity nucleate boiling. Adv. Space Res., 24(10): 1325-1330.

Ohta, H., 2003a. Review of reduced gravity boiling heat transfer: Japanese research. J. Jpn. Soc. Microgravity Appl., 20(4), 272–285.

Ohta, H., 2003b. Microgravity heat transfer in flow boiling. Adv. Heat Transfer, 37, 1–76.

Ohta, H., Kawasaki, K., Azuma, H., Yoda, S., Nakamura, T., 1999. On the heat transfer mechanisms in microgravity nucleate boiling. Adv. Space Res., 24(10), 1325–1330.

Straub, J., 2001. Boiling heat transfer and bubble dynamics in microgravity. Adv. Heat Transfer, 35, 57–172.

Sun, K.H., Lienhard, J.H., 1970. The Peak Pool Boiling Heat Flux on Horizontal Cylinders, Int. J. Heat Mass Transfer, 13: 1425-1439.

Wan, S.X., Zhao, J.F., 2008. Pool boiling in microgravity: recent results and perspectives for the project DEPA-SJ10. Microgravity Sci. Tech., 20(3-4), 219-224.

Wan, S.X., Zhao, J.F., Liu, G., Li, B., Hu, W.R., 2003. TCPB device: description and preliminary ground experimental results. In: 54th Int. Astronautical Cong., Sep. 29– Oct 3, Bremen, Germany.

Xue, Y.F., Zhao, J.F., Wei, J.J., Li, J., Guo, G., Wan, S.X., 2010. Experimental Study of FC-72 Pool Boiling on Smooth Silicon Chip in Short-Term Microgravity. Submitted to J. Heat Transfer Trans. ASME.

Yan, N., 2007. Experimental study on pool boiling heat transfer in microgravity. M. Sc. Thesis, Institute of Mechanics, Chinese Academy of Sciences, Beijing, China.

You, S.M., Hong, Y.S., O'Connor, J.P., 1994. The Onset of Film Boiling on Small Cylinders: Local Dryout and Hydrodynamic Critical Heat Flux Mechanisms, Int. J. Heat Mass Transfer, 37: 2561-2569.

Wei, J.J., Zhao, J.F., Yuan, M.Z., Xue, Y.F., 2009. Boiling heat transfer enhancement by using micro-pin-finned surface for electronics cooling. Microgravity Sci. Tech., 21(Suppl. 1): S159 – S173.

Wei, J.J., Xue, Y.F., Zhao, J.F., Li, J., 2010. High efficiency of heat transfer of nucleate pool boiling on micro-pin-finned surface in microgravity. submitted to Chin. Phys. Lett.

Zhao, J.F., Wan, S.X., Liu, G., Hu, W.R., 2004. Subcooled pool boiling in microgravity: results of drop tower testing. In: 7th Drop Tower Days, Sep. 12–25, Bremen, Germany.

Zhao, J.F., Liu, G., Li, Z.D., Wan, S.X., 2007. Bubble behaviors in nucleate pool boiling on thin wires in microgravity. In: 6th Int. Conf. Multiphase Flow, July 9–13, Leipzig, Germany.

Zhao, J.F., Liu, G., Wan, S.X., Yan, N., 2008. Bubble dynamics in nucleate pool boiling on thin wires in microgravity. Microgravity Sci. Tech., 20(2), 81-89.

Zhao, J.F., Li, J., Yan, N., Wang, S.F., 2009a. Bubble behavior and heat transfer in quasi-steady pool boiling in microgravity. Microgravity Sci. Tech., 21(Suppl. 1): S175 – S183.

Zhao, J.F., Lu, Y.H., Li, J., 2009b. CHF of pool boiling on microwires. ASME 2009 2nd Micro/Nanoscale Heat Mass Transfer Int. Conf., December 18-21, 2009, Shanghai, China.

Zhao, J.F., Lu, Y.H., Li, J.,2009c. CHF on cylinders– revisit of influences of subcooling and cylinder diameter. ECI Int. Conf. on Boiling Heat Transfer, May 3-7, 2009, Florianópolis, Brazil.

Zhao, J.F., Wan, S.X., Liu, G., Yan, N., Hu, W.R., 2009d. Subcooled pool boiling on thin wire in microgravity. Acta Astronautica, 64(2-3): 188 – 194.

Zhao, J.F., 2010. Two-phase flow and pool boiling heat transfer in microgravity. Int. J. Multiphase flow, 36(2): 135-143.

Zhao, J.F, Li, Z.D, Li, J. 2010. Numerical simulation of single bubble boiling in different gravity conditions. In: 8th Japan-China-Korea Workshop on Microgravity Sciences for Asian Microgravity Pre-Symposium, September 22 – 24, 2010, Akiu, Sendai, Japan.

9

Heat Transfer in Film Boiling of Flowing Water

Yuzhou Chen
China Institute of Atomic Energy
China

1. Introduction

Film boiling is a post critical heat flux (CHF) regime with such a high surface temperature, that the wall can not contact with the liquid, but is covered by the vapor and thus has relatively low heat transfer efficiency due to poor heat conductivity of the vapor. The film boiling is encountered in various practices, e.g., the metallurgy, the refrigeration, the chemical and power engineering, etc.. In a postulated break loss of coolant accident of nuclear reactors the uncovered core would experience this regime, and the maximum fuel temperature would be primarily dominated by the heat transfer of film boiling. Due to its significant importance to the applications the film boiling has received extensive investigations both experimentally and theoretically. It was one of three subjects in a coordinated research program on Thermal-hydraulic relationships for advanced water-cooled reactors, which was organized by the International Atomic Energy Agency (1994 – 1999). A comprehensive review on these investigations has been presented in the technical document (IAEA-TECDOC-1203, 2001)

In film boiling the heat is transferred from the wall to the vapor, then from the vapor to the liquid, characterized by non-equilibrium. The interaction between two phases dominates the vapor generation rate and the superheat, associated with extremely complicated characteristics. This presents a major challenge for the estimation of heat transfer because of less knowledge on the interfacial processes. In particular, due to the peculiar feature of the boiling curve it is difficult to establish the film boiling regime at stable condition in a heat flux controlled system by using a conventional experimental technique. As shown in Fig.1, the stable film boiling regime can only be maintained at a heat flux beyond the CHF, which associates with an excessively high surface temperature for water. But for a heat flux, q, below the CHF, the regime can not be maintained stably at the post-CHF region (F or T), but at the pre-CHF region (N).

The experimental data on film boiling were mostly obtained with refrigerant or cryogenic fluids, and the data of water were generally obtained in a temperature-controlled system or at transient condition with less accuracy. Since a so-called hot patch technique was developed for establishment of the stable film boiling regime (Groeneveld, 1974, Plummer, 1974, Groeneveld & Gardiner, 1978), a large number of experimental data have been obtained (Stawart & Groeneveld, 1981, Swinnerton et al., 1988, Mossad, 1988). Based on the data base various physical models have been proposed (Groeneveld & Snoek, 1984, Groeneveld, 1988, Mossad & Johannsen, 1989), and the tabular prediction methods have been developed for fully-developed film boiling heat transfer coefficients (Leung et al., 1997, Kirillov et al., 1996).

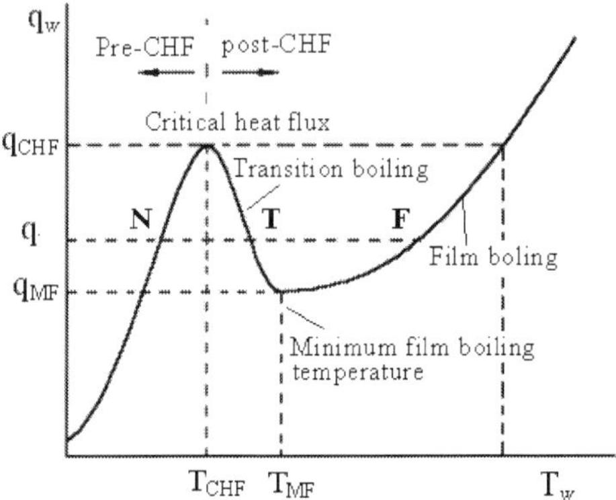

Fig. 1. Typical boiling curve

In 1984 a directly heated hot patch technique was applied by the authors to reach higher heat flux, enabling the steady-state experiment to cover extended range of conditions (Chen & Li, 1984). The results fill the gaps of data base, especially in the region of lower flow, where thermal non-equilibrium is significant, associated with much complicated parametric trends and strongly history-dependent features of the heat transfer coefficient (Chen, 1987, Chen et al., 1989, Chen & Chen, 1994). With these unique data the film boiling has been studied systematically and the prediction methods have been suggested, as will be shown in the following paragraphs.

2. Steady-state experimental technique

The hot patch technique is to supply separate power to a short section just ahead of the test section to reach CHF, preventing the rewetting front from moving forward. It was first used in freon and nitrogen experiments (Groeneveld, 1974, Plummer, 1974). To increase the power of hot patch for the experiment of water, it was improved by Groeneveld & Gardiner (1978), using a big copper cylinder equipped with a number of cartridge heaters.

To reach further high heat flux, a directly heated hot patch technique was applied by the authors (Chen & Li, 1984). As shown schematically in Fig.2, the test section included two portions, AB and BC, with each heated by a separate supply. The length of section AB was 10 - 25 mm. Near the end (B) the wall thickness was reduced locally, so that a heat flux peak can be created there by electric supply due to higher electric resistance.

During experiment, at first the inlet valve of the test section was closed, and the water circulation was established in a bypass at desired pressure, flow rate and temperature. The test section was then heated by switching on two supplies with it in empty of water. When the wall temperature reached above 500 °C, the flow was switched from the bypass to the test section. As the rewetting front moved upward the power to the upstream section was increased to reach CHF at the end (B), where the rewetting front was arrested without an excessive increase in the wall temperature as a result of axial heat conduction. In the same way, another rewetting front was arrested at the end of section BC by the upper hot patch. Therefore, the stable film boiling regime was maintained on the section BC with heat flux below the CHF. Shown in Fig.3 are the pictures of stable film boiling in an annulus for different water temperatures with the hot patch on and a reflooding transient with the hot patch off.

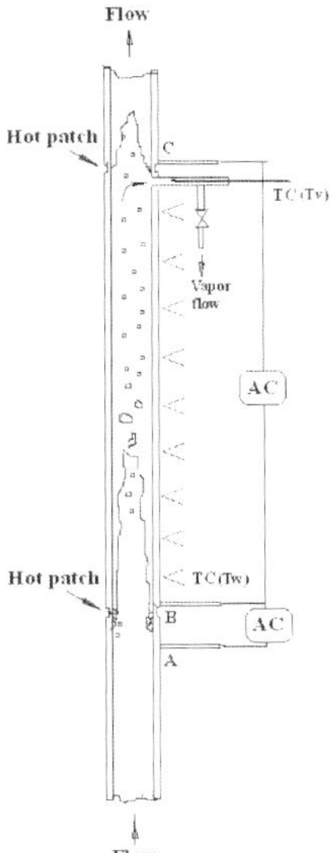

Fig. 2. Schematic of the test section with measurements of both the wall and vapor temperatures

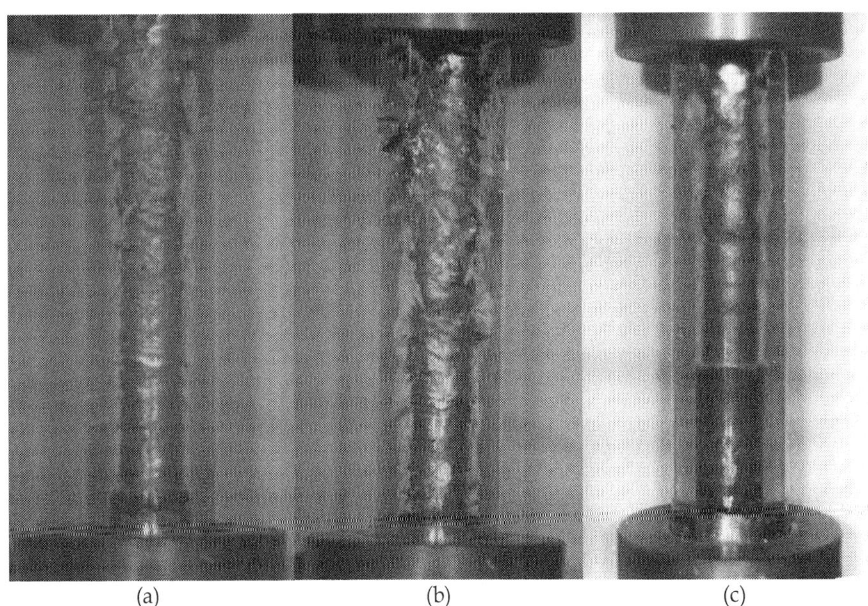

(a)   (b)   (c)

Fig. 3. Inverted annular film boiling in an annulus with water flowing upward (a) and (b): Stable regime (with the hot patch on), $T_{l,a} < T_{l,b}$, (c): Reflooding transient (with the hot patch off)

The steady-state film boiling experiments have been performed with water flowing upward in tubes of 6.7 – 20 mm in diameter and 0.15 – 2.6 m in length, covering the ranges of pressure of 0.1 – 6 MPa, mass flux of 23 – 1462 kg/m²s and inlet quality of -0.15 – 1.0.

3. Characteristics of the heat transfer in film boiling

The term "film boiling" was originally used for a post-CHF regime in a pool, characterized by the wall separated from the stagnant liquid by a continuous vapor film. It was then used in forced flow, though the flow pattern varied with the enthalpy in the channel. It includes two major regimes: 1) the inverted annular film boiling (IAFB), which occurs at subcooled or low quality condition, and 2) the dispersed flow film boiling (DFFB), which occurs at saturated condition with the void fraction larger than around 0.8. In IAFB the vapor film separates the wall from the continuous liquid core, in which some bubbles might be entrained for saturated condition. The DFFB is characterized by liquid droplets entrained in the continuous vapor flow. It can be resulted from break-up of the IAFB or from dryout of the liquid film in an annular flow. Fig.4 shows the film boiling regimes in a bottom reflooding transient at different flooding rates.

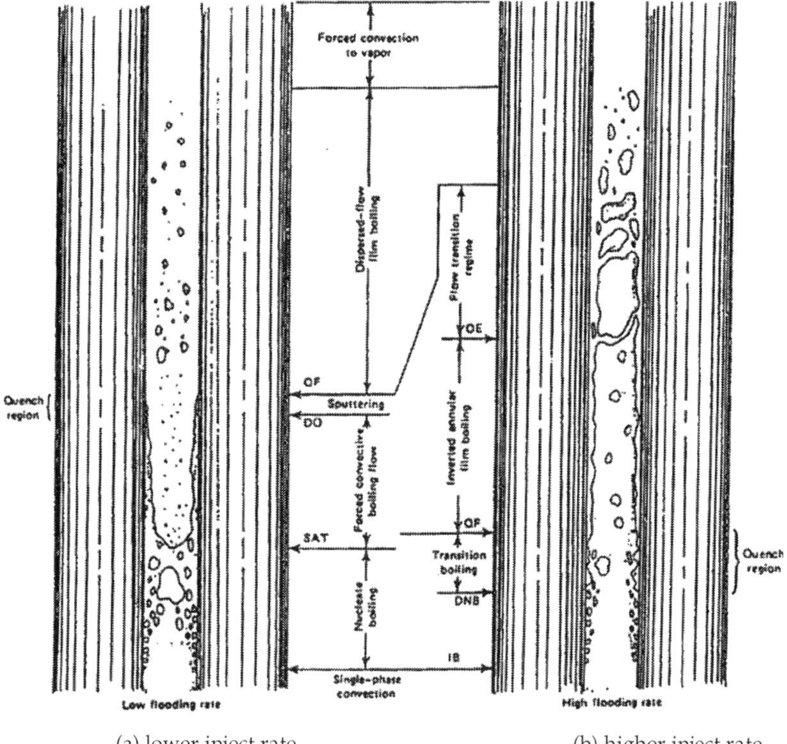

(a) lower inject rate   (b) higher inject rate

Fig. 4. Film boiling regimes during reflooding with different flooding rates (Arrieta & Yadigaroglu, 1978)

Typical experimental results are exemplified in Fig.5, where the heat transfer coefficient distributions in a tube for different inlet qualities are displayed by h (= $q_w/(T_w\text{-}T_s)$) versus x_E. For subcooled (run no.1) and low quality (run no. 2) inlet condition the post-CHF region initiates with IAFB followed by DFFB. While for relatively high inlet quality (run no. 3 and 4) the DFFB covers the whole post-CHF region. As seen, lower heat transfer coefficients are attained in the transition region.

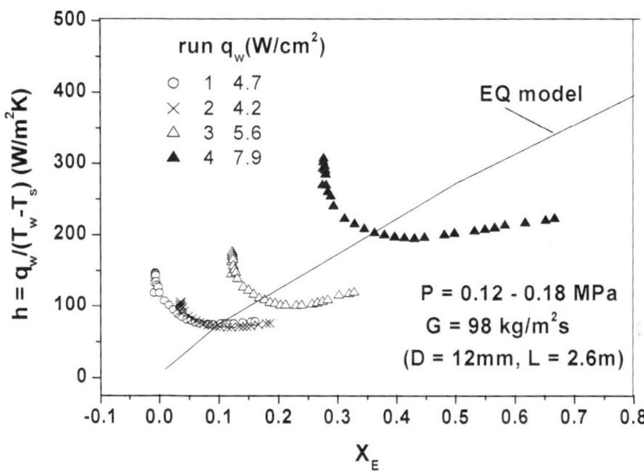

Fig. 5. Distributions of the heat transfer coefficients along the length for different inlet qualities

3.1 Inverted annular film boiling

In IAFB the heat is transferred by convection and radiation from the wall to the vapor, subsequently from the vapor to the interface with liquid. For subcooled condition it is then partially transferred to the liquid core. At the interface the vaporization takes place and the vapor generation rate is determined by the heat flux to the interface minus that to the liquid core. As the increase of vapor generation the vapor flow in the film may transit from laminar to turbulent. Furthermore, the interaction between two phases could result in interface oscillation, having enhancement effect on the heat exchange in both the vapor film and the liquid core.

3.1.1 Effects of the pressure, mass flux and subcooling

Fig.6 shows the distributions of heat transfer coefficients (h = $q_w/(T_w\text{-}T_s)$) under different conditions. For lower flow with higher subcooling the h decreases rapidly with distance, while as subcooling decreasing the h decreases, and the trend becomes mild (Fig.6(a)). For higher flow with higher subcooling a maximum h is attained at a few centimeters from the dryout point (Fig.6(b,c)). In this case the thickness of vapor film is very small, so the interface oscillation could lead to dry-collision between liquid and wall, resulting in a substantial increase in the h. For low inlet subcooling the variation of h along the length is not substantial (Fig.6(f)). This suggests that as the distance increases the negative effect of the increase in thickness and the positive effect of disturbance in the vapor film are comparable on the heat transfer.

At higher pressure the heat transfer coefficients are generally higher than those at lower pressure for low subcooling or saturation condition (Fig. 6(e, f). An opposite effect is observed for higher flow and higher subcooling (Fig. 6(d)). This can be explained in terms of the thickness of vapor film and the interface oscillation. Higher pressure corresponds to smaller volumetric vapor generation and thus smaller thickness of the film.

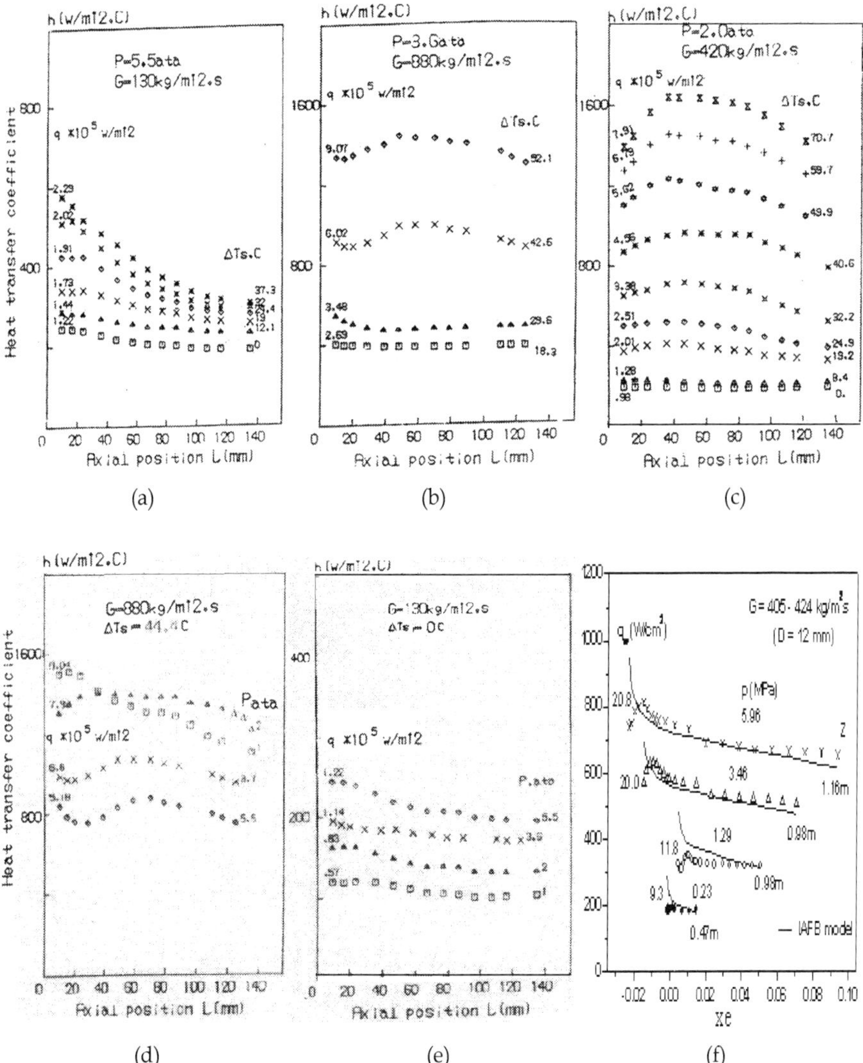

Fig. 6. Variation of the heat transfer coefficient in IAFB under different conditions (Chen, 1987)

It results in higher h for low subcooling or saturated condition. Nevertheless at higher flow and higher subcooling the film is very thin, and for lower pressure there could exist stronger interface oscillation, even dry-collision of liquid to wall, which has predominant effect on the heat exchange in both the vapor film and the liquid core. While for higher pressure this effect is less important due to less interface oscillation.

3.1.2 Effect of the preceding heating

To clarify the effect of preceding heating, an additional power supply was provided to a section of L = 225 mm immediately ahead of AB (with heat flux q_0). When the q_0 exceeded a value for the onset of boiling a substantial fall in the T_w was attained over the first about 100 mm for fixed p, G and ΔT_s at the dryout point, as shown in Fig.7 (Chen, 1987). In this case a bubble layer was produced upstream, which was determinant for the vapor flow rate and the interfacial oscillation over a certain length near the dryout point. For high subcooling the vapor film was very thing, and this effect could be more substantial. Nevertheless, at a q_0 without boiling the T_w near the dryout point was increased slightly. In this case a temperature profile was developed in the subcooled liquid core, which would result in lower heat transfer coefficient from the interface to liquid core, compared to that with uniform core temperature for the same average temperature.

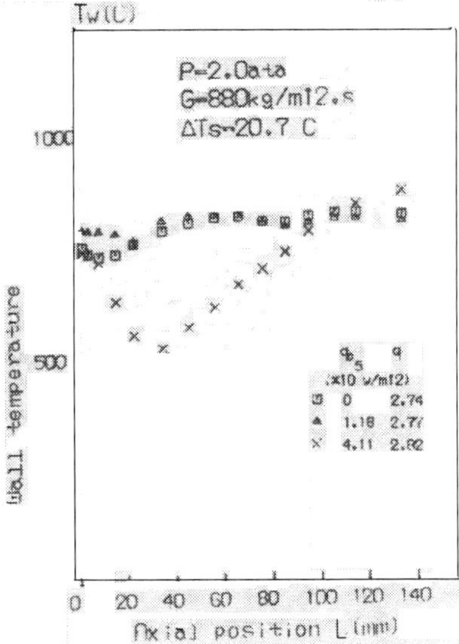

Fig. 7. Effect of the preceding heating power on the wall temperature (Chen, 1987)

3.2 Dispersed flow film boiling

In DFFB the heat is transferred from the wall to the vapor, then from the vapor to the liquid droplets entrained in the continuous vapor flow. The wall temperature is mainly dominated by the vapor convection heat transfer and the vapor temperature. The liquid droplets would induce some disturbance for the vapor convection, and the vapor-droplet interfacial heat transfer determines the vapor temperature. This effect is closely relative with the flow conditions, associated with complicated parametric trends of the wall temperature.

3.2.1 Effects of the pressure, mass flux and inlet quality

Typical distributions of the h (= $q_w/(T_w-T_s)$) along the length are shown in Fig.8. In general, as distance increases from the dryout point, at first the h decreases rapidly. For lower flow it decreases monotonously over the whole length, though the trend becomes milder downstream. For higher flow the h turns to increase after a certain distance. This behavior varies distinctly with pressure. At p < 0.2 MPa , for instance, the increase trend in the h is observed at mass flux below 300 kg/m²s, while for higher pressure it is attained at higher mass flux.

In addition to the local parameters, p, G and x_e, the inlet quality (at the dryout point) has a significant effect on the h. As seen, for the same pressure and mass flux with different inlet quality, different h may be attained at a fixed local x_e, and higher h corresponds to higher inlet quality, exhibiting a strongly history-dependent feature. This is understandable due to the fact that to reach a same x_e the flow with higher inlet quality subjects to less heat transfer and thus less superheat of vapor. At low flow this effect is so significant, that the fully-developed condition can not be reached even at L > 2 m or L/D > 200.

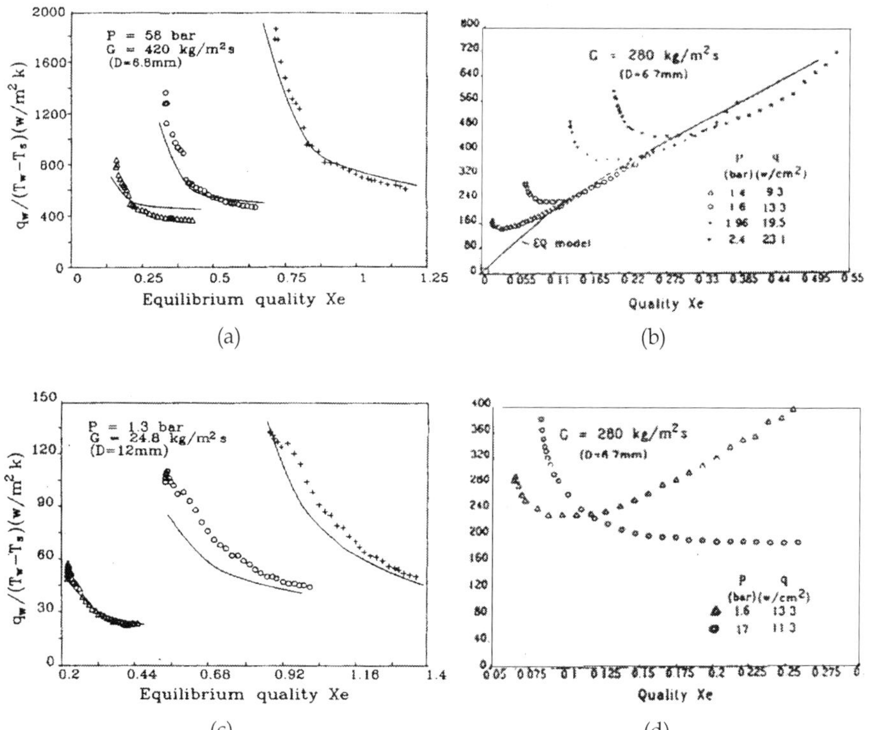

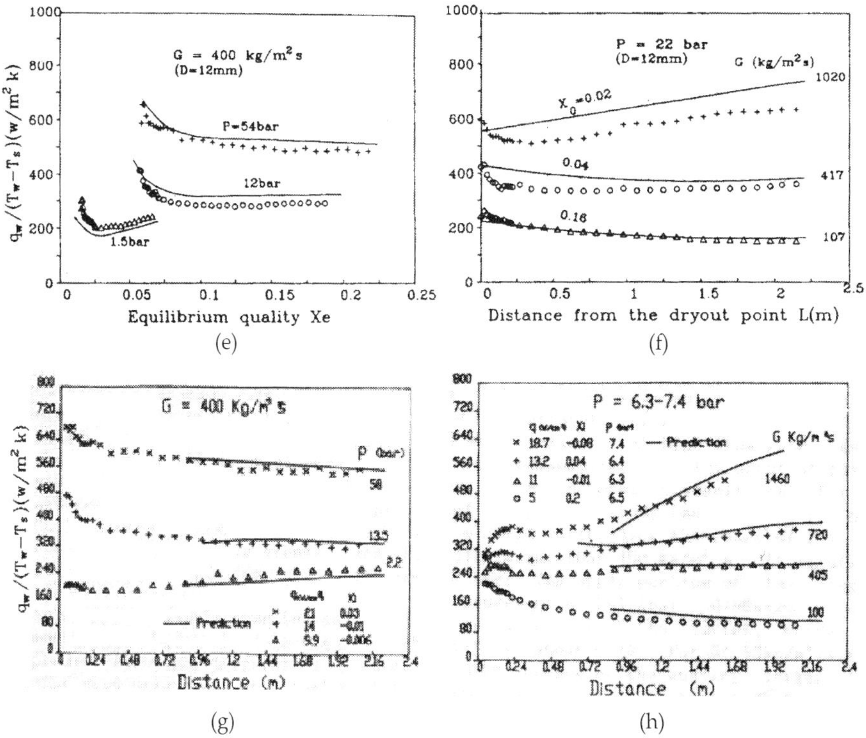

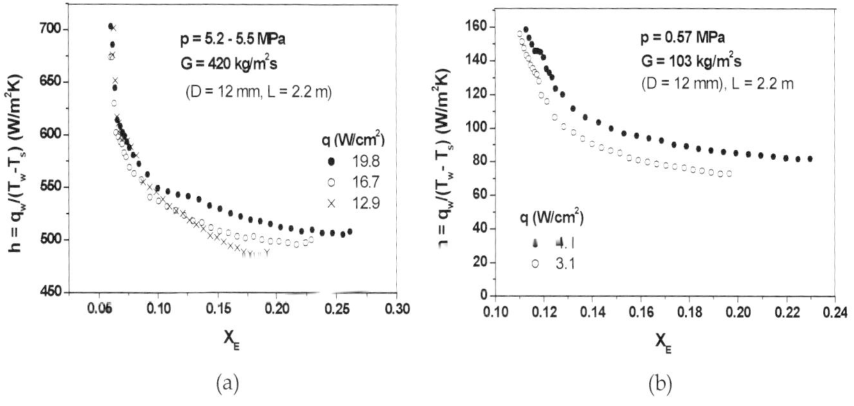

Fig. 8. Variations of heat transfer coefficient for different conditions in DFFB (− mechanistic model), (a-f): DFFB covering the whole post-CHF region;   (g,h): DFFB preceded by IAFB (Chen et al., 1991, 1992, 1994b)

Fig. 9. Effect of heat flux on the heat transfer coefficients in DFFB

3.2.2 Effects of other factors

The effect of heat flux on the h is shown in Fig.9. Higher heat flux corresponds to higher h. This is mainly attributed to the increase in the radiation heat transfer due to higher wall temperature at higher heat flux. Fig.10 shows the effect of diameter on the h. In general, smaller diameter corresponds to lower heat transfer coefficients over the downstream. It is expectable that for same heat flux and mass flux smaller diameter corresponds to greater increase rate of the enthalpy along the length, leading to stronger thermal non-equilibrium and thus lower h.

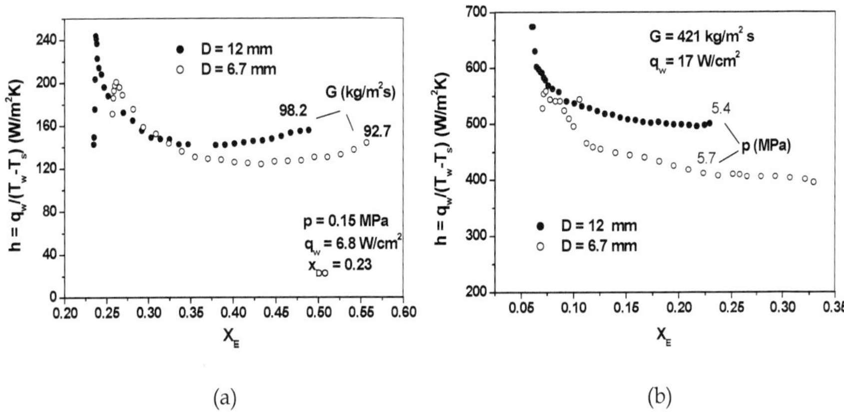

Fig. 10. Effect of diameter on the heat transfer coefficients in DFFB

3.2.3 Thermal non-equilibrium

The complicated parametric trends of the heat transfer in DFFB are closely related to the thermal non-equilibrium, which is determined by the fraction of total heat to the vapor for superheating. The following thermal non-equilibrium parameter was defined by Plummer et al. (1977),

$$K = \frac{x - x_0}{x_e - x_0} \tag{1}$$

with

$$x = x_e \left(1 + \frac{C_{pg}(T_v - T_s)}{h_{fg}} \right)^{-1}$$

where the T_v and T_s are the vapor temperature and saturation temperature, respectively, the x_0 is the quality at the dryout point, and the x and x_e are the local actual quality and equilibrium quality, respectively.

The case of K = 1 represents the thermal equilibrium, in which all the heat from the wall goes to liquid for evaporation and the vapor temperature keeps at constant (T_s), so the h increases along the length as the vapor flow rate increasing. The case of K = 0 represents that all the heat goes to the vapor for superheating without vapor generation, so the T_w increases as the T_v increasing and thus the h decreases monotonously. Fig.11 illustrates the substantial effect of the K on both the values and the trends of the heat transfer coefficient.

Using a technique to prevent the probe from striking by the liquid droplets and from the effect of radiation, the data of vapor superheat were successfully obtained in steady-state film boiling experiments near the exit of test section (Chen, 1992, Chen & Chen, 1994a). The values of K and ratio of $(T_v\text{-}T_s)/(T_w\text{-}T_s)$ were then evaluated from the vapor superheats measured at 2 m from the dryout point, as shown in Fig.12. For low X_0, the K decreases as X_0 increasing. At certain increased X_0 the trend becomes milder. It varies distinctly with pressure, and higher K is attained at lower pressure. The ratio $(T_v\text{-}T_s)/(T_w\text{-}T_s)$ decreases with mass flux. For G < 100 kg/m²s, the $(T_v\text{-}T_s)/(T_w\text{-}T_s)$ is larger than 0.5, suggesting a major contribution of the vapor superheat to the wall superheat. For G < 50 kg/m²s the thermal non-equilibrium is much significant, so that the T_v and T_w increase significantly along the length, and the h (= $q_w/(T_w\text{-}T_s)$) exhibits sharp decrease trend. The effects of various parameters on the thermal non-equilibrium can be explained in terms of droplet size and concentration, the vapor-droplet relative velocity and heat transfer coefficient, the properties, etc.. This is made clear in the analysis with the two-fluid mechanistic model.

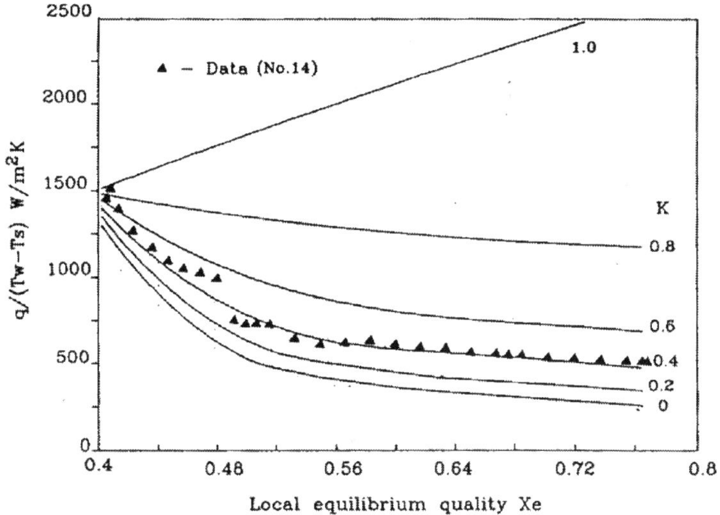

Fig. 11. Variations of the heat transfer coefficient along the length for different K (p = 5.8 MPa, G = 417 kg/m²s, x_{DO}=0.383, D=6.8mm) (Chen, et al., 1992)

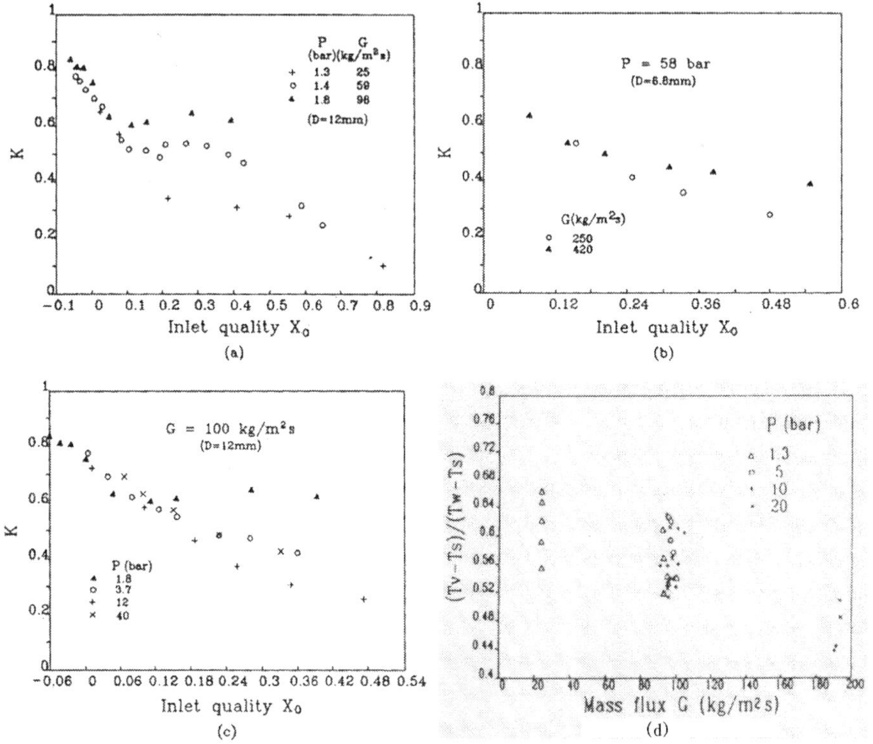

Fig. 12. Variations of the K with x_0 and $(T_v-T_s)/(T_w-T_s)$ with G for different conditions (Chen & Chen, 1994a, Chen, et al., 1992)

3.3 Minimum film boiling temperature

The minimum film boiling temperature, T_{min}, defines the boundary between the film boiling and the transition boiling, in which the wall contacts with the liquid intermittently and thus has much higher heat transfer coefficient than the film boiling. The collapse of film boiling could be resulted from the thermodynamic limit or the hydrodynamic instability. During a fast transient it could be thermodynamically controlled, while for low flow and low pressure it is likely to be hydraudynamically controlled. Six types of the film boiling termination mechanisms have been identified: (1) collapse of vapor film, (2) top flooding, (3) bottom flooding, (4) droplet cooling, (5) Leidenfrost boiling and (6) pool boiling. Significant discrepancies were found among the existing correlations of the T_{min}, and were attributed to the different types of the mechanism and scarcity of reliable data (Groeneveld & Snoek, 1984).

With the hot patch technique the minimum film boiling temperatures were measured in steady-state film boiling experiments by decreasing the power to the test section slowly with small steps until the collapse of film boiling occurred. The following empiric correlation was formulated from an experiment over the ranges of p = 115 – 6050 kPa, G = 53 – 1209 kg/m²s, x = -0.055 – 0.08 and ΔT_s = -35 – 25.1 K (Chen, 1989),

$$T_{min} = 363,6 + 38.37\ln p + 0.02844p - 3.86 \times 10^{-6} p^2 + a\Delta Ts \tag{2}$$

with

$$a = 17.1 / (3.3 + 0.0013p) \qquad \text{for } \Delta Ts > 0$$

and $\qquad\qquad\qquad a = 0 \qquad\qquad\qquad\qquad$ for $\Delta Ts \leq 0$

where the p is in kPa and the T_{min} and $\triangle T_s$ in K.

This correlation is in reasonable agreement with that derived from a similar experiment by Groeneveld and Steward (1982). It can be recommended for type (1-4) of film boiling termination.

4. Predictions for the heat transfer coefficients

As described above, the film boiling is characterized by non-equilibrium in both the velocity and temperature between phases, associated with extremely complicated parametric trends. The steady-state experimental data obtained in tube with flowing water were compared with the existing correlations, and significant discrepancies were observed between them, as shown in Fig.13. This result revealed the suspect of the correlations, and it was attributed to the lack of reliable data base and the difficulty in accounting for various physical mechanisms in a simple correlation (Stewart and Groeneveld, 1982, Groeneveld & Snoek, 1984). With steady-state technique the accuracy of the experimental data was improved substantially. As shown in Fig.14, the present steady-state data are in well agreement with those obtained by Swinnerton et al (1988) using indirectly heated hot patch technique for similar conditions.

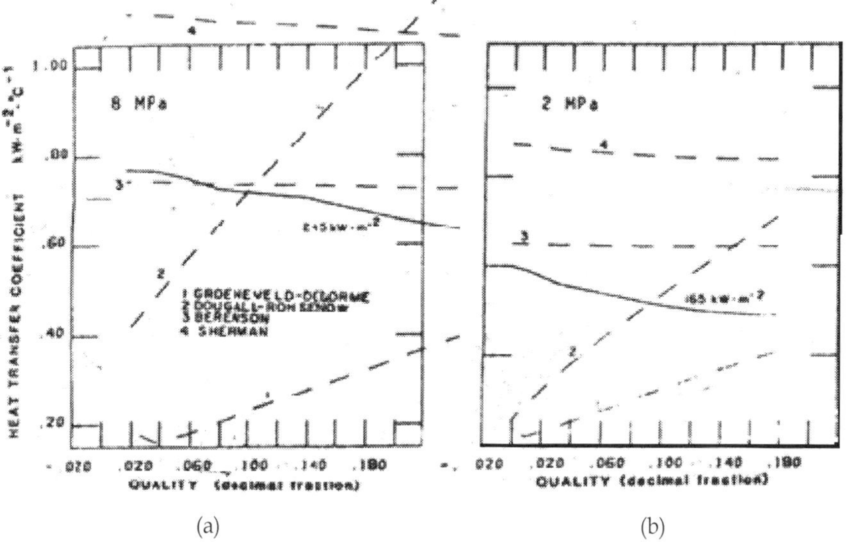

(a)   (b)

Fig. 13. Comparison of the steady-state experimental data of water with existing correlations (Stewart and Groeneveld, 1982)

To predict the non-equilibrium characteristics in film boiling the two-fluid models are favorable, and have been proposed by many investigators (Groeneveld, 1988, 1992, Mossad & Johannsen, 1989). The major challenge for these models is to simulate the interfacial heat and momentum exchanges. Due to less knowledge on these processes they were generally accounted by empiric or semi-empiric correlations. Therefore, the suitability of this kind of models is heavily determined by the ranges and the accuracy of data base. The following two-fluid models are developed based on the present experimental data.

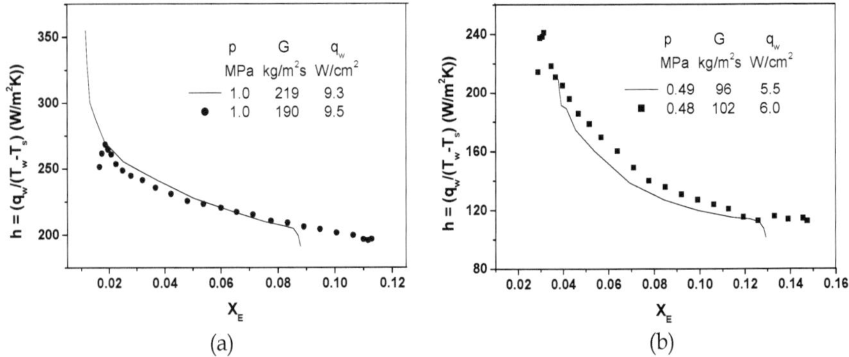

(a)   (b)

Fig. 14. Comparison of the results obtained in different steady-state experiments with water flowing in tube ($-$ Swinnerton, et al.D=9.75mm, ● Present experiment, D=12mm)

4.1 IAFB model

The schematic of flow structure in IAFB is shown in Fig.15, in which the vapor film is divided into two regions, A and B, bounded with the locus of the maximum velocity line.

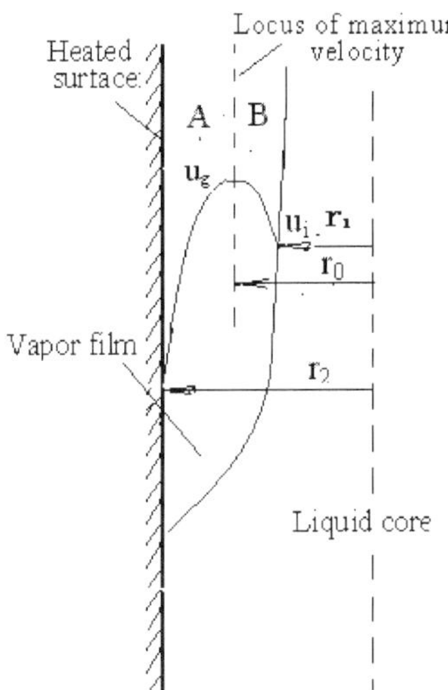

Fig. 15. Inverted annular film boiling mode

The assumptions are as follows:

- The pressure over a cross section is uniform.
- The kinetic energy and viscous dissipation and pressure loss due to acceleration are negligible.
- The properties of vapor are evaluated at $(T_w + T_s)/2$.
- The vapor-liquid interface is treated as smooth, and both the vapor and the liquid at the interface are at saturation.
- The velocity profile in the liquid core is uniform and equal to the vapor velocity at the interface.

In vapor film the force balance gives

$$\mu_v(1+\frac{\varepsilon_v}{v_v})\frac{du}{dr}+\frac{r^2-r_0^2}{2r}(\rho_v g+\frac{dp}{dz})=0 \qquad (3)$$

with boundary conditions as

$$u = 0 \ \text{at} \ r = r_2$$

and

$$u = u_i = \overline{u_l} \ \text{at} \ r = r_i$$

where $\overline{u_l}$ is the average velocity in the core.

Neglecting the weight of vapor, the integration of Eq.(3) gives, for region A

$$u_g = \frac{1}{2\mu_v(1+\varepsilon_v/v_v)}\frac{dp}{dz}\left[r_0^2\ln\frac{r}{r_2}+\frac{1}{2}(r_2^2-r^2)\right] \qquad (4)$$

and for region B

$$u_g = \frac{1}{2\mu_v(1+\varepsilon_v/v_v)}\frac{dp}{dz}\left[r_0^2\ln\frac{r}{r_1}-\frac{1}{2}(r^2-r_1^2)\right]+u_i \qquad (5)$$

with

$$\frac{dp}{dz}=\frac{r_1^2\rho_c g+(r_0^2-r_1^2)\rho_v g}{r_0^2} \qquad (6)$$

where ε_v is the eddy diffusivity in the vapor film, and the ρ_c is the average density of the core. Some vapor may be entrained in the core, and ρ_c is evaluated by

$$\rho_c = \alpha_c\rho_v+(1-\alpha_c)\rho_l$$

where α_c is the void fraction of the core. The following expression is attempted with adjustable factor c=40,

$$\alpha_c = \frac{1}{1+c\rho_v/\rho_l/x}$$

Assuming the momentum eddy diffusivity ε_v independent of r, by integrating Eq. (4) and (5) the vapor flow rate in the film is as

$$\Gamma = \frac{\pi \rho_v}{\mu_v (1 + \varepsilon_v / v_v)} \frac{dp}{dz} \left[\frac{r_0^4}{2} \ln \frac{r_2}{r_1} + \frac{1}{8}(r_2^4 - r_1^4) - \frac{1}{2} r_0^2 (r_2^2 - r_1^2) \right] + \pi u_i \rho_v (r_0^2 - r_1^2) \tag{7}$$

The wall heat flux is expressed as

$$q = q_{w-i,c} + q_{w-i,r} + q_v \tag{8}$$

where q_v is the heat for vapor superheating, $q_{w-i,c}$ and $q_{w-i,r}$ are the heat flux to the interface by convection and by radiation, respectively, evaluated by,

$$q_{w-i,c} = h_c (T_w - T_s) \tag{9}$$

with

$$h_c = \frac{k_v}{\delta} (1 + \frac{\varepsilon_v}{v_v} \frac{\mathrm{Pr}_v}{\mathrm{Pr}_t}) \tag{10}$$

and

$$q_{w-i,r} = 0.75 \frac{\sigma_{SB}}{(1 / \varepsilon_l + 1 / \varepsilon_w - 1)} (T_w^4 - T_s^4) \tag{11}$$

where δ is the thickness of film, and Pr_t is the turbulent Prandtl number.
The energy balance equation at the interface can be written as,

$$q_{w-i,c} + q_{w-i,r} = q_g + q_l$$

where q_l is the heat flux from the interface to the core, and q_g is the heat flux for evaporation, as

$$q_g = \frac{d\Gamma}{dz} h_{fg} / 2\pi r_1 \tag{12}$$

The mass equation is written as

$$\Gamma + \pi r_1^2 \overline{u} \, \rho_c = \pi r_2^2 G \tag{13}$$

From the authors experiments with saturated and low subcooling condition, the following empiric expression for ε_v is proposed with $\mathrm{Pr}_t = 1.0$.

$$\varepsilon_v / v_v = 0.011 \mathrm{Re}_v^{0.6} \mathrm{Re}_l^{0.3} \tag{14}$$

with

$$\mathrm{Re}_v = 2Gx(r_2 - r_1) / \mu_v$$

and

$$\mathrm{Re}_l = G(1 - x)D_1 / \mu_l$$

For the heat transfer from interface to subcooled liquid core the following correlation is available (Alalytis & Yadigaroglu 1987)

$$q_l = 0.06(k_l / D)\mathrm{Re}_l^{0.6}\mathrm{Pr}_l^{0.3}\left[\mu_v\rho_v / (\mu_l\rho_l)\right]^{-0.15}\mathrm{Pr}_v^{0.6}(Ts - T_f) \tag{15}$$

It should be noted that for subcooled condition the h near the dryout point is related to the condition at the dryout point, which is determined by the preceding heating, as described above. It is not simulated in the model, therefore, the calculation result could have appreciable uncertainty over a short length, especially for high subcooling.

For low subcooling and saturated condition, this model gives satisfactory calculations for the experimental data of p = 0.1 – 6 MPa, and G = 90 – 1462 kg/m²s, as exemplified in Fig.6(f).

4.2 DFFB model

The following two-fluid mechanistic model is based on the motion and energy equations, involving various equations for the wall-vapor-droplet heat transfer (Chen & Chen, 1994b). The heat from the wall, q_w, is transferred to the vapor by convection, q_c, and radiation, q_r, as

$$q_w = q_c + q_r \tag{16}$$

with

$$q_c = h_c(T_w - T_v) \tag{17}$$

and

$$q_r = \varepsilon\sigma_{SB}(T_w^4 - T_v^4) \tag{18}$$

where h_c is the convective heat transfer coefficient, σ_{SB} is the Boltzman's constant, and ε the emissivity.

The vapor temperature is evaluated by

$$T_v = T_s + (\frac{x_e}{x} - 1)\frac{h_{fg}}{C_{pv}} \tag{19}$$

where x and x_e are the actual quality and equilibrium quality, respectively.

From the heat balance equation, we have

$$\frac{dx}{dz} = \frac{6(1 - x)q_d^{''}}{u_l h_{fg}\rho_l\delta} \tag{20}$$

where u_l and δ are the droplet velocity and diameter, respectively. The heat flux to the droplet, $q_d^{''}$, includes the convective and radiative components, i.e.,

$$q_d^{''} = q_{d,c} + q_{d,r} \tag{21}$$

with

$$q_{d,r} = E\sigma_{SB}(T_w^4 - T_s^4) \tag{22}$$

and

$$E = \frac{2}{3}\varepsilon\frac{\delta}{(1 - \alpha)D}$$

The vapor to droplet convective heat transfer is evaluated by Frossling correlation, as

$$q_{d,c}'' = \frac{k_v}{\delta}(2 + 0.552(\frac{\rho_v(u_v - u_l)\delta}{\mu_v})^{0.5} \Pr_v^{0.33})(T_v - T_s) \tag{23}$$

The motion equation for the droplet gives

$$\rho_l u_l \frac{du_l}{dz} = \frac{3}{2}\frac{C_d}{\delta}\frac{\rho_v(u_v - u_l)^2}{2} - (\rho_l - \rho_v)g \tag{24}$$

where the drag coefficient, C_d, is evaluated by Ingebo's correlation, as

$$C_d = \frac{24(1 + 0.1\,\mathrm{Re}_d^{0.75})}{\mathrm{Re}_d} \quad \text{for} \quad \mathrm{Re}_d < 1000$$

with

$$\mathrm{Re}_d = \frac{\rho_v(u_v - u_l)\delta}{\mu_v}$$

and

$$C_d = 0.45 \text{ for } \mathrm{Re}_d \geq 1000$$

The wall to vapor convective heat transfer is predicted by the correlation derived from an author's convection heat transfer experiment in pure steam, multiplying an enhancement factor due to the disturbance induced by the droplets, as

$$h_c = h_0 F \tag{25}$$

with

$$h_0 = 0.0175\frac{k_{v,f}}{D}(\frac{\rho_{v,f}u_v D}{\mu_{v,f}})^{0.812} \Pr_{v,f}^{0.33}$$

The factor F would be related with pressure and quality. To fit the calculations with the experimental results, the following expression for F is proposed with p in MPa, as

$$F = 2.32(1 + 0.1p)e^{-12x} + 1$$

For the DFFB occurring from the dryout of liquid film of annular flow, the initial droplet diameter, δ_0, is evaluated by a modification of Nujiyama-Tanazawa equation, as

$$\delta_0 = 798\frac{\rho_v}{Gx_0}(\frac{\sigma}{\rho_l})^{0.5}(\frac{\mu_v}{Gx_0 D})^a \tag{26}$$

where

$$a = 0.63 - 0.1p \quad \text{for} \quad p < 1.3 MPa$$

and

$$a = 0.5 \quad \text{for} \quad p \geq 1.3 MPa$$

in which p is in MPa, δ and D in m, G in kg/m²s, μ in kg/ms, σ in kg/m, ρ_l and ρ_v in kg/m³.

The droplet has a maximum diameter, determined by the critical Weber number, We_c, as

$$\delta_m = \frac{We_c \sigma}{\rho_v (u_v - u_l)^2} \tag{27}$$

with $We_c = 5.0$.

For the DFFB occurring from the break-up of the IAFB, the initial diameter δ_0 is taken as δ_m. The onset of DFFB is defined by the criterion of droplet carryover (Yanomoto et al., 1987), as

$$J_{gc} = \frac{3.57}{\sqrt{\rho_g C_d}} (\sigma g (\rho_l - \rho_g))^{1/4} N_{\mu g}^{1/6} \tag{28}$$

with

$$N_{\mu g} = \frac{\mu_g}{\left[\rho_g \sigma \sqrt{\dfrac{\sigma}{g(\rho_l - \rho_g)}} \right]^{1/2}}$$

where J_{gc} is critical volumetric flux and $N_{\mu g}$ is viscosity number.

Assuming that the droplet break-up does not occur, so it varies with the actual quality, as

$$\delta = \delta_0 \left(\frac{1-x}{1-x_0} \right)^{1/3}$$

The present model gives satisfactory predictions for the DFFB data over the range of pressure of 0.1 – 6 MPa, mass flux of 23 – 1020 kg/m²s, as exemplified in Fig.8.

4.3 Tabular method

The film boiling models and phenomenological equations are time consuming, because they involve a large number of constitutive equations on the heat and momentum exchanges between phases and the evaluations of many properties. Furthermore, they are only applicable for the specific test condition of individual investigator. In recent years, the tabular method has been widely accepted due to its advantages of high accuracy, wide applicable range and convenience for application and updating.

4.3.1 fully-developed film boiling heat transfer coefficients

Based on over 15000 film boiling data points, Leung et al. (1997) have developed a look-up table for the fully-developed film boiling heat transfer coefficients in tubes with vertical upward flow. It contains a tabulation of normalized heat transfer coefficients at discrete local parameters of pressure (0.1 – 20 MPa in 14 steps), mass flux (0 – 7000 kg/m²s in 12 steps), quality (-0.2 – 1.2 in 11 steps) and heat flux (0.05 – 3 MW/m² in 9 steps). The agreements of the calculations of this table for surface temperatures with the data are 6.7% in r.m.s.

The range of flow conditions of the data base is presented in Fig.16. It shows significant gaps in the range of low flows and medium pressures, where the thermal non-equilibrium could be significant, and the heat transfer could be strongly history-dependent.

4.3.2 Thermal non-equilibrium parameter

For estimation of the heat transfer coefficient in the region of lower flow, a table for thermal non-equilibrium parameter has been proposed (Chen & Chen 1998). It formulates the non-equilibrium parameter, K_0, for tube of D = 8 mm at concrete values of the pressure p, mass flux G and quality at the dryout point x_0 (Tab.1). The values of K_0 were produced from the calculations of DFFB model, which was validated by the data of vapor temperature measured at the steady-state film boiling experiments.

The K is assumed to be related with the diameter, local quality and heat flux, as expressed by

$$K = K_0 F_d F_x F_q$$

where, F_d, F_x and F_q are the correction factors for the effects of diameter, local quality and heat flux, respectively.

For the present experimental results, the F_d is evaluated by

$$F_d = (\frac{D}{0.008})^{0.26}$$

The calculation indicates that from the dryout point the K increases rapidly, while after a short distance it does not vary appreciably. So it is simply accounted by a proper K_0 with $F_x = 1$. At present the expression for the F_q is not available and is also set to be 1.0.

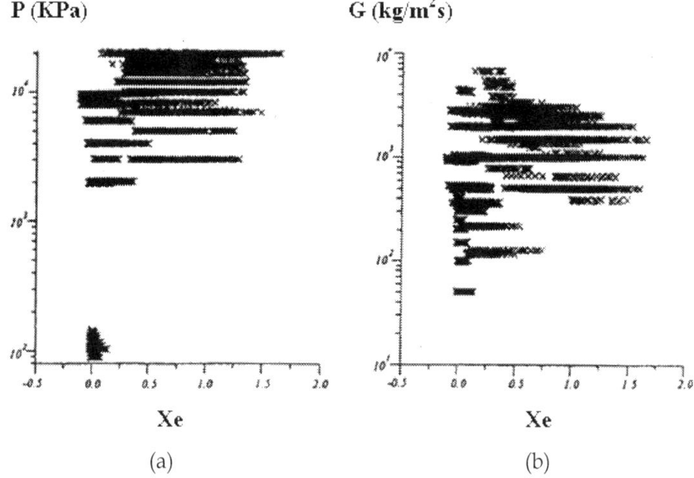

Fig. 16. Ranges of conditions of the data base for Leung's Look-up table (Leung, et al., 1997)

Having the value of K, the actual quality and the vapor temperature are evaluated by

$$K = \frac{x - x_0}{x_e - x_0}$$

and

$$T_v = T_s + (\frac{x_e}{x} - 1)\frac{h_{fg}}{C_{pg}}$$

Then, the wall temperature is calculated by

$$T_w = T_v + (q_w - q_r) / h_c$$

P (MPa)	G (kg/m²s)	X0							
		0.0	0.05	0.1	0.2	0.4	0.6	0.8	1.0
0.1	25	0.68	0.58	0.44	0.31	0.28	0.26	0.18	0.0
	50	0.69	0.59	0.45	0.42	0.48	0.36	0.25	0.0
	100	0.74	0.63	0.60	0.64	0.68	0.56	0.44	0.0
	200	0.78	0.73	0.75	0.78	0.77	0.74	0.60	0.0
	400	0.81	0.81	0.86	0.89	0.90	0.89	0.81	0.0
	600	0.82	0.88	0.92	0.94	0.96	0.95	0.90	0.0
	1000	0.83	0.92	0.96	0.98	0.98	0.98	0.97	0.0
	1500	0.84	0.96	0.98	0.99	0.99	0.99	0.98	0.0
0.3	25	0.72	0.60	0.45	0.40	0.31	0.17	0.10	0.0
	50	0.74	0.61	0.46	0.43	0.36	0.30	0.16	0.0
	100	0.76	0.64	0.54	0.57	0.52	0.47	0.33	0.0
	200	0.78	0.71	0.66	0.70	0.68	0.63	0.47	0.0
	400	0.80	0.78	0.79	0.81	0.83	0.75	0.63	0.0
	600	0.82	0.80	0.84	0.88	0.90	0.86	0.75	0.0
	1000	0.83	0.86	0.89	0.92	0.93	0.92	0.87	0.0
	1500	0.83	0.90	0.95	0.96	0.98	0.97	0.95	0.0
0.5	25	0.74	0.56	0.44	0.35	0.29	0.23	0.12	0.0
	50	0.76	0.58	0.45	0.38	0.33	0.22	0.14	0.0
	100	0.78	0.60	0.50	0.44	0.40	0.36	0.26	0.0
	200	0.79	0.62	0.63	0.64	0.61	0.56	0.40	0.0
	400	0.80	0.67	0.65	0.77	0.74	0.70	0.54	0.0
	600	0.81	0.80	0.81	0.83	0.83	0.80	0.67	0.0
	1000	0.82	0.84	0.86	0.89	0.90	0.88	0.78	0.0
	1500	0.82	0.87	0.92	0.94	0.95	0.94	0.89	0.0
1.0	25	0.76	0.62	0.53	0.42	0.25	0.10	0.08	0.0
	50	0.77	0.63	0.56	0.46	0.28	0.14	0.09	0.0
	100	0.79	0.66	0.58	0.53	0.33	0.22	0.13	0.0
	200	0.80	0.66	0.60	0.50	0.45	0.38	0.25	0.0
	400	0.80	0.68	0.66	0.65	0.62	0.56	0.40	0.0
	600	0.80	0.77	0.74	0.73	0.71	0.63	0.45	0.0
	1000	0.81	0.80	0.79	0.82	0.80	0.73	0.57	0.0
	1500	0.82	0.84	0.85	0.87	0.84	0.82	0.70	0.0
2.0	25	-	-	-	-	-	-	-	-
	50	0.81	0.68	0.65	0.54	-	-	-	-
	100	0.81	0.71	0.68	0.62	0.32	0.16	0.08	0.0
	200	0.81	0.69	0.65	0.53	0.35	0.28	0.16	0.0
	400	0.81	0.70	0.68	0.56	0.52	0.44	0.29	0.0
	600	0.82	0.73	0.71	0.64	0.61	0.53	0.36	0.0
	1000	0.83	0.74	0.74	0.74	0.71	0.65	0.47	0.0
	1500	0.83	0.81	0.80	0.80	0.80	0.71	0.55	0.0
4.0	25	-	-	-	-	-	-	-	-
	50	-	-	-	-	-	-	-	-
	100	0.81	0.72	0.62	0.45	0.30	0.21	0.08	0.0
	200	0.81	0.72	0.63	0.50	0.34	0.26	0.14	0.0
	400	0.82	0.72	0.63	0.55	0.49	0.41	0.26	0.0
	600	0.83	0.77	0.73	0.61	0.56	0.48	0.33	0.0
	1000	0.83	0.78	0.76	0.71	0.70	0.60	0.42	0.0
	1500	0.84	0.79	0.78	0.78	0.77	0.69	0.53	0.0
6.0	25	-	-	-	-	-	-	-	-
	50	-	-	-	-	-	-	-	-
	100	0.81	0.72	0.63	0.45	0.35	0.20	0.11	0.0
	200	0.82	0.73	0.64	0.52	0.42	0.25	0.14	0.0
	400	0.82	0.76	0.67	0.55	0.51	0.39	0.26	0.0
	600	0.83	0.77	0.70	0.60	0.58	0.48	0.31	0.0
	1000	0.84	0.78	0.74	0.69	0.68	0.61	0.43	0.0
	1500	0.84	0.80	0.77	0.76	0.74	0.68	0.53	0.0

Table 1. Table of K_0 with parameters p, G and x_0

where h_c is the convective heat transfer coefficient, and q_r is the radiative heat flux. They are evaluated by Eq.(25) and (18), respectively.

The present tabular method was based on 2192 experimental data in tubes of D = 6.7 and 12 mm for pressure of 0.1 –5.8 MPa, mass flux of 23 – 1462 kg/m²s and local equilibrium quality of 0 – 1.36. The conditions of the data base are shown in Fig. 17. The agreements of the calculations of wall temperatures with the data are 7.2% in r.m.s., as exemplified in Fig 18.

5. Conclusions

Using the directly heated hot patch technique the film boiling experiments have been performed at stead-state in tubes of 6.7 – 20 mm with water flowing upward, covering the ranges of pressure of 0.1 – 6 MPa, mass flux of 23 – 1462 kg/m²s and inlet quality of -0.15 – 1.0. The characteristics of film boiling have been investigated systematically, including the inverted annular film boiling, the dispersed flow film boiling and the minimum film boiling temperature. A great number of data of heat transfer coefficients, the minimum film boiling temperatures and the vapor superheats have been obtained. They fill the gaps of the data base and understanding in the regions of lower flow and medium pressure.

The film boiling is characterized by non-equilibrium between phases in both the velocity and temperature, associated with extremely complicated parametric trends and strongly history-dependent feature of the heat transfer coefficients. Two-fluid mechanistic models and tabular method have been proposed to predict the heat transfer coefficients satisfactorily for the inverted annular film boiling with saturated or low subcooling conditions and the dispersed flow film boiling.

The results show that the heat transfer coefficient near the dryout point is closely relative to the preceding heating. Therefore the IAFB model without encountering for this effect could have appreciable uncertainties over a short length for higher subcooling condition.

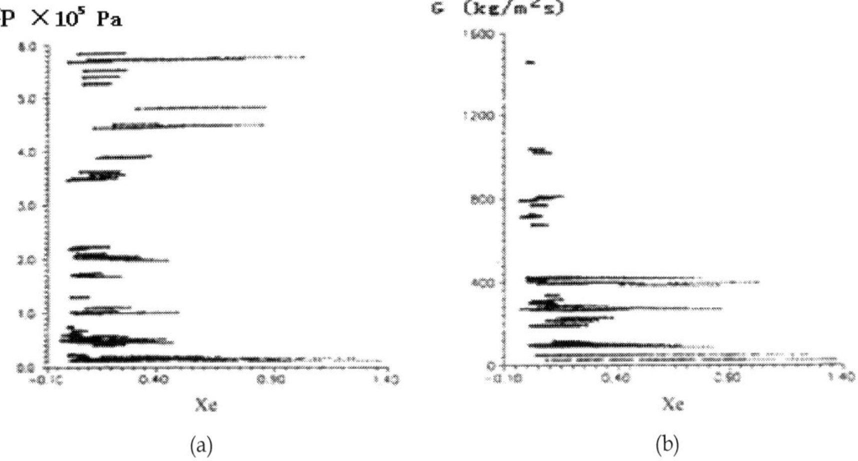

Fig. 17. Ranges of conditions of the present film boiling data for x_e > 0 (Chen & Chen 1998)

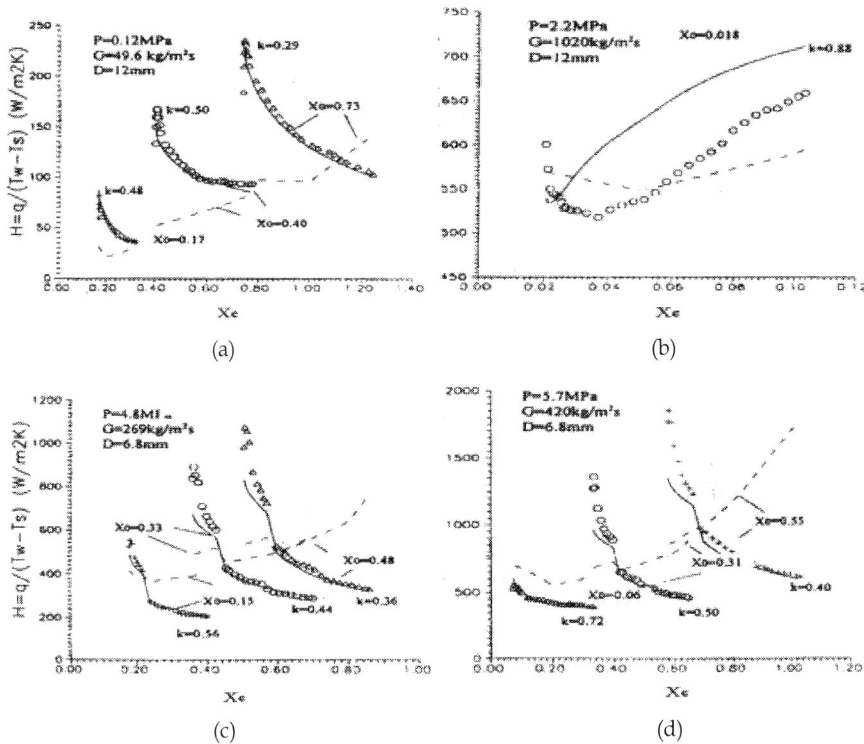

Fig. 18. Comparison of the calculation by tabular methods with experimental data (Chen & Chen, 1998) — Table of Non-equilibrium parameter,   --- Leung's Table ...
Note: in (c,d) two portions of boiling length were heated at different heat fluxes

6. Nomenclature

C_d	drag coefficient
C_p	specific heat
D	diameter
F	enhancement or correction factor
G	mass flux
h	heat transfer coefficient
h_{fg}	latent heat
K	heat conductivity
L	distance from the dryout point
p	pressure
q	heat flux
r	radius
T	temperature
u	velocity

x	quality
J_{gc}	critical volumetric flux
$N_{\mu g}$	viscosity number
Nu	Nusselt number
Pr	Prandtl number
Pr_t	turbulent Prandtl numbe
Re	Reynolds number
We	Weber number;
α	void fraction
δ	droplet diameter, thickness of film
μ	kinetic viscosity
ν	dynamic viscosity
ρ	density
ε	eddy diffusivity, emissivity
σ	surface tension
σ_{SB}	Stefan-Boltzman constant

Subscript

c	critical, convective, core
d	droplet
e	equilibrium
m	maximum
min	minimum
r	radiation
i	interface
l	liquid
s	saturation
v , g	vapor
w	wall
0, DO	inlet (at dryout point)

7. References

Analytis, G. Th. & Yadigaroglu, G. (1987), Analytical Modelling of Inverted Annular Film Boiling, *Nuclear Engineering and Design*, Vol. 99, PP 201 - 212

Arrieta, L. & Yadigaroglu, G. (1978), Analytical Model for Bottom Reflooding Heat Transfer in Light Water Reactor, *EPPI-NP 756*

Chen Y. & Li, J. (1984), Subcooled Flow Film Boiling of Water at Atmospheric Pressure, in *Two-phase Flow and Heat Transfer*, X. J. Chen and T. N. Veziroglu, PP 141 – 150, Hemisphere Pub. Co., ISBN 0-89116-432-4

Chen,Y. (1987), Experimental Study of Inverted Annular Flow Film Boiling Heat Transfer of Water，in *Heat Transfer Science and Technology*, Bu-Xuan Wang, PP 627 – 634, Hemisphere Pub. Co., ISBN 0-89116-571-1,

Chen, Y.; Wang, J.; Yang, M. & Fu, X. (1989a), Experimental Measurement of the Minimum Film Boiling Temperature for Flowing Water, in *Multiphase Flow and Heat Transfer*, Xue-Jun Chen, T. N. Veziroglu and C. L. Tien, Vol. 1, PP 393 - 400 17. Hemisphere Pub. Co., ISBN 1-56032-050-8

Chen, Y.; Cheng, P.; Wang, J. & Yang, M. (1989b), Experimental Studies of Subcooled and Low Quality of Film Boiling Heat Transfer of Water in Vertical Tubes at Moderate Pressure, *Proc. 4th Int. Topical Meeting on Nuclear Reactor Thermal-Hydraulics*, Vol. 2, PP 1106 - 1110

Chen Y., (1991), An Investigation of Dispersed Flow Film Boiling Heat Transfer of Water - Experiment and numerical analysis, *2d World Conf. on experimental fluid, heat transfer and thermodynamics, Yugoslavia*

Chen, Y.; Chen, H.& Zhu, Z. (1992), Post-Dryout Droplet Flow Heat Transfer-Measurements of both wall and vapor superheat at stable condition, in *Transport Phenomena Science and Technology*, Wang B. X. PP 319 – 324, Higher Education Press, ISBN 7-04-004122-7/TH324

Chen, Y & Chen, H. (1994a), An Experimental Investigation of Thermal Non-equilibrium in Dispersed Flow Film Boiling of Water, *Proc. Int. Conf. on New Trends in Nuclear System Thermal-Hydraulics*, Pisa Italy Vol.1 PP 31 - 37

Chen, Y. & Chen, H. (1994b), A Model of Dispersed Flow Film Boiling Heat Transfer of Water, *Proc. 10th Int. Heat Transfer Conf.* Brighton UK, Vol. 7, PP 419 – 424

Chen, Y & Chen, H. (1998), A Tabular Method for Prediction of the Heat Transfer during Saturated Film Boiling of Water in a Vertical Tube, , *Proc. 11th Int. Heat Transfer Conf.*, Kyongju, Korea, Vol. 2, PP 163 – 168

Groeneveld, D. C. (1974), Effect of a Heat Flux Spike on the Downstream Dryout Behavior, *J. of Heat Transfer* PP 121 – 125

Groeneveld, D. C. and Gardiner, R. M. (1978), A Method of Obtaining Flow Film Boiling Data for Subcooled Water, *Int. J. Heat Mass Transfer*, Vol. 21 PP 664 – 665

Groeneveld, D.C. & Steward, J. C. (1982), The Minimum Film Boiling Temperature for Water during Film Boiling Collapse, *Proc. 6th Int. Heat Transfer conf.* PP 393 – 398

Groeneveld, D.C. & Snoek, C. W. (1984), A Comprehansive Examination of Heat Transfer Correlations Suitable for Reactor Safety Analysis, *Multiphase Science and Technology*, Vol.2

Groeneveld, D.C. (1988), Recent Developments in Thermalhydraulic Prediction Methods, *Proc. Int. Topical Meeting on Thermalhydraulics of Nuclear Reactors*, Seoul

Groeneveld, D. C. (1992), A Review of Inverted Annular and Low Quality Film Boiling, in *Post-dryout Heat Transfer*, Hewitt, Delhaye, Zuber, CRC Press, PP 327 – 366

Kirillov, P. L., et al. (1996), The Look-up Table for Heat Transfer Coefficient in Post-Dryout Region for Water Flowing in Tubes, *FEI-2525*

Leung, L. K. H.; Hammouda, N. & Groeneveld, D. C. (1997), A look-up Table for Film Boiling Heat Transfer Coefficients in Tubes with Vertical Upward Flow, *Proc. 8th Int. Topical Meeting on Nuclear Reactor Thermal-Hydraulics*, Kyoto, Japan, Vol. 2, pp 671 – 678

Mossad, M. (1988), Sobcooled Film Boiling Heat Transfer to Flowing Water in Vertical Tube，Dr。-Ing. *Thesis*, Techn. Univ. Berlin, Berlin,

Mossad, M. & Johannsen, K. (1989), A new Correlation for Subcooled and Low Quality Film Boiling Heat Transfer of Water at Pressures from 0.1 to 8 MPa, *Proc. 4th Int. Topical Meeting on Nuclear Reactor Thermal-Hydraulics*, Vol. 2, PP 1111 - 1117

Plummer, D, N.; Iloeje, O. C.; Rohsenow, W. M.; Griffith, P. & Ganic, E. (1974), Post Critical Heat Transfer to Flowing Liquid in a Vertical Tube, MIT Dpt. Of Mech. Eng. Report 72718-91

Plummer, D. N.; Griffith, P. & Rohsenow, W. M. (1977), Post-Critical Heat Transfer to Flowing Liquid in a Vertical Tube, Trans. *ASME*, Vol. 4, N.3, PP 151 – 158

Stewart, J. C. & Groeneveld, D. C. (1982), Low Quality and Subcooled Film Boiling at Elevated Pressures, *Nucl. Eng. Design.* Vol. 67, PP 259 - 272

Swinnerton, D.; Hood, M. L. & Pearson, K.G. (1988), Steady State Post-Dryout at Low Quality and Medium Pressure –Data Report, *UKAEA AEEW-R 2267*

Thermohudraulic relationships for advanced water cooled reactors, (2001), *IAEA-TECDOC-1203*

Yonomoto, T.; Koizumi, Y. & Tasaka, K, (1987) Onset Criterion for Liquid Entrainment in Reflooding Phase of LOCA, *Nucl. Science and Technology*, Vol. 24, No. 10, pp 798-810

10

Two-Phase Flow Boiling Heat Transfer for Evaporative Refrigerants in Various Circular Minichannels

Jong-Taek Oh[1], Hoo-Kyu Oh[2] and Kwang-Il Choi[1]
[1]Department of Refrigeration and Air-Conditioning Engineering,
Chonnam National University,
San 96-1, Dunduk-dong, Yeosu, Chonnam 550-749
[2]Department of Refrigeration and Air Conditioning Engineering,
Pukyong National University,
100, Yongdang-dong, Nam-Ku, Busan 608-739,
Republic of Korea

1. Introduction

Global awareness of climate change has been raised in recent decades; many countries proposed to reduce carbon emissions and to control refrigerants based on the Montreal protocol. The ozone layer has become a concern for many researchers, focusing on reducing depletion. Refrigerant has been used for a long time and is still used as the principle material in refrigeration systems. The demand for new methods of refrigeration and air conditioning has promoted more effective and efficient refrigeration systems. Small refrigeration systems might be one of the solutions to reduce depletion of ozone layer indirectly.

Several studies and experimental results have been issued that discussed flow patterns in horizontal tubes, two-phase flow boiling heat transfer and pressure drop using refrigerant as the observed fluid. Two-phase flow boiling heat transfer pressure drop of refrigerants in minichannels has been researched for several decades. Only a few studies in the literature report on the two-phase flow heat transfer and pressure drop of refrigerants in minichannels. Compared with pure refrigerants in conventional channels, the flow boiling of refrigerants in minichannels has discrete characteristics due to the physical and chemical properties of the refrigerants and the dimensions of the minichannels.

The greatest advantages of the minichannels are their high heat transfer coefficients, significant decreases in the size of compact heat exchangers, and lower required fluid mass. A higher heat transfer in minichannels is due to large ratios of heat transfer surface to fluid flow volume and its properties. The decreasing size also allows heat exchangers to achieve significant weight reductions, lower fluid inventories, low capital and installation costs, and energy savings. Despite those advantages, pressure drop within minichannels is higher than that of conventional tube because of the increase of wall friction.

Chisholm (1967) proposed a theoretical basis for the Lockhart–Martinelli correlation for two-phase flow. The Friedel (1979) correlation was obtained by optimizing an equation for the two-phase frictional multiplier using a large measurement database. Many studies have developed pressure drop correlations on the basis of the Chisholm (1967) and Friedel (1979) correlations. Mishima and Hibiki (1996), Yu et al.(2002), and Kawahara et al. (2002) developed pressure drop correlations on the basis of the Chisholm (1967) correlation.

Chang et al. (2000), Chen et al. (2001), and Zhang and Webb (2001) developed pressure drop correlations on the basis of the Friedel (1979) correlation. Tran et al. (2000) measured the two-phase flow pressure drop with refrigerants R-134a, R-12, and R-113 in small round and rectangular channels. They modified Chisholm's (1983) correlation and proposed a new correlation.

This chapter reports on a study whose goals were to present pressure drop experimental data for the refrigerant R-22, and its alternatives, R-134a, R-410A, R-290, R-717 (NH_3) and R-744 (CO_2) which were measured in horizontal and local heat transfers during evaporation in smooth minichannels to establish a new correlation for heat exchangers with minichannel designs. The present experimental data for pressure drop were obtained by the previous experiments compared with existing two-phase pressure drop prediction methods, namely Beattie and Whalley (1982), Cicchitti et al (1960), McAdams (1954), Chang et al. (2000), Dukler et al. (1964), Friedel (1979), Chisholm (1983), Tran et al. (2000), Zhang and Webb (2001), Mishima and Hibiki (1996), Lockhart and Martinelli (1949), Shah (1988), Tran et al. (1996), Jung at al. (1989), Gungor and Winterton (1987), Takamatsu et al (1993), Kandlikar and Stainke (2003), Wattelet et al (1994), Chen (1966), Zhang et al. (2004), Chang and Ro (1996), Yu et al. (2002), Friedel (1979), Kawahara et al. (2002), Mishima (1983), Chisholm et al. (2000), Tran et al. (2000), Chen (2001), and Yoon et al. (2004). A new correlation for two-phase frictional pressure drop was developed on the basis of the Lockhart–Martinelli method using the present experimental data. In the present paper, heat fluxes to make the flow boiling heat transfer coefficients were electrically heated. The experimental results were compared with the predictions of seven existing heat transfer correlations, namely those reported by Mishima and Hibiki (1996), Friedel (1979), Chang at al. (2000), Lockhart and Martinelli (1949), Chisholm (1983), Zhang and Webb (2001), Chen (1966), Chen et al. (2001), Kawahara et al. (2002), Tran et al. (1996), Tran et al. (2000), Wattelet et al. (1993), Wattelet et al. (1994), Gungore-Winterton (1986), Gungore-Winterton (1989), Zhang et al (2004), Kandlikare-Steinke (1996), Kandlikare-Steinke (2003), Jung et al.(1989), Jung et al. (2004), Shah (1988), Gungore-Winterton (1987), Zhang et al (1987), and Takamatsu et al. (2003) was and were developed in this study based on superposition, due to the limitations in the correlation for forced convective boiling of refrigerants in small channels.

Compared with conventional channels, evaporation in small channels may provide a higher heat transfer coefficient due to their higher contact area per unit volume of fluid. In evaporation within small channels, as reported by Bao et al. (2000), Zhang et al. (2004), Kandlikare-Steinke (2003), Tran et al. (2000), Pettersen (2004), Park and Hrnjak (2007), Zhao et al. [7], Yun et al. (2005), Yoon et al. (2004), Pamitran et al. (2008), and K.-I. Choi (2009), the contribution of nucleate boiling is predominant and laminar flow appears.

The study that was done by A.S. Pamitran et al (2007) and [a/b]K.-I.Choi et al. (2007), yielded a basic understanding about predicting pressure drop and heat transfer coefficients during refrigerants evaporation in minichannels. The studies have also developed correlations and have been compared with other experimental correlations reported in much available literatures in the area of two-phase boiling heat transfer. The methods of creating correlations have good agreements with the experiment data gathered by using refrigerants as the working fluids.

2. Experimental aspects

2.1 Experimental apparatus and method

Experimental facility for inner diameter 3 and 1.5 mm

The experimental facilities are schematically shown in Fig. 1(a) and (b). The test facilities were constructed by A.S. Pamitran et al. (2007) and K.-I. Choi et al (2007) and consisted of a condenser, a subcooler, a receiver, a refrigerant pump, a mass flow meter, a preheater, and test sections. For the test with 3.0 and 1.5mm inner diameter tubes, a variable AC output motor controller was used to control the flow rate of the refrigerant. A Coriolis-type mass flow meter was installed in a horizontal layout for the test with 1.5 and 3.0mm inner diameter tubes. A preheater or a cooler was installed to control the vapor quality of the refrigerant by heating or condensing the refrigerant before it entered the test section. For evaporation at the test section, a pre-determined heat flux was applied from a variable A.C voltage controller. The vapor refrigerant from the test section was then condensed in the condenser and subcooler, and then the condensed refrigerant was supplied to the receiver.

The test section was made of stainless steel circular smooth tubes with inner tube diameters of 3.0 and, 1.5 mm. The rate of input electric potential E and current I were adjusted in order to control the input power and to determine the applied heat flux, which was measured by a standard multimeter. The test sections were uniformly and constantly heated by applying the electric current directly to their tube walls. The test sections were well insulated with foam and rubber; therefore, heating loss was ignored in the present study. The local saturation pressure of the refrigerant, which was used to determine the saturation temperature, was measured using bourdon tube type pressure gauges with a 0.005 MPa scale at the inlet and at the outlet of the test sections. Differential pressure was measured by the bourdon tube type pressure gauges and a differential pressure transducer. Circular sight glasses with the same inner tube diameter as the test section were installed at the inlet and outlet of the test section to visualize the flow. Each sight glass was held by flanges on both sides, as described in Fig. 1.

The temperature and flow rate measurements were recorded using the Darwin DAQ32 Plus logger R9.01 software program and version 2.41 of the Micro Motion ProLink Software package, respectively. The physical properties of the refrigerants were obtained from the REFPROP 8.0.

Experimental facility for inner diameter 0.5 mm

Another experimental facility was made for 0.5 mm inner diameter tubes; this facility is an open-loop system. This system allows the refrigerant flow from higher pressure containers to refrigerant receivers. This system used a needle valve to the control flow rate before entering the test section which is made of stainless steel. A weighing balance was used for the test with the 0.5 mm inner diameter tube to measure the refrigerant flow rate. Heating and measurements were similar to those on Fig. 1.

The experimental conditions used in the studies of A.S Pamitran et al and K.-I. Choi et al. are listed in Table 1. Five refrigerants are use as the working fluid in the experiments; the studies express the effects of dimensional factors that are represented by the inner diameter of the tubes. The mass flux effect was observed by configuring the velocity of the fluids and heat flux control by regulating the electrical heating.

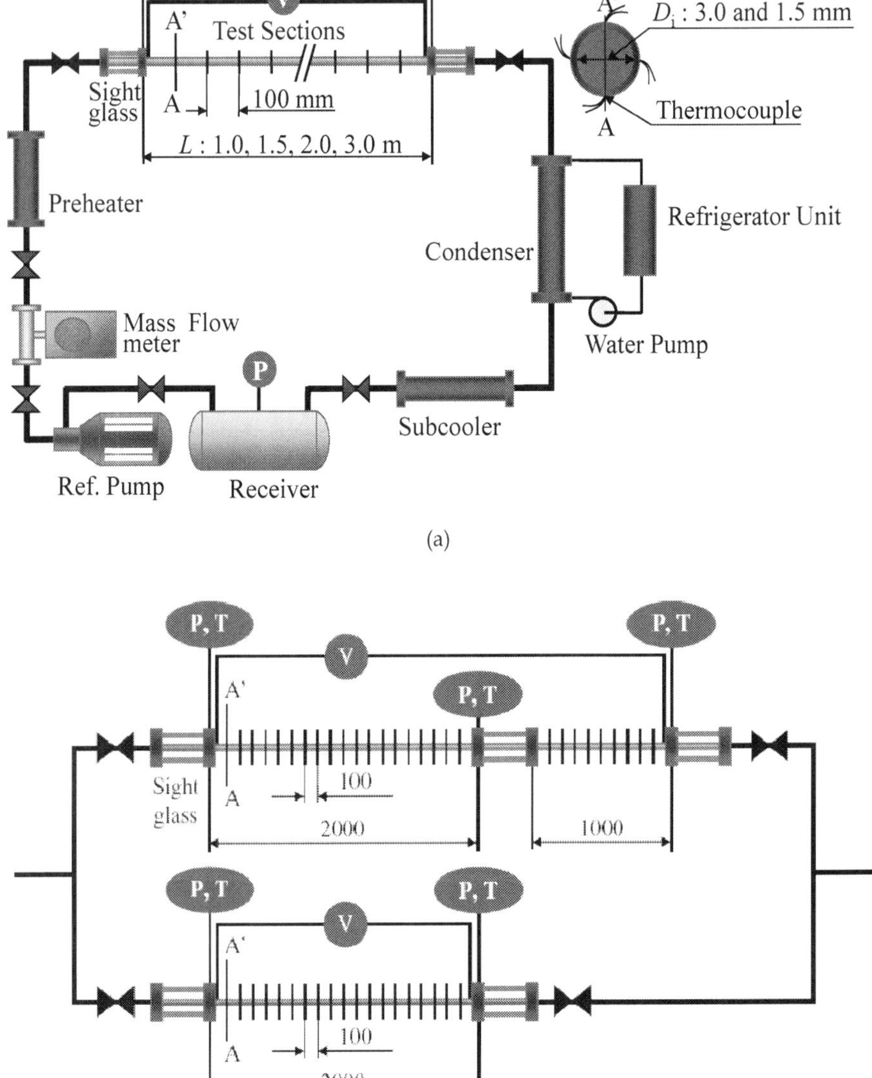

Fig. 1. Experimental test facility: (a) for test section with inner tube diameter of D_i=3.0mm and D_i=1.5 mm

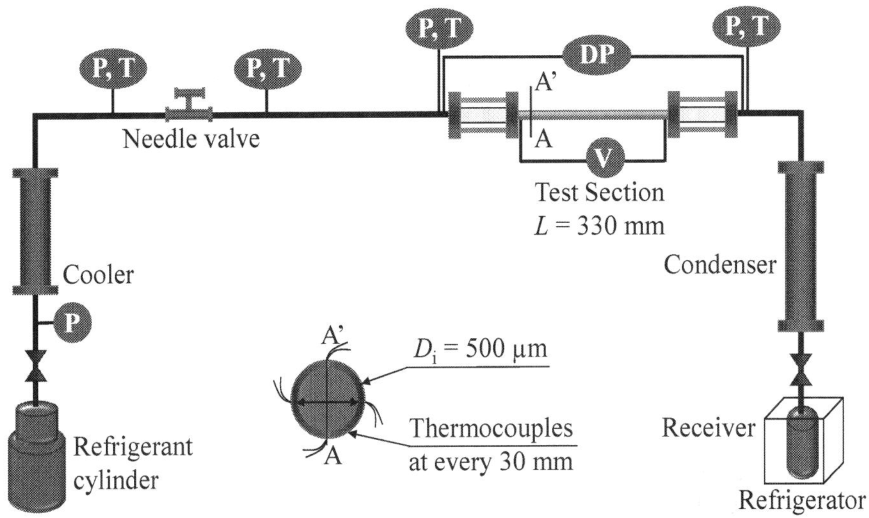

Fig. 2. For test section with inner tube diameter of D_i =0.5 mm

Test section	Quality	Working refrigerant	Inner diameter (mm)	Tube length (mm)	Mass flux (kg/(m²·s))	Heat flux (kW/m²)	Inlet Tsat (°C)
Horizontal circular smooth small tubes	up to 1.0	R-22	3.0	2000	400 – 600	20 – 40	10
			1.5	2000	300 – 600	10 – 20	10
		R-134a	3.0	2000	200 – 600	10 – 40	10
			1.5	2000	200 – 400	10	10
			0.5	330	100	5 – 20	6 – 10
		R-410A	3.0	3000	300 – 600	10 – 40	10 – 15
			1.5	1500	300 – 600	10 – 30	10 – 15
			0.5	330	70 – 400	5 – 20	1 – 11
		C_3H_8	3.0	2000	50 – 240	5 – 25	0 – 11
			1.5	2000	100 – 400	5 – 20	0 – 12
		CO_2	3.0	2000	200 – 600	20 – 30	1 – 10
			1.5	2000	300 – 600	10 – 30	0 – 11
		NH_3	3.0	2000	100 – 800	10 – 70	0 – 10
			1.5	2000	100 – 500	10 – 35	0 – 10

Table 1. Experiment conditions

The experimental uncertainty associated with all the parameters is tabulated in Table 2. The uncertainties were obtained using both random and systematic errors, and these uncertainty values changed according to the flow conditions; their minimum to maximum ranges are shown.

Parameters	Uncertainty (%)				
	CO_2	C_3H_8	R-410A	R-134a	R-22
T_{wi} (%)	±2.10 to ±4.56	±0.18 to ±5.58	±0.23 to ±6.53	±0.29 to ±3.92	±0.44 to ±3.90
P (kPa)	±2.5 kPa	±2.5 kPa	±2.5 kPa	±2.5 kPa	±2.5 kPa
G(%)	±1.85 to ±9.48	±3.24 to ±9.78	±1.84 to ±9.48	±1.85 to ±3.80	±1.84 to ±3.16
q(%)	±1.67 to ±2.70	±2.07 to ±3.58	±1.67 to ±3.20	±1.79 to ±2.59	±1.78 to ±2.26
x(%)	±1.79 to ±9.71	±4.27 to ±9.82	±1.78 to ±9.85	±2.23 to ±4.19	±1.88 to ±3.39
h(%)	±4.46 to ±8.23	±1.78 to ±9.89	±2.59 to ±10.33	±2.75 to ±9.24	±2.74 to ±9.07

Table 2. Summary of estimated uncertainty

3. Data reduction

3.1 Pressure drop

The saturation pressure at the initial point of saturation was determined by interpolating the measured pressure and the subcooled length. The experimental two-phase frictional pressure drop can be obtained by subtracting the calculated acceleration pressure drop from the measured pressure drop.

$$\left(\frac{dp}{dz}\right) = \left(\frac{dp}{dz}F\right) + \left(\frac{dp}{dz}a\right) + \left(\frac{dp}{dz}z\right) \tag{1}$$

$$-\left(\frac{dp}{dz}a\right) = G^2\frac{d}{dz}\left(\frac{x^2 v_g}{\alpha} + \frac{(1-x)^2 v_f}{(1-\alpha)}\right) \tag{2}$$

$$-\left(\frac{dp}{dz}z\right) = g\sin\theta\left[\frac{A_g}{A}\rho_g + \frac{A_f}{A}\rho_g\right] = g\sin\theta\left[\alpha\rho_g + (1-\alpha)\rho_f\right] \tag{3}$$

The equation for the friction pressure gradient for horizontal tubes should be reduced to the static head factor, therefore the equation is

$$\left(\frac{dp}{dz}F\right) = \left(\frac{dp}{dz}\right) - \left(\frac{dp}{dz}a\right) = \left(\frac{dp}{dz}\right) - G^2\frac{d}{dz}\left(\frac{x^2}{\alpha\rho_g} + \frac{(1-x)^2}{(1-\alpha)\rho_f}\right) \tag{4}$$

The predicted void fraction with Steiner (1993), CISE (Premoli et al., 1971) and Chisholm (1972) are compared in the present study, as is shown in Fig. 3. The absolute deviations of the Steiner (1993), CISE (Premoli et al., 1971) and Chisholm (1972) void fractions from the homogenous void fraction are 7.11%, 3.72% and 11.49%, respectively. The predicted pressure drop for the present experimental data using some previous pressure drop correlations with the Steiner (1993) void fraction showed a better prediction than that using the CISE (Premoli et al., 1971) and Chisholm (1972) void fractions. Therefore, the void fraction in the present study was obtained from the Steiner (1993) void fraction.

$$\alpha = \frac{x}{\rho_g}\left[(1+0.12(1-x))\left(\frac{x}{\rho_g}+\frac{(1-x)}{\rho_f}\right)+\left(\frac{1.18}{G^2}\right)\left(\frac{(1-x)}{\rho_f^{0.5}}\right)\left(g\sigma(\rho_f+\rho_g)\right)^{0.25}\right]^{-1} \quad (5)$$

The friction factor was determined from the measured pressure drop for a given mass flux by using the Fanning equation

$$f_{tp} = \frac{D\bar{\rho}}{2G^2}\left(-\frac{dp}{dz}F\right) \quad (6)$$

where the average density is calculated with the following equation:

$$\bar{\rho} = \alpha\rho_g + (1+\alpha)\rho_f \quad (7)$$

In order to obtain the two-phase frictional multiplier based on the pressure drop for the total flow assumed for the liquid, ϕ_{fo}^2, the calculated two-phase frictional pressure drop is divided by the calculated frictional two-phase pressure drop assuming the total flow to be liquid, as shown by

$$\phi_{fo}^2 = \frac{\left(-\frac{dp}{dz}F\right)_{tp}}{\left(-\frac{dp}{dz}F\right)_{fo}} = \frac{\left(-\frac{dp}{dz}F\right)_{tp}}{\left(-\frac{2f_{fo}G^2}{D\rho_f}F\right)} \quad (8)$$

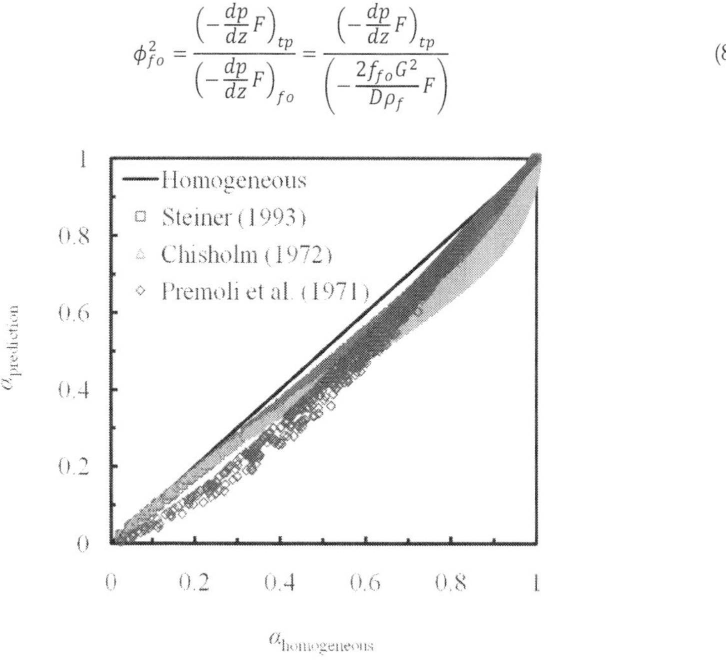

Fig. 3. Void fraction comparison

3.2 Heat transfer coefficient

The inside tube wall temperature, T_{wi} was the average temperature of the top, both right and left sides, and bottom wall temperatures, and was determined using steady-state one-dimensional radial conduction heat transfer through the wall with internal heat generation.

The quality, x, at the measurement locations, z, were determined based on the thermodynamic properties, namely,

$$x = \frac{i - i_f}{i_{fg}} \tag{9}$$

The refrigerant flow at the inlet of the test section was not completely saturated. Even though it was just short of being completely saturated, it was necessary to determine the subcooled length to ensure reduction data accuracy. The subcooled length was calculated using the following equation to determine the initial point of saturation.

$$z_{sc} = L\frac{i_f - i_{fi}}{\Delta i} = L\frac{i_f - i_{fi}}{(Q/W)} \tag{10}$$

The outlet mass quality was then determined using the following equation:

$$x = \frac{\Delta i + i_{fg} - i_f}{i_{fg}} \tag{11}$$

4. Results and discussion

The flow with heat addition, or adiabatic flow, is a coupled thermodynamic problem. On the one hand, heat transfer leads to a phase change, which leads to the change of phase distribution and flow pattern; on the other hand, it causes a change in the hydrodynamics, such as the pressure drop along the flow path, which affects the heat transfer characteristics. Furthermore, a single-component, two-phase flow in a tube can hardly become fully developed at low pressure because of the shape change in large bubbles and the inherent pressure change along the tube, which continually changes the state of the fluid and thereby changes the phase distribution and flow pattern. The current study presents the characteristics of a two-phase flow pattern and pressure drop in small tubes for some refrigerants.

4.1 Flow pattern

The hydrodynamic characteristic of two-phase flows, such as the pressure drop, void fraction, or velocity distribution, varies systematically with the observed flow pattern, just as in the case of a single-phase flow, whose behavior depends on whether the flow is in the laminar or turbulent regime. However, in contrast to a single-phase flow, liquid–vapor flows are difficult to describe by general principles, which could serve as a framework for solving practical work. The identification of a flow regime provides a picture of the phase boundaries.

The present experimental results were mapped on Wang et al. (1997) and Wojtan et al. (2005) flow pattern maps, which were developed for diabatic two-phase flows. The Wang et al. (1997) flow pattern map is a modified Baker (1954) map, developed using R-22, R-134a, and R-407C inside a 6.5mm horizontal smooth tube. Therefore, they claim that their flow pattern map is better for prediction of flow patterns in small tubes. The Wang et al. (1997) study showed that the flow transition for a mixture refrigerant showed a considerable delay compared with that of pure refrigerants. For R-410A, at the initial stage of evaporation, R-32 evaporated faster than R- 125. Therefore, R-32 increased the concentration of the vapor phase, and R-125 increased the concentration of the liquid phase throughout the evaporation process at the liquid–vapor interface. This resulted in a higher mean vapor velocity and a lower mean liquid velocity during evaporation. For the other working refrigerants, the physical properties of the refrigerant such as density, viscosity and surface tension have a strong effect on the flow pattern.

The predicted flow pattern for the selected current experimental data according to the existing flow pattern maps of Wojtan et al. (2005) and Wang et al. (1997) can be seen in Figs. 4(a) - (f) and 5(a) – (f), respectively. The Wang et al. (1997) map showed a better prediction of the flow pattern of the current experimental data for the beginning of annular flow than the Wojtan et al. (2005) map; however, the Wang et al. (1997) map could not show the prediction for dry-out conditions. The flow pattern prediction of the current test results for the all test conditions with the Wang et al. (1997) flow pattern map was for the intermittent, stratified wavy and annular flow. The stratified wavy flow appeared earlier for higher mass fluxes and its regime was longer for the low mass flux condition. The annular flow appeared earlier for higher mass fluxes.

The Wojtan et al. (2005) flow pattern map is a modified Kattan et al. (1998) map, developed using R-22 and R-410A inside a 13.6mm horizontal smooth tube. Kattan et al. (1998) used five refrigerants R-134a, R-123, R-402A, R-404A, and R-502 inside 12mm and 10.92mm (only for R-134a) tubes, which were heated by hot water flowing counter-currently to develop their flow pattern maps on the basis of the Steiner (1993) flow pattern map. The flow pattern prediction of the experimental results for all the test conditions with the Wojtan et al. (2005) flow pattern map was on intermittent, annular, dry-out, and mist flows. The flow pattern prediction, as shown in Fig. 4(b)–(c), showed that the mass flux, heat flux and inner diameter had an effect on the flow-pattern. Since the Wojtan et al. (2005) map was developed using a conventional tube, the flow pattern transition of the present experimental data showed a delay on this map. However, the Wojtan et al. (2005) map could predict the dry-out condition. The Wojtan et al. (2005) map predicts the dry-out condition of the R-410A experimental data better than those of the other working refrigerants. Overall, the Wang et al. (1997) flow pattern map provides a better flow pattern prediction for the current experimental results than the Wojtan et al. (2005) flow pattern map.

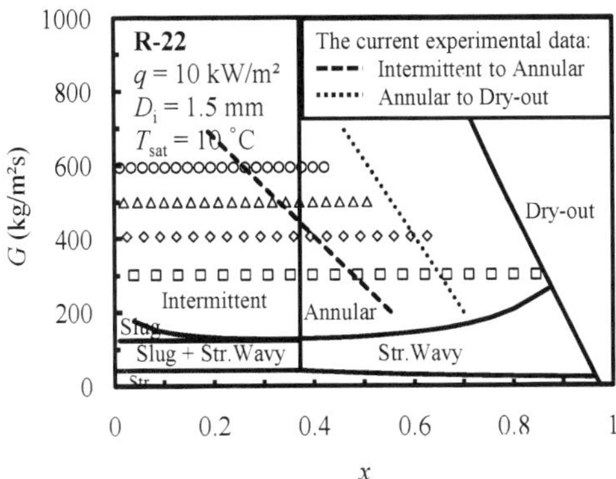

Fig. 4. (a) Present experimental results mapped on Wojtan et al. (2005) flow pattern map for R-22 at q = 10 kW/m², D_i = 1.5 mm, T_{sat} = 10°C

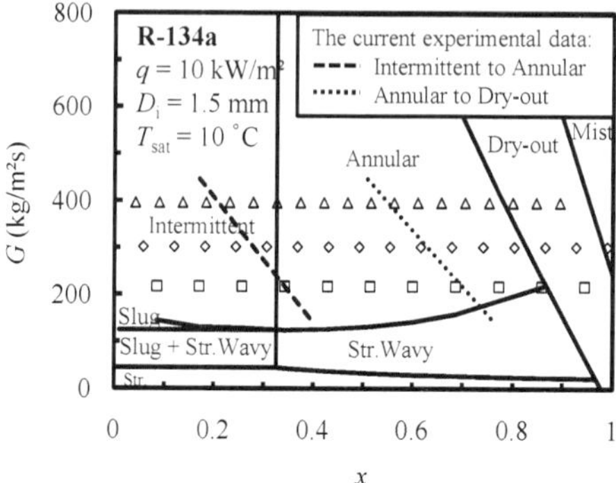

Fig. 4. (b) Present experimental results mapped on Wojtan et al. (2005) flow pattern map for R-134a at q = 10 kW/m², D_i = 1.5 mm, T_{sat} = 10°C

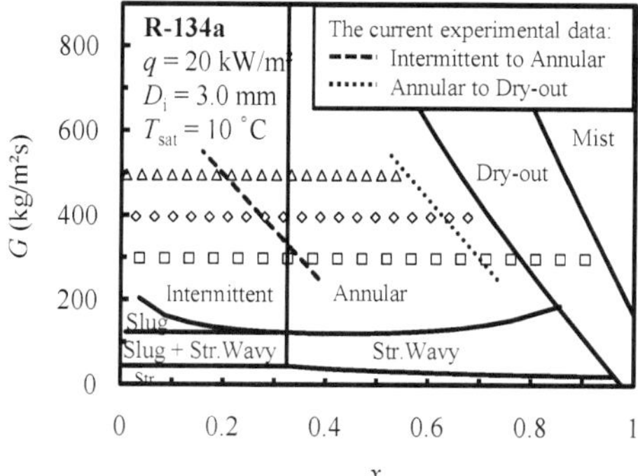

Fig. 4. (c) Present experimental results mapped on Wojtan et al. (2005) flow pattern map for R-134a at q = 20 kW/m², D_i = 3.0 mm, T_{sat} = 10°C

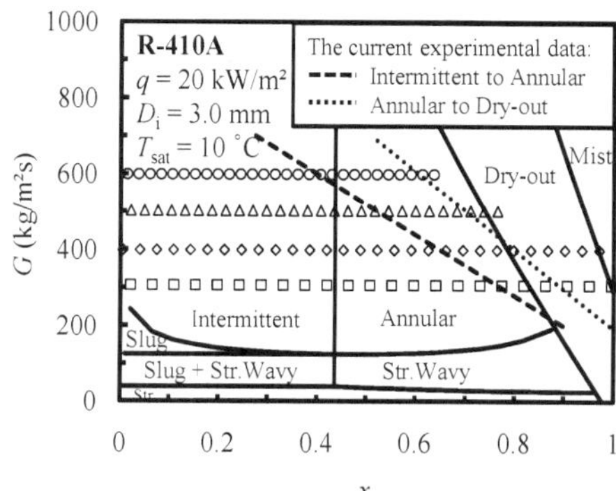

Fig. 4. (d) Present experimental results mapped on Wojtan et al. (2005) flow pattern map for R-410A at q = 20 kW/m², D_i = 3.0 mm, T_{sat} = 10°C

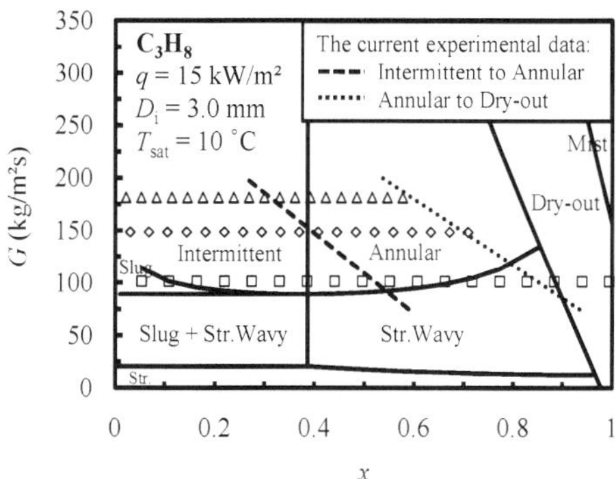

Fig. 4. (e) Present experimental results mapped on Wojtan et al. (2005) flow pattern map for
C_3H_8 at q = 15 kW/m², D_i = 3.0 mm, T_{sat} = 10°C

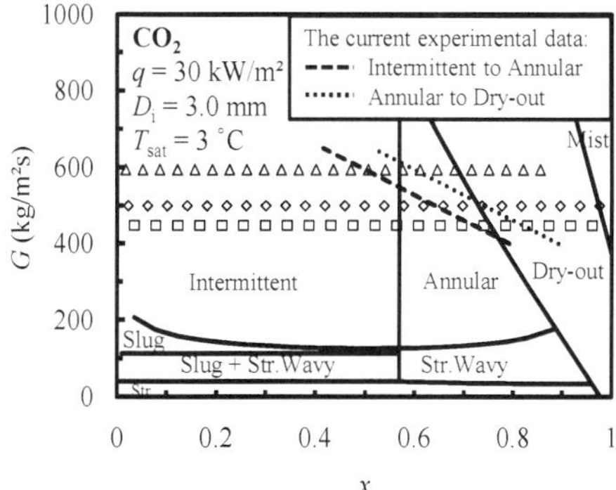

Fig. 4. (f) Present experimental results mapped on Wojtan et al. (2005) flow pattern map for
CO_2 at q = 30 kW/m², D_i = 3.0 mm, T_{sat} = 3°C

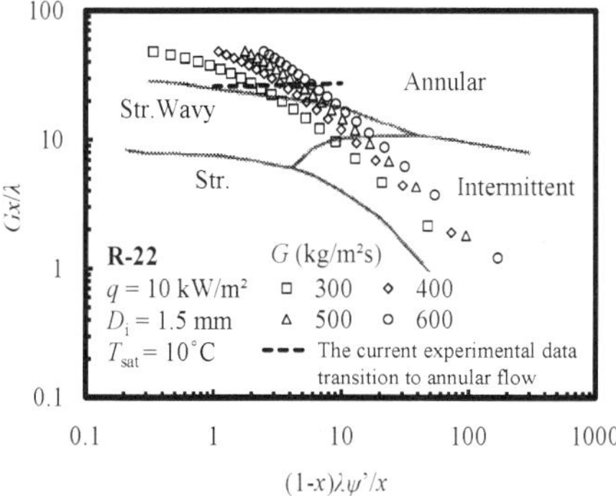

Fig. 5. (a) Present experimental results mapped on Wang et al. (1997) flow pattern map for
R-22 at q = 10 kW/m², D_i = 1.5 mm, T_{sat} = 10°C

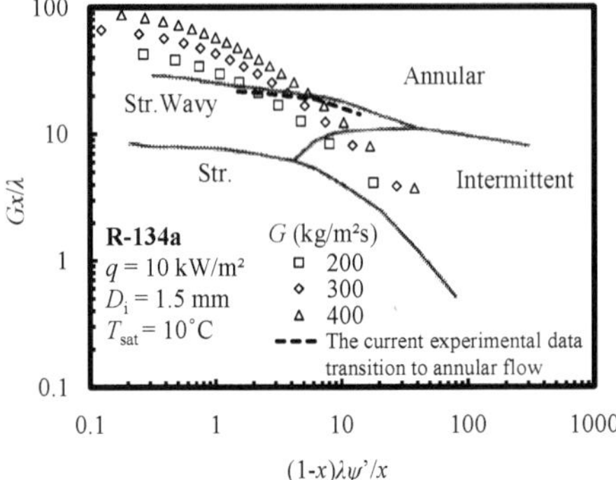

Fig. 5. (b) Present experimental results mapped on Wang et al. (1997) flow pattern map for R-134a at q = 10 kW/m², D_i = 1.5 mm, T_{sat} = 10°C

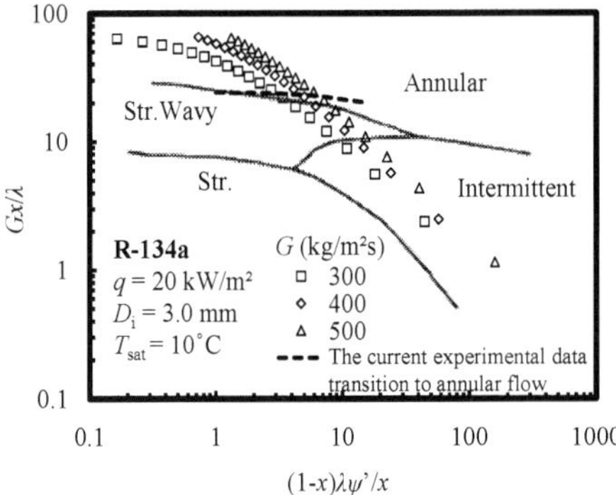

Fig. 5. (c) Present experimental results mapped on Wang et al. (1997) flow pattern map for R-134a at q = 20 kW/m², D_i = 3.0 mm, T_{sat} = 10°C

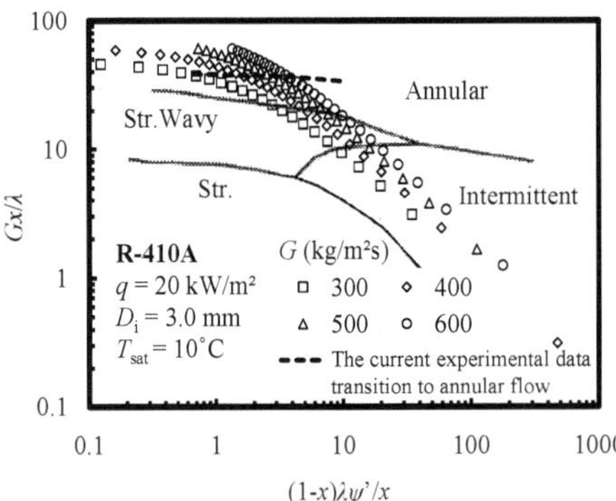

Fig. 5. (d) Present experimental results mapped on Wang et al. (1997) flow pattern map for R-410A at q = 20 kW/m², D_i = 3.0 mm, T_{sat} = 10°C

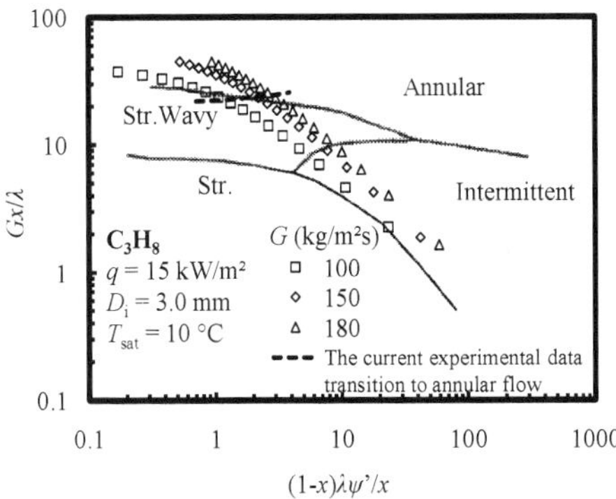

Fig. 5. (e) Present experimental results mapped on Wang et al. (1997) flow pattern map for C_3H_8 at q = 15 kW/m², D_i = 3.0 mm, T_{sat} = 10°C

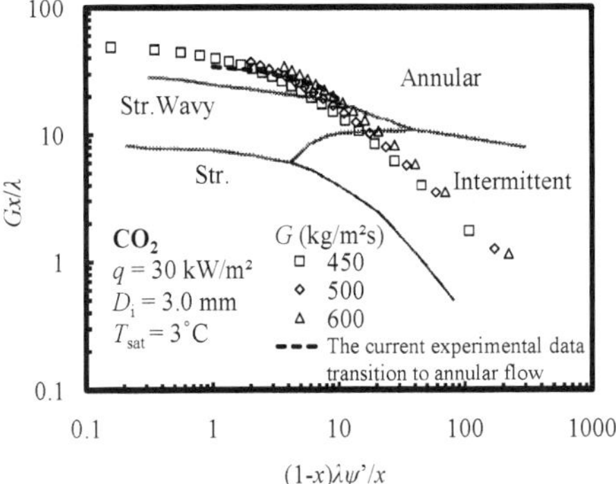

Fig. 5. (f) Present experimental results mapped on Wang et al. (1997) flow pattern map for CO_2 at q = 30 kW/m², D_i = 3.0 mm, T_{sat} = 3°C

4.2 Pressure drop

Fig. 6(a)-(f) shows that mass flux has a strong effect on the pressure drop. An increase in the mass flux results in a higher flow velocity, which increases the friction and acceleration pressure drops. This similar trend is shown by Zhao et al. (2000), Yoon et al. (2004), Park and Hrnjak (2007), Oh et al. (2008) and Cho and Kim (2007). Fig. 6 also illustrates that the pressure drop increases as the heat flux increases. It is presumed that the increasing heat flux results in a higher vaporization, which increases the average fluid vapor quality and flow velocity; this trend is similar to that shown by Zhao et al. (2000). The effect of the inner tube diameter on the pressure drop is also illustrated in Fig. 6(a)-(f). The pressure gradient in the 1.5 mm tube is higher than that in the 3.0mm tube. The explanation for this is that the smaller inner tube diameter results in a higher wall shear stress, wherein for a given temperature condition it results in a higher friction factor and flow velocity, and then results in both higher frictional and acceleration pressure drops.

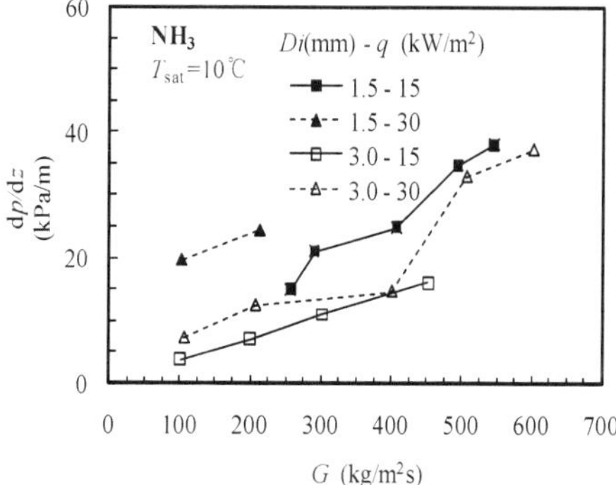

(a) Effect of mass flux, heat flux and inner diameter on pressure drop for NH_3 at T_{sat} = 10°C

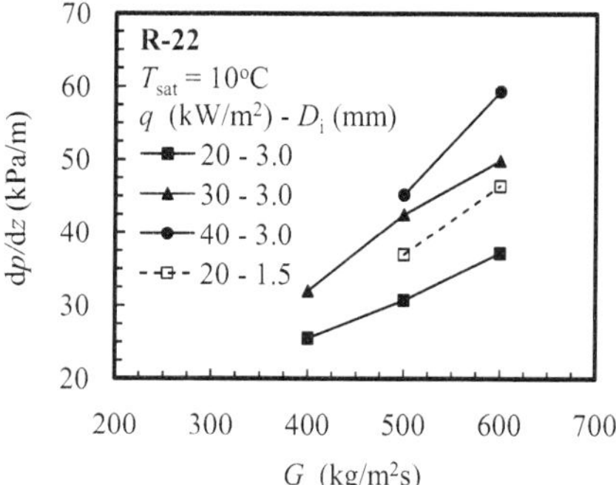

(b) Effect of mass flux, heat flux and inner tube diameter on pressure drop for R-22 at T_{sat} = 10°C

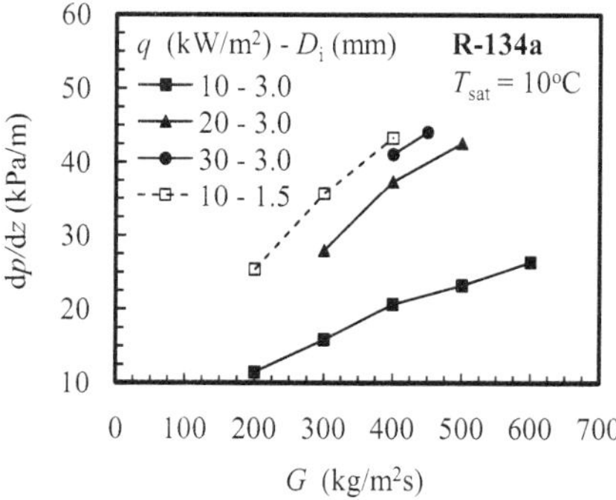

(c) Effect of mass flux, heat flux and inner tube diameter on pressure drop for R-134a at T_{sat} = 10°C

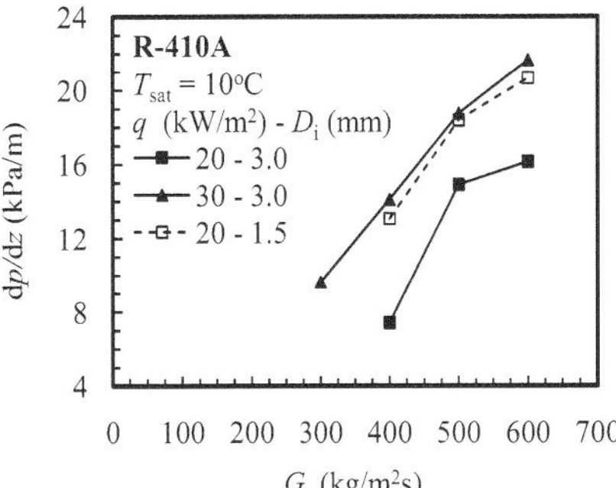

(d) Effect of mass flux, heat flux and inner tube diameter on pressure drop for R-410A at T_{sat} = 10°C

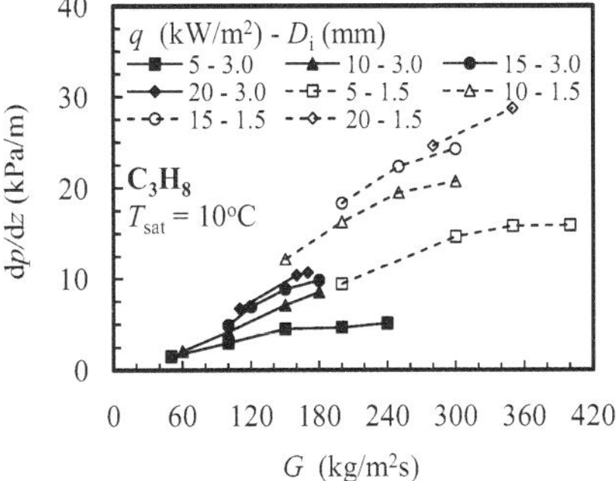

(e) Effect of mass flux, heat flux and inner tube diameter on pressure drop for C$_3$H$_8$ at T_{sat} = 10°C

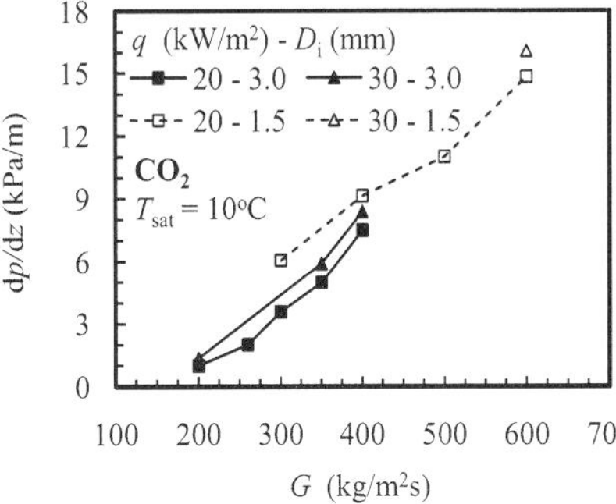

(f) Effect of mass flux, heat flux and inner tube diameter on pressure drop for CO$_2$ at T_{sat} = 10°C

Fig. 6. (a)-(f) The effects of mass flux, heat flux, inner tube diameter, and saturation temperature on pressure drop

Fig. 6 also depicts the effect of saturation temperature on pressure drop, where a lower saturation temperature results in a higher pressure drop. This can be explained by the effect of the physical properties of density and viscosity on the pressure drop at different temperatures. The liquid density, ρ_f, and liquid viscosity, μ_f, increase as the temperature decreases, whereas the vapor density, ρ_g, and vapor viscosity, μ_g, decrease as the temperature decreases. As the temperature decreases for a constant mass flux condition, the increasing liquid density and liquid viscosity result in a lower liquid velocity, whereas the decreasing vapor density and vapor viscosity are the result in a higher vapor velocity. It is clear that the pressure drop increases during evaporation, and this increasing of the pressure drop is higher when the saturation temperature is lower.

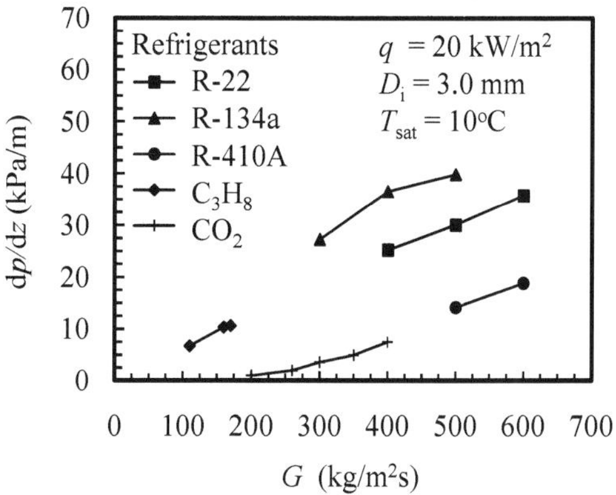

(a) Pressure drop comparison of the present working refrigerants at q = 20 kW/m², D_i = 3.0 mm, T_{sat} = 10°C

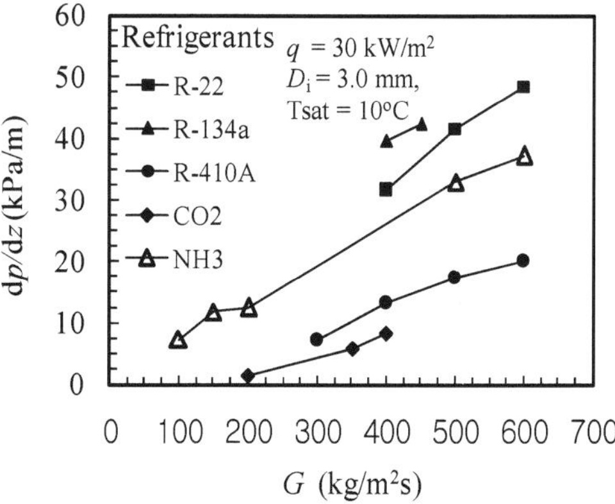

(b) Pressure drop comparison of the present working refrigerants at q = 30 kW/m², D_i = 3.0 mm, T_{sat} = 10°C

Fig. 7. Pressure drop comparison of the present working refrigerants

The pressure drop of the present working refrigerants is compared under several of the same experimental conditions. The pressure drop comparisons are illustrated in Fig. 7. The order of the pressure drop from the highest to the lowest is R-134a, R-22, NH_3, C_3H_8, R-410A, and CO_2. The pressure drop is strongly affected by the physical properties of the working fluid such as density, viscosity, surface tension and pressure. As shown in Table 3, the working refrigerant with a higher pressure drop has a higher density ratio ρ_f/ρ_g, viscosity ratio μ_f/μ_g, surface tension, and a lower pressure, generally. Therefore, it is clear that in the present comparison that CO_2 has the lowest pressure drop, making CO_2 an effective future environmental friendly refrigerant

Refrigerant	T (°C)	σ (10-3 N/m)	μ_f/μ_g	μ_g (10-6 Pa s)	μ_f (10-6 Pa s)	ρ_f/ρ_g	ρ_g (kg/m³)	ρ_f (kg/m³)	P (MPa)
R-22		10.22	16.36	11.96	195.7	43.27	28.82	1247	0.681
R-134a		10.14	21.42	11.15	238.8	62.33	20.23	1261	0.415
R-410A	10	7.16	11.36	12.91	146.6	27.07	41.74	1130	1.085
C_3H_8		8.85	13.96	8.15	113.8	37.32	13.8	515	0.636
CO_2		2.77	5.59	15.46	86.37	6.41	134.4	861.7	4.497
NH_3		29.59	16.35	9.36	153.03	128.32	4.89	623.64	0.615
R-22		10.95	17.62	11.73	206.7	50.99	24.79	1264	0.584
R-134a		10.84	23.25	10.94	254.4	74.61	17.13	1278	0.35
R-410A	5	7.9	12.38	12.6	156	32.21	35.73	1151	0.934
C_3H_8		9.48	15.03	7.97	119.8	43.57	11.98	522	0.551
CO_2		3.64	6.46	14.83	95.84	7.85	114.1	896.7	3.965
NH_3		31.24	17.51	9.21	161.23	153.52	4.115	631.66	0.515

Table 3. Physical properties of R-22, R-134a, R-410A, C_3H_8, CO_2 and NH_3 at 10 and 5°C

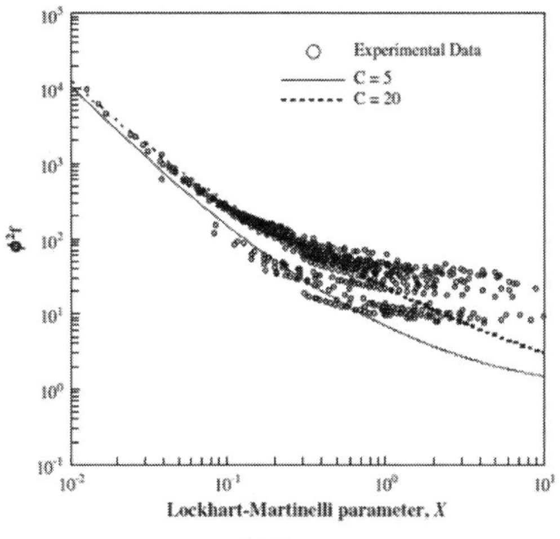

(a) Propane

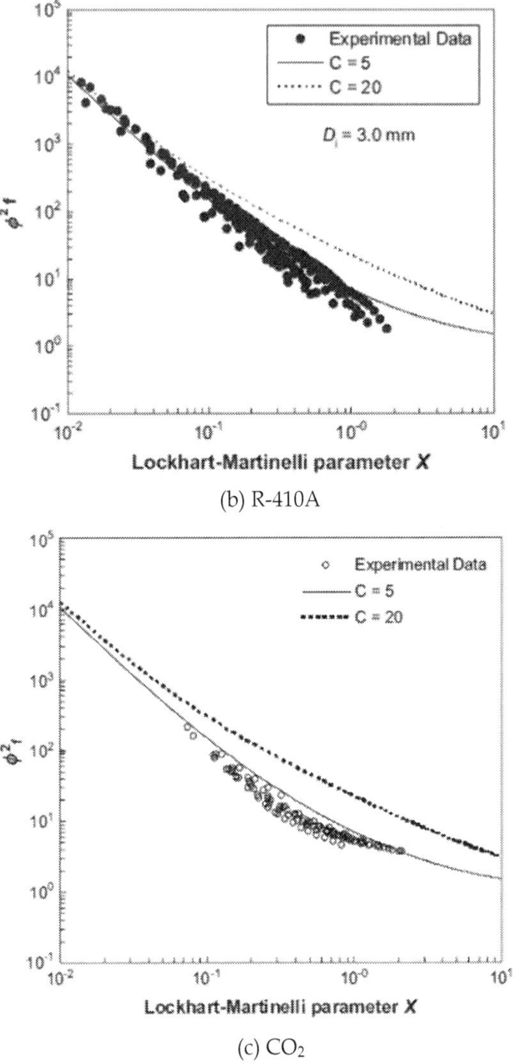

(b) R-410A

(c) CO_2

Fig. 8. Variation of the two-phase frictional multiplier data with the Lockhart–Martinelli parameter., (a) Propane, (b) R-410A and (c) CO_2

Fig. 8 (a)-(c) shows a comparison of the two-phase frictional multiplier data with the values predicted by the Lockhart–Martinelli correlation with C= 5 and C= 20, where the Cs are taken from Chisholm (1967). The figure shows that the present data are at the upper the baseline of C = 20, between the baseline of C = 20 and C = 5, and under the baseline of C = 5, which means that laminar, turbulent and the co-current laminar–turbulent flows exist in the present data.

	Previous correlations	Mean deviation(%)	Average deviation(%)
A.S. Pamitran et al. (2010)	Beattie and Whalley (1982)	23.35	15.91
	Cicchitti et al (1960)	23.51	2.9
	McAdams (1954)	25.25	21.75
	Chang et al. (2000)	26.41	22.65
	Dukler et al. (1964)	27.03	25.37
	Friedel (1979)	28.87	16.44
	Chisholm (1983)	34.65	3.89
	Tran et al. (2000)	42.48	33.05
	Zhang and Webb (2001)	49.04	22.4
	Mishima and Hibiki (1996)	50.71	39.89
	Lockhart and Martinelli (1949)	107	97.69
K.-I. Choi et al. (2009)	Mishima and Hibiki (1996)	35.37	25.08
	Friedel (1979)	38.79	21.91
	Chang at al. (2000)	38.86	21.99
	Lockhart and Martinelli (1949)	41.36	3.46
	Chisholm (1983)	45.24	18
	Zhang and Webb (2001)	46.82	7.58
	Chen et al. (2001)	63.67	63.6
	Kawahara et al. (2002)	73.04	72.92
	Tran et al. (2000)	77.99	35.6
	Homogeneous (Cicchitti et al., 1960)	48.67	-40.27
	Homogeneous (Beattie and Whalley, 1982)	54.68	-51.69
	Homogeneous (Mc Adam, 1954)	56.07	-53.73
	Homogeneous ((Dukler et al., 1964)	58.75	-57.32
K.-I. Choi et al. (2008)	Chang and Ro (1996)	19.57	7.43
	Homogeneous (Beattie and Whalley, 1982)	20.04	7.73
	Homogeneous (Mc Adam, 1954)	20.08	13.31
	Homogeneous (Cicchitti et al., 1960)	20.12	8.11
	Homogeneous ((Dukler et al., 1964)	21.87	18.08
	Yu et al. (2002)	27.14	17.88
	Chen et al(2001)	29.85	7.25
	*Chen et al(2001)	14.4	1.43
	Zhang and Webb (2001)	30.82	30.77
	Friedel (1979)	39.21	39.18
	Chang et al. (2000)	39.43	39.41
	Kawahara et al. (2002)	44.63	44.63
	Mishima (1983)	67.81	66.11
	Chisholm et al. (2000)	98.54	97.25
	Tran et al. (2000)	133.97	133.97
	Lockhart and Martinelli (1949)	145.44	144.58

	Previous correlations	Mean deviation(%)	Average deviation(%)
A.S. Pamitran et al. (2007)	Homogeneous (Dukler et al. (1964))	18.14	1.63
	Homogeneous (McAdams (1954))	18.33	0.53
	Homogeneous (Beattie and Whalley, 1982)	21.16	12.83
	Homogeneous (Cicchitti et al., 1960)	24.82	14.78
	Kawahara et al. (2002)	30.8	30.43
	Zhang and Webb (2001)	34.3	21.28
	Yu et al. (2002)	48.23	4.33
	Friedel (1979)	68.56	67.89
	Chang et al. (2000)	68.72	68.06
	Yoon et al. (2004)	101.18	93.59
	Chisholm (1983)	101.55	100.24
	Chen et al. (2001)	122.13	122.13
	Tran et al. (2000)	179.57	179.57

Table 4. Deviation of the pressure drop comparison with some existing correlations.

The current experimental two-phase pressure drop data were compared with the predictions from eleven existing correlations. The deviations from the correlation results from the previous experimental data are listed in Table 4, and the selected comparison figures are shown in figures 9(a)-(j). The pressure drop homogenous model of Beattie and Whalley (1982) provided the best prediction in the A.S. Pamitran et al. (2010) study among the other methods, yielding a mean deviation of around 23%. Mishima and Hibiki (1996) is the best correlation to predict the pressure drop in K.-I Choi et al. (2009) with a mean deviation about 35%. The other presented homogenous models also provided relatively good predictions.

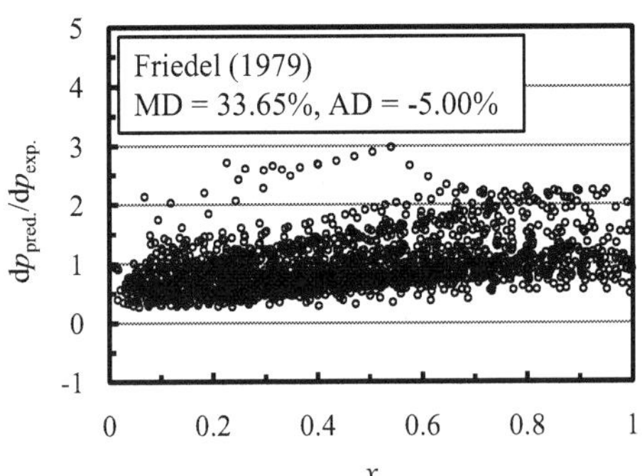

Fig. 9. (a) Comparison between the experimental data and the prediction pressure drop with Fridel (1979) correlation

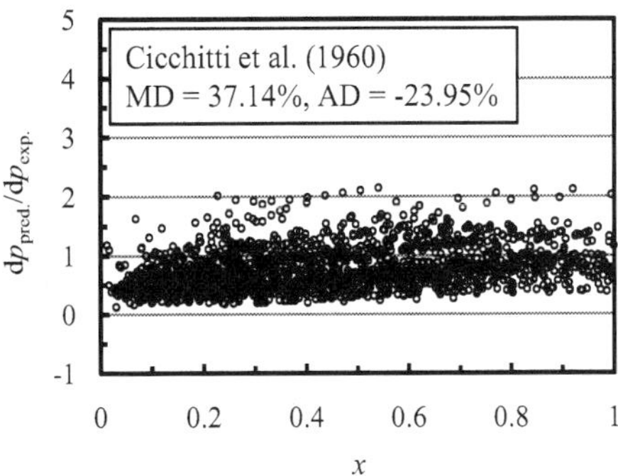

Fig. 9. (b) Comparison between the experimental data and the prediction pressure drop with Cicchitti et al (1960) correlation

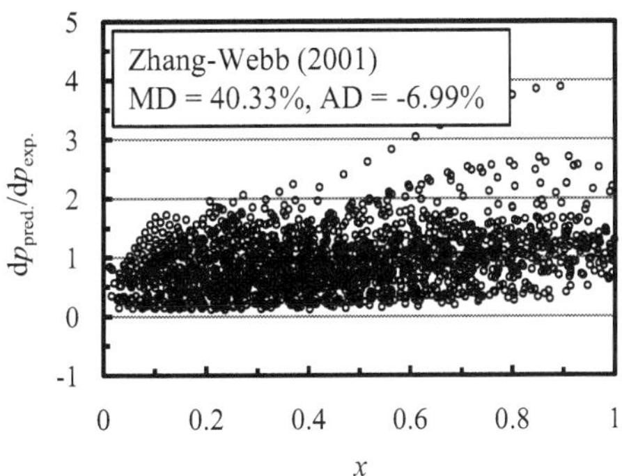

Fig. 9. (c) Comparison between the experimental data and the prediction pressure drop with Zang-Webb (2001) correlation

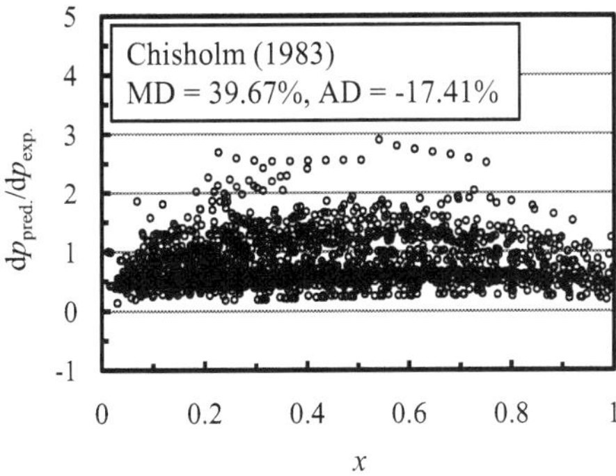

Fig. 9. (d) Comparison between the experimental data and the prediction pressure drop with Chisholm (1983) correlation

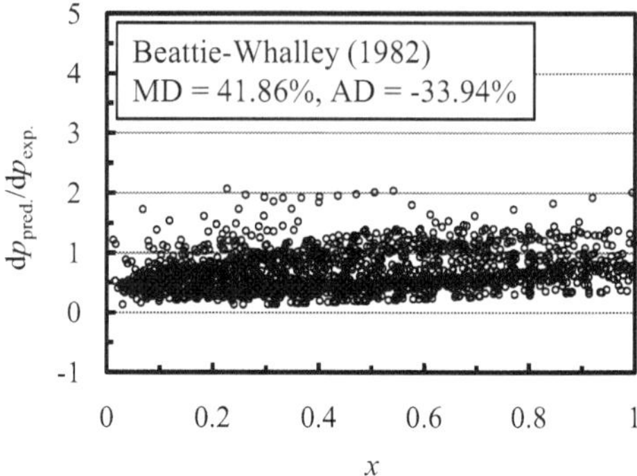

Fig. 9. (e) Comparison between the experimental data and the prediction pressure drop with Beattie-Whalley (1982) correlation

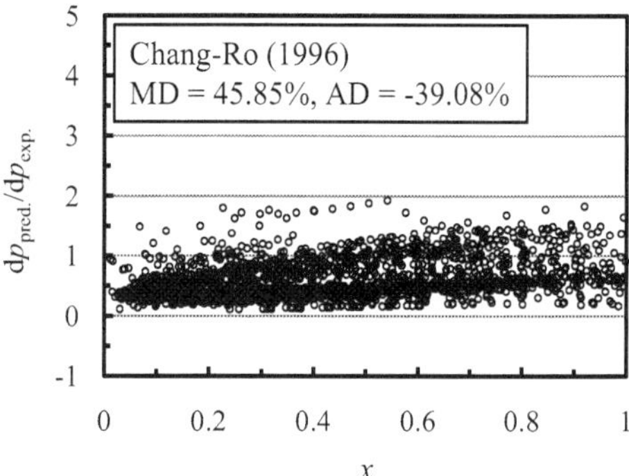

Fig. 9. (f) Comparison between the experimental data and the prediction pressure drop with Chang-Ro (1996) correlation

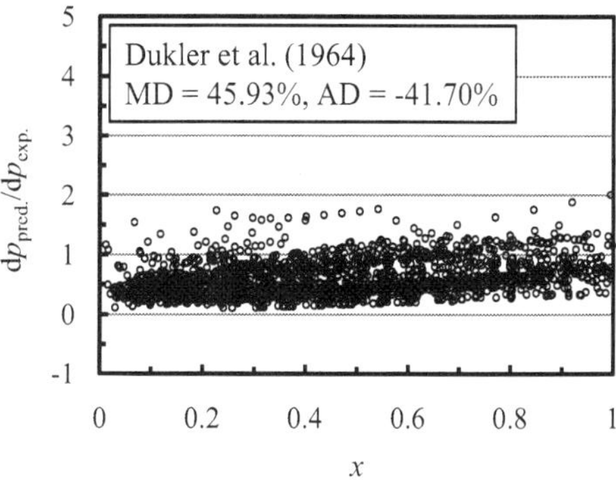

Fig. 9. (g) Comparison between the experimental data and the prediction pressure drop with Dukler et al. (1964) correlation

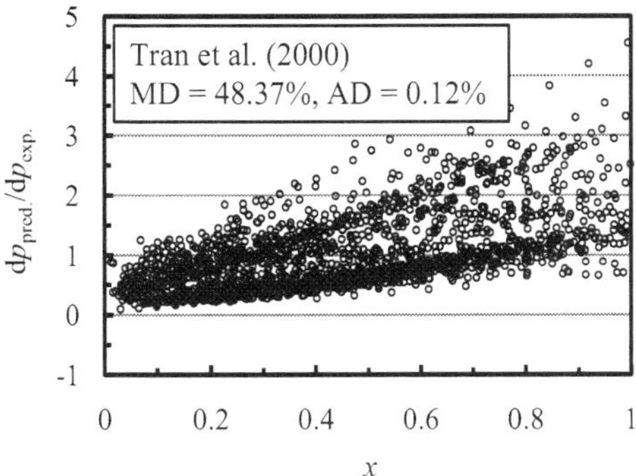

Fig. 9. (h) Comparison between the experimental data and the prediction pressure drop with
Trans et al. (2000) correlation

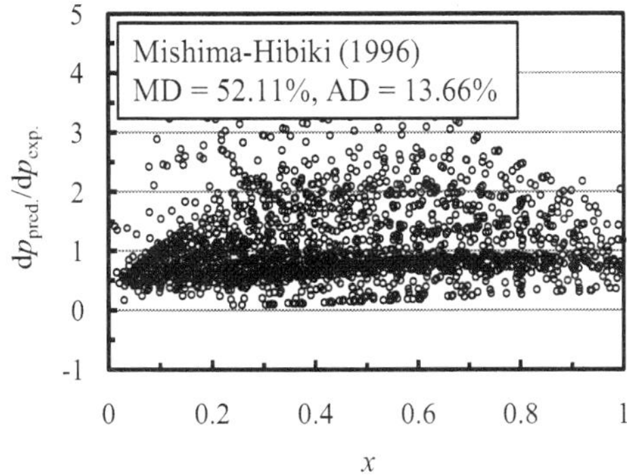

Fig. 9. (i) Comparison between the experimental data and the prediction pressure drop with
Mishima – Hibiki (1996) correlation

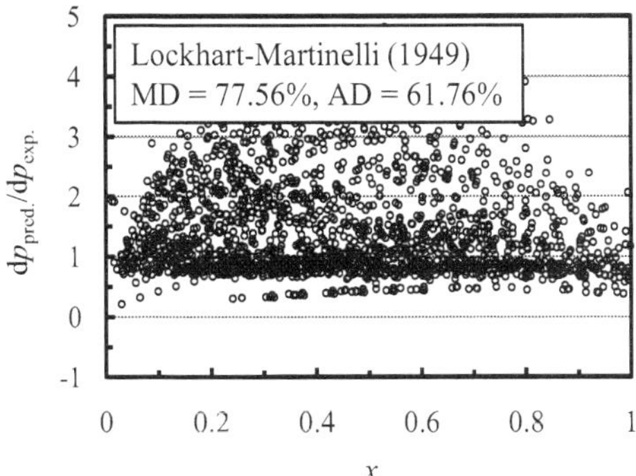

Fig. 9. (j) Comparison between the experimental data and the prediction pressure drop with
Lochart-Martinelli (1949) correlation

The pressure drop homogenous model of Beattie and Whalley (1982) provided the best prediction in the A.S. Pamitran et al. (2010) study among the other methods, yielding a mean deviation of around 23%. Mishima and Hibiki (1996) is the best correlation to predict the pressure drop in K.-I Choi et al. (2009) with a mean deviation about 35%. The other presented homogenous models also provided relatively good predictions. The homogenous model assumes that vapor and liquid velocities are equal. The McAdams et al. (1954) correlation was developed based on a homogeneous model. The homogeneous model is a simplification model; therefore, the model is not an empirical model. For some conditions, the homogeneous model can predict the pressure drop well, but not for other conditions; it depends on how close the conditions are to the assumptions of the homogenous model. In this study we develop a new correlation based on the present experimental data. The prediction by the Chang et al.'s (2000) and Friedel's (1979) models yielded a deviation of lower than 28% in every study. The Chang et al. (2000) model was developed with R-410A and air–water in a 5 mm smooth tube. Friedel's (1979) model was developed using a large database; it was valid for horizontal flow and vertical upward flow. The prediction with the Chisholm (1983) method showed a fair result with a mean deviation of lower than 40%. The remaining predictions of Tran et al. (2000), Zhang and Webb (2001), Lockhart and Martinelli (1949) provided a large mean deviation of more than 40% in every study as mentioned in Table 4. The correlations have failed to predict the experimental data of A.S Pamitran et al. (2010) and (2007) and K.-I. Choi et al. (2009) and (2008). Table 5 shows the previous correlations as comparisons, predicting the pressure drop by using the experimental data.

Previous Correlations	
Homogeneous [Beattie and Whalley (1982)]	$\bar{\mu} = \mu_f\left(1 - \dfrac{x\rho_f}{x\rho_f + (1-x)\rho_g}\right)\left(1 - \dfrac{2.5x\rho_f}{x\rho_f + (1-x)\rho_g}\right) + \mu_g\dfrac{x\rho_f}{x\rho_f + (1-x)\rho_g}$
Homogeneous [Cicchitti et al (1960)]	$\bar{\mu} = x\rho_g + (1-x)\rho_f$
Homogeneous [McAdams (1954)]	$\bar{\mu} = \left(\dfrac{x}{\mu_g}\dfrac{1-x}{\mu_f}\right)^{-1}$
Chang et al. (2000)	$\phi_{fo}^2 = (1-x)^2 + x^2\left[(\rho_f f_{go})/(\rho_g f_{fo})\right] + 3.24A_2A_3\mathrm{Fr}^{-0.045}\mathrm{We}^{-0.035+\Phi}$ $A_2 = x^{0.78}(1-x)^{0.224}$, $A_3 = (\rho_f/\rho_g)^{0.91}(\mu_g/\mu_f)^{0.19}\left[1-(\mu_g/\mu_f)\right]^{0.7}$ $\Phi = e^{-\left(\frac{D_{i,ref}}{D_i}\right)^{1.7}}\log\left(\dfrac{350}{\mathrm{We}\left(\frac{\mathrm{Re}_{fo}}{\mathrm{Re}_{fo,ref}}\right)^{1.3}}\right)$
Homogeneous [Dukler et al. (1964)]	$\bar{\mu} = \mu_f\left(1 - \dfrac{x\rho_f}{x\rho_f + (1-x)\rho_g}\right) + \mu_g\dfrac{x\rho_f}{x\rho_f + (1-x)\rho_g}$

Previous Correlations	
Friedel (1979)	$\phi_{fo}^2 = (1-x)^2 + x^2\left[(\rho_f f_{go})/(\rho_g f_{fo})\right] + 3.24 A_2 A_3 Fr^{-0.045} We^{-0.035}$ $A_2 = x^{0.78}(1-x)^{0.224}$, $A_3 = \left(\rho_f/\rho_g\right)^{0.91}\left(\mu_g/\mu_f\right)^{0.19}\left[1-\left(\mu_g/\mu_f\right)\right]^{0.7}$
Chisholm (1983)	$\phi_{fo}^2 = 1 + \left(\Gamma^2 - 1\right)\left[Bx^{0.875}(1-x)^{0.875} + x^{1.750}\right]$, where $\Gamma^2 = \dfrac{(dp/dz)_{go}}{(dp/dz)_{fo}}$ <table><tr><td>Γ</td><td>G (kg m^{-2} s^{-1})</td><td>B</td></tr><tr><td rowspan="3">≤ 9.5</td><td>≤ 500</td><td>4.8</td></tr><tr><td>$500 < G < 1900$</td><td>$2400/G$</td></tr><tr><td>≥ 1900</td><td>$55/G^{0.5}$</td></tr><tr><td rowspan="2">$9.5 < \Gamma < 28$</td><td>≤ 600</td><td>$520/(\Gamma G^{0.5})$</td></tr><tr><td>> 600</td><td>$21/\Gamma$</td></tr><tr><td>≥ 28</td><td>-</td><td>$\Gamma^2 G^{0.5}$</td></tr></table>
Tran et al. (2000)	$\phi_{fo}^2 = 1(4.3\Gamma^2 - 1)\left[N_{conf} x^{0.875} + x^{1.75}\right]$ $\Gamma^2 = \dfrac{(dp/dz)_{go}}{(dp/dz)_{fo}}$ and $N_{conf} = \dfrac{\left[\alpha/\left(g(\rho_f - \rho_g)\right)\right]^{0.5}}{D}$
Zhang and Webb (2001)	$\phi_{fo}^2 = (1-x)^2 + 2.87 x^2 \left(\dfrac{P}{P_c}\right)^{-1} + 1.68 x^{0.8}(1-x)^{0.25}\left(\dfrac{P}{P_c}\right)^{-1.64}$
Mishima and Hibiki (1996)	$\phi_f^2 = 1 - \dfrac{C}{X} + \dfrac{1}{X^2}$ where X $= \left(-\dfrac{dp}{dz}F\right)_f \Big/ \left(-\dfrac{dp}{dz}F\right)_g$ and $C = 21\left(1 - e^{-319\times 10^{-6}}\right)$
Lockhart and Martinelli (1949)	$\phi_f^2 = 1 - \dfrac{C}{X_{tt}} + \dfrac{1}{X_{tt}^2}$ $X_{tt} = \left(\dfrac{1-x}{x}\right)^{0.9}\left(\dfrac{\rho_g}{\rho_f}\right)^{0.5}\left(\dfrac{\mu_f}{\mu_g}\right)^{0.1}$ and C(turbulent − turbulent) $= 20$

Table 5. Some of previous correlations to predict pressure drop used by S. Pamitran et al. (2010) and (2007) and K.-I. Choi et al. (2009) and (2008)

4.3 Heat transfer coefficient

Fig. 10(a) shows the effect of mass flux on heat transfer coefficient. Mass flux has an insignificant effect on the heat transfer coefficient in the low quality region. The insignificant effect of mass flux on the heat transfer coefficient indicates that nucleate boiling heat transfer is predominant. Several previous studies performed by Kew and Cornwell (1997), Lazarek and Black (1982), Wambsganss et al. (1993), Tran et al. (1996) and Bao et al. (2000) used small tubes and showed that, in small channels, nucleate boiling tends to be predominant. The high amount of nucleate boiling heat transfer occurs because of the physical properties of the refrigerants, namely their surface tension and pressure, and the geometric effect of small channels. A higher mass flux corresponds to a higher heat transfer coefficient at moderate-high vapor quality, due to an increase in the convective boiling heat transfer contribution.

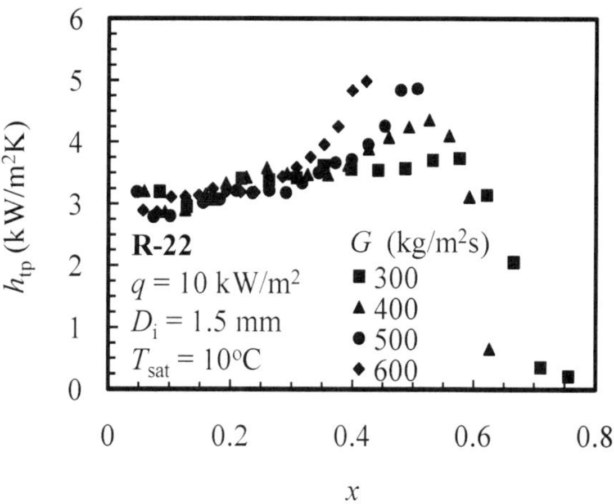

Fig. 10. (a) Effect of mass flux on heat transfer coefficient for R-22 at $q = 10$ kW/m^2, $D_i = 1.5$ mm, $T_{sat} = 10°C$

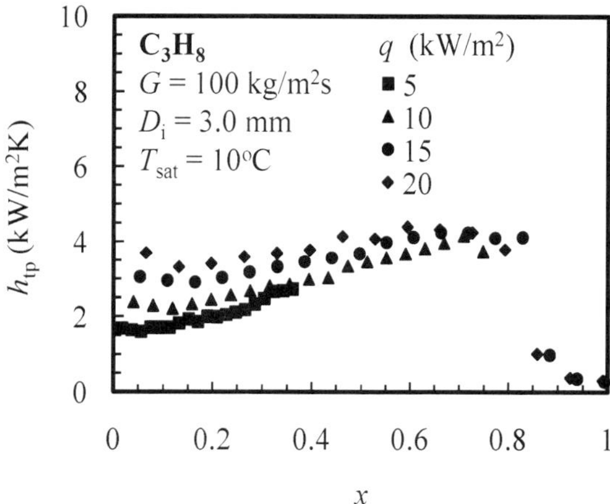

Fig. 10. (b) Effect of heat flux on heat transfer coefficient for C$_3$H$_8$ at $G = 100$ kg//(m^2·s), $D_i = 3.0$ mm, $T_{sat} = 10°C$

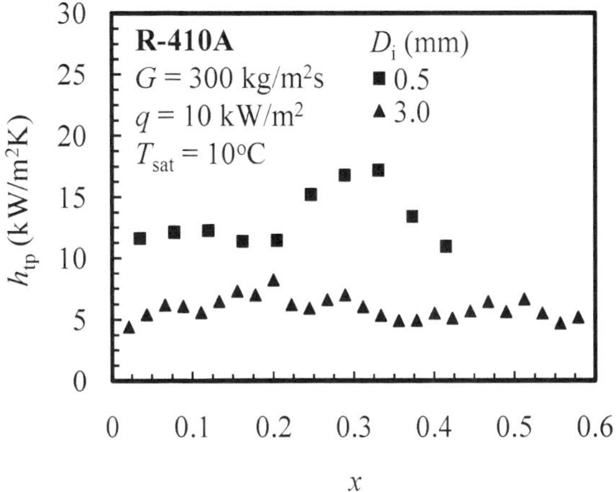

Fig. 10. (c) Effect of inner tube diameter on heat transfer coefficient for R-410A at
G = 300 kg//(m² ·s), q = 10 kW/m², T_{sat} = 10°C

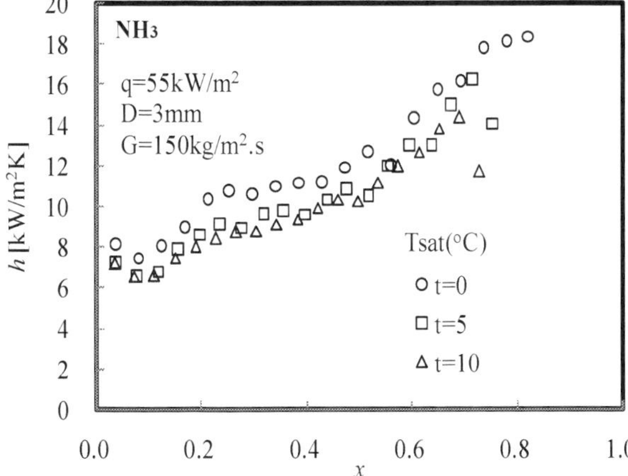

Fig. 10. (d) Effect of saturation temperature on heat transfer coefficient for NH₃ at
G = 150 kg//(m² ·s), q = 55 kW/m², D_i = 3.0 mm

In the high quality region, a drop in the heat transfer coefficient occurs at lower qualities for
a relatively higher mass flux. The steep decreasing of the heat transfer coefficient at high
qualities is due to the effect of a small diameter tube on the boiling flow pattern because
dry-patch occurs more easily in smaller diameter tubes and at a higher mass flux.

Fig. 10(b). depicts the dependence of heat flux on heat transfer coefficients in the low-
moderate quality region. The large effect of heat flux on the heat transfer coefficient shows
the domination of the nucleate boiling heat transfer contribution. At the higher quality
region nucleate boiling is suppressed or convective heat transfer contribution is
predominant; this is indicated by a low effect of heat flux on heat transfer coefficient.

Fig. 10 (c) shows that a smaller inner tube diameter has a higher heat transfer coefficient at low quality regions. This is due to a more active nucleate boiling in a smaller diameter tube. As the tube diameter gets smaller, the contact surface area for heat transfer increases. The more active nucleate boiling causes dry-patches to appear earlier. The quality for a rapid decrease in the heat transfer coefficient is lower for the smaller tube. It is supposed that the annular flow appears at a lower quality in the smaller tube and therefore, the dry-out quality is relatively lower for the smaller tube. The effect of saturation temperature on heat transfer coefficient is depicted in Fig. 10 (d). The heat transfer coefficient increases with an increase in saturation temperature, which is due to a larger effect from nucleate boiling. A higher saturation temperature provides a lower surface tension and higher pressure. The vapor formation in the boiling process explains that a lower surface tension and higher pressure provides a higher heat transfer coefficient.

	Previous Correlation	Deviation	
		Mean (%)	Average (%)
K.-I. Choi et al. (2009)	Shah (1988)	19.21	3.55
	Tran et al. (1996)	21.18	-6.15
	Jung at al. (1989)	26.05	23.38
	Gungor and Winterton (1987)	28.44	23.78
	Takamatsu et al (1993)	32.69	32.15
	Kandlikar and Stainke (2003)	33.84	24.41
	Wattelet et al (1994)	48.28	48.28
	Chen (1966)	50.82	18.74
	Zhang et al. (2004)	79.21	77.89
K.-I. Choi et al. (2007)	Wattelet et al.	0.06	-3.03
	Gungore-Winterton	17.48	1.59
	Zhang et al	19.83	-13.35
	Kandlikare-Steinke	21.12	-15.43
	Tran et al.	22.31	-17.37
	Jung et al.	29.1	-23.44
K.-I. Choi et al. (2007)	Wattelet et al.	19.09	12.22
	Jung et al.	23.48	9.53
	Kandlikare-Steinke	24.32	13.05
	Tran et al.	24.81	-10.51
	Shah	25.22	-12.29
	Gungore-Winterton	25.26	2.39
	Chen	36.13	-18.48
A.S. Pamitran et al (2007)	Zhang et al	27.45	-17.34
	Gungore-Winterton	31.07	-12.65
	Tran et al.	31.69	-28.52
	Takamatsu et al.	33.22	-20.65
	Jung et al.	33.39	-26.45
	Kandlikare-Steinke	38.59	-18.39
	Wattelet et al.	38.68	-31.48

Table 6. Mean deviation and average deviation calculated for the different pressure drop prediction methods

Fig. 11 (a) and 11 (b) show the comparisons of the heat transfer coefficients of R-22, R-134a, R-410A, C_3H_8 and CO_2 at some experimental conditions. The mean heat transfer coefficient ratio of R-22, R-134a, R-410A, C_3H_8 and CO_2 was approximately 1.0 : 0.8 : 1.8 : 0.7 : 2.0.

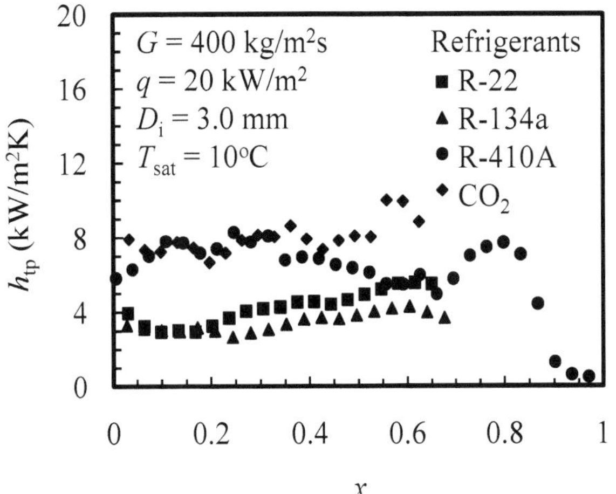

Fig. 11. (a) Heat transfer coefficient comparison of the present working refrigerants at G = 400 kg//(m² ·s), q = 20 kW/m², D_i = 3.0 mm, T_{sat} = 10 °C

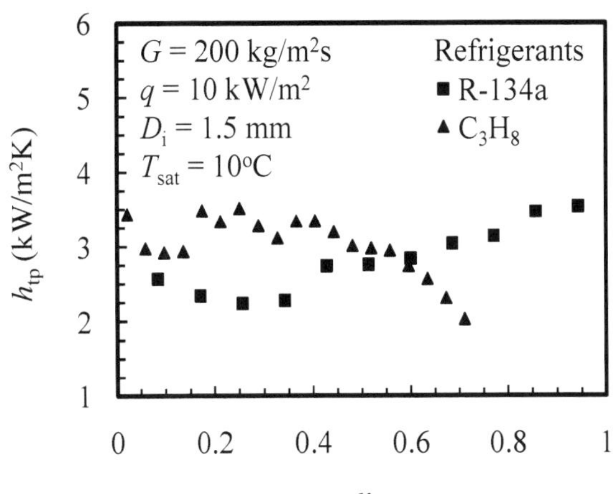

Fig. 11. (b) Heat transfer coefficient comparison of the present working refrigerants at G = 200 kg//(m² ·s), q = 10 kW/m², D_i = 1.5 mm, T_{sat} = 10°C

The heat transfer coefficient of CO_2 was higher than that of the other working refrigerants during evaporation under all test conditions. The higher heat transfer coefficient of CO_2 is believed to be due to its high boiling nucleation. The CO_2 has much lower surface tension and applies much higher pressure than the other working refrigerants. The heat transfer coefficients of R-22, R-134a and C_3H_8 are similar due to their similar physical properties. The comparisons of the physical properties of the present working refrigerants are given in Table 3. The CO_2 has a much lower viscosity ratio μ_f/μ_g than the other working refrigerants, which means that the liquid film of CO_2 can break more easily than those of the other refrigerants. The CO_2 has also a much lower density ratio ρ_f/ρ_g than the other working refrigerants, which leads to a lower vapor velocity, which in turn causes less suppression of nucleate boiling, as shown.

The heat transfer coefficients of the present study are compared with the results given by several correlations for boiling heat transfer coefficients as shown in Table 6. The Gungor-Winterton (1987), Jung et al. (1989), Shah (1988) and Tran et al. (1996) correlations provided better predictions, with mean deviations of lower than 30% in A.S. Pamitran et al (2010), than the other correlations. Wattelet et al (1994) successively predict the experiment data for [a,b]K.I. Choi et al (2007). The Gungor-Winterton (1987) correlation was a modification of the superposition model; it was developed using fluids in several small and conventional tubes under various test conditions. The Jung et al. (1989) correlation was developed with pure and mixture refrigerants in conventional channels; its F factor contributed a big calculation deviation with the current experimental data. The Shah (1988) correlation was developed using a large data set for conventional tubes. The prediction with the Shah (1988) correlation was fair under conditions at the low quality region. The Tran et al. (1996) correlation was developed for R-12 and R-113 in small tubes. The correlations of Chen (1966) and Wattelet et al. (1994), which were developed for large tubes, have a high prediction deviation in K.I Choi et al.'s (2009) experimental data. The correlations of Kandlikar (1990) and Zhang et al. (2004) were developed for small tubes; however, the correlations could not predict well the present experimental data. The correlations of Wattelet et al. (1994), Kandlikar (1990) and Zhang et al. (2004) showed a large deviation in the prediction of the CO_2 data. The Kandlikar (1990) correlation failed to predict the heat transfer coefficient at the high quality region.

Table 7 shows the previous correlation use by A.S. Pamitran et al. (2007) and K.I Choi (2009), (2007)

5. Development of a new correlation

5.1 Pressure drop

In this section, the effort of A.S. Pamitran et al and Choi et al. in developing a pressure drop correlation that is appropriate to use in specific refrigerants for the horizontal smooth minichannel is highlighted. The area average-gas fraction or void fraction denoted with α is defined as the ratio of gas phase cross-sectional area (A_g) to the total cross-sectional area. $A = A_g + A_f$

$$\alpha = \frac{A_g}{A} \tag{12}$$

In boiling and condensation it is often convenient to use the fraction of the total mass flow which is composed of vapor or liquid. The mass quality, x, is defined as

$$x = \frac{W_g}{W_g + W_f} = \frac{W_g}{W} \tag{13}$$

Heat Transfer Coefficient Correlations	
Gungor and Winterton (1987)	$h_{\text{tp}} = E.h_l + S.h_{\text{pb}}$ $E = \text{fn}(Bo, X_{\text{tt}})$ and $S = \text{fn}(Bo, X_{\text{tt}}, Re_l)$ $h_l = h_{\text{Dittus-Boelter}}$, $h_{\text{pb}} = h_{\text{Cooper}}$
Jung et al. (1989)	$h_{\text{tp}} = \dfrac{N}{C_{\text{un}}} h_{\text{un}} + C_{\text{me}} F_p h_{\text{lo}}$ $N = \text{fn}(X_{\text{tt}}, Bo)$ $C_{\text{un}} = \text{fn}(X, Y, p, p_{cmvc})$ $h_{\text{un}} = \dfrac{h_1 h_2}{(h_1 X_2 + h_2 X_1)}$, $h_1 \ and \ h_2 \ are \ h_{sa} = h_{Stephan-Abdesalam}$ $C_{\text{me}} = \text{fn}(X, Y)$, $F_p = \text{fn}(X_{\text{tt}})$ $h_{\text{lo}} = h_{\text{Dittus-Boelter}}$
Shah (1988)	For horizontal tube, $Fr_f > 0.04$ $N = C_o = \left(\dfrac{1-x}{x}\right)^{0.8} \left(\dfrac{\rho_g}{\rho_f}\right)^{0.5}$ $h_{\text{tp}} = max(h_{\text{nb}}, h_c)$
Tran et al. (1996)	$h_{\text{tp}} = 8.4 \times 10^2 (Bo^2 We_f)^{0.3} \left(\dfrac{\rho_f}{\rho_g}\right)^{-0.4}$
Chen (1966)	$h_{\text{tp}} = S.h_{\text{nb}} + F.h_{lo}$ $h_{\text{nb}} = h_{\text{Foster-Zuber}}$ $h_{\text{lo}} = h_{\text{Dittus-Boelter}}$ $F = \text{fn}(X_{\text{tt}})$ and $S = \text{fn}(Re_{\text{tp}})$
Wattelet et al.(1994)	$h_{\text{tp}} = [h_{\text{nb}}^n + h_{\text{cb}}^n]^{1/2}$, $n = 2.5$ $h_{\text{nb}} = h_{\text{Cooper}}$ $h_{\text{cb}} = h_{\text{Dittus-Boelter}} \times F \times R$ $R = \text{fn}(X_{\text{tt}})$ and $R = \text{fn}(Fr_1)$
Kandlikar and Steinke (2003)	$\dfrac{h_{\text{tp}}}{h_l} = D_1 Co^{D_2}(1-x)^{0.8}\text{fn}(Fr_{lo})$ $\qquad\qquad\qquad + D_3 Bo^{D_4}(1-x)^{0.8} Fn$ $\text{fn}(Fr_{lo}) = 1$
Zang et al. (2004)	$h_{\text{tp}} = S.h_{\text{nb}} + F.h_{\text{sp}}$ for horizontal circular channel $h_{\text{nb}} = h_{\text{Foster-Zuber}}$ $h_{\text{sp}} = max(h_{\text{Dittus-Boelter}})$, if $Re_f < 2300$ $h_{\text{sp}} = h_{\text{Dittus-Boelter}}$, if $Re_f \geq 2300$ $F = \text{fn}(\phi_f)$ and $S = \text{fn}(Re_f)$

Table 7. Some of previous correlations to predict heat transfer coefficient used by .S Pamitran et al. (2010) and (2007) and K.-I. Choi et al. (2009) and (2008)

The rate of mass flow divided by the flow area is given in the name 'mass velocity' and the symbol G.

$$G = \frac{W}{A} = \rho.u = \frac{u}{v} \tag{14}$$

The mean velocity of the liquid phase is shown by

$$u_f = \frac{G(1-x)}{\rho_f.(1-\alpha)} \tag{15}$$

The mean velocity of the gas phase is shown by

$$u_g = \frac{Gx}{\rho_g.\alpha} \tag{16}$$

The total pressure drop consists of friction, acceleration and static head components as illustrated in the Equation below.

$$\left(\frac{dp}{dz}\right) = \left(\frac{dp}{dz}F\right) + \left(\frac{dp}{dz}a\right) + \left(\frac{dp}{dz}z\right)$$

For flow boiling in a horizontal layout test section, the static head pressure drop is excluded. Therefore, the experimental two-phase frictional pressure drop can be obtained by subtracting the calculated acceleration pressure drop from the measured pressure drop, as shown in Eq. (2), and may then be expressed in terms of the single-phase pressure drop for the liquid phase, considered to exist in the tube, as below.

$$\left(\frac{dp}{dz}F\right) = \left(\frac{dp}{dz}F\right)_f \phi_f^2 \tag{17}$$

The liquid frictional pressure drop is calculated from the Fanning equation. The frictional pressure drop can then be rewritten as Eq. (19) below.

$$\left(\frac{dp}{dz}F\right)_f = \frac{2f_f G^2(1-x)^2}{\rho_f D} \tag{18}$$

The friction factor in Eq. (7) was obtained by considering the flow conditions of laminar–turbulent flows, where $f = 16\text{Re}^{-1}$ for Re < 2300 (laminar flow) and $f = 0.079\ \text{Re}^{-0.25}$ for Re > 3000 (turbulent flow). The laminar–turbulent transition Reynolds number was obtained from Yang and Lin (2007).

$$\left(\frac{dp}{dz}F\right) = \left[\frac{2f_f G^2(1-x)^2}{\rho_f D}\right] \phi_f^2 \tag{19}$$

A new modified pressure drop correlation was developed on the basis of the Lockhart–Martinelli method. The two-phase pressure drop of Lockhart–Martinelli consists of the following three terms: the liquid phase pressure drop, the interaction between the liquid phase and the vapor phase, and the vapor phase pressure drop. The relationship among these terms is expressed in Eq. (20).

$$\left(-\frac{dp}{dz}F\right)_{tp} = \left(-\frac{dp}{dz}F\right)_{f} + C\left[\left(-\frac{dp}{dz}F\right)_{f}\left(-\frac{dp}{dz}F\right)_{g}\right]^{1/2} + \left(-\frac{dp}{dz}F\right)_{g} \tag{20}$$

The two-phase frictional multiplier based on the pressure gradient for liquid alone flow, ϕ_f^2, is calculated by dividing Eq. (20) by the liquid phase pressure drop, as is shown in Eq. (21).

$$\phi_f^2 = \frac{\left(-\frac{dp}{dz}F\right)_{tp}}{\left(-\frac{dp}{dz}F\right)_{f}} = 1 + C\left[\frac{\left(-\frac{dp}{dz}F\right)_{g}}{\left(-\frac{dp}{dz}F\right)_{f}}\right]^{1/2} + \frac{\left(-\frac{dp}{dz}F\right)_{g}}{\left(-\frac{dp}{dz}F\right)_{f}} = 1 + \frac{C}{X} + \frac{1}{X^2} \tag{21}$$

The Martinelli parameter, X, is defined by the following equation:

$$X = \left[\frac{\left(-\frac{dp}{dz}F\right)_{f}}{\left(-\frac{dp}{dz}F\right)_{g}}\right]^{1/2} = \left[\frac{2f_f G^2(1-x)^2\rho_g/D}{2f_g G^2 x^2 \rho_f/D}\right]^{1/2} = \left(\frac{f_f}{f_g}\right)^{1/2}\left(\frac{1-x}{x}\right)\left(\frac{\rho_g}{\rho_f}\right)^{1/2} \tag{22}$$

The friction factor in Eq. (22) was obtained by considering the flow conditions of laminar (for Re < 1000, $f = 16\mathrm{Re}^{-1}$) and turbulent (for Re > 2000, $f = 0.079\mathrm{Re}^{-0.25}$). The pressure drop prediction method can be summarized as follows:

$$\left(\frac{dp}{dz}\right) = \left(\frac{dp}{dz}F\right) + \left(\frac{dp}{dz}a\right)$$

$$\left(\frac{dp}{dz}\right) = \left(\frac{2f_f G^2(1-x)^2}{\rho_f D}\right)\phi_f^2 + G^2\frac{d}{dz}\left(\frac{x^2}{\alpha\rho_g} + \frac{(1-x)^2}{(1-\alpha)\rho_f}\right)$$

$$\left(\frac{dp}{dz}\right) = \left(\frac{2f_f G^2(1-x)^2}{\rho_f D}\right)\left(1 + \frac{C}{X} + \frac{1}{X^2}\right) + G^2\frac{d}{dz}\left(\frac{x^2}{\alpha\rho_g} + \frac{(1-x)^2}{(1-\alpha)\rho_f}\right)$$

$$\left(\frac{dp}{dz}\right) = \left(\frac{2f_f G^2(1-x)^2}{\rho_f D}\right)\left(1 + \frac{C}{\left(\frac{f_f}{f_g}\right)^{1/2}\left(\frac{1-x}{x}\right)\left(\frac{\rho_g}{\rho_f}\right)^{1/2}} + \frac{1}{\left(\frac{f_f}{f_g}\right)\left(\frac{1-x}{x}\right)^2\left(\frac{\rho_g}{\rho_f}\right)}\right)$$
$$+ G^2\frac{d}{dz}\left(\frac{x^2}{\alpha\rho_g} + \frac{(1-x)^2}{(1-\alpha)\rho_f}\right) \tag{23}$$

The calculated factor C is obtained from Chisholm (1967). For the liquid-vapor flow condition of turbulent-turbulent (tt), laminar-turbulent (vt), turbulent-laminar (tv) and laminar-laminar (vv), the values of the Chisholm (1967) parameter, C, are 20, 12, 10, and 5, respectively. The value of C in this thesis is obtained by considering the flow conditions of laminar and turbulent with thresholds of Re=2300 and Re=3000 for the laminar and turbulent flows, respectively. The laminar-turbulent transition Reynolds number was obtained from Yang and Lin (2007). Fig. 12 shows a comparison of the two-phase frictional multiplier data with the values predicted by the Lockhart-Martinelli correlation with C=5 and C=20. The figure shows that the present data are located at mostly between the baseline of C=5 and C=20, which means that laminar, turbulent and the co-current laminar-turbulent flows exist in the present data. For liquid-vapor flow conditions, the present experimental data shows 5.43% laminar-laminar, 28.46% laminar-turbulent, 3.78% turbulent-laminar, and 62.33% turbulent-turbulent.

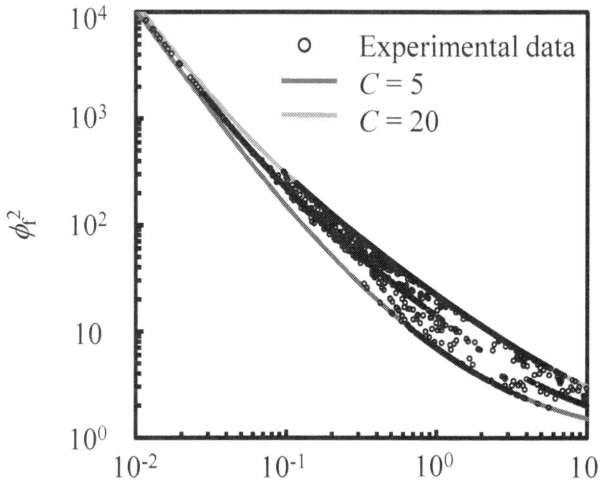

Lockhart-Martinelli parameter, X

Fig. 12. Variation of the two-phase frictional multiplier data with the Lockhart-Martinelli parameter

The experimental result showed that the pressure drop is a function of mass flux, inner tube diameter, surface tension, density and viscosity, therefore the factor C in Eq. (21) and (23) will be developed as a function of the two-phase Weber number, We_{tp}, and the two-phase Reynolds number, Re_{tp}.

$$C = \left(\phi_f^2 - 1 - \frac{1}{X^2} \right) X = fn\left(We_{tp}, Re_{tp} \right) \tag{24}$$

Where We_{tp} and Re_{tp} are defined as:

$$We_{tp} = \frac{G^2 D}{\bar{\rho}\sigma} \tag{25}$$

$$Re_{tp} = \frac{GD}{\bar{\mu}} \tag{26}$$

A new factor C was developed using the regression method, as is shown in the Equations in the table below.

A new pressure drop correlation was developed on the basis of the Lockhart–Martinelli method as a function of the Weber number, We_{tp}, and the Reynolds number, Re_{tp} by considering the laminar–turbulent flow conditions. Using a regression method with 812 data points, a new factor C was developed by A.S. Pamitran et al. (2010) with mean and average deviations of 21.66% and -2.47%, respectively, based on comparison as shown in Fig. 13.

	Correlation	Deviation
Pamitran, A.S., et al. (2010)	$C = 3 \times 10^{-3} \times Re_{tp}^{1.23} We_{tp}^{-0.433}$	MD = 21.66% AD = -2.47%
Choi, K.-I., et al. (2009)	$C = 1732.953 \times Re_{tp}^{-0.323} We_{tp}^{-0.24}$	MD = 10.84% AD = 1.08%
Choi, K.-I., et al. (2008)	$C = 5.5564 \times Re_{tp}^{0.2837} We_{tp}^{-0.288}$	MD = 4.02 AD = -0.14%
Pamitran, A.S., et al. (2008)	$C = 1.2897 \times 10^6 \times Re_{tp}^{0.5674} We_{tp}^{-3.3271}$	MD = 9.41% AD = -0.55%

Table 8. Equations developed by A.S. Pamitran et al. and K.I. Choi et al

This correlation will contribute to the design of heat exchangers with small tubes.

Fig. 14 illustrates the two-phase frictional multiplier comparison between the present experimental data and the prediction with the newly developed correlation. The comparison shows a mean deviation of 10.84% and an average deviation of 1.08%.

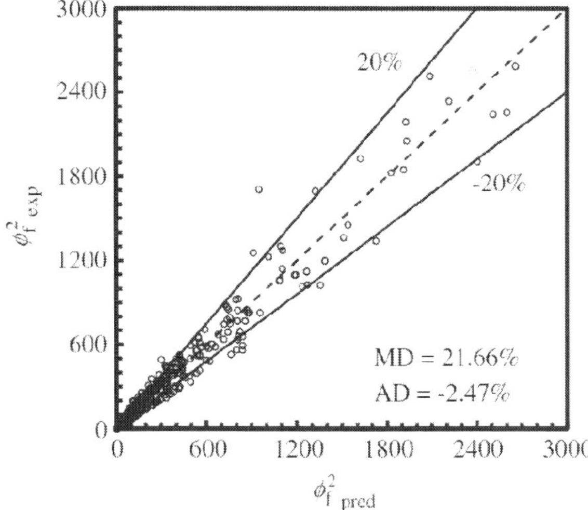

Fig. 13. Two-phase frictional multiplier comparison between the present experimental data ($\phi_{f\,exp}^2$) and the prediction with the newly developed correlation ($\phi_{f\,pred}^2$), A.S. Pamitran et al. (2010)

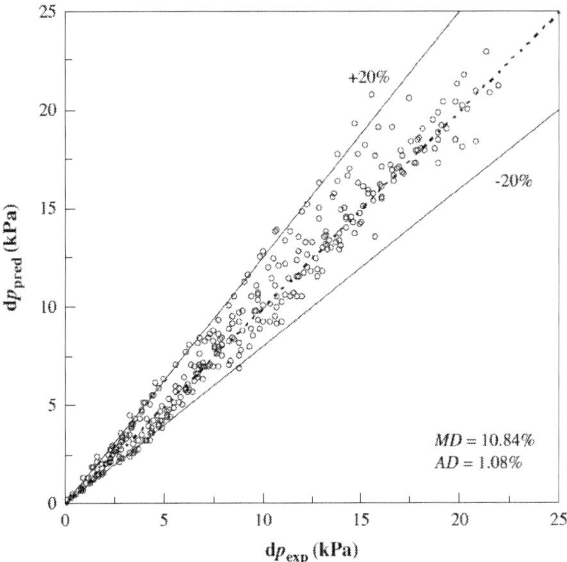

Fig. 14. Comparison of the experimental and predicted heat transfer coefficients using the new developed correlation. K.I. Choi et al (2009)

5.2 Heat transfer coefficient

It is well known that the flow boiling heat transfer is mainly governed by two important mechanisms, namely nucleate boiling and forced convective heat transfer.

$$h_{tp} = h_{nb} + h_c \qquad (27)$$

In two-phase flow boiling heat transfer, the nucleate boiling heat transfer contribution is suppressed by the two-phase flow. Therefore, the nucleate boiling heat transfer contribution may be correlated with a nucleate boiling suppression factor, S. Another contribution of convective heat transfer may be correlated with a liquid single phase heat transfer. The F factor is introduced as a convective two-phase multiplier to account for enhanced convective properties due to co-current flow of liquid and vapor. A superposition model of heat transfer coefficient may be written as follows:

$$h_{tp} = S.h_{nbc} + F.h_f \qquad (28)$$

The appearance of convective heat transfer for boiling in small channels occurs later than it does in large channels because of its high boiling nucleation.

The new heat transfer coefficient correlation in this study is developed by only using the experimental data prior to the dry-out. Chen (1966) introduced a multiplier factor, $F= fn(X_{tt})$, to account for the increase in the convective turbulence that is due to the presence of the vapor phase. The function should be physically evaluated again for flow boiling heat transfer in a minichannel that has a laminar flow condition, which is due to the small diameter effect. By considering the flow conditions (laminar or turbulent) in the Reynolds number factor, F, Zhang et al. (2004) introduced a relationship between the factor F and the two-phase frictional multiplier that is based on the pressure gradient for liquid alone flow, ϕ_f^2. This relationship is

$$F = fn\left(\phi_f^2\right) \tag{29}$$

Where ϕ_f^2 is a general form for four conditions according to Chisholm (1967), as is shown in Eq. (21). For the liquid–vapor flow condition of turbulent–turbulent (tt), laminar–turbulent (vt), turbulent–laminar (tv) and laminar–laminar (vv), the values of the Chisholm parameter, C, are 20, 12, 10, and 5, respectively. The value of C is found by an interpolation of the Chisholm parameter with thresholds of Re = 1000 and Re = 2000 for the laminar and turbulent flows, respectively.

The Martinelli parameter is defined as in equation (22) together with the Blasius equation of friction factors, f_f and f_g ; the Martinelli parameter can be rewritten as

$$\left(\frac{f_f}{f_g}\right)^{1/8}\left(\frac{1-x}{x}\right)^{7/8}\left(\frac{\rho_g}{\rho_f}\right)^{1/2} \tag{30}$$

There is an important effect of quality, density ratio, ρ_f/ρ_g, and the viscosity ratio, μ_f/μ_g, on heat transfer coefficient. The liquid heat transfer is defined by existing liquid heat transfer coefficient correlations by considering flow conditions of laminar and turbulent. For laminar flow, $Re_f < 2300$, where

$$Re_f = \frac{G(1-x)D}{\mu_f} \tag{31}$$

And the liquid heat transfer coefficient is obtained from the following correlation:

$$h_f = 4.36\frac{k_f}{D} \tag{32}$$

For flow with $3000 \leq Re_f \leq 10^4$, the liquid heat transfer coefficient is obtained from the Gnielinski (1976) correlation:

$$h_f = \frac{(Re_f-1000)Pr_f\left(\frac{f_f}{2}\right)\left(\frac{k_f}{D}\right)}{1 + 12.7\left(Pr_f^{2/3} - 1\right)\left(\frac{f_f}{2}\right)^{0.5}} \tag{33}$$

where the friction factor is calculated from (for Re < 1000, $f = 16Re^{-1}$) and turbulent (for Re > 2000, $f = 0.079Re^{-0.25}$). For flow with $2300 \leq Re_f \leq 3000$, the liquid heat transfer coefficient is calculated by interpolation. For turbulent flow with $10^4 \leq Re_f \leq 5\times10^6$, the liquid heat transfer coefficient is obtained from the Petukhov and Popov (1963) correlation:

$$h_f = \frac{Re_f Pr_f\left(\frac{f_f}{2}\right)\left(\frac{k_f}{D}\right)}{1 + 12.7\left(Pr_f^{2/3} - 1\right)\left(\frac{f_f}{2}\right)^{0.5}} \tag{34}$$

The Dittus Boelter (1930) correlation is used for turbulent flow with $Re_f \geq 5\times10^6$.

$$h_f = 0.023 \frac{k_f}{D} Re_f^{0.8} Pr_f^{0.4} = 0.23 \frac{k_f}{D} \left[\frac{G(1-x)D}{\mu_f} \right]^{0.8} \left[\frac{Cp_f \mu_f}{k_f} \right]^{0.4} \tag{35}$$

The F factor in this study is developed as a function of ϕ_f^2, $F = \mathrm{fn}(\phi_f^2)$, where ϕ_f^2 is obtained from Eqs. (21) – (22). The liquid heat transfer is defined by the Dittus Boelter correlation and a new factor F, as is shown in Fig. 15(a)-(d), is developed using a regression method.

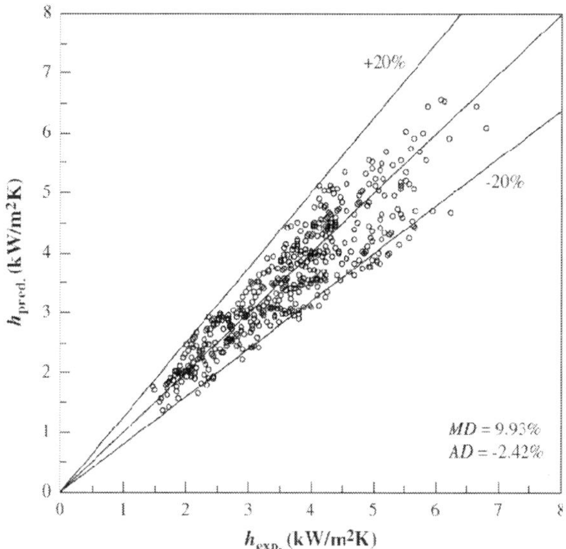

Fig. 15. (a) Comparison of the experimental and predicted heat transfer coefficients using the new developed correlation. K.I. Choi et al. (2009)

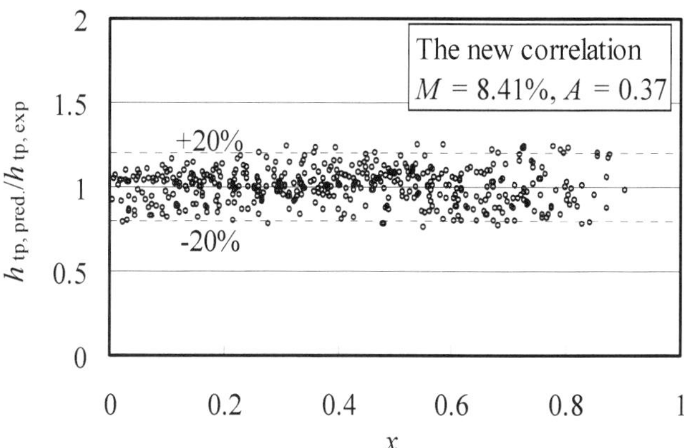

Fig. 15. (b) Diagram of the ratio of the experimental heat transfer coefficient, $h_{tp,exp}$, and the predicted heat transfer coefficient, $h_{tp,pred}$, vs. quality, x. [a]K.I. Choi et al. (2007)

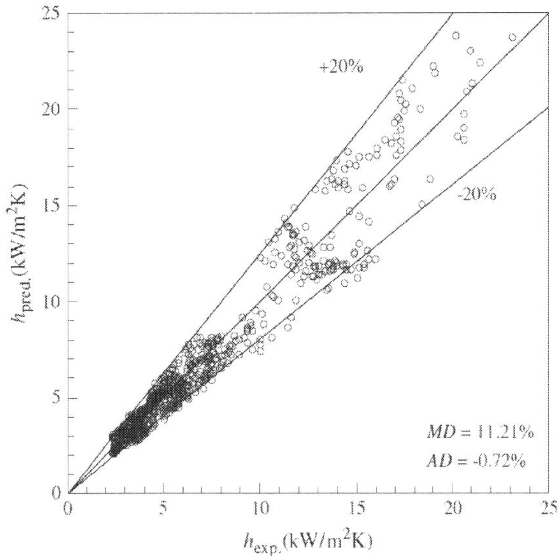

Fig. 15. (c) Diagram of the experimental heat transfer coefficient, $h_{\text{exp.}}$, vs prediction heat transfer coefficient, $h_{\text{pred.}}$. [b]K.I. Choi et al. (2007)

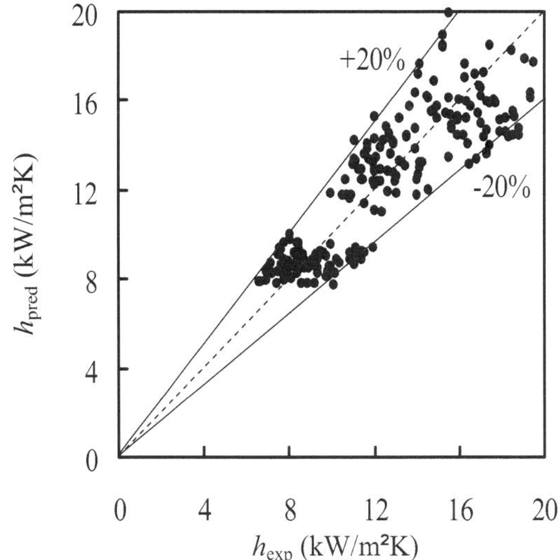

Fig. 15. (d) Diagram of experimental heat transfer coefficient h_{exp} vs prediction heat transfer coefficient h_{pred}. A.S. Pamitran et al (2007)

The prediction of the nucleate boiling heat transfer for the present experimental data used Cooper (1984), which is a pool boiling correlation developed based on an extensive study.

$$h = 55p_r^{0.12}(-0.4343 ln p_r)^{-0.55}M^{-0.5}q^{0.67} \tag{36}$$

Where the heat fluxes, q, is in Wm^{-2}. Kew and Cornwell (1997) and Jung et al. (2003) showed that the Cooper (1984) pool boiling correlation best predicted their experimental data. Chen (1966) defined the nucleate boiling suppression factor, S, as a ratio of the mean superheat, ΔT_e, to the wall superheat, ΔT_{sat}. Jung et al. (1989) proposed a convective boiling heat transfer multiplier factor, N, as a function of quality, heat flux and mass flow rate (represented by employing Xtt and Bo) to represent the strong effect of nucleate boiling in flow boiling as it is compared with that in nucleate pool boiling, h_{nbc}/h_{nb}. The Martinelli parameter, X_{tt}, is replaced by a two-phase frictional multiplier, ϕ_f^2, in order to consider laminar flow in minichannels. By using the experimental data of this study, a new nucleate boiling suppression factor, as a ratio of h_{nbc}/h_{nb}, is proposed as showen in table 9.

A.S. Pamitran et al. and K.I. Choi et al have developed a correlation based on the Zhang et al (2004) that modified the F factor and Chen (1966) introduced the suppression factor, S. The new heat transfer coefficient correlation is developed using a regression method with 461 data points for C_3H_8 in Choi, K.I et al (2009), 471 data points for CO_2 in [a]K.-I. Choi et al (2007), 681 data points for R-22, R-134a, and R-744 (CO_2) in [b]K.I. Choi et al.(2007), and 217 data points for R-410A in A.S. Pamitran et al. (2007).

The comparison of the experimental heat transfer coefficient, $h_{tp,exp}$, and the predicted heat transfer coefficient, $h_{tp,pred}$, for propane is illustrated in Fig. 15(a). The new correlation agrees closely for the comparison with a mean deviation of 9.93% and an average deviation of -2.42%. Fig. 15(b) illustrates CO_2 on the comparison with a mean deviation of 8.41% and an average deviation of 0.37%.

The comparison of the experimental heat transfer coefficient, h_{exp}, and the prediction heat transfer coefficient, h_{pred}, is shown in Fig. 15(c). The new correlation for three refrigerants R-22, R-134a and R-744 (CO_2) showed good agreement with a mean deviation of 11.21% and an average deviation of -0.72%. Fig 14 (d), the new correlation for 410A also show a suitable range of mean deviation, 11.20% and an average deviation 0.09%; Table 9 gives a summary of the new correlation.

	F factor	Supression factor	Deviation
Choi, K.-I. et al. (2009)	$F = MAX(0.5\phi_f, 1)$	$S = 181.458(\phi_f^2)^{0.002}Bo^{0.816}$	MD = 9.93% AD = 2,42%
[a]K.-I. Choi et al. (2007)	$F = 0.05\phi_f^2 + 0.95$	$S = 7.2694(\phi_f^2)^{0.0094}Bo^{0.2814}$	MD = 8.41% AD = 0.37%
[b]K.-I. Choi et al. (2007)	$F = 0.042\phi_f^2 + 0.958$	$S = 469.1689(\phi_f^2)^{-0.2093}Bo^{0.7402}$	Overall MD = 11.21% AD = -0,72%
A.S. Pamitran et al. (2007)	$F = 0.062\phi_f^2 + 0.938$	$S = 9.4626(\phi_f^2)^{-0.2747}Bo^{0.1285}$	MD= 11.20% AD = 0.09%

Table 9. Equations developed by A.S. Pamitran et al. and K.I. Choi et al

6. Concluding remarks

The pressure drop and heat transfer experiments in convective boiling performed with R-22, R-134a, R-410A, R-744 (CO_2), R-717 (NH_3) and R-290 (C_3H_8) in horizontal smooth minichannels. It can be explained, that, the pressure drop is higher at the conditions of higher mass and heat fluxes, and for the conditions of smaller inner tube diameter and lower saturation temperature. The experimental results showed that pressure drop is a function of mass flux, inner tube diameter, surface tension, density and viscosity. The new pressure drop correlations were developed on the basis of the Lockhart–Martinelli method as a function of the two-phase Reynolds number, Re_{tp}, and the two-phase Weber number, We_{tp}. The new factor C was developed using a regression method

Mass flux, heat flux, inner tube diameter and saturation temperature have an effect on the heat transfer coefficient. The heat transfer coefficient increases with a decreased inner tube diameter and with an increased saturation temperature. The geometric effect of the small tube must be considered to develop a new heat transfer coefficient correlation.

The two-phase flow pattern was mapped in the Wang et al (1997) and the Wotjan et al.(2005) flow pattern maps. The Wotjan et al. flow pattern map illustrates the formation of intermittent, annular, dry-out, and mist flows until the dry-out. The Wang et al (1997) flow pattern map illustrates the data start from annular flow, Str. wavy to intermittent flow.

Laminar flow appears for flow boiling in small channels, so the modified correlation of the multiplier factor for the convective boiling contribution, F, and the nucleate boiling suppression factor, S, is developed in the study using laminar and turbulent flows consideration. The new boiling heat transfer coefficient correlations that are based on a superposition model for refrigerants in minichannels were presented. The work of developing the new correlations used 2288 data points.

The documentation contained in this manuscript endeavors to perceive the two-phase flow boiling heat transfer and pressure drop in horizontal circular channels as the basic understanding for application in refrigeration fields or any concern that is related to two-phase flow. It is known that there is no single correlation that has the ability to predict accurately the two-phase flow heat transfer and pressure drop. Every single or mixed refrigerant could be had differently as a result of pressure drop correlation and heat transfer correlation. Therefore, the modification of the previous correlations or a new development in this matter is still open.

7. Nomenclature

A   Cross section area

AD   Average Deviation, $AD = \left(\dfrac{1}{n}\right)\sum_{1}^{n}\left(\left(dp_{pred} - dp_{exp}\right)\times 100 \big/ dp_{exp}\right)$ *for pressure drop or*

$$AD = \left(\dfrac{1}{n}\right)\sum_{1}^{n}\left(\left(h_{pred} - h_{exp}\right)\times 100 \big/ h_{exp}\right) \text{ for heat transfer coefficient}$$

Bo   Boiling number, $Bo = q/Gi_{fg}$

C Correction factor for two-phase pressure drop

c_p Specific heat capacity at constant pressure $(kJ/(kg \cdot K))$

D Diameter (m)

E Electric potential (V)

F Multiplier factor for convective heat transfer contribution

f Friction factor

G Mass flux $(kg/(m^2 \cdot s))$

g Acceleration due to gravity (m/s^2)

h Heat transfer coefficient $(kW/(m^2 \cdot K))$

I Electric current (A)

i Enthalpy (kJ/(kg·K))

k Thermal conductivity (kW/(m·K))

L Tube length (m)

M Molecular weight of the liquid

MD   Mean Deviation, $\mathrm{MD} = \left(\dfrac{1}{n}\right)\sum_{1}^{n}\left|\left(\left(\mathrm{d}p_{\mathrm{pred}} - \mathrm{d}p_{\mathrm{exp}}\right)\times 100\big/\mathrm{d}p_{\mathrm{exp}}\right)\right|$ *for pressure drop or*

$\mathrm{MD} = \left(\dfrac{1}{n}\right)\sum_{1}^{n}\left|\left(\left(h_{\mathrm{pred}} - h_{\mathrm{exp}}\right)\times 100\big/h_{\mathrm{exp}}\right)\right|$ *for heat transfer coefficient*

n Number of data

P Pressure (N/m²)

Pr   Prandtl number, $\mathrm{Pr} = c_{\mathrm{p}}\mu/k$

Q Electric power (kW)

q Heat flux (kW/m²)

$\dot{q}$ Heat generation (kW/m³)

R Electrical resistance (ohm)

Re   Reynolds number, $\mathrm{Re} = GD/\mu$

R_{p} Surface roughness parameter (µm)

S Suppression factor

T Temperature (K)

V Volume (m³)

W Mass flow rate (kg/s)

We Webber number,

X Lockhart-Martinelli parameter

x Mass quality

z Length (m)

Greek letters

λ Correction factor on Baker (1954) flow pattern map, $\lambda = \left[\left(\rho_{\mathrm{g}}/\rho_{\mathrm{A}}\right)\left(\rho_{\mathrm{f}}/\rho_{\mathrm{W}}\right)\right]^{1/2}$

a Void fraction

μ Viscosity (Pa·s)

ρ Density (kg/m³)

σ Surface tension (N/m)

ϕ^2 Two-phase frictional multiplier

ψ Correction factor on Wang et al.(1997) flow pattern map, $\psi = \left(\sigma_{\mathrm{W}}/\sigma\right)^{1/4}\left[\left(\mu_{\mathrm{f}}/\mu_{\mathrm{W}}\right)\left(\rho_{\mathrm{W}}/\rho_{\mathrm{f}}\right)^2\right]^{1/3}$

Gradients and differences

$(\mathrm{d}p/\mathrm{d}z)$ Pressure gradient (N/m²m)

$(\mathrm{d}p/\mathrm{d}z\ F)$ Pressure gradient due to friction (N/m²m)

$(\mathrm{d}p/\mathrm{d}z\ a)$ Pressure gradient due to acceleration (N/m²m)

$(\mathrm{d}p/\mathrm{d}z\ z)$ Pressure gradient due to static head (N/m²m)

Subscripts

A   Air

c   Convective

exp   Experimental value

f   Saturated liquid

g   Saturated vapor

i	Inner tube
fo	Liquid only
nb	Nucleate boiling
nbc	Nucleate boiling contribution
o	Outer tube
pb	Pool boiling
pred	Predicted value
red	Reduced
sat	Saturation
sc	Subcooled
t	Turbulent flow
tp	Two-phase
v	Laminar flow
W	Water
w	Wall

8. References

Ali, M. I., Sadatomi, M. and Kawaji, M., 1993, "Two-phase flow in narrow channels between two flat plates", Can. J. Chem. Eng. 71, pp. 657-666.

Baker, O., 1954, "Design of pipe lines for simultaneous flow of oil and gas", Oil and Gas J. July, 26.

Baroczy, C. J., 1963, "Correlation of liquid fraction in two-phase flow with application to liquid metals", NAA-SR-8171. Fluid Sci. 28;111-121.

Premoli, A., Francesco, D., Prina, A., 1971, "A dimensionless correlation for determining the density of two-phase mixtures", Lo Termotecnica 25;17-26.

Shah, M. M., 1988, "Chart correlation for saturated boiling heat transfer: equations and further study", ASHRAE Trans 2673;185-196.

Smith, S. L., 1969, "Void fractions in two phase flow: a correlation based upon an equal velocity head model", Proc. Inst. Mech. Engrs, London 184;647–657 Part 1.

Steiner, D., 1993, "Heat transfer to boiling saturated liquids", VDI-Wärmeatlas (VDI Heat Atlas), Verein Deutcher Ingenieure, ed., VDI-Gessellschaft Verfahrenstechnik und Chemieinge- nieurwesen (GCV), Düsseldorf, Germany, (J.W. Fullarton, translator). Bao, Z. Y., Fletcher, D. F., Haynes, B. S., 2000, "Flow boiling heat transfer of freon R11 and HCFC123 in narrow passages", Int. J. Heat and Mass Transfer 43;3347-3358.

Beattie, D. R. H. and Whalley, P. B., 1982, "A simple two-phase flow frictional pressure drop calculation method", Int. J. Multiphase Flow 8;83-87.

Chang, S. D. and Ro, S. T., 1996, "Pressure Drop of Pure HFC Refrigerants and Their Mixtures Flowing in Capillary Tubes", Int J. Multiphase Flow 22(3); 551-561.

Chang Y. J., Chiang S. K., Chung T. W., Wang C. C., 2000, "Two-phase frictional characteristics of R-410A and air-water in a 5 mm smooth tube", ASHRAE Trans; DA-00-11-3;792-797.

Chen I. Y., Yang K. S., Chang Y. J., Wang C. C., 2001, "Two-phase pressure drop of air-water and R-410A in small horizontal tubes", Int. J. Multiphase Flow 27;1293-1299.

Chen, J. C., 1966, "A correlation for boiling heat transfer to saturated fluids in convective flow", Industrial and Engineering Chemistry, Process Design and Development 5;322-329.

Chisholm, D., 1967, "A theoretical basis for the Lockhart-Martinelli correlation for two-phase flow", Int. J. Heat Mass Transfer 10;1767-1778.

Chisholm, D., 1972, "An equation for velocity ratio in two-phase flow", NEL Report 535.

Chisholm, D., 1968, "The influence of mass velocity on friction pressure gradients during steam-water flow", Paper 35 presented at Thermodynamics and Fluid Mechanics Convention, Institutes of Mechanical Engineers (Bristol), March.

Chisholm, D., 1983, "Two-phase flow in pipelines and heat exchangers", New York: Longman.

Chisholm, D. and Sutherland, L. A., 1969, "Predicted of pressure gradients in pipeline systems during two-phase flow", Paper 4 presented at Symposium on Fluid Mechanics and Measurements in Two-phase Flow Systems, Leeds, 24-25, September.

Cho, J. M. and Kim, M. S., 2007, "Experimental studies on the evaporative heat transfer and pressure drop of CO_2 in smooth and micro-fin tubes of the diameters of 5 and 9.52 mm", Int. J. Refrigeration 30;986-994.

Cicchitti, A., Lombardi, C., Silvestri, M., Solddaini, G., Zavalluilli, R., 1960, "Two-phase cooling experiments— Pressure drop, heat transfer and burnout measurement", Energia Nucl 7(6);407-425.

Choi K-I A.S. Pamitran, Chun-Young Oh, Jong-Taek Oh, 2007, "Boiling heat transfer of R-22, R-134a, and CO_2 in horizontal smooth minichannels", Int. J. of Refrigeration, Vol. 30, 1336-1346.

Choi K-I, A. S. Pamitran, Chun-Young Oh, Jong-Taek Oh, 2008 "Two-phase pressure drop of R-410A in horizontal smooth minichannels", Int. J. of Refrigeration, Vol. 31, 119-129.

Choi K-I. A.S. Pamitran, Jong-Taek Oh, 2007, "Two-phase flow heat transfer of CO_2 vaporization in smooth horizontal Minichannels", Int. J. of Refrigeration, vol. 30, 767-777.

Choi K-I. A.S. Pamitran, Jong-Taek Oh, Kiyoshi Saito, 2009, "Pressure drop and heat transfer during two-phase flow vaporization of propane in horizontal smooth minichannels", Int. J. of Refrigeration, vol. 32, 837-845.

Cooper, M. G., 1984, "Heat flow rates in saturated nucleate pool boiling–a wide-ranging examination using reduced properties", In: Advances in Heat Transfer. Academic Press 16;157-239.

Dittus, F. W. and Boelter, L. M. K., 1930, "Heat transfer in automobile radiators of the tubular type", University of California Publication in Engineering 2;443-461.

Dukler, A. E., Wicks, I. I. I. M., Cleveland, R. G., 1964, "Pressure drop and hold-up in two-phase flow", AIChE J. 10(1);38-51.

Friedel, L., 1979, "Improved friction pressure drop correlations for horizontal and vertical two-phase pipe flow", Presented at the European Two-phase Flow Group Meeting, Ispra, Italy, Paper E2, June.

Gnielinski, V., 1976, "New equations for heat and mass transfer in turbulent pipe and channel flow", International Chemical Engineering 16: 359-368

Grönnerud, R., 1979, "Investigation of Liquid Hold-Up, Flow-Resistance and Heat Transfer in Circulation Type Evaporators", Part IV: Two-phase flow resistance in boiling refrigerants, Annexe 1972-1, Bull. de l'Inst. du Froid, International Inst. of Refrigeration, Paris.

Gungor, K. E., Winterton, H. S., 1987, "Simplified General Correlation for Saturated Flow Boiling and Comparisons of Correlations with Data", Chem.Eng.Res 65; 148-156.

Jung, D., Kim, Y., Ko, Y., Song, K., 2003, "Nucleate boiling heat transfer coefficients of pure halogenated refrigerants", Int. J. Refrigeration 26;240-248.

Jung, D. S., McLinden, M., Radermacher, R., Didion, D., 1989, "A Study of Flow Boiling Heat Transfer with Refrigerant Mixtures", Int J. Mass Transfer 32(9);1751-1764.

Kandlikar, S. G., 1990, "A general correlation for saturated two-phase flow boiling heat transfer inside horizontal and vertical tubes", Journal of Heat Transfer 112;219-228.

Kandlikar, S. G., 2002, "Fundamental issues related to flow boiling in minichannels and microchannels", Experimental Thermal and Fluid Science 26;389-407.

Kattan N., 1996, "Contribution to the heat transfer analysis of substitute refrigerants in evaporator tubes with smooth or enhanced tube surfaces", PhD thesis No 1498, Swiss Federal Institute of Technology, Lausanne, Switzerland.

Kattan, N., Thome, J. R., Favrat, D., 1998, "Flow boiling in horizontal tubes: part 1 – development of a diabatic two-phase flow pattern map", J. Heat Transfer 120;140-147.

Kawahara, A., Chung, P. M. Y., Kawaji, M., 2002, "Investigation of two-phase flow pattern, void fraction and pressure drop in a microchannel", Int J. of Multiphase Flow 28;1411-1435.

Kew, P. A., Cornwell, K., 1997, "Correlations for the Prediction of Boiling Heat Transfer in Small-Diameter Channels", Applied Thermal Engineering 17(8-10);705-715.

Kuo, C. S. and Wang C. C., 1996, "In-tube evaporation of HCFC-22 in a 9.52 mm microfin/smooth tube", Int. J. Heat Mass Transfer 39;2559-2569.

Lazarek, G. M. and Black, S. H., 1982, "Evaporative heat transfer, pressure drop and critical heat flux in a small diameter vertical tube with R-113", Int. J. Heat Mass Transfer 25;945-960.

Lockhart, R. W. and Martinelli, R. C., 1949, "Proposed correlation of data for isothermal two-phase, two-component flow in pipes", Chem. Eng. Prog. 45;39-48.

McAdams, W. H., 1954, "Heat transmission", third ed. New York: McGraw-Hill.

Mishima, K. and Hibiki, T., 1996, "Some characteristics of air-water two-phase flow in small diameter vertical tubes", Int. J. Multiphase Flow 22;703-712.

Müller-Steinhagen, H. and Heck, K., 1986, "A simple friction pressure drop correlation for two-phase flow in pipes", Chemical Engineering and Processing 20(6);297-308

Oh, H. K., Ku, H. G., Roh, G. S., Son, C. H., Park, S. J., 2008, "Flow boiling heat transfer characteristics of carbon dioxide in a horizontal tube", Applied Thermal Engineering 28;1022-1030.

Ould Didi, M. B., Kattan, N., Thome, J. R., 2002, Prediction of Two-phase Pressure Gradients of Refrigerants in Horizontal Tubes, Int J. Refrig. 25;935-947.

Pamitran A.S, Kwang-Il Choi, Jong-Taek Oh, Hoo-Kyu Oh, 2008, "Two-phase pressure drop during CO2 vaporization in horizontal smooth minichannels" Int. J. of Refrigeration, Vol. 31, 1375–1383.

Pamitran A.S. Kwang-Il Choi, Jong-Taek Oh, Hoo-Kyu Oh, 2007, "Forced convective boiling heat transfer of R-410A in horizontal minichannels", Int. J. of Refrigeration, Vol. 30,155-165.

Pamitran A.S. Kwang-Il Choi, Jong-Taek Oh, Pega Hrnjak., 2010, "Characteristics of two-phase flow pattern transitions and pressure drop of five refrigerants in horizontal circular small tubes", Int. J. of Refrigeration, vol. 33, 578-588.

Park, C. Y. and Hrnjak, P. S., 2007, "CO_2 and R410A flow boiling heat transfer, pressure drop, and flow pattern at low temperatures in a horizontal smooth tube", Int. J. Refrigeration 30;166-178.

Peng, X. F. and Peterson, G. P., 1996, "Forced convective heat transfer of single-phase binary mixtures through microchannels", Experimental Thermal and Fluid Science 12;98-104.

Petukhov, B. S. and Popov, V. N., 1963, "Theoretical calculation of heat exchanger in turbulent flow in tubes of an incompressible fluid with variable physical properties", High Temp. 1(1);69-83.

Pettersen, J., 2004, "Flow vaporization of CO_2 in microchannels tubes", Exp. Therm.

Tran, T. N., Chyu, M. C., Wambsganss, M. W., France, D. M., 2000, "Two-phase pressure drop of refrigerants during flow boiling in small channels: An experimental investigation and correlation development", Int. J. Multiphase Flow 26;1739-1754.

Tran, T. N., Wambsganss, M. W., France, D. M., 1996, "Small circular- and rectangular-channel boiling with two refrigerants", Int. J. Multiphase Flow 22(3);485-498.

Wambsganss, M. W., France, D. M., Jendrzejczyk, J. A., Tran, T. N., 1993, "Boiling Heat Transfer in a Horizontal Small-diameter Tube", AMSE Trans 115;963-975.

Wang, C. C., Chiang, C. S., Lu, D. C., 1997, "Visual observation of two-phase flow pattern of R-22, R-134a, and R-407C in a 6.5-mm smooth tube", Exp. Therm. Fluid Sci. 15;395-405.

Wojtan, L., Ursenbacher, T., Thome, J. R., 2005, "Investigation of flow boiling in horizontal tubes: part I – a new diabatic two-phase flow pattern map", Int. J. Heat Mass Transfer 48;2955-2969.

Wattelat, J. P., Chato, J. C., Souza, A. L., Christoffersen, B. R., 1994, "Evaporative Characteristics of R-12, R-134a, and a Mixture at Low Mass Fluxes", ASHRAE Trans 94-2-1;603-615.

Yan, Y. Y., Lin, T. F., 1998, "Evaporation Heat Transfer and Pressure Drop of Refrigerant R-134a in a Small Pipe", Int J. of Heat and Mass Transfer 41;4183-4194.

Yang, C. Y. and Lin, T. Y., 2007, "Heat transfer characteristics of water flow in microtubes", Experimental Thermal and Fluid Science 32(2);432-439.

Yoon, S. H., Cho, E. S., Hwang, Y. W., Kim, M. S., Min, K., Kim, Y., 2004, "Characteristics of evaporative heat transfer and pressure drop of carbon dioxide and correlation development", Int. J. Refrigeration 27;111-119.

Yu, W., France, D. M., Wambsganss, M. W., Hull, J. R., 2002, "Two-phase Pressure Drop, Boiling Heat Transfer, and Critical Heat Flux to Water in a Small-diameter Horizontal Tube", Int J. of Multiphase Flow 28;927-941.

Yun, R., Kim, Y., Kim, M. S., 2005, "Convective boiling heat transfer characteristics of CO_2 in microchannels", Int. J. Heat. Mass Transfer 48;235-242.

Zhang, L., Hihara, E., Saito, T., Oh, J. T., 1997, "Boiling Heat Transfer of a Ternary Refrigerant Mixture inside a Horizontal Smooth Tube", Int J. Mass Transfer 40(9);2009-2017.

Zhang, M. and Webb, R. L., 2001, "Correlation of two-phase friction for refrigerants in small-diameter tubes", Experimental Thermal and Fluid Science 25;131-139.

Zhang, W., Hibiki, T., Mishima, K., 2004, "Correlation for flow boiling heat transfer in mini-channels", Int. J. Heat and Mass Transfer 47;5749-5763.

Zhao, Y., Molki, M., Ohadi, M. M., Dessiatoun, S. V., 2000, "Flow boiling of CO_2 in microchannels", ASHRAE Trans DA-00-2-1;437-445.

Zivi, S. M., 1964, "Estimation of Steady-State Steam Void-Fraction by Means of the Principle of Minimum Entropy Generation", J. Heat Transfer 86;247-252.

11

Comparison of the Effects of Air Flow and Product Arrangement on Freezing Process by Convective Heat Transfer Coefficient Measurement

Douglas Fernandes Barbin[1] and Vivaldo Silveira Junior[2]
[1]*University College Dublin*
[2]*University of Campinas*
[1]*Ireland*
[2]*Brazil*

1. Introduction

Since energy level could implement on the final cost of food products, the reduction of the freezing time is a major goal for industries and researchers. The product thermal load is generally the greatest of the demands on the refrigeration system, most of the heat removal occurring during the first hours of the process, mainly the freezing step.

Food products are predominantly frozen in air blast freezing tunnels. Air blast systems are based on fluid convection, where the cooling air flows around the product that must have the temperature reduced. Convection is the transfer of thermal energy by the movement of molecules from one part of the material to another. The rate of heat exchange between the bulk of a fluid and a known solid is given by (1):

$$\frac{dQ}{dt} = h_c A (T_b - T_\infty) \tag{1}$$

where Q is the amount of energy (J) drawn back per time t (s); h_c is the convection heat transfer coefficient (W/m^2 $^\circ$C); A is the heat transfer area (m^2); T_b is the product temperature ($^\circ$C) and T_∞ is the air temperature ($^\circ$C).

The convective heat flow is enhanced as the fluid motion increases. Air blast cooling processes are majorly ruled by the convective heat transfer, which relates the amount of transferred energy from the product surface to the cooling air. Wide variations in convective heat transfer coefficients may occur in different positions. Determination of this parameter can be a useful tool to measure the efficiency of the process.

In industrial situations, the freezing process of large amounts of food products may be very complex. Products to be cooled act like a barrier to air flow, and heat transfer can be compromised. A careful investigation of product characteristics and processing conditions must be conducted, as inadequate sample positioning and poor air circulation can cause longer cooling periods. If the system is not correctly designed, problems related to product or process qualities can appear, such as inadequate cooling or freezing rate, microbiological growth, alterations in consistency and changes in its composition caused by chemical reactions during the subsequent storage. There is also need to consider proper packaging materials and the possibility of the existence of voids of air bubbles inside the package, as well as improper arrangement of products, which could lead to poor heat transfer and ineffective temperature reduction.

To produce homogeneous heat transfer it is important to have proper sample arrangement inside the system, as well as adequate air flow. Therefore, exhausting air may represent an economical choice for air blast systems, since it minimizes air short-circuiting, leading to more uniform cooling. Several implementations of forced air devices to aid cooling and freezing process have been reported. The most common method considers the use of a forced-air freezing tunnel that enhances the air flow around the products. Hence, the investigations of objective methods to determine the best design for cooling and freezing systems have been recently reported.

Frequently air velocity and air temperature are the parameters which regard most attention from designers of air blast systems. However, it is also important to consider the ways this air will flow trough, as product distribution may directly affect the efficiency of cooling and freezing processes. Assessing the efficiency of freezing processes is a troublesome task. Recent researches have suggested that the convective coefficient, among other parameters, can be a useful tool for comparing different processing conditions.

Considering the complexity of the industrial process, the objective of this chapter is to discuss the influence of the main factors that affect the freezing time of food samples measuring the convective heat transfer coefficient. Results from comparative studies for different air flow and samples arrangement inside the system will be discussed, showing how slight changes to product arrangement and air flow orientation can affect the heat transfer.

2. Literature review

2.1 Cooling and freezing processes for food products

Regarding the rapid increase in frozen foods production and consumption, there is growing interest in determining food thermal properties during freezing process for the development of new systems and improvement of processes equipments (Scott et al. 1992).

Temperature reduction processes aims at reducing microbial growth and hence extending the shelf life of perishable food. Freezing and cooling processes are driven by the heat exchange between the product to be cooled and the cooling medium (Welti-Chanes et al., 2005). Some researchers have presented models that can be used to assess the cost of food freezing by different methods (Becker & Fricke, 1999; Chourot et al., 2003).

The processes of cooling and freezing of food are complex. Freezing food basically depends on the amount of water that is present in the food and will freeze during the process. Prior to freezing, sensible heat must be removed from food until reaching the freezing temperature. Supercooling occurs when temperature reaches values below the freezing point and crystal nucleation starts. In pure water, heat is released during nucleation, causing a rise in temperature to the initial freezing point, and the temperature remains constant until all the water is converted to ice. However, it decreases slightly in foods, due to the increasing concentration of solutes in the unfrozen water portion.

After that, during the phase change of water into ice, there is the removal of latent heat from the frozen product. It starts when most freezable water has been converted to ice, and ends when the temperature is reduced to storage temperature (Zaritzky, 2000).

Among most popular frozen food products are fruits and fruit pulps, normally used as raw material for processing industries as ice cream, yogurt, jams and others. Such products can be frozen in batch or continuous processing (Salvadori & Mascheroni, 1996). In order to make cooling processes financially affordable, studies based on information from manufacturers and users are necessary to design refrigeration equipment in accordance with the demanded application.

2.2 Heat transfer for cooling process

Cooling time is directly influenced by the ratio between the resistances to heat transfer inside and outside the sample. This ratio is called the Biot number, defined by Equation 1:

$$Bi = \frac{hL_c}{k_b} \tag{2}$$

where h is the convective heat transfer coefficient (W/m² °C), L_c is the characteristic length of the body (m), usually defined as the volume of the body divided by the surface area in contact with the cooling medium, and k_b is the thermal conductivity of the body (W/m °C). Values of Biot number close to zero (Bi → 0), imply that the heat conduction inside the body is much faster than the heat convection away from its surface, and temperature gradients are negligible within the body (Heldman, 1992). This condition allows for the applicability of certain methods of solving transient heat transfer problems. Within this condition, it can be assumed a lumped-capacitance model of transient heat transfer, leading to Newtonian cooling behaviour since the amount of thermal energy in the body is directly proportional to its temperature, which can be assumed uniform throughout the body. This leads to a simple first-order differential equation which describes heat transfer in these systems.

When the Biot number is very large (Bi → ∞), the internal resistance to heat transfer is much larger and it only can be assumed that the surface temperature is equal to the cooling medium, but not the interior of the body. For this situation, solutions of the equation of Fourier heat transfer are useful. When the Biot number is within the range of 0.1 <Bi <40, both internal and external resistance should be considered.

The factors that are inherent to the products such as thermal conductivity and diffusivity cannot be changed. Hence, the reduction of cooling and freezing time must be achieved by changing system variables such as temperature and velocity of the cooling air, and product arrangement

2.3 Parameters affecting cooling time

The cooling or freezing rate is among the most important parameters in designing freezing systems, as it can directly affect the quality of products. For industrial applications, it is the most essential parameter in the process when comparing different types of systems and equipment. Besides the characteristics of the products, the temperature removal time varies according to some parameters involved in the heat transfer process, such as size and areas of openings of the packaging and characteristics of the cooling medium. The cost of the cooling process is related to cooling rate, which is directly affected by the opening area of packaging for air circulation, bed depth, temperature and speed of the cooling air (Baird et al., 1988).

The freezing systems can be described according to different methods of heat removal. Among the most used methods are forced air freezers. This technique has several advantages, from ease of installation and operation and efficiency of the process to the variety of products that can be cooled in this type of equipments.

The selection of the best cooling method varies according to the desired application and depends on several factors, including the cooling rate required, subsequent storage conditions and costs of equipment and operation. Systems properly designed may increase efficiency and reduce the cost of operation (Talbot & Chau, 1998; ASHRAE, 2002).

Talbot & Fletcher (1996) compared the efficiency between an air blast system and a storage chamber. During the cooling process of grapes, there was a reduction of 6.7 ºC in one hour and 14.6 ºC after 2.5 hours, compared to a decrease of only 2 °C in one hour and 3.5 ºC in 2.5 hours in the storage chamber. Experiments carried out using a prototype portable forced-air device offered promising results (Barbin et al., 2009). The system was designed to be used inside cooling and freezing chambers, aiming to improve heat flow rates.

2.4 Air flow

Cooling time by forced air systems is determined by airflow and product thermal load, which affects the amount of energy to move the air around the product and inside the system. The most common industrial applications use direct cold air insufflations inside the system. The airflow varies according to the speed and amount of air flowing through products and its variation results in longer or shorter freezing time. A correct orientation of the air flow inside the equipment and around the product can significantly reduce processing times (Cortbaoui et al., 2006). Surface area of contact between products and cooling air and products arrangement are other parameters that affect forced air cooling (Baird et al., 1988; Fraser, 1998; Laguerre et al., 2006).

In industrial plants, air flow is highly turbulent due to the fans movement and to wakes originated from upstream obstacles. Resende & Silveira Jr. (2002b) showed that the air velocity in forced air tunnels are strongly influenced by any changes in the amount of product inside the system, causing the air to flow through preferential paths, leading to increased freezing time and poor heat transfer coefficients. Results show that variations in heat transfer coefficients may occur according to the product positioning inside the equipment. Vigneault et al. (2005) studied how gravity influences air circulation in horizontal air flow, showing that low levels of air flows may be more affected, causing temperature variation of the cooling air and reducing the flow to the upper chamber. This could be an important parameter to consider when designing cooling and freezing systems.

Exhausting air is more appropriate to avoid air to flow through preferential corridors, leading to more uniform heat exchange when compared to insufflations processes (Fraser, 1998).

Conventional cooling methods by forced air are an efficient alternative to removing the heat load of fruits and vegetables during post-harvest cooling. Air exhaustion is widely used for this purpose, as it improves the air distribution in the products surrounding. This system is usually used inside cooling chambers. With the fan in operation, it creates a low pressure region surrounding the products. The cooling air flows through this region between the small opening areas, reducing the product temperature (Talbot & Fletcher, 1996; Talbot & Chau, 1998; Fraser, 1998; Abrahao, 2008). The possibility of adapting a cold room for use as a system for forced air represents an economical advantage of this process (Talbot & Fletcher, 1996).

Baird et al. (1988) showed that the velocity of cooling air influences directly the operational cost of cooling systems, as it can change with the increase of air velocity in the system. The lowest costs were obtained with air velocities between 0.1 and 0.3 m/s. To study the cooling of plastic balls filled with a solution of carrageen, Allais et al. (2006) showed that increasing the speed of air flow, ranging from 0.25 m/s to 6 m/s, reduced the half-time cooling of samples from 800 s to 500 s. But this variation is exponential, and the reduction tends to be smaller from speed of 2 m/s. Results obtained by Vigneault et al. (2004a, b) for cooling process using forced air show that air flows above 2 l/s.kg and air velocities of insufflations higher than 0.5 m/s cause no influence on half-cooling time of the samples.

2.5 Instruments and methods for measurement of air flow velocity

There are several methods for measuring air flow velocity described in the literature, with different principles, and the accuracy of the sensors used in each of these techniques varies significantly, making them suitable for particular applications. Hot-wire anemometer is one of the most used instruments because of its wide applicability. Due to the small size and short response time, these instruments are suitable for detailed study of fluid flow, and are commonly used to measure the air flow in ventilation systems and air conditioning. The hot-wire anemometer measures the instantaneous velocity of fluids.

The core of the anemometer is an exposed very fine hot wire heated by a constant current up to some temperature above ambient. Air flowing past the wire has a cooling effect on the wire. By measuring the change in wire temperature under constant current, a relationship can be obtained between the resistance of the wire and the fluid flow velocity, as the electrical resistance of most metals is dependent upon the temperature of the metal. This kind of instrument has a fine spatial resolution compared to other measurement methods, and as such is employed for the detailed study of turbulent flows, or when rapid velocity fluctuations are of interest.

Resende & Silveira Jr. (2002b) and Nunes et al. (2003) suggest that several measurements in the cross-section of the air flow could lead to a more accurate determination of the air velocity. The average velocity is therefore used for determination of the air flow in the selected position, according to (3):

$$\dot{V} = \int_S \bar{v}\, dS \tag{3}$$

In this equation, V is the flow (m³/s), S is the total area (m²) and v is the velocity vector (m/s).

2.6 Packaging and storage

Packaging affects the heat transfer coefficients of food items in several ways. It is a barrier to the transfer of energy from the food by acting as insulation to the food item, thus lowering the heat transfer coefficient. Packaging may also create air-filled voids and bubbles around the food item which further insulates the food and lowers the heat transfer coefficient (Becker and Fricke, 2004). Results presented by Santos et al. (2008) showed that freezing process of meat in cardboard boxes is underrated and the processing time sometimes is not enough for all the samples to reach the desired temperature. Replacing cardboard boxes by metal perforated boxes produced a reduction of up to 45% at the freezing time for this product.

Becker and Fricke (2004) developed an iterative algorithm to estimate the surface heat transfer coefficients of irregularly shaped food items based upon their cooling curves, considering the density of the food item and the packaging. This algorithm extends to irregularly shaped food items existing techniques for the calculation of the surface heat transfer coefficient previously applicable to only regularly shaped food items, taking into account the concept of equivalent heat transfer dimensionality. In this method, the density used to calculate the heat transfer coefficient is affected by the packaging, as it is calculated from the mass of the food item plus the packaging and the outside dimensions of the package around the food item, generating results for the heat transfer coefficient for the food within its packaging.

An important parameter for improved performance of an air blast cooling system is the apertures and gaps that the packages and pallets must have to allow the circulation of the cold air through the packed product in order to achieve rapid and uniform heat transfer between the cooling air and the product (Vigneault et al. 2004a; Zou et al., 2006a, b).

Results obtained by Talbot & Fletcher (1996) and Abrahao (2008) showed the importance of proper cooling system design, proving that the larger the opening area in the packaging, the lower the requirement on refrigeration and air circulation systems to obtain a more uniform cooling rate. Meana et al. (2005) showed that the empty regions between the plastic containers that are used in the cooling of strawberries by forced air influence significantly the cooling time of the products. According to Baird et al. (1988), opening areas smaller than 10% of the total area of the box can significantly increase the cost of cooling processes. Castro et al. (2003) suggest that an opening area of 14% is appropriate for a rapid and uniform cooling process.

Large opening areas can lead to poorly designed boxes that are not suitable for industrial processing. The main goal is to get an optimal opening area of the boxes to enable a low freezing time without, however, affecting the mechanical structure of boxes.

2.7 Convective heat transfer coefficients (h_c)

Convective heat transfer is related to the amount of energy transferred from the product surface when it is in contact with the refrigerating fluid (Welty et al, 2000). Dincer (1995a) determined the experimental heat transfer coefficient with data obtained during forced air cooling of figs in air blast systems, with results varying from 21.1 to 32.1 Wm^{-2}°C^{-1} for air velocities of 1.1 to 2.5 ms^{-1}.

Experiments carried out in a forced air room with air velocities in the range of 1 to 2 ms^{-1} resulted in h_c values varying from 28 Wm^{-2}°C^{-1} up to 52 Wm^{-2}°C^{-1} for cylindrical products (cucumber) during cooling (Dincer and Genceli, 1994). Mohsenin (1980) obtained h_c values in the range of 20 to 35 Wm^{-2}°C^{-1} for forced air systems with air velocity from 1.5 to 5.0 ms^{-1}. Dussán Sarria et al. (2006) studied the influence of the air velocity in a cooling tunnel. According to the authors, air velocities greater than 2.0 ms^{-1} did not affect the convective coefficients (h_c), as results obtained were not greater than 23.8 Wm^{-2}°C^{-1}.

Considering the complexity of freezing processes and the recent results presented, many parameters influence the experimental results for heat transfer coefficients, thus varying according to the flow characteristics of the cooling medium and the products involved.

Accurate descriptions of the boundary conditions are rather difficult for industrial air blast systems, and software solutions such as CFD will not be effective in solving the momentum and heat transport equation without precise information (Mohamed, 2008). Regarding the wide range of convective heat transfer coefficients (h_c) reported, it is important to calculate this coefficient in order to understand different operating conditions of distinct cooling systems and compare to any new systems developed. Several methods for convective heat transfer measurements are reported. The most common are those involving temperature measurements in permanent and transient state (Cleland, 1990).

2.7.1 Temperature measurements in steady state

In this method, a constant thermal load is created in the system such as an electrical heating probe, for example. The coefficient of heat transfer can be calculated using the values of the surface area of the heating probe, the amount of energy added and the temperatures of the cooling medium and the probe. However, the temperature and velocity of the cooling medium should be kept constant, which is not an easy task in experimental conditions, limiting the use of this method.

2.7.2 Temperature measurements in transient state

Temperature measurement in transient state consists of a metallic test body with a known high thermal conductivity being used to minimize the temperature gradient during the heat exchange between the cooling medium and the product (Bi<0.1), allowing the test body to have an almost uniform temperature during the cooling process. When the internal resistance of the test body to heat transfer is neglected, an energy balance conducts to the convective heat transfer coefficient. By Newton's cooling law, the rate of heat transfer in a given volume of control is given by equation 1.
The variation of energy in a metal body with constant properties is given by the equation:

$$\frac{dQ}{dt} = \rho_m V c_{pm} \frac{dT}{dt} \tag{4}$$

where ρm is the density, V is the volume and cpm is the specific heat of the metallic body, respectively. Combining equations 1 and 4, integrating and adopting the initial boundary condition T $(t = 0) = T_i$, leads to the solution for the temperature variation as a function of time:

$$\frac{T_b - T_\infty}{T_i - T_\infty} = e^{\frac{-h_c A t}{\rho_m c_{pm} V}} \tag{5}$$

Equation 5 proves that the cooling process has an exponential behaviour, as verified by several authors for horticultural products (Mohsenin, 1980; Dincer, 1995a).
In practice, this method consists of using a test body made of some material with high thermal conductivity, so that tests are carried out without phase change and assuming the constant thermal properties within temperature variation range. Le Blanc et al. (1990a, b), Resende et al. (2002), Mohamed (2008) and Barbin et al. (2010) reported experiments using the described method for obtaining convective coefficients from the cooling curves obtained for a metallic test body, indicating the capability of the present method in handling complex boundary situation such as encountered in industrial systems. Results for convective heat transfer coefficients were reported by Barbin et al. (2010), comparing two air flow direction in the same equipment, concluding that this is a useful method for studying temperature reduction processes.
According to Resende et al. (2002), some points arise when using this method. If the test body consists of a metal block, there may be heat transfer through the edges of the material, affecting the values of h_c calculated. Furthermore, condensation can occur in experiments with cooling air, causing changes to the measurements. Thus, the positioning of the test body must

be carefully chosen in order to prevent any condensation of water during the tests, and the edges of the body or other parts that may interfere with the temperature measurements during the process must be perfectly insulated to avoid heat transfer through these regions.

3. Experiments

3.1 Samples for simulation of thermal load

Food model system with 15% (weight / weight of solution) of sucrose and 0.5% (weight / weight of solution) of carboxyl-methyl cellulose (Carbocel AM, Arinos, SP, Brazil) was packed in polyethylene bags (0.1 kg) with similar dimensions (0.095 m x 0.07 m x 0.015 m) to pulp fruit products in the market. The samples were stored in 35 plastic boxes (Figure 1a), with external dimensions of 0.6 m x 0.4 m x 0.12 m, which were stacked on a commercial pallet (1.00 m x 1.20 m, Figure 1b) and kept inside the freezing room. The boxes had an opening area of 21% of the total area, accounting for more than the minimum values recommended for proper air flow (Castro et al., 2003).

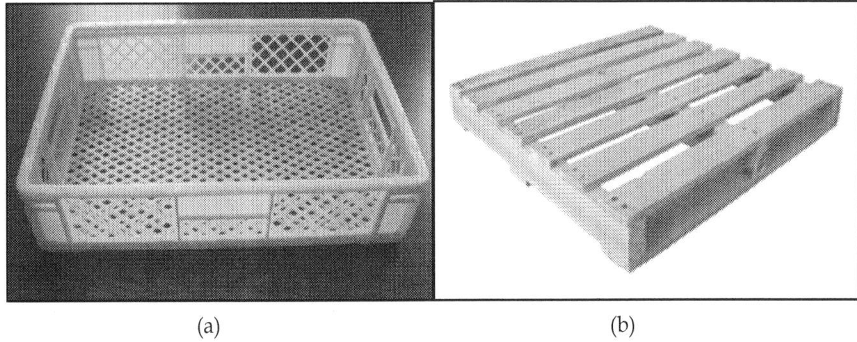

(a)   (b)

Fig. 1. (a) Plastic box for freezing products; (b) Commercial pallet

3.2 Product arrangement

Two arranges of samples were tested to determine the influence of opening areas to the refrigeration process. Using the industrial arrangement of samples, ninety six packages of sample were allocated in each box, in three layers (top, middle and bottom), with thirty two packs in each layer, corresponding to about 9.6 kg of product, similar to the amount used in the industrial process. This assembly is shown in Figure 2 (Arrangement 1). The boxes were piled in six layers with five boxes per layer, totalling thirty boxes, simulating a commercial assembly of a pallet used regularly in the process (Figure 3, Arrangement 1).

A second distribution of packages inside the boxes was tested, with larger distance between the packages inside the boxes in order to improve the circulation of air around the samples.

In this assembly, eighty four packs of sample were allocated in each box, within five layers three layers with twenty units, and two layers of twelve units, distanced from each other for air circulation, totaling 8.4 kg of product per box.

Figure 2 (Arrangement 2) shows the new distribution of packaging inside the boxes. The second arrangement had a smaller amount of samples in each box. Hence, it was added another layer of boxes in the system in order to have the same amount of product and the same thermal load for all the tests (Figure 3, Arrangement 2).

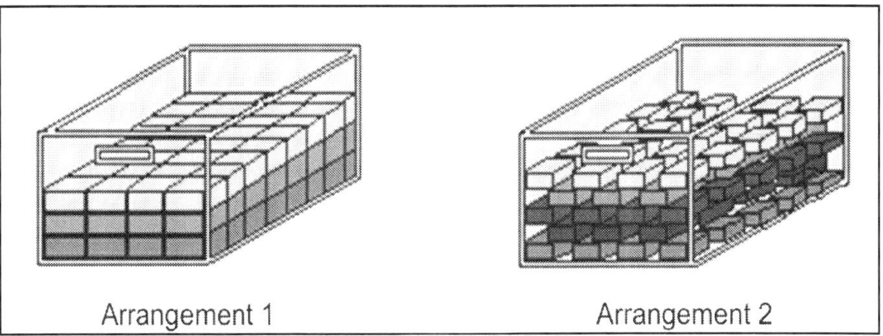

Fig. 2. Schematic diagram for sample distribution used for packaging boxes.

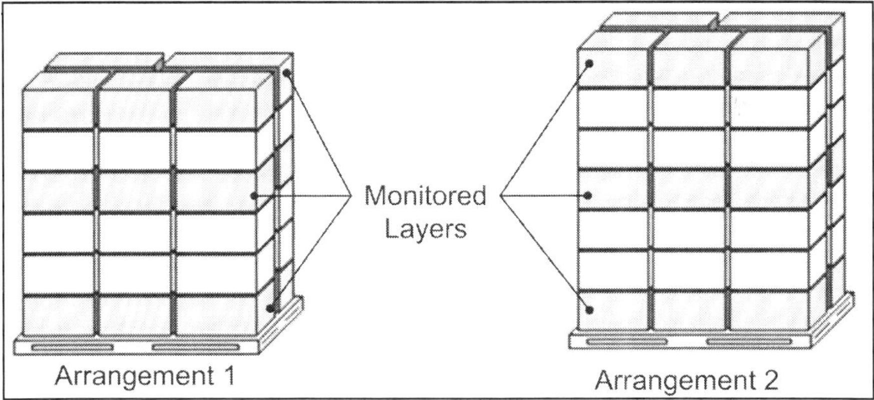

Fig. 3. Pallet with boxes and layers that were monitored

3.3 Temperature measurement

Insufflations and exhaustion air tests were run in triplicate. The velocity of the cooling air
was measured for comparison with the convective coefficients. Three layers of boxes had its
temperature monitored until the centre of the samples reached -18°C. Each of the monitored
layers had four thermocouples in the corners of the layers and one in the middle, as shown
in Figure 4. The thermocouples were inserted inside the samples in the plastic bags to
measure the samples temperatures variation during the freezing process.

The monitoring system used for temperature acquisition is composed by an automatic
channel selector system (Scanner 706, Keithley Instruments Inc.). Samples temperatures
were monitored using T-type thermocouples (copper–constantan). The thermocouples were
calibrated using a controlled temperature bath with a propylene glycol solution and a
standard thermometer as reference. Five different temperature values were chosen (-19 ºC, -
10 ºC, 0 ºC, 10 ºC and 20 ºC). The average temperatures measured in the water bath (10
measurements for each one of the five different chosen values) were plotted against the
corresponding thermocouple (mV) values (ASTM, 1989). The difference for the correlation
coefficient of the curve-fitted line (R^2) were not lower than 0.99.

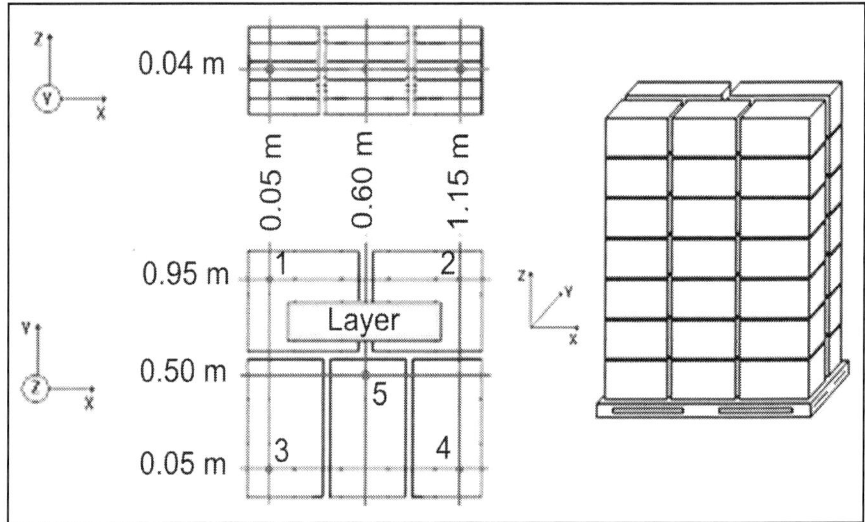

Fig. 4. Samples monitored in the layers of boxes

3.4 Portable forced air system

The forced air system was designed as described in Barbin et al. (2009), with a plastic sheet cover connected to a flexible duct and a fan that insufflates or exhausts the air inside the system. The plastic covers the boxes that contain the product, stacked on a commercial pallet. The portable tunnel fan used has axial airscrews with a tri-phase induction engine (Weg, Brazil, model 71586, 0.5 hp). The device was placed inside a freezing storage room (Recrusul, Brazil), with internal dimensions of 3 m x 3 m x 2.3 m (20.7m^3) and walls made of 0.01 m aluminium panels filled with expanded polystyrene as insulation.

The cooling process consists in circulating the internal air of the storage room through the boxes open spaces and around the product samples. In the exhaustion process, the system is connected to the fan suction, and the air flows from the lower part of the system to inside the boxes and through the fan back to the room. In the insufflations processes, the airflow is changed, blowing the cooling air from the room directly to the product. The forced air circulation is vertically oriented in both the exhaustion and the insufflations process. During exhaustion, it goes from the bottom to the top of the pallet; while in the blowing process, it goes from top to bottom (Figure 5).

3.5 Air flow measurement

A hot-wire anemometer (Tri-Sense, model EW-37000-00, Cole-Parmer Instrument Company, IL, USA) was used for measurements of air velocity. The sensor was inserted through openings in the air diffuser for measuring air velocity in different positions of the area normal to the air flow. The measurement points were aligned and positioned at regular distances. The sensor was introduced for measuring the air speed with different depth of insertion, providing fixed points in the surface area perpendicular to the airflow.

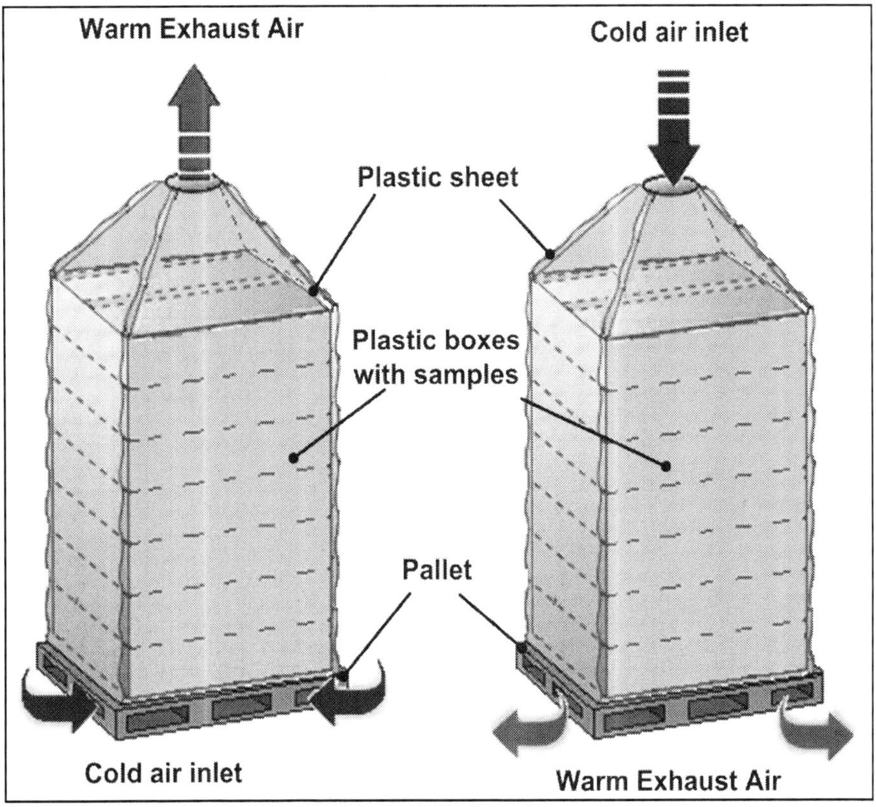

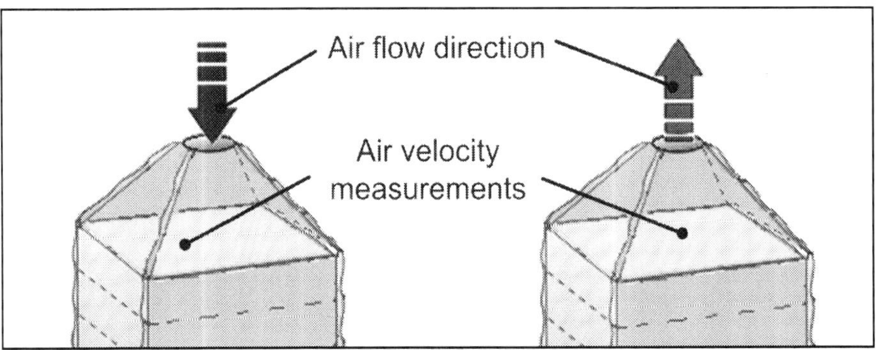

Fig. 5. Portable tunnel with boxes stacked on a commercial transport pallet covered with plastic, and air flow orientation during the exhaustion and insufflations processes.

Fig. 6. Surface area for air velocity measurements during insufflations and exhaustion processes.

Velocities were measured for comparison between the exhaustion and insufflations with the fan operating at steady state and no obstructions to the air flow. In this study, the area that the air flows through is a cross-section of the pallet represented by the five boxes. The greater the number of velocity measurements, the more accurate the result of air flow. Thus the new equation for calculating the flow in the tunnel is:

$$\dot{V} = \int_{x_0}^{x_1} \int_{y_0}^{y_1} \overline{v}(x,y)dydx \tag{6}$$

where x and y represent the coordinates of the cross-section perpendicular to the air flowing stream, comprising the dimensions of the surface formed by the boxes from the pallet.
Common approach is to measure the air velocities in several points of the flow and obtain one average result, for a more consistent representation of the profile of the flow and avoid the high variability of measurements.

3.6 Convective heat transfer measurement

The experiments for determining the convective heat transfer coefficients during freezing processes were carried out according to procedures described by Le Blanc et al. (1990a, b) and Resende et al. (2002), using a specimen of high thermal conductivity metal. The method consisted of measuring the temperature variation of a test body with high thermal conductivity during cooling. The high conductivity is necessary to minimize the temperature gradient formed during the heat transfer process between the sample and the cooling medium. The test body shown in Figure 4a is an aluminium brick with known dimensions (0.10 m x 0.07 m x 0.025 m), with perforations for insertion of thermocouples for temperature measurement. Empty spaces around the thermocouples were filled with thermal paste to prevent formation of air pockets within the holes that could affect the measurements. Aluminium thermo physical properties (as a metallic test body) used for the determination of the convective heat transfer coefficients at 20°C are: density (ρ_{Al}=2701.1 kgm^{-3}), specific heat (C_{pAl}=938.3 Jkg^{-1} °C^{-1}), thermal conductivity(k_{Al}=229 Wm^{-1} °C^{-1}) (Welty et al., 2000).

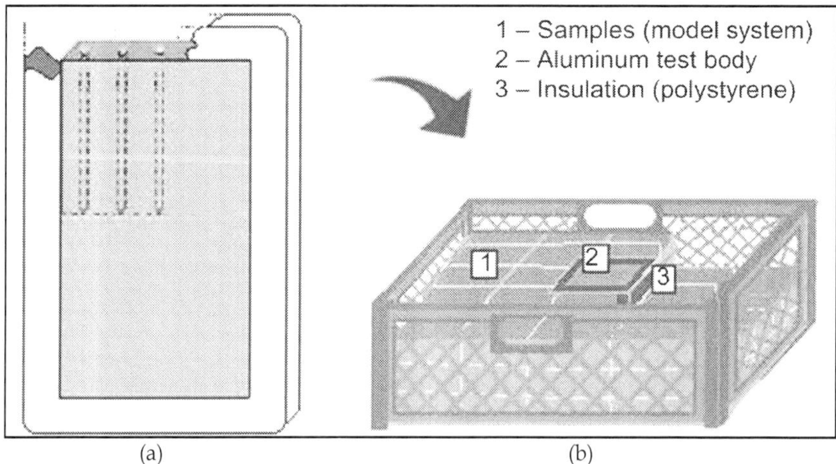

(a)   (b)

Fig. 7. (a) Aluminium test body insulation and (b) positioning inside the box.

Regarding heat flow analysis, polystyrene was used as insulation around the test body to keep only one surface exposed in contact with the cooling air. The test body had all of its edges insulated, except the upper face which was kept exposed to the cold air flow (Figure 4). This procedure was adopted to avoid edge effects and generate a one-dimensional heat flow. The test bodies were assembled inside the boxes and over the packages of samples, as shown in Figure 4b.

The fast cooling curve can be described by Equation 7, which is a simplification of Equation 6:

$$\frac{(T - T_\infty)}{(T_i - T_\infty)} = e^{S_2 t} \tag{7}$$

where T is the average value obtained from three thermocouple temperature measurements inside the test body during the cooling process, T_i is the initial temperature of the aluminium plate and T_∞ is the cooling medium temperature. The S_2 parameter represents the cooling coefficient, defined as the test body temperature change per time for each temperature degree difference between the product and the cooling medium, and is expressed in s^{-1}.

The values of S_2 were used for the calculation of the heat transfer coefficients for the period of sensible heat removal of the samples during the cooling process, as Equation 8:

$$h_c = \frac{-\rho_m V c_{pm}}{A} S_2 \tag{8}$$

Values for the average dimensionless temperature [(T - T1)/ (Ti - T1)] logarithm were calculated for every test body monitored during chilling period, with values obtained plotted versus cooling time. The angular coefficients (S_2) were calculated from these graphs. Convective heat transfer coefficients were obtained according to Equation 8, using temperature measurements of the 5 identified (T1 to T5) aluminium test bodies, distributed in layers 1, 3, 4, 5 and 6, respectively, including both the extreme layers (1 and 7) and the central layers (3, 4 and 5) (Barbin et al., 2010). With the new arrangement of samples inside the boxes, one layer of boxes was added to the pallet, namely layer number 7. In this new arrangement, the layer number 6 was not monitored. All the aluminium test bodies were positioned over the samples in the centre of the boxes along with the thermocouples identified with number 5 as last algorism (15, 35, 45, 55 and 75, Figure 3) and in contact with the cooling air (Figure 4b). Only the second layer in the first arrangement, and both second and sixth layers in the second arrangement, did not have a test body. The main objective was to analyze the air circulation in the central layers in comparison to the top and bottom layers, and to determine the local convection coefficient of heat transfer. After obtaining S_2 values (according to equation 7), the effective heat transfer coefficients were calculated for the samples in the sensible heat loss phase, as shown in Equation 8.

4. Experimental results

4.1 Determination of air flow velocity

The air flow in the region surrounding the product was investigated and results were previously reported by Barbin et al. (2009). These results are reproduced here for comparison with the heat transfer coefficients. Flow direction and turbulence intensity are generally not accurately known in practical situations. Furthermore, impending variation of heat transfer coefficient value for a product of certain size is primarily dependent on airflow

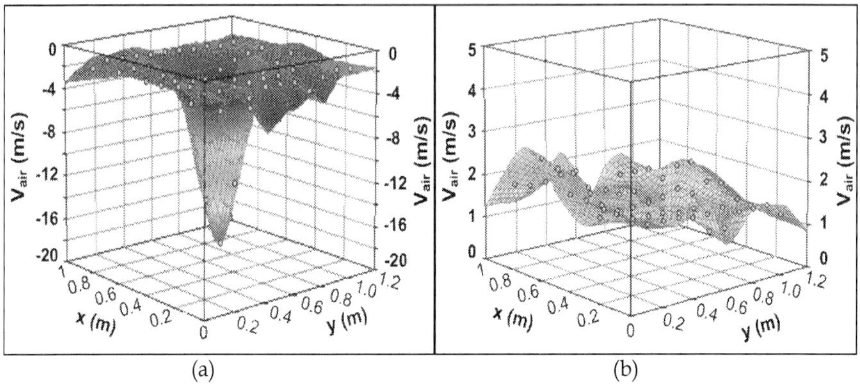

(a)　　　　　　　　　　　　　　　　　　(b)

Fig. 8. Results for air flow velocity obtained in the product surface for: (a) insufflation process, and (b) exhaustion process.

properties such as velocity and turbulence, and less affected by product shape and direction into the flow. Due to high turbulence of air flow in industrial applications, wind tunnel experiments are not useful to determine heat transfer values (Kondjoyan, 2006); hence, the importance in studying the airflow behaviour for the experiments.

The graphs shown in Figure 8 illustrate the results of measurements of air velocity for each process (Barbin et al., 2009). Figure 8a shows the experimental values for the air velocities measured during the process of insufflations, while Figure 8b shows the results for the exhaustion process. The dimensions x (m) and y (m) represent the surface area perpendicular to the air flow, where velocities were measured. The vertical axis shows the velocity results (m/s), containing a response surface for visualization of the airflow.

High values for the air velocity were observed in the insufflation process, reaching above 15 m/s in the central region of measurement, while in the edge and corner positions the velocity results are lower, ranging between 1 and 2 m/s. This difference or lack of uniformity in this distribution of velocities is not observed in the exhaustion process, as it can be seen in Figure 5. This phenomenon may be consequence of the fact that the exhaustion is performed more uniformly, causing the distribution and movement of air to be more even inside the tunnel. Another reason could be that the air had not enough space to expand properly along the short region between the duct exit and the uppermost layer of samples. This makes it a difficult task to overcome because it needs more space for the system according to the appropriate technique for the design of air blast systems (ASHRAE, 1977). However, given the dimensions of the chamber and set of boxes with the product, the assembly was done with the maximum extension possible to fit inside the cold room.

Another factor that may cause interference measures in this place of assembly is that the anemometer does not indicate the direction of air flow during the measurements. Table 1 shows the values obtained and the average velocity of air flow calculated for each flow.

Test	Average air velocity (m/s)	Air flow (m³/s.kg)
Insufflation	- 3.05 ± 0.2	3.7 x 10⁻³
Exhaustion	1.88 ± 0.2	2.3 x 10⁻³

Table 1. Average values obtained for air velocity and air flow during insufflation and exhaustion processes.

The negative sign for the value of the inflation rate represents the direction of flow opposite
to the exhaustion, as the z axis of the graph shown in Figure 8, and it did not affect air flow
calculations, since the absolute values of air velocity were used for both cases.

4.2 Freezing time

A freezing process without the portable tunnel was carried out as a reference test to be
compared to the experiments using the tunnel device. This reference freezing process
consisted in leaving the boxes inside the cold room until all of the samples reached the final
freezing temperature.

The average freezing time for the pallet with the industrial sample arrangement was 47
hours, for the reference test, without the use of the portable tunnel. It was also observed the
lack of uniformity of process during the freezing process, where the samples located in the
upper and lower layers reached -18 ° C in about 42 hours, while it took more than 52 hours
for all the samples in the central layer to reach the final freezing temperature.

The top layer of boxes (layer 6) was quickly frozen in the insufflation test, as all the samples
reached -18 ° C after 35 hours. The samples in the bottom and middle layers were frozen
after about 45 hours, showing that the cooling air was not flowing properly in the bottom
part of the pallet.

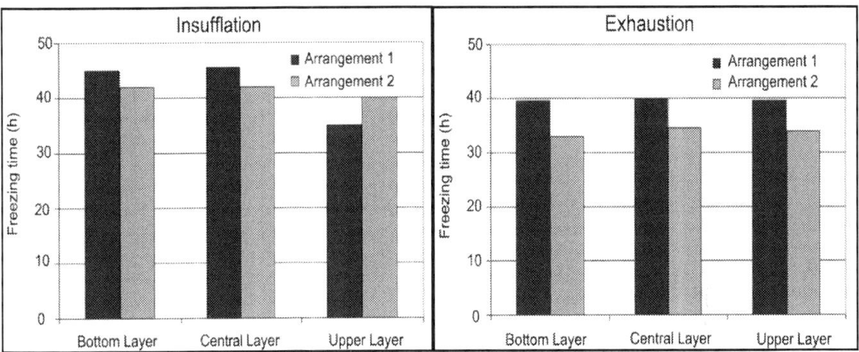

Fig. 9. Comparison of freezing time between two arrangements of samples for insufflation
and exhaustion processes.

During the exhaustion test it was observed a difference for the freezing time between the
central and peripheral layers of about 1 to 2 hours; however, all samples reached the
temperature of -18o C after 40 hours of processing.

Regarding the new arrangement of samples, the freezing time for the exhaustion process
was 38 hours. It took 43 hours for the samples to be frozen in the insufflation process. The
comparison between the freezing time for industrial and new arrangement can be seen in
Figure 9.

The heat transfer analysis comparing the air distribution performed shows that, although
showing values around 2 m/s for air velocity results, the exhaustion process reduced up to
14% of the freezing time compared to the blowing system, where air velocity were up to
19m/s.

The top layer of the assembly showed an increase in freezing time for the new configuration
of the samples for the testing of inflation (Figure 9). This may have occurred because the

industrial configuration might block air circulation for the other layers, reducing the efficiency of heat exchange between air and the other layers of product assembly.

4.3 Convective heat transfer coefficients

The lumped-capacitance method was applied to obtain the heat transfer coefficients for comparison between two different arrangements of products inside the cold chamber. Results of Table 2 show that there is an increase in the convective coefficient for the all the layers of boxes in the pallet, except the test body number 4, located on the second layer from the top.

In the insufflation process for the industrial arrangement, the results were in the range from 3 to 6 W/m^2°C for the three lower layers increasing to 15 W/m^2°C for the test body 4 and reaching 30 W/m^2°C the top layer. The values for the new proposed arrangement reached 10 W/m^2°C for the first two bodies and were greater than 30 W/m^2°C the top layer.

In the exhaustion process, results were nearly twice as high as the industrial process for the first two test bodies. The top layer had also higher values compared to the industrial arrange. For the central layer, results were similar, around 10 W/m^2°C. Only the test body in the second layer from the top showed higher results for the industrial arrange. Thus, results show that enlarging opening space for air flow increases the convective heat transfer, even if there is more product or layers in the pallet.

| | Convective coefficient (W/m^2°C) | | | |
| | Exhaustion | | Insufflation | |
Test body	Arrangement 1	Arrangement 2	Arrangement 1	Arrangement 2
T1 lower	5.79	10.74	3.58	9.90
T2	5.25	12.88	6.13	9.92
T3 middle	10.41	10.76	5.77	6.38
T4	14.97	11.06	15.41	13.04
T5 upper	5.92	7.73	29.46	31.72

Table 2. Results for convective coefficient measured between two arrangements of samples in the boxes and two different air flow orientation

The variability of the heat transfer coefficients with position inside a refrigerated room is a crucial aspect to be considered as this will directly affect cooling time. Considering that energy costs are the major expense associated with most refrigerated warehouses, this becomes critical to the managerial decisions, spanning from the initial investment to the long term running costs.

5. Conclusion

This study was initiated to resolve deficiencies in temperature reduction processes by investigating the convective heat transfer coefficient data for food cooling and freezing processes, endeavouring to assist the food refrigeration industry to improve heat transfer process efficiency with uncomplicated solutions. A literature examination was conducted to collect support information and compare with the results obtained.

Results have shown that increasing the cooling air velocity inside the system, or using more powerful equipments to introduce cooling air with lower temperatures is not the best way to improve cooling rates. Improved results can be achieved by applying simple changes to the process, such as rearranging the product or samples to be cooled or frozen and changing the direction of cooling air to flow around the product can save some time in freezing processes. On the other hand, the determination of convective heat transfer coefficient can help to determine how the cooling air is working on some specific region of the system or samples, leading to more precise ways to improve the cooling process.

The experimental data resulting from this project will be used by designers of cooling and freezing systems for foods. This information will make possible a more accurate determination of convective heat transfer coefficients inside these equipments, leading to more efficient systems and reduction in cooling and freezing times. Such information is essential in the venture and operation of cooling and freezing facilities and will be of immediate usefulness to engineers involved in the design and manoeuvre of such systems. Taking the above into consideration and incorporating these factors to the equipment's design will have a significant impact on energy savings.

6. References

Abrahao, R. F.; Correia, E.; Teruel, B. J. (2008). Computational simulation for the development of packages for bananas [Simulação computacional aplicada ao desenvolvimento de embalagens para bananas]. Revista Brasileira de Fruticultura, 30 (1). p. 79-87.

Allais, I.; Alvarez, G.; Flick, D. (2006). Modeling cooling kinetics of spheres during mist chilling. *Journal of Food Engineering*. V. 72, p. 197-209.

ASHRAE Handbook. *Fundamentals*. (1977) Atlanta, Georgia: American Society of Heating, Refrigerating and Air Conditioning Engineers, Inc. 1977.

ASHRAE. *Refrigeration*. (2002). Atlanta, Georgia: American Society of Heating, Refrigerating and Air Conditioning Engineers, Inc. 2002.

ASTM. (1989). *Manual on the use of thermocouples in temperature measurement*. American society for testing and materials.

Baird, C. D.; Gaffney, J. J.; Talbot, M. T. (1988). Design criteria for efficient and cost effective forced air cooling systems for fruits and vegetables. *ASHRAE Transactions*, Atlanta, v.94, p.1434-1453.

Barbin, D. F.; Neves Filho, L. C.; Sllveira Junior, V. (2009). Processo de congelamento em túnel portátil com convecção forçada por exaustão e insuflação para paletes. [Freezing process evaluation using a portable forced air system with air evacuation and air blowing in pallets]. *Ciencia e Tecnologia de Alimentos*, 29, (3), p.1-9.

Barbin, D. F.; Neves Filho, L. C.; Silveira Junior, V. (2010). Convective heat transfer coefficients evaluation for a portable forced air tunnel. *Applied Thermal Engineering*, 30. p.229–233.

Becker, B. R.; Fricke, B. A. (1999). Food thermophysical property models. *International Communications in Heat and Mass Transfer*, v.26, n.5, p.627-636.

Becker, B. R.; Fricke, B. A. (2004). Heat transfer coefficients for forced-air cooling and freezing of selected foods. *International Journal of Refrigeration*, v. 27, (5). p. 540-551.

Castro, L. R.; Vigneault, C.; Cortez, L. A. B. (2003). Container opening design for horticultural produce cooling efficiency. *International Journal of Food Agriculture and Environment*, v.2, n.1, p.135-140.

Chourot, J. M.; Macchi, H.; Fournaison, L.; Guilpart, J. (2003). Technical and economical model for the freezing cost comparison of immersion, cryomechanical and air blast freezing processes. *Energy Conversion and Management*, n.44, p.559-571.

Cleland, A. C. (1990). *Food Refrigeration Process. Analysis, Design and Simulation.* Elsevier Applied Science. London and New York. 1990. 284p.

Cortbaoui, P.; Goyette, B.; Gariepy, Y.; Charles, M. T.; Raghavan, V. G. S.; Vigneault, C. (2006). Forced air cooling system for Zea mays. *Journal of Food Agriculture and Environment*. V. 4, p. 100-104.

Dincer, I. (1995a). Thermal cooling data for figs exposed to air cooling. *International Communications Heat Mass Transfer*. v.22, n.4. p.559-566.

Dincer, I. (1995b). Transient heat transfer analysis in air cooling of individual spherical products. *Journal of Food Engineering*, v. 26, p. 453-467.

Dincer, I.; Genceli, F. (1994). Cooling process and heat transfer parameters of cylindrical products cooled both in water and air. International Journal Heat Transfer, v. 37, n. 4, p. 625-633.

Dussán Sarria, S.; Honório, S.L.; Nogueira, D.H. (2006). Precooling parameters of 'Roxo de Valinhos' figs (Ficus carica L.) packed in a carton box. *Fruits*. v. 61, Issue 6, Nov. 2006, p 401-406.

Fraser, H. (1998). Tunnel Forced-Air Coolers for Fresh Fruits & Vegetables. Ministry of Agriculture, Food and Rural Affairs, Government of Ontario, Canada, 1998. Available: <http://www.omafra.gov.on.ca/english/engineer/facts/98-031.htm>.

Heldman, D.R. Food Freezing. In: Heldman, D. R.; Lund, D. B. *Handbook of Food Engineering*. New York: Dekker, 1992. 277-315.

Kondjoyan, A. (2006). A review on surface heat and mass transfer coefficients during air chilling and storage of food products. *International Journal of Refrigeration*, v.29, p.863–875.

Laguerre, O.; Ben Amara, S.; Flick, D. Heat transfer between wall and packed bed crossed by low velocity airflow. *Applied Thermal Engineering*, v. 26, p. 1951-1960, 2006.

Le Blanc, D. I.; Kok, R.; Timbers, G. E. (1990a). Freezing of a parallelepiped food product. Part 1: Experimental determination. *International Journal of Refrigeration*, v.13, p.371-378.

Le Blanc, D. I.; Kok, R.; Timbers, G. E. (1990b). Freezing of a parallelepiped food product. Part 2: Comparison of Experimental and Calculated Results. *International Journal of Refrigeration*, v.13, p.379-392.

Meana, M. B.; Chau, K. V.; Emond, J. P.; Talbot, M. T. Forced-air cooling of strawberries in reusable plastic containers. *Proceedings of Florida State Horticultural Society*. v. 118, p. 379-382, 2005.

Mohamed, I. O. (2008). An inverse lumped capacitance method for determination of heat transfer coefficients for industrial air blast chillers. *Food Research International*, 41. p.404–410.

Mohsenin, N. N. (1980). *Thermal properties of foods and agricultural materials.* New York: Gordon and Breach. p. 198-224. 1980.

Nunes, M.; Nader, G.; Jesus, F. B. G.; Cardoso, M.; Jabardo, P. J. S.; Pereira, M. T. (2003). Estudo sobre técnicas de medição de vazão em bocas de insuflamento e de exaustão de ar. Metrologia-2003 – Metrologia para a Vida Sociedade Brasileira de Metrologia (SBM) Setembro 01-05, 2003, Recife, Pernambuco - BRASIL

Resende, J. V.; Silveira Jr, V. (2002a). Medidas da Condutividade Térmica Efetiva em Modelos de Polpas de Frutas no Estado Congelado. *Ciência e Tecnologia de Alimentos*, v.22, n.2, p.177-183.

Resende, J. V.; Silveira Jr. V. (2002b). Escoamento de ar através de embalagens de polpa de frutas em caixas comerciais: Efeitos sobre os perfis de velocidade em túneis de congelamento. *Ciência e Tecnologia de Alimentos*, v.22, n.2, p.184-191.

Resende, J. V.; Neves Filho, L. C.; Silveira Jr, V. (2002). Coeficientes de Transferência de Calor Efetivos no Congelamento com Ar Forçado de Modelos de Polpas de Frutas em Caixas Comerciais. *Brazilian Journal of Food Technology*, v.5, p.33-42.

Salvadori, V. O.; Mascheroni, R. H. (1996). Freezing of strawberry pulp in large containers: experimental determination and prediction of freezing times. *International Journal of Refrigeration*, v.19, n.2, p.87-94.

Santos, C. A.; Laurindo, J. B.; Silveira Junior, V.; Hense, H. (2008). Influence of secondary packing on the freezing time of chiken meat in air blast freezing tunnels [Influência da embalagem secundária sobre o tempo de congelamento de carne de frango em túneis de circulação de ar forçada]. *Ciencia e Tecnologia de Alimentos*, 28 (SUPPL.). p.252-258.

Scott, E. P.; Beck, J. V.; Heldman, D. R. (1992). Estimation of Time Variable Heat Transfer Coefficients in Frozen Foods during Storage. *Journal of Food Engineering*, v.15, p.99-121, 1992.

Talbot, M. T.; Chau, K. V. (1998). Precooling Strawberries. Agricultural and Biological Engineering Department, Florida Cooperative Extension Service, Institute of Food and Agricultural Sciences, University of Florida, pub.CIR942/AE136, 1998. Available in: <http://edis.at.ufl.edu/AE136>.

Talbot, M. T.; Fletcher, J. H. A (1996). Portable Demonstration Forced-Air Cooler. Agricultural and Biological Engineering Department, Florida Cooperative Extension Service, Institute of Food and Agricultural Sciences, University of Florida, pub.CIR1166/AE096, 1996. Available in:<http://edis.at.ufl.edu/AE096>.

Vigneault, C.; Goyette, B.; Markarian, N. R.; Hui, C. K. P.; Cote, S.; Charles, M. T.; Emond, J.-P. (2004a). Plastic container opening area for optimum hydrocooling. *Canadian Biosystems Engineering*, v.46, p.41-44.

Vigneault, C.; De Castro, L. R.; Gautron, G. (2004b). Effect of open handles on packages during precooling process of horticultural produce. *ASAE Annual International Meeting.* p. 6901-6908.

Vigneault, C.; De Castro, L. R.; Cortez, L. A. B. (2005). Effect of Gravity on Forced-air Precooling. *IASME Transactions* Vol. 2(3):459-463.

Welty, J. R.; Wicks, C. E.;, Wilson, R. E.; Rorrer, G. (2000). *Fundamentals of Momentum, Heat, and Mass Transfer.* 4th Edition, John Wiley & Sons: New York, 2000.

Welty-Chanes, J.; Vergara-Balderas, F.; Bermudez-Aguirre, D. (2005). Transport phenomena in food engineering: Basic concepts and advances. *Journal of Food Engineering* V. 67, (1-2), March 2005, Pages 113-128.

Zaritzky, N. E. (2000). Factors affecting the stability of frozen foods. In: KENNEDY, C. J. *Managing frozend foods.* Cambridge: Woodhead Publishing Limited, 2000.

Zou, Q; Opara, L. U.; Mckibbin, R. (2006a). A CFD modeling system for airflow and heat transfer in ventilated packaging for fresh foods: I. Initial analysis and development of mathematical models. *Journal of Food Engineering.* v. 77, Issue 4, p. 1037-1047.

Zou, Q; Opara, L. U.; Mckibbin, R. (2006b). A CFD modeling system for airflow and heat transfer in ventilated packaging for fresh foods: II. Computational solution, software development, and model testing. *Journal of Food Engineering*. v. 77, Issue 4, p. 1048-1058.

12

Marangoni Condensation Heat Transfer

Yoshio Utaka
Yokohama National University
Japan

1. Introduction

Marangoni condensation phenomena, which show the extremely high heat transfer coefficient in the dropwise condensation regime, occur due to surface tension instability of the condensate in the condensation of binary a vapor mixture of a positive system (i.e., one in which the surface tension of the mixture has a negative gradient with the mass fraction of the volatile component, such as water - ethanol and water - ammonium mixtures.) Marangoni dropwise condensation differ from so-called dropwise condensation which occurs only on the lyophobic surface and easily occurs on the wetting surface. This phenomenon was first reported by Mirkovich & Missen (1961) for a binary mixture of organic vapors. Ford & Missen (1968) demonstrated that the criterion for instability of a condensate liquid film. Fujii et al. (1993) experimentally investigated the condensation of water - ethanol mixtures in a horizontal tube and found that several different condensation modes such as dropwise and rivulets occur depending on concentration. Morrison & Deans (1997) measured the heat transfer characteristics of a water - ammonium vapor mixture and found that it exhibited enhanced heat transfer.

Utaka & Terachi (1995a, 1995b) measured the condensation characteristic curves and clarified that surface subcooling is one of the dominant factors in determining the condensate and heat transfer characteristics of Marangoni condensation although the effect of concentration of binary mixture was major factor deciding the condensation modes in most experimental reserches. Moreover, the effects of external conditions such as the vapor mass fraction (Utaka & Wang, 2004) and the vapor velocity (Utaka & Kobayashi, 2003) were investigated. Heat transfer was significantly enhanced at low mass fractions of ethanol in a water - ethanol mixture. Murase et al. (2007) studied Marangoni condensation of steam - ethanol mixtures using a horizontal condenser tube. Their results showed similar trends as those of (Utaka & Wang, 2004) for vertical surfaces.

On the other hand, the mechanisms of Marangoni condensation have also been studied. Hijikata et al. (1996) presented a theoretical drop growth mechanism for Marangoni dropwise condensation. They found that the Marangoni effect that occurs due to a difference in the surface tension plays a more important role than the absolute value of the surface tension in Marangoni condensation. Utaka et al. (1998) investigated the effect of the initial drop distance, which is the average distance of the initially formed drops grown from a thin flat condensate film that appears immediately after a drop departs. They clarified that the initial drop distance is closely related to the heat transfer characteristics of Marangoni condensation. Further, Utaka & Nishikawa (2003a, 2003b) measured the thickness of condensate films on the tracks of departing drops and between drops by applying the laser extinction method. They found that the condensate film thickness was approximately 1 µm and that it is closely related to the initial drop distance and the heat transfer characteristics.

In this paper, the mechanisms and the heat transfer characteristics of Marangoni condensation phenomena are described on the basis of those researches.

2. General description of mechanisms and characteristics of Marangoni condensation phenomena

2.1 Outline of mechanisms and characteristics of Marangoni condensation phenomena

The conditions determining the condensate modes in condensation of binary mixtures depend upon the phase equilibrium relation between liquid and vapor and the magnitude relation between surface tension of two liquids. Marangoni condensation appears typically for the case of so-called positive system, in which the surface tension of more volatile component σ_L is lower than that of non-volatile component σ_H. Such a instability in the system of liquid evaporation was shown by Hovestreijit (1963) for the first time. The mechanism of Marangoni condensation was explained by Fujii et al. (1993) from the similar point of view as shown in Fig. 1. The thicker condensate liquid (point B in Fig. 1) pull the thinner condensate (point A) due to the surface tension difference occurred by the distributions of surface temperature and concentration of liquid. As a result, the irregular condensate thickness augmented and the irregular modes of condensate such as dropwise appear by the surface instability. The phase equilibrium ralation and the variation of surface tension against mass fraction for water – ethanol binary mixture, which is major test material as a positive system in this study, is shown in Fig. 2.

The major dominant parameters in Marangoni condensation are the concentration of vapor and the surface subcooling of condensing surface. Althoug the concentration of vapor was a main factor determining the condensation mode, in investigating the heat transfer characteristics of Marangoni condensation, surface subcooling was found to be the fundamental factor controlling the condensate modes by Utaka-Terachi (1995a, 1995b). In other words, the change in heat transfer coefficient showed strong non-linearity with condensate mode transition because the Marangoni dropwise condensation formed in the vapor-side appears over the wide range of surface subcooling, in addition to the change in diffusion resistance, which is inherent in the condensation of binary mixtures. The features of the condensation characteristic curves are summarized schematically in Fig. 3, in which the variations of heat flux q and heat transfere coefficient α aginst surface subcooling ΔT are shown. Table 1 shows the characters in the change in heat transfer codfficient. Alphabetical symbols in the figure and table denote characteristic points in the condensation characteristic curves. Points B and D are the steep increase point of heat transfer and the maximum heat transfer point, respectively. The condensate shows the dropwise mode in a wide range of surface subcooling, encompassing B and D, and the dropwise mode appears when approaching point D from B. With further increase of the surface subcooling, the minimum point of the heat flux and the point of inflection of the heat transfer coefficient (E) appear as the end point of the transition region.

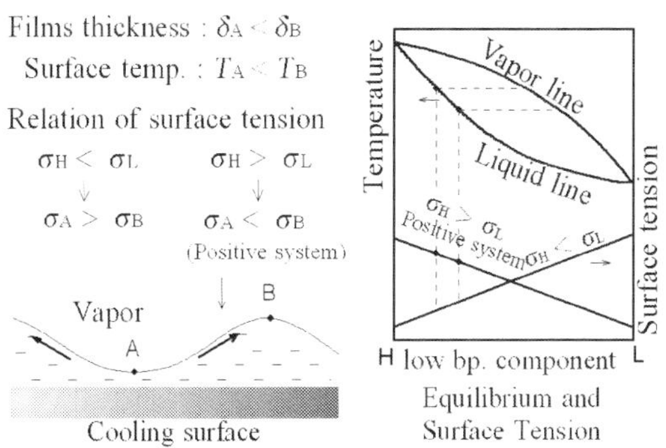

Fig. 1. Mechanisms of Marangoni condensation phenomena

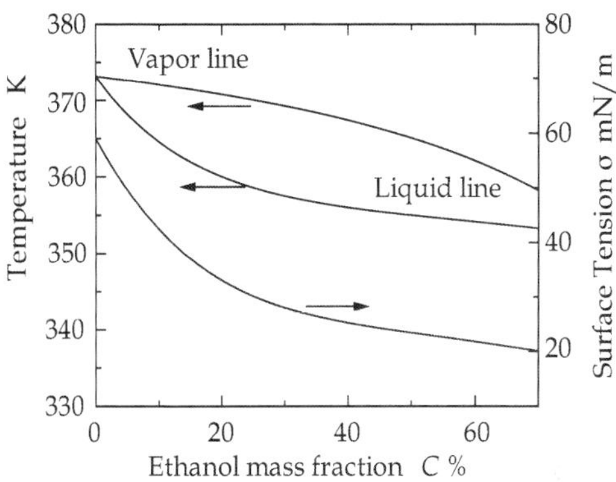

Fig. 2. Phase equilibrium relation and surface tension variation for water-ethanol mixtures

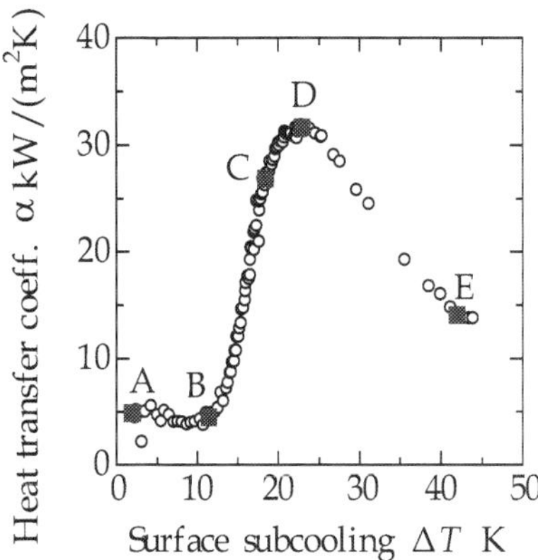

Fig. 3. Nature of condensation characteristic curves for Marangoni condensation

Domain	Characteristics
Smaller ΔT than steep increase pt. (B) (Vapor-side dominant region, A -B)	Small α Diffusion resistance dominant
Linear increase of α from pt. (B) to pt. (C) (Steep increase region, B -C)	Reduction of diffusion resistance Dropwise condensation
Departing from linear increase Maximum α pt. (D) Negative gradient region (Transition region, C -E)	Reduction of α
(Film region (E -))	Film condensation

Table 1. Nature of Marangoni condensation heat transfer

2.2 Marangoni dropwise condensation cycle

There are drop cycles in Marangoni dropwise condensation similar to dropwise condensation on a lyophobic surface. Since detailed aspects of condensate variation describe later in the relating sections, only the basic items are discussed here. First, just after sweeping by a departing drop in a typical condensation process, thin liquid condensate film remains. Next, the formation of initial drops commences along with the continuous condensation. Then, the drops grow with condensation and coalescences of drops. At the final stage of the cycle, some largest drops begin to depart due to the effect of external forces such as gravity and vapor flow and the new thin liquid film appears. Those cycles are repeated irregularly.

3. Experimental apparatus and methods common in measurements

Typical experimental apparatus and method common for Marangoni condensation experiments were shown next. A copper heat transfer block devised specifically for investigating phenomena with large heat flux and high heat transfer coefficients (Utaka & Kobayshi, 2003) was shown in Fig. 4. The heat transfer block having a cross-section of trapezoidal shape with notches was constructed in order to realize uniformity of surface temperature and large heat flux. The condensing surface had an area of 10 mm×20 mm. The copper block of one-dimensional rectangular column was also utilized for the cases realizing moderate heat flux. Oxidized titanium was applied to the condensing surface in order to achieve a wetting surface. In addition, impinging water jets from a bundle of thin tubes were used so as to provide high and uniform cooling intensity. A schematic diagram of the leak-tight experimental apparatus, intended to minimize the effects of non-condensing gas, is shown in Fig. 6 (Utaka & Wang, 2004). After passing through the condensing chamber (Fig. 5) in which the heat transfer block is placed, the vapor generated in the steam generator is condensed almost entirely in the auxiliary condenser. The condensate is returned to the vapor generator by the plunger pump via the flow measurement equipment. The vapor flow is in the same direction in which gravity acts, through a duct. Non-condensing gas is continuously extracted by the vacuum pump near the outlet of the auxiliary condenser. The inlet of the vacuum pump is cooled by an electronic cooler to maintain a constant concentration in the vapor mixture, by maintaining low vapor pressure. The loop was divided into a high-pressure part and a low-pressure part bounded by the pressure adjusting valve and the return pump. The vapor pressure of the high-pressure-side is maintained at approximately 1 kPa above atmospheric pressure. The concentration of non-condensing gas in the vapor mixture is measured before and after the experiment. Another heat transfer block shown in Fig. 6 for the vapor concentration measurement is attached in the condensing chamber located downstream of the main heat transfer block.

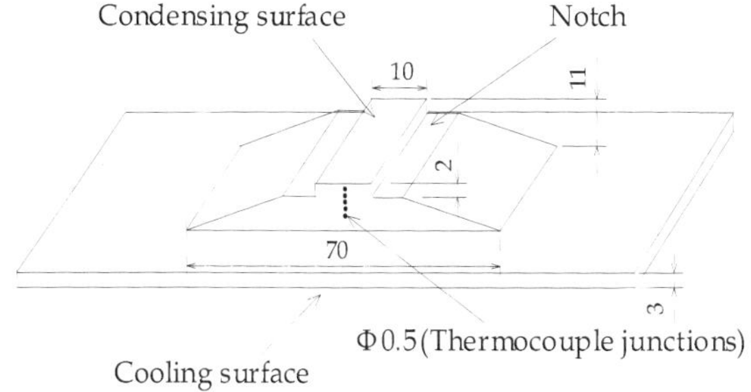

Fig. 4. Heat transfer block for large heat flux

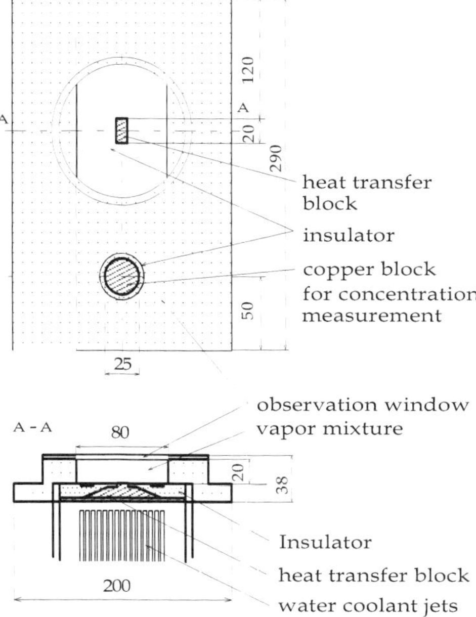

Fig. 5. Condensing chamber

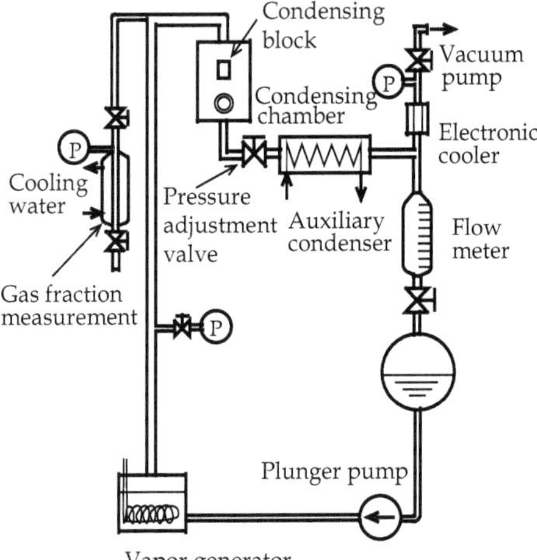

Fig. 6. Schematic diagram of experimental apparatus

After the vapor condition reaches the steady state, the condensation characteristic curves were measured continuously using a quasi-steady measurement in which the temperature of the cooling water was changed very slowly for a fixed concentration and fixed velocity of vapor. The aspect of condensate was observed and recorded through the glass window of the condensing chamber using a high speed camera to analyze the condensate characteristics.

4. Factors concerning mechanisms of Marangoni condensation

As described in Section 2, the iregurarity of condensate appears due to surface tenstion instability of the condensation system for binary mixtures of positive system in Marangoni condensation. Hijikata et al. (1996) performed the instability analyses and gave proof of Maragoni condensation phenomena theoretically. In this section, the dominant factors concerning the mecanism of heat transfer in Marangoni condensation is discussed.

4.1 Relation between initial drop distance and heat transfer
(a) Observation and measurements of drop formation
The observation and measurement of the process of formation and growth of condensate drops commenced after the sweeping action by the departing drops were carried out for four ethanol mass fractions of vapor c, i.e., $c = 0.07, 0.17, 0.37$ and 0.52 under constant vapor velocity at atmospheric pressure (Utaka, et al. 1998). Figure 7 (a) - (c) shows the photographs taken by the high speed camera. Figure 8 (a) shows the condensation heat transfer characteristic curves measured simultaneously with the taking photographs. The local maximum drops sizes increased with increasing distance from the up-flow side peripherals of the departing drops, which are shown as the largest drops in the images, because the time elapsed from the sweep motions by departing drops increased with the distance owing to the limited velocity of the drop departure. It is also seen that almost no droplets existed at the relatively thin belt-wise areas just next to the up-flow side peripherals of the departing drops. The small droplets appeared at the next areas. Hence, the initial drops form from the

thin liquid film and then grow by condensation and coalescences as reported in the observation of Hijikata et al. (1996) by making the uniform sweeping over the whole surface. It could be understood that the drop sizes nearest to the peripherals of the departing drops varied with the surface subcooling.

The initial drop distance d_i, which is the distance between adjoining drops forms after the sweeping, were adopted as parameters determining the characteristics against the variation of the ethanol concentration and the surface subcooling. Here, the reason for adopting the initial drop distance to decide the characteristics of drop formation is as follows. It is difficult to determine the minimum drop size because the unevenness of thin condensate film and drop shape are not distinguishable each other. On the contrary, since the initial drop distance does not change until an occurrence of a first coalescence, it can be defined certainly.

The variations of initial drop distance measured are shown in Fig. 8(b). Since the elapsed time after the end of sweep changes along the track of the departing drop as described above, the drop sizes varies along the departure direction. Therefore, the distance between the neighboring drop centers in the direction of the departing drop peripheral was adopted

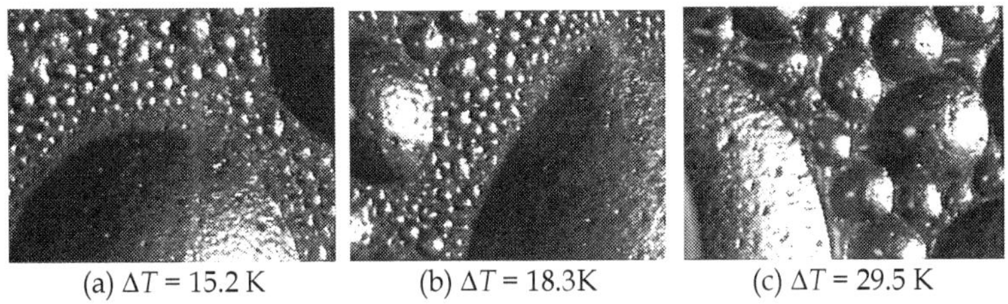

(a) ΔT = 15.2 K   (b) ΔT = 18.3K   (c) ΔT = 29.5 K

Fig. 7. Aspect of condensate (c =0.37) 0.2 mm

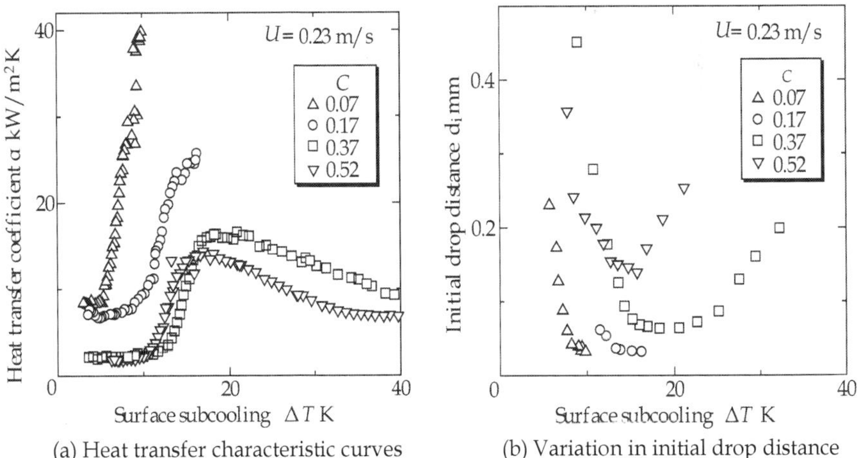

(a) Heat transfer characteristic curves   (b) Variation in initial drop distance

Fig. 8. Relation among aspect of condensation, heat transfer characteristic curve, and initial drop distance as the initial drop distance. As is evident from Fig. 8(b), the initial drop distances depend markedly on the surface subcooling in the Marangoni dropwise condensation. The U-shaped curves having minima were obtained for c = 0.37 and 0.52 since the measurement of the whole regions of the surface subcoolings were achieved. The curves until the minima were obtained for c = 0.07 and 0.17 as same subcooling regions as the characteristic curves. The ratio of least to greatest initial drop distances for each concentration changed widely and was in the range of 2 to 7. It is notable that the surface subcooling at the minimum of the initial drop distances nearly coincides with those at the maximum heat transfer coefficient for each concentration.

The smallest drops, which is nearly equal to the sizes of initial drop distances, were quite larger than those of dropwise condensation on a lyophobic surface and were 0.03 mm to 0.15 mm in this experimental range. Thus, the initial drop distance depended considerably on the ethanol concentration and surface subcooling. That is, the initial drop distance decreased with decreasing ethanol concentration and the maximum is about 5 times larger than the minimum. The minimum diameters of drops measured by Fujii et al (1993) were 0.2 mm and 0.05 mm for c = 0.48 and 0.34, respectively, and are ranged in those measurements. However, since their conditions of the surface subcooling were unknown as well as the concentrations are different from this measurement, it is difficult to get detailed comparison. Therefore, it indicates the possibility that the reduction of the thickness of the thin condensate film among drops together with the decrease of the initial drop distance and the approaching of the spherical shape is given by the increase of the driving force of the Marangoni effect.

4.2 Condensate film thickness
(a) Apparatus for measuring condensate film thickness
The entire experimental system, composed of the vapor loop, condensing chamber, cooling system, which are similar to those shown in Section 3, and laser extinction measurement system, is shown in Fig. 9 (Utaka & Nishikawa 2003a, 2003b).

Lambert's law (Eq. 1) was used for determining the condensate liquid thickness. Where, A is the laser exstinction coefficient, I_0 and I are the intensities of incident light and of transmission light, respectively, and δ is the optical path length.

$$\delta = -A^{-1} \times \log\left(I / I_0\right) \tag{1}$$

Since the extinction coefficient A is unknown and depends upon the mass fraction of the liquid mixture through which the laser beam passes, and was measured prior to the measurement of film thickness. The helium-neon laser beam, having 3.39 μm of 2.6 mm in diameter, is transformed to parallel light of 53 mm in diameter by the beam collimator lens. The beam is concentrated by passing through a condenser lens made from optical silicon. The beam waist is defined at the thin liquid layer of the mixture. The transmitted laser beam, which is partially absorbed by the test liquid enclosed in a quartz glass gap, is converged by the quartz glass lens. The transmitted light intensity I is measured by the detector (light reception area of 3×3 mm², response time of $1 \sim 3$ μs) made of lead selenide. The incident light intensity I_0 is measured under the same conditions of transmitted light intensity without the test liquid. The diameter of the laser beam is approximately 30 μm at the test liquid.

The aspect of the condensate was photographed by a high-speed digital camera synchronously with the condensate film thickness measurement in order to confirm correspondence with the condensate behavior.

(b) Condensate behavior and variation in condensate film thickness

Simultaneous measurement of condensate film thickness and observation of condensate behavior during the formation of the initial drops commenced after the departing drops swept by, under four ethanol vapor mass fractions c of 0.11, 0.21, 0.30, and 0.40, and a constant vapor velocity of 0.7 m/s. Figures 10 (a) and (b) show the images of condensate behavior and the variations of condensate thickness as measured from light transmittance for c = 0.30. Figure 10 (a) shows the aspect of condensate appearing after departing drops swept by, and figure 10 (b) shows variations in condensate thickness immediately after the departing drops swept by. Here, t = 0 was determined arbitrarily, and the variation in liquid thickness and the image coincide with each other. The relatively large dark region on the right-hand side of each image denotes the departing drop, and the light-colored area is thin condensate film at t = 0 ms. Condensate thickness is non-uniform; i.e., the droplets grew rapidly from a flat thin film that appeared after the departing drops swept by. The measuring position irradiated by the laser beam is highlighted as a circle measuring about 30 μm in diameter. After the measuring position was covered by the departing drop at t = 0 ms, the thin film observed as the minimum condensate thickness immediately after sweeping at t = 6 ms by the movement of the departing drop towards the lower right-hand corner. At t = 28 ms, the thickness was measured at the drop having diameter of less than 0.2 mm. Thus, good correspondence between the variation of measured value and the aspect of condensate is confirmed. In this case, the measured minimum condensate thickness was 1.3 μm. These results confirm that a thin liquid film having a thickness of around 1 μm remains after sweeping and that the heat transfer surface was always covered by liquid in Marangoni dropwise condensation.

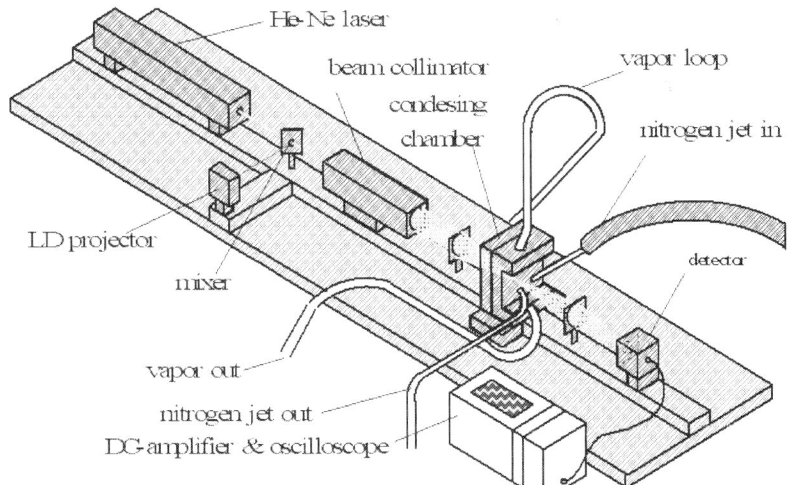

Fig. 9. Experimental apparatus for measuring condensate film thickness

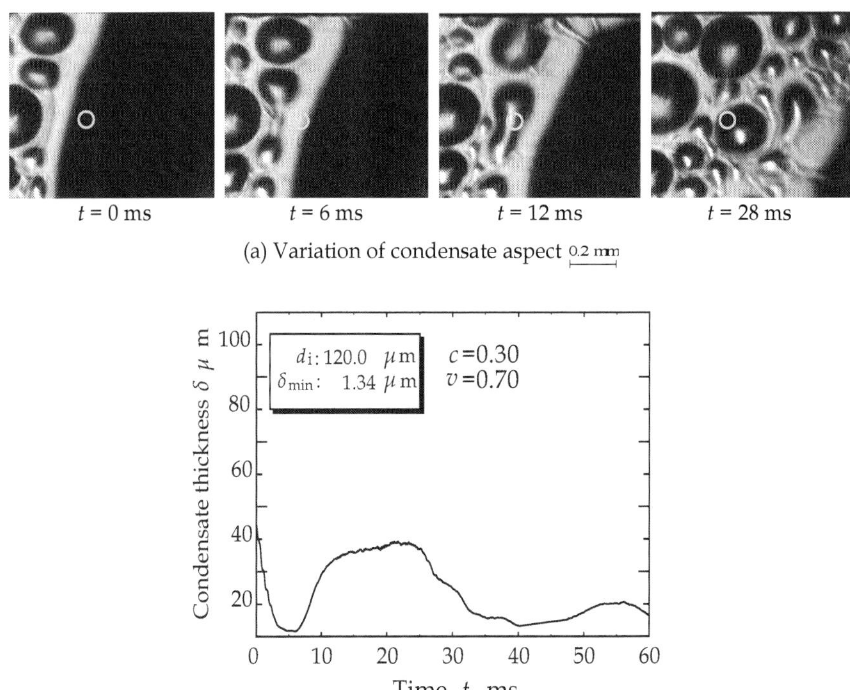

(b) Variation of condensate thickness

Fig. 10. Aspect of condensation and condensate thickness for $c = 0.30$ and $\delta_{min} = 1.34$ μm

Figures 11 (a),(b),(c), and (d) show the relations between the minimum condensate film thickness δ_{min} appearing immediately after departing drops swept by in the Marangoni condensation cycle and the initial condensate thickness measured from the simultaneously obtained images for four ethanol vapor concentrations, i.e.; c = 0.11, 0.21, 0.30, and 0.42. Initial drop distance was adopted as the averaged value for at least several drop departures. For all vapor concentrations, minimum condensate thickness decreased as initial drop distances decreased in association with increasing surface subcooling. For the higher ethanol vapor concentrations of 0.30 and 0.42 shown in Figs. 11 (c) and (d), respectively, the minimum drop distances decreased linearly with decreasing initial drop distance in association with increased surface subcooling. Afterwards, the minimum thickness became minimum. Both the initial drop distance and the minimum condensate thickness increased with further increase in cooling intensity. Since, as shown in section 4.1, the maximum heat transfer coefficient appeared in a region near the minimum initial drop distance, and the return portion of the curve; i.e., the region near the minimum initial drop distance and the smallest minimum condensate thickness, should correspond to the maximum heat transfer point in the condensation characteristic curve. The minimum condensate thickness for the vapor ethanol mass fractions of 0.30 and 0.42 was very thin; that is, 1.2 μm and 1.5 μm, respectively. Minimum condensate thickness showed a tendency to decrease with decreasing vapor ethanol concentration. The minimum values of minimum condensate thickness for the mass fraction of ethanol vapor of 0.11 and 0.21 were about 1 µm, because of the limit of cooling intensity. In the present study, larger surface subcooling in the thinner ethanol concentration region associated with very large heat flux was difficult to realize, because quartz glass and the jet of evaporated liquid nitrogen gas were used as a heat transfer surface and a coolant, respectively. Therefore, minimum condensate thickness could not be measured near the region of minimum value. Linear curves fitted by the least squares

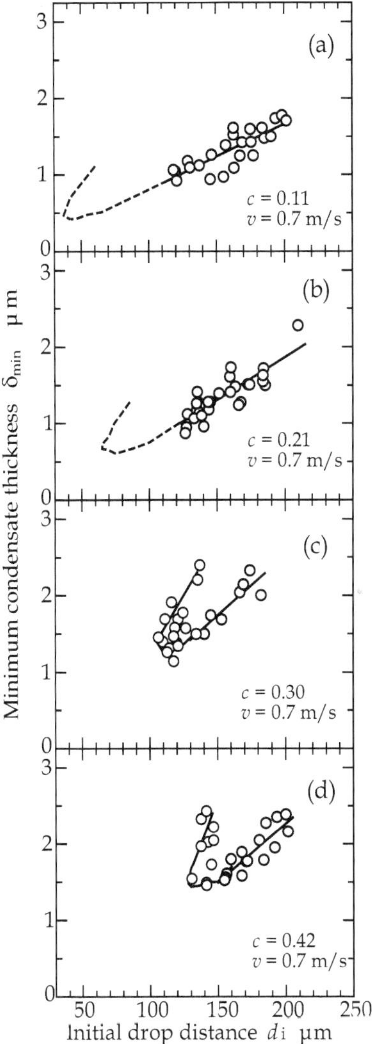

Fig. 11. Variation of minimum film thickness against initial drop distances after sweeping by departing drops

method to the data showing the linear changes had slopes of 0.0087, 0.011, 0.016, and 0.016 for ethanol mass fractions of 0.11, 0.21, 0.30, and 0.42, respectively, and tended to increase with ethanol concentration. The dotted lines in Figs. 11 (a) and (b) were predicted on the basis of the characteristic features of the curves shown in Figs. 11 (c) and (d) and the previous results shown in section 4.1. Although results do not lend themselves to detailed quantitative discussion, minimum thickness can be predicted to decrease with increasing ethanol concentration and to reach less than 0.5 μm.

The results of this section and section 4.1 clarified that close coincidence exists among the features of heat transfer variation in the condensation characteristic curves and their characteristic quantities, such as initial drop distance, departing drop diameter, and minimum condensate thickness, whose minimum is attained at the surface subcooling near maximum heat transfer coefficient. Similar features of these characteristic quantities in relation to ethanol concentration were also obtained. Therefore, the variations in heat transfer coefficient in the condensation characteristic curves were shown to correspond to the changes in the distributions of surface tension differences, which are the driving force of condensate irregularity in Marangoni condensation. Most likely, the maximum driving force appears in the vicinity of peak value of heat transfer coefficient and the thinning of the condensate thin film. Therefore, the heat transfer is enhanced near the maximum heat transfer coefficient point when the thinning of the condensate film with the augmentation of surface tension difference and the decrease in diffusion resistance, which is discussed in next section, occur.

4.3 Vapor-side diffusion characteristics

For Marangoni condensation, the condensate resistance based on the condensate aspect and the mass diffusion resistance in the vapor-side are important fundamentally in controlling the condensation characteristics. This phenomenon consists of the unsteady process whereby the condensate thickness and the vapor concentration change due to the variation of the condensate mode. An analysis of the vapor-side diffusion resistance is required in order to explain the detailed mechanism. The relation between the condensate aspect and the condensation characteristic was previously reported in former sections.

Here, a model approximating the process just after sweeping by a departing drop in a typical condensation process was considered along with the diffusion resistance (Wang & Utaka, 2005). Rapid condensation occurs with the beginning of sweeping of the heat transfer surface by a departing drop. The concentration distribution proceeds due to rapid condensation from the initial situation where the heat transfer surface is covered by a large drop with comparatively lower heat transfer and nearly uniform concentration distribution. By solving a one-dimensional unsteady diffusion equation derived from the idealized process described above numerically, the thermal conductance variation of the vapor-side for a process in which the condensation occurs swiftly from uniform concentration is investigated.

The vapor-liquid interface temperature is given as a boundary condition in order to consider only the conductance in the vapor phase. In addition, since the condensate film thickness was very thin, approximately 1 μm, as describe in seciton 4.2, in the surface subcooling range near the maximum heat transfer coefficient, the thermal resistance of the condensate film is small and the surface temperature of the condensate liquid is near that of the heat transfer surface. The basic equation, the boundary conditions, and the initial condition can be written as:

$$\frac{\partial C}{\partial t} + V\frac{\partial C}{\partial y} = D\frac{\partial^2 C}{\partial y^2} \tag{2}$$

$$y = 0 : C = C_I \tag{3}$$

$$(\rho D\frac{\partial C}{\partial y})_I = (C_{IL} - C_I)m \tag{4}$$

$$m = \rho V \tag{5}$$

$$y = \infty : C = C_\infty \tag{6}$$

$$t = 0 : C = C_\infty \tag{7}$$

where C is the ethanol mass fraction of the water-ethanol vapor mixture, D is the diffusivity between water and ethanol, V is the velocity of the vapor flowing into the heat transfer surface due to the condensation, ρ is the vapor density, m is the condensation rate, and C_I, C_{IL} and C_∞ are the ethanol concentrations of vapor and liquid in the vapor-liquid interface and the bulk vapor concentration, respectively. The thermal conductance H and the non-dimensional temperature difference ΔT^* in the vapor phase are defined respectively as:

$$H = Lm /(T_S - T_I) \tag{8}$$

$$\Delta T^* = (T_S - T_I)/(T_S - T_L) \tag{9}$$

where L is the latent heat of condensation, T_s and T_L are the dew point and boiling point tempetatures, respectively, corresponding to the vapor concentration C_∞ in the phase equilibrium diagram of water and ethanol, and T_I is the temperature of the vapor-liquid interface.

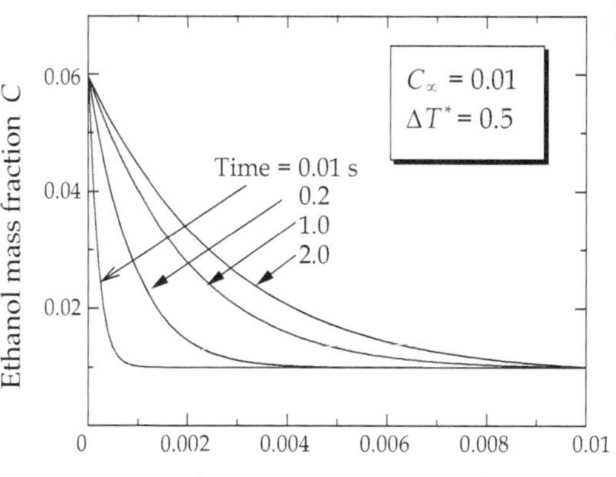

Fig. 12. Variation of distribution of mixture vapor concentration with time

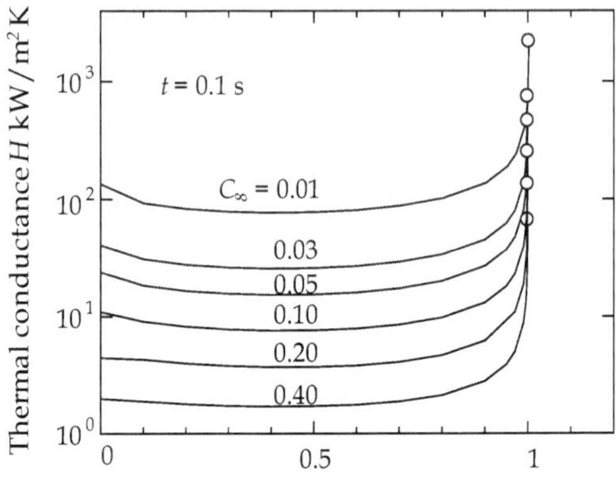

Fig. 13. Variation of thermal conductance with respect to non-dimensional temperature difference in the vapor layer ($t = 0.1$ s)

Figure 12 shows the time variation of ethanol mass fraction distribution in the vapor phase. It is seen that the thickness of different concentration region becomes thicker with time. Figure 13 shows the variation of thermal conductance with the non-dimensional temperature difference in the vapor phase for the different bulk vapor mass fractions, calculated when the time equals 0.1 seconds. Assuming a similar distribution of the condensate shape, the period of frequency between drop departures is inversely proportional to the heat flux. Therefore the characteristic time is dependent on that period. Also, the maximum heat flux of such a diffusion process is different for different vapor concentrations. However, for convenience, the thermal conductance at a fixed time is compared in order to examine the character of the variation in the condensation characteristic curve and its characteristic points. Since the sweeping period by the departing drop is approximately 0.1~0.2 s at the maximum heat transfer coefficient on the condensation characteristic curves, the calculation result for this time was adopted herein.

As a characteristic of the vapor-side diffusion resistance, the heat flux tends to increase rapidly for any vapor concentration when the non-dimensional temperature difference approaches 1.0 (i.e. the vapor-liquid interface temperature approaches the boiling point), because the compositions of the condensation and the bulk vapors also approach each other. The decrease in the diffusion resistance in the vapor-side due to the decrease in ethanol concentration improved the heat transfer performance. This is thought to be one of the reasons why a relatively high maximum heat transfer coefficient appears in the low-vapor-concentration region.

5. Heat transfer characteristics of Marangoni condensation

The factors affecting the characteristics of heat transfer in Marangoni condensation is investigated in this section. That is, the effects of given external conditions such as vapor concentration, vapor velocity and non-condensing gas concentration. Also, the effect of surface subcooling varying the cooling intensity in detail.

5.1 Effects of surface subcooling and vapor concentration

(a) Condensation characteristic curves

The condensation characteristic curves under atmospheric pressure and at vapor velocitie U of 1.5 m/s for a considerable range of ethanol concentrations are shown in Figs. 14 (a) and (b) (Utaka & Wang, 2004). Since film condensation appeared in all subcooling regions for pure water vapor, monotonical change, which is a feature of ordinary film condensation heat transfer, was exhibited. In contrast, a common trend was observed for ethanol vapor mass fractions larger than 0.05%. That is, the qualitative features of each curve showed a common characteristic change having heat flux and heat transfer coefficient maxima as shown in section 2.

From these characteristic curves, excellent heat transfer were confirmed, whereby the heat transfer coefficient maxima shifted to the smaller subcooling region and the values at the maxima were very high at low ethanol concentrations, with the exception of the extremely low ethanol concentrations of 0.05% and 0.1%. In addition, in the low ethanol concentration range, the dependencies of the condensation heat transfer characteristics on the vapor concentration and the surface subcooling become greater, and the gradient of the characteristic curves at the smaller subcooling range increases.

As indicated in a previous report (Utaka & Terachi, 1995a), the realization of excellent heat transfer characteristics is likely due to the following. Since the temperature difference between the boiling point line and dew point line decreases as the ethanol concentration of

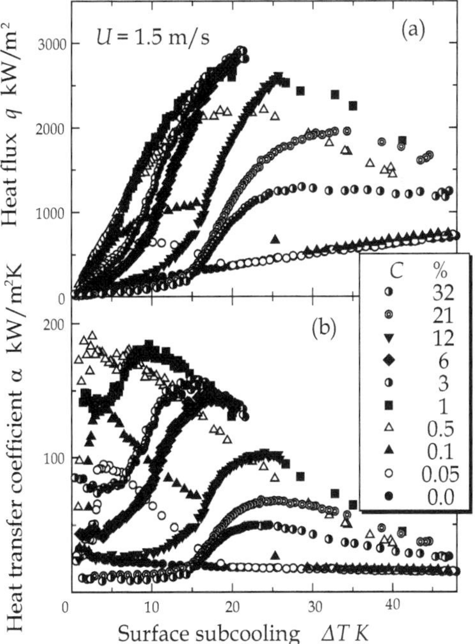

Fig. 14. Condensation heat transfer characteristic curves (U = 1.5 m/s)

the vapor mixture decreases, the subcooling region of condensation controlled by the diffusion resistance in the vapor-side becomes narrower due to the nature of vapor-liquid equilibrium relation. Moreover, since the difference between ethanol concentrations of the vapor in bulk and at the vapor-liquid interface is small and the variation of surface tension with respect to the concentration is large, higher heat transfer conductance is realized.

In order to evaluate the condensation characteristic curves quantitatively, we examined the heat transfer quantities at the commencement points of the steep increase in heat transfer, the maximum heat transfer coefficient and maximum heat flux, as well as the condensing surface subcooling. Figure 15 shows the variation of surface subcooling with respect to vapor ethanol concentration at the commencement points of the steep increase in heat flux and the heat transfer coefficient for vapor velocity of 0.4 and 1.5 m/s. The steep increase in heat transfer is confirmed to have begun approximately when the temperature of the heat transfer surface reached the dew point line temperature. The surface subcooling at the commencement point of the steep increase almost coincided with the temperature difference between the boiling point line and the dew point line in the vapor-liquid equilibrium relation. However, the surface subcooling at the steep increase point is slightly higher than the temperature difference between the dew point and the bubble point in the vapor-liquid phase equilibrium, and is almost constant in the low-concentration range of less than 3% ethanol.

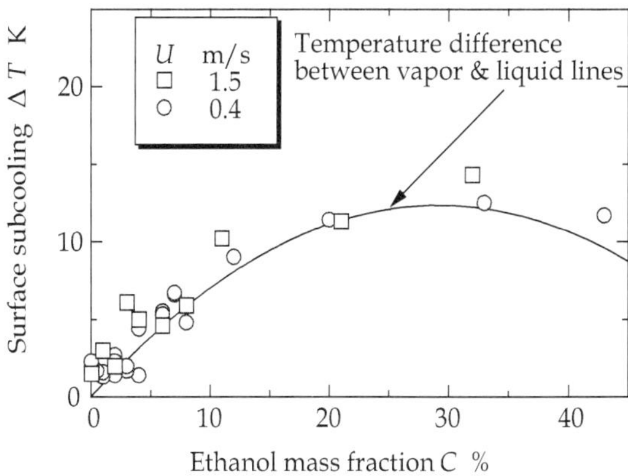

Fig. 15. Surface subcooling at the steep increase point of heat transfer

Figures 16 (a) and (b) show the changes in the maximum heat flux and heat transfer coefficient, respectively, with the condensing surface subcooling, the maximum values of which appear in the region of low ethanol concentration. When the vapor velocity was 0.4 m/s, a maximum heat flux of approximately 1.7 MW/m^2 appeared at an ethanol concentration of nearly $C = 6\%$. The surface subcooling at the maximum value of heat flux increased monotonically with respect to the vapor concentration, and the rate of increase was relatively high in the low-concentration range. When the vapor velocity was 1.5 m/s, a maximum heat flux could not be directly obtained due to insufficient cooling intensity and excessive heat flux. Assuming a similar trend in behavior as for the vapor velocity of 0.4 m/s, the maximum heat flux at vapor velocity of 1.5 m/s was estimated to be approximately

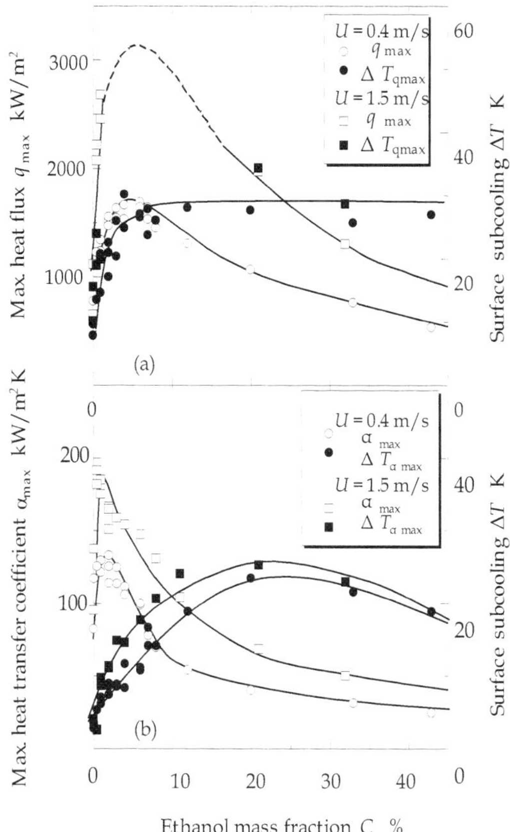

Fig. 16. Characteristics of maximum points of heat transfer (a) Maximum heat flux, (b) Maximum heat transfer coefficient

3 MW/m², as indicated by the dotted line in Fig. 15 (a). As shown in Fig. 15 (b), the maximum heat transfer coefficients appeared at an ethanol concentration of approximately C = 1%, and were 0.12 MW/m²K and 0.18 MW/m²K for vapor velocities of 0.4 m/s and 1.5 m/s, respectively. In addition, the subcooling at the maximum heat transfer coefficient showed some variation, with a peak that appeared at a vapor concentration of approximately 20%.

(b) Promotion of steam condensation by addition of ethanol

We examine the promotion of condensation heat transfer by adding ethanol to water. Figure 17 shows the variation of the condensation heat transfer coefficient of the vapor mixture normalized by that of pure steam with respect to the subcooling, for a vapor velocity of 0.4 m/s. For all vapor concentrations, the maximum shifted slightly from that of the characteristic curve toward larger subcooling. This is due to the lowering of the heat transfer coefficient for pure steam with the increase in subcooling. The condensation heat transfer was improved over almost the entire measured subcooling region for ethanol concentrations of less than approximately 6%. On the other hand, when the ethanol concentration of the vapor mixture was higher than 12%, the ratio of condensation heat transfer of the vapor mixture is smaller than that of steam in the low-subcooling region, but is higher in the subcooling region larger than that of the commencement point of steep increase. As discussed in section 3.4, the mutual relationship of the Marangoni driving force on the condensate and the diffusion resistance in the vapor phase in the condensation of binary vapor mixtures determines the condensation characteristics. The decrease in condensation heat transfer coefficient due to the diffusion resistance of the vapor-side is relatively small in the case of $C \leq 6\%$ for low ethanol concentration. In the case of $C \geq 12\%$, as the diffusion resistance is the controlling factor, the condensation heat transfer coefficient was reduced even in the dropwise mode.

Figure 18 shows the relationships between the ethanol concentration of the vapor mixture and the maximum value of the ratio of the heat transfer coefficient of the vapor mixture, α_{peak}, to steam, α_0, with respect to the surface subcooling. For an ethanol concentration of approximately 1%, the addition of ethanol improves the condensation heat transfer coefficient of the mixture, compared to steam, by a factor of approximately six at a vapor velocity of 0.4m/s, and by a factor of approximately eight at a vapor velocity of 1.5 m/s. In addition, for vapor concentrations of 0.05% and 0.1% (ethanol concentrations of the liquid mixture were 0.005% and 0.01%, respectively), which are very low, the ratios of the heat transfer coefficients reached 3.5~5.5, and the subcooling is in the relatively small temperature difference range of approximately 3~5 K. Thus, the addition of a very small amount of ethanol is expected to promote the heat transfer process for a comparatively small temperature difference and at sustained high heat transfer coefficient regardless of the nature of the condensation surface.

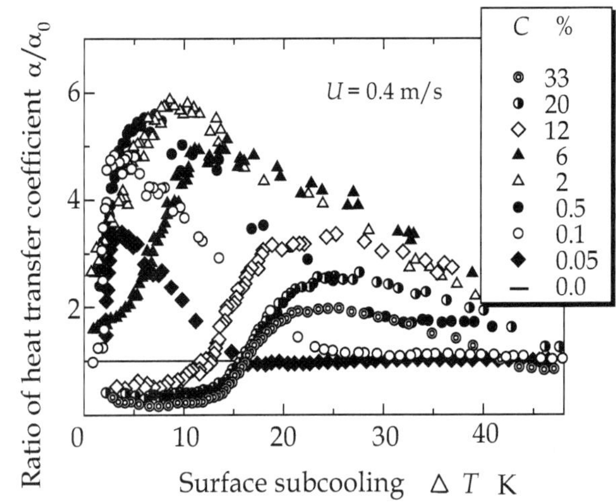

Fig. 17. Ratio of the condensation heat transfer coefficient of the mixture vapors to that of pure steam, with respect to surface subcooling

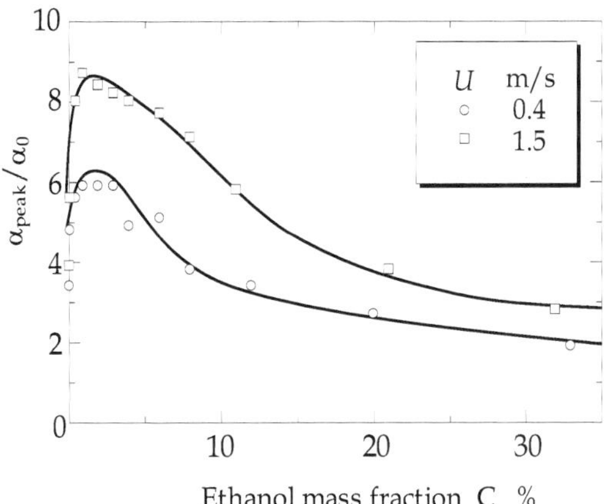

Fig. 18. Variation of the peak ratio of the condensation heat transfer coefficient of the mixture vapors to that of pure steam with respect to the ethanol concentration in the mixture vapors

5.2 Effect of vapor velocity

Condensation characteristic curves were measured for three ethanol mass fractions of water-ethanol vapor mixture c = 0.09, 0.32, 0.53. The results for each of the concentrations are shown in Figs. 19(a) and (b) for c = 0.09, 0.32 (Utaka & Kobayashi, 2003). The vapor velocity was varied from 0.3 to 45 m/s for c = 0.32 and 0.53 and from 0.3 to 68 m/s for c = 0.09. The region near the maximum heat flux could not be measured when the vapor velocity was over 18 m/s for c = 0.09 because of the limited cooling intensity. The condensation heat transfer coefficient increased remarkably throughout whole subcooling region with the increase in vapor velocity under all conditions. Moreover, the surface subcooling in which the maximum heat transfer coefficient appeared moved towards the smaller subcooling-side with the increase in vapor velocity. The maximal value of heat transfer coefficient is very high for the thinnest ethanol concentration of c = 0.09 and is about 200 kW/m2K for vapor velocity over 18 m/s. As describe before, thermal resistance of Marangoni condensation is composed of the resistance layers of diffusion in steam-side and the condensate. As shown in Table 1, with increasing surface subcooling, the dominant heat transfer resistances shifts from diffusion layer in the vapor phase to condensate layer. There are two effects of increase in vapor velocity, that is, weakening of the vapor-side resistance and condensate resistance due to the decrease in the largest droplet size in dropwise condensation region. Therefore, the characteristic measured from the above-mentioned experiment is related to these features. First, the increase in heat transfer coefficient with the vapor velocity is remarkable in the region where surface subcooling is relatively small. For example, the heat transfer coefficient increased by nearly tenfold at the steep increase point (the point B in Fig. 3) from 5 kW/m2K to 45 kW/m2K with the change in vapor velocity from 0.8 m/s to 45 m/s for c = 0.32. Therefore, the vapor velocity affects the heat transfer characteristic at the steep increase point which is mainly controlled by the diffusion resistance, because the effect of the vapor flow influences steam-side diffusion resistance. Next, the maximum heat trasfer coefficient increases nearly 2 times from 45 kW/m2K to 80 kW/m2K under the same condition. Similar tendencies are also found in the other two concentration conditions. In general, since the diffusion resistance in vapor layer is smaller than the condensate thermal resistance near the surface subcooling at the point of maximum heat transfer coefficient, and the condensate resistance is also smaller compared to that at smaller surface subcooling region because of the appearance of clear dropwise mode. Especially, when the vapor velocity increases, the decrease of condensate resistance caused by a reduction in the size of the largest drop on the heat transfer surface becomes remarkable and is similar to that in so-called dropwise condensation process on a lyophobic surface. With increased vapor velociity, maximum heat transfer coefficient point shifts to smaller subcooling. This is due to lower condensate surface temperature caused by reduced average condensate thickness.

Figures 20 (a) and (b) show the relations among the maximum heat transfer coefficient as a representative point showing the feature of condensation characteristic curve, the vapor velocity and the departing drop diameter. In this region, since thermal resistance of the condensate is the main factor over the vapor-side thermal resistance, it is worth comparing the effect of vapor velocity with the result (Tanasawa et al. 1973) of dropwise condensation on a lyophobic surface where the effect of drop conduction is important. While the relation between maximum heat transfer coefficient and vapor velocity in Marangoni condensation is similar for all of the conditions, those are greatly different from that of dropwise condensation. Thus, the difference in the dependecies of departing drop size to the vapor velocity observed between two phenomena is due not only to the droplet shape but also to the condition of drop peripheral. That is, while there is a three phase interface of solid, vapor and liquid in dropwise condensation on a lyophobic surface, a continuous condensate film exists among the drops so that there is no boundary between drop and condensate film in Marangoni dropwise condensation. Figure 20 (b) shows the change of the maximum heat transfer coefficient against the departing drop diameter. The line fitted by using least square method in the figure was applied to the above measuring points for vapor velocity greater than 0.8 m/s. Since each gradient of the curves almost agrees with that for dropwise condensation on a lyophobic surface, it can be seen that the effect of departing drop diameter on the heat transfer coefficient is similar to each other. Similar trends of dependencies on departing drop diameters both for the maximum heat transfer coefficient in Marangoni condensation and the heat transfer coefficient of dropwise condensation is considered here.

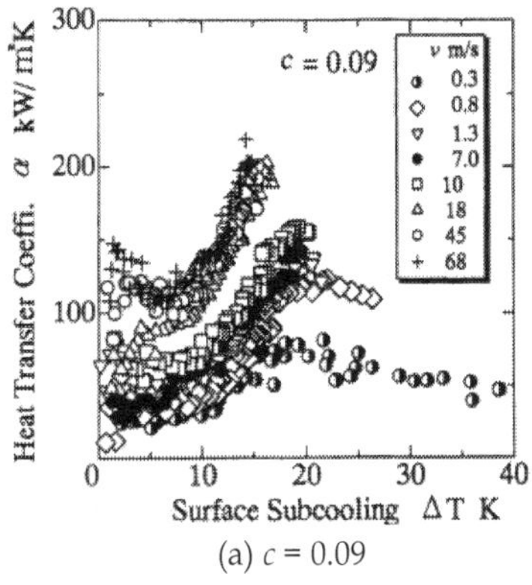

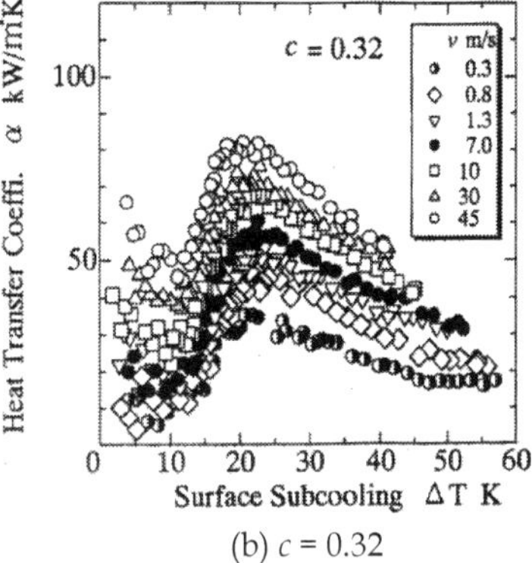

(a) c = 0.09   (b) c = 0.32

Fig. 19. Condensation characteristic curves

It has been clarified that comparatively large drops on condensing surface in dropwise condensation on a lyophobic surface works as a hindrance to heat transfer and that the major part of heat transfer occurs through small drops. Therefore, considering that the condensate resistance is dominant and that the dropwise mode appears at the point of maximum heat transfer coefficient, it can be understood why the exponents agree with one another because relatively large drops work as heat resistance and the thin condensate between the dorps are taking on the major heat transfer.

5.3 Effect of non-condensing gas

In order to realize the lowest non-condensing gas concentration, the leak-tight vapor loop, which is shown in Fig. 5, was first sealed off from the atmosphere (Wang & Utaka, 2005). In order to change the non-condensing gas concentration, a structure in which nitrogenous gas could be added to the vapor mixture as non-condensing gas was adopted. The amount and the stability of the added non-condensing gas were observed by measuring the flow of nitrogenous gas using a flow meter.

Figures 20 (a) and (b) shows the condensation characteristic curves of heat transfer coefficient under ethanol mass fractions of the vapor mixture of 0.01 and 0.45, respectvely. The condensation heat transfer performance improves with the decrease in the concentration of the non-condensing gas for vapor mixtures of all ethanol concentrations. In particular, the condensation heat transfer performance of the vapor mixture can be improved by further reducing the concentration of the non-condensing gas, even in cases of very small non-condensing gas concentration.

As shown in Fig. 3, Marangoni condensation characteristic curves can be divided into four domains: the domain controlled by the diffusion resistance of the vapor-side, the rapid decrease domain of the diffusion resistance, and the domains controlled by dropwise and filmwise condensations. For each domain, since the main factor governing the heat transfer is different, the degree of the effect of the non-condensing gas on the condensation heat transfer should also be different.

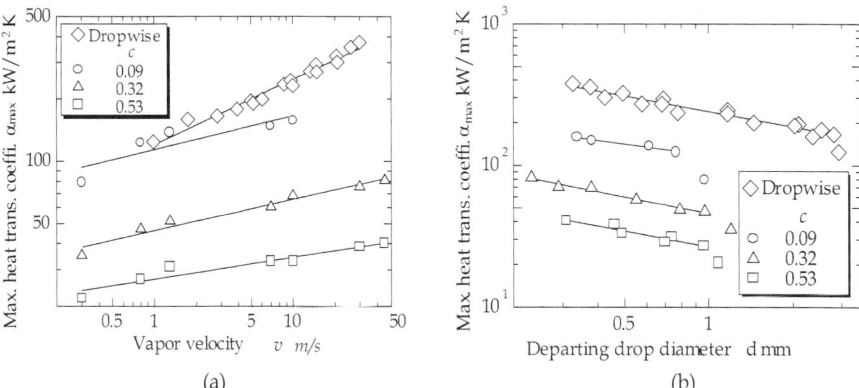

Fig. 20. Heat transfer characteristics at point of maximum heat transfer coefficient;
(a) Variation of maximum heat transfer coefficient against vapor velocity; (b) Variation of maximum heat transfer coefficient against departing drop diameter

First, in the domain controlled by the diffusion resistance of the vapor-side (domain A-B in Fig. 3), the diffusion resistance of the vapor-side is large and the amount of condensate is comparatively small due to the small heat flux. Therefore, the effect of the non-condensing gas is comparatively small. Particularly, for the case in which the vapor ethanol concentration is high (Fig. 21 (b)), this tendency is clear. Next, in the rapid decrease domain of the diffusion resistance (domain B-C), as the diffusion resistance of the vapor-side decreases rapidly while the condensate is maintained dropwise, the condensation heat flux increases suddenly. Corresponding to this change, the effect of the non-condensing gas increases suddenly. As seen in Fig. 21, with the increase of the non-condensing gas concentration, the condensation characteristic curve showed the tendency to decrease from the commencement point of the steep increase, and the drop became large with the increase in surface subcooling. Moreover, in the domain controlled by the dropwise condensation (domain C-E), since the condensate maintains the dropwise aspect and the diffusion resistance on the vapor-side is small, the heat transfer coefficient showed was high. Hence, the domain is a region in which the effect of non-condensing gas on Marangoni condensation appears easily. These figures show that the decrease in heat transfer coefficient or heat flux due to the existence of the non-condensable gas becomes remarkable around the peak value of the condensation characteristic curve, and is comparatively large. For example, in the case of the ethanol concentration in vapor mixture of $C_e = 0.01$, when the non-condensable gas concentration was increased from 15×10^{-6} to 494×10^{-6}, the peak value of the heat transfer coefficient was reduced by about 35% and the peak value of the heat flux was reduced by about 40%. Even at other vapor ethanol concentrations, the decrease of the peak value of the heat transfer coefficient or the heat flux due to the increase in the non-condensing gas concentration to the degree mentioned above is between approximately 30%-50%. The reasons for this are as follows. 1) In this domain, since the condensation heat flux is comparatively high, the non-condensing gas accumulates easily, therefore the diffusion resistance layer of the non-condensing gas form easily. 2) In this domain, both the heat transfer resistance in the condensate and the diffusion resistance due to the low-boiling-point component in the vapor mixture are small. In other words, the diffusion resistance caused by the non-condensing gas becomes relatively large in comparison with other heat

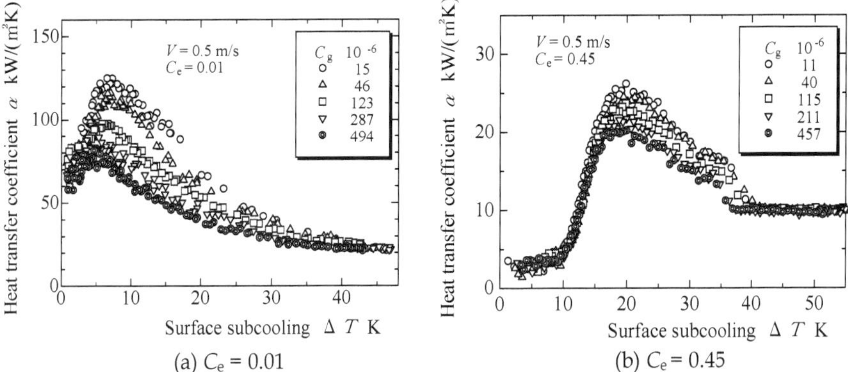

(a) $C_e = 0.01$ (b) $C_e = 0.45$

Fig. 21. Effect of non-condensing gas in Marangoni condensation curve transfer resistance in the condensation process. Furthermore, for this domain, as the driving force that makes the condensate irregular, the temperature difference formed on the surface of the condensate decreases again after an increase and gradually approaches zero (the condensate becomes filmwise). As a result, the variation of the heat transfer resistance in the condensate with respect to the surface subcooling having a tendency to decrease in the initial stage and increase again after reaching the minimum around the point of peak heat transfer coefficient. Hence, the effect of non-condensing gas on Marangoni condensation becomes greatest around the subcooling at which the peak heat transfer coefficient. Finally, in the domain controlled by the filmwise condensation (the area to the right of point E), as the condensate becomes filmwise, the diffusion resistance formed by the non-condensing gas and the low-boiling-point component in the vapor mixture is relatively small in the condensation process, the effect of the non-condensable gas is not readily apparent.

These figures also indicate that the effect of the non-condensing gas is different under different ethanol concentrations of vapor mixture. Specifically, in the case of the low ethanol concentration of vapor mixture, the heat flux characteristic curve of which has a strongly nonlinear shape, the nonlinearity became weak and the heat flux peak becomes indistinct with the increase in non-condensing gas concentration. The effect of the non-condensable gas becomes stronger because the diffusion resistance of the low-boiling-point component in the condensation process is small when the ethanol concentration of the vapor mixture is low. On the other hand, in the case of the larger ethanol concentration, the effect of non-condensing gas is weaker because of the existence of larger vapor-side heat transfer resistance due to much non-volatile (ethanol) concentration as seen in Fig. 21 (b).

6. Concluding remarks

The mechanisms and heat transfer characteristics of Marangoni condensation phenomena, which show the extremely high heat transfer coefficient in the dropwise condensation regime and occur due to surface tension instability of the condensate in the condensation of a binary vapor mixture of a positive system, are investigated mainly on the basis of the researches of author's group. That is, the general description of mechanisms and characteristics of Marangoni condensation phenomena, Marangoni dropwise condensation cycle, experimental apparatus and methods common in measurements, factors concerning the mechanisms of marangoni condensation (Relation between initial drop distance and heat transfer, condensate film thickness, and vapor-side diffusion characteristics), and the heat transfer characteristics of Marangoni condensation (effects of surface subcooling, vapor concentration, vapor velocity and non-condensing gas) were discussed.

7. References

Ford, J.D. and Missen, R.W. (1968). On the Conditions for Stability of Falling Films Subject to Surface Tension Disturbances; the Condensation of Binary Vapor, *Can. J. Chem. Eng.*, Vol. 48, pp. 309-312.

Fujii, T., Osa, N., Koyama, S. (1993). Free Convective Condensation of Binary Vapor Mixtures on a Smooth Horizontal Tube: Condensing Mode and Heat Transfer Coefficient of Condensate, *Proc. US Engineering Foundation Conference on Condensation and Condenser Design*, St. Augustine, Florida, ASME, pp. 171-182.

Hijikata, K., Fukasaku, Y., Nakabeppu, O. (1996). Theoretical and Experimental Studies on the Pseudo-Dropwise Condensation of a Binary Vapor Mixture, *Journal of Heat Transfer*, Vol. 118, pp. 140-147.

Hovestreijdt, J. (1963). The Influence of the Surface Tension Difference on the Boiling of Mixture, *Chem. Eng. Sci.*, Vol. 18, pp. 631-639.

Mirkovich, V.V. and Missen, R.W. (1961). Non-Filmwise Condensation of Binary Vapor of Miscible Liquids, *Can. J. Chem. Eng.*, Vol. 39, pp. 86-87.

Morrison, J.N.A. and Deans, J. (1997). Augmentation of Steam Condensation Heat Transfer by Addition of Ammonia, *International Journal of Heat and Mass Transfer*, Vol. 40, pp. 765-772.

Murase, T., Wang, H.S., Rose, J.W. (2007). Marangoni condensation of steam-ethanol mixtures on a horizontal tube, *International Journal of Heat and Mass Transfer*, Vol. 50, pp. 3774-3779.

Tanasawa, I., Ochiai, J., Utaka, Y. and Enya, S. (1976). Experimental Study on Dropwise Condensation Process (Effect of Departing Drop Size), *Trans. JSME*, Vol. 42, No. 361, pp. 2846-2853.

Utaka, Y. and Terachi, N. (1995a). Measurement of Condensation Characteristic Curves for Binary Mixture of Steam and Ethanol Vapor, *Heat Transfer-Japanese Research*, Vol. 24, pp. 57-67.

Utaka, Y. and Terachi, N. (1995b). Study on Condensation Heat Transfer for Steam-Ethanol Vapor Mixture (Relation between Condensation Characteristic Curve and Modes of Condensate), *Transactions of Japan Society of Mechanical Engineers*, Series B, Vol. 61, No. 588, pp. 3059-3065.

Utaka, Y., Kenmotsu, T., Yokoyama, S. (1998). Study on Marangoni Condensation (Measurement and Observation for Water and Ethanol Vapor Mixture), *Proceedings of 11th International Heat Transfer Conference*, Vol. 6, pp. 397-402.

Utaka, Y. and Kobayashi, H. (2003). Effect of Vapor Velocity on Condensation Heat Transfer for Water-Ethanol Binary Vapor Mixture, *Proceedings of 6th ASME-JSME Thermal Engineering Conference*.

Utaka, Y. and Nishikawa, T. (2003a). An Investigation of Liquid Film Thickness during solutal Marangoni Condensation Using a Laser Absorption Method (Absorption Property and Examination of Measuring Method) , *Heat Transfer – Asian Research*, Vol.32, No.8, pp.700-711.

Utaka, Y. and Nishikawa, T. (2003b). Measurement of Condensate Film Thickness for Solutal Marangoni Condensation Applying Laser Extinction Method, *Journal of Enhanced Heat Transfer*, Vol. 10, No. 1, pp. 119-129.

Utaka, Y. and Wang, S. (2004). Characteristic Curves and the Promotion Effect of Ethanol Addition on Steam Condensation Heat Transfer, *International Journal of Heat and Mass Transfer*, Vol. 47, pp. 4507-4516.

Wang, S. and Utaka, Y. (2005). An Experimental Study on the Effect of Non-condensable Gas for Solutal Marangoni Condensation Heat Transfer, *Experimental Heat Transfer*, Vol. 18, No. 2, pp. 61-79.

Part 3

Heat Transfer Phenomena and Its Assessment

13

Quantitative Visualization of Heat Transfer in Oscillatory and Pulsatile Flows

Cila Herman
Department of Mechanical Engineering, The Johns Hopkins University
USA

1. Introduction

Oscillatory and pulsatile flows arise in a variety of engineering applications as well as in nature. Typical examples include blood flow, breathing, flow in some pipe systems, acoustic systems, etc. Typical engineering applications include enhancement of heat transfer and species transport, chemical species separation, flow control, flow velocity measurement equipment calibration and biomedical applications. Very often these flows are accompanied by heat or mass transfer processes. During the past four decades numerous studies have addressed issues specific for **purely oscillatory** and **modulated or pulsatile flows** (oscillatory flow superimposed on a mean steady flow). Recent advances in the study of oscillatory flows were reviewed by Cooper et al. (1993) and Herman (2000). A better understanding of these flows and the accompanying heat transfer processes is essential for the proper design of equipment for such processes and physical situations. Both the experimental study and the computational modeling of oscillatory and pulsatile flows pose specific challenges, which will be addressed in this chapter, with the emphasis on quantitative experimental visualization using holographic interferometry.

Experimental visualization of oscillatory and pulsatile flows and heat transfer requires non-invasive measurement techniques, to avoid affecting the investigated process. In this chapter we discuss holographic interferometry (HI) as a powerful tool in the quantitative visualization of oscillatory and pulsatile flows and heat transfer. Two situations will be considered to demonstrate the applications of the method: (i) the study of **self-sustained oscillatory flows** and the accompanying heat transfer in grooved and communicating channels and the study of (ii) **oscillatory flow and heat transfer in the stack region of thermoacoustic refrigerators.**

In this chapter we introduce holographic interferometry as an experimental technique that simultaneously renders quantitative flow and heat transfer data. We demonstrate that for a certain class of problems HI is superior to conventional flow visualization techniques, such as tracer methods or dye injection, since it can provide not only qualitative but also quantitative insight into certain types of unsteady flows and it does not require the seeding of the flow. Several types of flows and heat transfer processes amenable for quantitative evaluation will be analyzed in the paper. We begin the discussion by introducing the experimental apparatus and technique, followed by the description of the investigated physical situation. Next, we present, based on three case studies, visualized temperature fields along with numerous examples of how quantitative data can be extracted from interferometric visualization images. Data reduction procedures, image processing tools, experimental uncertainties as well as advantages and limitations of the method are explained.

2. Real-time holographic interferometry for quantitative visualization of fluid flow and heat transfer

Holographic Interferometry (HI) is a well-established measurement and visualization technique widely used in engineering sciences (Vest, 1979; Hauf and Grigull, 1970). In transparent fluids it visualizes refractive index fields, which are related to fluid properties, such as temperature, pressure, species concentration, as well as density in compressible flows. Optical measurement techniques have virtually no "inertia", therefore they are ideal tools for investigation of high-speed, unsteady processes. The combination of HI and high-speed cinematography (that allows high spatial resolutions) is used in the present study to visualize oscillatory or pulsatile flows.

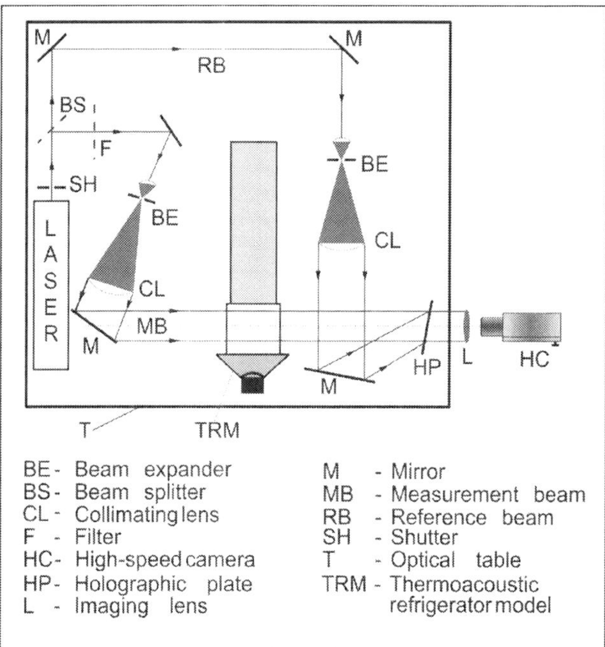

Fig. 1. Optical arrangement for holographic interferometry

2.1 Optical arrangement for holographic interferometry

Holographic interferometry uses light as information carrier to provide both qualitative (visual) insight and quantitative data on the investigated physical process. In convective heat transfer measurements the temperature fields in the thermal boundary layer above the heated surface and in the transparent working fluid are of particular interest. For more details about the technique, the reader is referred to the comprehensive literature on this topic. General information can be found in the publications such as those of Vest (1979) and Mayinger (1994). Specific information on the optical setup used in our studies can be found in the descriptions by Amon et al. (1992).

A standard **optical arrangement** for HI is presented in Fig. 1. The light source is a laser. In our research we used both a 25 mW Helium-Neon and a 1 W Argon-Ion laser. The laser power required for analyzing a particular physical process depends on the speed of the process, i.e. the highest frequency of oscillations in oscillatory and pulsatile flows, as well as the sensitivity of the film or digital sensor used to record the high-speed image sequence, in order to be able to resolve the smallest time scales of interest. The type and wavelength of the laser determine the choice of holographic and film materials for highest sensitivity, resolution and contrast, which is especially critical in high-speed applications.

For imaging by HI, the laser beam is divided into a reference beam, RB, and an object beam, OB, by means of a, usually variable, semitransparent mirror (beam splitter), BS, as shown in Fig. 1. Both beams are then expanded into parallel light bundles by a beam expander, BE, which consists of a microscope objective, a spatial filter and a collimating lens. The object beam passes through the test section, TS, with the phase object (representing the refractive index field to be visualized and related to temperature, concentration or density in the evaluation phase) and then falls on the holographic plate, H. The reference beam falls directly onto the holographic plate. The photograph of the optical arrangement for holographic interferometry in the Heat Transfer Lab of the Johns Hopkins University with the thermoacoustic refrigerator model mounted on the optical table is displayed in Fig. 2.

Fig. 2. Photograph of the optical arrangement for holographic interferometry at the Heat Transfer Lab of the Johns Hopkins University and the thermoacoustic refrigerator model mounted on the optical table.

2.2 Visualization of temperature fields: infinite and finite fringe field arrangements

HI allows the visualization and analysis of high-speed, transient phenomena by using the **real-time method**, which is a single exposure technique. The visualization is carried out in two steps. First, the reference state (usually with the fluid in the measurement volume at ambient temperature) is recorded on the holographic plate. Next, the holographic plate is developed, bleached, dried and exactly repositioned into a precision plate holder. In the second step, the reference state of the object under investigation is reconstructed by illuminating the holographic plate with the reference beam. At the same time, the investigated physical process is initiated (in our experiment the blocks in the wind tunnel are heated or the thermoacoustic refrigerator is activated). The heating causes the refractive index of the fluid in the measurement volume to change, and, consequently, this causes the object wave to experience a phase shift on its way through the test section. The difference between the reference state recorded earlier and the new state of the fluid in the measurement volume, i.e. the phase shift between reference and measurement beams, is visualized in the form of a macroscopic interference fringe pattern. This fringe pattern can be recorded with a photographic camera or a high-speed camera (when the process is unsteady).

If, during the measurement, the object wave is identical to the original state for which the reference hologram was recorded (the object is unheated, for example), no interference fringes will appear. This state can be adjusted before initiating the experiment, and the corresponding method of reconstruction is called the **infinite fringe field alignment.** The infinite fringe field alignment was used in all measurements reported in this paper. When the heat transfer process is initiated, the object wave passing through the test section becomes distorted, and behind the hologram the object and reference waves interact to form a macroscopic interference pattern. In our study we record this fringe pattern with a high-speed camera with speeds of up to 10,000 image frames per second. It is desirable to record around 10 images or more during one period of oscillations to achieve good reconstruction accuracy. The interferometric fringes obtained using the infinite fringe field alignment correspond to isotherms, and are suitable, apart from the fairly common temperature measurements, also for the quantitative visualization of fluid flow phenomena, which will be demonstrated in this chapter.

Another alignment of the optical equipment frequently used in interferometric measurements is the **finite fringe field alignment.** In this method a small tilt is applied to the mirror M in Fig. 1 that projects the reference beam onto the holographic plate. At ambient conditions this tilt will cause a regular, parallel macroscopic fringe pattern to form in the field of view. Our experience indicates that the finite fringe field alignment is less suitable for quantitative flow visualization, since the fringe patterns cannot be easily and intuitively related to the flow field. The finite fringe field alignment is frequently used when temperature gradients on the heated surface are measured (rather than temperatures). An example contrasting images obtained by the infinite and finite fringe field alignments is shown in Fig. 3. Both interferometric images visualize temperature fields around two heated stack plates in crossflow. The thermal boundary layers can be identified in both alignments, the fringes in the infinite fringe field alignment visualize the isotherms in the thermal boundary layer.

In heat transfer measurements, that were the original and primary objective of our investigations, high spatial resolutions are required to analyze the thin thermal boundary layers in the vicinity of the heated surface in forced convection, such as those shown in the bottom image of Fig. 3. In order to achieve sufficient accuracy in heat transfer measurements, the interferometric images in our experiments were recorded on 16 mm high-speed film first, then scanned, digitized with resolutions up to 2700 dpi, and finally evaluated quantitatively using digital image processing techniques. The cost of video equipment suitable for these high-speed heat transfer measurements would have been prohibitive in addition to unsatisfactory spatial resolution, since the resolution of images recorded by digital video cameras decreases with increasing recording speed. Depending on the thickness of the thermal boundary layer and the refractive index of the working fluid, tens to hundreds of fringes may need to be resolved accurately over a distance of few millimeters. In flow visualization experiments the spatial resolution is less critical than in temperature measurements, since the fringes relevant for the characterization of flow phenomena in the main channel and recirculating regions are wider than in the thermal boundary layer.

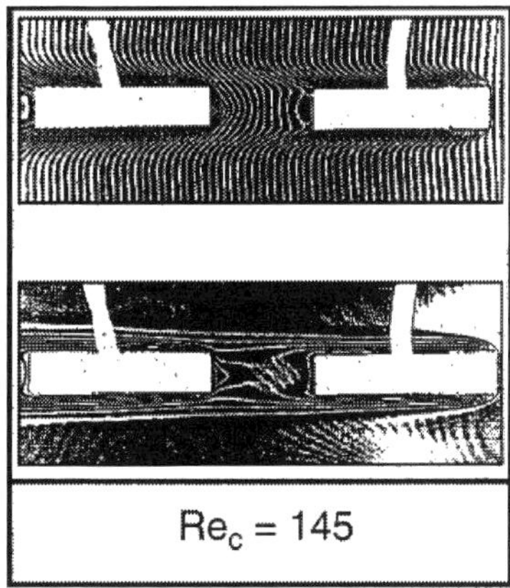

Fig. 3. Temperature fields around two heated plates in crossflow in a rectangular channel visualized by HI using the finite fringe field arrangement (top) and the infinite fringe field arrangement (bottom).

3. Physical situations

Oscillating flows can be classified according to the method used to generate the oscillations, flow geometry, role of compressibility, character of the undisturbed flow as well as other flow parameters that will influence the development of the flow field and the heat transfer process. The impact of self-sustained oscillation in grooved and communicating channels as well as the impact of acoustic oscillations on convective heat transfer and the role of compressibility in the stack of a thermoacoustic refrigerator will be addressed in this paper.

3.1 Self-sustained oscillatory flows in grooved and communicating channels

Self-sustained oscillatory flows in grooved and communicating channels were visualized in wind tunnels specially designed to allow accurate measurements by HI using air as the working fluid. The length of the path of light across the heated region is a critical design parameter that determines the number of fringes present in the interferometric image for a prescribed temperature difference.

During the past four decades the development of compact heat transfer surfaces has received considerable attention in the research community. It was found that oscillation of the driving flow is a promising approach to heat transfer augmentation (Ghaddar et al. 1986a and 1986b). Resonant heat transfer enhancement is a passive heat transfer enhancement technique, which is appropriate for systems with naturally occurring separated flows, such as the grooved and communicating channels shown in Fig. 4. Grooved channels are typically encountered in electronic cooling applications (the heated blocks represent electronic chips mounted on a printed circuit board) and communicating channels represent a model of the rectangular plate fin, offset-fin and offset strip-fin flow passages of compact heat exchangers as well as heat sinks used in electronic packaging solutions.

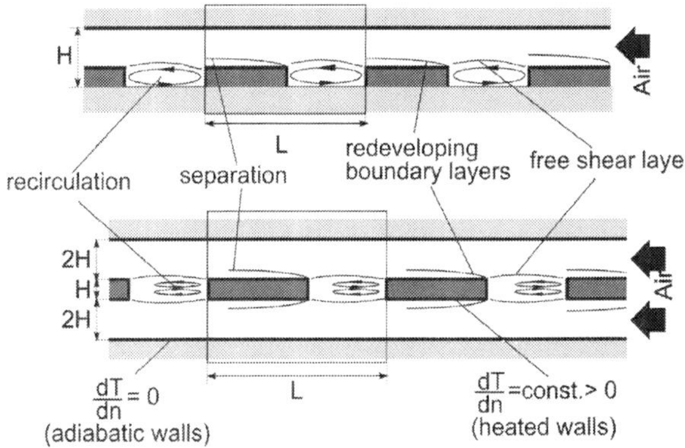

Fig. 4. Schematic of the geometries and physical situations for the study of self-sustained oscillatory flows in grooved (top) and communicating (bottom) channels

The enhanced surfaces we investigated involve the repeated formation and destruction of thin thermal boundary layers by interrupting the heat transfer surface in the streamwise direction, as shown in the schematic in Fig. 4. In addition to their practical significance, the two situations presented in Fig. 4 are examples of separated shear flows featuring complex interactions between separated vortices, free shear layers and wall bounded shear layers. In both channel geometries, two main flow regions can be identified: (i) the bulk flow in the main channel and (ii) the weak recirculating vortex flow in the groove or communicating region. They are separated by a free shear layer. In laminar, steady-state conditions there is virtually no exchange of fluid between these two regions. The results of Patera and Mikic (1986), Karniadakis et al. (1987), Greiner et al. (1990) and Greiner (1991) showed that self-sustained oscillations develop in such flow configurations at a relatively low Reynolds number in the transitional regime, and the interaction of separated flow with imposed unsteadiness leads to lateral convective motions that result in overall transport enhancement.

The heated blocks attached to the bottom of the grooved channel and the plates in the central plane of the communicating channels were heated electrically. The thermal boundary conditions on the surface of the heated blocks are described by constant heat flux, as indicated in Fig. 4. The top and bottom plane walls of the test section are manufactured of low thermal conductivity material to maintain approximately adiabatic thermal boundary conditions. Details on the experimental setup and instrumentation are available elsewhere (Amon et al. 1992; Farhanieh et al., 1993; Kang, 2002).

In both channels, above a critical Reynolds number and at sufficient downstream distance, a periodically, fully developed flow regime is established. This was the region of interest in our visualization experiments, since the instantaneous velocity and temperature fields repeat periodically in space. Therefore temperature fields were visualized in the region of the ninth heated block (there were 11 blocks in the test section), sufficiently far downstream from the channel entrance to satisfy the periodicity requirement. The channel height to spanwise dimension aspect ratio is selected to ensure that the flow and temperature fields investigated by HI are two-dimensional.

3.2 Oscillatory flow in a thermoacoustic refrigerator

HI can also be applied to visualize time dependent temperature distributions in oscillating flows with zero mean velocity. A need for such measurements arose in the investigations of heat transfer in thermoacoustic refrigerators. Thermoacoustic refrigeration is a new, environmentally safe refrigeration technique that was developed during the past two decades (Wheatley et al. 1983; Swift 1988). The schematic of a thermoacoustic refrigerator is presented in Fig. 5. The purpose of the acoustic driver is to generate an acoustic standing wave in the resonance tube. Thus, the working fluid in the resonance tube oscillates with zero mean velocity.

Over the past decades environmental concerns have become increasingly important in the design and development of energy conversion and refrigeration systems. Thermoacoustic energy conversion was introduced into engineering systems during the past four decades as a new, alternative, environmentally safe energy conversion technology. It uses noble gases and mixtures of noble gases as working fluids rather than hazardous refrigerants required for the vapor compression cycle. A thermoacoustic system can operate both as a prime mover/engine, when a temperature gradient and heat flow imposed across the stack leads to the generation of acoustic work/sound in the resonator. When reversing the thermodynamic cycle, the thermoacoustic system functions as a refrigerator: acoustic work is used to pump heat from the low temperature reservoir to release it into a higher temperature ambient. Heat transfer in the stack region of the thermoacoustic refrigerator was the focus of our visualization experiments.

The schematic of a half-wavelength thermoacoustic refrigerator is shown in Fig. 5. Energy transport in thermoacoustic systems is based on the thermoacoustic effect. Using an acoustic driver, the working fluid in the resonance tube is excited to generate an acoustic standing wave. When introducing a stack of plates of length Δx at a location specified by x_c into the acoustic field, a temperature difference ΔT develops along the stack plates. This temperature difference is caused by the thermoacoustic effect. The thermoacoustic effect is visualized in our study using high-speed holographic interferometry. In HI both temperature and pressure variations impact the refractive index and they are both present in our thermoacoustic system. Therefore, temperature variations need to be uncoupled from pressure variations in our evaluation process, to accurately quantitatively visualize the oscillating temperature fields around the stack plate.

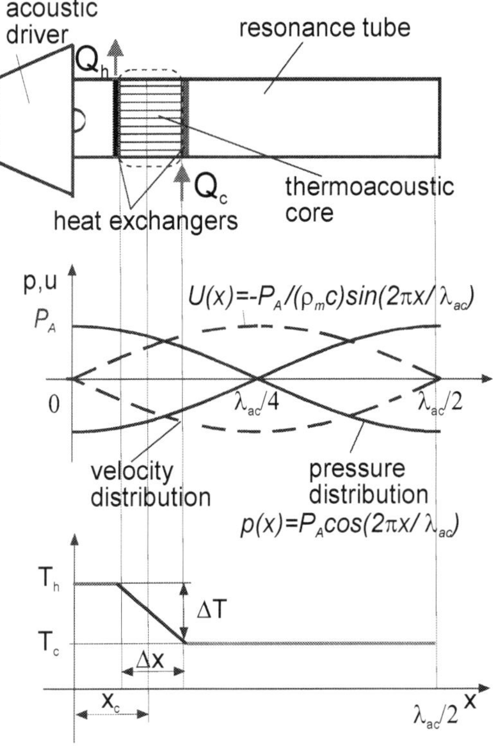

Fig. 5. Schematic of the thermoacoustic refrigerator, pressure and velocity distributions in the resonance tube and temperature distribution along the stack and in the resonator.

In the top portion of Fig. 5, a schematic of a thermoacoustic refrigerator is shown. The length of the resonance tube in the study corresponds to half the wavelength of the acoustic standing wave, $\lambda_{ac}/2$. The corresponding pressure and velocity distributions are displayed in the middle image in Fig. 5. A densely spaced stack of plates of length Δx is introduced at a location specified by the stack center position x_c into the acoustic field. During the operation of the refrigerator a temperature difference ΔT develops along the stack plates (bottom image in Fig. 5). By attaching heat exchangers on the cold and hot ends of the stack, heat Q_c can be removed from a low temperature reservoir, pumped along the stack plate to be delivered into the high temperature heat exchanger and ambient as Q_h. The temperature difference forming along the stack is caused by the thermoacoustic effect. This paper focuses on the visualization of the oscillating temperature fields in the thermoacoustic stack near the edge of the stack plates, which allows the visualization of the thermoacoustic effect.

The mechanism of thermoacoustic heat pumping (Swift, 1988) is illustrated in the schematic in Fig. 6, by considering the oscillation of a single gas parcel of the working fluid along a stack plate. The gas parcel begins the cycle at a temperature T. In the first step, the gas parcel moves to the left, towards the pressure antinode, the movement caused by the acoustic wave. During this displacement it experiences adiabatic compression, which causes its temperature to rise by two arbitrary units to T^{++}(Step 1 in Fig. 6). In this state the gas parcel is warmer than the stack plate and irreversible heat transfer from the parcel towards the stack plate takes place (Step 2 in Fig. 6). The resulting temperature of the gas parcel after this heat loss step is T^+ . On its way back to the initial location the gas parcel experiences adiabatic expansion and cools down by two arbitrary units, to the temperature $T-$ (Step 3 in Fig. 6). At this state the gas parcel is colder than the stack plate and irreversible heat transfer from the stack plate towards the gas parcel takes place (Step 4 in Fig. 6). After these four steps the gas parcel has completed one thermodynamic cycle and reached its initial location and temperature T. At this point the cycle can start again. In this paper we visualize the oscillating temperature distributions near the edge of two stack plates to visualize the thermoacoustic effect.

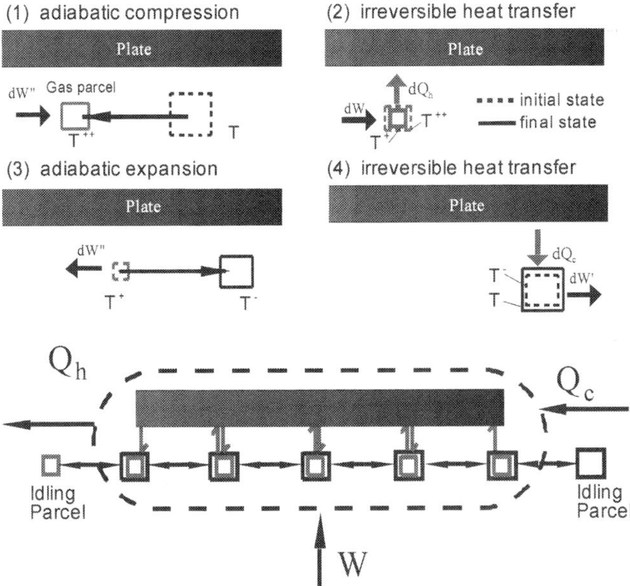

Fig. 6. Heat pumping cycle along a stack plate by considering the oscillation of one gas parcel. The four steps of the cycle are: 1: $T \rightarrow T^{++}$ adiabatic compression, 2: $T^{++} \rightarrow T^+$ irreversible heat transfer, 3: $T^+ \rightarrow T^-$ adiabatic expansion and 4: $T^- \rightarrow T$ irreversible heat transfer.

There are many gas parcels subjected to this thermodynamic cycle along each stack plate, and the heat that is delivered to the plate by one gas parcel is transported further by the adjacent parcel, as illustrated in the bottom portion of Fig. 6. The result of this transport is a temperature gradient developing along the stack plates. The described cycle can also be reversed by imposing a temperature gradient ΔT along the stack plates. In this situation the directions of irreversible heat transfer and work flux are reversed, and the thermoacoustic device operates as a prime mover, also known as the thermoacoustic engine. Thermoacoustic prime movers can be used to generate acoustic work that can drive a thermoacoustic refrigerator, a pulse tube or a Stirling refrigerator. The advantage of this solution is a refrigeration system that does not require moving parts (especially at the low temperature). The combination of HI and high-speed cinematography (that allows the high spatial resolutions needed in our study) is used to visualize the thermoacoustic effect at the edge of the thermoacoustic stack plates. In the thermoacoustic resonator the temperature fields oscillate with a frequency of 337Hz. Therefore sampling rates of the order of 5,000 frames per second were needed to accurately resolve the temporal evolution of the physical process. At the same time high spatial resolutions of up to 2700 dpi were realized.

Fig. 7. Thermoacoustic refrigerator model used in the visualization experiments assembled on the optical table in the Heat Transfer Lab of JHU with the components for holographic interferometry.

In Figure 7, the photograph of the thermoacoustic refrigerator model used in the visualization experiments described in this paper is shown. To investigate temperature fields in the region of the stack and heat exchangers using HI, we built a thermoacoustic refrigerator model with a transparent stack region that allows the irradiation of the measurement volume with laser light. The loudspeaker (Electro Voice EVM-10M) was used to generate an acoustic standing wave in the resonance tube. The 337Hz input signal for the loudspeaker was generated by an HP 8116A function generator and amplified using a Crown DC-300A Series II amplifier, before being led to the loudspeaker. A dynamic pressure transducer (Sensym SX01) was mounted at the entrance of the resonance tube to measure the acoustic pressures. At this location the transducer measures the dynamic peak pressure amplitude P_A. The drive ratio in the system is defined as the ratio of peak pressure amplitude P_A to the mean pressure p_m within the working fluid, $DR \equiv P_A/p_m$, and it is determined from this measurement. Experiments were conducted for drive ratios ranging from 1% to 3%. The length of the resonance tube (510mm) matches half the wavelength of the acoustic standing wave $\lambda_{ac}/2$. Visualization experiments were carried out on two stack plates. The dimensions of the stack plates are: plate spacing 3mm, stack length 76mm and stack center position 127mm. To capture details of the movements and to resolve the unsteady temperature fields as a function of time, we used a high-speed film camera, which is capable of recording speeds up to 10,000 picture frames per second. This speed corresponds to a temporal resolution of 0.1ms.

When applying HI to the visualization of temperature fields in a thermoacoustic refrigerator, the experimenter faces the problem that the changes in the refractive index cannot be directly related to temperature changes. This is the case because acoustic pressure variations cannot be neglected in the evaluation. Therefore, we developed a new interpretation and evaluation procedure for the interferometric fringe pattern that allows accurate measurements of oscillating temperature fields by accounting for the effect of periodic pressure variations. For a complete description of the unsteady temperature distributions it was also necessary to include frequency and phase measurements into the evaluation procedure.

4. Reconstruction of temperature fields in the presence of pressure variations

The feature that makes HI a powerful measurement tool is the possibility to detect optical path length differences $\Delta\Phi$ between an object wave and a reference wave in the nanometer range. These path length differences $\Delta\Phi$ are the multiple S of the wavelength λ of laser light, and they can be visualized in form of interference fringes. The interference fringes can then, in the case of a transparent phase object *(Vest, 1979)*, be related to a difference Δn in the refractive index along the optical paths of the object and reference waves as follows

$$\Delta\Phi = S \cdot \lambda = \Delta\, n \cdot L \,. \tag{1}$$

Equation (1) holds when the refractive index along the direction of transillumination, along the phase object of the optical path length L, can be considered constant. In conventional applications of HI the difference Δn in the refractive index is easily related to a single field variable of interest. Such field variables are density or pressure in the case of aerodynamic applications, or temperature and concentration in heat and mass transfer applications of HI, respectively. Most applications of high-speed HI reported in the literature were limited to physical situations in which the difference Δn in the refractive index is caused by a single field variable.

In many heat and mass transfer processes, such as chemical reactions, the experimenter is interested in simultaneously measuring temperature T and species concentration C. In such a situation the difference $\Delta n(T,C)$ in the refractive index depends on two field variables, temperature as well as concentration, and Equation (1) can be written as *(Panknin, 1972)*

$$S \cdot \lambda = L \cdot \left[\frac{\partial\, n}{\partial\, T}\bigg|_{C} \cdot \Delta\, T + \frac{\partial\, n}{\partial\, C}\bigg|_{T} \cdot \Delta\, C \right]. \tag{2}$$

Equation (2) represents a more general form of fringe interpretation in the study of phase objects. The difficulty the experimenter is facing in the quantitative evaluation is that interference fringes cannot be easily interpreted as isotherms or lines of constant concentration. In order to determine both field variables, temperature and concentration, in such a situation, *Panknin (1977)* took advantage of the fact that the difference Δn in the refractive index also depends on the wavelength λ of the laser light. Therefore he used two lasers operating at two different wavelengths λ_i in his experiments, and applied Equation (2) two times to resolve the interference fringes for temperature as well as concentration.

In the case of acoustically driven flow in our thermoacoustic system we are facing a similar situation, since, in addition to the oscillating temperature field, periodic pressure variations are also present in the fluid. Consequently, the difference $\Delta n(T,p)$ in the refractive index depends on both field variables, temperature as well as pressure, and Equation (1) can be written as

$$S \cdot \lambda = L \cdot \left[\frac{\partial n}{\partial T}\bigg|_p \cdot \Delta T + \frac{\partial n}{\partial p}\bigg|_T \cdot \Delta p \right].$$

(3)

Again, Equation (3) shows that the interference fringes cannot directly be interpreted as isotherms or isobars. In the paper the impact of periodic pressure variations on the fringe interpretation is discussed first, and an evaluation formula that expands the applicability of HI to temperature measurements in the presence of pressure variations is introduced. This evaluation formula reduces to its conventional form when the pressure of the measurement state equals the pressure of the reference state. The drive ratio, DR, the ratio of the peak pressure amplitude to the mean pressure in the working fluid

$$DR = \frac{P_A}{p_m},$$

(4)

is the parameter describing the magnitude of the periodic pressure fluctuations. From Equation (3), it follows that additional information regarding the pressure variations is necessary to be able to resolve the interference fringes for temperature. This information can be derived from the acoustic field. We demonstrated that the interference fringes can be approximated as "quasi"-isotherms because the temperature of one interference fringe can vary, depending on the magnitude of the pressure variations, by up to 18K in the present study. We also demonstrate that the conventional evaluation formula can be applied only in cases when the experimenter is interested in time averaged temperature distributions. Furthermore, an analytical error function was derived and applied to correct the unsteady temperature distribution determined with the conventional evaluation formula. The result of this correction shows good agreement with a simple theoretical model.

The reconstruction of temperature fields in the presence of pressure variations involves a series of steps. After expanding the difference in the refractive index Δn appearing in Equation (1), we obtain the equation of ideal interferometry *(Vest, 1979; Hauf and Grigull, 1970)* in the form

$$S(x,y,t) \cdot \lambda = (n(x,y,t) - n_\infty) \cdot L,$$

(5)

that relates the interference order $S(x,y,t)$ to the refractive index field $n(x,y,t)$ of the measurement state. We note that these parameters are not only functions of the two spatial coordinates x and y, as in conventional applications of Equation (5), but also time dependent quantities, since we are investigating an unsteady process. The wavelength λ, the refractive index n_∞ of the reference state and the spanwise dimension (optical pathlength) L, are known constants, and therefore the refractive index field can be reconstructed from the interference fringe pattern. Equation (5) implies an averaging of the refractive index in the spanwise direction L (along the light beam). Therefore, the experimental setup has been designed to maintain the refractive index along the spanwise direction constant, and thus allow the experimenter to deal with a simplified 2D model of the physical process. To simplify the present discussion, we also neglect refraction of the laser beam and assume that the beam passes along a straight line through the test section. For gases with a refractive index $n \approx 1$, such as air, the Gladstone-Dale equation

$$\bar{r}(\lambda) = \frac{2}{3} \frac{1}{\rho\,(x,y,t)} (n(x,y,t) - 1)\,, \tag{6}$$

can be applied to relate the refractive index field $n(x,y,t)$ to the density $\rho\,(x,y,t)$ of the working fluid. The specific refractivity $\bar{r}(\lambda)$ appearing in Equation (6) is a material specific constant for a given wavelength λ of laser light. When comparing the measurement state to the reference state, the Gladstone-Dale equation can be written as

$$\frac{n(x,y,t) - 1}{n_\infty - 1} = \frac{\rho\,(x,y,t)}{\rho_\infty}\,. \tag{7}$$

Substituting Equation (7) into (5) we can relate the density field $\rho\,(x,y,t)$ of the measurement state to the interference order $S(x,y,t)$ as

$$\frac{\rho\,(x,y,t)}{\rho_\infty} = \frac{S(x,y,t)\cdot\lambda}{L\cdot(n_\infty - 1)} + 1\,. \tag{8}$$

The goal of this derivation is to find a relationship between the temperature field $T(x,y,t)$, describing the measurement state, and the interference order $S(x,y,t)$. Therefore, we substitute the ideal gas law in the form

$$\frac{\rho\,(x,y,t)}{\rho_\infty} = \frac{p(x,y,t)}{p_\infty}\cdot\frac{T_\infty}{T(x,y,t)}\,, \tag{9}$$

into Equation (8) to eliminate density and obtain the following evaluation formula

$$T(x,y,t) = \frac{T_\infty}{1 + a(T_\infty)\cdot S(x,y,t)}\frac{p(x,y,t)}{p_\infty}\,, \tag{10}$$

describing the temperature field as a function of the interference order $S(x,y,t)$ as well as the pressure $p(x,y,t)$. The evaluation constant $a(T_\infty)$ in Equation (10) is defined as

$$a(T_\infty) \equiv \frac{\lambda}{L\cdot(n_\infty - 1)} = \frac{2}{3}\frac{\lambda}{L}\frac{R\cdot T_\infty}{\bar{r}(\lambda)\cdot p_\infty}\,. \tag{11}$$

Equation (10) incorporates the expansion of the conventional evaluation formula for HI, when the changes in the refractive index are caused not only by temperature changes, but also by pressure variations. Since it has been shown that the temperature field $T(x,y,t)$ under these conditions is a function $f(S,p)$ of interference order $S(x,y,t)$ and pressure $p(x,y,t)$, it is easily understood that the interference fringes cannot be directly related to isotherms. Consequently, in order to apply Equation (10) to quantitatively reconstruct the temperature field $T(x,y,t)$, information on the interference order $S(x,y,t)$ and the measurement state's pressure $p(x,y,t)$ is required. The interference order $S(x,y,t)$ can be determined from the recorded interferometric images, which is not the case for the measurement state's pressure $p(x,y,t)$.

In conventional applications of HI this complication is avoided by conducting the experiments such, that the pressure p_∞ of the reference state equals the pressure $p(x,y,t)$ of the measurement state. Therefore, in such cases the pressure ratio $p(x,y,t)/p_\infty$ equals unity, and Equation (10) reduces to the evaluation formula used in conventional applications of HI in heat transfer measurements *(Hauf and Grigull, 1970)*. The infinite fringe field alignment offers an additional advantage in the application discussed in the present paper: images obtained through these experiments also contain important information about the flow field. Thus, qualitative and quantitative information about the temperature field as well as the flow field is obtained simultaneously.

In the case of an acoustically driven flow it is naturally not possible to maintain the pressure $p(x,y,t)$ of the measurement state constant. Therefore, we acquire the additional information needed to determine the temperature field $T(x,y,t)$, from the acoustic field. For an acoustic field the pressure variations can be expressed as

$$p(x,y,t) = p_m(x,y) + \delta\, p(x,y,t) . \tag{12}$$

Substituting Equation (12) into (10) and taking advantage of the fact that the mean pressure p_m of the acoustic field in our experiment corresponds to the pressure p_∞ of the reference state, we obtain

$$T(x,y,t) = \frac{T_\infty}{1 + a(T_\infty) \cdot S(x,y,t)}\left(1 + \frac{\delta\, p(x,y,t)}{p_m}\right) \tag{13}$$

In Equation (13) the term $\delta\, p(x,y,t)/p_m$ describes the periodic pressure fluctuations. For an acoustic field this pressure term is generally at least one to two orders of magnitude smaller than unity (in our measurements $[\delta\, p/p_m]_{max} = 0.03$ for a drive ratio of 3%). Thus, one is tempted to neglect this pressure term and treat the measurement state as having the constant pressure p_∞ of the reference state, to arrive at the conventional evaluation formula

$$T(x,y,t) = \frac{T_\infty}{1 + a(T_\infty) \cdot S(x,y,t)} . \tag{14}$$

However, such an approximation may or may not have a significant impact on the temperature measurements, depending on the measurement parameter the experimenter is interested in.

In acoustically driven flow the working fluid is adiabatically compressed and expanded. These compression and expansion cycles cause small temperature fluctuations $\delta\, T(x,y,t)$

around a mean temperature $T_m(x,y)$. Consequently, we can describe the temperature field $T(x,y,t)$, similar to the way we have done it with the pressure, as a linear combination of the mean temperature distribution $T_m(x,y)$ and a small temperature fluctuation $\delta\, T(x,y,t)$. Furthermore, we can assume a harmonic time dependence for both temperature and pressure fluctuations, such that the temperature field can be written as

$$T(x,y,t) = T_m(x,y) + T_A(x,y) \cdot e^{i\,\omega\, t} \tag{15}$$

and the pressure as

$$p(x,y,t) = p_m + p_A(x,y) \cdot e^{i\,\omega\, t} . \tag{16}$$

Next we will apply these considerations to emphasize two important aspects in the application of Equation (14) to temperature measurements in the presence of pressure variations: (i) the impact on fringe interpretation and (ii) the impact on measurements of the temperature field $T(x,y,t)$.

The advantage of Equation (14) when compared to Equation (13) is its simplicity: the temperature field $T(x,y,t)$ is described as a function $f(S)$ of the interference order $S(x,y,t)$ only. Consequently, we can assign a constant temperature T_{S_0} to each interference order S_0, and interpret the interference fringes as isotherms. In order to quantify the influence of pressure variations, we will now focus our attention on the temperature T_{S_0} that is assigned to an interference order S_0 through Equation (14). If we substitute the small pressure fluctuations of Equation (16) into Equation (13) and consider the interference order S_0 of one interference fringe to be constant, we obtain

$$T_{S_0}(x,y,t) = \frac{T_\infty}{1 + a(T_\infty) \cdot S_0}\left(1 + \frac{p_A(x,y)\,e^{i\omega\, t}}{p_m}\right). \tag{17}$$

Equation (17) confirms the conclusion that one interference fringe cannot be directly related to an isotherm, because its temperature is a function of the spatially as well as time dependent pressure fluctuation. However, if we average with respect to time over one period of oscillations we obtain

$$\overline{T_{S_0}} = \frac{1}{\tau}\int_0^\tau \frac{T_\infty}{1 + a(T_\infty) \cdot S_0}\left(1 + \frac{p_A(x,y)\,e^{i\omega\, t}}{p_m}\right)dt = \frac{T_\infty}{1 + a(T_\infty) \cdot S_0} . \tag{18}$$

Equation (18) represents the conventional evaluation formula identical to Equation (14) for one particular interference fringe with the interference order S_0. Thus, we can conclude that using Equation (14) in the presence of periodic pressure variations, can be interpreted as approximating the temperature of one interference fringe with its time averaged temperature value. Since in this case we are dealing with approximated, time averaged isotherms, we will call them "quasi"-isotherms. To emphasize this fact, let us consider the temperature changes an interference fringe experiences in a typical experiment of the present study. For this purpose let us assume that we can assign a time averaged temperature $\overline{T_{S_0}} = 300K$ to one particular interference fringe and that the experiment was conducted at a drive ratio of 3%. Thus, the pressure term will be $\delta p / p_m = 0.03 \cdot e^{i\,\omega t}$.

Substituting these values into Equation (17), we will find that the temperature of this interference fringe can oscillate by $\pm 9K$ around its time averaged value of 300K.

5. Visualized and reconstructed temperature fields

5.1 Grooved channel

In the grooved channel the flow velocity is described by the Reynolds number defined as $\mathrm{Re}_g \equiv \dfrac{u_m D_h}{\nu}$, where u_m is the mean air flow velocity averaged over a cross section at the channel inlet (corresponds to the dimensions of the cross section in the groove region). The critical Reynolds number for the onset of oscillations was found to be approximately $\mathrm{Re}_{g,crit}$ = 1320, a flow rate significantly below the value characteristic for the onset of turbulence. The temperature oscillations that mirror the flow structure confirm the existence of a natural frequency in the investigated channel geometry. After the onset of oscillations, significant mixing between groove and bulk flows is initiated and it contributes to heat transfer enhancement. Typical histories of temperature fields corresponding to one period of oscillations at two representative Reynolds numbers, Re_g = 1580 and 2890, are shown in Fig. 8.

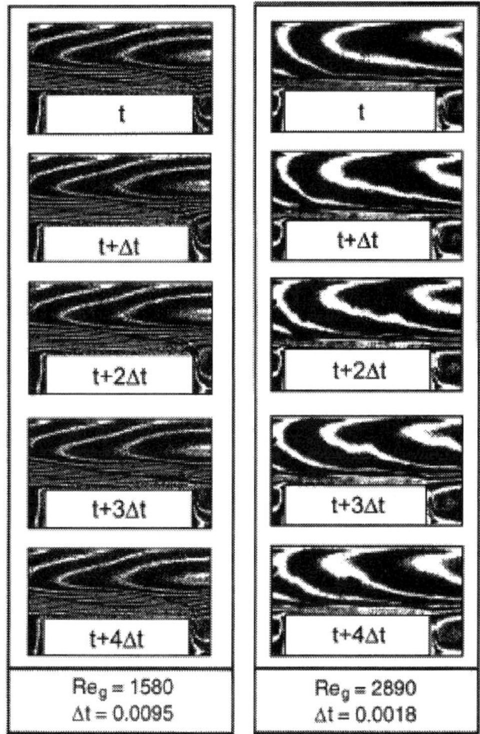

Fig. 8. Temperature fields in the grooved channel during a cycle of self-sustained oscillations at Re_g = 1580 and Re_g = 2890.

In the interferometric images shown in Fig. 8 two dominant oscillatory features were observed. The first feature is characterized by traveling waves of different wavelengths and amplitudes in the main channel, visualized by moving isotherms, that are continuously being swept downstream. Typically, for Re_g = 1580, the waves are comprised of several isotherms assembled in a stack and these structures are characteristic of the Tollmien-Schlichting waves in the main channel, which are activated by the Kelvin-Helmholtz instabilities of the free shear layer spanning the groove. The speed of a traveling wave can be determined by measuring the time it takes for the wave peak to traverse a known distance. The frequency of traveling waves is the inverse of the average wave period defined as the time required for two consecutive wave peaks to cross a fixed location during a given time interval. The second characteristic oscillatory feature is observed at a location close to the leading edge of the heated block. The horizontal isotherm separating the main channel and groove regions remains relatively motionless for low Reynolds numbers. As the Reynolds number increases, the isotherm starts "whipping" up and down, which is indicative of vertical velocity components.

For Re_g = 1580 the dominant frequency of the oscillatory whip was 29 Hz and corresponded well to the frequency of the traveling waves in the main channel, 26 Hz. At this Reynolds number two full waves, easily identified in Figure 8 (left), spanned one geometric periodicity length. The data regarding wave characteristics, such as wavelength propagation speed and flow oscillations, obtained for two Reynolds numbers, Re_g = 1580 and Re_g = 2370, and the number of images evaluated are summarized in Table 1.

PARAMETER	Re_g= 1580		Re_g= 2370	
	Main channel	Groove lip	Main channel	Groove lip
Waves/Period, L / λ_{TS}	1.85	N/A	2.94	N/A
Mean wave speed (m/s)	0.71	N/A	1.76	N/A
Dominant frequencies (Hz)	26	29	125	44, 67
Average period (s)	0.039	0.035	0.01	0.023
No. of periods analyzed	9	6	161	161
Total images sampled	350	208	1582	3754

Table 1. Wavelengths, oscillation frequencies and speeds of traveling waves in the grooved channel at Re_g = 1580 and Re_g = 2370

Increasing the Reynolds number from 1580 to 2370 results in an increase of the frequency of the oscillatory whip and a much more pronounced increase of frequency of the propagating waves. On the average, at Re_g = 2370, the geometric periodicity contained three Tollmien-Schlichting waves. These three waves are also present at Re_g = 2890, but they are not as obvious or easy to recognize in Fig. 8 (right). For Re_g = 2370, the measured frequencies of the main channel wave activity and oscillating isotherm show more scatter than was the case for the lower Reynolds number. The dominant frequencies indicated in Table 1 have different definitions for Re_g = 1580 and Re_g = 2370. For Re_g = 1580 the flow exhibits a more ordered behavior and the dominant frequency is the average of all the frequencies recorded. The number of image frames that needs to be analyzed to determine frequencies and speeds of these two characteristic features depends primarily on the Reynolds number of the flow and on the appropriate recording speed for that particular Reynolds number. For low Re_g, when the flow is ordered and well behaved, several hundred images were analyzed. For Re_g greater than 2000, several thousand images were processed sequentially in order to obtain an acceptable sample for quantification of the disorderly flow behavior. These numbers are indicated in Table 1.

From images obtained by holographic interferometry, such as those shown in Fig. 8, temperature fields can be reconstructed using the approach described in Section 4, with the algorithm for situations without the pressure fluctuations. Reconstructed temperature distributions in the basic grooved channel (left) and the grooved channel enhanced with curved vanes near the trailing edge of the heated block (right) for $Re_g = 530$ and $Re_g = 1580$ are displayed in Fig. 9. From the temperature distributions we can determine temperature gradients near the heated surface, find the values of the local Nusselt numbers along the surface and identify regions with low heat transfer rates and high heat transfer rates for different time instants during an oscillation cycle. Therefore we can analyze both spatial and time dependences of temperature and flow distributions as well as heat transfer. This information allows understanding and quantifying the mechanisms responsible for the heat transfer enhancement and optimizing the geometry of the grooved channel for maximum heat transfer.

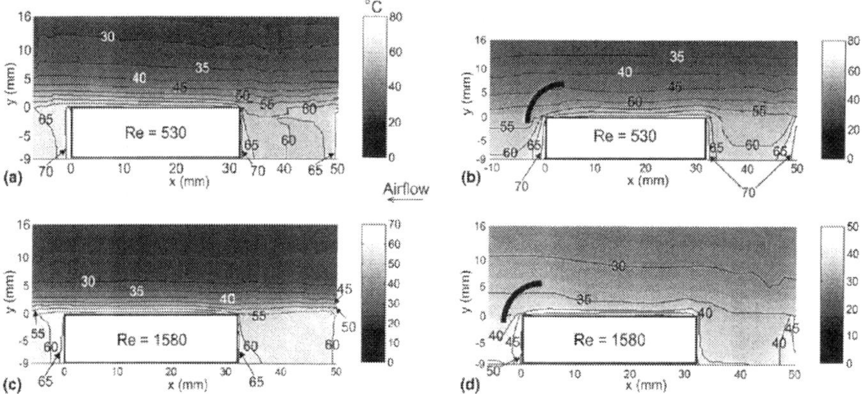

Fig. 9. Temperature fields measured using HI in the basic grooved channel (left) and the grooved channel enhanced with curved vanes near the trailing edge of the heated block (right) for $Re_g = 530$ and $Re_g = 1580$.

5.2 Communicating channels

The Reynolds number in the communicating channels is defined as $Re_c \equiv \dfrac{3}{2}\dfrac{u_m H}{\nu}$, where the characteristic length scale H in this definition represents the half-height of the main channel. A temperature distribution obtained by HI for the steady-state regime is displayed in Fig. 10. The critical Reynolds number for the onset of oscillations was experimentally determined to be around 200 and this result is in good agreement with data obtained numerically (Amon et al. 1992). At Reynolds numbers above the critical value for the onset of oscillations the vortices in the communicating region are unsettled and the steady state of the flow is disrupted. The vortex configuration becomes unstable, and vortices are ejected alternately to the top and bottom channels, thus inducing mixing between the vortex and bulk flows.

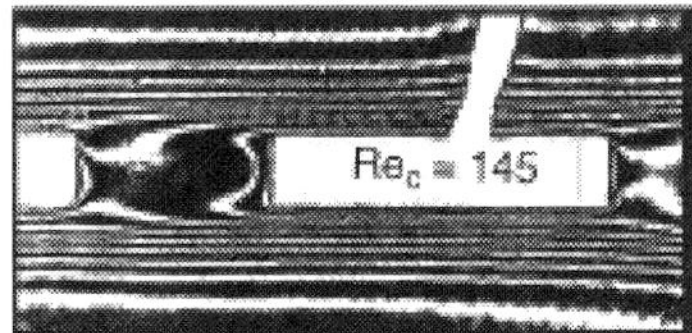

Fig. 10. Instantaneous temperature fields visualized by holographic interferometry in the form of isotherms in the communicating channels at Re$_c$ = 145, for the steady-state situation.

By analyzing sequences of interferometric images recorded by high-speed camera, different oscillatory regimes and varying oscillatory amplitudes were detected in the communicating channels. Two characteristic flow situations are illustrated schematically in Fig. 11, together with the corresponding temperature fields. One can notice the presence of a) four and b) three traveling waves over the double periodicity length in these images. As the flow structure in the communicating region is in agreement with the traveling wave structure in the main channels, the vortical structures in two successive communicating regions are either a) identical or b) antisymmetric, respectively.

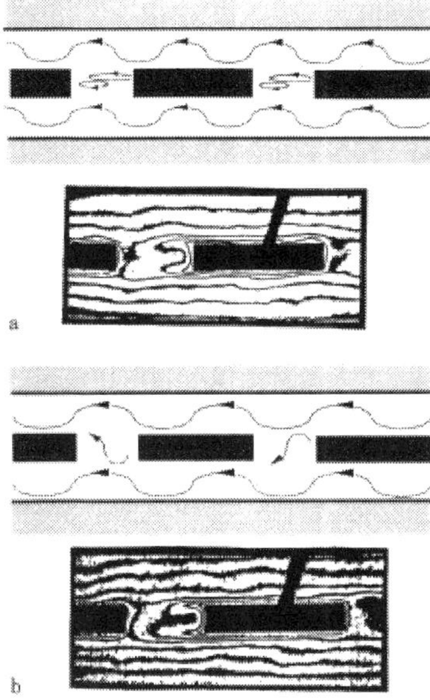

Fig. 11. Oscillations modes in two geometric periodicity lengths for a) four waves and c) three waves in the investigated region.

The studies on communicating channels led to interesting discoveries regarding flow instabilities: different oscillatory regimes were detected and amplitudes of oscillations varied significantly at the same flow velocity and during the same experimental run. These phenomena, captured by visualizing temperature fields, are illustrated in Figure 12, showing the history of temperature fields during one period of oscillations for the two different oscillatory regimes developing at Re_c = 493. In the first set of interferometric images displayed in Figure 12 (left), hardly any vortex activity can be observed in the communicating region, indicating reduced lateral mixing. The second sequence of interferometric images in Fig. 12 (right) was recorded during a time interval of the same length as the one corresponding to the sequence shown on the left hand side. The pattern of oscillations has changed, and the set of interferometric images displayed in Fig. 12 (right) shows intensive vortex activity in the communicating region, indicating that vortices ejected into the top and bottom channels improve lateral mixing. The intensity of waviness in the main channels does not change significantly.

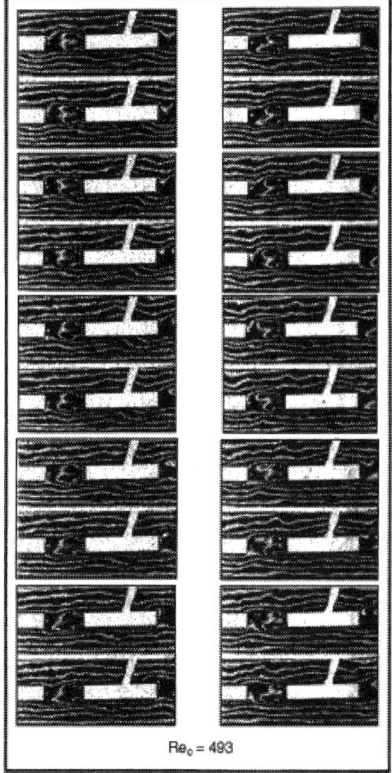

Fig. 12. Time evolution of temperature fields corresponding to two oscillatory regimes visualized by holographic interferometry in the communicating channels at Re_c = 493 .

As discussed in Section 2, in order to demonstrate the advantages of using the infinite fringe field alignment in quantitative flow visualization, in Fig. 3 we compared the characteristic features of the fringe patterns developed along the first two heated plates in the communicating channels. These interferometric images were obtained using the finite fringe field alignment (top) and the infinite fringe field alignment (bottom). The initially imposed pattern for the finite fringe field alignment consisted of thin vertical fringes. The temperature distribution obtained by the infinite fringe field alignment allows good qualitative insight into the flow structures in the visualized region, as one can, for example, easily identify the redeveloping thermal and viscous boundary layers along the plates. Wide isotherms develop in the wake downstream of the plate. Fig. 3 clearly demonstrates the advantages of the infinite fringe field alignment in analyzing wave structures in the main channel flow. These cannot be identified in images obtained by the finite fringe field alignment. In order to quantify the oscillatory flow in the vortex region of the communicating channels, we can measure the locations of specific isotherms as a function of time for a sequence of images, such as those in Fig. 12, for a series of Reynolds numbers using digital image processing.

5.3 Thermoacoustic refrigerator

The temperature fields in the stack region of the thermoacoustic refrigerator model, visualized by HI, were analyzed and reconstructed quantitatively by first identifying a sequence of interferometric images that describe a complete period of acoustic oscillations in the movie segment of interest. A sequence of 8 interferometric images, representative of one period of acoustic oscillations for a drive ratio of 1%, is presented in Fig. 13.

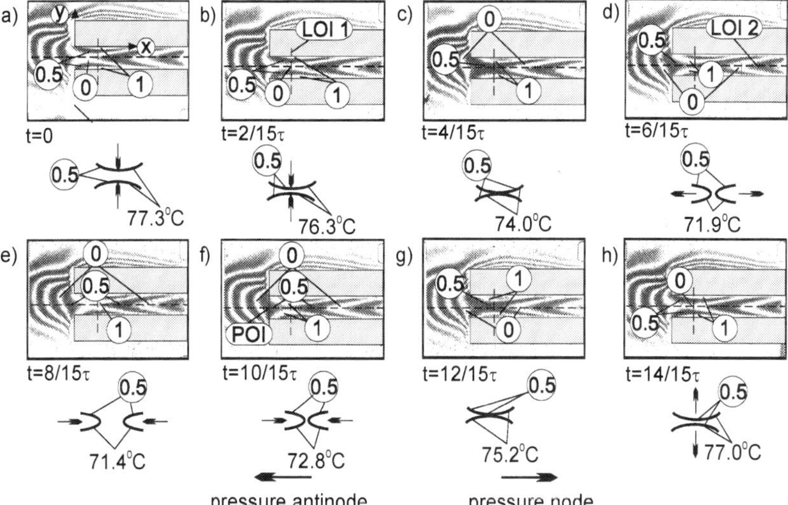

Fig. 13. Oscillating temperature fields around two stack plates of the thermoacoustic refrigerator at a drive ratio DR=1%. Images were recorded at a rate of 5,000 picture frames per second. Fringes of the order 0, 0.5 and 1 are indicated in the images and the measured temperature of the fringe of the order of 0.5 is displayed under the image. It should be noted that the temperature of this fringe changes periodically with the change of the acoustic pressure.

The changes the fringe pattern undergoes during one period of oscillations are analyzed by following the motion of the fringes tagged with the order 0.5. The changes of temperature of the fringe tagged with interference order 0.5 from 77.3°C to 71.4°C, obtained by taking advantage of the concept of quasi isotherms, are indicated in the schematic for the time instants illustrated in Figure 13.

Prior to the quantitative evaluation, an image specific coordinate system was defined, and image processing as well as curve fit algorithms were applied to determine the spatially and temporally continuous interference order $S(x,y,t)$. Details regarding the image processing procedure are available in the literature (Wetzel, 1998) and were also discussed in Section 4. From the continuous interference order, the time averaged interference order $S_m(x,y)$ and the interference order amplitude $S_A(x,y)$ were recovered. The latter two quantities were then used to determine the time averaged temperature fields $T_m(x,y)$, shown in Fig. 14.

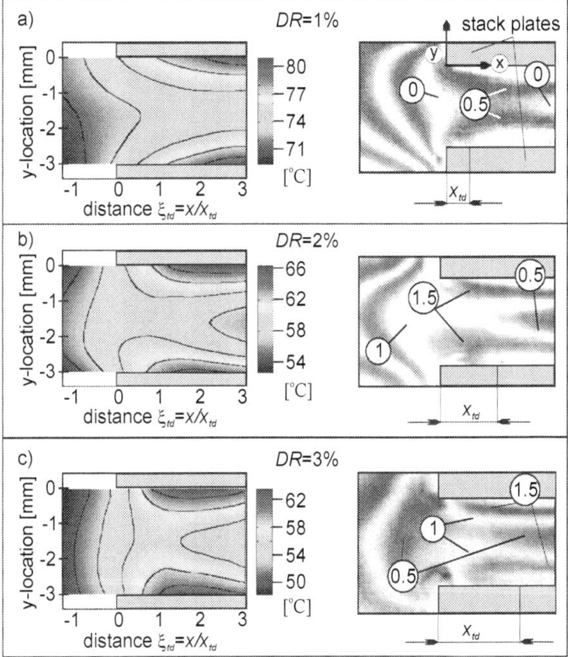

Fig. 14. Time-averaged visualized temperature fields (left: color coded reconstructions from interferometric fringe patterns and right: corresponding interferometric fringe patterns) at the hot end of the two stack plates for drive ratios DR of 1%, 2% and 3%.

An examination of the time averaged temperature field and interferometric image in Figure 14a shows that for the drive ratio DR=1% the working fluid along the axis of the channel is colder than at the stack plates. This trend is confirmed by considering the interferometric image on the right hand side of Figure 14a. This image demonstrates that the colder fringe of order 0 bends over the warmer fringe of order 0.5. Such a temperature distribution and fringe pattern indicate that heat is being transferred from the stack plates to the working fluid. This behavior is expected, since heat is generated in the stack plates by resistive heaters in addition to the thermoacoustic heating.

The shape of the isotherms in Figures 14b and c, representing the time averaged temperature fields for the drive ratios $DR=2\%$ and 3% becomes more complicated than for $DR=1\%$, with the shape of the isotherms reflecting vortex shedding in the region $0 \le \xi \le 1$, the edge of the stack plate. The temperature of the working fluid is higher than that of the lower stack plate close to the edge of the plate ($\xi \sim 0$). Both fringe pattern and temperature distribution indicate that heat is transferred into the lower stack plate at $\xi \sim 0$.

Time dependent temperature distributions $T(x,y,t)$ in the upper half of the channel between the two investigated stack plates for the drive ratio $DR=3\%$ are shown in Figure 15.

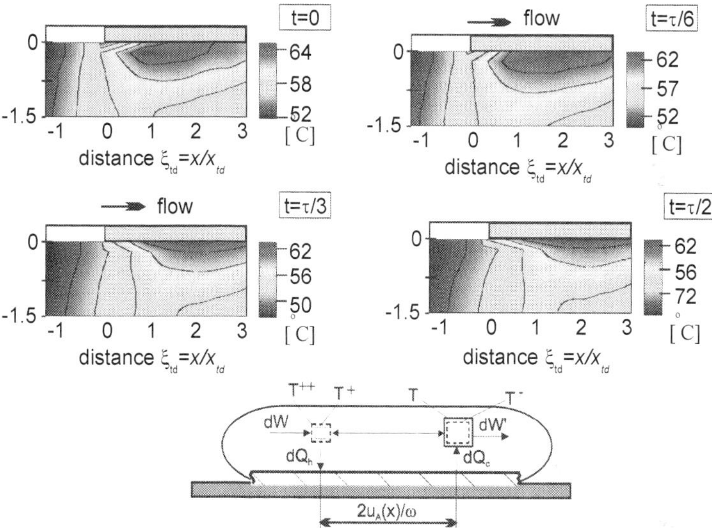

Fig. 15. Visualization of the thermoacoustic effect through the reconstructed temperature distributions during one half of the acoustic cycle. The first image illustrates full compression and the last one full expansion with the associated reversal of the direction of the heat flow, which is the characteristic of the thermoacoustic effect.

The flow direction is also indicated in Figure 15, and the length of the arrow is proportional to the flow velocity. The instanteneous temperature distributions correspond to four time instants within one half of the acoustic cycle, $t = 0$, $\tau/6$, $\tau/3$ and $\tau/2$, steps 2 through 4, as described by the gas parcel model (Swift, 1988). At t=0 (top left) the working fluid is fully compressed and hotter than the stack plate in the region $0 < \xi < 3$. This is indicative of heat being transferred into the stack plate. This time instant therefore corresponds to the second step in the gas parcel model with the working fluid at the temperature T^{++} (Figure 6). As time progresses, at $t = \tau/6$ and $\tau/3$ (top right and bottom left), the working fluid gradually expands and cools down to the temperature T^-. This time sequence corresponds to the third step in the gas parcel model. The fourth step in the gas parcel model is the last temperature field image (bottom right) in Figure 15. At this time instant the working fluid is fully expanded, it is colder than the stack plate, and heat is being transferred from the stack plate to the working fluid, illustrated as dQ_c. During the second half of the cycle the working fluid is being compressed, and the temperature distribution goes through the phases displayed in Figure 15 in reversed order. The temperature gradient changes sign and therefore heat is being transferred from the stack plate into the working fluid at $\xi = 0$.

6. Conclusions

The results reported in this paper demonstrate that HI, apart from the measurement of unsteady temperature distributions and local heat transfer, also allows the quantitative study of flow characteristics for certain classes of oscillatory flows coupled with heat transfer, and it can be recommended in situations when quantitative local velocity data are not required. We can conclude that through visualization of temperature fields we can indirectly study flow instabilities and resonant modes, as well as measure wavelength, propagation speed and frequency of traveling waves. HI allows the visual identification of separated flow regions, and provides limited amount of detail regarding the recirculation region. However, it should also be pointed out that the method does not replace the widely accepted, nonintrusive and quantitative velocity measurement techniques, such as LDA or PIV, rather it complements them.

The advantage of the approach introduced in the paper over "classical" visualization methods (such as the use of tracers that allows qualitative insight only) is that quantitative information on the structure of flow and temperature fields as well as heat transfer is obtained simultaneously, using the same experimental setup and during the same experimental run, thus yielding consistent flow and heat transfer data. This feature makes the technique particularly attractive for applications such as the development of flow control strategies leading to heat transfer enhancement. Using temperature as tracer offers the additional advantage that a quantitative analysis of high-speed thermofluid processes becomes possible, since optical measurement techniques are nonintrusive, and they, due to the high speed of light, have virtually no inertia. In the study of complex flow situations and flow instabilities, the investigated process can be very sensitive to disturbances, and the injection of dye or tracer particles can interfere with the flow, causing it to switch to another oscillatory regime due to perturbations. With HI, the contamination of the test section by tracers is avoided, thus allowing longer experimental runs and analyses of different steady states and transients during a single experimental run.

The effort involved in the study of high-speed, unsteady flows, in the measurement of phase shift and the extraction of quantitative information from the obtained images, is significant. When analyzing oscillatory flows, the required recording frequencies are usually at least one order of magnitude higher than the frequency of the physical process. Due to the high sampling rates, thousands of images are generated within seconds, and these images are then successively processed to extract the required quantitative information. Algorithms for a completely automated analysis of images recorded by holography, PIV and HPIV are available. Algorithms that allow an automated analysis of images generated by HI have been described in the literature, however, at this point of time the issue has not yet been adequately resolved. This is especially true for unsteady processes. Therefore the quantitative evaluation of interferometric fringe patterns still poses a challenge and limits the applications of HI.

7. Acknowledgments

The research reported in the paper was supported by the National Science Foundation and the Office of Naval Research. Dr. Martin Wetzel and Dr. Eric Kang were instrumental in conducting the experiments as well as in the data analyis.

8. References

Amon, C. H.; Majumdar, D.; Herman, C. V.; Mayinger, F.; Mikic, B. B.; Sekulic, D. P. (1992) *Numerical and experimental studies of self-sustained oscillatory flows in communicating channels*, Int. J. Heat and Mass Transfer, Vol. 35, 3115-3129

Cooper, W. L.; Yang, K. T.; Nee, V. W. (1993) *Fluid mechanics of oscillatory and modulated flows and associated applications in heat and mass transfer - A review*, J. of Energy, Heat and Mass Transfer, Vol. 15, 1-19

Farhanieh, B.; Herman, C.; Sunden, B. (1993) *Numerical and experimental analysis of laminar fluid flow and forced convection heat transfer in a grooved duct*, Int. J. Heat and Mass Transfer, Vol. 36, No. 6, 1609-1617

Ghaddar, N. K.; Korczak, K. Z.; Mikic, B. B.; Patera, A. T. (1986a) *Numerical investigation of incompressible flow in grooved channels. Part 1. Stability and self-sustained oscillations*, J. Fluid Mech., Vol. 163, 99-127

Ghaddar, N. K.; Magen, M.; Mikic, B. B.; Patera, A. T. (1986b) *Numerical investigation of incompressible flow in grooved channels*. Part 2. Resonance and oscillatory heat transfer enhancement, J. Fluid Mech., Vol. 168, 541-567

Greiner, M.; Chen, R.-F.; Wirtz, R. A. (1990) *Heat transfer augmentation through wall-shape-induced flow destabilization*, ASME Journal of Heat Transfer, Vol. 112, May 1990, 336-341

Greiner, M. (1991) *An experimental investigation of resonant heat transfer enhancement in grooved channels*, Int. J. Heat Mass Transfer, Vol. 34, No. 6, 1383-1391

Hauf, W.; Grigull, U. (1970) *Optical methods in heat transfer*, in Advances in Heat Transfer, Vol. 6, Academic Press, New York

Herman, C., 2000, *The impact of flow oscillations on convective heat transfer*, Annual Review of Heat Transfer, Editor: C.-L. Tien, Vol. XI, Chapter 8, pp. 495-562, invited contribution.

Kang, E. (2002) *Experimental investigation of heat transfer enhancement in a grooved channel*, Dissertation, Johns Hopkins University, Baltimore, MD, USA.

Karniadakis, G. E.; Mikic, B. B.; Patera, A. T. (1987), *Heat transfer enhancement by flow destabilization: application to the cooling of chips*, Proc. Int. Symposium on Cooling Technology for Electronic Equipment, 587-610

Mayinger, F. Editor (1994) *Optical measurements - Techniques and applications*, Springer Verlag, Berlin

Patera, A. T.; Mikic, B. B. (1986), *Exploiting hydrodynamic instabilities. resonant heat transfer enhancement*, Int. J. Heat and Mass Transfer, Vol. 29, No. 8, 1127-1138

Panknin, W., 1977, *Eine holographische Zweiwellenlangen-Interferometrie zur Mesung uberlagerter Temperatur- und Konzentrationsgrenzschichten*, Dissertation, University of Hannover, Germany.

Swift, G.W. (1988) *Thermoacoustic Engines*, J. Acoust. Soc. Am. 84(4), Oct., 1145-1180

Vest, C. M. (1979) *Holographic interferometry*, John Wiley & Sons, New York

Wetzel, M. (1998) *Experimental investigation of a single plate thermoacoustic refrigerator,* Dissertation, Johns Hopkins University, Baltimore, MD, USA.

Wheatley, J.C.; Hofler, T.; Swift, G.W.; Migliori, A. (1983) *An intrinsically irreversible thermoacoustic heat engine,* J. Acoust. Soc. Am., 74 (1), July, 153-170

14

Application of Mass/Heat Transfer Analogy in the Investigation of Convective Heat Transfer in Stationary and Rotating Short Minichannels

Joanna Wilk
Rzeszów University of Technology
Poland

1. Introduction

Convective heat transfer during laminar fluid flow through channels is an important technological problem. The literature provides much research data on heat transfer phenomena occurring during fluid flow through channels of different dimensions and shapes based on theoretical analysis, numerical calculations and experimental investigations. Theoretical analysis has attempted to solve the problem of convective heat transfer in channels of basic shape in simplified conditions of fluid flow. Laminar convective heat transfer in pipes where the fluid velocity profile was parabolic was described by Graetz (Graetz, 1885). His solution to this classical problem was further developed by Sellars, Tribus and Klein (Sellars et al., 1956). Alternatively, Levěque (Levěque, 1928) investigated heat transfer in the entrance region of pipes where hydraulic stabilisation occurs. All these works offer an analytical solution to the problem. Numerical investigation of heat transfer in the entrance region of pipes was initiated by Kays (Kays, 1955) who provided the results of numerical calculations for three types of gas flow conditions: uniform wall temperature, uniform heat flux and uniform difference between wall and fluid temperature. On the other hand, the work of Sider and Tite (Sider and Tite, 1936) is an example of an early experimental investigation where an empirical formula for the heat transfer coefficient calculations, regardless of the fluid being heated or cooled, is provided.

The above-mentioned works were followed by further research on the laminar convective fluid flow in channels. Particularly, heat transfer in channels of small hydraulic diameters was extensively investigated. It turned out that heat transfer and fluid flow in small diameter channels often differed from those in channels of conventional dimensions. Mathematical equations describing heat transfer cannot always be applied to mini- or micro-channels. Hence some researchers (Adams et al., 1998), (Tso and Mahulikar, 2000), (Owhaib and Palm, 2004), (Lelea et al., 2004), (Celata et al., 2006), (Kandlikar et al., 2006), (Yang and Lin, 2007), (Yarin et al., 2009) examined liquid and gas convective heat transfer in circular mini- and micro-channels using experimental methods, mostly the thermal balance method. However, the surface and fluid temperature measurements using this method are difficult to obtain due to the small size of the channels tested. Application of mass transfer investigations and the mass/heat transfer analogy makes it possible to avoid the problem as it excludes temperature measurements. In this chapter application of the mass/heat transfer analogy in the study of heat transfer in short minichannels is described. The electrochemical technique is employed in measuring the mass transfer coefficients.

2. Mass/heat transfer analogy

The heat transfer coefficient is most often determined by quite intricate experiments based on the thermal balance method which requires complex instruments and sometimes difficult measurements. An alternative method of obtaining it is to measure it by applying the mass/heat transfer analogy. There is an exact analogy between the mass and heat transport processes so mass transfer results may be converted to heat transfer results. This depends on the similarity of the equations describing the heat and mass change processes. If the boundary conditions are the same for a given geometry, the differential equations have the same solution. For the mass transfer related to a forced convection, the solution to the problem is given by the general correlation

$$Sh = Sh(Re,Sc) , \qquad (1)$$

analogously, for heat transfer

$$Nu = Nu(Re,Pr) , \qquad (2)$$

where: Nu – Nusselt number, $h{\cdot}d_h/k$; Re – Reynolds number, $w{\cdot}d_h/v$;
Pr – Prandtl number, v/a; Sc – Schmidt number, v/D;
Sh – Sherwood number, $h_D{\cdot}d_h/D$; a – thermal diffusivity $[m^2/s]$;
d_h – hydraulic diameter $[m]$; D – mass diffusion coefficient $[m^2/s]$;
h – heat transfer coefficient $[W/(m^2K)]$; h_D – mass transfer coefficient $[m/s]$;
k – thermal conductivity $[W/(mK)]$; w – mean fluid velocity $[m/s]$;
v - kinematic viscosity $[m^2/s]$.

In the case when the forced convective heat transfer occurs during rotation, the Rossby number has to be introduced into Eqs (1) and (2), thus

$$Sh = Sh(Re,Ro,Sc) \qquad (3)$$

and

$$Nu = Nu(Re,Ro,Pr) , \qquad (4)$$

where: Ro – Rossby (rotation) number, $\omega{\cdot}d_h/w$; ω - angular velocity $[rad/s]$.
The results of the experimental investigation of the mass change processes may be correlated in the empirical equation

$$Sh = cRe^pSc^q , \qquad (5)$$

where c, p, q – empirical constants.
Taking into account the mass/heat transfer analogy, the results of the heat transfer processes may be correlated in the form

$$Nu = cRe^pPr^q . \qquad (6)$$

The analogy requires that the Sc and Pr numbers be equal. However, the similarity of the fluid properties expressed by the equal Schmidt and Prandtl numbers is very difficult to obtain. Nevertheless, the research data (Goldstein & Cho, 1995) show a good agreement of the mass transfer experimental results with heat transfer results for different Sc and Pr numbers.

Based on equations (5) and (6) the mass transfer measurement results may be converted to the heat transfer formula

$$\frac{Nu}{Sh} = \left(\frac{Pr}{Sc}\right)^q .$$

(7)

In the Chilton-Colburn analogy, the experimentally obtained exponent q = 1/3 (Chilton & Colburn, 1934). The authors present the analogy in the form:

$$j_M = St_M Sc^{2/3}$$

(8)

$$j_H = St_H Pr^{2/3}$$

(9)

$$j_M = j_H$$

(10)

where: j_H, j_M – Chilton-Colburn coefficients for heat and mass transfer,
    St_H – Stanton number for heat transfer, $h/(c_p \cdot \rho \cdot w)$,
    St_M – Stanton number for mass transfer, h_D/w,
    c_p – specific heat capacity [J/(kgK)],
    ρ - density [kg/m³].
According to this description some of the researchers (Bieniasz and Wilk, 1995), (Bieniasz 1998), (Wilk, 2004), (Bieniasz, 2010) present the results of the mass transfer measurement as

$$j_M = pRe^q .$$

(11)

When rotary conditions are involved, the analogy between mass forces is maintained by Rossby number equality. In this case the Rossby number can occur as a parameter in the equations describing mass or heat transfer. Bieniasz (Bieniasz, 2010) proposes the following correlation describing mass transfer in rotary conditions:

$$j_M = p_1 Re^{q_1} + p_2 Re^{q_1} Ro ,$$

(12)

where: p_1, p_2, q_1, q_2 – empirical constants.

3. Experimental technique of the mass transfer coefficient measurement

The limiting current method and naphthalene sublimation are basic experimental techniques which may be used for mass transfer coefficient measurements.
Using naphthalene sublimation involves the following stages (Goldstein and Cho, 1995):
* preparing the coat test models with naphthalene,
* measuring the initial naphthalene surface profile or weight,
* conducting the experiment with a naphthalene-coated model,

- measuring the naphthalene surface profile or weight after the mass change experiment,
- calculating the mass transfer coefficient. The coefficient is given by the equation

$$h_D = \frac{\Delta m}{\left(\rho_{VN,w} - \rho_{VN,f}\right)\tau} , \tag{13}$$

where: Δm – surface mass decrement of naphthalene [kg/m²],

   $\rho_{VN,w}$ – naphthalene vapour density on the surface [kg/m³],

   $\rho_{VN,,f}$ - naphthalene vapour density in the fluid [kg/m³],

   τ - time of the experiment [s].

Recently, naphthalene sublimation has been often used (Szewczyk, 2002), (Kim and Song, 2003), (Hong and Song, 2007), mainly because mass transfer coefficients can be determined by this method with high accuracy. Moreover, based on the mass transfer coefficients obtained, and using mass/heat transfer, reliable results for heat transfer coefficients can be achieved.

However, naphthalene sublimation cannot be applied in all flow and geometric conditions (Goldstein and Cho, 1995). Another technique used to determine heat transfer coefficients by mass/heat transfer analogy is the limiting current method. This method involves observing the controlled ion diffusion at one of the electrodes, usually the cathode. Once the external voltage is applied to the electrodes which are immersed in the electrolyte, electric current arises in the external circuit. According to Faraday's law, the magnitude, I, of the current generated is given by

$$I = iA = nFAN , \tag{14}$$

where: A – surface area of the cathode [m²],

   F – Faraday constant [96 493 A s/kmol],

   I – current in the circuit [A],

   i – current density [A/m²],

   N – molar flux density [kmol/(m² s],

   n – valence charge of reacting ions.

In this process ion transport is caused by convection, migration and diffusion. The convection element is small. If it is ignored, the error is not greater than 0.3% (Bieniasz, 2005). Reduction of the migration element can be achieved through adding a background electrolyte to the electrolyte investigated. If only the diffusion process is taken into account, the molar flux density according to Fick's law is given by:

$$N = -D\frac{dC}{dy} , \tag{15}$$

where: dC/dy – gradient of the reacting ions concentration,

   C – ions concentration [kmol/m³],

   y – normal coordinate [m].

Then, based on the Nernst model (linear dependence of ion concentration vs the distance from the electrode surface in the diffusion layer – Fig.1), one may write:

$$\frac{dC}{dy} = \frac{C_b - C_w}{\delta} , \tag{16}$$

$$\delta = D/h_D \; ,\tag{17}$$

where: C_b – bulk ions concentration [kmol/m³],
             C_w – ions concentration at the electrode surface [kmol/m³],
             δ - mean thickness of Nernst diffusion layer [m].

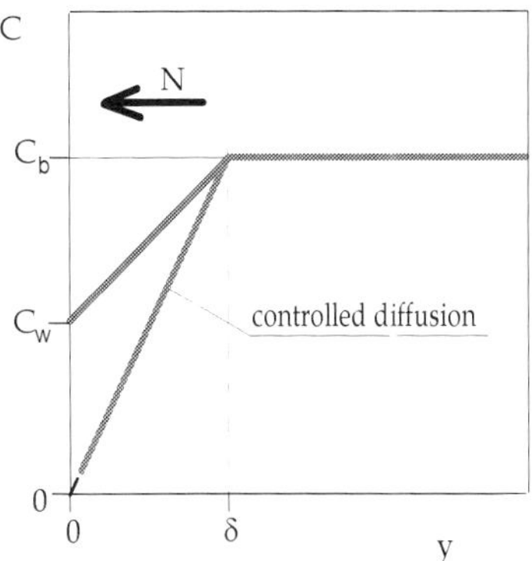

Fig. 1. Ion concentration distribution in the electrolyte at the cathode surface

On the basis of the equations (14) – (17) one can obtain

$$i = nFh_D\left(C_b\text{-}C_w\right).\tag{18}$$

It is impossible to calculate the mass transfer coefficient from equation (18) because C_w is practically non-measurable. In the limiting current method an increasing current is caused to flow across the electrodes by increasing the applied voltage until a characteristic point is reached – point A in Fig. 2. If the anode surface is much bigger than the cathode surface, a further increase in the applied voltage will not lead to increased current intensity – segment AB in Fig.2. This is the limiting current plateau. Under these conditions the ion concentration at the cathode surface C_w approaches zero. Based on the measurement of the limiting current I_p, , the mass transfer coefficient can be calculated from the equation

$$h_D = \frac{I_p}{nFAC_b}\; .\tag{19}$$

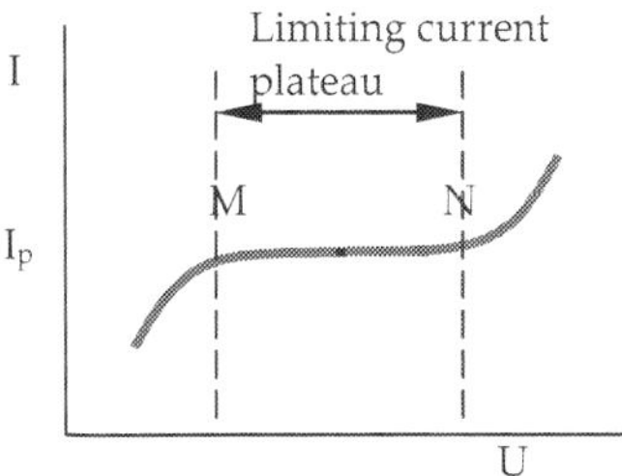

Fig. 2. Typical polarization curve

The polarization curve in Fig. 2 shows three zones: zone →A where the primary reaction is under mass transport and electron transfer control, plateau zone A↔B where the primary reaction is controlled by mass transport–controlled convective-diffusion, and zone B← where a secondary reaction occurs at the same time as the primary reaction. When employing the limiting current method it is important to fulfil the following   conditions (Szántó et al., 2008):

- the electrode should be carefully polished before each experiment,
- the electrode should be activated before the experiment,
- the background electrolyte should provide stable limiting current measurement over a long time,
- the electrolytes should be freshly prepared before each experiment and investigations should be performed in the absence of direct sunlight.

4. Experimental investigations

4.1 Properties of the electrochemical system applied

Measurements of the mass transfer coefficients were performed using the ferrocyanide/ferricyanide redox couple at the surface of nickel electrodes immersed in an aqueous solution of equimolar quantities of $K_3Fe(CN)_6$ and $K_4Fe(CN)_6$. A molar solution of sodium hydroxide NaOH was applied as the background electrolyte.

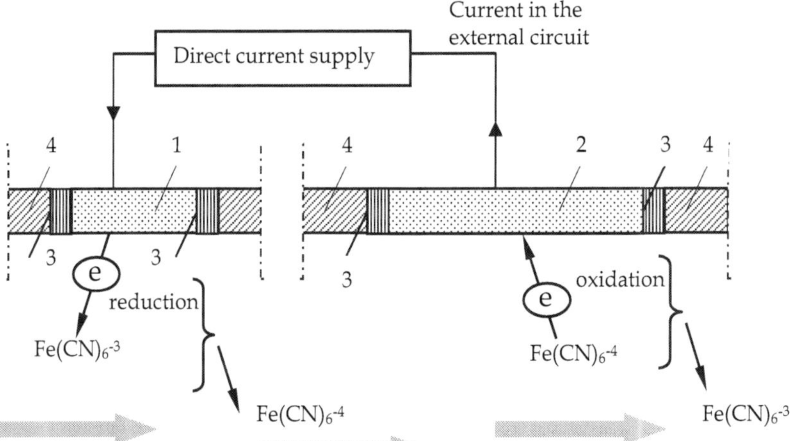

Electrolyte - aqueous solution of equimolar quantities of $K_3Fe(CN)_6$ and $K_4Fe(CN)_6$. and molar solution of sodium hydroxide NaOH; 1 – nickel cathode; 2 – nickel anode; 3 – electric insulation; 4 – construction element

Fig. 3. Scheme of the electrochemical process.

The oxidation-reduction process under the convective-diffusion controlled conditions is written as:

$$Fe(CN)_6^{-4} \xleftrightarrow{\text{ox red}} Fe(CN)_6^{-3} + e \ . \tag{20}$$

Besides the redox couple described by eq. (20), an unwanted redox couple shown in the equation below may take place:

$$O_2+2H_2O+4e \leftrightarrow 4OH^- .\tag{21}$$

In this case the limiting current is the sum of the limiting current of the reduced $Fe(CN)^3$ ions and oxygen. It was therefore necessary to eliminate the dissolved oxygen from the electrolyte by washing it with nitrogen.

Fig.3. shows the electrochemical system used in the experiment. The physical properties of the electrolyte at 25°C were as follows: $D = 6.71×10^{-10}m^2/s$, $v = 1.145×10^{-6}m^2/s$ (Wilk, 2004). The value of ferricyanide ion concentration was measured using iodometric titration. The concentration of fericyanide ions changed over time (Szanto et al., 2008), its sample value during measurements being $3.52×10^{-3}kmol/m^3$.

4.2 Experimental rig

The measurements of limiting current were performed using a universal rig. Its electrolyte as well as its nitrogen and electrical measurement system are shown in Fig.4. The anode was located behind the cathode in the direction of the electrolyte flow. Because the purpose of the experiment was to achieve controlled diffusion at the cathode, the necessary condition anode surface >> cathode surface had to be fulfilled.

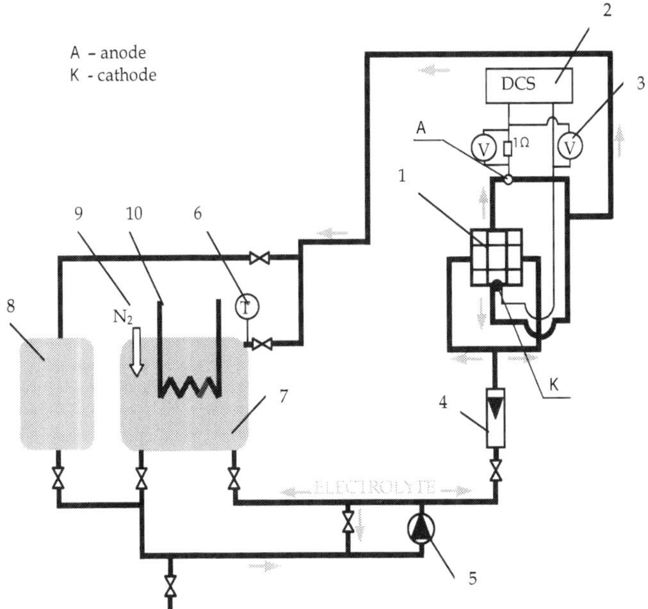

1 – test section, 2 – direct current supply, 3 – voltage measurement, 4 – flow rate measurement, 5 – pump, 6 – temperature measurement, 7 – main tank of electrolyte, 8 – tank for activation, 9 – nitrogen bubbling, 10 – preheater.

Fig. 4. Scheme of the experimental rig.

The measurement stand made it possible to carry out the following stages of the experiment: cathode activation, stabilisation of the electrolyte temperature at 25°C and its measurement, release of oxygen from the electrolyte by nitrogen bubbling, step alteration and measurement of the external voltage applied to the electrodes and measurement of the electric current in the external circuit. The test section mounted on the rig enabled measurements to be made during rotation.

5. Convective mass transfer in circular minichannels

5.1 Stationary conditions

The experiment was performed in circular minichannels of d=1.5 mm inner diameter and of L=15 mm length. Measurements were made using a PVC ring with 600 radially drilled identical minichannels. The ring used for the measurements was part of the test section prepared and used previously in investigations on the rotor of a high-speed regenerator (Bieniasz and Wilk, 1995), (Wilk 2004), (Bieniasz, 2009), (Bieniasz, 2010). Nickel cathodes were mounted as the inner surfaces of three minichannels. A scheme of the ring with the minichannels and electrolyte flow direction is shown in Fig. 5.

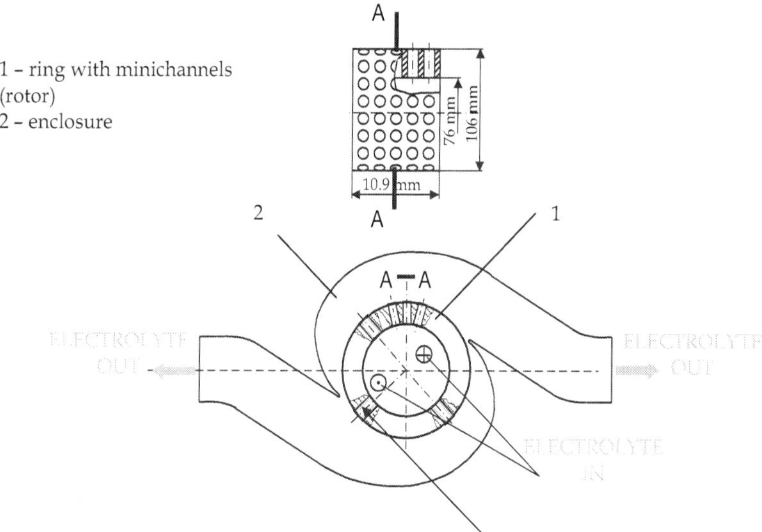

Fig. 5. Simplified scheme of part of the test section for the investigations of the mass transfer coefficient in short circular minichannels.

Before the experiment the inner surfaces of the minichannels were polished using a greasy diamond abrasive compound. After polishing the surface roughness was about 0.012μm so the investigated minichannels were considered smooth (Kandlicar et al., 2001).

As a result of the experiment, linear sweep voltammograms of ferricyanide ion reduction at the cathode were obtained. Examples of the voltammograms for one of the investigated cathodes (inner surface of the minichannel) with the Reynolds number as a parameter are shown in Fig. 6.

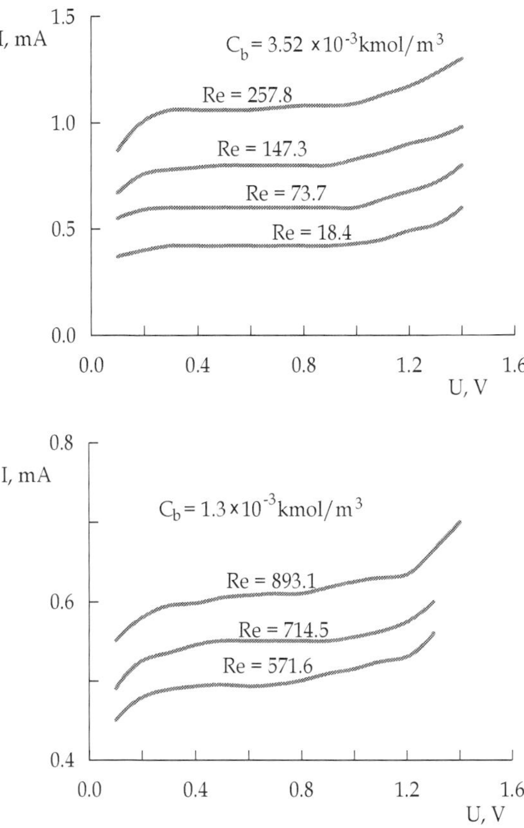

Fig. 6. Examples of the limiting current plateau at different Reynolds numbers and different bulk ion concentrations

The values of the Reynolds numbers were calculated based on the flow rate measured. The volumetric flow rate through a single minichannel was calculated as the ratio of the total measured rate to the number of minichannels. The assumption about an identical flow rate in all the minichannels was based on a theoretical analysis of the flow through the rotor (ie the ring with minichannels) (Bieniasz, 2009) and the assumption that all the channels had identical inner diameters.

As the final result of the experiment the mean mass transfer coefficient at the inner surface of the circular short minichannel was calculated from eq. (19). The necessary limiting current plateau was obtained as the average value of the values I_p measured for all the cathodes investigated. The plot of the mean mass transfer coefficient vs. the Reynolds number is shown in Fig. 7.

The results were compared with those from the literature data concerning the mass transfer coefficients in mini or microchannels obtained using the electrochemical limiting current method.

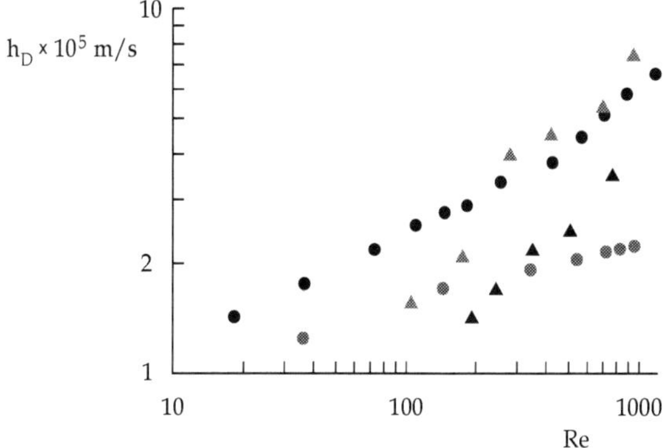

• - this study, d = 1.5 mm, L = 15 mm; • - (Sara et al., 2009), circular microchannel, d = 0.205 mm, L = 20.22 mm; ▲- (Bieniasz and Wilk, 1995), irregular curved minichannel, d_h = 2.46 mm, L = 15.8 mm; ▲- (Bieniasz and Wilk, 1995), irregular curved minichannel, d_h = 1.34 mm, L = 15.8 mm

Fig. 7. Results of mass transfer coefficient measurements

Application of the limiting current method made it possible to perform preliminary investigations of the mass transfer coefficient distribution along the short minichannel. Additional cathodes were used in the measurements. They were of nickel and of length 1/4L and 3/4L. The shorter cathodes were located as inner surfaces of the minichannels. Examples of voltammograms for the 3/4L length cathode are shown in Fig. 8.

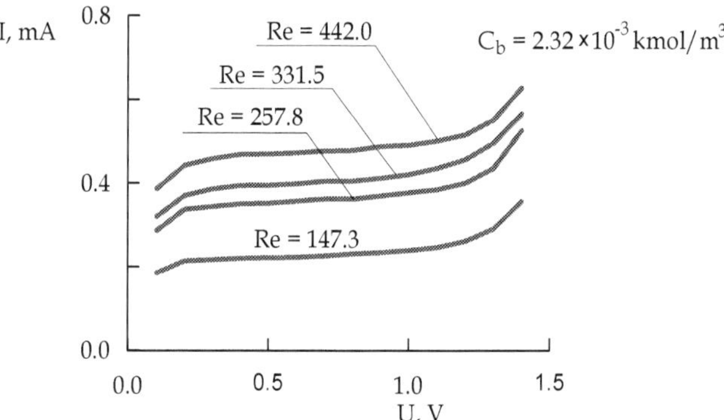

Fig. 8. Examples of the limiting current plateau for the 3/4L length cathode

The distribution of the mean mass transfer coefficient along the length of the channel for different Reynolds numbers is shown in Fig. 9.

The measurements obtained were compared with values given by the extension of the Graetz-Leveque solution for mass transfer (Acosta et al., 1985) which are given by

$$Sh = 1.85 \left(ReSc \right)^{1/3} \left(\frac{d_h}{x} \right)^{1/3} , \qquad (22)$$

where: x – distance from the mass transfer leading edge.

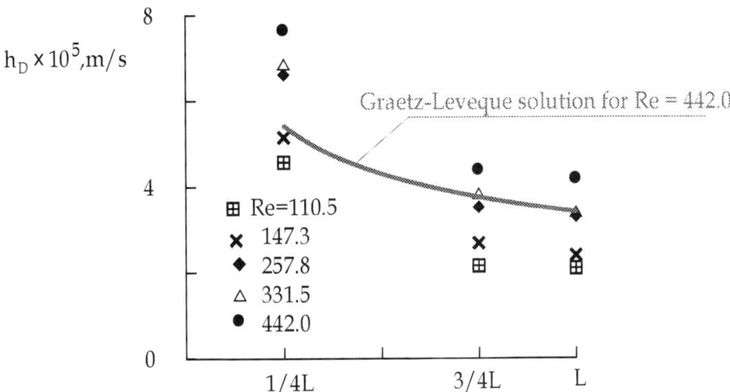

Fig. 9. Variation of mean mass transfer coefficient with the length of the minichannel

5.2 Rotary conditions

Experimental investigations of the mass transfer coefficients in the rotating minichannels were performed using the rig shown in Fig. 4. Additionally, the drive system of the tested ring was applied. The rotor was driven by a three-phase electric motor. The rotational speed was controlled by a speed indicator. A simplified scheme of part of the experimental rig with the ring drive system is shown in Fig. 10.

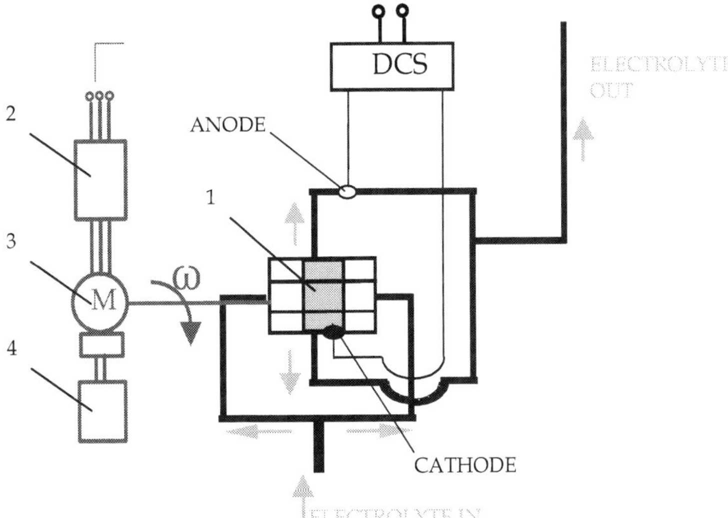

1 – test section with the ring with minichannels, 2 – inverter, 3 – driving motor, 4 – angular rate sensor.

Fig. 10. Scheme of the rotary part of the rig.

The rotational speed $\dot{n}$ of the minichannels was obtained based on an assumed value 0.1 of the Rossby number. The value of $\dot{n}$ was calculated from the formula:

$$\dot{n} = \frac{\nu \, \text{Re} \, \text{Ro}}{2\pi \, \text{d}^2}.$$ (23)

The values of the rotational speed of the ring with minichannels varied between 100 and 534 revolutions per minute.

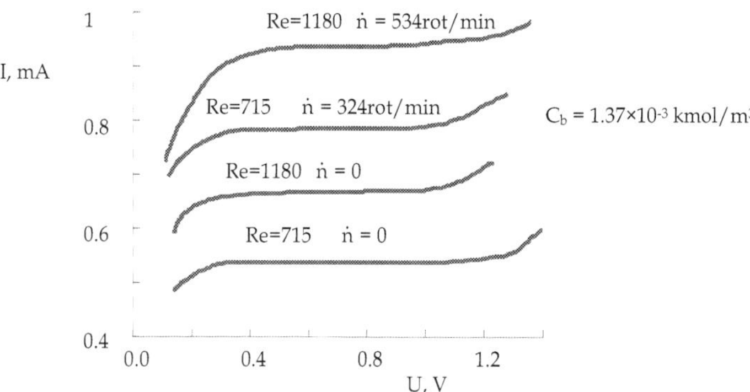

Fig. 11. Influence of the rotation on the limiting current.

Rotation caused intensification of the mass transfer process in the minichannel. The limiting current increased with the increase in the minichannel rotational speed. Examples of the voltammograms are shown in Fig. 11.

Finally from the measurements the dimensionless mass transfer coefficients (Sherwood numbers) were calculated. The results are shown in Fig. 12, where the Rossby number (rotation number) occurs as a parameter.

Based on Eqs (9) and (11) the results were obtained in the forms:

$$j_M = 0.31 \, \text{Re}^{-0.48}$$ (24)

for Ro = 0, and

$$j_M = 1.93 \, \text{Re}^{0.67}$$ (25)

for Ro = 0.1.

The electrochemical results were compared with the correlations described by Bieniasz (Bieniasz, 2010) for rotating short curved minichannels of cross-section varying in shape and surface area along the axis, namely

$$j_M = 0.282 \, \text{Re}^{-0.437} + 2.48 \, \text{Re}^{-0.732} \, \text{Ro}$$ (26)

and

$$j_M = 0.351 \, \text{Re}^{-0.418} + 2.20 \, \text{Re}^{-0.703} \, \text{Ro}.$$ (27)

Bieniasz gave two correlations (26) and (27) depending on the kind of baffle applied in the test section (Bieniasz, 2010). The comparison is shown in Fig. 13.

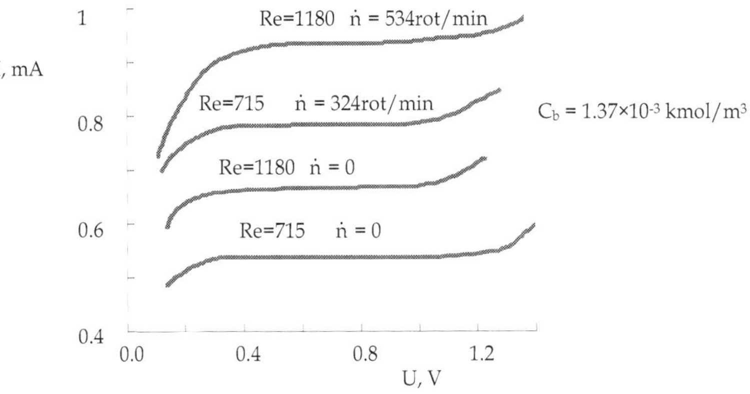

Fig. 12. Influence of rotation on the mass transfer in the circular short minichannel.

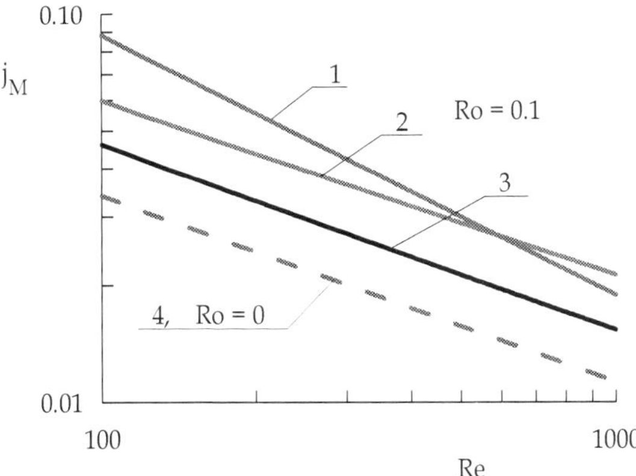

1 – circular minichannels, d = 1.5mm, Eq.(25), 2 – curved minichannels, d_h = 2.36mm, Eq.(27), 3 – as previously, Eq.(26), 4 – circular minichannels, d = 1mm, stationary conditions, Eq.(24).

Fig. 13. Chilton-Colburn mass transfer coefficients in short rotating minichannels,

6. Experimental uncertainties of the major parameters

The average relative uncertainties of the complex quantities $y = y(x_1, x_2,...,x_i)$ were calculated according to the general relation:

$$\frac{\Delta y}{y} = \left[\frac{\sum_{i=1}^{n}\left(\frac{\partial y}{\partial x_i} \Delta x_i \right)}{y^2} \right]^{\frac{1}{2}} \tag{28}$$

where: Δx_i – the mean uncertainties of the partial measurements

Based on Eqs (19) and (28) the relative uncertainty of the mass transfer coefficient measurement is given by

$$\left|\frac{\Delta h_D}{h_D}\right| = \left[\left(\frac{\Delta I_p}{I_p}\right)^2 + \left(\frac{\Delta A}{A}\right)^2 + \left(\frac{\Delta C_b}{C_b}\right)^2\right]^{\frac{1}{2}}.$$
(29)

The limiting current I_p measurement was made by means of a digital millivoltmeter, so $\left|\frac{\Delta I_p}{I_p}\right|$ = 0.1% is the uncertainty resulting from the degree of accuracy of the measuring instrument (measurement of the voltage drop at standard resistance).

The average uncertainty in the determination of the cathode surface area $\left|\frac{\Delta A}{A}\right|$ was related to the minichannel inner diameter and length measurements. It was estimated to be 3.4%.

For iodometric titration the following data was necessary in order to obtain $\left|\frac{\Delta C_b}{C_b}\right|$: solution normality of $Na_2S_2O_3$ N = 1 ± 0.002, volume of electrolyte sample V = (25 ± 0.1) ml and uncertainty of titrant added volume measurement ΔV_t = 0.005 ml. Based on this data the value of $\left|\frac{\Delta C_b}{C_b}\right|$ was calculated as 1.4%.

The relative uncertainty of the mass transfer coefficient measurement was calculated according to Eq.(25) and was 3.7%.

7. Convective heat transfer in circular minichannels

On the basis of the mass transfer coefficient measurements and mass/heat transfer analogy described in section 2, some correlations describing the convective heat transfer in stationary and rotating circular minichannels were obtained.
The equation:

$$Nu = 0.257Re^{0.52}$$
(30)

describes the dependence of the mean Nusselt number vs Reynolds number for stationary conditions in the Reynolds number range 250 to 1200.
The similar correlation

$$Nu = 1.714Re^{0.33}$$
(31)

concerns the convective heat transfer in a rotating minichannel. The range of Reynolds numbers was the same as in Eq.(30). Eq.(31) is valid for the Rossby number 0.1.
A different correlation was obtained for stationary conditions where convective heat transfer in a short circular minichannel takes place at low Reynolds numbers. It was

$$Nu = 1.067e^{0.52}Pr^{1/3}.$$
(32)

Correlation (32) is valid for d/L = 0.1 and for Reynolds numbers ranging from 20 to 250. The extension of Eq.(32) by the term $(d/L)^r$ and the assumption of the power r = 1/3 lead to the following form of Eq.(32):

$$Nu = 2.3 Re^{0.52} \left(Pr \frac{d}{L} \right)^{1/3}. \tag{33}$$

Relationship (32) in comparison with the thermal balance results (Celata et al., 2006) is shown in Fig. 14.

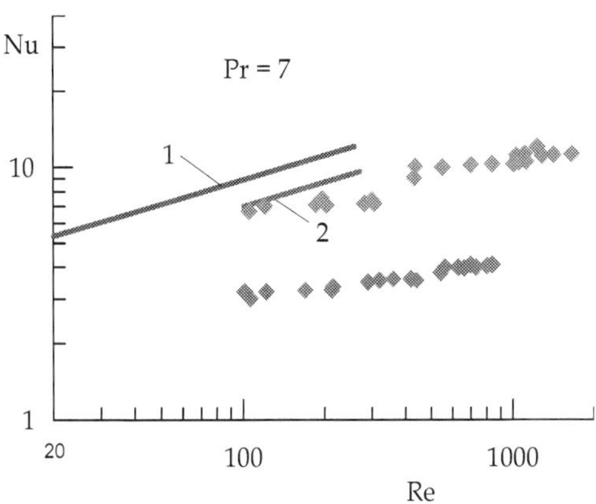

1 – heat/mass transfer analogy results, mean Nu number, d/L = 0.1; 2 – graphical presentation of the Eq.(33), d/L = 0.048; ◆◆◆ experimental thermal balance results, local Nusselt number, d/L = 0.048 (Celata et al., 2006); ◆◆◆ as previously, d/L = 0.016 (Celata et al., 2006)

Fig. 14. Heat/mass transfer analogy results for a short minichannel in comparison with the thermal balance results.

8. Summary

In this chapter the possibility of applying the mass/heat transfer analogy to the investigation of convective laminar heat transfer in rotating and stationary short minichannels was explored. The author has provided a general form of the mass/heat transfer analogy, assumptions of the processes, as well as the dimensionless numbers and equations describing the analogy. The limiting current method used for mass transfer coefficient determination was described. Some limiting current voltammograms which formed the basis for mass transfer coefficient calculations were provided. The use of an electrochemical technique and the Chilton-Colburn analogy produced correlations describing heat transfer processes in short minichannels.

Results of the mass transfer coefficient uncertainty calculations were also presented.

An important problem in the application of the mass/heat transfer analogy is determination of the analogy uncertainty. In order to estimate the uncertainty of the heat transfer coefficient resulting from application of the mass/heat transfer analogy, a comparison of experimental heat tests, mass transfer experiments and theoretical analysis in defined cases should be performed. This problem was described in papers by (Wilk, 2004) and (Lucas & Davies, 1970). Taking into account all the facts discussed in these works, one may tentatively conclude that the mass/heat transfer analogy uncertainty is of the same order of magnitude as the uncertainty of the heat transfer coefficient determined by means of a specific measuring technique.

9. References

Acosta R.E., Muller R.H. & Tobias C.W. (1985). Transport processes in narrow (capillary) channels. *American Institute of Chemical Engineers Journal*, 31, 3, 473-482, ISSN: 0001-1541

Adams T.M, Abdel-Khalik S.I, Jeter S.M. & Qureshi Z.H. (1998). An experimental investigation of single-phase forced convection in microchannels. *International Journal of Heat and Mass Transfer*, 41, 6-7, 851-857, ISSN:0017-9310

Bieniasz, B. & Wilk, J. (1995). Forced convection mass/heat transfer coefficient at the surface of the rotor of the sucking and forcing regenerative exchanger. *International Journal of Heat and Mass Transfer*, 38, 1, 1823-1830, ISSN:0017-9310

Bieniasz, B., Kiedrzyński K., Smusz R. & Wilk, J. (1997). Effect of positioning the axis of a lamellar rotor of a sucking and forcing regenerative exchanger on the intensity of convective mass/heat transfer. *International Journal of Heat and Mass Transfer*, 40, 14, 3275-3282, ISSN: 0017-9310

Bieniasz, B. (1998). Short ducts consisting of cylindrical segments and their convective mass/heat transfer, pressure drop and performance analysis. *International Journal of Heat and Mass Transfer*, 41, 3, 501-511, ISSN: 0017-9310

Bieniasz B. (2005). Convective mass/heat transfer for sheet rotors of the rotary regenerator. Publishing House of Rzeszów Univeersity of Technology, ISBN: 83-7199-348-X, Rzeszów, Poland

Bieniasz, B. (2009). Static research of flow in rotor channels of the regenerator. *International Journal of Heat and Mass Transfer*, 52, 25-26, 6050-6058, ISSN:0017-9310

Bieniasz, B. (2010). Convectional mass/heat transfer in a rotary regenerator rotor consisted of the corrugated sheets. *International Journal of Heat and Mass Transfer*, 53, 15-16, 3166-3174, ISSN: 0017-9310

Celata G.P., Cumo M., Marcowi V.McPhail S.J. & Zummo G. (2006). Microtube liquid single-phase heat transfer in laminar flow. *International Journal of Heat and Mass Transfer*, 49, 19-20, 3538-3546, ISSN: 0017-9310

Chilton T.H. & Colburn A.P. (1934). Mass transfer (absorption) coefficients prediction from data on heat transfer and fluid friction. *Industrial and Engineering Chemistry* 26, 11, 1183-1187, ISSN: 0888-5885

Goldstein R.J. & Cho H.H. (1995). A review of mass transfer measurements using naphthalene sublimation. *Experimental Thermal and Fluid Science*. 10, 4, 416-434, ISSN: 0894-1777

Graetz L. (1885). Über die Wärmeleitfähigkeit von Flüssigkeiten, *Annalen der Physik*, p.337

Hong K. & Song T-H. (2007). Development of optical naphthalene sublimation method. *International Journal of Heat and Mass Transfer*, 50, 19-20, 3890-3898-511, ISSN: 0017-9310

Kandlikar S.G., Joshi S. & Tian S. (2001) Effect of channel roughness on heat transfer and fluid flow characteristics at low Reynolds numbers in small diameter tubes. *Proceedings of 35th National Heat Transfer Conference*, Anaheim, California, Paper 12134

Kandlikar S.G., Garimella S., Li D., Colin S., & King M.R. (2006). *Heat Transfer and Fluid Flow in Minichannels and Microchannels*, Elsevier, ISBN: 0-0804-4527-6, Kidlington, Oxford

Kays W.M. (1955). Numerical solution for laminar flow heat transfer in circular tubes. *Transactions of the ASME Journal of Heat Transfer*, 77, 1265-1272

Kim J-Y. & Song T-H. (2003). Effect of tube alignment on the heat/mass transfer from a plate fin and two-tube assembly: naphthalene sublimation results. *International Journal of Heat and Mass Transfer*, 46 , 16, 3051-3059, ISSN: 0017-9310

Lelea D., Nishio K. & Takano K. (2004). The experimental research on microtube heat transfer and fluid flow of distilled water. *International Journal of Heat and Mass Transfer*, 47, 12-13, 2817-2830, ISSN: 0017-9310

Levĕqe M.A. (1928). Les lois de la transmission de la chaleur par convection. *Annales des Mines* 13, 201-299, 305-362, 381-415

Lucas D. M. & Davies R. M. (1970). Mass transfer modelling techniques in the prediction of convective heat transfer coefficients in industrial heating processes. *Proceedings of Fourth International Heat Transfer Conference*, Paris, Versailles, vol. VII, Paper M T 1.2

Owhaib W. & Palm B. (2004). Experimental investigation of single-phase convective heat transfer in circular microchannels. *Experimental Thermal and Fluid Science*. 28, 2-3, 105-110, ISSN: 0894-1777

Sara O.N., Barlay Ergu Ő., Arzutug M.E. & Yapıcı S. (2009). Experimental study of laminar forced convective mass transfer and pressure drop in microtubes. *International Journal of Thermal Sciences*, 48, 10, 1894-1900, ISSN: 1290-0729

Sellars J.R., Tribus M. & Klein J.S. (1956). Heat transfer to laminar flow in a round tube or flat conduit – the Graetz problem extended. *Transactions of the ASME Journal of Heat Transfer*, 78, 441-448

Sider E.N. & Tate G.E. (1936). Heat transfer and pressure drop of liquids in tubes. *Industrial and Engineering Chemistry*, 28, 1429-1436, ISSN: 0888-5885

Szănto D. A., Cleghorn S., Ponce – de - León C. & Walsh F. C. (2008). The limiting current for reduction of ferricyanide ion at nickel: The importance of experimental conditions. *American Institute of Chemical Engineers Journal*, 54, 3, 802-810, ISSN: 0001-1541

Szewczyk M. (2002). *The influence of the mutual displacement of the rib turbulators in the narrow channel on the intensity of the convective mass/heat transfer*. Doctor thesis. Rzeszów, Poland

Tso C.P. & Mahulikar S.P. (2000). Experimental verification of the role of Brinkman number in microchannels using local parameters. *International Journal of Heat and Mass Transfer*, 43, 10, 1837-1849, ISSN: 0017-9310

Wilk J. (2004). Mass/heat transfer coefficient in the radially rotating circular channels of the rotor of the high-speed heat regenerator. *International Journal of Heat and Mass Transfer*, 47, 8-9, 1979-1988, ISSN: 0017-9310

Wilk J. (2009). Experimental investigation of convective mass/heat transfer in short minichannel at low Reynolds numbers. *Experimental Thermal and Fluid Science*. 33, 2, 267-272, ISSN: 0894-1777

Yang C.Y. & Lin T.Y. (2007). Heat transfer characteristics of water flow in microtubes. *Experimental Thermal and Fluid Science*. 32, 2, 432-439, ISSN: 0894-1777

Yarin L.P., Mosyak A. & Hetsroni G. (2009). *Fluid Flow, Heat Transfer and Boiling in Micro-Channels*. Springer, ISBN: 978-3-540-78754-9, Berlin Heidelberg

15

Heat Transfer Enhancement for Weakly Oscillating Flows

Efrén M. Benavides
Universidad Politécnica de Madrid
Spain

1. Introduction

Heat exchangers that work with an oscillatory fluid flow exhibit a heat transfer dependence on the oscillation parameters, and hence such devices can modify its range of applicability and its performances by changing the oscillation parameters. This capability of oscillating flows to modulate the heat transfer process of some devices makes them interesting for some applications. Among the current devices that use oscillating flows are thermoacoustic engines and refrigerators (Backhaus & Swift, 2000; Gardner & Swift, 2003; Ueda et al., 2004), oscillatory flow reactors (Lee et al., 2001; Harvey et al., 2003), and, in general, any kind of heat exchangers with a periodical mass flow rate (Benavides, 2009). The theoretical characterization of the heat transfer process in such devices, where an oscillating flow is imposed over a stationary flow, is necessary either because it appears in a natural way such as in some kind of thermoacoustic devices (Benavides, 2006, 2007) or well because it is forced by pulsating the flow with vibrating or moving parts placed far enough in the upstream or downstream path (Wakeland & Keolian, 2004a, 2004b). Since the heat exchanged depends on the heat fluxes, an interesting way to modify these fluxes could be to change the mean velocity of the fluid just as it has been experimentally corroborated in baffled pipes (Mackley & Stonestreet, 1995). Other interesting way of changing the flux of heat is to change the amplitude and the frequency of an imposed pulsation. Yu et al. (2004) summarize this situation by classifying previous works into four categories according to the conclusion being reached: (a) pulsation enhances heat transfer (Mackley & Stonestreet, 1995; Faghri et al., 1979) (b) pulsation deteriorates heat transfer (Hemida et al., 2002) (c) pulsation does not affect heat transfer (Yu et al., 2004), and (d) heat transfer enhancement or deterioration may occur depending on the flow parameters (Cho & Hyun, 1990). Although, the main conclusion reached by Yu et al. (2004) was that pulsation neither enhances nor deteriorates heat flow and that the same result was found by Chattopadhyay et al. (2006), who showed that pulsation has no effect on the time-averaged heat transfer along straight channels, other results show that forced flow pulsation enhances convective mixing and affects Nusselt number (Kim & Kang, 1998; Velazquez et al., 2007). These works show how the heat transfer reaches its maximum for a specific frequency of oscillation and decreases for both higher and lower values of frequency. This disparity of results motivated a theoretical study (Benavides, 2009), based on a second order expansion of the integral equations that governs the oscillating flow, which predicts the possibility of having heat transfer enhancement or deterioration as a function of the frequency of the oscillation. It is interesting to note that this feature is present even in the case of having flows where the amplitude of the velocity oscillation is small when it is compared with the time-averaged velocity inside the device. Although the model proposed in that article is highly simplified, it fixes the basic mechanisms leading to a heat transfer enhancement when a pulsating flow is present and hence it gives a common explanation to all the aforementioned results. Here, we discuss this model with the purpose of *i*) obtaining a final formula able to fit the frequency response, and *ii*) showing that a correct measurement of the heat transferred requires a dynamic characterization of the outlet mass flow rate and temperature.

2. Problem formulation

2.1 Characterization of the heat transfer coefficient for weakly oscillating flows

For an incompressible flow under the assumption that the properties of the fluid are temperature independent, the dimensionless form of the conservation laws are given by the following partial derivative equations (Baehr & Stephan, 2006):

$$\sum_{i=1}^{3} \partial_i v_i = 0 \tag{1}$$

$$\mathrm{St}\,\partial_0 v_j + \sum_{i=1}^{3} v_i \partial_i v_j = -\partial_j p + \frac{1}{\mathrm{Re}} \sum_{i=1}^{3} \partial_i \partial_i v_j \tag{2}$$

$$\mathrm{St}\,\partial_0 \theta + \sum_{i=1}^{3} v_i \partial_i \theta = \frac{1}{\mathrm{Pe}} \sum_{i=1}^{3} \partial_i \partial_i \theta + \frac{\mathrm{Ec}}{\mathrm{Re}} \sum_{i=1}^{3}\sum_{j=1}^{3} \partial_j v_i (\partial_i v_j + \partial_j v_i) \tag{3}$$

In these equations, non-dimensional variables are related to dimensional ones by means of the followings relations (dimensional variables are on the left-hand side of the equalities):

$$\omega t = \xi_0 \;\; ; \;\; \frac{x_i}{L} = \xi_i \;\; ; \;\; \frac{u_i}{U} = v_i \;\; ; \;\; \frac{P}{\frac{1}{2}\rho U^2} = p \;\; ; \;\; \frac{T_W - T}{T_W - T_L} = \theta(T) \tag{4}$$

$$\omega^{-1}\frac{\partial}{\partial t} = \partial_0 \;\; ; \;\; L\frac{\partial}{\partial x_i} = \partial_i \tag{5}$$

$$\frac{\omega L}{U} = \mathrm{St} \;\; ; \;\; \frac{\rho L U}{\mu} = \mathrm{Re} \;\; ; \;\; \frac{\rho L U c}{k} = \mathrm{Pe} \;\; ; \;\; \frac{U^2}{c(T_W - T_L)} = \mathrm{Ec} \;\; ; \;\; \frac{c\mu}{k} = \frac{\mathrm{Re}}{\mathrm{Pe}} = \mathrm{Pr} \tag{6}$$

Here, t is the time, x_i are the spatial coordinates, u_i are the fluid velocities, and P and T are, respectively, the pressure and the temperature of the fluid at time t and position x_i; ω is the angular frequency of the oscillation, L is a characteristic length of the device, U is a characteristic velocity, T_W is the temperature of the hottest surface of the device, T_L is the inlet fluid temperature; ρ, μ, k, and c are, respectively, the density, the viscosity, the thermal conductivity, and the specific heat capacity of the fluid. Note that the dimensionless numbers St, Re, Pe, and Ec in Eqs. (6) are evaluated as characteristic constant numbers, and hence do not change with the oscillation. Eckert number, Ec, does not depend on the size of the device (i.e., on the characteristic length, L). Besides, it only affects Eq. (3), which is the one governing the temperature field, and only has to be taken into account when friction gives rise to a noticeable warming of the fluid, which is not the case because the kinetic energy is considered to be negligible when compared with the increment of the internal energy due to heat transfer. The order of magnitude of the third term, compared with the heat conduction term in Eq. (3), is given by the dimensionless number:

$$\frac{\mathrm{Pe}\,\mathrm{Ec}}{\mathrm{Re}} = \frac{\mu U^2}{k(T_W - T_L)} = \mathrm{Pr}\frac{U^2}{c(T_W - T_L)} \tag{7}$$

This expression allows defining the following characteristic velocity, whose values at different temperatures are given for water and air in Tables 1 and 2 respectively:

$$U_c = \sqrt{\frac{c(T_W - T_L)}{\mathrm{Pr}}} \tag{8}$$

T	ρ	c	μ	k	Pr	U_c $T_W\text{-}T_L$=1°C	U_c $T_W\text{-}T_L$=10°C
°C	kg m⁻³	J kg⁻¹ K⁻¹	Pa s	W K⁻¹ m⁻¹		m s⁻¹	m s⁻¹
10	999.70	4192.1	1307·10⁻⁶	0.5800	9.447	21.2	66.6
50	988.03	4180.6	547.0·10⁻⁶	0.6435	3.554	34.3	108
90	965.35	4205.0	314.5·10⁻⁶	0.6753	1.958	46.3	147

Table 1. Properties of liquid water at three different temperatures (elaborated from Lide, 2004)

T	ρ	c	μ	k	Pr	U_c $T_W\text{-}T_L$=1°C	U_c $T_W\text{-}T_L$=10°C
K	kg m⁻³	J kg⁻¹ K⁻¹	Pa s	W K⁻¹ m⁻¹		m s⁻¹	m s⁻¹
200	1.746	1007	13.3·10⁻⁶	0.0184	0.728	37.2	117.6
300	1.161	1007	18.6·10⁻⁶	0.0262	0.715	37.5	118.7
500	0.696	1030	27.1·10⁻⁶	0.0397	0.703	38.3	121.0

Table 2. Properties of air (1 bar) at three different temperatures (elaborated from Lide, 2004)

Equations (7) and (8) state that the last term in Eq. (3) can be neglected when $U<<U_c$. If this term is drooped, an error in the calculated temperature appears. Values collected in Table 1 show that the error is of the order of 1°C for water and a characteristic velocity near 20 m/s (a similar error is obtained for air with a characteristic velocity near 38 m/s). If the characteristic velocity is increased, the error grows: for water at a characteristic velocity near 67 m/s, the order of magnitude of the error is 10°C. When the characteristic velocities inside the device are much less than those given by Tables 1 and 2, the error in the temperature due to neglect this term is much less than 1°C (or 10°C). This means an error less than 1% (or 10%) for water suffering a heating in the range 10-90°C. Note that the errors due to consider that the fluid properties are independent of temperature are greater than these (Table 1 shows that the thermal conductivity of water suffers a variation of 15% in the rage 10-90°C). Thus, the energy equation can be substituted by:

$$\text{St}\,\partial_0\theta + \sum_{i=1}^{3} \upsilon_i\partial_i\theta = \frac{1}{\text{Pe}}\sum_{i=1}^{3}\partial_i\partial_i\theta \qquad (9)$$

The remaining numbers, St, Re, and Pe, depend on the size of the device, L. The Strouhal number, St, typically is near one in these applications, and hence the complete device cannot be treated as stationary. However, the Reynolds number, Re, is normally much greater than one. Since, as can be seen in Tables 1 and 2, the Prandtl number, Pr, is not much less than one (for air it is near 0.7 and for water lies in the range 2-10), the Péclet number, Pe, is also much greater than one. This means that the viscosity and the thermal conductivity lead to negligible effects in regions with characteristic lengths of the order of L. However, their effects are stronger in regions with a characteristic length much less than L. In order to see this, it is convenient to introduce these numbers as a function of a characteristic length, l_c, as:

$$\text{St}(l_c) = \frac{\omega l_c}{U} \quad ; \quad \text{Re}(l_c) = \frac{\rho l_c U}{\mu} \quad ; \quad \text{Pe}(l_c) = \text{Re}(l_c)\text{Pr} \qquad (10)$$

Indeed, these numbers fix the thickness of the thermal and viscous convective boundary layers. We define the characteristic thickness of the viscous boundary layer as $\text{Re}(\delta_\mu)=1$, which leads to $\delta_\mu=\mu/(\rho U)$. In the same way, the characteristic thickness of the thermal convective boundary layer is defined by $\text{Pe}(\delta_k)=1$, which leads to $\delta_k=k/(\rho c U)$. The conditions $\text{Re}(L)\gg1$ and $\text{Pe}(L)\gg1$ lead to $\delta_\mu\ll L$ and $\delta_k\ll L$, respectively. Therefore, the characteristic thicknesses of the boundary layers are much smaller than the characteristic length of the device. Although, the Strouhal number is typically near one in the device scale, i.e., $\text{St}(L)\sim1$, the Strouhal number in the boundary layers is much less than one, i.e., $\text{St}(\delta_\mu)\ll1$ and $\text{St}(\delta_k)\ll1$. Table 3 collects the different orders of magnitude of the dimensionless numbers $\text{St}(l_c)$, $\text{Re}(l_c)$, and $\text{Pe}(l_c)$, for the three different length scales.

l_c	$\text{St}(l_c)$	$\text{Re}(l_c)$	$\text{Pe}(l_c)$
L	1	$\gg1$	$\gg1$
δ_μ	$\ll1$	1	Pr
δ_k	$\ll1$	Pr⁻¹	1

Table 3. Orders of magnitude of the dimensionless numbers St, Re, and Pe, as a function of the length scale.

As a consequence of the results shown in the first column of Table 3, the boundary layers can be modeled as quasi-steady. Additionally, the dominant effects in the heat transfer forced by the wall temperature can be restricted to those that are important in a region near the wall that accomplishes $\text{Pe}(l_c)\sim1$. The energy equation can be rewritten for this layer by substituting the characteristic length L wih $\delta_k\ll L$, so that, near the wall, $\text{St}(\delta_k)\sim0(\delta_k/L)\ll1$. The same happens with the viscous boundary layer. This fact allows the removal of the temporal derivatives from the momentum and energy equations that describe the behaviour of the fluid near the wall. Therefore, although the complete device is not stationary, the thermal boundary layers can be modeled as quasi-steady, and Eqs. (2) and (9) can be reduced to:

$$\sum_{i=1}^{3}\upsilon_i\partial_i\upsilon_j = -\partial_j p + \text{Pr}\sum_{i=1}^{3}\partial_i\partial_i\upsilon_j \qquad (11)$$

$$\sum_{i=1}^{3} \upsilon_i \partial_i \theta = \sum_{i=1}^{3} \partial_i \partial_i \theta \qquad (12)$$

Let us introduce a local frame of reference in the wall which has the component $i=1$ parallel to the velocity, the component $i=2$ perpendicular to the surface wall, and $i=3$ parallel to the surface wall and perpendicular to the velocity. In this situation, the thermal conduction in the direction of the flow ($i=1$) can be neglected as well as all the derivatives in the third component ($i=3$), and hence, the equations are:

$$\upsilon_1 \partial_1 \upsilon_1 = -\partial_1 p + \mathrm{Pr}\, \partial_2 \partial_2 \upsilon_1 \qquad (13)$$

$$\upsilon_1 \partial_1 \theta = \partial_2 \partial_2 \theta \qquad (14)$$

Equation (14) states that, near the wall, the convective transport of energy is as important as the conductive transport of heat. Eq. (14) also states that the convective term is proportional to the velocity. As a result, the higher the velocity, the higher the convective term. However, the presence of the wall and the viscosity reduces the velocity inside the thermal boundary layer to zero: $\upsilon_i(\xi_2=0)=0$. Besides, Eq. (13) states that the higher the Prandtl number, the higher the reduction of velocity. It is interesting to study the following limits.

a.   Pr tends to infinity. This is the situation for water at 10°C whose Prandtl number is near ten (see Table 1). When the Prandtl number tends to infinity, the viscosity dominates the velocity inside the thermal boundary layer. This result is obtained from retaining in Eq. (13) the last term as the dominant one: $\partial_2 \partial_2 \upsilon_1 = 0 \Rightarrow \upsilon_1 = C\xi_2$. Here C is a constant imposed by the velocity outside of the thermal boundary layer.

b.   Pr tends to zero. A negligible Prandtl number produces the higher velocity inside the thermal boundary layer. Although this extreme case is not representative of common fluids, it may be useful for obtaining an upper bound to the maximum heat transfer (Benavides, 2009). This result is obtained from neglecting in Eq. (13) the last term: $\upsilon_1 \partial_1 \upsilon_1 = -\partial_1 p$. The vanishing viscosity allows to approach the velocity near the wall by a parallel flux ($i=1$) which does not depend on the transversal coordinate ($i=2$).

The heat transferred over a square region of the wall with an area $L_1 L_3$, which accomplishes $L \gg L_1 \sim L_3 \gg \max(\delta_\mu, \delta_k)$, amounts to:

$$\int_0^{L_1 L_3} -k \left.\frac{\partial T}{\partial x_2}\right|_{x_2=0} dx_1 dx_3 = kL_3 (T_W - T_L) \int_0^{\infty} \left.\frac{\partial \theta}{\partial \xi_2}\right|_{\xi_2=0} d\xi_1 \qquad (15)$$

When the Prandtl number is near one, such as the Prandtl number of air or water at higher temperatures, the velocity in the thermal boundary layer is neither linear nor constant with the transversal component. In general, Eqs. (13) and (14) must be integrated with the boundary conditions $\theta(\xi_2=0)=\upsilon_i(\xi_2=0)=0$, $\theta(\xi_2\to\infty)=(T_W\text{-}T_{ext})/(T_W\text{-}T_L)$, and $\upsilon_i(\xi_2\to\infty)=\upsilon_{ext}$ being T_{ext} and υ_{ext} the solution obtained from Eqs. (1), (2) and (9) with Re=∞ and Pe=∞. The condition $\theta(\xi_2\to\infty)=(T_W\text{-}T_{ext})/(T_W\text{-}T_L)$ indicates that it is convenient to rescale the function θ as $(T_W\text{-}T_I)\theta/(T_W\text{-}T_{ext})=\theta_b$, which has to accomplish the boundary conditions $\theta_b(\xi_2=0)=0$, $\theta_b(\xi_2\to\infty)=1$. This change of variable does not affect the form of Eqs. (13) and (14) so that the function θ_b only can depend on the Prandtl number and on the value of υ_{ext}. This change of variable allows rewriting the above expression as:

$$\int_0^{L_1 L_3} -k\frac{\partial T}{\partial x_2}\Bigg|_{x_2=0} dx_1 dx_3 = kL_3(T_{1V} - T_{ext})\int_0^{\infty}\frac{\partial \theta_b}{\partial \xi_2}\Bigg|_{\xi_2=0} d\xi_1 = kL_3(T_W - T_{ext})f(\text{Pr},\upsilon_{ext}) \tag{16}$$

Note that, in general, the external solution depends on the position of the surface where the region is located. Although it has been assumed that this dependence is negligible when it is compared with the transversal variation through the boundary layer, it can give a significant contribution when it is evaluated along the full wall. The convective heat transfer coefficient for the full device is defined as:

$$\int_0^{S_W} -k\frac{\partial T}{\partial x_2}\Bigg|_{\substack{solid \\ surface}} dS_W = \int_0^{S_W} kf(\text{Pr},\upsilon_{ext})(T_{1V} - T_{ext})dS_W \tag{17}$$

It is convenient to introduce a spatial-averaged temperature $T_P(t)$ and a spatial-averaged heat transfer coefficient $h(t)$, by means of the following definition:

$$\int_0^{S_W} kf(\text{Pr},\upsilon_{ext})(T_{1V} - T_{ext})dS_W = S_W h(t)(T_W - T_P(t)) \tag{18}$$

Obviously, Eq. (18) states that the heat transfer coefficient depends on the convention used for defining T_P (Baehr & Stephan, 2006). This question is also important for oscillating flows, and hence it has been discussed in the literature (Faghri et al., 1979; Guo & Sung, 1997; Benavides, 2009). Note that Eq. (18) states that the final outcome is given by the product of T_P and h: the heat transfer coefficient is used only as an intermediate function. In section 2.3 (see Eq. 35), the logarithmic mean temperature will be used to model T_P. This assumption will be discussed later. As Eq. (18) indicates, functions $T_P(t)$ and $h(t)$ depend on the Prandtl number, and on the external solution which also depends on the Strouhal number and on the boundary conditions. For oscillating flows, the effect of the pulsation on the velocity near the wall can be described by a temporal evolution of the external solution. This external solution depends both, on the time-averaged mass flow rate and on the oscillatory component of the mass flow rate. It is convenient to define the oscillating component of the mass flow rate as:

$$\frac{G(t)}{\overline{G}} - 1 = \varepsilon g_1(t) + \varepsilon^2 g_2(t) + 0(\varepsilon^3) \tag{19}$$

where $0<\varepsilon<<1$ is a dimensionless number that assures a weakly oscillating flow, and $g_1(t)$ and $g_2(t)$ are two dimensionless functions taking into account the effects of the oscillation on the mass flow rate. Here, $G(t)$ is the instantaneous mass flow rate and $\overline{G}$ is the stationary mass flow rate. (Through the chapter, a bar above the symbols means stationary solution, i.e., the solution for $\varepsilon=0$.) The stationary mass flow rate $\overline{G}$ induces a heat transfer process which can be obtained from a characteristic velocity U, and a characteristic cross sectional area A, which define a Reynolds number given by the expression:

$$\text{Re} = \frac{\rho L U}{\mu} = \frac{L\overline{G}}{\mu A} \tag{20}$$

Since the oscillatory part of the mass flow rate modifies the velocities v_{ext} and the temperatures T_{ext}, it also modifies the heat transfer process. Let $\wp$ be the set of parameters that are necessary to define the function $G/\bar{G}-1$ for any value of time; for example, parameters that might belong to the set $\wp$ are the relative amplitude of the oscillation or its phase. Although parameter ε could belong to set $\wp$, due to its importance, it will be explicitly written as an argument. Thus, in general, the functions $T_P(t)$ and $h(t)$ depend also on Prandtl, Reynolds and Strouhal numbers and on the set of parameters defining the oscillation, i.e., $T_P(t;Pr,Re,St,\wp,\varepsilon)$ and $h(t;Pr,Re,St,\wp,\varepsilon)$. Finding a solution of the heat transfer problem requires a model for both, T_P and h. This will be done in following sections. For weakly oscillating flows, it is $\varepsilon \ll 1$ and the effect of the pulsation on the heat transfer coefficient can be described by:

$$h\left(t;\Pr,\mathrm{Re},\mathrm{St},\wp,\varepsilon\right)=h\left(t;\Pr,\mathrm{Re},\mathrm{St},\wp,0\right)\left[1+h_1\left(t;\Pr,\mathrm{Re},\mathrm{St}\right)\varepsilon+h_2\left(t;\Pr,\mathrm{Re},\mathrm{St}\right)\varepsilon^2+0\left(\varepsilon^3\right)\right] \quad (21)$$

Finally, we will assume that in absence of oscillation of the mass-flow rate there is not any self-oscillating behaviour, i.e., if $\varepsilon=0$, the solution does not depend on time. Under this condition, the leading term in the above expression is the convective heat transfer coefficient without oscillation, $\bar{h}$. Thus, the above expression can be rewritten as:

$$\frac{h\left(t;\Pr,\mathrm{Re},\mathrm{St},\wp,\varepsilon\right)}{\bar{h}\left(\Pr,\mathrm{Re}\right)}-1=h_1\left(t;\Pr,\mathrm{Re},\mathrm{St},\wp\right)\varepsilon+h_2\left(t;\Pr,\mathrm{Re},\mathrm{St},\wp\right)\varepsilon^2+0\left(\varepsilon^3\right) \quad (22)$$

2.2 Estimation of heat transfer coefficient for weakly oscillating flows

For stationary flows the heat transfer coefficient is obtained by relating Nusselt, Reynolds and Prandtl numbers. A frequently used expression for this function is (Baehr & Stephan, 2006):

$$\mathrm{Nu} = K\,\mathrm{Re}^\alpha\,\Pr^\beta \quad (23)$$

where K, α and β are dimensionless constants that depend on the geometry and on the type of flow. Here, the Nusselt number is defined as usual, $\mathrm{Nu}=\bar{h}L/k$, where L is a characteristic length associated with the device geometry and k is the fluid thermal conductivity at the representative temperature in that region. From this expression, we can obtain the heat transfer coefficient for stationary flows as:

$$\bar{h}=\frac{K}{L}\mathrm{Re}^\alpha\,\Pr^\beta \quad (24)$$

Because the Reynolds number depends linearly on the velocity, any variation of the velocity must induce a variation of the heat transfer coefficient. This fact can be used to estimate the value of function h in Eq. (18) by means of $\bar{h}$. The following expression can be used as a first attempt:

$$h\left(t;\Pr,\mathrm{Re},\mathrm{St},\wp,\varepsilon\right)\cong\bar{h}\left(\Pr,\frac{LG(t;\wp,\varepsilon)}{\mu A}\right)=\bar{h}\left(\Pr,\mathrm{Re}+\mathrm{Re}\left(\frac{G(t;\wp,\varepsilon)}{\bar{G}}-1\right)\right) \quad (25)$$

For weakly oscillating flows it is $G/\bar{G}-1 \ll 1$, and hence, the expression above can be expanded as:

$$\bar{h}\left(\Pr,\mathrm{Re}+\mathrm{Re}\left(\frac{G}{\bar{G}}-1\right)\right) = $$
$$\bar{h}\left(\Pr,\mathrm{Re}\right)+\mathrm{Re}\left.\frac{\partial \bar{h}}{\partial \mathrm{Re}}\right|_{G=\bar{G}}\left(\frac{G}{\bar{G}}-1\right)+\frac{\mathrm{Re}^2}{2}\left.\frac{\partial^2 \bar{h}}{\partial \mathrm{Re}^2}\right|_{G=\bar{G}}\left(\frac{G}{\bar{G}}-1\right)^2+0\left(\frac{G}{\bar{G}}-1\right)^3 \qquad (26)$$

The comparison between Eq. (26) and Eq. (22) allows writing an estimation of h_1 and h_2 as:

$$h_1\left(t;\Pr,\mathrm{Re},\mathrm{St},\wp\right)\varepsilon \cong \frac{\mathrm{Re}}{\bar{h}\left(\Pr,\mathrm{Re}\right)}\left.\frac{\partial \bar{h}}{\partial \mathrm{Re}}\right|_{G=\bar{G}}\left(\frac{G(t;\wp,\varepsilon)}{\bar{G}}-1\right) \qquad (27)$$

$$h_2\left(t;\Pr,\mathrm{Re},\mathrm{St},\wp\right)\varepsilon^2 \cong \frac{\mathrm{Re}^2}{2\bar{h}\left(\Pr,\mathrm{Re}\right)}\left.\frac{\partial^2 \bar{h}}{\partial \mathrm{Re}^2}\right|_{G=\bar{G}}\left(\frac{G(t;\wp,\varepsilon)}{\bar{G}}-1\right)^2 \qquad (28)$$

From Eq. (24), we can estimate the value of functions h_1 and h_2 in Eq. (21) as:

$$h_1 \cong \alpha g_1(t;\wp) \qquad (29)$$

$$h_2 \cong \alpha g_2(t;\wp)+\frac{\alpha(\alpha-1)}{2}g_1(t;\wp)^2 \qquad (30)$$

This approximation was explored by Benavides (2009) with a reasonable succes for different configurations. For example, one interesting case is the incompressible laminar flow in circular tubes, that assumes all the properties to be constant and both hydrodynamic and thermally fully developed, where it is known (Baehr & Stephan, 2006) that the Nusselt number in this case does not depend on the Reynolds and Prandtl numbers, i.e., $\alpha=\beta=0$. Additionally, it has been shown (Chattopadhyay et al., 2006) that a circular tube in the laminar regime under pulsating flow conditions does not have any oscillation of the local Nusselt number if the flow is both thermally and hydrodynamically developed. The present model explains this behavior: the stationary solution for a constant wall temperature Nu=3.66 (Baehr & Stephan, 2006) shows that in this case α is zero and hence Eqs. (18), (20), (29) and (30) avoid any fluctuation of the heat transfer coefficient. However, Chattopadhyay et al. (2006) also found that the Nusselt number varies in time in the near-entry region of the pipe. The explanation comes again from Eqs. (29) and (30) by taking into account that the thermal entry flow with a fully developed velocity profile has a behavior given by the Léveque solution Nu~Re[0.33] (Baehr & Stephan, 2006), i.e., $\alpha=0.33$, and hence the expansion given by Eq. (18), together with Eqs. (29) and (30), retains the pulsation effects.

2.3 Integral formulation of the energy conservation for weakly oscillating flows

The model given by Eq. (18) allows calculating the heat transferred if the spatial-averaged temperature T_P and the heat transfer coefficient h are known. The heat transfer coefficient can be estimated by means of Eqs. (22), (24), (29) and (30). This section will introduce the estimation of T_P as a function of time. For this purpose, the physics involved will be simplified to retain only the most important aspects for the dynamical response. Let us define V as the volume of fluid inside the device where the unsteady temperature field T and the three-dimensional unsteady velocity field u_i are being studied; let be $S(V)$ the surface that limits the volume V. The Reynolds' transport theorem applied to the energy balance leads to:

$$\frac{dQ}{dt} = \int_V \frac{\partial}{\partial t}\left(\rho c T + \rho \sum_{i=1}^{3}\frac{u_i^2}{2}\right)d^3x + \oint_{S(V)}\left(cT + \frac{P}{\rho} + \sum_{i=1}^{3}\frac{u_i^2}{2}\right)\rho\sum_{i=1}^{3}u_i \cdot dA^i \tag{31}$$

The term on the left-hand side of the equality represents the heat flux and dA is a surface element vector. Since $u_i^2/2cT\ll1$ holds due to $U^2\ll U_c^2<2cT$ [see Eq. (8) and Tables 1 and 2], the kinetic energy can be neglected when it is compared with the internal energy of the fluid. Additionally, for an ideal gas, which satisfies $P=\rho R T$, the specific heat at constant pressure $c_p=c+R$ appears; for an incompressible flow, the following inequality holds $P/\rho cT\ll1$ and it can be considered $c_p=c$. These considerations lead to:

$$\frac{dQ}{dt} = \int_V \frac{\partial}{\partial t}(\rho c T)d^3x + Gc_pT_H - Gc_pT_L \tag{32}$$

Here, $\rho c_p T_L$ and $\rho c_p T_H$ are the spatial-averaged specific enthalpies (specific internal energies for a liquid) over the inlet and exit ports respectively. If there is not any source of mechanical work in the volume V, the surface $S(V)$ does not change with time and hence the temporal derivative and the spatial integral commutate:

$$\frac{dQ}{dt} = \frac{d}{dt}\int_V \rho c T d^3x + Gc_pT_H - Gc_pT_L \tag{33}$$

Although the fluid temperature changes over the device, the integral can be removed by defining $\rho c T_p$ as the spatial-averaged specific internal energy over the volume V, and hence the energy balance equation leads to:

$$\frac{dQ}{dt} = V\frac{d}{dt}\left[\rho c T_P(t)\right] + Gc_p\left[T_H - T_L\right] \tag{34}$$

Finally, the spatial-averaged specific internal energy will be modelled by considering that the spatial averaged density does not change over time, and that the spatial-averaged temperature T_p follows the logarithmic mean temperature (Baehr & Stephan, 2006) given by (this assumption will be discussed later):

$$T_P = T_W - \frac{T_H - T_L}{\ln\dfrac{T_W - T_L}{T_W - T_H}} \tag{35}$$

Taking into account Eq. (20), the heat flow amounts to:

$$\frac{dQ}{dt} = hS_W(T_W - T_P) \tag{36}$$

Here h is the spatial-averaged heat transfer coefficient defined by Eqs. (18) and (22), and S_W is the wall area at the temperature T_W. The energy balance comes from Eqs. (34) and (36) with the spatial-averaged temperature defined by Eq. (35):

$$hS_W(T_W - T_P) = \rho V c \frac{dT_P}{dt} + Gc_p(T_H - T_L) \tag{37}$$

The temporal response of the device is obtained by solving Eqs. (35) and (37) for T_P and T_W. For studying the solution, it is convenient to use a dimensionless form of these equations and variables. For this purpose, let us define a dimensionless measure of the heating efficiency as:

$$\eta(T_H) = \frac{T_H - T_L}{T_W - T_L} \tag{38}$$

The dimensionless coefficient that takes into account the temperature inside the device is:

$$\theta(T_P) = \frac{T_W - T_P}{T_W - T_L} \tag{39}$$

The dimensionless coefficient that takes into account the outlet temperature is:

$$\theta(T_H) = \frac{T_W - T_H}{T_W - T_L} = 1 - \eta(T_H) \tag{40}$$

These coefficients transform Eq. (35), which defines the logarithmic temperature, into:

$$\ln\left(1 - \eta(T_H)\right) + \frac{\eta(T_H)}{\theta(T_P)} = 0 \tag{41}$$

and the energy balance into:

$$\frac{h}{\overline{h}}\theta(T_P) + \frac{\rho V c}{\overline{h}S_W}\frac{d\theta(T_P)}{dt} = \frac{\overline{G}c}{\overline{h}S_W}\frac{G}{\overline{G}}\eta(T_H) \tag{42}$$

This differential equation let us define a characteristic time t_c (see Eq. 43) and a characteristic dimensionless number χ (see Eq. 44) which depends on the stationary solution.

$$t_c = \frac{\rho V c}{\overline{h}S_W} \quad ; \quad t = t_c\tau \tag{43}$$

$$\chi = \frac{\overline{h}S_W}{\overline{G}c_p} \tag{44}$$

As a consequence of the above definitions, the behaviour of the device is characterized by the following two dimensionless equations:

$$\ln\left(1 - \eta(T_H(\tau))\right) + \frac{\eta(T_H(\tau))}{\theta(T_P(\tau))} = 0 \tag{45}$$

$$\frac{h(\tau)}{\bar{h}}\theta(T_P(\tau)) + \frac{d\theta(T_P(\tau))}{d\tau} = \frac{1}{\chi}\frac{G(\tau)}{\bar{G}}\eta(T_H(\tau)) \tag{46}$$

2.4 Non-oscillating solution

The condition $\varepsilon=0$ retains only the leading terms of the solution, which can be identified with the stationary part of the problem. Therefore, when we are interested on the stationary solution, Eqs. (45) and (46) admit the following simplification:

$$\ln\left(1-\eta(\bar{T}_H)\right) + \frac{\eta(\bar{T}_H)}{\theta(\bar{T}_P)} = 0 \tag{47}$$

$$\frac{\eta(\bar{T}_H)}{\theta(\bar{T}_P)} = \chi \tag{48}$$

The stationary solution obtained from both equations is:

$$\eta(\bar{T}_H) = 1 - e^{-\chi} \tag{49}$$

$$\theta(\bar{T}_P) = \frac{1-e^{-\chi}}{\chi} \tag{50}$$

$$\theta(\bar{T}_H) = 1 - \eta(\bar{T}_H) = e^{-\chi} \tag{51}$$

For the subsequent discussions, it is convenient to define the dimensionless number Z as:

$$\frac{\theta(\bar{T}_P)}{\theta(\bar{T}_H)} = \frac{1-e^{-\chi}}{\chi e^{-\chi}} = 1 + \frac{1}{Z} \tag{52}$$

$$Z = \frac{\chi e^{-\chi}}{1-(1+\chi)e^{-\chi}} \tag{53}$$

Since it is $Z>0$ for any value of $\chi>0$, the above expressions lead to the following inequalities concerning the temperatures involved: $T_L < \bar{T}_P < \bar{T}_H < T_W$.

2.5 Second-order expansion of the oscillating solution

The stationary solution found in the previous section allows rewriting Eqs. (45) and (46) in the following form:

$$\ln\left(1-\frac{\eta(T_H(\tau))}{\eta(\bar{T}_H)}(1-e^{-\chi})\right) + \frac{\eta(T_H(\tau))}{\eta(\bar{T}_H)}\frac{\theta(\bar{T}_P)}{\theta(T_P(\tau))}\chi = 0 \tag{54}$$

$$\frac{h(\tau)}{\bar{h}}\frac{\theta(T_P(\tau))}{\theta(\bar{T}_P)} + \frac{d}{d\tau}\frac{\theta(T_P(\tau))}{\theta(\bar{T}_P)} = \frac{G(\tau)}{\bar{G}}\frac{\eta(T_H(\tau))}{\eta(\bar{T}_H)} \tag{55}$$

In order to solve these equations, it is necessary to know $G/\bar{G}$ and $h/\bar{h}$. Here, we will use the model for weakly oscillating flows introduced in Section 2.1 as:

$$\frac{G}{\overline{G}} = 1 + g_1\varepsilon + g_2\varepsilon^2 + 0(\varepsilon^3) \tag{56}$$

$$\frac{h}{\overline{h}} = 1 + h_1\varepsilon + h_2\varepsilon^2 + 0(\varepsilon^3) \tag{57}$$

The oscillatory solution can be expanded as a series of the small parameter ε as:

$$\frac{\theta(T_P)}{\theta(\overline{T}_P)} = 1 + p\varepsilon + q\varepsilon^2 + 0(\varepsilon^3) \tag{58}$$

$$\frac{\eta(T_H)}{\eta(\overline{T}_H)} = 1 + r\varepsilon + s\varepsilon^2 + 0(\varepsilon^3) \tag{59}$$

where p, q, r, and s can be found by substituting expressions (56) to (59) in Eqs. (54) and (55). The first and second order terms of these equations lead to:

$$p = -\frac{r}{Z} \tag{60}$$

$$q = -Y\frac{r^2}{Z^2} - \frac{s}{Z} \tag{61}$$

$$\frac{dp}{d\tau} = r - p + g_1 - h_1 \tag{62}$$

$$\frac{dq}{d\tau} = s - q + g_1 r - h_1 p + g_2 - h_2 \tag{63}$$

$$Y = \frac{(2 + \chi)e^{-2\chi} - 4e^{-\chi} + 2 - \chi}{2\left[1 - (1 + \chi)e^{-\chi}\right]^2} \tag{64}$$

Therefore, when the fluctuation is taken into account, the ratio (58) is expressed as:

$$\frac{\theta(T_P)}{\theta(\overline{T}_P)} = 1 - \frac{r}{Z}\varepsilon - \left(Y\frac{r^2}{Z^2} + \frac{s}{Z}\right)\varepsilon^2 + 0(\varepsilon^3) \tag{65}$$

The substitution of p and q, given by Eqs. (60) and (61), in Eqs. (62) and (63) yields to:

$$\frac{dr}{d\tau} + (Z + 1)r = Z(h_1 - g_1) \tag{66}$$

$$\frac{ds}{d\tau} + (Z + 1)s = \left(2 + \frac{1}{Z}\right)Yr^2 - (Z - 2Y)g_1 r - (1 + 2Y)h_1 r + Z(h_2 - g_2) \tag{67}$$

This system of differential equations determines r and s when functions g_1, g_2, h_1, and h_2 are given as data. From Eqs. (60) and (61) functions p and q are also determined. Without any loss of generality for the purpose of the paper, g_1, h_1, p, q, r and s, are oscillating functions that accomplishes:

$$\int_0^{\frac{2\pi}{\omega t_c}} g_1 d\tau = \int_0^{\frac{2\pi}{\omega t_c}} h_1 d\tau = 0 \tag{68}$$

$$\int_0^{\frac{2\pi}{\omega t_c}} \frac{dp}{d\tau} d\tau = \int_0^{\frac{2\pi}{\omega t_c}} \frac{dq}{d\tau} d\tau = \int_0^{\frac{2\pi}{\omega t_c}} \frac{dr}{d\tau} d\tau = \int_0^{\frac{2\pi}{\omega t_c}} \frac{ds}{d\tau} d\tau = 0 \tag{69}$$

As a consequence, Eq. (66) states that the mean value of function r is zero:

$$\int_0^{\frac{2\pi}{\omega t_c}} r d\tau = 0 \tag{70}$$

Contrarily, the term g_1^2 in Eq. (30) shows that the mean value of function h_2 can be distinct of zero:

$$\int_0^{\frac{2\pi}{\omega t_c}} h_2 d\tau \neq 0 \tag{71}$$

Therefore, function s exhibits a mean value given by Eq. (67) as:

$$\int_0^{\frac{2\pi}{\omega t_c}} s d\tau = \frac{2Z+1}{Z(Z+1)} Y \int_0^{\frac{2\pi}{\omega t_c}} r^2 d\tau - \frac{Z-2Y}{Z+1} \int_0^{\frac{2\pi}{\omega t_c}} g_1 r d\tau - \frac{1+2Y}{Z+1} \int_0^{\frac{2\pi}{\omega t_c}} h_1 r d\tau + \frac{Z}{Z+1} \int_0^{\frac{2\pi}{\omega t_c}} (h_2 - g_2) d\tau \tag{72}$$

Multiplying Eq. (66) by r and taking the mean value of the resulting expression allow expressing the mean value of r^2 as a function of the mean value of $g_1 r$ and $h_1 r$ as:

$$\int_0^{\frac{2\pi}{\omega t_c}} r^2 d\tau = \frac{Z}{Z+1} \int_0^{\frac{2\pi}{\omega t_c}} (h_1 - g_1) r d\tau \tag{73}$$

Thus, Eq. (72) simplifies to:

$$\int_0^{\frac{2\pi}{\omega t_c}} s d\tau = -\frac{Z(Z+1)-Y}{(Z+1)^2} \int_0^{\frac{2\pi}{\omega t_c}} g_1 r d\tau - \frac{Z+Y+1}{(Z+1)^2} \int_0^{\frac{2\pi}{\omega t_c}} h_1 r d\tau + \frac{Z}{Z+1} \int_0^{\frac{2\pi}{\omega t_c}} (h_2 - g_2) d\tau \tag{74}$$

This interesting result shows that the oscillation can modify the time-averaged value of the outlet temperature. Thus, it is plausible to expect that a thermocouple placed on the outlet stream measures this deviation. This solution also shows that this effect is of order ε^2. However, although the time-averaged outlet temperature is related to the heat flux, they are not exactly the same. The heat transferred per unit of time is obtained from Eq. (34) as:

$$\frac{dQ}{dt} = \rho c V \frac{dT_P}{dt} + Gc_p(T_H - T_L) \tag{75}$$

The dimensionless form of this equation is:

$$\frac{\dfrac{dQ}{dt}}{\overline{h}S_W(T_W - \overline{T}_P)} = -\frac{d}{d\tau}\frac{\theta(T_P)}{\theta(\overline{T}_P)} + \frac{G}{\overline{G}}\frac{\eta(T_H)}{\eta(\overline{T}_H)} \tag{76}$$

The solution given by Eqs. (56) to (59) leads to:

$$\frac{\dfrac{dQ}{dt}}{\overline{h}S_W(T_W - \overline{T}_P)} = 1 + \left(r + g_1 - \frac{dp}{d\tau}\right)\varepsilon + \left(s + g_1 r + g_2 - \frac{dq}{d\tau}\right)\varepsilon^2 \tag{77}$$

Remembering Eqs. (68) to (70), the time-averaged flux of heat evolves to:

$$\frac{\omega t_c}{2\pi}\int_0^{\frac{2\pi}{\omega t_c}}\frac{dQ/dt}{\overline{h}S_W(T_W - \overline{T}_P)}d\tau - 1 = \varepsilon^2 \frac{\omega t_c}{2\pi}\int_0^{\frac{2\pi}{\omega t_c}}(s + g_1 r + g_2)d\tau \tag{78}$$

This expression shows that the terms $g_1 r$ and g_2 under the second integral of Eq. (78) introduce a bias respect to the time-averaged outlet temperature. The solution given by Eq. (74) lets us write the right-hand side of this equation as:

$$\int_0^{\frac{2\pi}{\omega t_c}}(s + g_1 r + g_2)d\tau = \frac{Z+Y+1}{(Z+1)^2}\int_0^{\frac{2\pi}{\omega t_c}}(g_1 - h_1)r d\tau + \frac{Z}{Z+1}\int_0^{\frac{2\pi}{\omega t_c}}h_2 d\tau + \frac{1}{Z+1}\int_0^{\frac{2\pi}{\omega t_c}}g_2 d\tau \tag{79}$$

Equations (74) and (79) differ only in the coefficients that precede the integrals on the right-hand side of both equations. The value of g_1-h_1 can be obtained from Eq. (73), what simplifies the above expression to:

$$\int_0^{\frac{2\pi}{\omega t_c}}(s + g_1 r + g_2)d\tau = -\frac{Z+Y+1}{Z(Z+1)}\int_0^{\frac{2\pi}{\omega t_c}}r^2 d\tau + \frac{Z}{Z+1}\int_0^{\frac{2\pi}{\omega t_c}}h_2 d\tau + \frac{1}{Z+1}\int_0^{\frac{2\pi}{\omega t_c}}g_2 d\tau \tag{80}$$

3. Discussion

In the discussion carried out by Benavides (2009), in order to overcome the problems derived from how the time-averaged temperature is defined (Guo & Suang, 1997), the heat flux was estimated as the time-averaged increment of the specific enthalpy along the device, i.e., as the value of s. Although, this approximation is interesting because the time-averaged outlet temperature is near the one that a thermocouple would measure in the exit port of an experimental arrangement, Eq. (78) shows that the constant part of s is not equivalent to the time-averaged heat flux under oscillating conditions. Indeed, the presence of the term $g_1 r$ modifies averaged value and hence, it also modifies the prediction that the model produces.

Equation (80), which takes into account this term, shows that this term removes completely the possibility of having an enhancement of the heat flux due to the first order terms g_1 and h_1 because the first term on the right-hand side of this equation is always negative. Depending on the sign of the integrals of h_2 and g_2, the heat transfer could suffer enhancement or deterioration. However, note that following the estimation given by Eq. (30) this term is proportional to $\alpha(\alpha-1)/2$, which depends on the level of turbulence in the device and which is maximum for $\alpha=1/2$ (i.e., for laminar flow), proportional to $Z/(Z+1)$ which is maximum for $\chi\to0$ (i.e., for low efficiency devices – see the asymptotic behaviour of function Z in Table 4 at the end of Section 3), and proportional to the mean value of g_1^2, which is ½ for a sine law of maximum amplitude without reversing flow. Therefore, the maximum attainable discouragement predicted by this term is near $\alpha(\alpha-1)/4<1/16\sim6\%$. The other term that can produce deterioration is the one with the time-averaged value of g_2, which produces deterioration of heat transfer in the presence of pulsating flow if the pulsation significantly increments the losses through the channel (Kim and Kang, 1998, reported that, in a 2D channel with two heated blocks working with a pulsating flow of air, the force over the blocks is typically amplified by the pulsating flow; the amplitude of the instantaneous friction factor increases dramatically, easily reaching 50 times the stationary value). Nevertheless, this term is proportional to $1/(Z+1)$ which tends to zero when $\chi\to0$ and hence it becomes negligible in this limit. It is interesting to note that the limit $\chi\to0$ can also be obtained by substituting the logarithmic definition of the averaged temperature [see Eq. (35)] by the condition $T_P=T_L$ (i.e., by $\theta(T_P)=1$). The solution for this definition of the spatial-averaged temperature inside the device comes from substituting Eq. (55) with $\theta(T_P)/\theta(\overline{T}_P)=1$ and hence, the solution is $Z=\infty$, $Y=0$, $p=q=0$, $r=h_1$-g_1, $s=g_1^2$-$g_1h_1+h_2$-g_2 and $s+g_1r+g_2=h_2$. The last expression shows that the dominant term is the one generated by the second order term of the heat transfer coefficient. Thus, in this limit, the enhancement of the heat transfer requires a positive value of h_2 which is incompatible with the estimation given by Eq. (30), what shows that this estimation has a restricted range of validity. Eq. (35) can also be substituted by $T_P=T_H$ (i.e., by $\theta(T_P)+\eta(T_H)=1$). In this case Eq. (54) must be substituted by $\theta(T_P)/\theta(\overline{T}_P)+\chi\eta(T_H)/\eta(\overline{T}_H)=1+\chi$ and the entire solution remains valid for $Z=1/\chi$ and $Y=0$. It is possible to see that the behaviour of this approach is qualitatively equal to the obtained with the logarithmic definition. Indeed, depending on the chosen values of Z and Y, the solutions in Eqs. (74) and (80) are valid for different definitions of the spatial-averaged temperature T_P. This holds while the relationship between temperatures remains algebraic, which means that the response of both temperatures to the external mass flow rate pulsation is instantaneous. A more detailed model would need a differential equation where the relationship between temperatures included temporal delays; then, the expected solution would depend not only on the value of r^2 but on the value of g_1r and h_1r. Note that Eq. (74) retains these terms for the time-averaged outlet temperature. The approach in the next section is based on retaining only one fundamental harmonic and shows that the outlet temperature may suffer an enhancement when the relative phases between g_1, h_1 and r are correctly chosen.

3.1 Single-harmonic approach

Let us characterise the oscillation by means of only one harmonic. Thus the solution can be written in the following way:

$$g_1 = \frac{g_{1+}e^{j\omega t_c \tau} + g_{1-}e^{-j\omega t_c \tau}}{2} \tag{81}$$

$$h_1 = \frac{h_{1+}e^{j\omega t_c \tau} + h_{1-}e^{-j\omega t_c \tau}}{2} \tag{82}$$

$$r = \frac{r_+e^{j\omega t_c \tau} + r_-e^{-j\omega t_c \tau}}{2} \tag{83}$$

where subscripts + and − indicates two conjugate complex functions that do not depend on time, and j is the imaginary unit. Then, the solution is obtained from Eqs. (66) and (74) as:

$$r_+ = \frac{Z}{Z+1+j\omega t_c}(h_{1+} - g_{1+}) \tag{84}$$

$$\frac{\omega t_c}{2\pi}\int_0^{\frac{2\pi}{\omega t_c}} s\, d\tau = -\frac{Z(Z+1)-Y}{(Z+1)^2}\frac{g_{1+}r_- + g_{1-}r_+}{4} - \frac{Z+Y+1}{(Z+1)^2}\frac{h_{1+}r_- + h_{1-}r_+}{4}$$
$$+\frac{Z}{Z+1}\frac{\omega t_c}{2\pi}\int_0^{\frac{2\pi}{\omega t_c}}(h_2 - g_2)d\tau \tag{85}$$

When Eq. (84) is used to eliminate the function r in Eq. (85), the final solution for the mean value of s arises as:

$$\frac{\omega t_c}{2\pi}\int_0^{\frac{2\pi}{\omega t_c}} s\, d\tau = \left[-C_1|h_{1+}|^2 + C_2|g_{1+}|^2 - C_3|h_{1+}||g_{1+}|\cos\left(\varphi_{h_{1-}} - \varphi_{g_{1-}}\right)\right]\frac{1}{1+\left(\dfrac{\omega t_c}{Z+1}\right)^2}$$
$$-C_4|h_{1+}||g_{1+}|\sin\left(\varphi_{h_{1-}} - \varphi_{g_{1-}}\right)\frac{\dfrac{\omega t_c}{Z+1}}{1+\left(\dfrac{\omega t_c}{Z+1}\right)^2} + 2C_4\frac{\omega t_c}{2\pi}\int_0^{\frac{2\pi}{\omega t_c}}(h_2 - g_2)d\tau \tag{86}$$

$$C_1 = \frac{Z(Z+Y+1)}{2(Z+1)^3} \quad C_2 = \frac{Z^2(Z+1)-YZ}{2(Z+1)^3} \quad C_3 = C_2 - C_1 \quad C_4 = C_2 + C_1$$

From Eq. (79), the heat transferred amounts to:

$$\frac{\omega t_c}{2\pi}\int_0^{\frac{2\pi}{\omega t_c}}\left(s + g_1 r + g_2\right)d\tau = -C_1\frac{1}{1+\left(\dfrac{\omega t_c}{Z+1}\right)^2}\left[|h_{1+}|^2 + |g_{1+}|^2 - 2|h_{1+}||g_{1+}|\cos\left(\varphi_{h_{1-}} - \varphi_{g_{1-}}\right)\right]$$
$$+2C_4\frac{\omega t_c}{2\pi}\int_0^{\frac{2\pi}{\omega t_c}}h_2\, d\tau + \frac{1}{Z+1}\frac{\omega t_c}{2\pi}\int_0^{\frac{2\pi}{\omega t_c}}g_2\, d\tau \tag{87}$$

3.2 Predicted dependence on the Strouhal number

Expression (86) shows that the time-averaged outlet temperature depends explicitly on the angular frequency as:

$$F = \frac{\Gamma_0 + \Gamma_1 \dfrac{\omega t_c}{Z+1} + \Gamma_2 \left(\dfrac{\omega t_c}{Z+1}\right)^2}{1 + \left(\dfrac{\omega t_c}{Z+1}\right)^2} \tag{88}$$

Here, Γ_0, Γ_1, and Γ_2 are functions that can be obtained from comparing Eqs. (88) and (86). Since these functions depend on the value adopted for h_{1+} and h_2, Eq. (21) states that these functions depend also on the Strouhal number. Note that the Strouhal number and the angular frequency are related to each other by means of the following expression:

$$\mathrm{St} = \frac{\omega L}{U} = \frac{\omega \rho L A}{\overline{G}} = \omega t_c \chi \frac{LA}{V} \frac{c_p}{c} \tag{89}$$

Therefore, function F in Eq. (88) has an implicit dependence on the Strouhal number due to the coefficients Γ_0, Γ_1, and Γ_2, and an explicit dependence due to ω. The influence of the Strouhal number on the coefficients Γ_0, Γ_1, and Γ_2 comes through the effect that produces on the external solution previously discussed in section 2.1. Therefore, in those devices where the external solution does not suffer a significant change due to the variations of the Strouhal number, these coefficients can be considered constant values. In this case, Eq. (89) allows us to rewrite expression (88) as a function that depends on the Strouhal number and on four constants A, B, C and D:

$$F = \frac{A + B\,\mathrm{St} + C\,\mathrm{St}^2}{1 + D\,\mathrm{St}^2} \tag{90}$$

This function does not exhibit any resonant behaviour since all the terms in the denominator are always positive quantities. However, this function presents an extreme at the angular frequency:

$$\frac{\omega t_c}{Z+1} = \frac{\Gamma_2 - \Gamma_0}{\Gamma_1} \pm \sqrt{\left(\frac{\Gamma_2 - \Gamma_0}{\Gamma_1}\right)^2 + 1} \quad \text{or} \quad \mathrm{St} = \frac{C}{BD} - \frac{A}{B} \pm \sqrt{\left(\frac{C}{BD} - \frac{A}{B}\right)^2 + \frac{1}{D}} \tag{91}$$

Since this extreme disappears for those devices where Γ_1 vanishes, the above expression is able to reproduce all the results found in the current scientific literature. For example, for very low frequencies, Eq. (90) can produce the enhancement or deterioration depending on the sign of the coefficient Γ_0 which changes, for example, with the relative phase between the dimensionless heat transfer coefficient, h_1, and the dimensionless mass flow rate coefficient g_1. The same can happen at very high frequencies depending on the sign of Γ_2. Thus, although this expression has been obtained under the assumption of several restrictive hypotheses, it retains an influence on the Strouhal number that reflects the actual behaviour of some heat exchangers. For example, a backward facing step like the one presented in Fig. 1 exhibits a heat transfer enhancement, which was calculated by Velazquez et al. (2007) for several Strouhal numbers. For this configuration, they reported that the maximum heat transfer is 42% higher than in the steady case. The method of least square allows fitting the simplified model in Eq. (90) giving the following values for the four constants: $A=0$, $B=0.8453$, $C=0.1997$ and $D=1.561$. (The constant $A=0$ imposes that the solution at zero frequency must coincide with the steady one.) The result of the fitting and the numerical data are plotted in Fig. 2. Discrepancies between the numerical calculations and the approach given in this section can be due to the neglected terms, to the harmonics of order greater than one which could be present, and to the fact of having neglected the influence of the Strouhal number on the amplitudes and phases that drive the coefficients A, B and C. Extended discussions about the estimation of these coefficients from physical considerations and about the behaviour of other configurations are given by Benavides (2009).

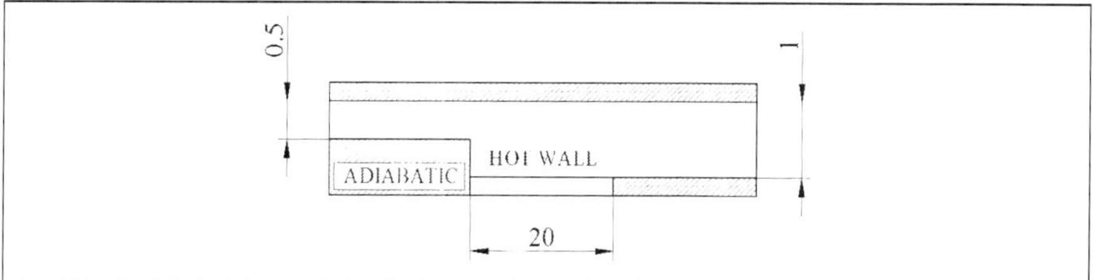

Fig. 1. Dimensionless definition of the 2-D backwards facing step (not at scale). The outlet height of this device is L=450 μm. It works with ideal water ($k=0.598$ W K^{-1}m^{-1}, $\mu=10^{-3}$ Pa s, $c=4180$ J kg^{-1}K^{-1} and $\rho=998$ kg m^{-3}) in an operational point identified by Nu=5.18 and Re=100. These values give the following non-dimensional characterization of the stationary heat flow: $\chi=0.074$, $\eta(\overline{T}_H)=0.071$, $\theta(\overline{T}_P)=0.96$, $Z=26$ and $Y=0.34$.

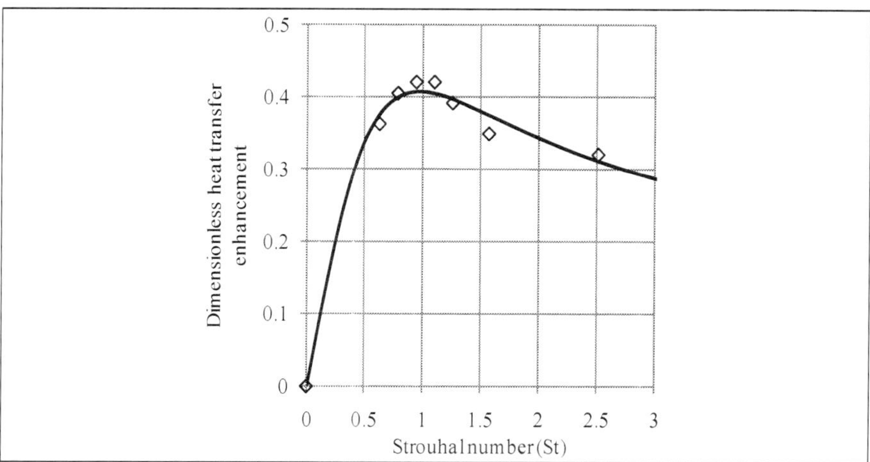

Fig. 2. Set of data points (diamond points) obtained from direct numerical simulation (Velazquez et al., 2007) for different values of the frequency for the device whose scheme is given in Fig. 1. The result of fitting these data points with the model given by Eq. (90) (solid line) gives $A=0$, $B=0.8453$, $C=0.1997$ and $D=1.561$.

Equation (87) can also be represented by Eq. (88). However, for this equation, $\Gamma_1=0$ always holds. This means that the result expressed by Eq. (87) does not have extremes. Besides, the condition $\Gamma_0=0$ removes any effect at zero frequency. Finally, in case of considering Γ_2 constant with the frequency, the model would follow the expression:

$$F = \frac{C St^2}{1 + D St^2} \qquad (92)$$

Figure 3 presents a qualitative drawing of Eqs. (90) and (92). Note that the condition $\Gamma_0=0$ must only be satisfied for zero frequency. Besides, as it has been discussed above, one possible reason for losing the term proportional to Γ_1 is the algebraic relationship for defining the spatial-averaged temperature. Therefore, Eq. (92) is a very restrictive case of Eq. (90). Besides, coefficients in Eqs. (90) and (92) should depend on Strouhal number with a mathematical law fixed by the device geometry; since Eq. (90) has more degrees of freedom that Eq. (90) it will be a better choice for fitting empirical laws derived from different test configurations. Note that, due to the inertial terms, it is plausible to expect a negligible response for higher frequencies: this feature is not described by Eq. (90) except for the case $C=0$. Since the constant C is required to model a flat asymptotic behaviour for high Strouhal numbers, an external model to characterize the dependency of the coefficients A, B and C is required. The correction due to this missing model is qualitatively represented as the dashed lines of Fig. 3.

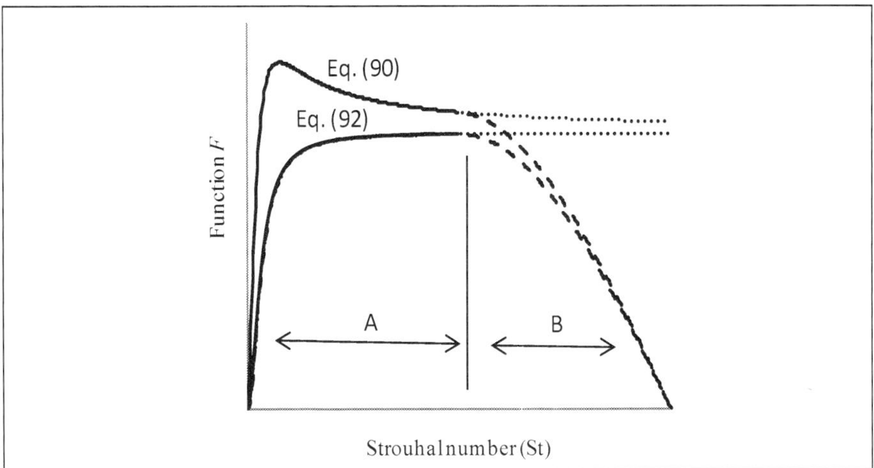

Fig. 3. Solid and dot curves are the solutions obtained from Eqs. (90) (upper line) and (92) with the same values of constants A, B, C and D (this qualitative drawing can be reproduced by choosing the values 0, 0.8, 0.5 and 0.5 for A, B, C and D). In the range of low Strouhal numbers, labelled as region A in the figure, it is assumed that constants A, B, C and D do not depend on the Strouhal number; in the region of higher Strouhal numbers, labelled as region B, the effect of the Strouhal number on the constants could be significant, and hence dot curves represents the behaviour predicted by the model for high Strouhal numbers while dashed curves represents the plausible decay due to the variation of constants A, B, C and D with the Strouhal number.

As expected, the model does not predict any change respect to the steady state when χ tends to infinity. The proof of this statement comes from the asymptotic behaviour of the coefficients C_1, C_2, C_3 and C_4, that tends to zero exponentially when χ tends to infinity. The situation described by $\chi \to \infty$ represents a very effective device where the steady heat transfer coefficient is so high that the outlet temperature matches the hottest wall temperature. Therefore, the enhancement predicted by this model cannot be obtained for those devices whose efficiency is high or, in other words, the enhancement only is present for low efficiency devices. Normally, actual devices have small values of χ, typically $\chi<0.1$, and hence the efficiency $\eta(\overline{T}_H)$ of the device is close to zero. The asymptotic behaviour of the relevant functions and coefficients for both limits, low and high values of χ, are collected in Table 4.

$\eta(\overline{T}_H) = 1 - e^{-\chi}$	$\chi - \dfrac{\chi^2}{2} + 0(\chi^3) << 1$	$0(1)$
$\theta(\overline{T}_P) = \dfrac{1 - e^{-\chi}}{\chi}$	$1 - \dfrac{\chi}{2} + \dfrac{\chi^2}{6} + 0(\chi^3) < 1$	$0\left(\dfrac{1}{\chi}\right)$
$\theta(\overline{T}_H) = e^{-\chi}$	$1 - \chi + \dfrac{\chi^2}{2} + 0(\chi^3) < 1$	$0\left(\dfrac{1}{e^{\chi}}\right)$
$\dfrac{\theta(\overline{T}_P)}{\theta(\overline{T}_H)} = 1 + \dfrac{1}{Z} = \dfrac{1 - e^{-\chi}}{\chi e^{-\chi}}$	$1 + \dfrac{\chi}{2} + \dfrac{\chi^2}{6} + 0(\chi^3) > 1$	$0\left(\dfrac{e^{\chi}}{\chi}\right)$
$Z = \dfrac{\chi e^{-\chi}}{1 - (1 + \chi)e^{-\chi}}$	$\dfrac{2}{\chi} - \dfrac{2}{3} + \dfrac{\chi}{18} + \dfrac{\chi^2}{270} + 0(\chi^3) >> 1$	$0\left(\dfrac{\chi}{e^{\chi}}\right)$
$Y = \dfrac{(2 + \chi)e^{-2\chi} - 4e^{-\chi} + 2 - \chi}{2\left[1 - (1 + \chi)e^{-\chi}\right]^2}$	$\dfrac{1}{3} + \dfrac{\chi}{9} + \dfrac{\chi^2}{45} + 0(\chi^3)$	$0(\chi)$
$C_1 = \dfrac{Z(Z + Y + 1)}{2(Z + 1)^3}$	$\dfrac{\chi}{4} - \dfrac{\chi^2}{8} + 0(\chi^3)$	$0\left(\dfrac{\chi^2}{e^{\chi}}\right)$
$C_2 = \dfrac{Z^2(Z + 1) - ZY}{2(Z + 1)^3}$	$\dfrac{1}{2} - \dfrac{\chi}{2} + \dfrac{\chi^2}{6} + 0(\chi^3)$	$0\left(\dfrac{\chi^2}{e^{\chi}}\right)$
$C_3 = \dfrac{Z^3 - Z(2Y + 1)}{2(Z + 1)^3}$	$\dfrac{1}{2} - \dfrac{3}{4}\chi + \dfrac{7}{24}\chi^2 + 0(\chi^3)$	$0\left(\dfrac{\chi^2}{e^{\chi}}\right)$
$C_4 = \dfrac{Z}{2(Z + 1)}$	$\dfrac{1}{2} - \dfrac{\chi}{4} + \dfrac{\chi^2}{24} + 0(\chi^3)$	$0\left(\dfrac{\chi}{e^{\chi}}\right)$

Table 4. Limits of the relevant functions and constants for $\chi \to 0$ and $\chi \to \infty$.

Finally, the model predicts that the form of measuring the heat flux is completely decisive when oscillating flows are present: in Eq. (90), which is a simplification of Eq. (86), the heat flux is obtained from averaging the outlet temperature or the specific enthalpy; in Eq. (92), which is a simplification of Eq. (87), it is obtained from averaging the enthalpy. As explained, both descriptions lead to the difference between the solid lines in region A of Fig. 3. While the averaged outlet temperature is closer to the response of a thermocouple placed in the outlet flow, the determination of the heat flux requires two time recordings: the model described by Eqs. (78) and (87) shows that the measurement setup should include the instantaneous value of the temperature and the instantaneous value of the mass flow rate. For this reason, in an experimental setup where the heat flux in the wall is only measured by a time-averaged value obtained from a final thermocouple, the heat transfer enhancement of the device could be overestimated. For example, for $\chi \to 0$, coefficients C_1, C_2, C_3 and C_4 tend to 0, $\frac{1}{2}$, $\frac{1}{2}$, $\frac{1}{2}$ respectively, this means that the model predicts a response of order unity for the three terms involving St^0, St and St^2 in the outlet temperature given by Eq. (86) whereas the only term of order unity in Eq. (87) is the one associated to St^0.

4. Conclusion

The proposed model applies to channels with different solid structures on the inside under the assumptions: *i*) the amplitude of the oscillation of the mass flow rate is small when it is compared with the averaged mass flow rate; *ii*) the spatial-averaged temperature inside the device and the outlet temperature are related to each other by means of an algebraic equation; *iii*) there is only one dominant harmonic; and *iv*) the effect of the Strouhal number on the coefficients that determine the expansions is negligible. The work shows that, under these assumptions, the outlet temperature is determined as a function of the frequency of the oscillation. This dependence on the Strouhal number is characterised by the ratio of two polynomials of second order where the linear term of the polynomial in the denominator does not appears. As a consequence, the time-averaged outlet temperature can exhibit a maximum response for a Strouhal number determined by the model. The model also predicts that, under the aforementioned assumptions, the time-averaged heat flux does not exhibit such a maximum; however, it is suggested that this feature probably is due to the assumption *ii*) that inhibits any temporal delay between temperatures. Besides, such as the derivation of the model establishes, the coefficients that determine the expansion depend on the Strouhal number: this implies that the model fixed by the ratio of two polynomials should be modified. The greater the effects of the Strouhal number on the coefficients, the greater the discrepancies. In particular, it has been discussed that due to the inertial terms, it is plausible to expect that the coefficients tend to zero for high Strouhal numbers, and hence the heat transfer enhancement will tend to zero when the Strouhal number tends to infinity. This fact introduces an effect highly dependent on the device geometry and produces a maximum response for a given Strouhal number that cannot be calculated with the present model. The relative importance of these two physical phenomena establishes the final response for a given device. However, it has been shown that the formula obtained from the model fits well the numerical results obtained for a 2-D backward facing step with the hot surface parallel to the flow in an adiabatic channel. Thus, the presented model reveals a mathematical law that can be used to fit several empirical results and to collect them under a similar dimensionless law. A final important conclusion is that those testing procedures where the heat flux is estimated by a time-averaged measurement of the outlet temperature might overestimate the heat transfer enhancement. A full characterization of the heat transfer requires recording the temporal evolution of both, the outlet temperature and the mass flow rate. Further studies are required to remove assumption *ii*), which imposes severe limitations to the results, and to incorporate the effect of the external solution on the dimensionless coefficients, which are very sensitive to the geometry and to the flow patterns in the channel, especially when stagnant or recirculation regions are involved.

5. References

Backhaus, S. & Swift, G.W. (2000). A thermoacoustic-Stirling heat engine: Detailed study, *J. Acoust. Soc. Am.*, Vol. 107, No. 6, (June 2000) (3148-3166), ISSN

Baehr, H. D. & Stephan, K. (2006). *Heat and Mass Transfer*, Springer-Verlag Berlin Heidelberg

Benavides, E.M. (2006). An analytical model of self-starting thermoacoustic engines, *Journal of Applied Physics*, Vol. 99 (2006)

Benavides, E.M. (2007). Thermoacoustic nanotechnology: Derivation of a lower limit to the minimum reachable size, *Journal of Applied Physics*, Vol. 101, No. (2007)

Benavides, E.M. (2009). Heat Transfer Enhancement by Using Pulsating Flows, *Journal of Applied Physics*, Vol. 101, No. (2007)

Chattopadhyay, H.; Durst, F. & Ray, S. (2006). Analysis of heat transfer in simultaneously developing pulsating laminar flow in a pipe with constant wall temperature, *Int. Commun Heat Mass Transfer*, Vol. 33 (2006) (475-481)

Cho, H.W. & Hyun, J.M. (1990). Numerical solution of pulsating flow and heat transfer characteristic in a pipe, *Int. J. Heat Flow*, Vol. 11, No. 4 (1990) (321-330)

Faghri, M.; Javadani, K. & Faghri, A. (1979). Heat Transfer with laminar pulsating flow in a pipe, *Lett. Heat Mass Transfer* , Vol. 6 (1979) (259-270)

Gardner, D.L. & Swift, G.W. (2003). A cascade thermoacoustic engine, *J. Acoust. Soc. Am.*, Vol. 114, No. 4, (October 2003), (1905-1919)

Guo, Z. & Sung, H.J. (1997). Analysis of the Nusselt number in pulsating pipe flow, *Int. J. Heat Mass Transfer*, Vol. 40, No. 10 (1997) (2486-2489)

Harvey, A.P.; Mackley, M.R. & Seliger, T. (2003). Process intensification of biodiesel production using a continuous oscillatory flow reactor, *Journal of Chemical Technology & Biotechnology*, Vol. 78, No. 2-3 (2003) (338-341)

Hemida, H. N.; Sabry, M.N.; Abdel-Rahim, A. & Mansour, H. (2002). Theoretical analysis of heat transfer in laminar pulsating flow, *Int. J. Heat Mass Transfer*, Vol. 45 (2002) (1767-1780)

Kim, S.Y. & Kang, B.H. Forced convection heat transfer from two heated blocks in pulsating channel flow, *Int. J. Heat Mass Transfer*, Vol 41, No.3 (1998) (625-634)

Lee, C.T.; Mackley, M.R.; Stonestreet, P. & Middelberg, A.P.J. (2001). Protein refolding in an oscillatory flow reactor, *Biotechnology Letters*, Vol. 23, No. 22 (2001) (1899-1901)

Lide, D.R. (2004). *CRC handbook of chemistry and physics: a ready-reference book of chemical and physical data : 2004-2005*, CRC Press Boca Raton, ISBN, Florida

Mackley, M.R. & Stonestreet, P. (1995). Heat transfer and associated energy dissipation for oscillatory flow in baffles tubes, *Chem. Sci. Eng.*, Vol. 50, No. 14 (1995) (2211-2224)

Ueda, Y.; Biwa, T.; Mizutani, U. & Yazaki, T. (2004). Experimental studies of a thermoacoustic Stirling prime mover and its application to a cooler, *J. Acoust. Soc. Am.* , Vol. 115, No. 3, (March 2004), (1134-1141)

Velazquez, A.; Arias, J.R. & Mendez, B. (2007). Laminar heat transfer enhancement downstream of a backward facing step by using a pulsating flow, *Heat Mass Transfer* (2007)

Wakeland, R. S. & Keolian, R.M. (2004a). Calculated effects of pressure-driven temperature oscillations on heat exchangers in thermoacoustic devices with and without a stack, *J. Acoust Soc. Am.* , Vol. 116, No. 1, (July 2004), (294-302)

Wakeland, R.S. & Keolian, R.M. (2004b). Efectiveness of parallel-plate heat exchangers in thermoacoustic devices, *J. Acoust. Am.*, Vol. 115, No. 6, (June 2004), (2873-2886)

Xiongwei N. & Pereira N. E. (2000). Parameters affecting fluid dispersion in a continuous oscillatory baffled tube, *AIChE Journal*, Vol. 46, No. 1 (2000) (37-45)

Yu, J.C.; Li, Z.X. & Zhao, T.S. (2004). An analytical study of pulsating laminar heat convection in a circular tube with constant heat flux, *Int. J. Heat Mass Transfer*, Vol. 47 (2004) (5297-5301)

16

Flow Patterns, Pressure Drops and Other Related Topics of Two-phase Gas-liquid Flow in Microgravity

Jian-Fu ZHAO
*Key Laboratory of Microgravity (National Microgravity Laboratory)/CAS;
Institute of Mechanics, Chinese Academy of Sciences (CAS)
China*

1. Introduction

Two-phase gas-liquid systems have wide applications both on Earth and in space. On Earth, they occur in a variety of process equipment, such as petroleum production facilities, condensers and re-boilers, power systems and core cooling of nuclear power plants during emergency operation. The potential space applications include active thermal control system, power cycle, storage and transfer of cryogenic fluids, and so on. Reliable design of such systems requires a thorough understanding of the mechanism of two-phase flow, such as the phase distributions (flow patterns), pressure drops and heat transfer coefficients at different gas and liquid flow rates. Among them, flow patterns will play an important role because of strong influence of the phase distributions on pressure drops and heat transfer coefficients and thus attract more attentions of the academic and technical communities all over the world.

With the aid of numerous meticulous experiments, our present knowledge on two-phase gas-liquid systems has been built. It is, however, far from complete due to the complicate influence of gravity which is a dominant factor in normal gravity. Gravity strongly affects many phenomena of two-phase gas-liquid systems by creating forces in the systems that drive motions, shape boundaries, and compress fluids. Furthermore, the presence of gravity can mask effects that ever present but comparatively small. Depending on the flow orientation and the phase velocities, gravity can significantly alter the flow patterns, and hence the pressure drops and heat transfer rates associated the flow. Advances in the understanding of two-phase flow have been greatly hindered by masking effect of gravity on the flow. Therefore, the microgravity researches will be conductive to revealing of the mechanism underlying the phenomena, and then developing of more mechanistic models for the two-phase flow and heat transfer both on Earth and in space.

Research on two-phase gas-liquid flow in microgravity has a history of more than 50 years with a short pause in the 1970s and has been advanced with the development of various microgravity facilities and with increased experimental opportunities, especially in the last two decades. Due to much strict restriction on flight chance, weight and size of the test facility, power supply, and so on, studies on two-phase gas-liquid flow in microgravity are much limited. Furthermore, few experiments have been conducted in real microgravity environment. Most data are obtained from experiments performed in short-term reduced gravity with a relatively large value and equivalent pulsation of the residual gravity aboard parabolic airplanes. However, some advances, particularly on flow patterns of two-phase flow in microgravity have been made. On the progress in this field, many comprehensive reviews and monographs are available now. Among many others, Hewitt (1996), McQuillen et al. (1998), Gabriel (2007) and Zhao (2010) summarized the experimental and theoretical works all over the world. The first two papers provided the status of this field at the beginning of our research, while the last one presented a brief review of our researches in the past years.

Since the middle of 1990's, a series of microgravity research projects on two-phase gas-liquid flow in microgravity have been conducted in the National Microgravity Laboratory/CAS (NMLC). These activities cover experimental and theoretical studies on classification of flow patterns, transition criteria between different flow patterns, and pressure drops of two-phase gas-liquid flows in pipelines, as well as some other related topics such as the characteristics of phase distribution inside flow fields and their influence on the performance of direct methanol fuel cells and proton exchange membrane fuel cells. In the following sections, the results obtained in these researches will be presented.

2. Two-phase gas-liquid flows in pipelines in microgravity

2.1 Microgravity flow pattern transition models

Our first activity in this field is re-analyzing the existing database and models/correlations in the literature up to 1998 on flow patterns of two-phase gas-liquid flows in straight, circular pipe in microgravity (Zhao, 1999). Major attentions focus on the classification of the flow patterns and the development of flow pattern maps of adiabatic two-phase gas-liquid flows in pipelines in microgravity. Because the influence of buoyancy is removed or weakened strongly, two-phase flows in microgravity are believed inherently simpler than those in normal gravity. For example, although other flow patterns appearing essentially in the transition zones are classified by some researchers, annular, slug, and bubble flows are usually considered as the major two-phase flow patterns in straight pipes in microgravity. Thus, there are two major transitions need to be modeled in microgravity.

For predicting the slug-to-annular flow transition in microgravity, the void fraction matched model proposed by Dukler et al. (1988) and modified by Colin et al. (1991) and Bousman (1995) is commonly used. For the case of turbulent liquid and gas phases, this model can be rewritten as following (Zhao, 2000)

$$C_0 = \left\{ \alpha + \left[\zeta \Phi (1-\alpha)^2 \right]^{5/9} \alpha^{-1/3} \right\}^{-1} \tag{1}$$

As shown in Fig.1, it is possible that there will be 0, 1, or 2 solutions in this model according to the different values of the phase distribution parameter C_0 in the drift-flux model and the material parameter $\zeta = (\rho_G/\rho_L)(v_G/v_L)^{1/5}$ (here ρ and v denote the density and viscosity respectively, while the subscripts G and L denote the gas and liquid phase respectively). The solutions will alter if different correlations for interfacial friction factor Φ, such as those proposed by Wallis (1969) and Chen et al. (1991), are used. Furthermore, the suggestion of Dukler (1989) that the slug-to-annular flow transition will take place at the solution of smaller void fraction α is not a feasible criterion. For example, such a solution may locate in bubble flow regime for the case of $\zeta = 0.001$, which approximates to the experimental condition of Dukler et al. (1988).

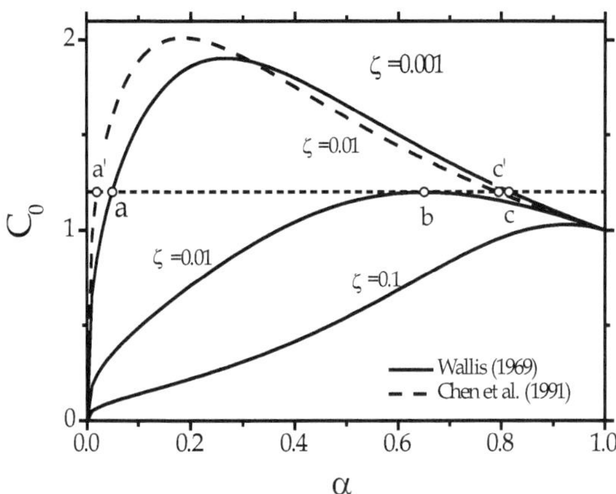

Fig. 1. Characteristics of the solution of the void fraction matched model for the slug-to-annular flow transition in microgravity (Zhao, 2000).

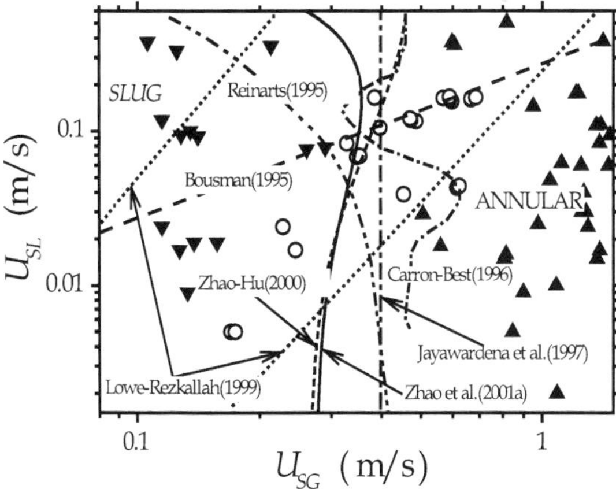

Fig. 2. Comparisons of the semi-theoretical Weber number model with the experimental data of Reinarts (1993) and other commonly used models. Symbols ▲, ▼, and ○ denote slug, annular, and transitional flows, respectively (Zhao, 2010).

A semi-theoretical Weber number model was developed firstly by Zhao & Hu (2000), and later modified by Zhao et al. (2001a), which was based on the balance between the impulsive force due to the gas inertia and the surface tension force near the slug-to-annular flow transition in microgravity. This model can be written in the following dimensionless form

$$\left\{ \begin{array}{l} We_{SG} \cong \dfrac{\rho_G U_{SG}^2 D}{\sigma} = \dfrac{\kappa\psi C_0 \left(1-\alpha\right)\alpha^{3/2}}{C_0 - 1} \\[4mm] q \cong \dfrac{U_{SL}}{U_{SG}} = \dfrac{1 - C_0\alpha}{C_0\alpha} \end{array} \right. \tag{2}$$

where D, σ, USG, and USL denote respectively the pipe diameter, the surface tension, the superficial velocities of the gas and liquid phases, ψ denotes the geometrical corrector of the pipe cross-section (*e.g.* $\psi = 4$ for a circular pipe while $\psi = 2\sqrt{\pi}$ for a square one), while κ is an empirical parameter with an order of 1. As shown in Fig.2, the model can provide an improvement of the accuracy in comparison with others. It was also proved to be accurate over a rather wide range of working fluids, tube diameters, and experimental methods including both flight experiments and ground simulated tests such as capillary gas-liquid experiments and equi-density, or neutral buoyancy, immiscible liquid-liquid experiments on the ground.

For predicting the bubble-to-slug flow transition in microgravity, the drift-flux model (Dukler et al., 1988; Colin et al., 1996) was commonly used in the literature.

$$q = \frac{1 - X_{cr}}{X_{cr}} \tag{3}$$

$$X_{cr} \cong C_0\alpha_{cr} = \left\{ \begin{array}{ll} 0.54, & \left(Su < 1.5\times 10^6\right) \\[2mm] 0.24, & \left(Su > 1.7\times 10^6\right) \end{array} \right. \tag{4}$$

where α_{cr} denotes the critical void fraction corresponding to the bubble-to-slug transition, while $Su \cong \sigma D\rho_L / \mu_L^2$ is the liquid Suratman number. The empirical model proposed by Jayawardena et al. (1997) can also be re-written in the same form with the following transition quantity Xcr

$$X_{cr} = K_1 \frac{v_G}{v_L} \bigg/ \left(K_1 \frac{v_G}{v_L} + Su^{2/3} \right) \tag{5}$$

It was, however, found that there exists an obvious difference between bubble flows in mini-scale channels in normal gravity and those in normal channels in microgravity. It may arise from the difference of the relative bubble initial size in the two cases. A Monte Carlo method was then used to simulate the influence of the initial bubble size d_b on the bubble-to-slug flow transition based on the bubble coalescence mechanism (Zhao, 2005). It was found that the dimensionless rate of collision is a universal function of the dimensionless bubble diameter d_b/D, and that the bubble initial size can affect the bubble-to-slug flow transition when its dimensionless value locates in the range from 0.03 to 0.4. Assuming the transition void fraction α_{cr} depends only on the dimensionless collision rate, the following correlation was obtained for the critical void fraction in this range,

$$\alpha_{cr} = 0.60 - 2.32 d_b/D \tag{6}$$

which agreed well with the experimental data (Fig. 3).

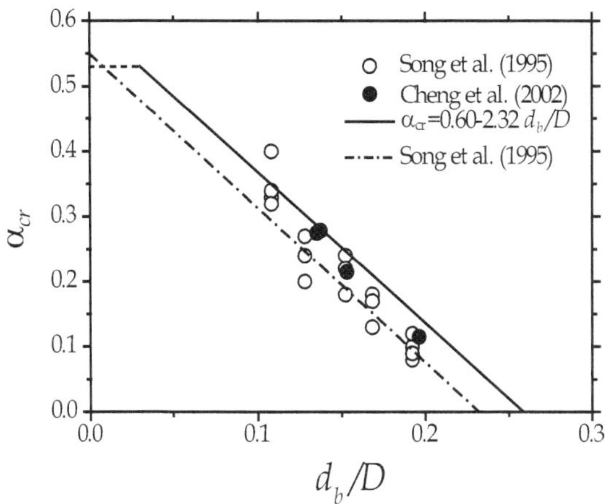

Fig. 3. The influence of bubble initial size on the transition void fraction for the bubble-to-slug flow transition (Zhao, 2005).

2.2 Two-phase flow pattern maps in microgravity

Another major activity of us is to perform experiments on two-phase flow both in microgravity environment aboard the Russian space station Mir and IL-76 parabolic airplane and in simulated microgravity conditions using capillary gas-liquid experiments on the ground. Several sets of data on two-phase flow patterns have been collected from these experiments, which are helpful both to understand the mechanism underlying the phenomena and to design two-phase systems operating on board spacecrafts and/or satellites.

Collaborating with researchers from the Keldysh Research Center of Russia, an experimental study was conducted on two-phase flow patterns in a circular pipe with an inside diameter of 10 mm and a length of 356 mm aboard the Russian space station Mir in August 1999 (Zhao et al., 2001b, c), which is the first opportunity and also the sole one up to now to have data from an experiment conducted in a long-term, steady microgravity environment. Air was used as the gas phase, while carbogal, an odorless, colorless and non-toxic liquid with a small contact angle (0~7°) with the test tube, was used as the liquid phase. Bubble, slug, and annular flow were observed as in other researches. Fine dispersed bubble flow was observed at higher liquid superficial velocity, while a wider region of slug-annular flow was also observed at moderate gas superficial velocity. A new region of annular flow with smooth interface at much lower liquid superficial velocity was discovered firstly, which can not be interpreted presently (Fig. 4).

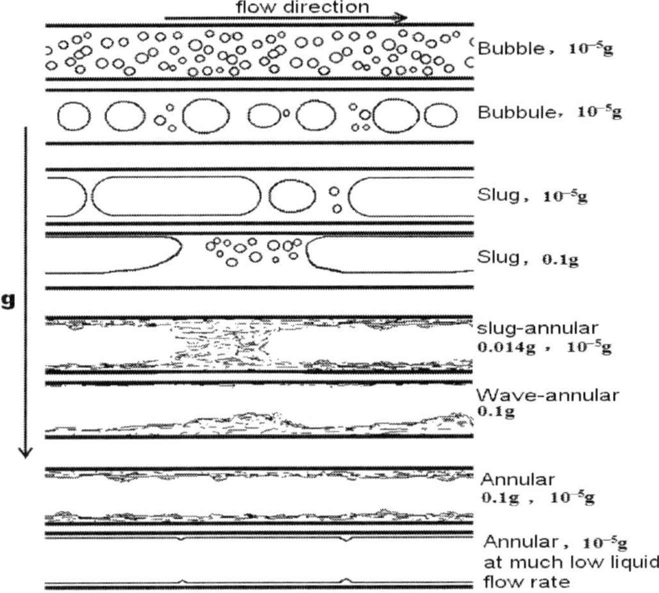

Fig. 4. Typical flow patterns in different gravity conditions aboard the space station Mir (Zhao et al., 2001b,c).

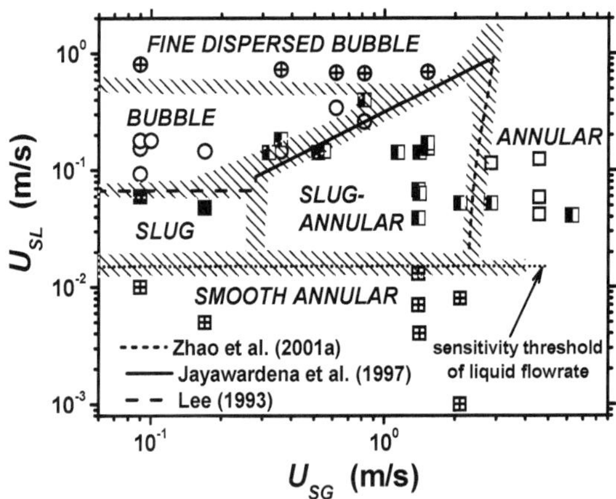

Fig. 5. Flow pattern map of two-phase flow in microgravity aboard the space station Mir (Zhao et al., 2001b).

With detailed comparisons between the reliable data obtained aboard the space station Mir and the predictions of the models proposed in the literature, it was indicated that the observed flow patterns at low gas superficial velocity should be considered to be developing ones due to the small length-to-diameter ratio. The entrance effects were much weak on the flow pattern transitions at moderate and high gas superficial velocity. A comprehensive flow pattern map (Fig.5) was provided according to the results in the background microgravity environment aboard the space station Mir. Data in partial gravity conditions provided by rotating the experimental facility with constant velocities aboard the space station Mir were also reported by Zhao et al. (2004b).

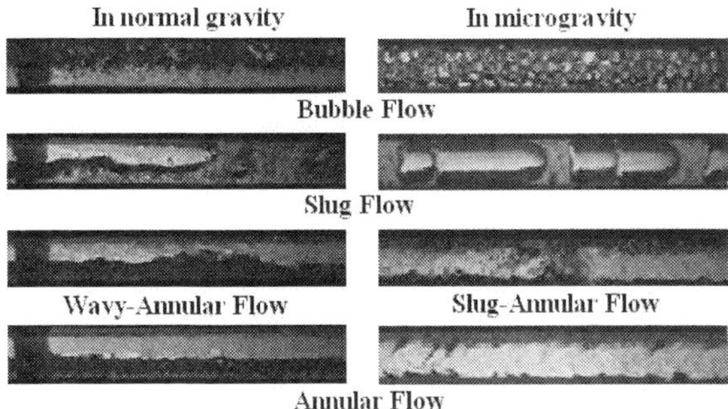

Fig. 6. Typical flow patterns in a square channel in different gravity conditions (Zhao et al., 2001a).

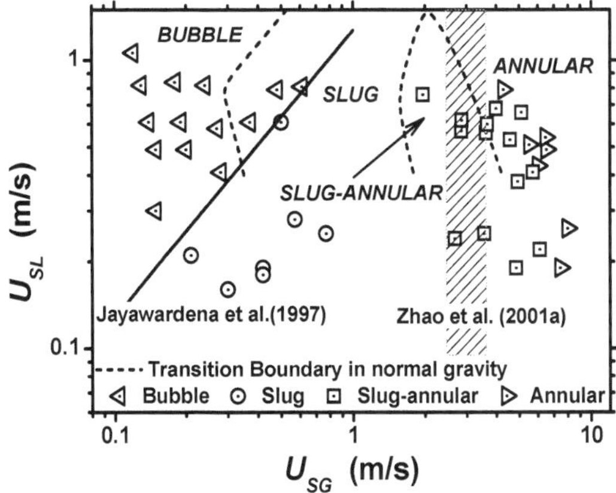

Fig. 7. Flow pattern map of water-air flow in a square channel in microgravity (Zhao et al., 2001a).

Another experimental research was conducted on two-phase water-air flow patterns in a 12×12 mm2 square channel aboard the Russian IL-76 reduced gravity airplane in July 1999 (Zhao et al., 2001a). Bubble, slug, slug-annular, and annular flows were observed in different gravity conditions (Fig. 6). The flow pattern map obtained in microgravity was shown in Fig. 7, while that obtained from the ground control tests in normal gravity was also shown for comparison. If the slug-annular flow is considered as no one of the major flow patterns but a transitional one between the slug and annular flows, the semi-theoretical Weber number model of Zhao & Hu (2000) with the improvement about the shape influence can predict well the slug-to-annular flow transition both in normal and in microgravity. The prediction by Jayawardena et al. (1997) was also in reasonable agreement with the observed boundary between bubble and slug flows in microgravity. Furthermore, it was found that the transition void fraction for the bubble-to-slug flow transition with large Froude number in normal gravity is in direct proportion to the gas relative area in the channel cross-section and its counterpart in microgravity.

Two-phase flow patterns in a 90° bend in microgravity were analyzed by Zhao & Gabriel (2004). The experimental data were obtained by the Microgravity Research Group at the University of Saskatchewan, Canada. Three major flow patterns, namely slug, slug-annular, and annular flows, were observed in this study (Fig. 8). The transitions between adjoining flow patterns were found to be more or less the same as those in straight pipes, and can be predicted satisfactorily by the Weber number models (Zhao & Hu, 2000; Lowe & Rezkallah, 1999). Attention was also paid to the difference of the flow structure between single- and two-phase flows. The bullet-shaped bubbles in slug flow and the gas core in annular flow usually exhibited imposed rotation, which could be inferred from the obvious striations on the gas-liquid interface spreading from the inside to the outside at an acute angle to the forward direction. It may arise from the secondary liquid flow in the bend, which is modified by the presence of the gas phase. This information will be valuable for more sophisticated modeling of two-phase flow in bends in the future.

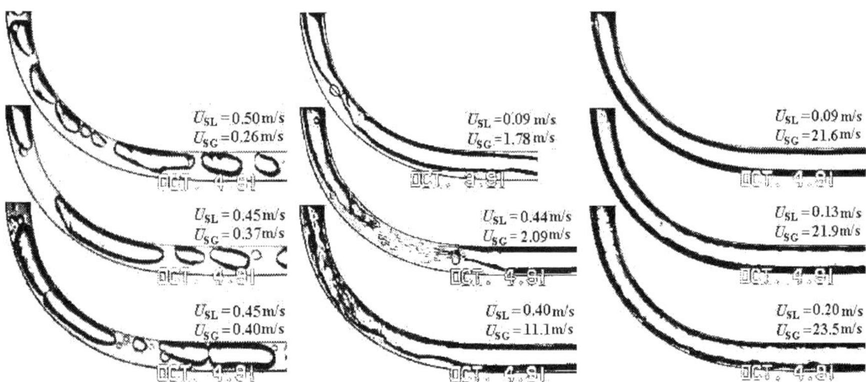

Fig. 8. Flow patterns in 90° bend in microgravity (Zhao & Gabriel, 2004).

Mini-scale modeling was also used to simulate the behavior of two-phase flow in microgravity (Zhao et al., 2004a). A 1×1 mm^2 square mini-channel was used. A mixer with four 0.7-mm holes perpendicular to the channel axis was located before the channel. The experimental data were compared with other data in mini-channels reported in literature, and also compared with those in normal channel in microgravity, in which the Bond number had the same order of magnitude. The transition to annular flow was consistent in all cases (Fig. 9). Comparing with the prediction of the empirical relation of Jayawardena et al. (1997), a much smaller value of the transition void fraction was obtained for the bubble-to-slug flow transition in mini-channels. Obvious difference was found between bubble flow in mini-scale experiments and that in microgravity experiments (Fig. 10). As discussed above, it may arise from the difference of the relative bubble initial size in the two cases. Thus, the mini-scale modeling can be used to anticipate the behavior of two-phase flows with high flow rates through normal size channels in microgravity, while it can not be an effective way for simulating the behavior of microgravity two-phase bubble flows.

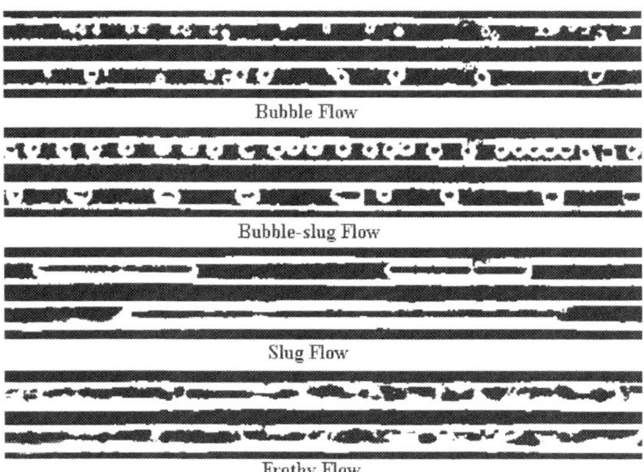

Fig. 9. Typical flow patterns of water-air flow in a 1×1 mm² capillary square pipe on the ground (Zhao et al., 2004a).

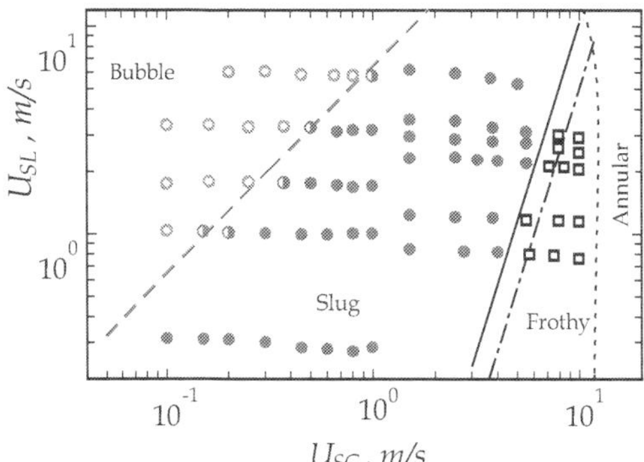

Fig. 10. Flow pattern map of water-air flow in a 1×1 mm² capillary square pipe on the ground (Zhao et al., 2004a).

2.3 Pressure drop of two-phase flow in microgravity

In the experiment aboard the Russian IL-76 reduced gravity airplane in July 1999, pressure drops of two-phase flow in microgravity were also measured and compared with some commonly used correlations in the literature (Zhao et al., 2001d), such as the homogenous model, the Lockhart-Martinelli-Chisholm model, and the Friedel model. It was found that much large differences exist between the experimental data and the predictions. Among these models, the Friedel model provided a relative good agreement with the experimental data (Fig. 11). A more accurate model should be developed based on a more physical analysis of flow characteristics and a large empirical database developed with the aid of numerous meticulous experiments both in normal and reduced gravity.

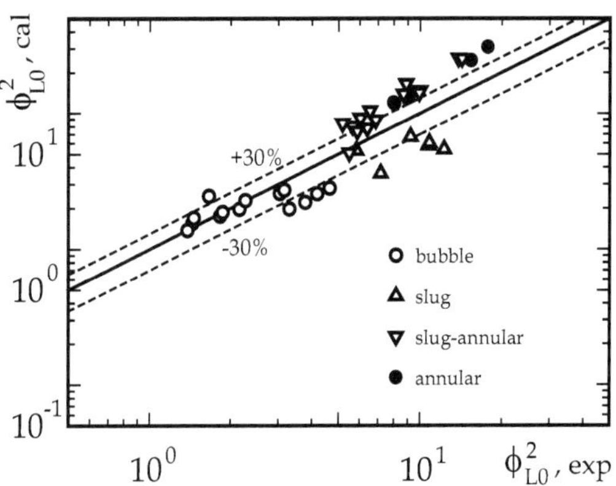

Fig. 11. Comparison of two-phase pressure drop in microgravity with Friedel's correlation (Zhao et l., 2001d).

Based on the analysis of the flow structure of two-phase bubble flow in microgravity, it was suggested by Zhao et al. (2002) that the friction factor and the Reynolds number in this case should be defined based on the mixture velocity Um and the properties of the liquid phase, namely

$$f_{TP} = \left[(dp/dz)_F D \right] / \left(2\rho_L U_m^2 \right) \text{ and } Re_{TP} = \rho_L U_m D / \mu_L \tag{7}$$

respectively. A semi-theoretical relationship, *i.e.*

$$f_{TP} = A Re_{TP}^{-1} \tag{8}$$

was also proposed, in which the parameter A is dependent on the Reynolds number and should be determined empirically. Comparing the present data in the square channel and those collected by Zhao & Rezkallah (1995) and Bousman (1995) in circular pipes, little influence of the cross-sectional shape was found (Fig. 12). Constant values of the parameter A, namely $A = 35$ for $Re_{TP} < 3000$ and $A = 120$ for $Re_{TP} > 4000$, were obtained. It was indicated that there exists a transition of flow structure in the range of $3000 < Re_{TP} < 4000$, similar to the laminar-to-turbulent transition in single-phase pipe flow. A further comparison, however, with the data obtained by Colin (1990) showed that an exponent for the Reynolds number between 0 and -1 should be more suitable for the case of large Reynolds number.

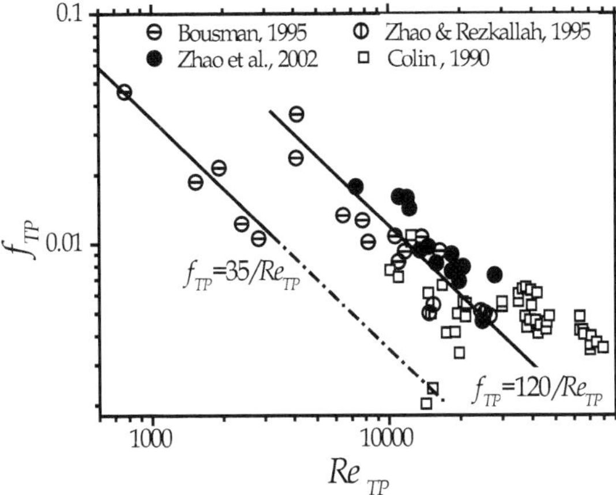

Fig. 12. Pressure drop of two-phase bubble flow in microgravity (Zhao, 2010).

3. Phase distribution and performance of fuel cells in microgravity

Electrolysis cells and fuel cells will be major parts of regenerative environmental control and life support systems (ECLSS) for a long-term manned space flight mission. Electrolysis/fuel cells may be operated in different gravity condition during different flight stage. Two-phase flow inside cells may exhibit different characteristics and then affect their performances, even cause some severe problems of safety. Thus, study the phase distribution and corresponding electric performance of electrolysis/fuel cells plays an important role for the design of such systems related to space applications.

Collaborating with Profs. H. Guo and C. F. Ma at Beijing University of Technology, a serial of experiment projects have been proposed to study the phase distribution and corresponding electric performance of electrolysis/fuel cells in different gravity conditions in our group by utilizing the drop tower Beijing to provide a short-term microgravity environment.

At first, a small liquid fed direct methanol fuel cells (DMFC) was used as a simplified model. A commercially available MEA was sandwiched between two graphite bipolar plates with sealing gasket. The MEA with an active area of 5.0×5.0 cm^2 consists of a Nafion 117 membrane and two carbon cloth gas diffusion layers. The catalyst loading was 4 mg/cm^2 Pt/Ru on the anode side and 4 mg/cm^2 Pt on the cathode side. Electric resistors were employed as external circuit load. Both the anode and the cathode bipolar plates were made of graphite and consisted of channel area, extension area and location pinholes. Extension area of both the anode and the cathode plates was used as current connector. A single serpentine channel, which had rectangular cross section of 2.0 mm in depth and 2.5 mm in width, was fluted in cathode flow field. The rib width is 2.0 mm and the channel length was 566 mm. In order to preheat the fuel cell, a heater was stuck to the cathode bipolar plate. The channel area of anode bipolar plate had 2 manifolds and 11 parallel straight channels, whose length was 48.0 mm. Each channel had rectangle cross-section with depth of 2.0 mm and width of 2.5 mm. The width of rib, which was between two adjacent channels, was 2.0 mm. For the purpose of visual observation, the end plate of the anode side was made of transparent polycarbonate (PC). A high-speed video camera (VITcam CTC) with a CCTV C-mount lens (SE2514, AVENIR) was employed to capture two-phase flow images in the anode flow field. A shutter speed of 3996 μs, a recording speed of 250 frames/s and a resolution of 1280×1024 pixels were set to visualize and record two-phase flow in the anode flow field.

Oxygen gas with purity of 99.999%, without humidification, was used as oxidant reactant. The oxygen gas flow rate was controlled by a mass flow controller (Cole Parmer, CZ-32907-67) at constant flow rate of 400 mL/min. The prepared methanol solution was stored in a storage bag and was driven by a peristaltic pump and sent to a liquid flow meter (Cole Parmer, CZ-32908-43). The oxygen gas and the methanol solution were heated up before flowing into the anode channels. The produced mixtures from DMFC were sent to two separate containers.

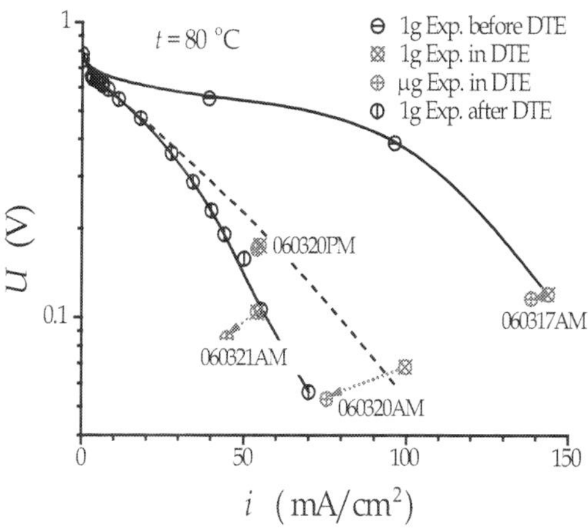

Fig. 13. Influence of gravity on the power performance of DMFC (Wan et al., 2006).

Fig. 13 shows the preliminary results using gold-plating stainless steel as both the anode and the cathode bipolar plates (Wan et al., 2006). In despite of the deterioration of performance of the fuel cell, it is very evident that the cell performance falls more strongly with the degree of concentration polarization deepening.

After re-design of the anode and the cathode bipolar plates, an in-situ visualization of two-phase flow inside anode flow bed of a small liquid fed direct methanol fuel cells in normal and reduced gravity has been conducted in a drop tower Beijing. The experimental results indicated that when the fuel cell orientation is vertical, two-phase flow pattern in anode channels can evolve from bubbly flow in normal gravity into slug flow in microgravity (Fig. 14). In normal gravity environment, the gravitational buoyancy is the principal detaching force. The carbon dioxide bubbles were produced uniformly with tiny shape in normal gravity before the release of the drop tower. The diameter of most bubbles, which were detached from the MEA surface, ranged from 0.05 to 0.3 mm in our experiments (Fig. 14a, corresponding to 40 ms before the release). After detaching from the MEA surface, the carbon dioxide bubbles moved fast at a speed above 100 mm/s. Considering that the mean velocity of liquid at the entry of channel was 3.03 mm/s, which was calculated from the inlet flow rate of methanol solution, the speed of bubbles removal was quite fast because of buoyant lift force. Big bubbles with fast velocity would push the small ones anterior. When the bubbles collided with each other, coalescence took place and it was a dominative way of bubbles growth. The typical flow pattern in the anode flow channels in normal gravity was bubbly flow.

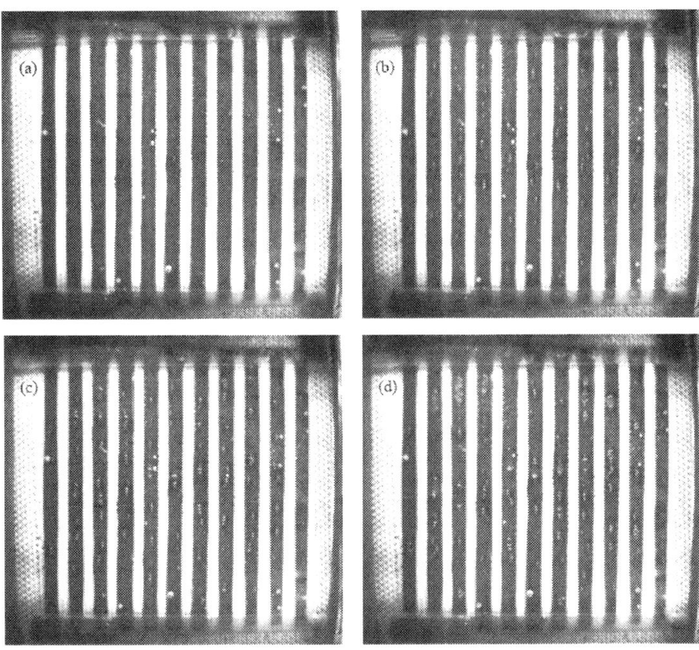

Fig. 14. Gas-liquid two-phase flow pattern in vertical parallel channels of anode bipolar plate of DMFC in different gravity (Guo et al., 2009).

In microgravity, the carbon dioxide bubbles could not get away from the MEA surface in time (Fig. 14b~d, corresponding to 1, 2 and 3 s after the release, respectively). At the beginning, the bubbles accreted on the wall of carbon cloth surface. Then, the bubbles on the surface grew gradually because of producing carbon dioxide by anode electrochemical reaction. The longer the time was, the bigger the bubble was. Furthermore, the gravity affects not only detaching diameter, but also bubbles rising velocity. The in-situ observation showed that once the capsule was released, the bubbles move was slowed down immediately. Bubbles, which were detached from MEA, almost suspended in methanol solution. The average rising velocities of bubbles in channels are near to the mean velocity of liquid, which was obviously slower than those in normal gravity because buoyancy lift was very weak and the bubbles removal was governed by viscous drag of fluid in the reduced gravity. Some bubbles coalesced with each other and formed larger bubbles. Those large bubbles decreased the effective area of fuel mass transfer and hence the DMFC performance deterioration took place. The gravitational effect on power performance of DMFC is considerable when the concentration polarization is dominant in fuel cells operation. The higher the current density is, the bigger the effect of gravity is. Increasing methanol feeding molarity is conducive to weaken the effect of gravity on performance of liquid fed direct methanol fuel cells. Increasing feeding flow rate of methanol solution from 6 to 15 ml/min could reduce the size of carbon dioxide bubbles. But the influence of gravity still exists (Ye et al., 2010).

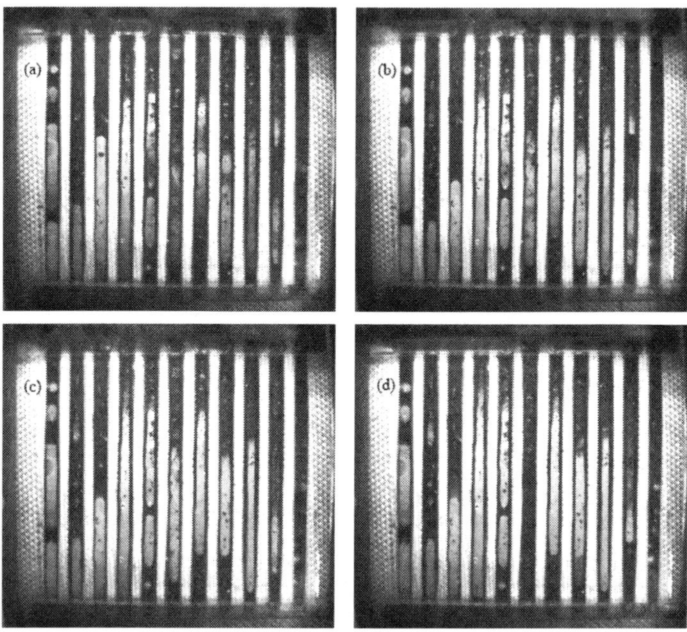

Fig. 15. Gas-liquid two-phase flow pattern in horizontal parallel channels of anode bipolar plate of DMFC in different gravity (Guo et al., 2009).

When the fuel cell orientation is horizontal, the typical flow pattern is the elongated slug flow. The slug flow in the reduced gravity has almost the same characteristic with that in normal gravity (Fig. 15). It implies that the effect of gravity on two-phase flow is small and the bubbles removal is governed by viscous drag. Like in the condition of vertical orientation of fuel cell, once the gas slugs or even gas columns occupy channels, the performance of liquid fed direct methanol fuel cells will fall rapidly. The phenomena infer that in long-term microgravity condition, flow bed, fuel cell orientation and operation condition should be optimized to ensure timely discharge carbon dioxide bubbles and avoid concentration polarization.

A compact transparent proton exchange membrane fuel cell (PEMFC) with a single serpentine channel in graphite cathode flow field, which had a square cross section of 2.0×2.0 mm^2 and a rib width of 2.0 mm, was also designed and tested in short-term microgravity environment in the drop tower Beijing. Hydrogen and oxygen gases with purity of 99.999%, without humidification, were used as fuel and oxidant reactant, respectively. The experimental facility was similar with that for DMFC. Its detail can be found in Liu (2008).

It was found that the accumulated liquid water in the vertical parts of flow channel for the vertical orientation configuration can be removed easily by the reactant gas in microgravity environment comparing with in normal gravity. The PEMFC performance was then enhanced dramatically in microgravity because of the flooded areas in the flow channel before the release of the drop capsule was exposed to the reactant gas again. However, for the horizontal orientation configuration with the lower outlet, liquid water produced in flow channel can move along the bottom of the channel in normal gravity and then flow freely out off the channel. Then little liquid water was found and water columns to pinch off the flow channel were difficult to be formed in normal gravity. On the contrary, the liquid water formed in microgravity was prone to stay in the flow channel, and the departure diameter of water droplets increased. Therefore, the PEMFC performance was deteriorated due to liquid water flooding in the flow channel. The influence of gravity on the characteristics of phase distribution and performance of PEMFC increases with the increase of the current, and/or increase with the decrease of the cell temperature.

4. Further researches on two-phase flow in microgravity in china

Several new projects for two-phase flow in microgravity have been proposed to study pressure drop in in-tube condensation, flow boiling heat transfer enhancement of micro-pin-finned surface, membrane separation of two-phase air-water mixture, two-phase flows inside fuel cells and electrolysis cells, and so on. These projects will be helpful for the development of space systems involving two-phase flow phenomena, as well as for the improvement of understanding of such phenomena themselves.

5. Conclusion

Two-phase gas-liquid systems have wide applications both on Earth and in space. Gravity strongly affects many phenomena of two-phase gas-liquid systems. It can significantly alter the flow patterns, and hence the pressure drops and heat transfer rates associated the flow. Advances in the understanding of two-phase flow and heat transfer have been greatly hindered by masking effect of gravity on the flow. Therefore, the microgravity researches will be conductive to revealing of the mechanism underlying the phenomena, and then developing of more mechanistic models for the two-phase flow and heat transfer both on Earth and in space.

The present chapter summarizes a series of microgravity researches on two-phase gas-liquid flow in microgravity conducted in the National Microgravity Laboratory/CAS (NMLC) since the middle of 1990's, which included ground-based tests, flight experiments, and theoretical analyses. In the present chapter, the major results obtained in these researches will be presented and analyzed.

Up to now, the sole flow pattern map of two-phase gas-liquid flow in long-term, steady microgravity was obtained in the experiments aboard the Russian space station Mir, which is intended to become a powerful aid for further investigation and development of two-phase systems for space applications. Flow pattern map of two-phase air-water flow through a square channel in reduced gravity was obtained in the experiments aboard IL-76 parabolic airplane, too. Mini-scale modeling was also used to simulate the behavior of microgravity two-phase flow on the ground. The criteria of gravity-independence of two-phase gas-liquid flow were proposed based on experimental observations and theoretical analyses. A semi-theoretical Weber number model was proposed to predict the slug-to-annular flow transition of two-phase gas-liquid flows in microgravity, while the influence of the initial bubble size on the bubble-to-slug flow transition was investigated numerically using the Monte Carlo method.

Pressure drops of two-phase flow through a square channel in reduced gravity were also measured experimentally, which were used to validate the common used correlations for microgravity applications. It was found that much large differences exist between the experimental data and the predictions. Among these models, the Friedel model provided a relative good agreement with the experimental data. A new correlation for bubbly flow in microgravity was proposed successfully based on its characteristics, which indicates that there may exist a transition of flow structure in the range of two-phase Reynolds number from 3000 to 4000, which is similar to the laminar-to-turbulent transition in single-phase pipe flow.

In-situ visualizations of two-phase gas-liquid flow inside fuel cells (DMFC and PEMFC) in different gravity conditions have carried out utilizing the drop tower Beijing. The gravity influence of the cells performance, namely deterioration or enhancement, depends upon the operation conditions. It also infers form the short-term microgravity experiments utilizing the drop tower Beijing that space experiments with long-term microgravity environment are needed.

6. Acknowledgement

The studies presented here were supported financially by the National Natural Science Foundation of China (19789201, 10202025, 10432060, 50406010, 50976006), the Ministry of Science and Technology of China (95-Yu-34), the Chinese Academy of Sciences (KJCX2-SW-L05), and the Chinese National Space Agency. The author really appreciates Prof. W. R. Hu, Prof. J. C. Xie, Mr. S. X. Wan, Mr. M. G. Wei, and all research fellows who have contributed to the success of these studies. The author also wishes to acknowledge the fruitful discussion and collaboration with Prof. K. S. Gabriel (UOIT, Canada), and Profs. H. Guo and C.F. Ma (Beijing University of Technology, China).

7. References

Bousman, W.S., 1995. Studies of two-phase gas-liquid flow in microgravity. Ph.D. thesis, Univ. of Houston, TX.

Carron, I., Best, F., 1996. Microgravity gas/liquid flow regime maps: can we compute them from first principles. In: AIChE Heat Transfer Symp., Nat. Heat Transfer Conf., August, Houston, TX.

Chen, I., Downing, R., Keshock, E., Al-Sharif, M., 1991. Measurements and correlation of two-phase pressure drop under microgravity conditions. J. Thermophy., 5, 514–523.

Cheng, H., Hills, J.H., Azzopardi, B.J., 2002. Effects of initial bubble size on flow pattern transition in a 28.9 mm diameter column. Int. J. Multiphase Flow, 28(7), 1047–1062.

Colin, C., 1990. Ecoulements diphasiques à bubbles et à poches en micropesanteur. Thesis, Institut de Mécanique des Fluides de Toulouse.

Colin, C., Fabre J., Dukler A.E., 1991. Gas-liquid flow at microgravity conditions-I. Dispersed bubble and slug flow. Int. J. Multiphase Flow, 17(4), 533–544.

Colin, C., Fabre, J., McQuillen, J., 1996. Bubble and slug flow at microgravity conditions: state of knowledge and open questions. Chem. Eng. Comm., 141/142, 155–173.

Dukler, A.E., Fabre, J.A., McQuillen, J.B., Vernon, R., 1988. Gas-liquid flow at microgravity conditions: flow patterns and their transitions. Int. J. Multiphase Flow, 14(4), 389–400.

Dukler, A.E., 1989. Response. Int. J. Multiphase Flow, 15(4), 677.

Gabriel, K.S., 2007. Microgravity Two-phase Flow and Heat Transfer. Springer.

Guo, H., Zhao, J.F., Ye, F., Wu, F., Lv, C.P., Ma, C.F., 2008. Two-phase flow and performance of fuel cell in short-term microgravity condition. Microgravity Sci. Tech., 20(3-4): 265-270.

Guo, H., Wu, F., Ye, F., Zhao, J.F., Wan, S.X., Lv, C.P., Ma, C.F., 2009. Two-phase flow in anode flow field of a small direct methanol fuel cell in different gravities. Sci. China E-Tech Sci, 52(6): 1576 – 1582.

Hewitt, G.F., 1996. Multiphase flow: the gravity of the situation. In: 3rd Microgravity Fluid Physics Conf., July 13–15, Cleveland, Ohio, USA.

Jayawardena, S.S., Balakotaiah, V., Witte, L.C., 1997. Flow pattern transition maps for microgravity two-phase flows. AIChE J., 43(6), 1637–1640.

Lee, J., 1993. Scaling analysis of gas-liquid two-phase flow pattern in microgravity. In: 31st Aerospace Sci. Meeting Exhibit, Jan. 11–14, Reno, NV.

Liu, X., 2008. Two-phase flow dynamic characteristics in flow field of proton exchange membrane fuel cells under micro-gravity conditions. Ph.D. thesis, Beijing University of Technology.

Lowe, D.C., Rezkallah, K.S., 1999. Flow regime identification in microgravity two-phase flows using void fraction signals. Int. J. Multiphase Flow, 25, 433–457.

McQuillen, J., Colin, C., Fabre, J., 1998. Ground-based gas-liquid flow research in microgravity conditions: state of knowledge. Space Forum, 3, 165–457.

Reinarts, T.R., 1993. Adiabatic two phase flow regime data and modeling for zero and reduced (horizontal flow) acceleration fields. Ph.D. thesis, Texas A&M Univ., TX.

Reinarts, T.R., 1995. Slug to annular flow regime transition modeling for two-phase flow in a zero gravity environment. In: Proc. 30th Int. Energy Conversion Eng. Conf., July 30–August 4, Orlando, FL.

Song, C.H., No, H.C., Chung, M.K., 1995. Investigation of bubble flow developments and its transition based on the instability of void fraction waves. Int. J. multiphase Flow, 21(3), 381–404.

Wallis, G.B., 1969. One-dimensional Two-phase Flow. McGraw-Hill Book Company, New York.

Wan, S.X., Zhao, J.F., Wei, M.G., Guo, H., Lv, C.P., Wu, F., Ye, F., Ma, C.F., 2006. Two-phase flow and power performance of DMFC in variable gravity. 3rd Germany-China Workshop on Microgravity & Space Life Sciences, October 8 - 11, 2006, Berlin, Germany.

Ye, F., Wu, F., Zhao, J.F., Guo, H., Wan, S.X., Lv, C.P., Ma, C.F., 2010. Experimental Investigation of Performance of a Miniature Direct Methanol Fuel Cell in Short-Term Microgravity. Microgravity Sci. Tech., 22(3): 347-352.

Zhao J.F., 1999. A review of two-phase gas-liquid flow patterns under microgravity conditions. Adv. Mech., 29(3): 369-382.

Zhao J.F., 2000. On the void fraction matched model for the slug-to-annular transition at microgravity. J. Basic Sci. Eng., 8(4): 394-397.

Zhao, J.F., 2005. Influence of bubble initial size on bubble-to-slug transition. J. Eng. Thermophy., 26(5), 793–795.

Zhao, J.F., 2010. Two-phase flow and pool boiling heat transfer in microgravity. Int. J. Multiphase flow, 36(2): 135-143.

Zhao, J.F., Gabriel, K.S., 2004. Two-phase flow patterns in a 90° bend at microgravity. Acta Mech. Sinica, 20(3), 206–211.

Zhao, J.F., Hu W.R., 2000. Slug to annular flow transition of microgravity two-phase flow. Int. J. Multiphase Flow, 26(8), 1295–1304.

Zhao, J.F., Xie, J.C., Lin, H., Hu, W.R., 2001a. Experimental study on two-phase gas-liquid flow patterns at normal and reduced gravity conditions. Sci. China E, 44(5), 553–560.

Zhao, J.F., Xie, J.C., Lin, H., Hu, W.R., Ivanov, A.V., Belyeav, A.Yu., 2001b. Microgravity experiments of two-phase flow patterns aboard Mir space station. Acta Mech. Sinica, 17(2), 151–159.

Zhao, J.F., Xie, J.C., Lin, H., Hu, W.R., Ivanov, A.V., Belyeav, A.Yu., 2001c. Experimental studies on two-phase flow patterns aboard the Mir space station. Int. J. Multiphase Flow, 27, 1931–1944.

Zhao, J.F., Xie, J.C., Lin, H., Hu, W.R., Lv, C.M., Zhang, Y.H., 2001d. Experimental study on pressure drop of two-phase gas-liquid flow at microgravity conditions. J. Basic Sci. Eng., 9(4), 373–380.

Zhao, J.F., Xie, J.C., Lin, H., Hu, W.R., 2002. Pressure drop of bubbly two-phase flow through a square channel at reduced gravity. Adv. Space Res., 29(4), 681–686.

Zhao, J.F., Liu, G., Li, B., 2004a. Two-phase flow patterns in a square micro-channel. J. Thermal Sci., 13(2), 174–178.

Zhao, J.F., Xie, J.C., Lin, H., Hu, W.R., Ivanov, A.V., Belyeav, A.Yu., 2004b. Study on two-phase gas-liquid flow patterns at partial gravity conditions. J. Eng. Thermophy., 25(1), 85–87.

Zhao, L., Rezkallah, K.S., 1995. Pressure drop in gas-liquid flow at microgravity conditions. Int J Multiphase Flow, 21(5), 837–849.

17

Heat Transfer and Its Assessment

Heinz Herwig and Tammo Wenterodt
Hamburg University of Technology
Germany

1. Introduction

Somebody, interested in heat transfer and therefore reading one of the many books about this subject might be confronted with the following statement after the heat transfer coefficient $h = \dot{q}_\mathrm{w}/\Delta T$ has been introduced:

"The heat transfer coefficient h is a measure for the quality of the transfer process."

That sounds reasonable to our *somebody* though for somebody else (the authors of this chapter) there are two minor and one major concerns about this statement. They are:

1. Heat cannot be transferred since it is a process quantity.

2. The coefficient h is not a nondimensional quantity what it better should be.

3. What is the meaning of "quality"?

The major concern actually is the last one and it will be the crucial question that is raised and answered in the following. Around that question there are, however, some further aspects that should be discussed. Two of them are the first two in the above list of concerns.

The heat transfer coefficient h is typically used in single phase convective heat transfer problems. This is a wide and important field of heat transfer in general. That is why the problem of heat transfer assessment will be discussed for this kind of "conduction based heat transfer" in the following sections 2 to 4. In section 5 extensions to the overall heat transfer through a wall, heat transfer with phase change and the fundamentally different "radiation based heat transfer" will be discussed.

2 The "quality" of heat transfer

2.1 Preliminary considerations

What is commonly named *heat transfer* is a process by which energy is transferred across a certain system boundary in a particular way. This special kind of energy transfer is characterised by two crucial aspects:

– The transfer process is initiated and determined by temperature gradients in the vicinity of the system boundary.

– As a consequence of this transfer process there is a change of entropy on both sides of the system boundary. That change can be interpreted as a transfer of entropy linked to the energy transfer. It thus is always in the same direction. The strengths of both transfer processes are not in a fixed proportion, but depend on the temperature level.

According to these considerations the phrase "heat transfer" actually should be replaced by "energy transfer in the form of heat". Since, however, "heat transfer" is established worldwide in the community we also use this phrase, but as a substitute for "energy transfer in the form of heat".

It is worth noting that from a thermodynamics point of view *heat* is one of only two ways in which energy can be transferred across a system boundary. The alternative way is *work*. This other kind of energy transfer is not caused by temperature gradients and is not accompanied by entropy changes. For further details see Moran & Shapiro (2003); Baehr & Kabelac (2009); Herwig & Kautz (2007), for example.

The energy transferred in the form of heat from a thermodynamics point of view is *internal energy* stored in the material by various mechanisms on the molecular level (translation, vibration, rotation of molecules). The macroscopic view on this internal energy can not only identify its amount (in Joule) but also its "usefulness". The thermodynamic term for that is its amount of *exergy*. Here exergy, also called *available work*, is the maximum theoretical work obtainable from the energy (here: internal energy) interacting with the environment to equilibrium.

According to this concept energy can be subdivided in two parts: exergy and anergy. Here anergy is energy which is not exergy (and thus "not useful"). If, however, energy has a certain value (its amount of exergy) a crucial question with respect to a transfer of energy (heat transfer) is that about the *devaluation of the energy in the transfer process*.

2.2 Energy devaluation in a heat transfer process

What happens to the energy in a heat transfer process can best be analysed on the background of the second law of thermodynamics. This kind of analysis which considers the entropy, its transfer as well as its generation is called second law analysis (SLA). In a heat transfer situation the entropy S is involved twofold:

- Entropy is transferred over the system boundary together with the transferred energy. In a thermodynamically reversible process entropy is transferred only and no entropy is generated. This infinitesimal transfer rate is

$$\mathrm{d}\dot{S} = \partial \dot{Q}_{\mathrm{rev}} / T \tag{1}$$

Here $\partial \dot{Q}_{\mathrm{rev}}$ is an infinitesimal heat flux, T the thermodynamic temperature at which it occurs and $\mathrm{d}\dot{S}$ the rate by which the entropy in the system is changed due to the heat transfer. Such a reversible heat transfer only occurs when there are no temperature gradients involved. Therefore (1) is either the ideal situation of a real transfer process (with ΔT as operating temperature difference) in the limit $\Delta T \to 0$ or that part of a real heat transfer process without the entropy generation due to local temperature gradients.

- Entropy is generated in the system wherever temperature gradients $\partial T / \partial n$ occur. The generation rate per volume ($'''$) is, see Bejan (1982); Herwig & Kock (2007),

$$\dot{S}_{\mathrm{C}}''' = \frac{k}{T^2} \left(\frac{\partial T}{\partial n} \right)^2 \tag{2}$$

Here k is the thermal conductivity and n a coordinate in the direction of the temperature gradient vector.

Entropy generation always means *loss of exergy*. According to the so-called Gouy–Stodola theorem, see Bejan (1982), the exergy loss rate per volume due to heat conduction is

$$\dot{E}_{\mathrm{LC}}''' = T_0 \, \dot{S}_{\mathrm{C}}''' \tag{3}$$

with $\dot{S}_{\mathrm{C}}'''$ from (2) and T_0 as the temperature of the environment.

The devaluation of energy that is transferred in the form of heat thus can be determined by integrating the local exergy loss rate (3) over the volume of the system under consideration, as will be shown in sec. 4.2.

2.3 Energetic and exergetic quality of heat transfer

Since heat transfer is caused by temperature gradients, or (integrated over a finite distance) by temperature differences ΔT there are two questions about the "quality" of a transfer process:

- How much energy can be transferred in a certain situation with ΔT as the operating temperature difference? This is called the *energetic quality* of the transfer process.

- How much is the energy devaluated in a certain transfer situation with ΔT as the operating temperature difference? This is called the *exergetic quality* of the transfer process.

Obviously two parameters are needed to answer both questions. If, however, there is only the heat transfer coefficient h or its nondimensional counterpart, the Nußelt number Nu, not all information about the heat transfer process is available as will be demonstrated hereafter.

3. Heat Transfer Assessment

3.1 The energetic quality of heat transfer

The energetic quality was introduced (sec. 2.3) as the answer to the question "How much energy can be transferred in a certain situation with ΔT as the operating temperature difference?" This amount of energy is finite (and not infinite) only because it occurs in a real process subject to losses. Finite values of the heat transfer coefficient h nevertheless are not a direct measure for these losses.

Though h, defined as

$$h \equiv \frac{\dot{q}_w}{\Delta T} \qquad ; [h] = \mathrm{W/m^2K} \tag{4}$$

is frequently used and widely accepted, there are several aspects of it that can be critically discussed:

- When h should characterise the energetic quality it should be a fixed value for a specific heat transfer situation. This especially referres to the strength of heat transfer, i.e. h should not depend on ΔT. There are, however, situations in which this is not the case, like for natural convection in general and for radiative heat transfer with large values of ΔT, see Herwig (1997) for details.

- Instead of h, its reciprocal $1/h = \Delta T/\dot{q}_w$ would be more appropriate. Then finite values (and not zero) for h^{-1} would be due to a *resistance* which a heat flux $\dot{Q}_w = \dot{q}_w A$ encounters on the heat transfer area A. This is in analogy to the resistance $R = \Delta U/I$ that an electrical current I ($I = iA$; i: current density, A: cross section) encounters with a voltage ΔU. By this analogy U corresponds to T and I to $\dot{Q}_w$.

- Since h is part of the Nußelt number

$$\mathrm{Nu} \equiv \frac{\dot{q}_w L}{k\,\Delta T} = h \cdot \frac{L}{k} \qquad ; [\mathrm{Nu}] = 1 \tag{5}$$

it is often assumed to be equivalent to this nondimensional group ($[\mathrm{Nu}] = 1$). The Nußelt number, however, as a result of a systematic dimensional analysis process, is a more significant parameter. It characterises a heat transfer situation irrespective of its geometrical size L and the thermal conductivity k of the fluid involved.

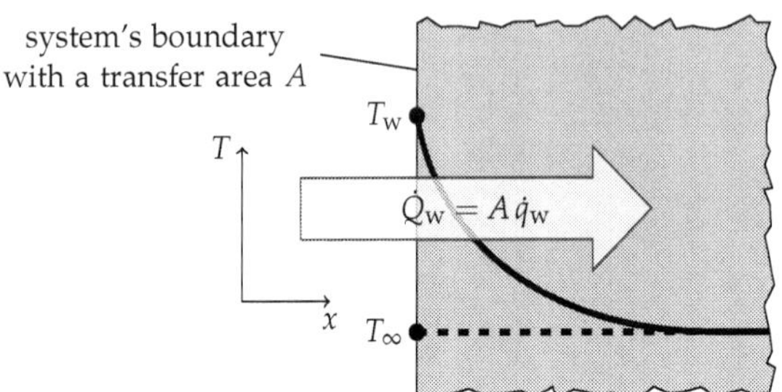

Fig. 1. One dimensional heat transfer at a system boundary ($\dot{q}_{\mathrm{w}} = $ const)

Whenever general statements about a certain heat transfer situation are the objective the Nußelt number is the preferred parameter.

3.2 The exergetic quality of heat transfer

The exergetic quality was introduced (sec. 2.3) as the answer to the question "how much is the energy devaluated in a certain transfer situation with ΔT as the operating temperature difference?" This devaluation is directly measured by the exergy loss rate based on the relation (3).

For a heat transfer situation as sketched in Fig. 1 which is characterised by Nu with respect to its energetic behavior a second parameter is now introduced and named *exergy loss number*. Its definition is

$$\mathrm{N}^{\mathrm{E}} \equiv \frac{\dot{E}_{\mathrm{LC}}}{\dot{E}} = \frac{T_0\,\dot{S}_{\mathrm{C}}}{\eta_{\mathrm{C}}\,\dot{Q}_{\mathrm{w}}} \qquad \text{(exergy loss number)} \qquad (6)$$

with $\dot{E}_{\mathrm{LC}}$ and $\dot{S}_{\mathrm{C}}$ as integration of the local rates $\dot{E}'''_{\mathrm{LC}}$ and $\dot{S}'''_{\mathrm{C}}$ and $\dot{E}$ as the exergy fraction of the heat flux $\dot{Q}_{\mathrm{w}}$. This exergy fraction corresponds to the transferred heat flux $\dot{Q}_{\mathrm{w}}$ multiplied by the *Carnot factor*

$$\eta_{\mathrm{C}} = 1 - \frac{T_0}{T_{\mathrm{w}}} \qquad (7)$$

The crucial quantity in N^{E} according to (6) is the overall entropy generation rate due to heat conduction, $\dot{S}_{\mathrm{C}}$. In a complex temperature field it emerges through a field integration of $\dot{S}'''_{\mathrm{C}}$, see section 4.2 below. If, however, there is a one-dimensional heat transfer perpendicular to the system boundary as sketched in Fig. 1, the situation is different.

Then only the two temperature levels T_{w} and T_∞ count, and $\dot{S}_{\mathrm{C}}$ is the difference of the transfer rates $\dot{Q}_{\mathrm{w}}/T_{\mathrm{w}}$ and $\dot{Q}_{\mathrm{w}}/T_\infty$, cf. (1) for incremental parts, i.e.

$$\dot{S}_{\mathrm{C}} = \dot{Q}_{\mathrm{w}} \left(\frac{1}{T_\infty} - \frac{1}{T_{\mathrm{w}}} \right) \qquad (8)$$

With this $\dot{S}_{\mathrm{C}}$ the exergy loss number for a one-dimensional heat transfer is

$$\mathrm{N}^{\mathrm{E}} \equiv \frac{\dot{E}_{\mathrm{LC}}}{\dot{E}} = \frac{T_0}{T_\infty} \cdot \frac{T_{\mathrm{w}} - T_\infty}{T_{\mathrm{w}} - T_0} \qquad \text{(one-dimensional exergy loss number)} \qquad (9)$$

3.3 An example

As a simple example, where N^{E} according to (9) holds, the fully developed pipe flow with heat transfer will be considered. What usually can be found with respect to its heat transfer

cycle / fluid	Nu	$\dfrac{\dot{q}_w}{W/m^2}$	$\dfrac{L}{m}$	$\dfrac{k}{W/mK}$	$\dfrac{T_0}{K}$	$\dfrac{T_w}{K}$	$\dfrac{T_w - T_\infty}{K}$	N^E
SPC / water	100	10^3	0.1	0.1	300	900	10	0.006
ORC / ammonia	100	10^3	0.1	0.038	300	400	26	0.3

Table 1. Heat transfer with $Nu = 100$ in two different power cycles

performance is a Nußelt number correlation $Nu = Nu(Re)$. This, however, is only the energetic part of the performance and N^E according to (9) should also be considered.

To be specific, it is assumed that a heat transfer situation with $Nu = 100$, $\dot{q}_w = 10^3\,W/m^2$ and $L = 0.1\,m$ occurs in two different power cycles. One is a steam power cycle (SPC) with water as the working fluid and a temperature level for heat transfer $T_w = 900\,K$. The alternative is an ORC cycle with ammonia (NH_3) as working fluid and a temperature level $T_w = 400\,K$. When in both cycles $Nu = 100$ with the same values for $\dot{q}_w$ and L holds the temperature difference ΔT is larger by a factor 2.6 for ammonia compared to water. This is due to the different values of thermal conductivity of water (at $T = 900\,K$ and $p = 250\,bar$) and ammonia (at $T = 400\,K$ and $p = 25\,bar$), assuming typical values for the temperature and pressure levels in both cycles.

Table 1 collects all data of this example including the exergy loss number according to (9). For the ORC-cycle this number is 0.3 and this is 50 times that for the SPC-cycle. Note, that an amount of 0.6% and 30% less exergy after the heat transfer in a power cycle means: that amount of available work is lost for a conversion into mechanical energy at the turbine of the cycle.

Here the devaluation of the transferred energy obviously is an important aspect of the process. This devaluation cannot be quantified by the Nußelt number though its finite (and not infinite) value is due to the fact that losses occur in a real transfer process. The exact amount of the losses is given by the exergy loss number N^E.

4. Complex convective heat transfer problems

Heat transfer often occurs as *convective heat transfer*, i.e. influenced and supported by a fluid flow. Especially when the flow is turbulent there is a strong impact on the heat transfer performance. This is due to the strong effect turbulent fluctuations have on the transport of internal energy. As a modeling strategy a so-called turbulent or effective thermal conductivity can be defined which often is a magnitude larger than the molecular conductivity k, see Munson et al. (2009); Herwig (2006) for details.

Increasing the flow rate almost always increases the heat transfer intensity. This is reflected by the increasing Nußelt numbers $Nu(Re)$. There is, however, a prize to pay: it turns out that not only the exergy losses due to the conduction of heat, $\dot{S}_C$ in (6), have to be accounted for, but that also the exergy losses due to the dissipation of mechanical energy in the fluid flow must be considered. Only when both losses are examined and accounted for together, one can answer the question whether an increase in the flow rate is beneficial for the transfer process as a whole.

4.1 Fluid flow assessment

Before the heat transfer process as a whole is considered we want to address the losses in a flow field. Again these losses are exergy losses accompanied by entropy generation. The common way to characterise a flow with respect to its losses is to define a *friction factor* for a flow in pipes and channels, for example. For external flows it would be a *drag coefficient*. Both parameters are finite values (and not zero) due to the fact that losses occur. Again the question arises, whether these parameters are an immediate measure of the exergy losses.

It turns out, that the friction factor f, which will be considered in the following, strictly speaking is a parameter that assesses the *energetic quality* of the flow. It is the answer to the question "How much mechanical energy (measured by the total head of the flow) can be transferred along the pipe or channel in a certain situation with Δp as the operating pressure difference?"

In analogy to the heat transfer process the second question is "How much is the mechanical energy (measured by the total head) devaluated in a certain situation with Δp as the operating pressure difference?" This is about the *exergetic quality* of the flow

Again in analogy to the heat transfer process we define two assessment parameters:

– Head loss coefficient K

$$K = f \frac{L}{D_\mathrm{h}} \equiv \frac{2\,\varphi}{u_\mathrm{m}^2} \tag{10}$$

with φ as specific dissipation of mechanical energy and u_m as the mean velocity in the cross section of the pipe or channel. This parameter is frequently used in fluid mechanics. By introducing K instead of f alone, a pipe or channel (of length L and with a hydraulic diameter D_h) is treated as a conduit component like bends, trijunction, diffusers etc.

– Exergy loss coefficient K^E

$$K^E \equiv \frac{T_0}{T_\infty} K = \frac{\dot{E}_{\mathrm{LD}}}{\dot{m}\, u_\mathrm{m}^2/2} \tag{11}$$

with T_0 as the temperature of the environment and T_∞ as that temperature level on which the flow occurs. The exergy loss rate due to dissipation $\dot{E}_{\mathrm{LD}}$ is the integrated local value $\dot{E}_{\mathrm{LD}}''' = T_0\,\dot{S}_\mathrm{D}'''$ with $\dot{S}_\mathrm{D}'''$ defined later, cf. (3) for the heat transfer counter parts.

The background of both definitions can again be analysed by looking at the entropy and its generation in the flow field. Here the specific dissipation φ and the entropy generation due to dissipation, $\dot{S}_\mathrm{D}$, are closely related, though not the same. The relation is, see Herwig et al. (2010), for example

$$\dot{m}\,\varphi = T_\infty \dot{S}_\mathrm{D} \tag{12}$$

Note, that $T_0\,\dot{S}_\mathrm{D}$ and not $T_\infty \dot{S}_\mathrm{D}$ is the exergy loss rate according to the Gouy–Stodola theorem, cf. (3). That is why K is the energetic, but not the exergetic assessment parameter.

For simple flow geometries such as straight pipes, bends, etc. (10) and (12) can be combined with

$$\dot{S}_\mathrm{D} = K \frac{u_\mathrm{m}^2\,\dot{m}}{2\,T_\infty} \tag{13}$$

as the entropy generation rate based on emperically determined correlations for K-values. For complex flow situations the entropy generation rate due to the dissipation of mechanical energy

$$\dot{S}_\mathrm{D} = \int \dot{S}_\mathrm{D}''' dV \tag{14}$$

and thus the dissipation rate $\dot{m}\,\varphi$ according to (12) can be determined by accounting for the local entropy generation rate per volume ($'''$)

$$\dot{S}_{\mathrm{D}}''' = \frac{\mu}{T}\left(2\left[\left(\frac{\partial u}{\partial x}\right)^2 + \left(\frac{\partial v}{\partial y}\right)^2 + \left(\frac{\partial w}{\partial z}\right)^2\right]\right.$$
$$\left. + \left(\frac{\partial u}{\partial y} + \frac{\partial v}{\partial x}\right)^2 + \left(\frac{\partial u}{\partial z} + \frac{\partial w}{\partial x}\right)^2 + \left(\frac{\partial v}{\partial z} + \frac{\partial w}{\partial y}\right)^2\right) \quad (15)$$

For details of the derivation see Herwig & Kock (2007); Herwig et al. (2010). When the flow is non-isothermal, T in (15) is different at different locations. As long as temperature variations are small compared to T this effect is also small and can be neglected as a first approximation. Then a unique temperature T_∞ appears in (12).

Equation (15) can be immediately used for the determination of $\dot{S}_{\mathrm{D}}'''$ when the flow is laminar or for a turbulent flow in a DNS-approach. In a RANS approach $\dot{S}_{\mathrm{D}}'''$ is split into $\dot{S}_{\overline{\mathrm{D}}}'''$ and $\dot{S}_{\mathrm{D}'}'''$ like all other variables with

$$\dot{S}_{\mathrm{D}}''' = \dot{S}_{\overline{\mathrm{D}}}''' + \dot{S}_{\mathrm{D}'}''' \quad (16)$$

and

$$\dot{S}_{\overline{\mathrm{D}}}''' = \frac{\mu}{\overline{T}}\left(2\left[\left(\frac{\partial \overline{u}}{\partial x}\right)^2 + \left(\frac{\partial \overline{v}}{\partial y}\right)^2 + \left(\frac{\partial \overline{w}}{\partial z}\right)^2\right]\right.$$
$$\left. + \left(\frac{\partial \overline{u}}{\partial y} + \frac{\partial \overline{v}}{\partial x}\right)^2 + \left(\frac{\partial \overline{u}}{\partial z} + \frac{\partial \overline{w}}{\partial x}\right)^2 + \left(\frac{\partial \overline{v}}{\partial z} + \frac{\partial \overline{w}}{\partial y}\right)^2\right) \quad (17)$$

$$\dot{S}_{\mathrm{D}'}''' = \frac{\mu}{\overline{T}}\left(2\left[\overline{\left(\frac{\partial u'}{\partial x}\right)^2} + \overline{\left(\frac{\partial v'}{\partial y}\right)^2} + \overline{\left(\frac{\partial w'}{\partial z}\right)^2}\right]\right.$$
$$\left. + \overline{\left(\frac{\partial u'}{\partial y} + \frac{\partial v'}{\partial x}\right)^2} + \overline{\left(\frac{\partial u'}{\partial z} + \frac{\partial w'}{\partial x}\right)^2} + \overline{\left(\frac{\partial v'}{\partial z} + \frac{\partial w'}{\partial y}\right)^2}\right) \quad (18)$$

Only $\dot{S}_{\overline{\mathrm{D}}}'''$ can be determined directly once a CFD-solution of the flow field exists. The fluctuating part $\dot{S}_{\mathrm{D}'}'''$ must be subject to turbulence modeling. For example, $\dot{S}_{\mathrm{D}'}'''$ can be linked to the turbulent dissipation rate ε which is known when a k–ε-model is used (but which is also part of almost all other models) by

$$\dot{S}_{\mathrm{D}'}''' = \frac{\varrho\,\varepsilon}{\overline{T}} \quad (19)$$

For details, again see Herwig & Kock (2007).

This approach can for example be used to determine the friction factor of a pipe with a special roughness type, called Loewenherz-thread roughness, by integrating the entropy generation rate as shown in Fig. 2, see Herwig et al. (2008) for details. The dark lines are numerical results based on (17)-(19). They compare very well with experimental results from Schiller (1923) and show that the classical Moody chart is not even a moderately good approximation for this kind of rough pipes.

Once the head loss coefficient K is known it is an easy though important step to find K^E according to (11). Only K^E is a direct measure for the devaluation of the transferred energy in terms of lost available work (or exergy).

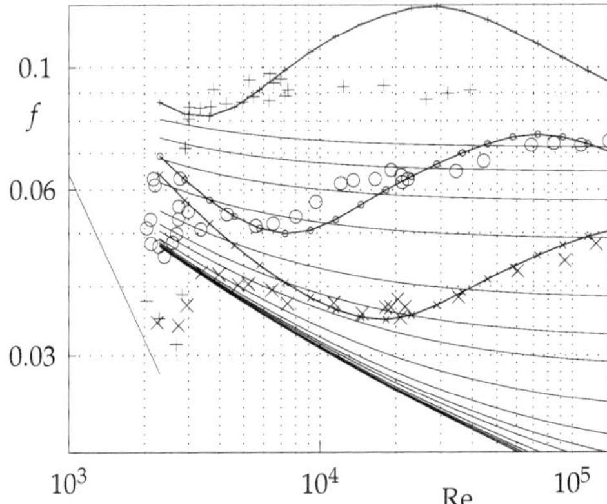

Fig. 2. Friction factor for a pipe with a Loewenherz-thread with different roughness heights; dark lines: numerical integration; large symbols: experiments; faint lines: Moody chart

As an example, a pipe with a certain head loss coefficient K is part of a power cycle. It is operated on different temperature levels when it is in a steam power cycle (SPC, temperature $T_\infty = 900\,\mathrm{K}$) or in an organic Rankine cycle (ORC, temperature $T_\infty = 400\,\mathrm{K}$). With $T_0 = 300\,\mathrm{K}$ according to (11)

– $K^E = 0.33 \cdot K$ for an SPC

– $K^E = 0.75 \cdot K$ for an ORC

The losses of exergy (available work) in an ORC-cycle are more than twice that for the same flow situation (and head loss coefficient K) in an SPC-cycle.

4.2 Convective heat transfer assessment

For a complete assessment of a convective heat transfer situation the *energetic* part is accounted for by the Nußelt number (with no need to look at the head loss coefficient). The *exergetic* part, however, requires the combined consideration of losses in the temperature and the flow field in order to determine the overall reduction of available work caused by the (convective) heat transfer.

This, however, is straight forward within the SLA-analysis. An *overall exergy loss number* N^E now is defined as

$$N^E \equiv \frac{\dot{E}_{LC} + \dot{E}_{LD}}{\dot{E}} = \frac{T_0(\dot{S}_C + \dot{S}_D)}{\eta_C\,\dot{Q}_w} \tag{20}$$

It referres the overall exergy loss $\dot{E}_{LC} + \dot{E}_{LD}$ in a convective heat transfer situation to the exergy fraction of the transferred energy, $\eta_C\,\dot{Q}_w$.

With the help of (20) it can be decided whether the increase in the Nußelt number by a certain technique to improve the heat transfer, like adding turbulence promoters, roughening of the wall or simply increasing the flow rate, is beneficial from the perspective of exergy conservation. When N^E is decreased, more available work is left after the transfer than without the change made in order to improve the heat transfer.

Since a device with a small number N^E obviously is more efficient than one with a larger N^E we introduce an *overall efficiency factor* as

$$\eta^E \equiv 1 - N^E = \frac{\dot{E} - (\dot{E}_{LC} + \dot{E}_{LD})}{\dot{E}} \tag{21}$$

When η^E with N^E should be applied to a complex convective heat transfer situation, $\dot{S}_C$ cannot be determined like in (9) which only holds for a one-dimensional case. Then $\dot{S}_C$ is found from

$$\dot{S}_C = \int \dot{S}_C''' dV \tag{22}$$

with $\dot{S}_C'''$ in Cartesian coordinates, cf. (2),

$$\dot{S}_C''' = \frac{k}{T^2}\left(\left(\frac{\partial T}{\partial x}\right)^2 + \left(\frac{\partial T}{\partial y}\right)^2 + \left(\frac{\partial T}{\partial z}\right)^2\right) \tag{23}$$

or after the turbulence splitting according to the RANS-approach

$$\dot{S}_C''' = \dot{S}_{\overline{C}}''' + \dot{S}_{C'}''' \tag{24}$$

with

$$\dot{S}_{\overline{C}}''' = \frac{k}{\overline{T}^2}\left(\left(\frac{\partial \overline{T}}{\partial x}\right)^2 + \left(\frac{\partial \overline{T}}{\partial y}\right)^2 + \left(\frac{\partial \overline{T}}{\partial z}\right)^2\right) \tag{25}$$

$$\dot{S}_{C'}''' = \frac{k}{\overline{T}^2}\left(\overline{\left(\frac{\partial T'}{\partial x}\right)^2} + \overline{\left(\frac{\partial T'}{\partial y}\right)^2} + \overline{\left(\frac{\partial T'}{\partial z}\right)^2}\right) \tag{26}$$

analogous to the determination of $\dot{S}_D$ in (14)-(18). Like $\dot{S}_{D'}'''$ in (19), $\dot{S}_{C'}'''$ has to be modeled, for example by (α, α_t: molecular and turbulent thermal diffusivities)

$$\dot{S}_{C'}''' = \frac{\alpha_t}{\alpha}\,\dot{S}_{\overline{C}}''' \tag{27}$$

described in Herwig & Kock (2007).

4.3 An example

As an example a counter flow plate heat exchanger is analysed with respect to its heat transfer performance. For that purpose a special 2D-geometry is chosen which corresponds to the geometric situation in a stack of sinusodially formed plates shown in Fig. 3. One element of the cold part is "cut out" as the numerical solution domain, assuming periodic boundary conditions (Fig. 3 b). The boundary conditions are non-slip for the flow and temperature boundary conditions for the heat transfer. Basically a temperature rise $T_2 - T_1$ is set which is assumed to be linearly distributed between the cross sections ① and ②. With $T_2 - T_1$ set, the overall heat flux into the solution domain is also prescribed. The questions to be answered now are

- What is the Nußelt number $Nu = Nu(Re)$

- What is the overall exergy loss rate $\dot{E}_L = \dot{E}_{LD} + \dot{E}_{LC}$ and thus the efficiency factor η^E?

- Is there a maximum of η^E with respect to the Reynolds number?

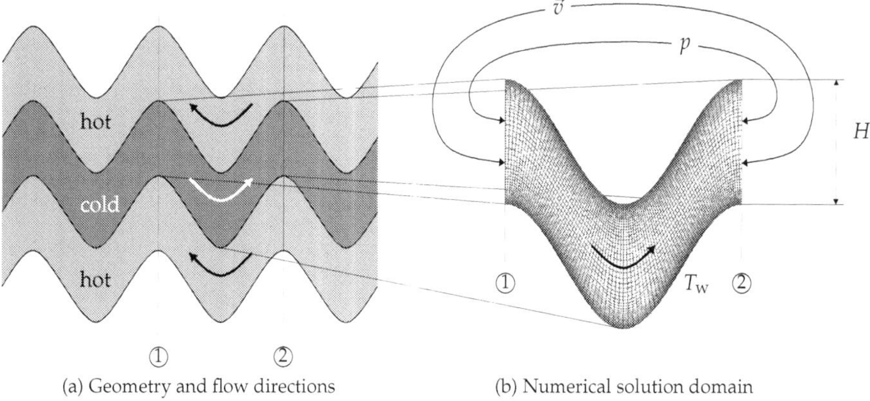

(a) Geometry and flow directions   (b) Numerical solution domain

Fig. 3. 2D-sinusodial plate arrangement in a plate heat exchanger

The flow part of the entropy generation $\dot{E}_{LD}/\dot{E}$ will increase when the Reynolds number gets higher whereas the heat transfer part $\dot{E}_{LC}/\dot{E}$ will decrease due to the favorable influence of a stronger convection. The overall effect $\dot{E}_L/\dot{E}$ is expected to show a minimum which corresponds to a maximum in η^E.

Details of the numerical approach, using the k–ω-SST turbulence model will not be given here (since we want to concentrate on the assessment strategy), they can be found in Redecker (2010). Figure 4 shows the qualitative distribution of the time averaged velocity $\vec{\bar{v}} = \bar{u},\bar{v}$, temperature $\bar{T}$ and the local entropy generation rate $\dot{S}''' = \dot{S}'''_D + \dot{S}'''_C$.

The overall heat transfer performance is shown in Fig. 5. The energetic part in terms of $Nu(Re)$ is shown in Fig. 5(a), the exergetic part in terms of η^E in Fig. 5(b). The Reynolds number $Re = 1995$, see Fig. 5(b), turns out to be the optimal Reynolds number with respect to the loss of available work, since η^E has its maximum for this parameter value.

When this heat exchanger element is analysed in the "conventional way" without recourse to the second law of thermodynamics, one would for example apply an often used *thermo-hydraulic performance parameter*

$$\eta \equiv \left(\frac{St}{St_0} \right) \left(\frac{f}{f_0} \right)^{-\frac{1}{3}} \tag{28}$$

introduced in Gee & Webb (1980) with $St = Nu/RePr$ and $f = K \cdot D_h/L$. This parameter η is used to compare a certain convective heat transfer device (with St_0, f_0) to modified versions (with St, f) and then to decide which of the modifications are beneficial. There is, however, no clear physical interpretation for η, except for the presumption that $\eta > 1$ corresponds to an improved situation.

Applied to the present example, η according to Fig. 6 results without an indication of an optimum within the Reynolds number range shown here. With the combination of St and f as two very different parameters in the definition of η there is no clear physical meaning to be recognised.

We therefore suggest to use η^E (or N^E) in addition to Nu in order to get a complete assessment of convective heat transfer situations.

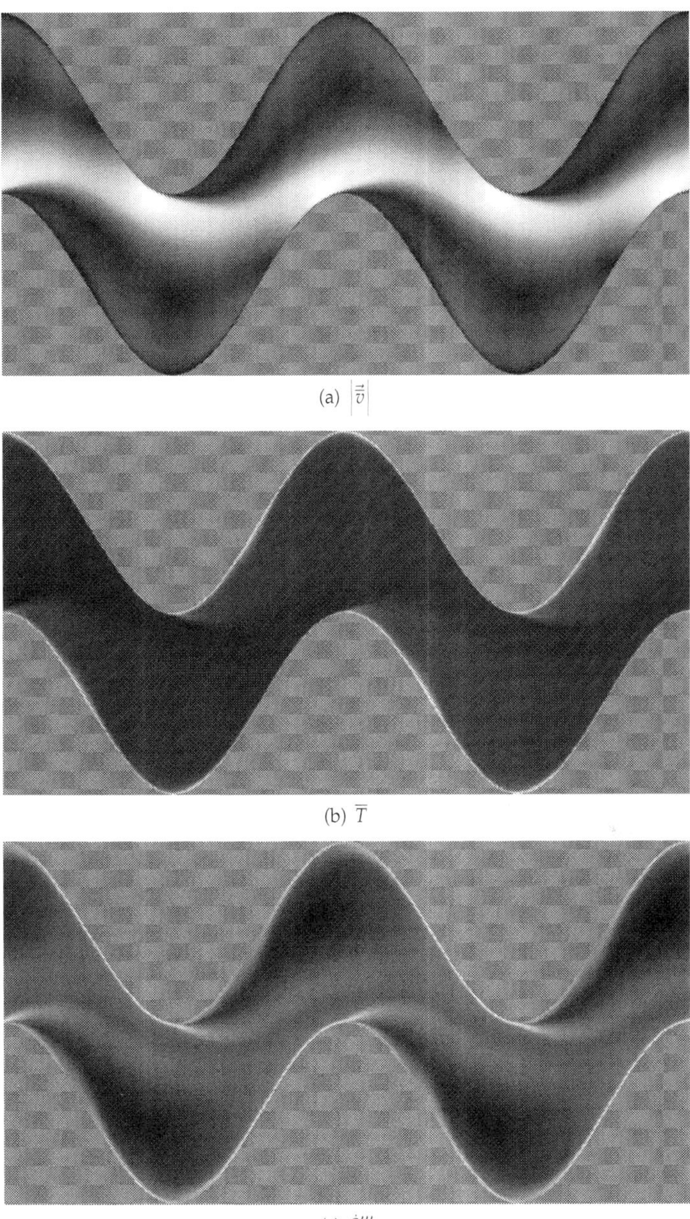

(a) $\left|\vec{\vec{v}}\right|$

(b) $\overline{T}$

(c) $\dot{S}'''$

Fig. 4. Numerical results in twice the solution domain, see Fig. 3(b); Re $= 1995$; light: high values, dark: low values

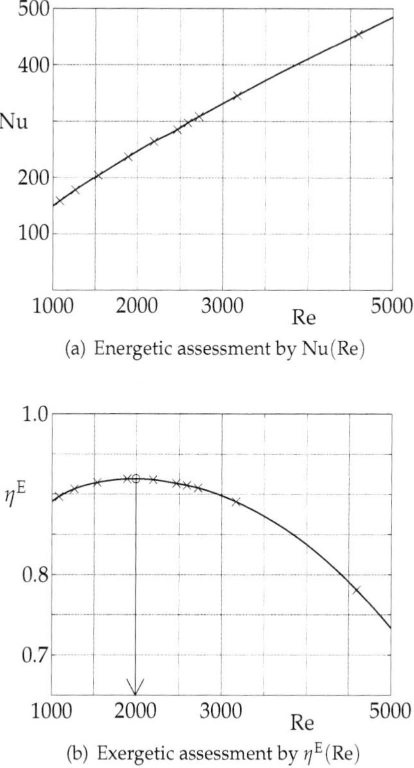

(a) Energetic assessment by $Nu(Re)$

(b) Exergetic assessment by $\eta^E(Re)$

Fig. 5. Heat transfer performance of a plate heat exchanger element

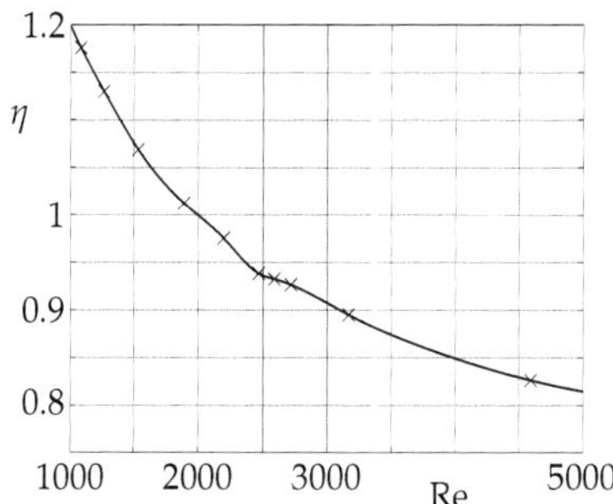

Fig. 6. Heat transfer performance of a plate heat exchanger element, see Fig. 5 for comparison; η: thermo-hydraulic performance parameter (28)

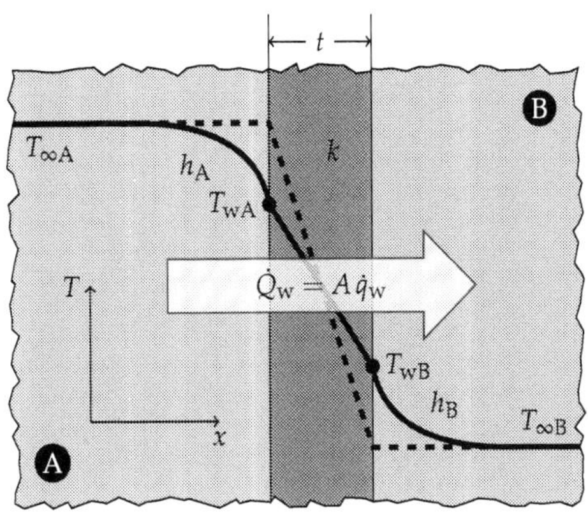

Fig. 7. Overall heat transfer through a wall of thickness t with a thermal conductivity k

5. Further heat transfer problems and their assessment

So far the general idea of a combined energetic and exergetic assessment of heat transfer situations has been discussed for single phase convective heat transfer processes. Some further aspects will be addressed in the following subsections.

5.1 Overall heat transfer through a wall

In Fig. 7 the overall heat transfer through a wall between two systems A and B is sketched. In terms of heat transfer resistances the overall resistance is the sum of two convective and one conductive parts, i.e. $h_A^{-1} + k\Delta T_w/t + h_B^{-1}$ with $\Delta T_w = T_{wA} - T_{wB}$. Since these details are not of interest an overall heat transfer coefficient

$$U = \frac{\dot{q}_w}{\Delta T_\infty} \qquad \text{with } \Delta T_\infty = T_{\infty A} - T_{\infty B} \tag{29}$$

is introduced, see for example Incropera et al. (2006). This is the *energetic* aspect of the assessment.

The *exergetic* aspect can again be accounted for by the exergy loss number N^E according to (6) but now with

$$\dot{S}_C = \dot{Q}_w \left(\frac{1}{T_{\infty B}} - \frac{1}{T_{\infty A}} \right) \tag{30}$$

so that

$$\mathrm{N}^E \equiv \frac{\dot{E}_{LC}}{\dot{E}} = \frac{T_0}{T_{\infty B}} \cdot \frac{T_{\infty A} - T_{\infty B}}{T_{\infty A} - T_0} \tag{31}$$

Here the whole process is assumed to be conductive in nature by replacing the real temperature distribution by one which is decreasing in the wall only (broken line in Fig. 7). Then the entropy generation occurs in the wall only. This simplyfied model is appropriate since it encounters the same devaluation of the transferred energy as the real case. The only thing that counts is the temperature drop from $T_{\infty A}$ to $T_{\infty B}$.

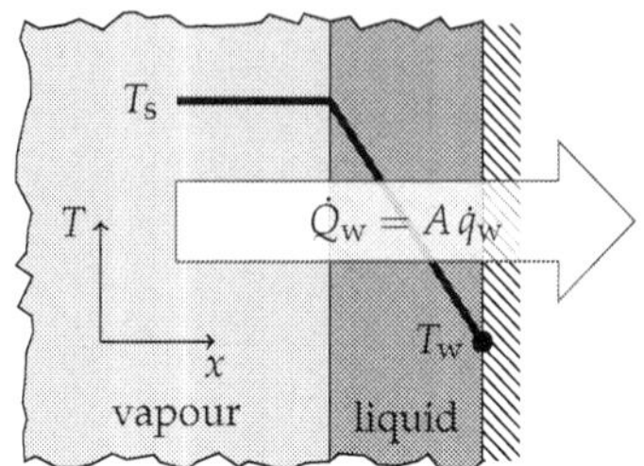

Fig. 8. Heat transfer at a vertical wall with film condensation at a position z away from the leading edge

5.2 Heat transfer with phase change

When close to a heated or cooled wall boiling or condensation occurs the heat transfer mechanism is still "conduction based" since the energy enters or leaves the system by conduction over the system boundary.

In Fig. 8 the situation is sketched for film condensation as an example, showing the local parameters at a certain position z from the leading edge of a vertical wall.

The energetic assessment is by the local heat transfer coefficient

$$h = \frac{\dot{q}_w}{\Delta T_s} \qquad \text{with } \Delta T_s = T_s - T_w \tag{32}$$

Here ΔT_s is the subcooling of the wall (beyond the saturation temperature T_s at the prevailing pressure). The exergetic assessment is possible again by the exergy loss number according to (6), now with

$$\dot{S}_C = \dot{Q}_w \left(\frac{1}{T_w} - \frac{1}{T_s} \right) \tag{33}$$

so that

$$N^E \equiv \frac{\dot{E}_{LC}}{\dot{E}} = \frac{T_0}{T_w} \cdot \frac{T_s - T_w}{T_s - T_0} \tag{34}$$

This simple model assumes that there is no overheating in the vapour phase. An extension to account for this effect would be straight forward, however.

5.3 Radiation based heat transfer

When heat transfer is by radiation the situation is very different compared to the "conduction based" heat transfer processes discussed so far. Energy transport occurs in an electromagnetic field and no longer by molecular interaction of adjacent molecules. For an effective non-zero transport of energy in the form of heat again temperature differences of the two surfaces between which this transport occurs is necessary. When these temperatures are T_{wA} and T_{wB} in a special situation, the overall heat transfer density is (σ: Stefan–Boltzmann constant)

$$\dot{q}_w = F_{AB}\,\sigma\,(T_{wB}^4 - T_{wA}^4) \tag{35}$$

The special situation behind (35) is, that both surfaces are "black surfaces", i.e. they are ideal radiators, see Incropera et al. (2006); Herwig & Moschallski (2009) for more details. Only then both surfaces interact alone with their geometrical orientation towards each other accounted for by the view factor F_{AB}. When $(T_{wB} - T_{wA})/T_{wA} \ll 1$ one approximately gets

$$T_{wB}^4 - T_{wA}^4 = 4T_{wA}^3(T_{wB} - T_{wA}) + \ldots \tag{36}$$

so that again a heat transfer coefficient

$$h = \frac{\dot{q}_w}{\Delta T} = F_{AB}\,\sigma\,4T_{wA}^3 \qquad \text{with } \Delta T = T_{wB} - T_{wA} \tag{37}$$

can be introduced for the energetic assessment.

The exergetic assessment, however, is not easily achieved. The entropy and also the entropy generation depend on various properties of the radiation, like the degree of polarisation and, more general, the deviation from the ideal black body radiation. Also, since radiation heat transfer is a mutual process between two surfaces which both are absorbers and emitters of radiation at the same time also the radiation properties of the counter surface have to be taken into account properly, see Kabelac (1994) for details. Therefore exergy loss numbers N^E according to (6) would only hold in very special situations which are of very limited interest for practical applications.

In this field more research and modelling with relevance for practical applications is needed.

6. Conclusion

The assessment of heat transfer processes is a crucial issue in the development of improved heat transfer devices. Especially nowadays where people are more and more concerned about the efficient use of energy the devaluation of the energy in a heat transfer process is of great importance. To account for this aspect of heat transfer one has to consider the second law of thermodynamics as has been shown in this chapter.

7. References

Baehr, H. & Kabelac, S. (2009). *Thermodynamik*, 14th edn, Springer Verlag, Berlin, Heidelberg, New York.

Bejan, A. (1982). *Entropy generation through heat and fluid flow*, John Wiley & Sons, New York.

Gee, D. & Webb, R. (1980). Forced convection heat transfer in helically rib-roughened tubes, *Int. J. Heat Mass Transfer* 23: 1127–1136.

Herwig, H. (1997). Kritische Anmerkungen zu einem weitverbreiteten Konzept: der Wärmeübergangskoeffizient α, *Forschung im Ingenieurwesen* 63(1–2): 13–17.

Herwig, H. (2006). *Strömungsmechanik*, 2nd edn, Springer.

Herwig, H., Gloss, D. & Wenterodt, T. (2008). A new approach to understanding and modelling the influence of wall roughness on friction factors for pipe and channel flows, *J. Fluid Mech.* 613: 35–53.

Herwig, H. & Kautz, C. (2007). *Technische Thermodynamik*, Pearson Studium, München.

Herwig, H. & Kock, F. (2007). Direct and indirect methods of calculating entropy generation rates in turbulent convective heat transfer problems, *Heat and Mass Transfer* 43: 207–215.

Herwig, H. & Moschallski, A. (2009). *Wärmeübertragung*, 2nd edn, Vieweg + Teubner.

Herwig, H., Schmandt, B. & Uth, M.-F. (2010). Loss coefficients in laminar flows: indispensible for the design of microflow systems, *Proc. ICNMM2010*, number ICNMM2010-30166, Montreal.

Incropera, F., DeWitt, D., Bergmann, T. & Lavine, A. (2006). *Fundamentals of heat and mass transfer*, 6th edn, John Wiley & Sons, New York.

Kabelac, S. (1994). *Thermodynamik der Strahlung*, Vieweg + Teubner.

Moran, H. & Shapiro, H. (2003). *Fundamentals of engineering thermodynamics*, 5th edn, John Wiley & Sons, New York.

Munson, B., Young, D., Okiishi, T. & Huebsch, W. (2009). *Fundamentals of Fluid Mechanics*, 6th edn, John Wiley & Sons, Inc., New York.

Redecker, C. (2010). Report of the Institute of Thermo-Fluid Dynamics, TUHH. available on request.

Schiller, L. (1923). Über den Strömungswiderstand von Rohren verschiedenen Querschnitts- und Rauhigkeitsgrades, *ZAMM* 3: 2–13.

18

Heat Transfer Phenomena in Laminar Wavy Falling Films: Thermal Entry Length, Thermal-Capillary Metastable Structures, Thermal-Capillary Breakdown

Viacheslav V. Lel and Reinhold Kneer
Institute of Heat and Mass Transfer, RWTH Aachen University,
Germany

1. Introduction

Liquid film flows are widely used in industrial applications and for example offer solutions for problems associated with the cooling of mechanical parts and electrical components where, for an optimum design of industrial devices, detailed knowledge about the different heat transfer phenomena in falling films is required. This chapter focuses on the three dominating aspects, such as

- formation and development of quasi-regular metastable structures within the residual layer between large waves;
- thermal-capillary breakdown of laminar-wavy falling film and
- thermal entry length.

1.1 Thermal-capillary metastable structures and thermal-capillary breakdown

Typically, a decrease in film thickness leads to an increase of the heat and mass transfer. However, very thin films are prone to flow breakdowns, which cause dry spot formation. In extreme cases, the appearance of dry spots and, as a consequence, a drastic reduction of the heat transfer at the solid liquid interface may lead to thermal destruction of the carrier surface.

There are several mechanisms leading to the appearance of dry spots: evaporation of liquid within the residual layer, i.e. between large waves, film separation due to interaction with the surrounding gas flow, a boiling crisis in the liquid film, a thermo-diffusive and thermo-capillary breakdown of the film due to concentration gradients or a temperature gradient on the film surface. For the thermo-capillary caused flow of fluid near the surface in falling films a detailed understanding is missing so far.

Some insight into thermo-capillary phenomena have been provided by the experimental study of (Kabov & Chinnov, 1998 and Kabov et al., 2001). These studies comprised investigations of slightly inclined heating surfaces in the small Reynolds number range where inertia can be neglected and as a consequence the free surface developed a two-dimensional stationary bump above the heater due to thermo-capillary Marangoni effects only. At a critical value of the heat flux from the heater to the liquid the bump developed an instability in the transverse direction that took the form of rivulets at the downstream edge of the bump. The first experiments relative to Marangoni instabilities were shown in the work by (Slattery & Stuckey, 1932) by draining a liquid film. Later, (Ludviksson & Lightfoot, 1968) formulated a hydrodynamic stability analysis of the problem, presented in (Slattery & Stuckey, 1932).

Jet formation in laminar-wavy flow was visualised recently by (Chinnov & Kabov, 2003). In this case the falling film is wavy through the hydrodynamic instability. Kalliadasis and co-workers (see for example Trevelyan & Kalliadasis, 2004; Scheid et al., 2005; Ruyer-Quil et al., 2005) in their theoretical works showed that at small Reynolds numbers increasing the Marangoni number leads to larger amplitudes and speads of the solitary pulses since the two instability modes (Kapiza and Marangoni) reinforce each other. On the other hand, in the region of large Reynolds numbers the destabilising Marangoni forces are weaker than the dominant inertia forces. Investigations (Pavlenko & Lel, 1997; Pavlenko et al., 2002) also revealed the presence of regular structures in the form of alternating jets and dry spots in the flow of saturated laminar-wavy liquid films with intense evaporation or boiling.

For the study of thermo-capillary phenomena in falling liquid films information about the surface temperature and the film thickness field is needed. The surface temperature distribution can only be found using non-invasive techniques. (Yüksel & Schlünder, 1988) used a pyrometer for the investigation of the temperature of the falling film's surface. This point technique requires a complex mechanical periphery for the determination of a surface temperature field.

This can be achieved more easily using a focal-plane-array IR camera (Klein et al., 2005). This approach has been pursued in this study. In addition, the camera used here attains a frame rate of up to 2,000 frames per second depending on the size of the array (Al-Sibai et al., 2003). Therefore, the surface temperature distribution of the entire film within the test section can be measured at the same time with a high temporal and spatial resolution. The disadvantage of the camera is that the detector chip requires a more complicated calibration procedure than a single point pyrometer and thus the pyrometer is better suited if absolute temperature measurements are required.

The thickness distribution of three-dimensional waves in laminar falling films was investigated by (Adomeit & Renz, 2000) using a fluorescence technique. A small amount of sodium-fluorescine ($C_{20}H_{10}Na_2O_5$) was added to the film as a fluorescent tracer. The tracer irradiates a bright orange light when exposed to UV-light. The emitted fluorescent light was recorded by a CCD camera equipped with a UV-filter and the brightness distribution was converted into the film thickness field using a calibration procedure. The error of this technique is approximately 0.02 mm, while the film thickness lies in the range of 0.1 up to 2.0 mm. Its disadvantage is the complicated calibration procedure and the dependence of the emitted light on the temperature. This can be avoided when studying falling films using excited waves (Alekseenko et al., 1994). In this case point thickness measurements can be transformed into field data, because the consecutive waves arise in the same places.

1.2 Thermal entry length of falling films

Only few publications relating to the thermal entry length of falling films are available. (Mitrovic, 1988) and (Nakoryakov & Grigorijewa, 1980) determine the thermal entry length of smooth laminar falling films defining the thermal entry length L_δ as the average distance between the beginning of a heated plate and the point where the thermal boundary layer arrives at the film surface. This approach was also adopted here. The results of (Mitrovic, 1988) and (Nakoryakov & Grigorijewa, 1980) yield a similar correlation for L_δ:

$$L_\delta = cRe_0^{\frac{4}{3}}Pr_0\left(\frac{v^2}{g}\right)^{\frac{1}{3}} \tag{1}$$

differing only in the numerical value of the constant c. (Mitrovic, 1988) found a value of c = 0.0974, whereas in (Nakoryakov & Grigorijewa, 1980) has a value of c = 0.06. Here the following symbols are used Pr = v/a (Prandtl number), Re = $\dot{V}/v$B (Reynolds number), $\dot{V}$/B (volume flow rate per test section width), a (thermal diffusivity), v (kinematic viscosity), g (gravitational acceleration).

When comparing laminar-wavy film flow with smooth laminar film flow, the influence of the Reynolds number could change since an influence of the waves can be assumed. Also the film thickness in the residual layer of laminar wavy films is smaller than the one of laminar-waveless films at the same Reynolds number.

Furthermore, the solutions derived by (Mitrovic, 1988) and (Nakoryakov & Grigorijewa, 1980) presume constant material properties which do not comply with changing temperatures in the flow, since due to the heating of the fluid the Prandtl numbers at inflow (Pr_0) and wall (Pr_W) temperature are different. With the temperature increase of the fluid its Prandtl number decreases and thereby the thermal entry length decreases as well. Thus, the ratio Pr_0/Pr_W should be added to the thermal entry length correlation.

The thermal entry length can also be referred to as the length L_α between the beginning of the heated section and the point from which on the heat transfer coefficient α remains constant. According to (Mitrovic, 1988), the ratio L_α/L_δ is about 2.2 for laminar flow. This means that the heat transfer coefficient takes much longer to reach a constant level than the thermal boundary layer to grow up to the film thickness. The time averaged ratio L_α/L_δ should be similar in laminar-wavy flow. Unfortunately, the local heat transfer coefficient could not be measured in the present set-up and therefore this value could not be confirmed.

Infrared thermography makes it possible to measure the surface temperature of the film flow without disturbing it, provided the film is opaque (Kabov et al., 1995). In (Kabov et al., 1996) an IR technique is used to determine the thermal entry length for a smooth film and it is qualitative shown that this length decreases with the heat flux because of Marangoni convection in the direction opposite to the main flow. Presented in Chap. 2.4 experimental data of the thermal entry length of laminar wavy falling films were obtained by means of infrared thermography too.

2. Experimental examination

2.1 Experimental setup

The closed-loop test facility is shown in Fig. 1. Silicone fluid is supplied from a liquid distributor through an adjustable gap onto the 150 mm wide test section. The test section consists of a vertical polyvinylchloride plate with a copper plate (130 x 70 mm² or 130 x 140 mm²) mounted in the hydrodynamically established region 330 mm downstream of the distribution gap. At the end of the test section the liquid film flows into a reservoir. The fluid is circulated by an adjustable piston pump. The flow rate is measured by a positive displacement flow meter. The experiments were carried out in a Reynolds number range between 2 and 39. The variation of Reynolds numbers was limited by the pump capacity. The entrance temperature of the liquid T_1 is checked by a thermocouple in the liquid distributor and is kept constant using a heat exchanger in the lower reservoir. Also, for safety reasons the liquid temperature T_2 is monitored with a thermocouple at a vertical position parallel to the measurement area. In order to avoid disturbances of the film flow this thermocouple is displaced sideways. The temperatures T_4 within the heater is measured at eight measurement points at distances of 6 and 10 mm from the surface. To excite regular waves on the liquid surface a loudspeaker above the upper reservoir is used.

The heat flux is regulated by adjusting the voltage supplied to heating cartridges in the copper plate. The average heat flux $\dot{q}''$ in the experiments was set between 0 and 3.1×10^4 W m^{-2}. An error less than 3% was assessed taking into account that heat losses through the isolation can be neglected. Silicone oils (Polydimethylsiloxane [DMS-T]) of different viscosity were used as test fluids to allow for a variation of the Prandtl number Pr between 57 and 167. The material properties of the different liquids are presented in Table 1.

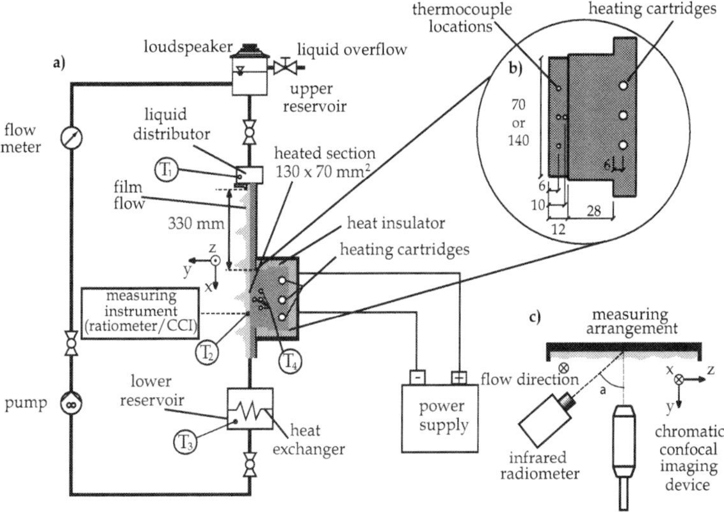

Fig. 1. a) Sketch of the test facility, b) heater, c) position of the measurement devices (Lel et al., 2008)

For description of the kinematic viscosity of DMS-T11 up to 100°C the Vogel-Fulcher-Tammann-equation was used, (Dietze, personal communication).

$$\eta[\text{Pa} \cdot \text{s}] = \frac{\exp(-3.748 + \frac{1848.1}{18 + T[K]})}{1000} \, . \tag{2a}$$

For the dynamic viscosity of DMS-T05 and DMS-T12 the linear dependencies were used:

$$\nu\left[\frac{\text{m}^2}{\text{s}}\right] = 6.62 \cdot 10^{-6} \left[\frac{\text{m}^2}{\text{s}}\right] - 7.37 \cdot 10^{-8} \left[\frac{\text{m}^2}{\text{s}°\text{C}}\right] \cdot T[°\text{C}] \quad \text{for DMS-T05} \tag{2b}$$

$$v\left[\frac{m^2}{s}\right] = 23.01 \cdot 10^{-6}\left[\frac{m^2}{s}\right] - 26.04 \cdot 10^{-8}\left[\frac{m^2}{s^\circ C}\right] \cdot T[^\circ C] \quad \text{for DMS–T12} \qquad (2c)$$

Experiments have been carried out at atmospheric pressure and ambient temperature.

	DMS-T05	DMS-T11	DMS-T12
ρ (kg m^{-3})	918	938	950
σ (N m^{-1})	0.0197	0.0201	0.0206
$d\sigma/dT$ (N K^{-1}m^{-1}) (10^{-5})	-6.7	-6.7	-6.7
c_p (J kg^{-1} K^{-1})	1,546	1,546	1,546
λ (W m^{-1}K^{-1})	0.119	0.136	0.145
ρ_{IR} (9.0 < λ <9.5 µm) (-)	0.049	0.052	0.055
Pr = v/a (-)	57	93	167
Ka = $(\sigma^3 \rho)/(g \eta^4)$ (-)	1,935,716	176,435	14,037

Table 1. Physical properties for silicone fluids Polydimethylsiloxane (DMS-T) at T = 25°C (Lel et al., 2007a)

2.2 Infrared radiometer

For the time-resolved recording of the temperature field an IR-camera CEDIP JADE 3 LW with a 320 x 240 HgCdTe focal-plane-array has been used. The camera operates in the long wavelength region between 7.7 and 9.5 µm. Using an additional filter which blocks all radiation below a wavelength of 9.0 µm, it is possible to measure the temperature in a very thin layer of the surface of the film (Al-Sibai et al., 2003). According to measurements conducted at the Institute for Organic Chemistry at RWTH Aachen University (Fig. 2a) the average absorption coefficient of the fluids at wavelengths between 9.0 and 9.5 µm is between $\alpha_{IR} = 3 \times 10^5$ and $\alpha_{IR} = 4 \times 10^5$ m^{-1}. Average reflection coefficients in the relevant range of wavelengths are included in Table 1.

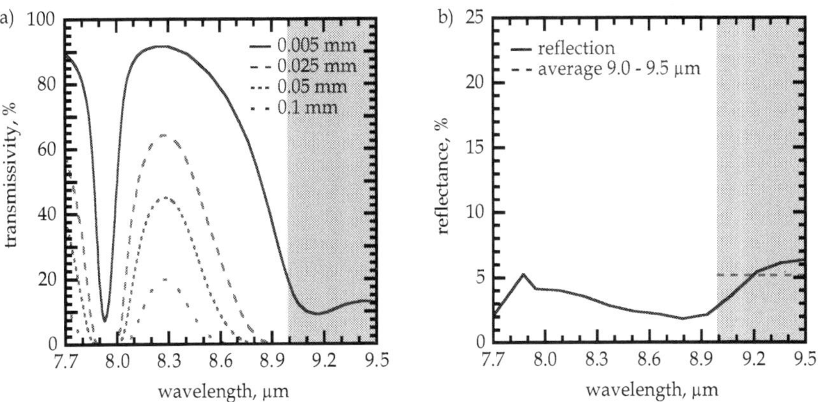

Fig. 2. Optical properties of the silicone fluid DMS-T11. a) Spectral transmissivity for layers of different thicknesses (Bettray, 2002), b) spectral reflectance (Mayerlen & Tacke, 2002)

Knowing the absorption coefficient allows to estimate that 99% of the radiation is absorben in a thin layer at the surface, having a thickness between 10 and 15 μm. The spectral reflectance of the silicone fluid was measured at another research institute (Mayerlen & Tacke, 2002), whose results are presented in Fig. 2b. The average reflection coefficient in the range of wavelengths between 9.0 and 9.5 μm is $\rho_{IR} = 0.052$.

The sensitivity (NETD) of this camera is about 25 mK. The integration time of the IR-camera varies depending on the range of working temperatures. For the given range of temperatures it was set to 500 μs. For a given set of parameters the film flow was recorded over period of 5 s at a frame rate of 200 Hz. The surface area recorded was 70 x 108 mm or 140 x 216 mm. This is equivalent to a resolution of 0.33 x 0.33 mm^2 or 0.66 x 0.66 mm^2 per pixel accordingly.

2.3 Chromatic confocal imaging

The film thickness was measured using a chromatic confocal imaging (CCI) technique (STIL CHR-450, see Cohen-Sabban et al., 2001). Its measuring principle is shown in Fig. 3a. Light from a polychromatic source passes through a semi-transparent mirror. Due to chromatic aberration the lens splits the white light into a continuum of monochromatic images with focal points at different distances. The measuring range (the distance between the focal points of the shortest and the longest used wavelength) is 2.7 mm. The light is reflected from the measured surface and conducted to the spectrometer. A spatial filter eliminates the wavelengths which are not in focus on the measured surface. The wavelength for which the spectrometer detects the highest intensity is the one in focus on the object. By means of a calibration curve this wavelength is converted to a position in the measuring range. This technique has been extended in (Lel et al., 2005) for measuring the thickness of falling liquid films for adiabatic fluid flow. The film surface reflection signal (the wavelength λ_{FS}, see Fig. 3b) is about ten times weaker than the one from the wall surface (the wavelength λ_{WS}, see Fig. 3b) and thus is often equal to the noise level, when the film surface is not parallel to the wall. Therefore the film thickness measurement refers only to the signal reflected from the wall. For the determination of the film thickness δ_r (Fig. 3b) the wall position y_3 has to be known before the start of the film thickness measurement. During the measurements only position y_2 is obtained. Then:

$$\delta_W = y_2 - y_3 \tag{3}$$

and

$$\delta_r = \frac{\delta_W n(\lambda_{WS})}{n(\lambda_{WS}) - 1} \tag{4}$$

Where $n(\lambda_{WS})$ is the refractive index fort the wavelength λ_{WS}, reflected at the wall surfaces.

In the case of heat transfer between surface and film, the heated copper plate expands in an uncontrolled manner and consequently the original position y_3 is displaced to position y_{3^*}. To determine y_{3^*}, the signal from the spectrometer, which is obtained at the times, when film and wall surfaces are parallel (usually in the residual layer between large waves), is analysed for two maxima λ_{FS} and λ_{WS} and the corresponding positions y_1 and y_2 within the measuring range of the optical sensor are calculated, Fig. 3b. The real film thickness and the exact position of the wall y_{3^*} then can be evaluated using the optical law:

$$\delta_r = \delta_{opt} n(\lambda_{WS}) ,\qquad(5)$$

and therefore

$$y_{3^*} = (y_2 - y_1) n(\lambda_{W.S}) + y_1 ,\qquad(6)$$

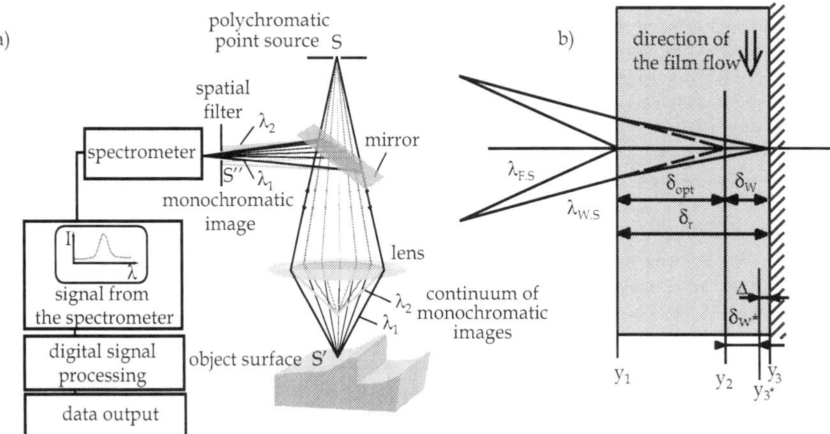

Fig. 3. a) Chromatic confocal imaging principle (Lel et al., 2005), b) light beam path through a falling film (Lel et al., 2008)

For the determination of δ_W, the position y_{3^*} has to be inserted into Eq. (3) instead of y_3. The procedure of refractive index $n(\lambda_{WS})$ determination is shown in Fig. 4. From the position within measuring range y_2 the wavelength λ_{WS} is determined and converted to the refractive index with the help of the dependencies shown in Fig. 4b.

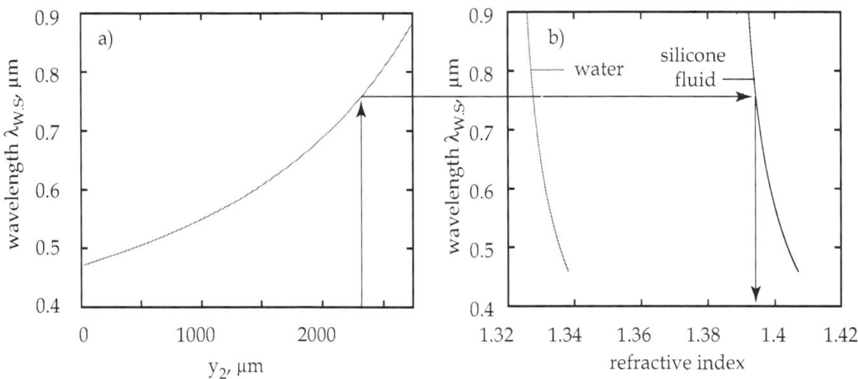

Fig. 4. The procedure of refractive index determination. a) Dependence between wavelength λ_{WS} and position within measuring range, b) Dependence between wavelength λ_{WS} and refractive index for silicone fluid and water (Lel et al., 2008)

2.4 Thermal entry length evaluation

The procedure of evaluating the thermal entry length from the infrared frames is shown on the basis of the example in Fig. 5. There the nondimensional temperature of the film surface according to Mitrović (1988) is shown in Fig. 5.b):

$$\Theta = \frac{T_W - T}{T_W - T_0} \tag{7}$$

Two horizontal stripes, one at the top and one at the bottom of the heated section can be seen (on the left and right sides in the image). The respective edges are shown as start and end markers. These stripes are spanned over the film so that the beginning and the end of the heated surface can be determined in the infrared recordings. The distance between both corresponds to the length of the heated section.

To determine the entry length, a slice A-A with a width of three pixels in the centre of the images was investigated. This procedure allows to average out faulty pixels within the detector array and to reduce noise from the pixels. Within the slice the inflow temperature for the film was identified by averaging over the first 10 pixels over all frames of each film.

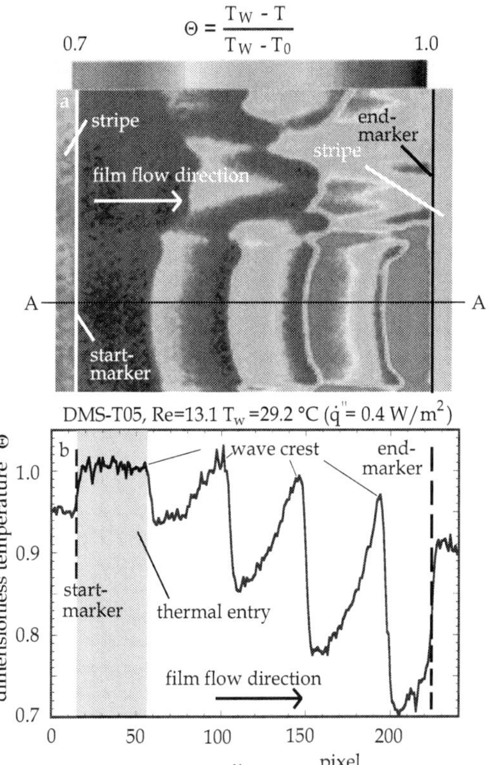

Fig. 5a), 5b) Nondimensional temperature Θ at the film surface (Lel et al., 2007b)

The graph of the nondimensional temperature distribution of the slice A-A, which is marked in Fig. 5a), can be seen in Fig. 5b). Clearly visible are the two stripes that are spanned over the film (the start and the end marker) in the beginning and in the end of the graph. The inflow temperature can be distinguished as the plateau at $\Theta = 1$, which varies in length over time. This plateau is also the thermal inflow length as it is defined for the present work. The noise in the data is smaller than the difference to the threshold value, $\Delta\Theta < 0.01$, which is equivalent to about 0.1 K for the example shown.

The first steep decline in the nondimensional temperature marks the front of a cool wave crest. Further downstream the residual layer is already heated up. In the direction of the flow the residual layer becomes cooler again until another wave crest is reached. This cooling is probably caused by the increasing thickness of the film. Measurements of the film thickness endorse this assumption. The second wave crest is again at inflow temperature. From the next wave crest on, an increase in the temperature can be determined.

At the certain momentary residual layer thickness the thermal boundary layer can grow to film thickness before a wave appears. In this case the current entry length is smaller than the distance between two large waves.

3. Regular structures within the residual layer

In this part the results of the visualization of the regular structures within the residual layer by means of infrared thermography and CCI technique are presented.

Fig. 6 shows a typical IR picture of quasi-regular metastable structures within the residual layer between large waves of laminar-wavy falling films. These structures propagate in flow direction (x-direction), whereas in cross direction (y-direction) the alternating nature of the structures, showing low and high surface temperatures can be determined. The distance between the two hot stripes is shown in Fig. 6 as Λ. As shown in a recent publication (Lel et al., 2007a), no dependence of the mean distance Λ between two hot stripes on the liquid flow rate has been found. This result differs from the conclusion of (Kabov et al., 2004). However the analysis of experimental data of (Kabov et al., 2004) shows, that almost all data for laminar-wavy films have same order of magnitude as the ones of (Lel et al., 2007a). Only the data for very low Reynolds Number (near to case of purely laminar film) have different values of transverse sizes of regular structures Λ.

There are supposably two different mechanisms causing the development of regular structures. The first mechanism is the development of a large bump, whose instability then leads to the regular structures, with the wavelength being a function of Reynolds number. This phenomenon has been described in detail by Kabov and co-workers (see for example, (Kabov & Chinnov, 1998) and (Kabov et al., 2004). The second case relates to the information of regular structures at the horizontal intersection points of larger, parabolic shaped wave fronts. The distance between these points was found (Lel et al., 2007a) to be in the same order as the critical length of the Rayleigh-Taylor instability and independent of the Reynolds number.

The stripes of high surface temperatures lie behind the intersection of the parabola-shaped waves and the temperature difference between the areas with the high and low temperatures grow until the arrival of the next wave front. The temporal evolution of the regular structure's "head" between two parabola-shaped waves is shown in Fig. 6b. From this series it can be seen, that until a time of 0.025 s, the top of the hot stripes (A) has no pronounced characteristic. From time 0.025 to 0.05 s the "head" of the regular structures rounds (B). Later it transforms to the mushroom form (C).

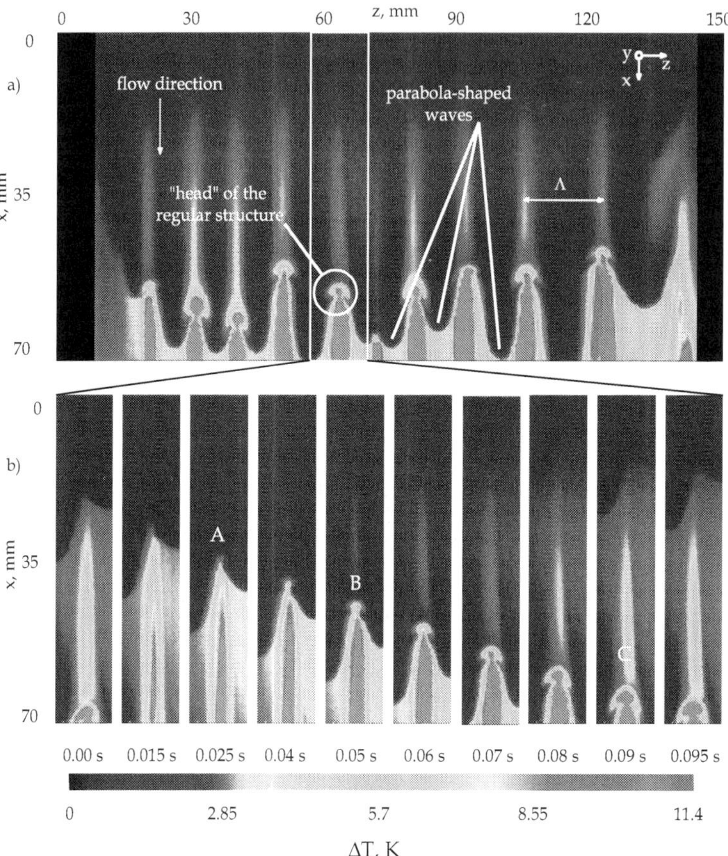

Fig. 6. a) Typical IR picture of quasi-regular metastable structures within the residual layer between large waves of laminar-wavy falling films, b) temporal evolution of the regular structure's "head" between two large parabola-shaped waves (zone marked in a), Re = 15, $\dot{q}\,''$ = 2.2 x 10⁴ W/m² (Lel et al., 2008)

The structure's "head" effect could be explained through interference of waves at the point of intersection of two large parabola-shaped waves. The visualisation of falling films in the work of (Alekseenko et al., 1994) and (Scheid et al., 2006) shows, that capillary waves in front of large waves generate an interference pattern. Also from a recent work by (Lel et al., 2005) has become clear, that the point with the minimal film thickness is always found directly in front of large waves. Now, if interference effects exist in the area of intersection of two large parabola-shaped waves, then these effects are strongest directly in front of the large waves and their intersections. Thus, close to the position of minimum film thickness, maximum film thickness gradients and the temperature gradients have to be expected. There, Marangoni's forces and the forces, induced by surface curvature (Laplace pressure), exhibit maximum values.

The effect of the "mushroom form" is not completely clear. Probably in the spanwise direction (z-direction) only the viscous force is a counterforce to the Marangoni force and the force, inducted by surface curvature. In contrast to this, in flow direction (x-direction) the influence of gravitational and inertial forces could play a certain role. Therefore the "heads" of the regular structures change their form from round (B) to the mushroom (C).

Fig. 7 and Fig. 8 show some results of the visualisation of quasi-regular metastable structures within the residual layer between large waves of laminar-wavy falling films at Reynolds numbers of approximately Re = 12 and 18 as a function of heat flux. In the first rows, IR pictures of the surface are presented. The measurements without heat flux show a uniform surface temperature (upper Fig. 7a.1). In the lower Fig. 7a.1 and in Fig. 8a.1 IR pictures taken without black shielding are shown. These allow obtaining the information about the wave front shapes from reflection of the ambience radiation. This information was used for the film surface imaging, see (Lel et al., 2008).

In the second row of Fig. 7 and Fig. 8 the film thickness fields are presented. In the third row film thickness and surface temperature profiles in the residual layer (directly in front of the large waves) as a function of the horizontal position are shown. At the positions in the z direction, where intersections of the parabola-shaped waves are located, the local film thickness of the residual layer has a minimum in both cases. At the same time a temperature distribution emerges whose extrema are distributed inversely. The differences between maxima and minima of both, film thickness and temperature distribution, increase in parallel with increasing heat flux [column (a)-(c)]. This observation confirms the assumption of the thermo-capillary nature of structures within the residual layer between large waves. Probably the nonuniformity of the surface temperature leads to a nonuniformity of surface tension and as a consequence there is some mass transfer in z direction. This development can be seen more clearly on the right hand side of the test section where the positions of the parabola-shaped waves were more stable. Here, even the small stripes of lower temperatures in the middle of the hot zones can be associated with small crests in the residual layer (Fig. 7b zone A).

The spatially resolved averaged maximum, mean and residual layer film thickness are shown in the fourth row of Fig. 7 and Fig. 8. It can be seen from the pictures a.4 in both figures that the mean film thickness correlates very well with the Nusselt film thickness for laminar flows. However, there is some uneven distribution in the z direction possibly concerned with the external excitation of the film. With increasing heat flux there is a decrease of the mean film thickness. On the one hand this could be explained by the reduction of viscosity with increasing temperature, on the other hand cross flow into the large waves and their acceleration could take place. In the position where the parabola-shaped waves were more stable (right hand side of Fig. 7), the amplitude of the mean film thickness distribution rises with an increase of the specific heat transfer, too. The maximum and residual layer thickness decrease with increasing of the heat flux. The distribution of the former becomes more scattered.

Further analysis of the film thickness fields of Fig. 7 and the temperature as a function of time at the same measurement positions yields the corresponding flow wise profiles of film thickness and surface temperature at different horizontal positions (from the middle of large waves — row one, to the middle of the hot regular structures — row four) and heat fluxes (from the heat flux equal to zero — column one, to the heat flux equal to 2.2×10^4 W/m^2), which are presented in Fig. 9. To ease orientation cut-outs from the film thickness fields of Fig. 7 are presented in each picture and the horizontal position is indicated by a vertical line.

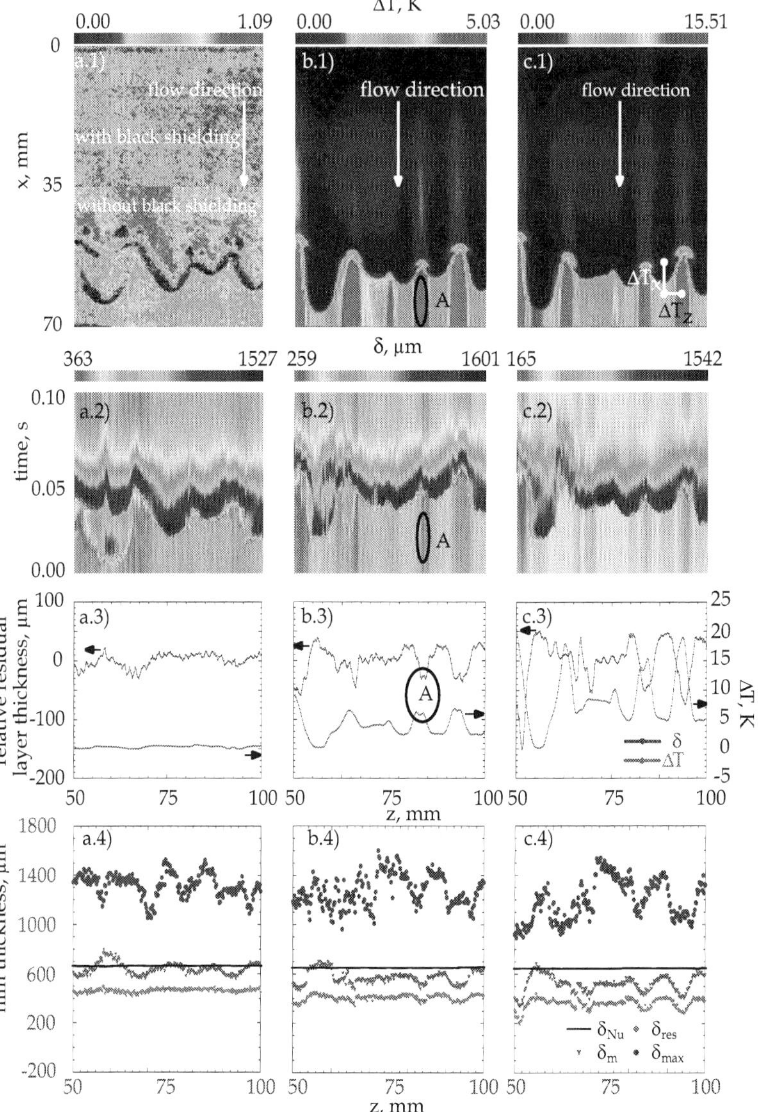

Fig. 7. Regular metastable structures within the residual layer. a) Re = 12.0, $\dot{q}'' = 0$ W/m^2, b) Re = 11.7, $\dot{q}'' = 1.22 \times 10^4$ W/m^2, c) Re = 12.0, $\dot{q}'' = 2.22 \times 10^4$ W/m^2, 1) surface temperature field, IR picture, linear scale, 2) film thickness field, logarithmic scale, 3) film thickness and temperature profile in the residual layer 2 mm ahead of the large waves' front, 4) maximum (δ_{max}), mean (δ_m), residual layer (δ_{res}) and Nusselt film thickness (δ_{Nu}), A: zone of double structure. (Lel et al., 2008)

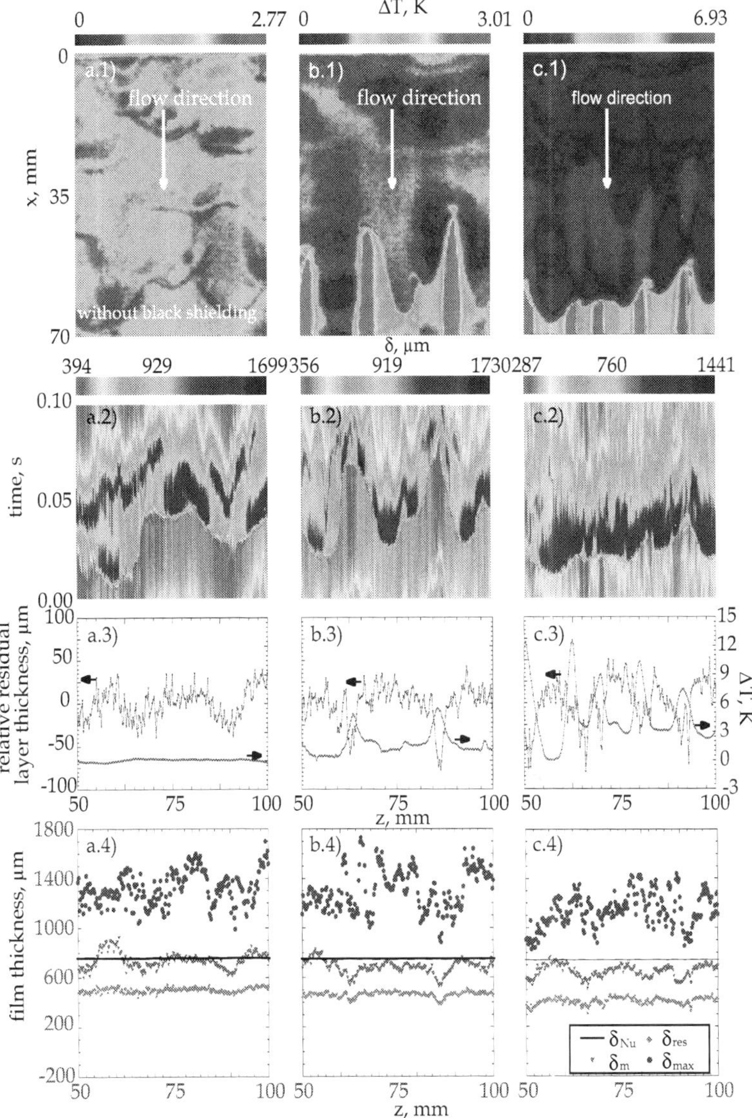

Fig. 8. Regular, metastable structures within the residual layer. a) Re = 17.9, $\dot{q}'' = 0$ W/m², b) Re = 18.0, $\dot{q}'' = 1.04 \times 10^4$ W/m², c) Re = 18.0, $\dot{q}'' = 2.23 \times 10^4$ W/m², 1) surface temperature field, IR picture, linear scale, 2) film thickness field, logarithmic scale, 3) film thickness and temperature profile in the residual layer 2 mm ahead of the large waves' front, 4 maximum (δ_{max}), mean (δ_m), residual layer (δ_{res}) and Nusselt film thickness (δ_{Nu}). (Lel et al., 2008)

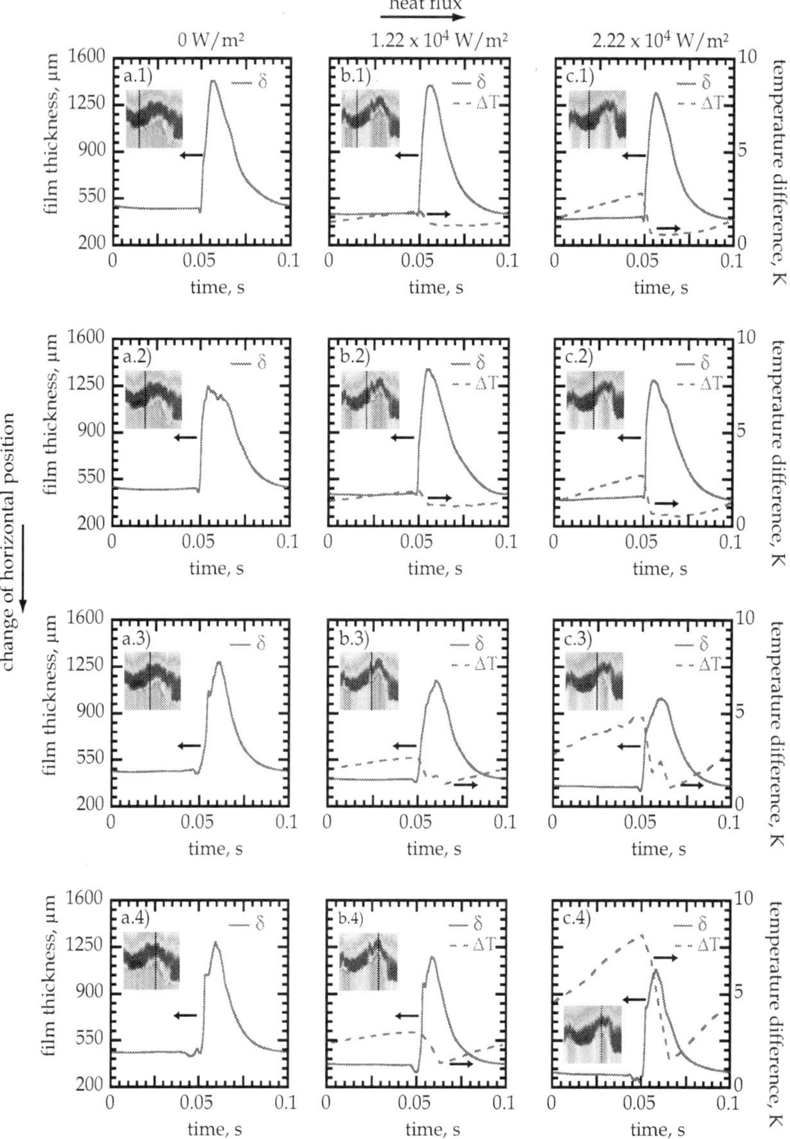

Fig. 9. The film thickness and the temperature profiles in the flow direction as a function of horizontal position and heat flux. a) Re = 12.0, $\dot{q}'' = 0$ W/m², b) Re = 11.7, $\dot{q}'' = 1.22 \times 10^4$ W/m², c) Re = 12.0, $\dot{q}'' = 2.22 \times 10^4$ W/m², 1) in the middle of the large waves, 2) at the side of the large waves, 3) at the side of the hot regular structures, 4) in the middle of the hot regular structures. (Lel et al., 2008)

It can be seen that the difference between maximum and minimum temperatures increases, with increasing heat flux as well as with a change of the horizontal position from the middle of large waves to the middle of the hot regular structures.

In Fig. 10 the non-dimensional film thickness as a function of the heat flux is presented. The different symbols indicate different Reynolds numbers. Open symbols refer to a non-dimensionalisation via the Nusselt film thickness at the inflow temperature, filled symbols indicate the usage of the local temperature. It can be seen that the film thickness decreases with an increase of the heat flux. This can be explained by the decrease in fluid viscosity. However, in this case the ratio of mean film thickness to the Nusselt film thickness based on the local temperature $\delta_m/\delta_{Nu,local}$ should be independent from the heat flux, which obviously is not the case. It can be seen that the non-dimensional film thickness also decreases with increasing heat flux for all Reynolds numbers. Where the local Nusselt film thickness is calculated as:

$$\delta_{NU,local} = \left(\frac{3\dot{V}}{Wg\nu(T_{loc})} \right)^{\frac{1}{3}} \tag{8}$$

where W is the width of test section.

The local temperature can be calculated from the equation:

$$T_{loc} = T_{in} + \frac{\dot{q}''WL}{\rho\dot{V}c_p} \tag{9}$$

where L is the distance between the beginning of the heated section and the measuring point, T_{loc} is the local fluid temperature at the position L, T_{in} is the inflow fluid temperature.

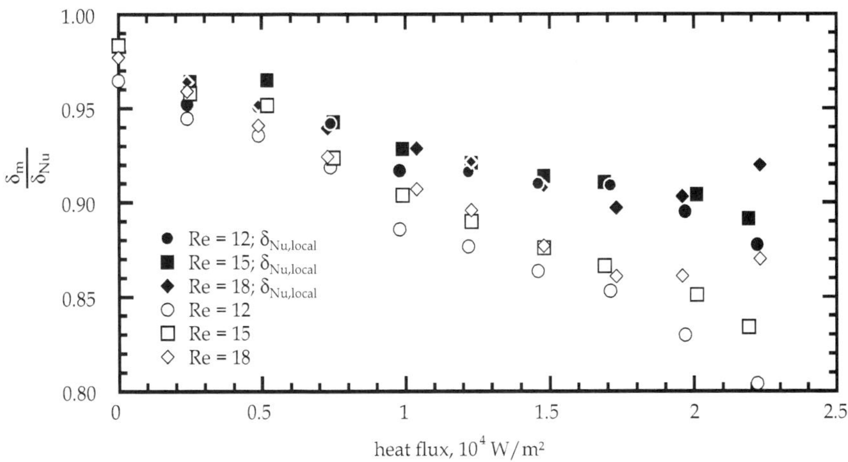

Fig. 10. Dimensionless film thickness, measured 60 mm downstream of the start of the heated section and averaged over the horizontal positions, as a function of heat flux, open symbols divided by Nusselt film thickness at the inflow temperature, filled symbols divided by local Nusselt film thickness. (Lel et al., 2008).

This effect of the film thickness decrease and consequently the increase of the mean film velocity can be explained on the basis of two phenomena. First, calculating the local temperature T_{loc} from Eq. (9), does not take into account the temperature gradient in the falling film in y direction. Close to the wall the viscosity is reduced due to higher temperatures. Therefore, the upper layers of the film move faster in flow-wise direction. Another possible reason is the Marangoni effect in flow-wise direction.

In the case of Marangoni flows being present, the fluid is transported from the residual layer into the large waves under the influence of the nonuniformity of the surface tension in flow direction. As a result the large waves transport more mass and the velocity of the large waves is increased. This assumption qualitatively correlates with the results from (Scheid et al., 2005), that the Marangoni forces could lead to the large amplitudes and speeds of the solitary waves. To confirm the latter considerations more detailed experimental information about mass, momentum and energy transport in the falling film is needed.

Upfront calculations have shown that the saturation pressure of silicone fluid in the air at room temperature is $p_{sat} < 5 \times 10^{-5}$ Pa, and therefore the mass flux of evaporation at room temperature is $\dot{m}'' < 2 \times 10^{-10}$ kg/m² s, thus, evaporation can be neglected.

4. Thermal-capillary breakdown of laminar-wavy falling film

4.1 Model of the thermal-capillary breakdown of laminar-wavy falling film

In this part a model of thermal-capillary breakdown of a liquid film and dry spot formation on the basis of a simplified force balance considering thermal-capillary forces in the residual layer is suggested.

In analogy to (Bohn & Davis, 1993) the force balance in the y-z plane within the residual layer is considered, Fig. 11. Regular structures are formed in flow direction. Neglecting friction between the fluid and the surrounding gas phase as well as inertial effects in the fluid the balance between the forces of surface tension and the tangential stress becomes:

$$\eta \frac{\partial u_z}{\partial y} = \frac{d\sigma}{dz} = \frac{d\sigma}{dT} \frac{\partial T}{\partial z} . \tag{10}$$

Approximating Eq. (10) by a finite difference equation the characteristic "thermal-capillary" velocity can be expressed as

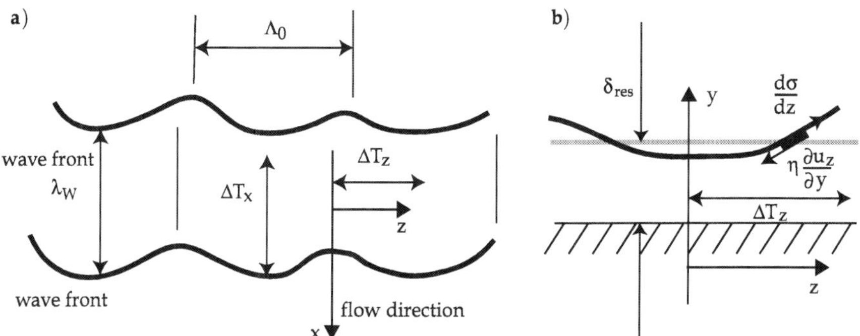

Fig. 11. a) Schematic pattern of regular structures within the residual layer; b) Thermal-capillary breakdown within the residual layer (Lel et al., 2007a)

$$U_{TC} = \left|\frac{d\sigma}{dT}\right| \frac{\Delta T_z}{\eta} \frac{\delta_{res}}{l_z} , \tag{11}$$

where ΔT_z is the characteristic temperature drop along l_z. Contrary to the model suggested in (Bohn & Davis, 1993) l_z is not the characteristic length scale but equal to one half of the distance between two structures as determined in experiments:

$$l_z = \frac{\Lambda_0}{2} . \tag{12}$$

In analogy to (Ganchev, 1984) dry spots may appear in the film when the time required for thermal-capillary breakdown in the residual layer and the period between the passing of two consecutive wave crests are of same order:

$$t_{TC} = t_W , \quad \text{thereby} \quad t_{TC} = \frac{l_z}{U_{TC}} , \quad t_W = \frac{1}{f_W} = \frac{\lambda_W}{c_W} , \tag{13}$$

and f_W is he dominating frequency of large wave propagation, λ_W is the length of large waves and c_W is the phase velocity of large waves.

Fig. 12 shows the energy balance inside an infinitely small element in a film in the x-y plane. Neglecting the heat conduction in x direction and the convective heat transfer in y direction we obtain the eqation:

$$\rho c_p u_x \frac{\partial T}{\partial x} = \lambda \frac{\partial^2 T}{\partial y^2} = -\frac{\partial \dot{q}''}{\partial y} . \tag{14}$$

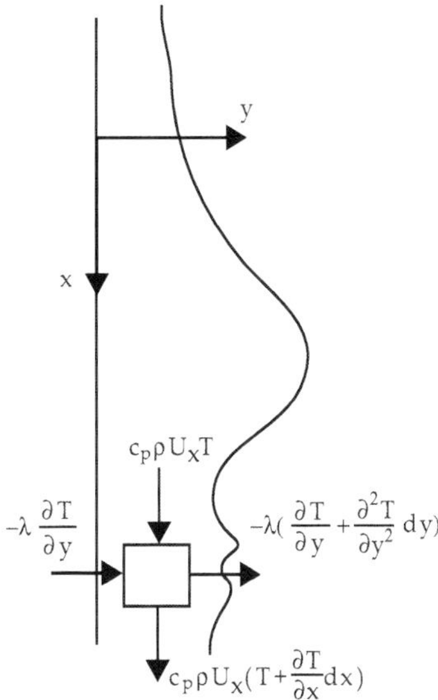

Fig. 12. Energy balance in the residual layer (Lel et al., 2007a)

Approximating this equation by a finite differential equation the typical temperature drop over $l_x \equiv 0.9\lambda_W$ can be determined:

$$\Delta T_x = \frac{\dot{q}''_W}{\delta_{res}} \frac{0.9\lambda_W}{\rho c_p U_{res}} \, ,$$

(15)

Where $\dot{q}''_W$ is the heat flux at the wall, U_{res} is the average liquid velocity in the residual layer, ρ is the liquid density and c_p is the specific heat. The remaining ten percent of l_x correspond to the length of the wave front. The material properties are taken at inflow temperatures. From the thermographic pictures as presented in Fig. 7.c.1, the temperature difference in x and y direction can be evaluated. Thereby a proportionality can be determined.

$$\Delta T_z = k \cdot \Delta T_x \, .$$

(16)

As for current data the constant of proportionality is in the range between $1 < k < 5$. For further considerations a value of $k = 3$ was assumed.

$$\Delta T_z = 3\Delta T_x \, .$$

(17)

Now, substituting expressions (16) and (15) into (11), we obtain:

$$U_{TC} = \left|\frac{d\sigma}{dT}\right| \frac{1}{\eta} \frac{1}{l_z} \frac{\dot{q}''_W}{\rho c_p U_{res}} \frac{0.9\lambda_W}{3}$$

(18)

and (18) into (13):

$$\frac{1}{l_z} \left|\frac{d\sigma}{dT}\right| \frac{1}{\eta} \frac{1}{l_z} \frac{\dot{q}''_W}{\rho c_p U_{res}} \frac{0.9\lambda_W}{3} = \frac{c_W}{\lambda_W} \, .$$

(19)

Solving Eq. (19) for $\dot{q}''_W$:

$$\dot{q}''_W = \frac{1}{0.9} \frac{c_W}{\lambda_W^2} \frac{\rho c_p U_{res}}{\left|\frac{d\sigma}{dT}\right|} \frac{\eta l_z^2}{3} = \frac{1}{0.9} \frac{f_W^2}{c_W} \frac{\rho c_p U_{res}}{\left|\frac{d\sigma}{dT}\right|} \eta \frac{\Lambda_0^2}{4} \frac{1}{3} \, .$$

(20)

According to expression (20) the critical heat flux depends on the liquid properties, the frequency of large waves and the typical transverse size of regular structures. If the following dimensionless numbers are used:

$$Ma_q = \frac{\dot{q}''_W \left|\frac{d\sigma}{dT}\right| \left(\frac{v^2}{g}\right)^{\frac{2}{3}}}{\lambda\rho v^2} \, , \quad Pr = \frac{v}{a} \quad \text{and} \quad K_\Lambda = \left(\frac{f_W \Lambda_0}{(vg)^{\frac{1}{3}}}\right)^2 \, .$$

(21)

Equation (21) can be presented in the form:

$$\frac{Ma_q}{Pr K_\Lambda} = \frac{1}{0.9} \frac{U_{res}}{c_W} \frac{1}{4} \frac{1}{3} = \frac{1}{10.8} \frac{U_{res}}{c_W} \, .$$

(22)

Therefore, in Fig. 13 the experimental data for the dimensionless critical heat flux (Ma_q/PrK_Λ) are presented as a function of the Reynolds number. Only data which were obtained on excited falling films (with the help of the loud speaker) were used, allowing to keep the major frequency f_W at a constant value. As can be seen Eq. (22) depends on the relation between the mean velocity of the residual layer and the mean velocity of large waves. In the literature many different analytical and empirical equations can be found for these velocities as functions of the Reynolds number. For example in (Brauner & Maron, 1983) a physical model for the falling film is presented. In this case the ratio is constant:

$$\frac{U_{res}}{c_W} = 0.091 \ . \tag{23}$$

Therefore the combination of dimensionless parameters from (22) is constant, too:

$$\frac{Ma_q}{PrK_\Lambda} = 8.43 \times 10^{-3} \ . \tag{24}$$

Equation (24) is in the same order for the dimensionless critical heat flux as the experimental data, but the trend of the latter has a different inclination, see Fig. 13.
In (Al-Sibai, 2004) the same silicone oils were used as in the current experiments. Therefore a better comparability could be given for dependencies from (Al-Sibai, 2004) as for other correlations from literature. Since the thickness of the residual layer is relatively small the Nusselt formula for laminar flow can be used:

$$U_{res} = \frac{g\delta_{res}^2}{3\nu} \ . \tag{25}$$

In (Al-Sibai, 2004) an equation for the residual layer thickness can be found:

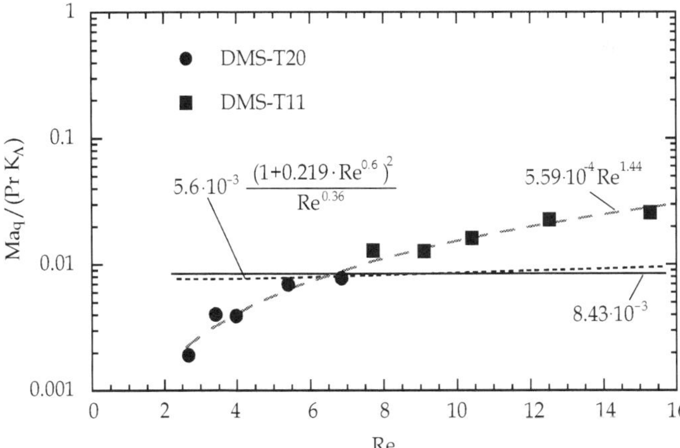

Fig. 13. The dependence of dimensionless parameter $Ma_q/(PrK_\Lambda)$ versus Reynolds-number. *Points* experimental data (Lel et al., 2007a)

$$\delta_{res} = \left(1 + 0.219 Re^{0.6}\right)\left(\frac{v^2}{g}\right)^{\frac{1}{3}}.$$ (26)

Substituting (26) into (25):

$$U_{res} = \frac{1}{3}g^{\frac{1}{3}}v^{\frac{1}{3}}\left(1 + 0.219 Re^{0.6}\right)^2$$ (27)

and another equation from (Al-Sibai, 2004) giving the velocity of large waves against the Reynolds number:

$$c_W = 5.516 g^{\frac{1}{3}}v^{\frac{1}{3}}Re^{0.36}.$$ (28)

From Eq. (22) the following non-dimensional expression can be derived with substitutions (27) and (28):

$$\frac{Ma_q}{PrK_\Lambda} = 5.6 \times 10^{-3}\frac{\left(1 + 0.219 Re^{0.6}\right)^2}{Re^{0.36}}.$$ (29)

The comparison of dependence (29) with experimental data gives a good agreement in the order of magnitude but a difference in the inclination, see Fig. 13. For a Reynolds number range Re < 3 the dimensionless parameter (Ma_q/PrK_Λ) according to Eq. (29) even decreases, but the experimental data show another tendency.

This disagreement between experimental data and theory can be ascribed to the uncertainty of the proportionality factor in Eq. (16) as describes above.

As can be seen from (29):

$$Ma_q = f\left(Re, Pr, K_\Lambda\right).$$ (30)

This approximation found from experimental data analysis is:

$$Ma_q = 5.59 \times 10^{-4} Re^{1.44} PrK_\Lambda.$$ (31)

In order to verify and consolidate this theory the range of Reynolds number should be increased. An elongating of heat section will allows the observation of further development of regular structures.

For the experiments without activated loud speaker the wavelength has to be determined by measuring the oscillations of the film surface and using the major frequency for the parameter K_Λ.

A comparison of experimental data with other dependencies from the literature is shown in the next part.

4.2 Comparing of experimental data with other approachs

In this part different approaches for the determination of the critical dimensionless heat flux are presented and compared with experimental data.

Experimental data for laminar-wavy and turbulent films were described in (Gimbutis, 1988) by the following empirical dependencies:

$$\mathrm{Ma}_q = 0.522\mathrm{Re}^{0.4}\left(1 + 0.12\left(\frac{\mathrm{Re}}{250}\right)^{4.5}\right)^{0.5} \quad \text{for } L \le 1 \text{ m.} \tag{32}$$

For $100 < \mathrm{Re} < 200$ in (Gimbutis, 1988) the scattering of data was up to 50 %, for $\mathrm{Re} < 100$ no experimental data have been recorded. It can be seen in Fig. 14 that this dependence suggests lower values than the current experimental data. The difference can be explained by the fact that in (Gimbutis, 1988) the experimental data were obtained only for a water film flow with a relatively long heated section. In this case evaporation effects and thus a shift in the thermophysical properties could have appeared.

In (Kabov, 2000) the empirical dependence of the critical Marangoni number on the Reynolds number for a shorter heated section (6.5 mm length along the flow) for laminar waveless falling films was obtained:

$$\mathrm{Ma}_q = 8.14\mathrm{Re}^{0.98} . \tag{33}$$

In this case the length of the heated section is in the same order of magnitude as the thermal entry length (Kabov, 2000). Therefore this curve indicates higher values than our experimental data.

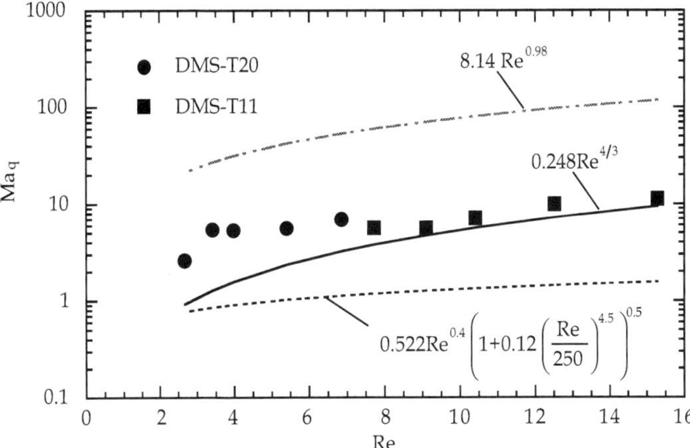

Fig. 14. The dependence of dimensionless parameter Ma_q versus the Reynolds-number. *Points* experimental data (Lel et al., 2007a)

It was shown in (Ito et al., 1995) that for the 2D case the modified critical Marangoni number is constant:

$$\mathrm{Ma}_c = \frac{\dot{q}''_w \left|\dfrac{d\sigma}{dT}\right|}{\lambda\rho U_s^2} = 0.23 . \tag{34}$$

With the film surface velocity based on Nusselt's film theory $U_s = (g/2\nu)\delta_m^2$, expression (34) can be transformed into:

$$Ma_q = 0.248 Re^{4/3} .$$

(35)

It can be seen, that only dependence (35) is in the same order of magnitude as our experimental data.

Other dimensionless parameters for generalisation of experimental data were used in (Bohn & Davis, 1993) and (Zaitsev et al., 2004). In (Bohn & Davis, 1993) the data for dimensionless breakdown heat flux is approximated in the form:

$$\frac{\dot{q}_w'' \left|\dfrac{d\sigma}{dT}\right|}{\rho^{1/3} c_p \eta^{5/3} g^{2/3}} = 4.78 \times 10^5 Re^{1.43} .$$

(36)

With elementary transformations (36) can be transformed into:

$$\frac{Ma_q}{Pr} = 4.78 \times 10^5 Re^{1.43} .$$

(37)

Fig. 15 shows that (37) again leads to lower values than our experimental data. Here, as in case of Eq. (32), evaporation effects could have appeared, because this dependence was obtained for water and for a 30 % glycerol-water solution at a 2.5 m long test section for Re > 959.

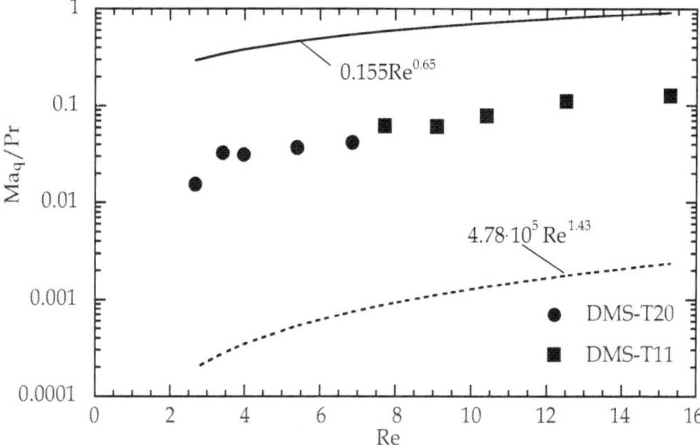

Fig. 15. The dependence of dimensionless parameter Ma_q/Pr versus the Reynolds-number. *Points* experimental data (Lel et al., 2007a)

A generalisation for water and an aqueous solution of alcohol is presented in (Zaitsev et al., 2004):

$$\frac{Ma_q}{Pr} = 0.155 Re^{0.65} .$$

(38)

This correlation leads to results which exceed current data by more than one order of magnitude. This can be partially explained by the fact that dependence (38) was obtained for stable dry spots, whereas the new data was recorded for the formation of local instable dry spots.

5. Thermal entry length

In this part experimental data for the thermal entry length with the correlation from the literature are compared and a new correlation, included dimensionless parameters incorporated several physical effects, is presented.

A comparison of experimental data for the thermal entry length with correlations for laminar flow by (Mitrovic, 1988) and (Nakoryakov & Grigorijewa, 1980), is shown in Fig. 16. Whereas for very low Reynolds numbers ($Re_0 < 3$) and heat fluxes the experimental data correlate satisfactorily with these dependencies, at larger Reynolds numbers and heat fluxes the experimental values lie under the ones obtained through correlations.

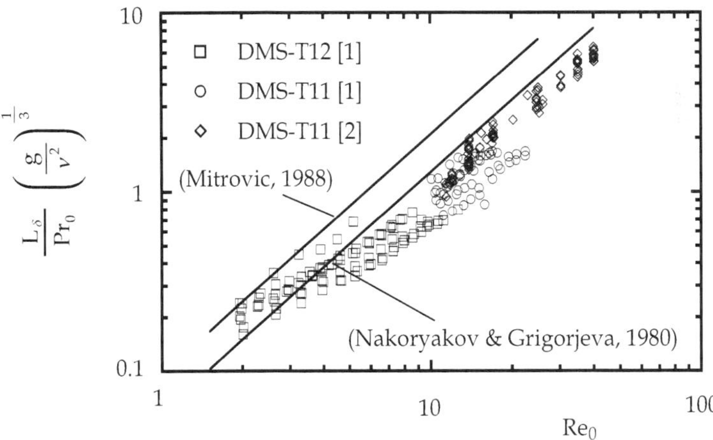

Fig. 16. Experimental data compared with solutions for smooth laminar falling films. [1] – experimental data by (Lel et al., 2007b); [2] – experimental data by (Lel et al., 2009).

Therefore, the experimental data were used in order to found a empirical dependency which describes the dimensionless entry length and attempts to incorporate several effects: i) the effect of nonlinear changing material properties due to temperature changes and the effects of ii) surface tension and iii) waves:

$$L_\delta = aRe_0^b Pr_0 Ka_0^{0.0606} \left(\frac{Pr_0}{Pr_W} \right)^{-0.29} \left(\frac{v^2}{g} \right)^{\frac{1}{3}} \tag{39}$$

$$\left. \begin{array}{l} a = 0.8367 \\ b = 0.718 \end{array} \right\} \text{for } Re < 8$$

$$\left. \begin{array}{l} a = 0.022 \\ b = 1.36 \end{array} \right\} \text{for } Re > 8$$

In Fig. 17 the comparison of experimental data with a correlation for $100<Pr_0<180$ and $2<Re_0<40$ is presented.

Here it is significant, that the differences of Eq. (1) and Eq. (39) are the additional terms involving Pr_0/Pr_W and the Kapitza number $Ka_0 = (\sigma^3\rho/g\eta^4)$.

(Brauer, 1956) found that the Kapitza number has an influence on the development of the waves at the film surface. He defined the point of instability $Re_i = 0.72Ka_0^{\frac{1}{10}}$, at which sinusoidal waves become instable. Therefore, the Kapitza number has also to be implemented into correlations (39).

The relation Pr_0/Pr_W between the Prandtl number at the inflow and at the wall temperature has to be added into the dependence, in order to take into account the dependency of the viscosity of the fluid on the temperature.

The effect of the decrease of the thermal entry length flux because of Marangoni convection, described in (Kabov et al., 1996), is subject to debate. We assume that in this case the influence of the waves on the thermal entry length play a dominant role. This question stays unsettled and should be investigated in future.

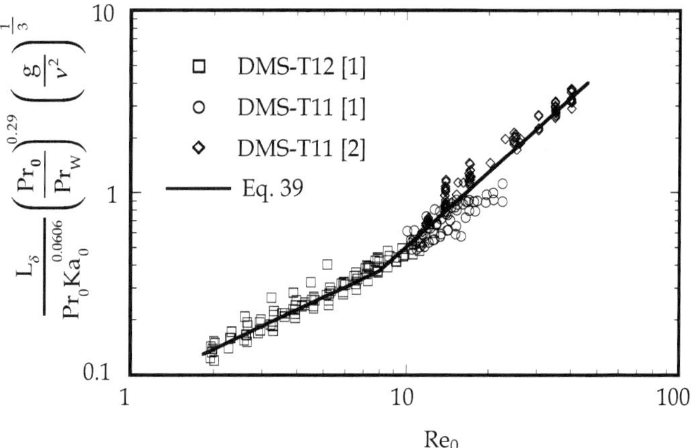

Fig. 17. Comparison of correlation (5) for thermal entry length for laminar wavy films with the experimental data. [1] – experimental data by (Lel et al., 2007b); [2] – experimental data by (Lel et al., 2009).

6. Conclusion

The results of the experimental investigation of different physical effects of heat and mass transfer in falling films were discussed in this chapter.

At first the visualization of quasi-regular metastable structures within the residual layer between large waves of laminar-wavy falling films were presented. To obtain a relation between the surface temperature and film thickness fields, infrared thermography and the chromatic confocal imaging technique were used.

By comparing the temperature and film thickness fields, the assumption of the thermo-capillary nature (Marangoni effect) of regular structures within the residual layer has been confirmed. An increase in local surface temperature leads to a decrease in local film thickness. The evolution of the regular structure's "head" between two large parabolic shaped waves over time was presented.

The decrease of the mean film thickness could be explained by a reduction of the viscosity and a cross flow into the faster moving large waves. Both effects cause a higher film velocity.

The results obtained are important for the investigation of the dependency between wave characteristics and local heat transfer, the conditions of "dry spot" appearance and the development of crisis modes in laminar-wavy falling films.

A model of thermal-capillary breakdown of a liquid film and dry spot formation is suggested on the basis of a simplified force balance considering thermal-capillary forces in the residual layer. It is shown that the critical heat flux depends on half the distance between two hot structures, because the fluid within the residual layer is transferred from hot structures to the cold areas in between them. It also depends on the main frequency of large waves, the Prandtl number, the heat conductivity, the liquid density and the change in surface tension in dependence on temperature. The model is also presented in a dimensionless form.

The investigations of the thermal entry length of laminar wavy falling films by means of infrared thermography are shown. Good qualitative agreement with previous works on laminar and laminar wavy film flow was found at low Reynolds numbers. However, with increasing Reynolds numbers and heat fluxes, these correlations describe the thermal entry length inadequately. The correlation established for laminar flow was extended in order to include the effect of temperature-dependent non-linear material properties as well as for the effects of surface tension and waves.

7. Acknowledgements

This work was financially supported by the "Deutsche Forschungsgemeinschaft" (DFG KN 764/3-1). The authors thank the student coworkers and colleagues A. Kellermann, Dr. H. Stadler, Dr. G. Dietze, M. Baltzer, Dr. F. Al-Sibai, M. Allekotte for the help in the preparation of this chapter.

8. References

Adomeit, P. & Renz, U. (2000). Hydrodynamics of three-dimensional waves in laminar falling films, *Int J Mult Flow*, Vol. 26, pp. 1183-1208

Alekseenko, S.V.; Nakoryakov, V.E. & Pokusaev, B.G. (1994). *Wave flow of liquid films*, Fukano, T. (Ed.), p. 313, Begell House, Inc., New York

Al-Sibai, F. (2004). Experimentelle Untersuchung der Strömungscharakteristik und der Wärmeübertragung bei welligen Rieselfilmen. *Thesis of Dr.-Ing. Degree*, Lehrstuhl für Wärme- und Stoffübertragung, RWTH Aachen

Al-Sibai, F.; Leefken, A.; Lel, V.V. & Renz, U. (2003). Measurement of transport phenomena in thin wavy film. *Fortschritt-Berichte VDI 817 Verfahrenstechnik Reihe 3*, pp. 1-15

Bohn, M.S. & Davis, S.H. (1993). Thermo-capillary breakdown of falling liquid film at high Reynolds number. *Int J Heat Mass Transf*, Vol. 7, pp. 1875-1881

Bettray, W. (2002). FTIR measurements of spectral transmissivity for layers of silicone fluids with different thicknesses, *Technical Report of the Institute for Organic Chemistry at the RWTH Aachen University*, Germany

Brauer, H. (1956). Strömung und Wärmeübergang bei Rieselfilmen, *VDI Forschungsheft 457*

Brauner, N. & Maron, D.M. (1983). Modeling of wavy flow in inclined thin films. *Chem Eng Sci*, Vol 38, No. 5, pp. 775-788

Chinnov, E.A. & Kabov, O.A. (2003). Jet formation in gravitational flow of a heated wavy liquid film, *J Appl Mech Tech Phys*, Vol. 44, No. 5, pp. 708-715

Cohen-Sabban, J.; Crepin, P.-J. & Gaillard-Groleas, J. (2001). Quasi confocal extended field surface sensing, *Processing of Optical Metrology for the Semicon*, Optical and Data Storage Industries SPIE's 46th Annual Meeting, San Diego, August 2-3, CA, USA

Ganchev, B.G. (1984). Hydrodynamic and heat transfer processes at downflows of film and two-phase gas-liquid flows (in Russian), *Thesis of Doctor's Degree in Phys-Math. Sci.*, Moscow

Gimbutis, G. (1988). *Heat transfer at gravitation flow of a liquid film* (in Russian), *Mokslas*, Vilnius

Ito, A.; Masunaga, N. & Baba, K. (1995). Marangoni effects on wave structure and liquid film breakdown along a heated vertical tube, In: *Advances in multiphase flow*, Serizawa, A.; Fukano, T.; Bataille, J. (Ed.), pp. 255-265, Elsevier, Amsterdam

Kabov, O.A. (2000). Breakdown of a liquid film flowing over the surface with a local heat source. *Thermophys Aeromech*, Vol. 7, No.4, pp. 513-520

Kabov, O.A. & Chinnov, E.A. (1998). Hydrodynamics and heat transfer in evaporating thin liquid layer flowing on surface with local heat source. In: *Proceedings of 11th international heat transfer conference*, Vol. 2, pp. 273-278, Kyondju, Korea, August 23-28

Kabov, O.A.; Chinnov, E.A. & Legros, J.C. (2004). Three-dimensional deformations in non-uniformly heated falling liquid film at small and moderate Reynolds numbers, In: *2nd International Berlin Workshop – IBW2 on Transport Phenomena with Moving Boundaries*, Schindler, F.-P. (Ed.), pp. 62-80, Düsseldorf: VDI Verlag, VDI Reihe 3, No. 817, 9-10 October, Germany

Kabov, O.A.; Diatlov A.V. & Marchuk, I.V. (1995). Heat transfer from a vertical heat source to falling liquid film, In: *G.P. Celata and R.K. Shah, Editors, Proceeding of First Internat. Symp. on Two-Phase Flow Modeling and Experimentation*, Vol.1, pp. 203–210, Rome, Italy, 1996

Kabov, O.A.; Marchuk, I. V. & Chupin, V.M. (1996). Thermal imaging study of the liquid film flowing on vertical surface with local heat source. *Russ. J. Eng. Thermophys*, Vol. 6, No. 2, pp. 104-138

Kabov, O.A.; Legros, J.C.; Marchuk IV & Scheid, B. (2001). Deformation of free surface in a moving locally-heated thin liquid layer. *Fluid Dyn*, Vol. 36, No. 3, pp. 521-528

Klein, D.; Hetsroni, G. & Mosyak, A. (2005). Heat transfer characteristics of water and APG surfactant solution slow in a micro-channel heat sink, *Int J Mult Flow*, Vol. 31, No. 4, pp.393-415

Lel, V.V.; Al-Sibai, F.; Leefken, A. & Renz, U. (2005). Local thickness and wave velocity measurement of wavy falling liquid films with chromatic confocal imaging method and a fluorescence intensity technique, *Exp Fluids*, Vol. 39, pp. 856-864

Lel, V.V.; Al-Sibai, F. & Kneer, R. (2009). Thermal entry length and heat transfer phenomena in laminar wavy falling films, *Microgravity Sci. Technol*, Vol. 21 (Suppl 1), pp. 215-220

Lel, V.V.; Dietze, G.F.; Stadler, H.; Al-Sibai, F. & Kneer, R. (2007b). Investigation of the Thermal Entry Length in Laminar Wavy Falling Films, *Microgravity sci. technol.*, Vol. XIX-3/4, pp. 66-68

Lel, V.V.; Kellermann, A.; Dietze, G.; Kneer, R. & Pavlenko, A.N. (2008). Investigations of the Marangoni effect on the regular structures in heated wavy liquid films, *Exp Fluids*, Vol.44, pp. 341-354

Lel, V.V.; Stadler, H.; Pavlenko, A. & Kneer, R. (2007a). Evolution of metastable quasi-regular structures in heated way liquid films, *J Heat Mass Transf*, Vol. 43, No. 11, pp. 1121-1132

Ludviksson, V. & Lightfoot, E.N. (1968). Hydrodynamic stability of Marangoni films, *AIChE J*, Vol. 14, No. 4, pp. 620-626

Mayerlen, W. & Tacke, M. (2002). Measurements of spectral reflectance of silicone fluids, *Technical report of FGAN-FOM*, Forschungsinstitut für Optronik und Mustererkennung, Germany

Mitrović, J. (1988). Der Wärmeaustausch am Berieselungskühler. *Brennstoff-Wärme-Kraft*, Vol. 40, No. 6, pp. 243-249

Nakoryakov V.E. & Grigorjeva, N.I. (1980). Calculation of heat and mass transfer in nonisothermal absorption on the initial portion of downflowing films. *Theoretical Foundations of Chemical Engineering*, Vol. 14, pp. 305-309

Pavlenko, A.N. & Lel, V.V. (1997). Heat transfer and crisis phenomena in falling films of cryogenic liquid, *Russ J Eng Thermophys*, Vol. 3-4, No. 7, pp. 177-210

Pavlenko, A.N.; Lel, V.V.; Serov, A.F.; Nazarov, A.D. & Matsekh, A.M. (2002). Wave amplitude growth and heat transfer in falling intensively evaporating liquid film, *J Eng Thermophys*, Vol. 11, No. 1, pp. 7-43

Ruyer-Quil, C.; Scheid, B.; Kalliadasis, S.; Velarde, M.G. & Zeytounian, R.K. (2005). Thermocapillary long waves in a liquid film flow. Part 1. Low dimensional formulation, *J Fluid Mech*, Vol. 538, pp. 199-222

Scheid, B.; Ruyer-Quil, C.; Kalliadasis, S.; Velarde, M.G. & Zeytounian, R.K. (2005). Thermocapillary long waves in a liquid film flow. Part 2. Linear stability and nonlinear waves, *J Fluid Mech*, Vol. 538, pp. 223-244

Scheid, B.; Ruyer-Quil, C. & Manneville, P. (2006). Wave patterns in film flows. Modeling and three-dimensional waves, *J Fluid Mech*, Vol. 562, pp. 183-222

Slattery, J. & Stuckey, E.L.; Trans R Soc Can Sect, Vol. XXVI, pp. 131

Trevelyan, P.M.J. & Kalliadasis, S. (2004). Wave dynamics on a thin liquid film falling down a heated wall, *J Eng Maths*, Vol. 50, pp. 177-208

Yükscl, M.K. & Schlünder, E.U. (1988). Wärme- und Stoffübertragung bei der Nichtisothermen Rieselfilmabsorption, *Wärme- und Stoffübertragung*, Vol. 22, pp. 209-218

Zaitsev, D.V.; Kabov, O.A.; Cheverda, V.V. & Bufetov, N.S. (2004). The effect of wave formation and wetting angle on the thermocapillary breakdown of a falling liquid film. *High Temperature*, Vol. 42, No. 3, pp. 450-456

19

Heat Transfer to Fluids at Supercritical Pressures

Igor Pioro and Sarah Mokry
University of Ontario Institute of Technology
Canada

1. Introduction

Prior to a general discussion on parametric trends in heat transfer to supercritical fluids, it is important to define special terms and expressions used at these conditions. Therefore, general definitions of selected terms and expressions, related to heat transfer to fluids at critical and supercritical pressures, are listed below. For better understanding of these terms and expressions a graph is shown in Fig. 1. General definitions of selected terms and expressions related to critical and supercritical regions are listed in the Chapter "Thermophysical Properties at Critical and Supercritical Conditions".

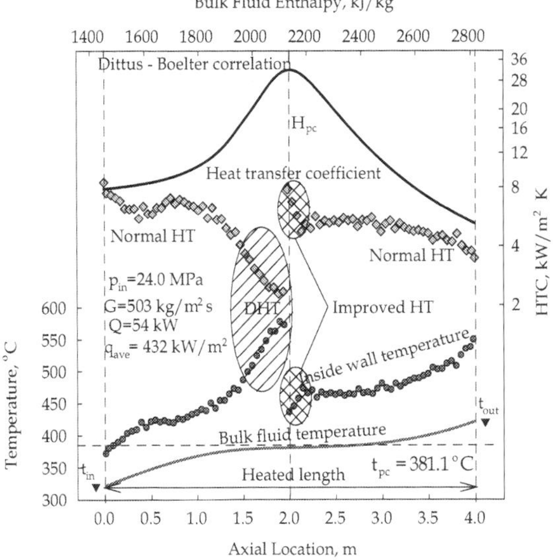

Fig. 1. Temperature and heat transfer coefficient profiles along heated length of vertical circular tube (Kirillov et al., 2003): Water, D=10 mm and L_h=4 m.

General definitions of selected terms and expressions related to heat transfer at critical and supercritical pressures

Deteriorated Heat Transfer (DHT) is characterized with lower values of the wall heat transfer coefficient compared to those at the normal heat transfer; and hence has higher values of wall temperature within some part of a test section or within the entire test section.

Improved Heat Transfer (IHT) is characterized with higher values of the wall heat transfer coefficient compared to those at the normal heat transfer; and hence lower values of wall temperature within some part of a test section or within the entire test section. In our opinion, the improved heat-transfer regime or mode includes peaks or "humps" in the heat transfer coefficient near the critical or pseudocritical points.

Normal Heat Transfer (NHT) can be characterized in general with wall heat transfer coefficients similar to those of subcritical convective heat transfer far from the critical or pseudocritical regions, when are calculated according to the conventional single-phase Dittus-Boelter-type correlations: $\mathbf{Nu} = 0.0023\ \mathbf{Re}^{0.8}\mathbf{Pr}^{0.4}$.

Pseudo-boiling is a physical phenomenon similar to subcritical pressure nucleate boiling, which may appear at supercritical pressures. Due to heating of supercritical fluid with a bulk-fluid temperature below the pseudocritical temperature (high-density fluid, i.e., "liquid"), some layers near a heating surface may attain temperatures above the pseudocritical temperature (low-density fluid, i.e., "gas") (for specifics of thermophysical properties, see Chapter "Thermophysical Properties at Critical and Supercritical Conditions"). This low-density "gas" leaves the heating surface in the form of variable density (bubble) volumes. During the pseudo-boiling, the wall heat transfer coefficient usually increases (improved heat-transfer regime).

Pseudo-film boiling is a physical phenomenon similar to subcritical-pressure film boiling, which may appear at supercritical pressures. At pseudo-film boiling, a low-density fluid (a fluid at temperatures above the pseudocritical temperature, i.e., "gas") prevents a high-density fluid (a fluid at temperatures below the pseudocritical temperature, i.e., "liquid") from contacting ("rewetting") a heated surface (for specifics of thermophysical properties, see Chapter "Thermophysical Properties at Critical and Supercritical Conditions"). Pseudo-film boiling leads to the deteriorated heat-transfer regime.

Water is the most widely used coolant or working fluid at supercritical pressures. The largest application of supercritical water is in supercritical "steam" generators and turbines, which are widely used in the power industry worldwide (Pioro and Duffey, 2007). Currently, upper limits of pressures and temperatures used in the power industry are about 30 – 35 MPa and 600 – 625°C, respectively. New direction in supercritical-water application in the power industry is a development of SuperCritical Water-cooled nuclear Reactor (SCWR) concepts, as part of the Generation-IV International Forum (GIF) initiative. However, other areas of using supercritical water exist (Pioro and Duffey, 2007).

Supercritical carbon dioxide was mostly used as a modelling fluid instead of water due to significantly lower critical parameters (for details, see Chapter "Thermophysical Properties at Critical and Supercritical Conditions"). However, currently new areas of using supercritical carbon dioxide as a coolant or working fluid have been emerged (Pioro and Duffey, 2007).

The third supercritical fluid used in some special technical applications is helium (Pioro and Duffey, 2007). Supercritical helium is used in cooling coils of superconducting electromagnets, superconducting electronics and power-transmission equipment.

Also, refrigerant R-134a is being considered as a perspective modelling fluid due to its lower critical parameters compared to those of water (Pioro and Duffey, 2007).

Experiments at supercritical pressures are very expensive and require sophisticated equipment and measuring techniques. Therefore, some of these studies (for example, heat transfer in bundles) are proprietary and hence, were not published in the open literature.

The majority of studies (Pioro and Duffey, 2007) deal with heat transfer and hydraulic resistance of working fluids, mainly water, carbon dioxide and helium, in circular bare tubes. In addition to these fluids, forced- and free-convection heat-transfer experiments were conducted at supercritical pressures, using liquefied gases such as air, argon, hydrogen; nitrogen, nitrogen tetra-oxide, oxygen and sulphur hexafluoride; alcohols such as ethanol and methanol; hydrocarbons such as n-heptane, n-hexane, di-iso-propyl-cyclo-hexane, n-octane, iso-butane, iso-pentane and n-pentane; aromatic hydrocarbons such as benzene and toluene, and poly-methyl-phenyl-siloxane; hydrocarbon coolants such as kerosene, TS-1 and RG-1, jet propulsion fuels RT and T-6; and refrigerants.

A limited number of studies were devoted to heat transfer and pressure drop in annuli, rectangular-shaped channels and bundles.

Accounting that supercritical water and carbon dioxide are the most widely used fluids and that the majority of experiments were performed in circular tubes, specifics of heat transfer and pressure drop, including generalized correlations, will be discussed in this chapter based on these conditions[1].

Specifics of thermophysical properties at critical and supercritical pressures for these fluids are discussed in the Chapter "Thermophysical Properties at Critical and Supercritical Conditions" and Pioro and Duffey (2007).

2. Convective heat transfer to fluids at supercritical pressures: Specifics of supercritical heat transfer

All[2] primary sources of heat-transfer experimental data for water and carbon dioxide flowing inside circular tubes at supercritical pressures are listed in Pioro and Duffey (2007).

In general, three major heat-transfer regimes (for their definitions, see above) can be noticed at critical and supercritical pressures (for details, see Figs. 1 and 2):

1.   Normal heat transfer;
2.   Improved heat transfer; and
3.   Deteriorated heat transfer.

Also, two special phenomena (for their definitions, see above) may appear along a heated surface:

1.   pseudo-boiling;
2.   pseudo-film boiling.

---

[1]Specifics of heat transfer and pressure drop at other conditions and/or for other fluids are discussed in Pioro and Duffey (2007).

[2] "All" means all sources found by the authors from a total of 650 references dated mainly from 1950 till beginning of 2006.

These heat-transfer regimes and special phenomena appear to be due to significant variations of thermophysical properties near the critical and pseudocritical points (see Fig. 3) and due to operating conditions.

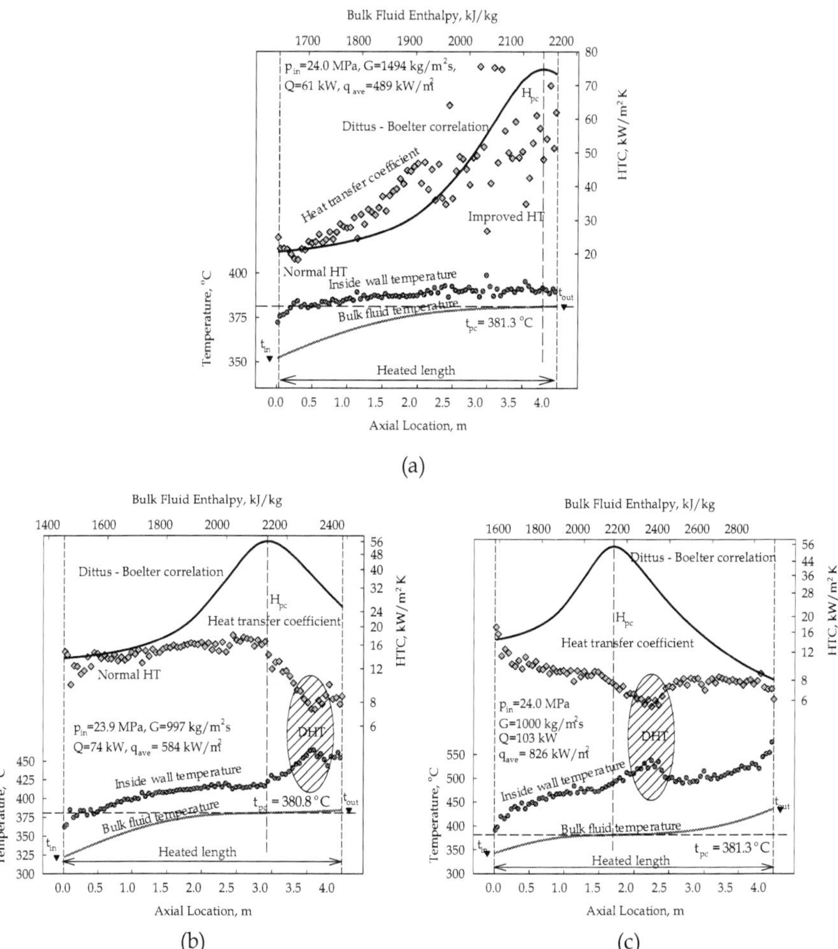

(a)

(b)

(c)

Fig. 2. Temperature and heat transfer coefficient profiles along heated length of vertical circular tube (Kirillov et al. 2003): Water, D=10 mm and L_h=4 m.

Therefore, the following cases can be distinguished at critical and supercritical pressures (for details, see Figs. 1 and 2):

a.   Wall and bulk-fluid temperatures are below a pseudocritical temperature within a part or the entire heated channel;

b.   Wall temperature is above and bulk-fluid temperature is below a pseudocritical temperature within a part or the entire heated channel;

c. Wall temperature and bulk fluid temperature is above a pseudocritical temperature within a part or the entire heated channel;
d. High heat fluxes;
e. Entrance region;
f. Upward and downward flows;
g. Horizontal flows;
h. Effect of gravitational forces at lower mass fluxes; etc.

All these cases can affect the supercritical heat transfer.

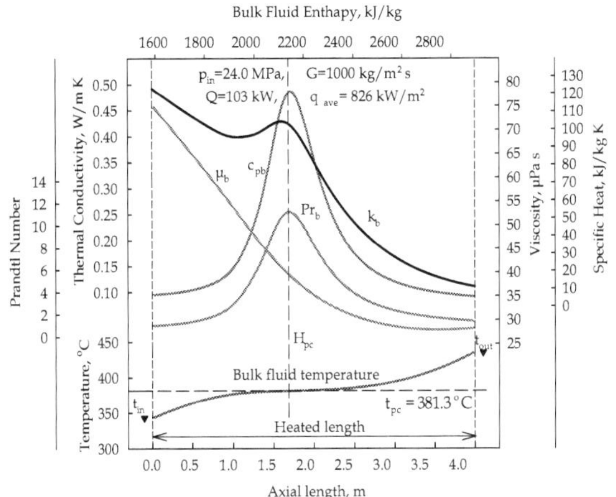

Fig. 3. Temperature and thermophysical properties profiles along heated length of vertical circular tube (operating conditions in this figure correspond to those in Fig. 2c): Water, D=10 mm and L_h=4 m; thermophysical properties based on bulk-fluid temperature.

3. Parametric trends

3.1 General heat transfer

As it was mentioned above, some researchers suggested that variations in thermophysical properties near critical and pseudocritical points resulted in the maximum value of Heat Transfer Coefficient (HTC). Thus, Yamagata et al. (1972) found that for water flowing in vertical and horizontal tubes, the HTC increases significantly within the pseudocritical region (Fig. 4). The magnitude of the peak in the HTC decreases with increasing heat flux and pressure. The maximum HTC values correspond to a bulk-fluid enthalpy, which is slightly less than the pseudocritical bulk-fluid enthalpy.

Results of Styrikovich et al. (1967) are shown in Fig. 5. Improved and deteriorated heat-transfer regimes as well as a peak ("hump") in HTC near the pseudocritical point are clearly shown in this figure. The deteriorated heat-transfer regime appears within the middle part of the test section at a heat flux of about 640 kW/m², and it may exist together with the improved heat-transfer regime at certain conditions (also see Fig. 1). With the further heat-flux increase, the improved heat-transfer regime is eventually replaced with that of deteriorated heat transfer.

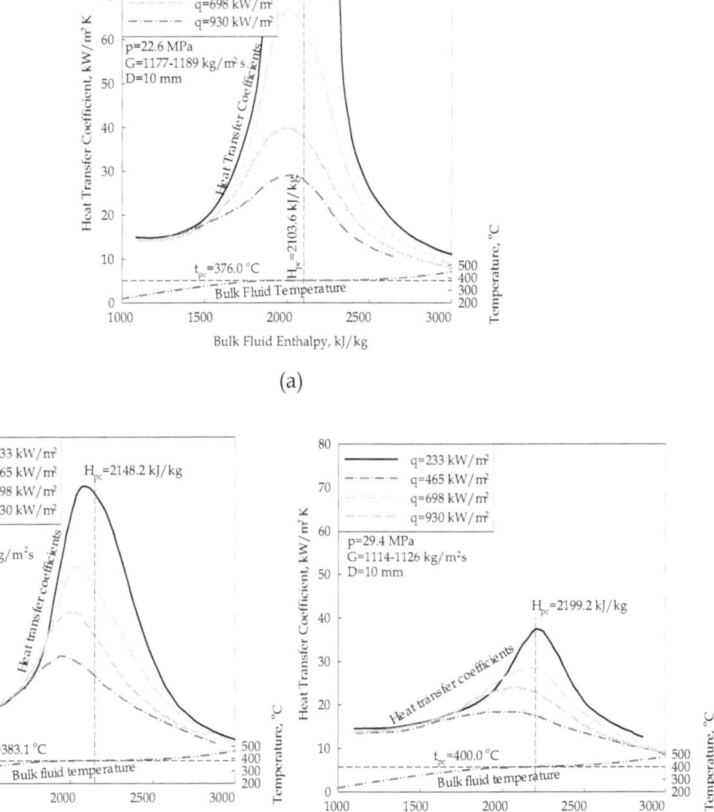

(a)

(b)

(c)

Fig. 4. Heat transfer coefficient vs. bulk-fluid enthalpy in vertical tube with upward flow at various pressures (Yamagata et al., 1972): Water – (a) p=22.6 MPa; (b) p=24.5 MPa; and (c) p=29.4 MPa.

Vikhrev et al. (1971, 1967) found that at a mass flux of 495 kg/m²s, two types of deteriorated heat transfer existed (Fig. 6): The first type appeared within the entrance region of the tube L / D < 40 – 60; and the second type appeared at any section of the tube, but only within a certain enthalpy range. In general, the deteriorated heat transfer occurred at high heat fluxes.

The first type of deteriorated heat transfer observed was due to the flow structure within the entrance region of the tube. However, this type of deteriorated heat transfer occurred mainly at low mass fluxes and at high heat fluxes (Fig. 6a,b) and eventually disappeared at high mass fluxes (Fig. 6c,d).

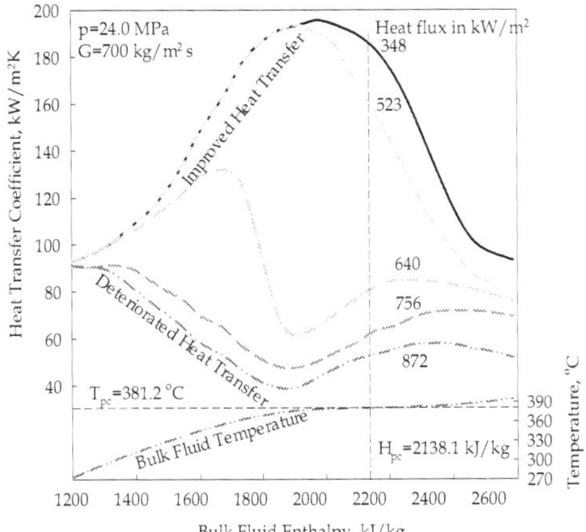

Fig. 5. Variations in heat transfer coefficient values of water flowing in tube (Styrikovich et al., 1967).

The second type of deteriorated heat transfer occurred when the wall temperature exceeded the pseudocritical temperature (Fig. 6). According to Vikhrev et al. (1967), the deteriorated heat transfer appeared when $q / G > 0.4$ kJ/kg (where q is in kW/m² and G is in kg/m²s). This value is close to that suggested by Styrikovich et al. (1967) ($q / G > 0.49$ kJ/kg). However, the above-mentioned definitions of two types of deteriorated heat transfer are not enough for their clear identification.

3.2 Pseudo-boiling and pseudo-film boiling phenomena

Ackerman (1970) investigated heat transfer to water at supercritical pressures flowing in smooth vertical tubes with and without internal ribs within a wide range of pressures, mass fluxes, heat fluxes and diameters. He found that pseudo-boiling phenomenon could occur at supercritical pressures. The pseudo-boiling phenomenon is thought to be due to large differences in fluid density below the pseudocritical point (high-density fluid, i.e., "liquid") and beyond (low-density fluid, i.e., "gas"). This heat-transfer phenomenon was affected with pressure, bulk-fluid temperature, mass flux, heat flux and tube diameter.

The process of pseudo-film boiling (i.e., low-density fluid prevents high-density fluid from "rewetting" a heated surface) is similar to film boiling, which occurs at subcritical pressures. Pseudo-film boiling leads to the deteriorated heat transfer. However, the pseudo-film boiling phenomenon may not be the only reason for deteriorated heat transfer. Ackerman noted that unpredictable heat-transfer performance was sometimes observed when the pseudocritical temperature of the fluid was between the bulk-fluid temperature and the heated surface temperature.

Kafengaus (1986, 1975), while analyzing data of various fluids (water, ethyl and methyl alcohols, heptane, etc.), suggested a mechanism for "pseudo-boiling" that accompanies heat transfer to liquids flowing in small-diameter tubes at supercritical pressures. The onset of pseudo-boiling was assumed to be associated with the breakdown of a low-density wall layer that was present at an above-pseudocritical temperature, and with the entrainment of individual volumes of the low-density fluid into the cooler (below pseudocritical temperature) core of the high-density flow, where these low-density volumes collapse with the generation of pressure pulses. At certain conditions, the frequency of these pulses can coincide with the frequency of the fluid column in the tube, resulting in resonance and in a rapid rise in the amplitude of pressure fluctuations. This theory was supported with experimental results.

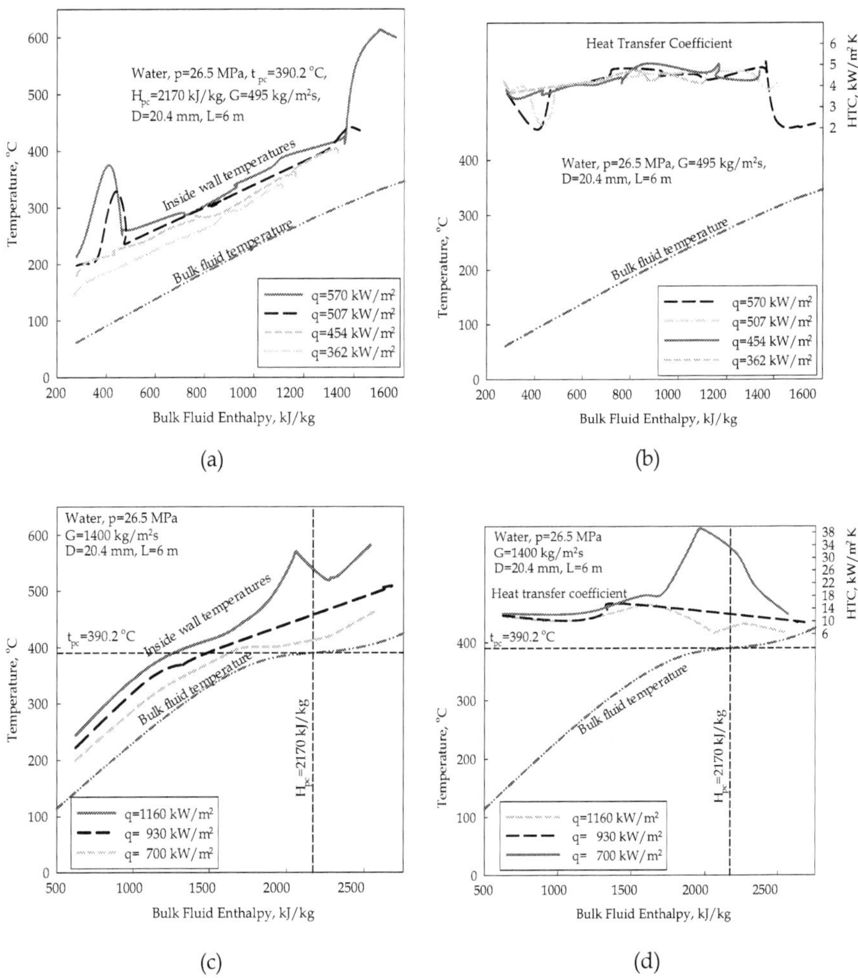

(a)   (b)

(c)   (d)

Fig. 6. Temperature profiles (a) and (c) and HTC values (b) and (d) along heated length of a vertical tube (Vikhrev et al., 1967): HTC values were calculated by the authors of the current chapter using the data from the corresponding figure; several test series were combined in each curve in figures (c) and (d).

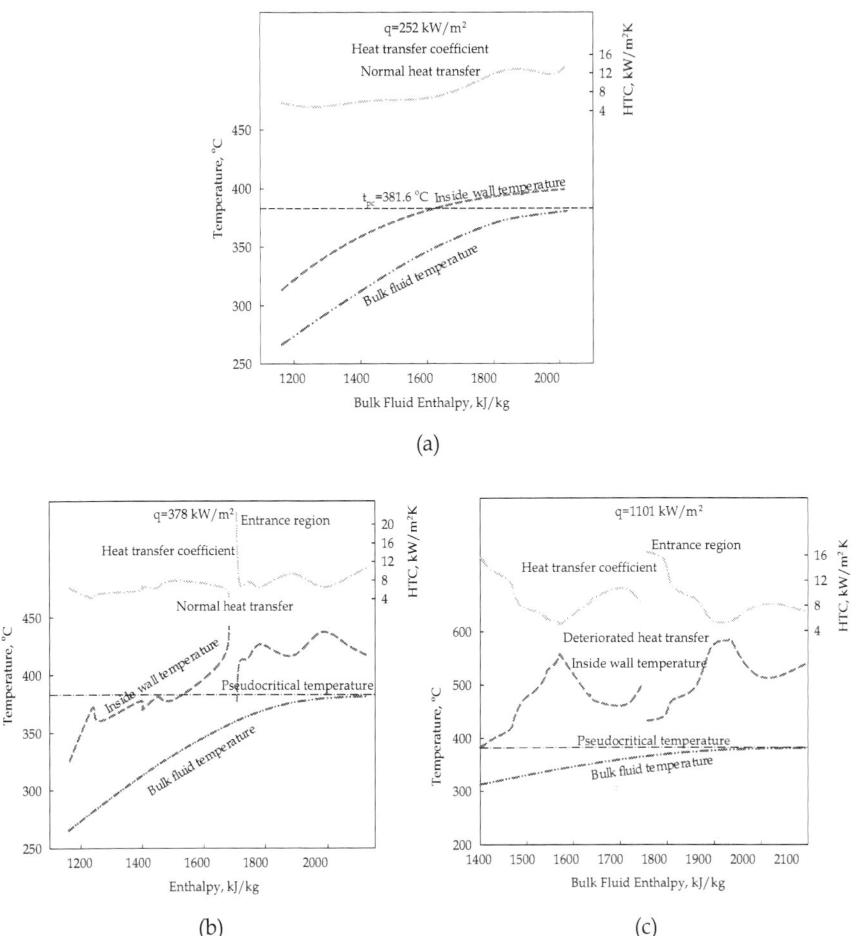

Fig. 7. Temperature and heat transfer coefficient profiles along 38.1-mm ID smooth vertical tube at different mass fluxes (Lee and Haller, 1974): Water, p=24.1 MPa, and H_{pc}=2140 kJ/kg; (a) G=542 kg/m²s, (b) G=542 kg/m²s, and (c) G=1627 kg/m²s; HTC values were calculated by the authors of the current chapter using data from the corresponding figure; several test series were combined in each curve.

3.3 Horizontal flows

All[3] primary sources of experimental data for heat transfer to water and carbon dioxide flowing in horizontal test sections are listed in Pioro and Duffey (2007).

---

[3] "All" means all sources found by the authors from a total of 650 references dated mainly from 1950 till beginning of 2006.

Krasyakova et al. (1967) found that in a horizontal tube, in addition to the effects of non-isothermal flow that is relevant to a vertical tube, the effect of gravitational forces is important. The latter effect leads to the appearance of temperature differences between the lower and upper parts of the tube. These temperature differences depend on flow enthalpy, mass flux and heat flux. A temperature difference in a tube cross section was found at $G = 300 - 1000$ kg/m^{2}s and within the investigated range of enthalpies ($H_b = 840 - 2520$ kJ/kg). The temperature difference was directly proportional to increases in heat-flux values. The effect of mass flux on the temperature difference is the opposite, i.e., with increase in mass flux the temperature difference decreases. Deteriorated heat transfer was also observed in a horizontal tube. However, the temperature profile for a horizontal tube at locations of deteriorated heat transfer differs from that for a vertical tube, being smoother for a horizontal tube compared to that of a vertical tube with a higher temperature increase on the upper part of the tube than on the lower part.

3.4 Heat-transfer enhancement

Similar to subcritical pressures, turbulization of flow usually leads to heat-transfer enhancement at supercritical pressures.

Shiralkar and Griffith (1970) determined both theoretically (for supercritical water) and experimentally (for supercritical carbon dioxide) the limits for safe operation, in terms of the maximum heat flux for a particular mass flux. Their experiments with a twisted tape inserted inside a test section showed that heat transfer was improved by this method. Also, they found that at high heat fluxes deteriorated heat transfer occurred when the bulk-fluid temperature was below and the wall temperature was above the pseudocritical temperature. Findings of Lee and Haller (1974) are shown in Fig. 7. They combined several test series into one graph. Due to the deteriorated heat-transfer region at the tube exit (one set of data) and the entrance effect in another set of data, experimental curves discontinue (see Fig. 7b,c). In general, they found heat flux and tube diameter to be the important parameters affecting minimum mass-flux limits to prevent pseudo-film boiling. Multi-lead ribbed tubes were found to be effective in preventing pseudo-film boiling.

3.5 Heat transfer in bundles

SCWRs will be cooled with a light-water coolant at a pressure of about 25 MPa and within a range of temperatures from 280 - 350°C to 550 - 625°C (inlet and outlet temperatures). Performing experiments at these conditions and bundle flow geometry is very complicated and expensive task. Therefore, currently preliminary experiments are performed in modelling fluids such as carbon dioxide and Freons (Richards et al., 2010). Their thermophysical properties are well known within a wide range of conditions, including the supercritical-pressure region (for details, see in Pioro and Duffey (2007) and in Chapter "Thermophysical Properties at Critical and Supercritical Conditions").

Experimental data obtained in a bare bundle with 7 circular elements, installed in a hexagonal flow channel located inside a ceramic insert surrounded by a pressure tube (Fig. 8) and cooled with R-12, are shown in Fig. 9 for reference purposes. The bundle has a 6 + 1 bare-element arrangement with each element being held at the ends to eliminate the use of spacers. Each of the 7 heating elements has a 9.5-mm outer diameter, and they are spaced one from another with a pitch of 11.29 mm. The total flow area is 374.0 mm^2, wetted perimeter – 318.7 mm, and hydraulic-equivalent diameter – 4.69 mm.

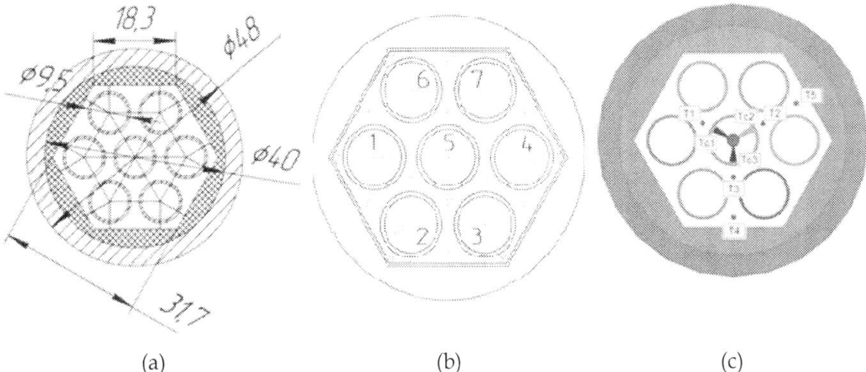

Fig. 8. Flow-channel cross sections: (a) with dimensions; (b) with elements numbering, and (c) with thermocouple layout.

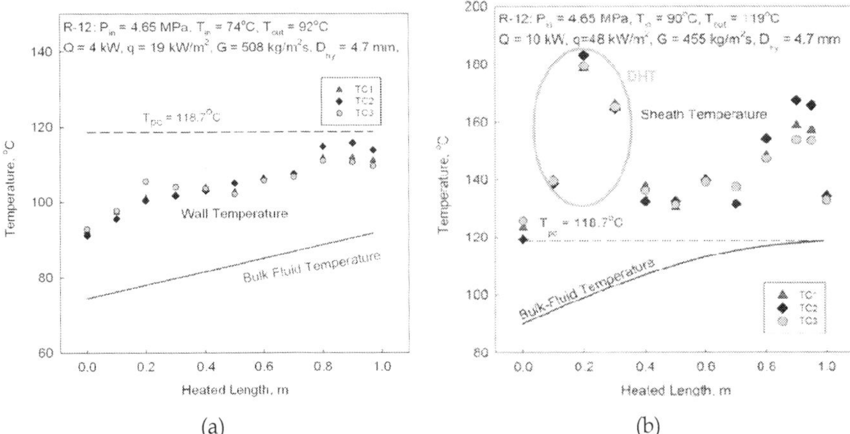

Fig. 9. Bulk-fluid and sheath-temperature profiles along bundle heated length: (a) normal heat-transfer regime; and (b) normal and deteriorated heat-transfer regimes.

The main test-section components are cylindrical heated elements installed tightly in the vertical hexagonal shell (downward flow). The entire internal setup is contained by a cylindrical 40×4 mm pressure tube with welded flanges at the edges that form the upper (inlet) chamber and lower (outlet) chamber, with a total heated length of 1000 mm. Four thermocouples installed into the top and bottom chambers were used to measure Freon-12 inlet and outlet temperatures. Basic parameters of the experimental setup are listed in Table 1.

The experiments showed that at certain operating conditions the deteriorated heat-transfer regime is possible not only in bare tubes, but also in "bare" bundles. This is the important statement, because previously deteriorated heat-transfer regimes have not been encountered in supercritical water-cooled bundles with helical fins (Pioro and Duffey, 2007).

Pressure	Up to 5.0 MPa (equivalent to 25.5 MPa for water)
Temperature of Freon-12	Up to 120°C (400°C heating elements)
Maximum flow rate	20 + 20 m³/h
Maximum pump pressure head	1.0 + 1.0 MPa
Experimental test-section power	Up to 1 MW
Experimental test-section height	Up to 8 m
Data Acquisition System (DAS)	Up to 256 channels

Table 1. Main parameters of 7-element bare bundle cooled with R-12.

4. Practical prediction methods for convection heat transfer at supercritical pressures

4.1 Circular vertical tubes

Unfortunately, satisfactory analytical methods have not yet been developed due to the difficulty in dealing with steep property variations, especially, in turbulent flows and at high heat fluxes. Therefore, generalized correlations based on experimental data are used for HTC calculations at supercritical pressures.

There are a lot of various correlations for convection heat transfer in circular tubes at supercritical pressures (for details, see in Pioro and Duffey (2007)). However, an analysis of these correlations showed that they are more or less accurate only within a particular dataset, which was used to derive the correlation, but show a significant deviation in predicting other experimental data. Therefore, only selected correlations are listed below.

In general, many of these correlations are based on the conventional Dittus-Boelter-type correlation (see Eq. (1)) in which the regular specific heat is replaced with the cross-section averaged specific heat within the range of $(T_w - T_b)$; $\left(\dfrac{H_w - H_b}{T_w - T_b} \right)$, J/kg K (see Fig. 8). Also,

additional terms, such as: $\left(\dfrac{k_b}{k_w} \right)^k$; $\left(\dfrac{\mu_b}{\mu_w} \right)^m$; $\left(\dfrac{\rho_b}{\rho_w} \right)^n$; etc., can be added into correlations to account for significant variations in thermophysical properties within a cross section, due to a non-uniform temperature profile, i.e., due to heat flux.

It should be noted that usually generalized correlations, which contain fluid properties at the wall temperature, require iterations to be solved, because there are two unknowns: 1) HTC and 2) the corresponding wall temperature. Therefore, the initial wall-temperature value at which fluid properties will be estimated should be "guessed" to start iterations.

The most widely used heat-transfer correlation at subcritical pressures for forced convection is the Dittus-Boelter (1930) correlation (Pioro and Duffey, 2007). In 1942, McAdams proposed to use the Dittus-Boelter correlation in the following form, for forced-convective heat transfer in turbulent flows at subcritical pressures:

$$Nu_b = 0.0243 \, Re_b^{0.8} Pr_b^{0.4} . \tag{1}$$

However, it was noted that Eq. (1) might produce unrealistic results within some flow conditions (see Figs. 1 and 2), especially, near the critical and pseudocritical points, because it is very sensitive to properties variations.

In general, experimental heat transfer coefficient values show just a moderate increase within the pseudocritical region. This increase depends on flow conditions and heat flux:

higher heat flux – less increase. Thus, the bulk-fluid temperature might not be the best characteristic temperature at which all thermophysical properties should be evaluated. Therefore, the cross-sectional averaged Prandtl number (see below), which accounts for thermophysical properties variations within a cross section due to heat flux, was proposed to be used in many supercritical heat-transfer correlations instead of the regular Prandtl number. Nevertheless, this classical correlation (Eq. (1)) was used extensively as a basis for various supercritical heat-transfer correlations.

In 1964, Bishop et al. conducted experiments in supercritical water flowing upward inside bare tubes and annuli within the following range of operating parameters: P=22.8 – 27.6 MPa, T_b= 282 – 527°C, G = 651 – 3662 kg/m²s and q = 0.31 – 3.46 MW/m². Their data for heat transfer in tubes were generalized using the following correlation with a fit of ±15%:

$$\mathrm{Nu}_b = 0.0069\,\mathrm{Re}_b^{0.9}\,\overline{\mathrm{Pr}}_b^{0.66}\left(\frac{\rho_w}{\rho_b}\right)^{0.43}\left(1+2.4\frac{D}{x}\right). \tag{2}$$

Equation (2) uses the cross-sectional averaged Prandtl number, and the last term in the correlation: $(1+2.4\,D/x)$, accounts for the entrance-region effect. However, in the present comparison, the Bishop et al. correlation was used without the entrance-region term as the other correlations (see Eqs. (1), (3) and (4)).

In 1965, Swenson et al. found that conventional correlations, which use a bulk-fluid temperature as a basis for calculating the majority of thermophysical properties, were not always accurate. They have suggested the following correlation in which the majority of thermophysical properties are based on a wall temperature:

$$\mathrm{Nu}_w = 0.00459\,\mathrm{Re}_w^{0.923}\,\overline{\mathrm{Pr}}_w^{0.613}\left(\frac{\rho_w}{\rho_b}\right)^{0.231}. \tag{3}$$

Equation (3) was obtained within the following range: pressure 22.8 – 41.4 MPa, bulk-fluid temperature 75 – 576°C, wall temperature 93 – 649°C and mass flux 542 – 2150 kg/m²s; and predicts experimental data within ±15%.

In 2002, Jackson modified the original correlation of Krasnoshchekov et al. from 1967 for forced-convective heat transfer in water and carbon dioxide at supercritical pressures, to employ the Dittus-Boelter-type form for $\mathbf{Nu}_0$ as the following:

$$\mathrm{Nu}_b = 0.0183\,\mathrm{Re}_b^{0.82}\mathrm{Pr}_b^{0.5}\left(\frac{\rho_w}{\rho_b}\right)^{0.3}\left(\frac{\overline{c_p}}{c_{pb}}\right)^n, \tag{4}$$

where the exponent n is defined as following:

$$n = 0.4 \quad \text{for } T_b < T_w < T_{pc} \text{ and for } 1.2\,T_{pc} < T_b < T_w;$$

$$n = 0.4 + 0.2\left(\frac{T_w}{T_{pc}}-1\right) \text{ for } T_b < T_{pc} < T_w; \text{ and}$$

$$n = 0.4 + 0.2\left(\frac{T_w}{T_{pc}}-1\right)\left[1-5\left(\frac{T_b}{T_{pc}}-1\right)\right] \text{ for } T_{pc} < T_b < 1.2\,T_{pc} \text{ and } T_b < T_w.$$

An analysis performed by Pioro and Duffey (2007) showed that the two following correlations: 1) Bishop et al. (1964) and 2) Swenson et al. (1965); were obtained within the same range of operating conditions as those for SCWRs.

The majority of empirical correlations were proposed in the 1960s – 1970s, when experimental techniques were not at the same level (i.e., advanced level) as they are today. Also, thermophysical properties of water have been updated since that time (for example, a peak in thermal conductivity in critical and pseudocritical points within a range of pressures from 22.1 to 25 MPa was not officially recognized until the 1990s).

Therefore, recently a new or an updated correlation, based on a new set of heat-transfer data and the latest thermophysical properties of water (NIST, 2007) within the SCWRs operating range, was developed and evaluated (Mokry et al., 2009):

$$\mathrm{Nu}_b = 0.0061 \, \mathrm{Re}_b^{0.904} \overline{\mathrm{Pr}}_b^{-0.684} \left(\frac{\rho_w}{\rho_b} \right)^{0.564}. \tag{5}$$

Figure 10 shows scatter plots of experimental HTC values versus calculated HTC values according to Eq. (5), and calculated and experimental values for wall temperatures. Both plots lie along a 45-degree straight line with an experimental data spread of ±25% for the HTC values and ±15% for the wall temperatures.

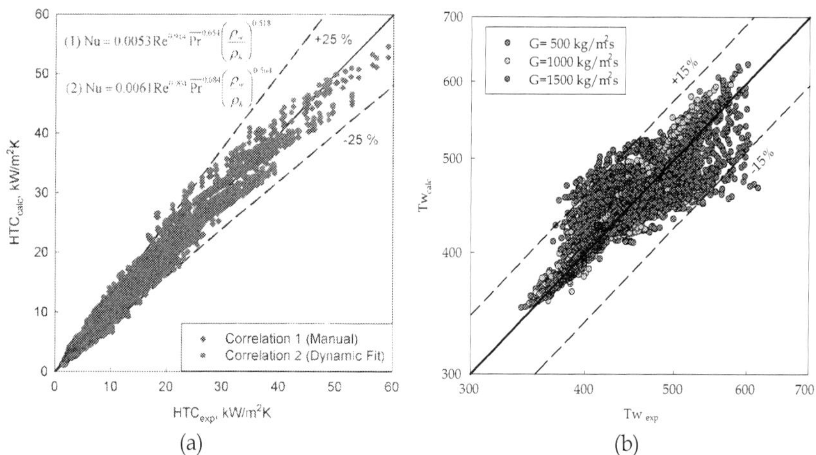

(a)   (b)

Fig. 10. Comparison of data fit through Eq. (5) with experimental data: (a) for HTC and (b) for wall temperature.

Figures 11 and 12 show a comparison of Eq. (5) with the experimental data. Figure 13 shows a comparison between experimentally obtained HTC and wall-temperature values and those calculated with FLUENT CFD code and Eq. (5).

It should be noted that all heat-transfer correlations presented in this chapter are intended only for the normal and improved heat-transfer regimes.

The following empirical correlation was proposed for calculating the minimum heat flux at which the deteriorated heat-transfer regime appears:

$$q_{dht} = -58.97 + 0.745 \cdot G, \, \mathrm{kW/m^2}. \tag{6}$$

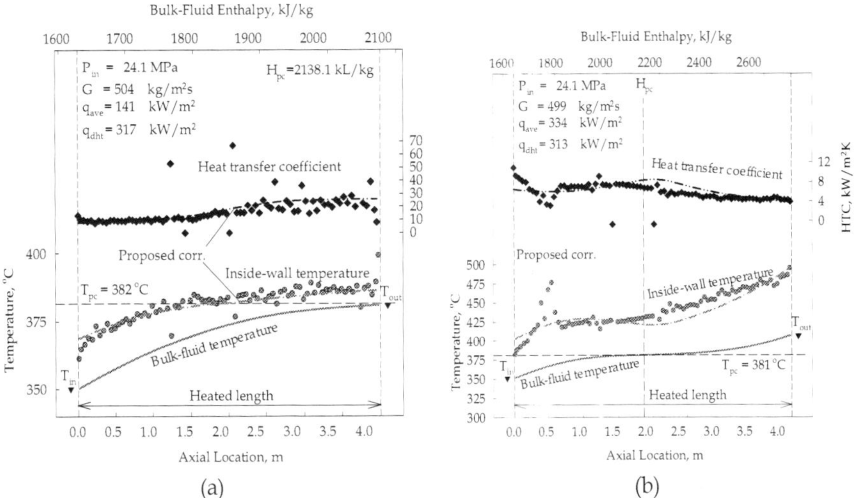

(a)   (b)

Fig. 11. Temperature and HTC profiles at various heat fluxes along 4-m circular tube (D=10 mm): P_{in}=24.1 MPa and G=500 kg/m²s; "proposed correlation" – Eq. (5).

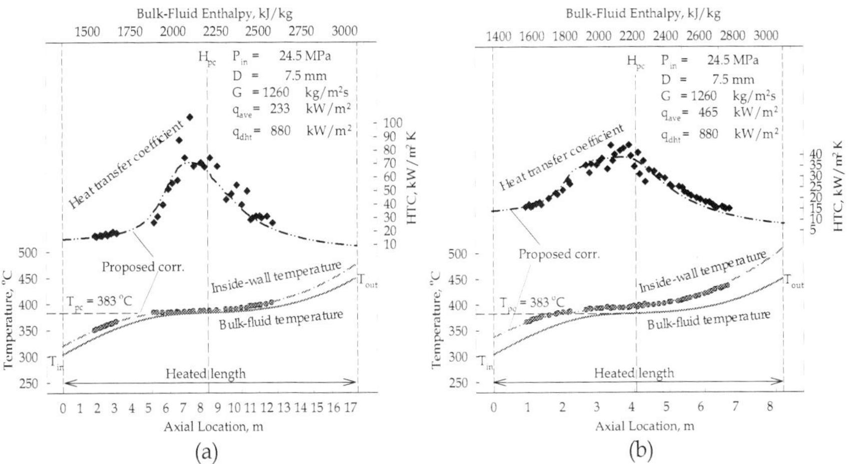

(a)   (b)

Fig. 12. Temperature and HTC profiles along circular tube at various heat fluxes: Nominal operating conditions – P_{in}=24.5 MPa and D=7.5 mm (Yamagata et al., 1972); "proposed correlation" – Eq. (5).

Figures 11 – 13 show that the latest correlation (Eq. (5)) closely represents experimental data and follows trends closely even within the pseudocritical range. CFD codes are nice and a modern approach. However, not all turbulent models are applicable to heat transfer at supercritical pressures, plus these codes should be tuned first on the basis of experimental data and after that used in similar calculations.

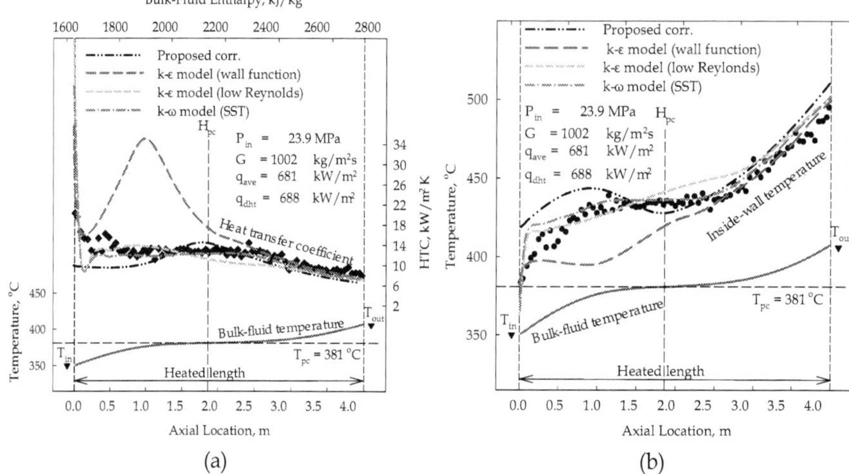

Fig. 13. Comparison of HTC and wall temperature values calculated with proposed correlation (Eq. (5)) and FLUENT CFD-code (Vanyukova et al., 2009) with experimental data along 4-m circular tube (D=10 mm): P_{in}=23.9 MPa and G=1000 kg/m²s.

Correlation*	Supercritical Region				Region	
	Liquid-Like		Gas-Like		Critical or Pseudocritical	
	Errors, %					
	Average	RMS	Average	RMS	Average	RMS
Bishop et al. (1965)	6.3	24.2	5.2	18.4	20.9	28.9
Swenson et al. (1965)	1.5	25.2	-15.9	20.4	5.1	23.0
Krasnoshchekov et al. (1967)	15.2	33.7	-33.6	35.8	25.2	61.6
Watts & Chou (1982)	4.0	25.0	-9.7	20.8	5.5	24.0
Chou (1982)	5.5	23.1	5.7	22.2	16.5	28.4
Griem (1996)	1.7	23.2	4.1	22.8	2.7	31.1
Jackson (2002)	13.5	30.1	11.5	28.7	22.0	40.6
Mokry et al. (2009)	-3.9	**21.3**	-8.5	**16.5**	**-2.3**	**17.0**
Kuang et al. (2008)	-6.6	23.7	**2.9**	19.2	-9.0	24.1
Cheng et al. (2009)	**1.3**	25.6	**2.9**	28.8	14.9	90.6
Hadaller & Benerjee (1969)	7.6	30.5	10.7	20.5	-	-
Sieder & Tate (1936)	20.8	37.3	93.2	133.6	-	-
Dittus & Boelter (1930)	32.5	46.7	87.7	131.0	-	-
Gnielinski (1976)	42.5	57.6	106.3	153.3	-	-

In bold – the minimum values.

* many of these correlations can be found in Pioro and Duffey (2007).

Table 2. Overall weighted average and RMS errors within three supercritical sub-regions (Zahlan et al., 2010).

A recent study was conducted by Zahlan et al. (2010) in order to develop a heat-transfer look-up table for the critical/supercritical pressures. An extensive literature review was conducted, which included 28 datasets and 6663 trans-critical heat-transfer data. Tables 2 and 3 list results of this study in the form of the overall-weighted average and Root-Mean-Square (RMS) errors: (a) Within three supercritical sub-regions for many heat-transfer correlations, including those discussed in this chapter (Table 2); and (b) For subcritical liquid and superheated steam (Table 3). In their conclusions, Zahlan et al. (2010) determined that within the supercritical region the latest correlation by Mokry et al. (Eq. (5)) showed the best prediction for the data within all three sub-regions investigated. Also, the Mokry et al. correlation showed quite good predictions for subcritical liquid and superheated steam compared to other several correlations.

Correlation	Subcritical liquid		Superheated steam	
	Error, %			
	Average	RMS	Average	RMS
Sieder & Tate (1936)	27.6	37.4	83.8	137.8
Gnielinski (1976)	-4.3	**18.3**	80.3	130.2
Hadaller & Banerjee (1969)	27.3	35.9	19.1	34.4
Dittus & Boelter (1930)	10.4	22.5	75.3	127.3
Mokry et al. (2009)	**-1.1**	19.2	**-4.8**	**19.6**

In **bold** – the minimum values.

Table 3. Overall average and RMS error within subcritical region (Zahlan et al., 2010).

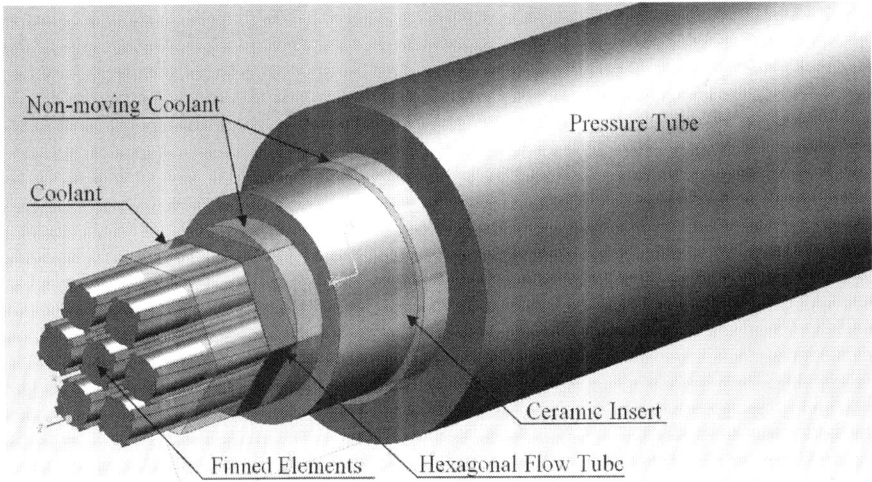

Fig. 14. Tested 7-element helically-finned bundle cooled with supercritical water and heated with electrical current (drawing prepared by W. Peiman, UOIT).

4.2 Bundles

As it was mentioned above, experiments in bundles cooled with supercritical water are very complicated and expensive. Therefore, only one empirical correlation is known so far in the open literature which predicts heat transfer coefficients in a special bundle design (Fig. 14). This correlation was developed by Dyadyakin and Popov (1977), who performed experiments in a tight 7-rod bundle with helical fins cooled with supercritical water. They have correlated their data for the local heat transfer coefficients as:

$$\mathrm{Nu_x} = 0.021\, \mathrm{Re_x^{0.8}}\, \overline{\mathrm{Pr}}_x^{0.7} \left(\frac{\rho_w}{\rho_b}\right)_x^{0.45} \left(\frac{\mu_b}{\mu_{in}}\right)_x^{0.2} \left(\frac{\rho_b}{\rho_{in}}\right)_x^{0.1} \left(1 + 2.5\,\frac{D_{hy}}{x}\right), \tag{9}$$

where x is the axial location along the heated length in meters, and D_{hy} is the hydraulic-equivalent diameter (equals 4 times the flow area divided by the wetted perimeter) in meters. This correlation fits the data (504 points) to within ±20%. The maximum deviation of the experimental data from the correlating curve corresponds to points with small temperature differences between the wall temperature and bulk temperature. Sixteen experimental points had deviations from the correlation within ±30%.

5. Hydraulic resistance

In general, the total pressure drop for forced convection flow inside a test section, installed in a closed-loop system, can be calculated according to the following expression:

$$\Delta p = \sum \Delta p_{fr} + \sum \Delta p_l + \sum \Delta p_{ac} + \sum \Delta p_g, \tag{10}$$

where Δp is the total pressure drop, Pa.
Δp_{fr} is the pressure drop due to frictional resistance (Pa), which defined as

$$\Delta p_{fr} = \left(\xi_{fr}\,\frac{L}{D}\,\frac{\rho\,u^2}{2}\right) = \left(\xi_{fr}\,\frac{L}{D}\,\frac{G^2}{2\rho}\right), \tag{11}$$

where ξ_{fr} is the frictional coefficient, which can be obtained from appropriate correlations for different flow geometries. For smooth circular tubes ξ_{fr} is as follows (Filonenko, 1954)

$$\xi_{fr} = \left(\frac{1}{\left(1.82 \log_{10} \mathrm{Re_b} - 1.64\right)^2}\right). \tag{12}$$

Equation (12) is valid within a range of **Re** = $4 \cdot 10^3 - 10^{12}$.
Usually, thermophysical properties and the Reynolds number in Eqs. (11) and (12) are based on arithmetic average of inlet and outlet values.
Δp_l is the pressure drop due to local flow obstruction (Pa), which is defined as

$$\Delta p_l = \left(\xi_l\,\frac{\rho\,u^2}{2}\right) = \left(\xi_l\,\frac{G^2}{2\rho}\right), \tag{13}$$

where ξ_ℓ is the local resistance coefficient, which can be obtained from appropriate correlations for different flow obstructions.

Δp_{ac} is the pressure drop due to acceleration of flow (Pa) defined as

$$\Delta p_{ac} = \left(\rho_{out}\, u_{out}^2 - \rho_{in}\, u_{in}^2 \right) = G^2 \left(\frac{1}{\rho_{out}} - \frac{1}{\rho_{in}} \right). \tag{14}$$

Δp_g is the pressure drop due to gravity (Pa) defined as

$$\Delta p_g = \pm g \left(\frac{\rho_{out} + \rho_{in}}{2} \right) L \sin\theta, \tag{15}$$

where θ is the test-section inclination angle to the horizontal plane, sign "+" is for the upward flow and sign "–" is for the downward flow. The arithmetic average value of densities can be used only for short sections in the case of strongly non-linear dependency of the density versus temperature. Therefore, in long test sections at high heat fluxes and within the critical and pseudocritical regions, the integral value of densities should be used (see Eq. (16)).

Ornatskiy et al. (1980) and Razumovskiy (2003) proposed to calculate Δp_g at supercritical pressures as the following:

$$\Delta p_g = \pm g \left(\frac{H_{out}\, \rho_{out} + H_{in}\, \rho_{in}}{H_{out} + H_{in}} \right) L \sin\theta. \tag{16}$$

In general, Equation (10) is applicable for subcritical and supercritical pressures. However, adjustment of this expression to conditions of supercritical pressures, with single-phase dense gas and significant variations in thermophysical properties near the critical and pseudocritical points, was the major task for the researchers and scientists.

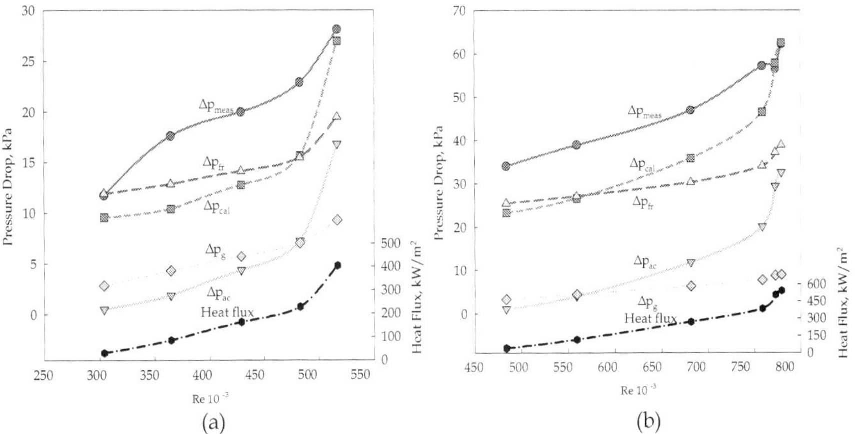

Fig. 15. Effect of Reynolds number on total pressure drop (measured and calculated) and its components (calculated values) in supercritical carbon dioxide flowing in vertical circular tube: p_{out}=8.8 MPa; (a) G=2040 kg/m²s, t_{in}=32°C; and (b) G=3040 kg/m²s, t_{in}=31°C.

In general, two major approaches to solve this problem were taken: an analytical approach (including numerical approach) and an experimental (empirical) approach.

Unfortunately, satisfactory analytical and numerical methods have not yet been developed, due to the difficulty in dealing with the steep property variations, especially in turbulent flows and at high heat fluxes. Therefore, empirical correlations are usually used.

For reference purposes, selected results obtained at Chalk River Laboratories (Pioro and Duffey, 2007; Pioro et al. 2004) are shown in Fig. 15. In these experiments, the local pressure drop due to obstructions along the heated length was 0, because of a smooth test section. Therefore, the measured pressure drop consists only of three components:

$$\Delta p_{meas} = \Delta p_{fr} + \Delta p_{ac} + \Delta p_{g}. \tag{17}$$

Other details of pressure drop at supercritical pressures are listed in Pioro and Duffey (2007).

Another important issue at supercritical and subcritical pressures is uncertainties of measured and calculated parameters. Pioro and Duffey (2007) dedicated a separate Appendix D to this important issue in their book.

6. Nomenclature

A	flow area, m^2
c_p	specific heat at constant pressure, J/kg K
$\bar{c_p}$	averaged specific heat within the range of $(t_w - t_b)$; $\left(\dfrac{H_w - H_b}{T_w - T_b}\right)$, J/kg K
D	inside diameter, m
G	mass flux, kg/m²s; $\left(\dfrac{m}{A_{fl}}\right)$
g	gravitational acceleration, m/s²
H	specific enthalpy, J/kg
h	heat transfer coefficient, W/m²K
k	thermal conductivity, W/m K
L	heated length, m
m	mass-flow rate, kg/s; (ρV)
P, p	pressure, MPa
Q	heat-transfer rate, W
q	heat flux, W/m²; $\left(\dfrac{Q}{A_h}\right)$
T, t	temperature, °C
u	axial velocity, m/s
V	volume-flow rate, m³/kg
x	axial coordinate, m

Greek letters

a	thermal diffusivity, m²/s; $\left(\dfrac{k}{c_p \rho}\right)$

Δ	difference
θ	test-section inclination angle, degree
μ	dynamic viscosity, Pa s
ξ	friction coefficient
ρ	density, kg/m^3
υ	kinematic viscosity, m^2/s

Non-dimensional numbers

Nu	Nusselt number; $\left(\dfrac{h\,D}{k}\right)$
Pr	Prandtl number; $\left(\dfrac{\mu\,c_p}{k}\right)=\left(\dfrac{\upsilon}{\alpha}\right)$
$\overline{\text{Pr}}$	averaged Prandtl number within the range of $(t_w - t_b)$; $\left(\dfrac{\mu\,\bar{c}_p}{k}\right)$
Re	Reynolds number; $\left(\dfrac{G\,D}{\mu}\right)$

Symbols with an overbar at the top denote average or mean values (e.g., $\overline{\text{Nu}}$ denotes average (mean) Nusselt number).

Subscripts or superscripts

ac	acceleration
ave	average
b	bulk
cal	calculated
cr	critical
dht	deteriorated heat transfer
exp	experimental
fl	flow
fr	friction
g	gravitational
h	heated
hy	hydraulic-equivalent
in	inlet
ℓ	local
meas	measured
out	outlet or outside
pc	pseudocritical
w	wall

Abbreviations and acronyms widely used in the text

DHT	Deteriorated Heat Transfer
GIF	Generation-IV International Forum
HT	Heat Transfer
HTC	Heat Transfer Coefficient
ID	Inside Diameter
IHT	Improved Heat Transfer
NHT	Normal Heat Transfer

NIST   National Institute of Standards and Technology (USA)
SCWR   SuperCritical Water-cooled Reactor

7. Reference

Ackerman, J.W., 1970. Pseudoboiling heat transfer to supercritical pressure water in smooth and ribbed tubes, Journal of Heat Transfer, Transactions of the ASME, 92 (3), pp. 490–498, (Paper No. 69-WA/HT-2, pp. 1–8).

Bishop, A.A., Sandberg, R.O., and Tong, L.S., 1965, Forced Convection Heat Transfer to Water at Near-critical Temperatures and Supercritical Pressures, *A.I.Ch.E.-I.Chem.E Symposium Series No.* 2, pp. 77–85.

Bishop, A.A., Sandberg, R.O. and Tong, L.S., 1964. Forced convection heat transfer to water at near-critical temperatures and super-critical pressures, Report WCAP-2056, Westinghouse Electric Corporation, Atomic Power Division, Pittsburgh, PA, USA, December, 85 pages.

Cheng, X., Yang, Y.H., and Huang, S.F., 2009, A Simple Heat Transfer Correlation for SC Fluid Flow in Circular Tubes, (NURETH-13), Kanazawa City, Ishikawa Prefecture, Japan, September 27-October 2.

Dittus, F.W. and Boelter, L.M.K., 1930. University of California, Berkeley, Publications on Engineering, Vol. 2, p. 443.

Dyadyakin, B.V. and Popov, A.S., 1977. Heat transfer and thermal resistance of tight seven-rod bundle, cooled with water flow at supercritical pressures, (In Russian), Transactions of VTI (Труды ВТИ), No. 11, pp. 244–253.

Filonenko, G.K., 1954. Hydraulic resistance of pipelines, (In Russian), Thermal Engineering, No. 4, pp. 40–44.

Gnielinski, V., 1976, New Equation for Heat and Mass Transfer in Turbulent Pipe and Channel Flow, *Intern. Chem. Eng.*, Vol. 16, No. 2, pp. 359-366.

Griem, H., 1996, A New Procedure for the at Near-and Supercritical Prediction Pressure of Forced Convection Heat Transfer, *Heat Mass Trans.*, Vol. 3, pp. 301–305.

Hadaller, G. and Banerjee, S., 1969, Heat Transfer to Superheated Steam in Round Tubes, AECL Report.

Jackson, J.D., 2002. Consideration of the heat transfer properties of supercritical pressure water in connection with the cooling of advanced nuclear reactors, Proceedings of the 13th Pacific Basin Nuclear Conference, Shenzhen City, China, October 21–25.

Kafengauz, N.L., 1986. About some peculiarities in fluid behaviour at supercritical pressure in conditions of intensive heat transfer, Applied Thermal Sciences, (Промышленная Теплотехника, стр. 6–10), 8 (5), pp. 26–28.

Kafengaus, N.L., 1975. The mechanism of pseudoboiling, Heat Transfer-Soviet Research, 7 (4), pp. 94–100.

Kirillov, P.L., Lozhkin, V.V. and Smirnov, A.M., 2003. Investigation of borders of deteriorated regimes of a channel at supercritical pressures, (In Russian), State Scientific Center of Russian Federation Institute of Physics and Power Engineering by the name of A.I. Leypunskiy, FEI-2988, Obninsk, Russia, 20 pages.

Krasnoshchekov, E.A., Protopopov, V.S., Van, F. and Kuraeva, I.V., 1967. Experimental investigation of heat transfer for carbon dioxide in the supercritical region, Proceedings of the 2nd All-Soviet Union Conference on Heat and Mass Transfer,

Minsk, Belarus', May, 1964, Published as Rand Report R-451-PR, Edited by C. Gazley, Jr., J.P. Hartnett and E.R.C. Ecker, Vol. 1, pp. 26–35.

Krasyakova, L.Yu., Raykin, Ya.M., Belyakov, I.I. et al., 1967. Investigation of temperature regime of heated tubes at supercritical pressure, (In Russian), Soviet Energy Technology (Энергомашиностроение), No. 1, pp. 1–4.

Kuang, B., Zhang, Y., and Cheng, X., 2008, A New, Wide-Ranged Heat Transfer Correlation of Water at Supercritical Pressures in Vertical Upward Ducts, NUTHOS-7, Seoul, Korea, October 5-9.

Lee, R.A. and Haller, K.H., 1974. Supercritical water heat transfer developments and applications, Proc. 5th International Heat Transfer Conference, Tokyo, Japan, September 3–7, Vol. IV, Paper No. B7.7, pp. 335–339.

McAdams, W.H., 1942. Heat Transmission, 2nd edition, McGraw-Hill, New York, NY, USA, 459 pages.

Mokry, S., Farah, A., King, K., Gupta, S., Pioro, I. and Kirillov, P., 2009. Development of Supercritical Water Heat-Transfer Correlation for Vertical Bare Tubes, Proceedings of the Nuclear Energy for New Europe 2009 International Conference, Bled, Slovenia, 2009 September 14 - 17, Paper #210, 13 pages.

National Institute of Standards and Technology, 2007. NIST Reference Fluid Thermodynamic and Transport Properties-REFPROP. NIST Standard Reference Database 23, Ver. 8.0. Boulder, CO, U.S.: Department of Commerce.

Ornatskiy, A.P., Dashkiev, Yu.G. and Perkov, V.G., 1980. Supercritical Steam Generators, (In Russian), VyshchaShkola Publishing House, Kiev, Ukraine, 287 pages.

Pioro, I., Duffey, R. and Dumouchel, T., 2004. Hydraulic resistance of fluids flowing in channels at supercritical pressures (survey), Nuclear Engineering and Design, 231 (2), pp. 187–197.

Pioro, I.L. and Duffey, R.B., 2007. Heat Transfer and Hydraulic Resistance at Supercritical Pressures in Power Engineering Applications, ASME Press, New York, NY, USA, 328 pages.

Razumovskiy, V.G., 2003. Private communications, State Technical University "KPI", Kiev, Ukraine.

Richards, G., Milner, A., Pascoe, C. et al., 2010. Heat Transfer in a Vertical 7-Element Bundle Cooled with Supercritical Freon-12, Proc. 2nd Canada-China Joint Workshop on Supercritical Water-Cooled Reactors (CCSC-2010), Toronto, Ontario, Canada, April 25-28, 10 pages.

Seider, N. M., and Tate, G. E., 1936, Heat Transfer and Pressure Drop of Liquids in Tubes, *Ind. Eng. Chem.* 28 (12), pp. 1429–1435.

Shiralkar, B.S. and Griffith, P., 1970. The effect of swirl, inlet conditions, flow direction, and tube diameter on the heat transfer to fluids at supercritical pressure, Journal of Heat Transfer, Transactions of the ASME, 92 (3), August, pp. 465–474.

Swenson,H.S., Carver, J.R. and Kakarala, C.R., 1965.Heat transfer to supercritical water in smooth-bore tubes, Journal of Heat Transfer, Transactions of the ASME, Series C, 87 (4), 1965, pp. 477–484.

Styrikovich, M.A., Margulova, T.Kh. and Miropol'skii, Z.L., 1967. Problems in the development of designs of supercritical boilers, Thermal Engineering (Теплоэнергетика, стр. 4–7), 14 (6), pp. 5–9.

Vanyukova, G.V., Kuznetsov, Yu.N., Loninov, A.Ya., Papandin, M.V., Smirnov, V.P. and Pioro, I.L., 2009. Application of CFD-Code to Calculations of Heat Transfer in a Fuel Bundle of SCW Pressure-Channel Reactor, Proc. 4th Int. Symp.on Supercritical Water-Cooled Reactors, March 8-11, Heidelberg, Germany, Paper No. 28, 9 pages.

Vikhrev, Yu.V., Barulin, Yu.D. and Kon'kov, A.S., 1967. A study of heat transfer in vertical tubes at supercritical pressures, Thermal Engineering (Теплоэнергетика, стр. 80–82), 14 (9), pp. 116–119.

Vikhrev, Yu.V.,Kon'kov, A.S., Lokshin, V.A. et al., 1971. Temperature regime of steam generating tubes at supercritical pressure, (In Russian), Transactions of the IVth All-Union Conference on Heat Transfer and Hydraulics at Movement of Two-Phase Flow inside Elements of Power Engineering Machines and Apparatuses, Leningrad, Russia, pp. 21–40.

Watts, M. J., and Chou, C-T., 1982, Mixed Convection Heat Transfer to Supercritical Pressure Water, Proc. 7th International Heat Transfer Conference, Munich, Germany, pp. 495–500.

Yamagata, K., Nishikawa, K., Hasegawa, S. et al., 1972. Forced convective heat transfer to supercritical water flowing in tubes, International Journal of Heat & Mass Transfer, 15 (12), pp. 2575–2593.

Zahlan, H., Groeneveld, D., & Tavoularis, S. (April 25-28, 2010). Look-Up Table for Trans-Critical Heat Transfer. The 2nd Canada-China Joint Workshop on Supercritical Water-Cooled Reactors (CCSC-2010). Toronto, Ontario, Canada: Canadian Nuclear Society.

20

Fouling of Heat Transfer Surfaces

Mostafa M. Awad
Mansoura University, Faculty of Engineering, Mech. Power Eng. Dept.,
Egypt

1. Introduction

Fouling is generally defined as *the accumulation and formation of unwanted materials on the surfaces of processing equipment,* which can seriously deteriorate the capacity of the surface to transfer heat under the temperature difference conditions for which it was designed. Fouling of heat transfer surfaces is one of the most important problems in heat transfer equipment. Fouling is an extremely complex phenomenon. Fundamentally, fouling may be characterized as a combined, unsteady state, momentum, mass and heat transfer problem with chemical, solubility, corrosion and biological processes may also taking place. It has been described as *the major unresolved problem in heat transfer*[1].

According to many [1-3], fouling can occur on any fluid-solid surface and have other adverse effects besides reduction of heat transfer. It has been recognized as a nearly universal problem in design and operation, and it affects the operation of equipment in two ways: Firstly, the fouling layer has a low thermal conductivity. This increases the resistance to heat transfer and reduces the effectiveness of heat exchangers. Secondly, as deposition occurs, the cross sectional area is reduced, which causes an increase in pressure drop across the apparatus.

In industry, fouling of heat transfer surfaces has always been a recognized phenomenon, although poorly understood. Fouling of heat transfer surfaces occurs in most chemical and process industries, including oil refineries, pulp and paper manufacturing, polymer and fiber production, desalination, food processing, dairy industries, power generation and energy recovery. By many, fouling is considered the single most unknown factor in the design of heat exchangers. This situation exists despite the wealth of operating experience accumulated over the years and accumulation of the fouling literature. This lake of understanding almost reflects the complex nature of the phenomena by which fouling occurs in industrial equipment. The wide range of the process streams and operating conditions present in industry tends to make most fouling situations unique, thus rendering a general analysis of the problem difficult.

In general, the ability to transfer heat efficiently remains a central feature of many industrial processes. As a consequence much attention has been paid to improving the understanding of heat transfer mechanisms and the development of suitable correlations and techniques that may be applied to the design of heat exchangers. On the other hand relatively little consideration has been given to the problem of surface fouling in heat exchangers. The

---

[1] The unresolved problems in heat transfer are: flow induced tube vibration, fouling, mixture boiling, flow distribution in two-phase flow and detailed turbulence flow modelling.

principal purpose of this chapter is to provide some insight into the problem of fouling from a scientific and technological standpoint. A better understanding of the problem and of the mechanisms that lead to the accumulation of deposits on surfaces will provide opportunities to reduce or even eliminate the problem in certain situations.

Fouling can occur as a result of the fluids being handled and their constituents in combination with the operating conditions such as temperature and velocity. Almost any solid or semi solid material can become a heat exchanger foulant, but some materials that are commonly encountered in industrial operations as foulants include:

Inorganic materials

    Airborne dusts and grit

    Waterborne mud and silts

    Calcium and magnesium salts

    Iron oxide

Organic materials

    Biological substances, e.g. bacteria, fungi and algae

    Oils, waxes and greases

    Heavy organic deposits, e.g. polymers, tars

    Carbon

Energy conservation is often a factor in the economics of a particular process. At the same time in relation to the remainder of the process equipment, the proportion of capital that is required to install the exchangers is relatively low. It is probably for this reason that heat exchanger fouling has been neglected as most fouling problems are unique to a particular process and heat exchanger design. The problem of heat exchanger fouling therefore represents a challenge to designers, technologists and scientists, not only in terms of heat transfer technology but also in the wider aspects of economics and environmental acceptability and the human dimension.

2. Types of fouling

Many types of fouling can occur on the heat transfer surfaces. The generally favored scheme for the classification of the heat transfer fouling is based on the different physical and chemical processes involved. Nevertheless, it is convenient to classify the fouling main types as:

1. *Particulate fouling*: It is the deposition of suspended particles in the process streams onto the heat transfer surfaces. If the settling occurs due to gravity as well as other deposition mechanisms, the resulting particulate fouling is called "sedimentation" fouling. Hence, particulate fouling may be defined as the accumulation of particles from heat exchanger working fluids (liquids and/or gaseous suspensions) on the heat transfer surface. Most often, this type of fouling involves deposition of corrosion products dispersed in fluids, clay and mineral particles in river water, suspended solids in cooling water, soot particles of incomplete combustion, magnetic particles in economizers, deposition of salts in desalination systems, deposition of dust particles in air coolers, particulates partially present in fire-side (gas-side) fouling of boilers, and so on. The particulate fouling is influenced by the following factors: concentration of suspended particles, fluid flow velocity, temperature conditions on the fouled surface (heated or nonheated), and heat flux at the heat transfer surface.

2. *Crystallization or precipitation fouling*: It is the crystallization of dissolved salts from saturated solutions, due to solubility changes with temperature, and subsequent precipitation onto the heat transfer surface. It generally occurs with aqueous solutions and other liquids of soluble salts which are either being heated or cooled. The deposition of inverse solubility salts on heated surfaces, usually called "scaling" and its deposited layer is hard and tenacious. The deposition of normal solubility salts on cooled surfaces, usually has porous and mushy deposited layers and it is called "sludge", "softscale", or "powdery deposit". Precipitation/crystallization fouling is common when untreated water, seawater, geothermal water, brine, aqueous solutions of caustic soda, and other salts are used in heat exchangers. The most important phenomena involved with this type of fouling include crystal growth during precipitation require formation of a primary nucleus. The mechanism controlling that process is nucleation, as a rule heterogeneous in the presence of impurities and on the heat transfer surface.

3. *Chemical reaction fouling*: The deposition in this case is the result of one or more chemical reactions between reactants contained in the flowing fluid in which the surface material *itself* is not a reactant or participant. However, the heat transfer surface may act as a catalyst as in cracking, coking, polymerization, and autoxidation. Thermal instabilities of chemical species, such as asphaltenes and proteins, can also induce fouling precursors. This fouling occurs over a wide temperature range from ambient to over 1000°C but is more pronounced at higher temperatures. The mechanism of this type of fouling is a consequence of an unwanted chemical reaction that takes place during the heat transfer process. Chemical reaction fouling is found in many applications of process industry, such as petrochemical industries, oil refining, vapor-phase pyrolysis, cooling of gas and oils, polymerization of process monomers, and so on. Furthermore, fouling of heat transfer surface by biological fluids may involve complex heterogeneous chemical reactions and physicochemical processes. The deposits from chemical reaction fouling may promote corrosion at the surface if the formation of the protective oxide layer is inhibited.

4. *Corrosion fouling*: It involves a chemical or electrochemical reaction between the heat transfer surface *itself* and the fluid stream to produce corrosion products which, in turn, change the surface thermal characteristics and foul it. Corrosion may cause fouling in two ways. First, corrosion products can accumulate and adhere to the surface providing resistance to heat transfer. Second, corrosion products may be transported as particulate materials from the corrosion site and be deposited as particulate fouling on the heat transfer surface in another site of the system. For example, fouling on the water side of boilers may be caused by corrosion products that originate in the condenser or feedtrain. Corrosion fouling is prevalent in many applications where chemical reaction fouling takes place and the protective oxide layer is not formed on the surface. It is of significant importance in the design of the boiler and condenser of a fossil fuel–fired power plant.

5. *Biological fouling*: It is the attachment and growth of macroorganisms and /or microorganisms and their products on the heat transfer surface. It is usually called *"Biofouling"*, and it is generally a problem in water streams. In general, biological fouling can be divided into two main subtypes of fouling: microbial and macrobial. Microbial fouling is accumulation of microorganisms such as algae, fungi, yeasts, bacteria, and molds, and macrobial fouling represents accumulation of macroorganisms such as clams, barnacles, mussels, and vegetation as found in seawater or estuarine cooling water. Microbial fouling precedes macrobial deposition as a rule and may be considered of primary interest. Biological fouling is generally in the form of a biofilm or a slime layer on the surface that is uneven, filamentous, and deformable but difficult to remove. Although biological fouling could occur in suitable liquid streams, it is generally associated with open recirculation or once-through systems with cooling water. Biological fouling may promote corrosion fouling under the slime layer.

6. *Solidification or freezing fouling*: It is the freezing of a pure liquid or a higher melting point components of a multicomponent solution onto a subcooled surfaces. Separation of waxes from hot streams when it come to contact with cooled surfaces, formation of ice on a heat transfer surface during chilled water production or cooling of moist air, deposits formed in phenol coolers, and deposits formed during cooling of mixtures of substances such as paraffin are some examples of solidification fouling. This fouling mechanism occurs at low temperatures, usually ambient and below depending on local pressure conditions. The main factors affecting solidification fouling are mass flow rate of the working fluid, temperature and crystallization conditions, surface conditions, and concentration of the solid precursor in the fluid.

It should be noted that, in many applications, where more than one fouling mechanism is present, the fouling problem becomes very complex with their synergistic effects. It is obvious that one cannot talk about a single, unified theory to model the fouling process wherein not only the foregoing six types of fouling mechanisms are identified, but in many processes more than one fouling mechanism exists with synergistic effects.

3. Fouling processes

The overall fouling process is usually considered to be the net result of two simultaneous sub-processes; a deposition process and a removal (reentrainment) process. A schematic representation of fouling process is given in Fig. (1). All sub-processes can be summarzed as:
- Formation of foulant materials in the bulk of the fluid.
- Transport of foulant materials to the deposit-fluid interface.
- Attachment/ formation reaction at the deposit-fluid interface.
- Removal of the fouling deposit (spalling or sloughing of the deposit layer).
- Transport from the deposit-fluid interface to the bulk of the fluid.

A schematic diagram for the fouling processes is shown in Fig. (2). It must be noted that, some of these sub-processes may not be applicable in certain fouling situations such as corrosion fouling.

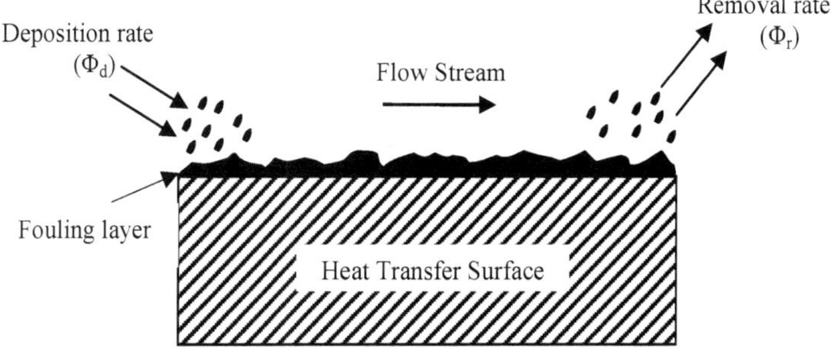

Fig. 1. Fouling processes

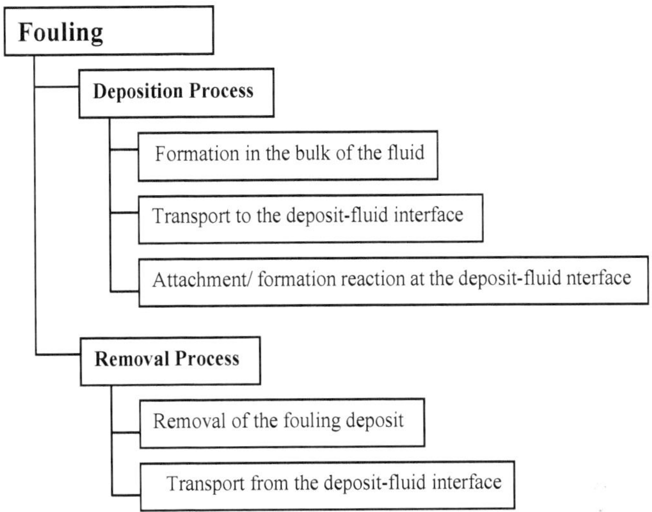

Fig. 2. Schematic diagram for the fouling processes

In another way, three basic stages may be visualized in relation to deposition on surfaces from a moving fluid. They are:

1. The diffusional transport of the foulant or its precursors across the boundary layers adjacent to the solid surface within the flowing fluid.
2. The adhesion of the deposit to the surface and to itself.
3. The transport of material away from the surface.

The sum of these basic components represents the growth of the deposit on the surface.

In mathematical terms the rate of deposit growth (fouling resistance or fouling factor, R_f) may be regarded as the difference between the deposition and removal rates as:

$$R_f = \Phi_d - \Phi_r \tag{1}$$

where Φ_d and Φ_r are the rates of deposition and removal respectively.

The fouling factor, R_f, as well as the deposition rate, Φ_d, and the removal rate, Φ_r, can be expressed in the units of thermal resistance as $m^2 \cdot K/W$ or in the units of the rate of thickness change as m/s or units of mass change as $kg/ m^2 \cdot s$.

4. Deposition and removal mechanisms

From the empirical evidence involving various fouling mechanisms discussed in Section 2, it is clear that virtually all these mechanisms are characterized by a similar sequence of events. The successive events occurring in most cases are illustrated in Fig. (2). These events govern the overall fouling process and determine its ultimate impact on heat exchanger performance. In some cases, certain events dominate the fouling process, and they have a direct effect on the type of fouling to be sustained. The main five events can be summarized briefly as following:

1-Formation of foulant materials in the bulk of the fluid or initiation of the fouling, the first event in the fouling process, is preceded by a delay period or induction period, t_d as shown in Fig. (3), the basic mechanism involved during this period is heterogeneous nucleation, and t_d is shorter with a higher nucleation rate. The factors affecting t_d are temperature, fluid velocity, composition of the fouling stream, and nature and condition of the heat exchanger surface. Low-energy surfaces (unwettable) exhibit longer induction periods than those of high-energy surfaces (wettable). In crystallization fouling, t_d tends to decrease with increasing degree of supersaturation. In chemical reaction fouling, t_d appears to decrease with increasing surface temperature. In all fouling mechanisms, t_d decreases as the surface roughness increases due to available suitable sites for nucleation, adsorption, and adhesion.

2-Transport of species means transfer of the fouling species itself from the bulk of the fluid to the heat transfer surface. Transport of species is the best understood of all sequential events. Transport of species takes place through the action of one or more of the following mechanisms:

- *Diffusion*: involves mass transfer of the fouling constituents from the flowing fluid toward the heat transfer surface due to the concentration difference between the bulk of the fluid and the fluid adjacent to the surface.
- *Electrophoresis*: under the action of electric forces, fouling particles carrying an electric charge may move toward or away from a charged surface depending on the polarity of the surface and the particles. Deposition due to electrophoresis increases with decreasing electrical conductivity of the fluid, increasing fluid temperature, and increasing fluid velocity. It also depends on the pH of the solution. Surface forces such as London–van der Waals and electric double layer interaction forces are usually responsible for electrophoretic effects.
- *Thermophoresis*: a phenomenon whereby a "thermal force" moves fine particles in the direction of negative temperature gradient, from a hot zone to a cold zone. Thus, a high-temperature gradient near a hot wall will prevent particles from depositing, but the same absolute value of the gradient near a cold wall will promote particle deposition. The thermophoretic effect is larger for gases than for liquids.
- *Diffusiophoresis*: involves condensation of gaseous streams onto a surface.
- *Sedimentation*: involves the deposition of particulate matters such as rust particles, clay, and dust on the surface due to the action of gravity. For sedimentation to occur, the downward gravitational force must be greater than the upward drag force. Sedimentation is important for large particles and low fluid velocities. It is frequently observed in cooling tower waters and other industrial processes where rust and dust particles may act as catalysts and/or enter complex reactions.
- *Inertial impaction*: a phenomenon whereby "large" particles can have sufficient inertia that they are unable to follow fluid streamlines and as a result, deposit on the surface.
- *Turbulent downsweeps*: since the viscous sublayer in a turbulent boundary layer is not truly steady, the fluid is being transported toward the surface by turbulent downsweeps. These may be thought of as suction areas of measurable strength distributed randomly all over the surface.

3-Attachment of the fouling species to the surface involves both physical and chemical processes, and it is not well understood. Three interrelated factors play a crucial role in the attachment process: surface conditions, surface forces, and sticking probability. It is the combined and simultaneous action of these factors that largely accounts for the event of attachment.

- *Surface properties*: The properties of surface conditions important for attachment are the surface free energy, wettability (contact angle, spreadability), and heat of immersion. Wettability and heat of immersion increase as the difference between the surface free energy of the wall and the adjacent fluid layer increases. Unwettable or low-energy surfaces have longer induction periods than wettable or high-energy surfaces, and suffer less from deposition (such as polymer and ceramic coatings). Surface roughness increases the effective contact area of a surface and provides suitable sites for nucleation and promotes initiation of fouling. Hence, roughness increases the wettability of wettable surfaces and decreases the unwettability of the unwettable ones.

- *Surface forces*: The most important one is the London–van der Waals force, which describes the intermolecular attraction between nonpolar molecules and is always attractive. The electric double layer interaction force can be attractive or repulsive. Viscous hydrodynamic force influences the attachment of a particle moving to the wall, which increases as it moves normal to the plain surface.

- *Sticking probability*: represents the fraction of particles that reach the wall and stay there before any reentrainment occurs. It is a useful statistical concept devised to analyze and explain the complicated event of attachment.

4-Removal of the fouling deposits from the surface may or may not occur simultaneously with deposition. Removal occurs due to the single or simultaneous action of the following mechanisms; shear forces, turbulent bursts, re-solution, and erosion.

- *Shear forces* result from the action of the shear stress exerted by the flowing fluid on the depositing layer. As the fouling deposit builds up, the cross-sectional area for flow decreases, thus causing an increase in the average velocity of the fluid for a constant mass flow rate and increasing the shear stress. Fresh deposits will form only if the deposit bond resistance is greater than the prevailing shear forces at the solid–fluid interface.

- *Randomly distributed* (about less than 0.5% at any instant of time) periodic turbulent bursts act as miniature tornadoes lifting deposited material from the surface. By continuity, these fluid bursts are compensated for by gentler fluid back sweeps, which promote deposition.

- *Re-solution*: The removal of the deposits by re-solution is related directly to the solubility of the material deposited. Since the fouling deposit is presumably insoluble at the time of its formation, dissolution will occur only if there is a change in the properties of the deposit, or in the flowing fluid, or in both, due to local changes in temperature, velocity, alkalinity, and other operational variables. For example, sufficiently high or low temperatures could kill a biological deposit, thus weakening its attachment to a surface and causing sloughing or re-solution. The removal of corrosion deposits in power-generating systems is done by re-solution at low alkalinity. Re-solution is associated with the removal of material in ionic or molecular form.

- *Erosion* is closely identified with the overall removal process. It is highly dependent on the shear strength of the foulant and on the steepness and length of the sloping heat exchanger surfaces, if any. Erosion is associated with the removal of material in particulate form. The removal mechanism becomes largely ineffective if the fouling layer is composed of well-crystallized pure material (strong formations); but it is very effective if it is composed of a large variety of salts each having different crystal properties.

5- Transport from the deposit-fluid interface to the bulk of the fluid, once the deposits are sloughed, it may/may not transported from the deposit-fluid interface to the bulk of the fluid. This depend on the mass and volume of the sloughed piece and on the hydrodynamic forces of the flowing fluid. If the sloughed piece is larg enough, it may moved on the surface and depoited on another site on the system such as some corrosion products. All deposits which removed due to erosion effect will be transported to the bulk of the fluid. The removal process in not complete without this action. The important parameter affecting the deposit sloughing is the aging of deposits in which it may strengthen or weaken the fouling deposits.

5. Fouling curves

The overall process of fouling is indicated by the fouling factor, R_f (fouling resistance) which is measured either by a test section or evaluated from the decreased capacity of an operating heat exchanger. The representation of various modes of fouling with reference to time is known as a fouling curve (fouling factor-time curve). Typical fouling curves are shown in Fig. (3).

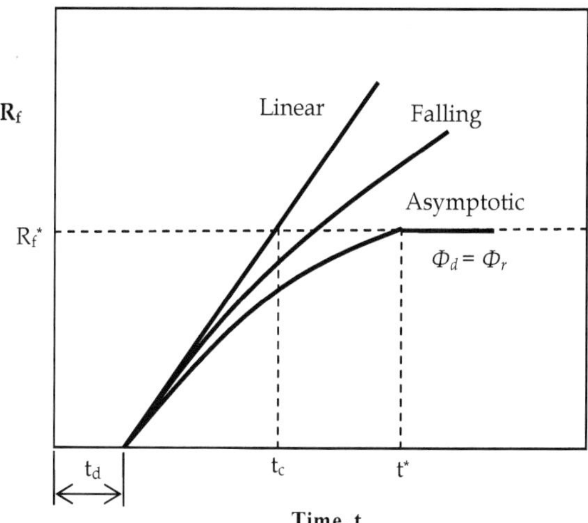

Fig. 3. Fouling Curves

The delay time, t_d indicates that an initial period of time can elapse where no fouling occurs. The value of t_d is not predictable, but for a given surface and system, it appears to be somewhat random in nature or having a normal distribution about some mean value or at least dependent upon some frequency factors. After clean the fouled surfaces and reused them, the delay time, t_d is usually shorter than that of the new surfaces when are used for the first time. It must be noted that, the nature of fouling factor-time curve is not a function of t_d. The most important fouling curves are:

- *Linear fouling curve* is indicative of either a constant deposition rate, Φ_d with removal rate, Φ_r being negligible (i.e. Φ_d = constant, $\Phi_r \approx 0$) or the difference between Φ_d and Φ_r

is constant (i.e. $\Phi_d - \Phi_r$ = constant). In this mode, the mass of deposits increases gradually with time and it has a straight line relationship of the form ($R_f = at$) where "a" is the slope of the line.

- *Falling rate fouling curve* results from either decreasing deposition rate, Φ_d with removal rate, Φ_r being constant or decreasing deposition rate, Φ_d and increasing removal rate, Φ_r. In this mode, the mass of deposit increases with time but not linearly and does not reach the steady state of asymptotic value.

- *Asymptotic fouling curve* is indicative of a constant deposition rate, Φ_d and the removal rate, Φ_r being directly proportional to the deposit thickness until $\Phi_d = \Phi_r$ at the asymptote. In this mode, the rate of fouling gradually falls with time, so that eventually a steady state is reached when there is no net increase of deposition on the surface and there is a possibility of continued operation of the equipments without additional fouling. In practical industrial situations, the asymptote may be reached and the asymptotic fouling factor, R^*_f is obtained in a matter of minutes or it may take weeks or months to occur depending on the operating conditions. The general equation describing this behavior is given in equation (4). This mode is the most important one in which it is widely existed in the industrial applications. The pure particulate fouling is one of this type.

For all fouling modes, the amount of material deposited per unit area, m_f is related to the fouling resistance (R_f), the density of the foulant (ρ_f), the thermal conductivity (λ_f) and the thickness of the deposit (x_f) by the following equation:

$$m_f = \rho_f x_f = \rho_f \lambda_f R_f \tag{2}$$

where

$$R_f = \frac{x_f}{\lambda_f} \tag{3}$$

(values of thermal conductivities for some foulants are given in table 1).

Foulant	Thermal conductivity (W/mK)
Alumina	0.42
Biofilm (effectively water)	0.6
Carbon	1.6
Calcium sulphate	0.74
Calcium carbonate	2.19
Magnesium carbonate	0.43
Titanium oxide	8.0
Wax	0.24

Table 1. Thermal conductivities of some foulants [2]

It should be noted that, the curves represented in Fig. (3) are ideal ones while in the industrial situations, ideality may not be achieved. A closer representation of asymptotic fouling practical curve might be as shown in Fig. (4). The "saw tooth" effect is the result of partial removal of some deposit due to "spalling" or "sloughing" to be followed for a short time by a rapid build up of deposit. The average curve (represented by the dashed line) can be seen to represent the ideal asymptotic curve on Fig. (3). Similar effects of partial removal and deposition may be experienced with the other types of foulin curves.

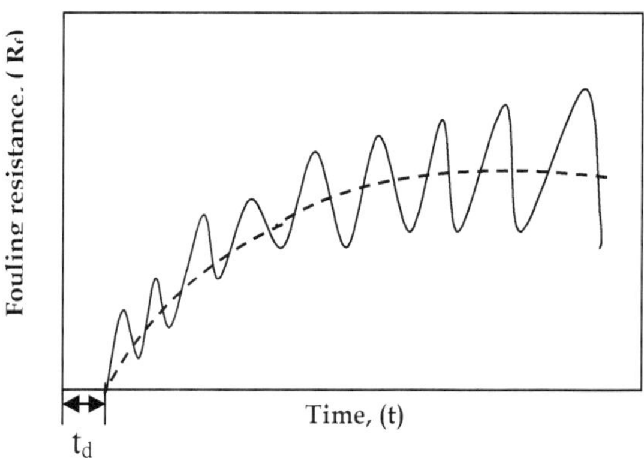

Fig. 4. Practical fouling curve

6. Cost of fouling

Fouling affects both capital and operating costs of heat exchangers. The extra surface area required due to fouling in the design of heat exchangers, can be quite substantial. Attempts have been made to make estimates of the overall costs of fouling in terms of particular processes or in particular countries. Reliable knowledge of fouling economics is important when evaluating the cost efficiency of various mitigation strategies. The total fouling-related costs can be broken down into four main areas:
4. Higher capital expenditures for oversized plants which includes excess surface area (10-50%), costs for extra space, increased transport and installation costs.
5. Energy losses due to the decrease in thermal efficiency and increase in the pressure drop.
6. Production losses during planned and unplanned plant shutdowns for fouling cleaning.
7. Maintenance including cleaning of heat transfer equipment and use of antifoulants.
The loss of heat transfer efficiency usually means that somewhere else in the system, additional energy is required to make up for the short fall. The increased pressure drop through a heat exchanger represents an increase in the pumping energy required to maintain the same flow rate. The fouling resistance used in any design brings about 50% increase in the surface area over that required if there is no fouling. The need for additional maintenance as a result of fouling may be manifested in different ways. In general, any extensive fouling means that the heat exchanger will have to be cleaned on a regular basis to restore the loss of its heat transfer capacity. According to Pritchard [4], the total heat exchanger fouling costs for highly industrialized countries are about 0.25% of the countries' Gross National Product (GNP). Table (2) shows the annual costs of fouling in some different countries based on 1992 estimation.

Country	Fouling Costs (million $)	Fouling Cost/GNP %
US	14175	0.25
UK	2500	0.25
Germany	4875	0.25
France	2400	0.25
Japan	10000	0.25
Australia	463	0.15
New Zealand	64.5	0.15

Table 2. Annual costs of fouling in some countries (1992 estimation) [5].

From this table, it is clear that fouling costs are substantial and any reduction in these costs would be a welcome contribution to profitability and competitiveness. The frequency of cleaning will of course depend upon the severity of the fouling problem and may range between one weak and one year or longer. Frequent cleaning involving repeated dismantling and reassembly will inevitably result in damage to the heat exchanger at a lesser or greater degree, which could shorten the useful life of the equipment. Fouling can be very costly in refinery and petrochemical plants since it increases fuel usage, results in interrupted operation and production losses, and increases maintenance costs.

Increased Capital Investment

In order to make allowance for potential fouling the area for a given heat transfer surface is larger than for clean conditions. To accommodate the fouling-related drop in heat transfer capacity, the tubular exchangers are generally designed with 20-50% excess surface, where the compact heat exchangers are designed with 15-25% excess surface. In addition to the actual size of the heat exchanger other increased capital costs are likely. For instance where it is anticipated that a particular heat exchanger is likely to suffer severe or difficult fouling, provision for off-line cleaning will be required. The location of the heat exchanger for easy access for cleaning may require additional pipe work and larger pumps compared with a similar heat exchanger operating with little or no fouling placed at a more convenient location. Furthermore if the problem of fouling is thought to be excessive it might be necessary to install a standby exchanger, with all the associated pipe work foundations and supports, so that one heat exchanger can be operated while the other is being cleaned and serviced.

Under these circumstances the additional capital cost is likely to more than double and with allowances for heavy deposits the final cost could be 4 - 8 times the cost of the corresponding exchanger running in a clean condition. Additional capital costs may be considered for on-line cleaning such as the Taprogge system (see sec. 12) or other systems. It has to be said however, that on-line cleaning can be very effective and that the additional capital cost can often be justified in terms of reduced operating costs. Furthermore the way in which the additional area is accommodated, can affect the rate of fouling. For instance if the additional area results say, in reduced velocities, the fouling rate may be higher than anticipated and the value of the additional area may be largely offset by the effects of heavy deposits. The indiscriminate use of excess surface area for instance, can lead to high capital costs, especially where exotic and expensive materials of construction are required.

Additional Operating Costs

The presence of fouling on the surface of heat exchangers decreases the ability of the unit to transfer heat. Due to this decrement in the exchanger thermal capacity, neither the hot stream nor the cold stream will approach its target temperature. To compensate this shortage in the heat flow, either additional cooling utility or additional heating utility is required. On the other hand, the presence of deposits on the surface of heat exchangers increases the pressure drop and to recover this increment, an additional pumping work is required and hence a greater pumping cost. Also the fouling may be the cause of additional maintenance costs. The more obvious result of course, is the need to clean the heat exchanger to return it to efficient operation. Not only will this involve labour costs but it may require large quantities of cleaning chemicals and there may be effluent problems to be overcome that add to the cost. If the cleaning agents are hazardous or toxic, elaborate safety precautions with attendant costs, may be required.

The frequent need to dismantle and clean a heat exchanger can affect the continued integrity of the equipment, i.e. components in shell and tube exchangers such as baffles and tubes may be damaged or the gaskets and plates in plate heat exchangers may become faulty. The damage may also aggravate the fouling problem by causing restrictions to flow and upsetting the required temperature distribution.

Loss of Production

The need to restore flow and heat exchanger efficiency will necessitate cleaning. On a planned basis the interruptions to production may be minimized but even so if the remainder of the plant is operating correctly then this will constitute a loss of output that, if the remainder of the equipment is running to capacity still represents a loss of profit and a reduced contribution to the overall costs of the particular site. The consequences of enforced shutdown due to the effects of fouling are of course much more expensive in terms of output. Much depends on recognition of the potential fouling at the design stage so that a proper allowance is made to accommodate a satisfactory cleaning cycle. When the seriousness of a fouling problem goes unrecognized during design then unscheduled or even emergency shutdown, may be necessary. Production time lost through the need to clean a heat exchanger can never be recovered and it could in certain situations, mean the difference between profit and loss.

The Cost of Remedial Action

If the fouling problem cannot be relieved by the use of additives it may be necessary to make modifications to the plant. Modification to allow on-line cleaning of a heat exchanger can represent a considerable capital investment. Before capital can be committed in this way, some assessment of the effectiveness of the modification must be made. In some examples of severe fouling problems the decision is straightforward, and a pay back time of less than a year could be anticipated. In other examples the decision is more complex and the financial risks involved in making the modification will have to be addressed. A number of contributions to the cost of fouling have been identified, however some of the costs will remain hidden. Although the cost of cleaning and loss of production may be recognized and properly assessed, some of the associated costs may not be attributed directly to the fouling problem. For instance the cost of additional maintenance of ancillary equipment such as pumps and pipework, will usually be lost in the overall maintenance charges.

7. Parameters affecting fouling

The fouling process is a dynamic and unsteady one in which many operational and design variables have been identified as having most pronounced and well defined effects on fouling. These variables are reviewed in principle to clarify the fouling problems and because the designer has an influence on their modification. Those parameters include the fluid flow velocity, the fluid properties, the surface temperature, the surface geometry, the surface material, the surface roughness, the suspended particles concentration and properties, …….etc. According to many investigators, the most important parameters are:

1. Fluid flow velocity

The flow velocity has a strong effect on the fouling rate where it has direct effects on both of the deposition and removal rates through the hydrodynamic effects such as the eddies and shear stress at the surface. On the other hand, the flow velocity has indirect effects on deposit strength (ψ), the mass transfer coefficient (k_m), and the stickability (P). It is well established that, increasing the flow velocity tends to increase the thermal performance of the exchanger and decrease the fouling rate. Uniform and constant flow of process fluids past the heat transfer surface favors less fouling. Foulants suspended in the process fluids will deposit in low-velocity regions, particularly where the velocity changes quickly, as in heat exchanger water boxes and on the shell side. Higher shear stress promotes dislodging of deposits from surfaces. Maintain relatively uniform velocities across the heat exchanger to reduce the incidence of sedimentation and accumulation of deposits.

2. Surface temperature

The effect of surface temperature on the fouling rate has been mentioned in several studies. These studies indicated that the role of surface temperature is not well defined. The literatures show that, *"increase surface temperature may increase, decrease, or has no effect on the fouling rates"*. This variation in behavior does indicate the importance to improve our understanding about the effect of surface temperature on the fouling process,
A good practical rule to follow is to expect more fouling as the temperature rises. This is due to a "baking on" effect, scaling tendencies, increased corrosion rate, faster reactions, crystal formation and polymerization, and loss in activity by some antifoulants [6]. Lower temperatures produce slower fouling buildup, and usually deposits that are easily removable [7]. However, for some process fluids, low surface temperature promotes crystallization and solidification fouling. To overcome these problems, there is an *optimum surface temperature* which better to use for each situation. For cooling water with a potential to scaling, the desired maximum surface temperature is about 60°C. Biological fouling is a strong function of temperature. At higher temperatures, chemical and enzyme reactions proceed at a higher rate with a consequent increase in cell growth rate [8]. According to Mukherjee [8], for any biological organism, there is a temperature below which reproduction and growth rate are arrested and a temperature above which the organism becomes damaged or killed. If, however, the temperature rises to an even higher level, some heat sensitive cells may die.

3. Surface material

The selection of surface material is significant to deal with corrosion fouling. Carbon steel is corrosive but least expensive. Copper exhibits biocidal effects in water. However, its use is limited in certain applications: (1) Copper is attacked by biological organisms including sulfate-reducing bacteria; this increases fouling. (2) Copper alloys are prohibited in high-pressure steam power plant heat exchangers, since the corrosion deposits of copper alloys are transported and deposited in high-pressure steam generators and subsequently block the turbine blades. (3) Environmental protection limits the use of copper in river, lake, and ocean waters, since copper is poisonous to aquatic life. Noncorrosive materials such as titanium and nickel will prevent corrosion, but they are expensive and have no biocidal effects. Glass, graphite, and teflon tubes often resist fouling and/or improve cleaning but they have low thermal conductivity. Although the construction material is more important to resist fouling, surface treatment by plastics, vitreous enamel, glass, and some polymers will minimize the accumulation of deposits.

4. Surface Roughness

The surface roughness is supposed to have the following effects: (1) The provision of "nucleation sites" that encourage the laying down of the initial deposits. (2) The creation of turbulence effects within the flowing fluid and, probably, instabilities in the viscous sublayer. Better surface finish has been shown to influence the delay of fouling and ease cleaning. Similarly, non-wetting surfaces delay fouling. Rough surfaces encourage particulate deposition and provide a good chance for deposit sticking. After the initiation of fouling, the persistence of the roughness effects will be more a function of the deposit itself. Even smooth surfaces may become rough in due course due to scale formation, formation of corrosion products, or erosion.

5. Fluid Properties

The fluid propensity for fouling is depending on its properties such as viscosity and density. The viscosity is playing an important rule for the sublayer thickness where the deposition process is taking place. On the other side the viscosity and density have a strong effect on the sheer stress which is the key element in the removal process.

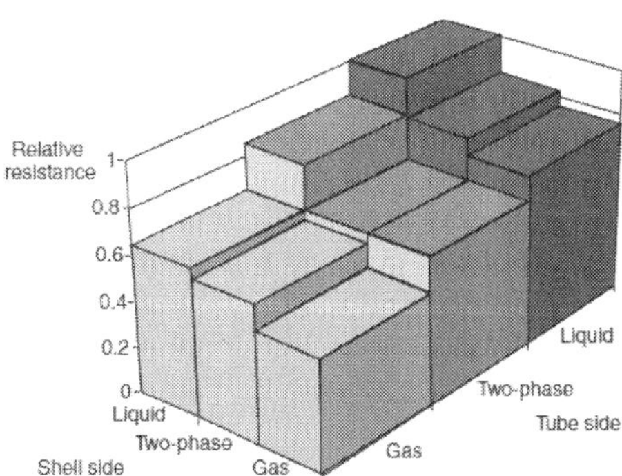

Fig. 5. Effect of the flow fluid type on the fouling

To show the effect of the flow fluid type on the fouling resistance, Chenoweth [7} collected data from over 700 shell and-tube heat exchangers. These data of combined shell- and tube-side fouling resistances (by summing each side entry), have been compiled and divided into nine combinations of liquid, two-phase, and gas on each fluid side regardless of the applications. The arithmetic average of total R_f of each two-fluid combination value has been taken and analyzed. The results are presented in Fig. (5) with ordinate ranges between 0 and

1.0. From this figure, it is clear that the maximum value is 1.0, that is due to liquid-liquid heat exchanger, where the minimum value is 0.5 which belong to gas-gas heat exchanger. If liquid is on the shell side and gas on the tube side, the relative fouling resistance is 0.65. However, if liquid is on the tube side and gas on the shell side, it is 0.75. Since many process industry applications deal with liquids that are dirtier than gases, the general practice is to specify larger fouling resistances for liquids compared to those for the gases. Also, if fouling is anticipated on the liquid side of a liquid–gas exchanger, it is generally placed in the tubes for cleaning purposes spite a larger fouling resistance is specified. These trends are clear from the figure. It should again be emphasized that Fig. (5) indicates the current practice and has no scientific basis. Specification of larger fouling resistances for liquids (which have higher heat transfer coefficients than those of gases) has even more impact on the surface area requirement for liquid–liquid exchangers than for gas–gas exchangers.

6. Impurities and Suspended Solids

Seldom are fluids pure. Intrusion of minute amounts of impurities can initiate or substantially increase fouling. They can either deposit as a fouling layer or acts as catalysts to the fouling processes [6]. For example, chemical reaction fouling or polymerization of refinery hydrocarbon streams is due to oxygen ingress and/or trace elements such as Va and Mo. In crystallization fouling, the presence of small particles of impurities may initiate the deposition process by seeding. The properties of the impurities form the basis of many antifoulant chemicals. Sometimes impurities such as sand or other suspended particles in cooling water may have a scouring action, which will reduce or remove deposits [9].
Suspended solids promote particulate fouling by sedimentation or settling under gravitation onto the heat transfer surfaces. Since particulate fouling is velocity dependent, prevention is achieved if stagnant areas are avoided. For water, high velocities (above 1 m/s) help prevent particulate fouling. Often it is economical to install an upstream filtration.

7. Heat Transfer Process

The fouling resistances for the same fluid can be considerably different depending upon whether heat is being transferred through sensible heating or cooling, boiling, or condensing.

8. Design Considerations

Equipment design can contribute to increase or decrease fouling. Heat exchanger tubes that extend beyond tube sheet, for example, can cause rapid fouling. Some fouling aspects must be considered through out the equipment design such as:
1.   Placing the More Fouling Fluid on the Tube Side
As a general guideline, the fouling fluid is preferably placed on the tube side for ease of cleaning. Also, there is less probability for low-velocity or stagnant regions on the tube side.
2.   Shell-Side Flow Velocities
Velocities are generally lower on the shell side than on the tube side, less uniform throughout the bundle, and limited by flow-induced vibration. Zero-or low-velocity regions on the shell side serve as ideal locations for the accumulation of foulants. If fouling is expected on the shell side, then attention should be paid to the selection of baffle design. Segmental baffles have the tendency for poor flow distribution if spacing or baffle cut ratio is not in correct proportions. Too low or too high a ratio results in an unfavorable flow regime that favors fouling.

3.   Low-Finned Tube Heat Exchanger

There is a general apprehension that low Reynolds number flow heat exchangers with low-finned tubes will be more susceptible to fouling than plain tubes. Fouling is of little concern for finned surfaces operating with moderately clean gases. Fin type does not affect the fouling rate, but the fouling pattern is affected for waste heat recovery exchangers. Plain and serrated fin modules with identical densities and heights have the same fouling thickness increases in the same period of time.

4.   Gasketed Plate Heat Exchangers

High turbulence, absence of stagnant areas, uniform fluid flow, and the smooth plate surface reduce fouling and the need for frequent cleaning. Hence the fouling factors required in plate heat exchangers are normally 10-25% of those used in shell and tube heat exchangers.

5.   Spiral Plate Exchangers

High turbulence and scrubbing action minimize fouling on the spiral plate exchanger. This permits the use of low fouling factors.

6.   Seasonal temperature changes

When cooling tower water is used as coolant, considerations are to be given for winter conditions where the ambient temperature may be near zero or below zero on the Celsius scale. The increased temperature driving force during the cold season contributes to more substantial overdesign and hence over performance problems, unless a control mechanism has been instituted to vary the water/air flow rate as per the ambient temperature. Also the bulk temperature of the cooling water that used in power condensers is changed seasonally. This change influences the fouling rate to some extent.

8. Fouling measurements and monitoring

The fouling resistances can be measured either experimentally or analytically. The main measuring methods include;

1-Direct weighing; the simplest method for assessing the extent of deposition on test surfaces in the laboratory is by direct weighing. The method requires an accurate balance so that relatively small changes in deposit mass may be detected. It may be necessary to use thin walled tube to reduce the tare mass so as to increase the accuracy of the method.

2-Thickness measurement; In many examples of fouling the thickness of the deposit is relatively small, perhaps less than *50 μm,* so that direct measurement is not easy to obtain. A relatively simple technique provided there is reasonable access to the deposit, is to measure the thickness. Using a removable coupon or plate the thickness of a hard deposit such as a scale, may be made by the use of a micrometer or travelling microscope. For a deformable deposit containing a large proportion of water, e.g. a biofilm it is possible to use an electrical conductivity technique

3-Heat transfer measurements; In this method, the fouling resistance can be determined from the changes in heat transfer during the deposition process. The basis for subsequent operations will be Equation (14). The data may be reported in terms of changes in overall heat transfer coefficient. A major assumption in this method is that the presence of the deposit does not affect the hydrodynamics of the flowing fluid. However, in the first stages of deposition, the surface of the deposit is usually rougher than the metal surface so that the turbulence within the fluid is greater than when it is flowing over a smooth surface. As a result the fouling resistance calculated from the data will be lower than if the increased level of turbulence had been taken into account. It is possible that the increased turbulence offsets the thermal resistance of the deposit and negative values of thermal resistance will be calculated.

4-Pressure drop; As an alternative to direct heat transfer measurements it is possible to use changes in pressure drop brought about by the presence of the deposit. The pressure drop is increased for a given flow rate by virtue of the reduced flow area in the fouled condition and the rough character of the deposit. The shape of the curve relating pressure drop with time will in general, follow an asymptotic shape so that the time to reach the asymptotic fouling resistance may be determined. The method is often combined with the direct measurement of thickness of the deposit layer. Changes in friction factor may also be used as an indication of fouling of a flow channel.

5-Other techniques for fouling assessment; In terms of their effect on heat exchanger performance the measurement of heat transfer reduction or increase in pressure drop provide a direct indication. The simple methods of measuring deposit thickness described earlier are useful, but in general they require that the experiment is terminated so as to provide access to the test sections. Ideally non-intrusive techniques would allow deposition to continue while the experimental conditions are maintained without disturbance. Such techniques include the use of radioactive tracers and optical methods. Laser techniques can be used to investigate the accumulation and removal of deposits. Also, infra red systems are used to investigate the development and removal of biofilms from tubular test sections. Microscopic examination of deposits may provide some further evidence of the mechanisms of fouling, but this is generally a "back up" system rather than to give quantitative data.

Gas-Side Fouling Measuring Devices

The gas-side fouling measuring devices can be classified into five groups: heat flux meters, mass accumulation probes, optical devices, deposition probes, and acid condensation probes. A heat flux meter uses the local heat transfer per unit area to monitor the fouling. The decrease in heat flux as a function of time is thus a measure of the fouling buildup. A mass accumulation device measures the fouling deposit under controlled conditions. Optical measuring devices use optical method to determine the deposition rate. Deposition probes are used to measure the deposit thickness. Acid condensation probes are used to collect liquid acid that accumulates on a surface that is at a temperature below the acid dew point of the gas stream.

Instruments for Monitoring of Fouling

Instruments have been developed to monitor conditions on a tube surface to indicate accumulation of fouling deposits and, in some cases, to indicate the effect on heat exchanger performance. The following is a summary of the different fouling monitors [10, 11]:

1. Removable sections of the fouled surface, which may be used for microscopic examination, mass measurements, and chemical and biological analysis of the deposits.

2. Increase in pressure drop across the heat exchanger length. This method provides a measure of fluid frictional resistance, which usually increases with buildup of fouling deposits. This device is relatively inexpensive and is easy to operate.

3. Thermal resistance monitors, which are used to determine the effect of the deposit on overall heat transfer resistance. The thermal method of monitoring has the advantage over the others of giving directly information that is required for predicting or assessing heat transfer performance.

9. Performance data analysis

As mentioned above, fouling has many effects on the heat exchanger perfornance. It decreases the exchanger thermal capacity and increases the pressure drop through the exchanger as shown in Fig. (6). From the figure it is clear that the total thermal resistance to heat transfer is decreased during the first stages of fouling due to the surface roughness resulting from initial deposition. After that and with deposits building up, the thermal resistace returns to increase again.

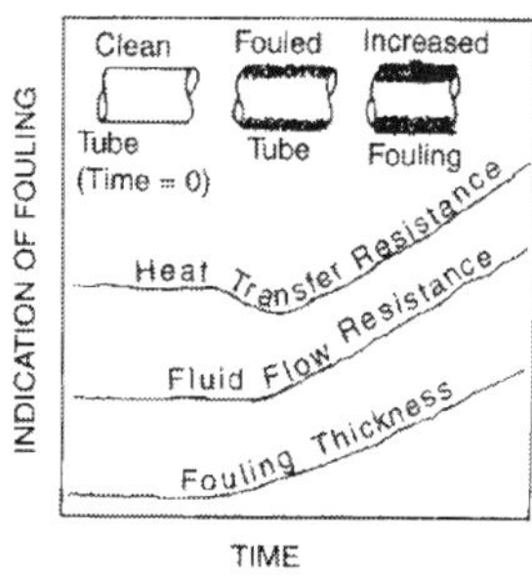

Fig. 6. Fouling effects on exchanger performance

In order to model and predict the industrial processes fouling problems it is first necessary to understand what is happening and what are the causes and effects of fouling. To achieve this, it is necessary to carefully examine and evaluate all the data and operating conditions at various plants in order to understand what the variables which are effective on fouling and what are the mechanisms of such phenomena. The objective of these efforts will be always to minimize the fouling clean-up / remediation shut-down frequency of the plants and to reduce the cost by making the minimum modification in the processes.

The possibility of whether the fouling material is a part of the feed to the system or it is a product of reaction / aggregation / flocculation in the system must be clarified. The role of various operating conditions in the system on fouling (pressures, temperatures, compositions, flow rates, etc. and their variations) must be understood and quantified. Only with appropriate modeling considering all the possible driving forces and mechanisms of fouling one may be able to predict the nature of fouling in each case and develop mitigation techniques to combat that.

The available fouling history data would be useful to test the packages which will be developed. Considering the diversity of the data, care must be taken in their analysis for any universality conclusions. However, in order to make comparisons between fouling data from various plants and test the accuracy of the developed packages, it will be necessary to acquire the compositions data of the feed in each plant as well as characteristics and conditions of operations of the process system used in those plants. Only then one can test the accuracy of the models developed and understand why in one case there is fouling and no fouling in another case.

Empirical data for fouling resistances have been obtained over many decades by industry since its first compilation by TEMA in 1941 for shell-and-tube heat exchangers. TEMA fouling resistances [12] are supposed to be representative values, asymptotic values, or those manifested just before cleaning to be performed.

It should be reiterated that the recommended fouling resistances are believed to represent typical fouling resistances for design. Consequently, sound engineering judgment has to be made for each selection of fouling resistances, keeping in mind that actual values of fouling resistances in any application can be either higher or lower than the resistances calculated. Finally, it must be clear that fouling resistances, although recommended following the empirical data and a sound model, are still constant, independent of time, while fouling is a transient phenomenon. Hence, the value of R_f selected represents a correct value only at one specific time in the exchanger operation. Therefore, it needs to be emphasized that the tables may not provide the applicable values for a particular design. They are only intended to provide guidance when values from direct experience are unavailable. With the use of finite fouling resistance, the overall U value is reduced, resulting in a larger surface area requirement, larger flow area, and reduced flow velocity which inevitably results in increased fouling. Thus, allowing more surface area for fouling in a clean exchanger may accelerate fouling initially.

Typical fouling resistances are roughly 10 times lower in plate heat exchangers (PHEs) than in shell-and-tube heat exchangers (TEMA values), (see Table 3).

Process Fluid	R_f, (m² · K/kW)	
	PHEs	TEMA
Soft water	0.018	0.18–0.35
Cooling tower water	0.044	0.18–0.35
Seawater	0.026	0.18–0.35
River water	0.044	0.35–0.53
Lube oil	0.053	0.36
Organic solvents	0.018–0.053	0.36
Steam (oil bearing)	0.009	0.18

Table 3. Liquid-Side Fouling Resistances for PHEs vs. TEMA Values (from Ref.13)

10. Fouling models

Fouling is usually considered to be the net result of two simultaneous processes: a deposition process and a removal process. A schematic representation of fouling process is given in Fig. (1). Mathematically, the net rate of fouling can be expressed as the difference between the deposition and removal rates as given in equation (1). Many attempts have been made to model the fouling process. One of the earliest models of fouling was that by Kern and Seaton [14]. In this model, it was assumed that the rate of deposition mass, $\dot{m}_d$, remained constant with time t but that the rate of removal mass, $\dot{m}_r$, was proportional to the accumulated mass, m_f, and therefore increased with time to approach $\dot{m}_d$ asymptotically. Thus

Rate of accumulation = Rate of deposition – Rate of removal

$$dm_f/dt = \dot{m}_f = \dot{m}_d - \dot{m}_r \tag{4}$$

then integration of Eqn. (4) from the initial condition $m_f = 0$ at $t = 0$ gives

$$m_f = m_f^*(1 - e^{-\beta t}) \tag{5}$$

where m_{f^*} is the asymptotic value of m_f and $\beta = 1/t_c$. The time constant t_c represents the average residence time for an element of fouling material at the heat transfer surface. Referring to Eqn. (2), Eqn. (5) can be expressed in terms of fouling resistance R_f at time t in terms of the asymptotic value R^*_f by

$$R_f = R^*_f(1 - e^{-\beta t}) \tag{6}$$

It is obvious that the real solution would be to find expressions for R^*_f and t_c as a function of variables affecting the fouling process.

The purpose of any fouling model is to assist the designer or indeed the operator of heat exchangers, to make an assessment of the impact of fouling on heat exchanger performance given certain operating conditions. Ideally a mathematical interpretation of Eqn. (6) would provide the basis for such an assessment but the inclusion of an extensive set of conditions into one mathematical model would be at best, difficult and even impossible.

Modeling efforts to produce a mathematical model for fouling process have been based on the general material balance given in Eqn. (4) and centered on evaluating the functions $\dot{m}_d$ and $\dot{m}_r$ for specific fouling situations, some of these models are:

Watkinson Model:

Watkinson [15] reported the effect of fluid velocity on the asymptotic fouling resistance in three cases as;

1. Calcium carbonate scaling (with constant surface temperature and constant composition)

$$R^*_f = 0.101/(v^{1.33} \cdot D^{0.23}) \tag{7}$$

2. Gas oil fouling (with constant heat flux)

$$R^*_f = 0.55/v^2 \tag{8}$$

3. Sand deposition from water (with constant heat flux)

$$R^*_f = 0.015/v^{1.2} \tag{9}$$

   where;
   R^*_f the asymptotic fouling resistance
   v the fluid velocity
   D the tube diameter

Taborek, et al. Model:

Taborek, et al [16] introduced a water characterization factor to the deposition term to account for the effect of water quality. The deposition term, also involves two processes; (1) Diffusion of the potential depositing substance to the surface and (2) Bonding at the surface. They expressed the deposition rate in an arrhenius type equation as the following:

$$\Phi_d = k_1 P_d \Omega^n \exp(\frac{-Ea}{R_g T_s}) \tag{10}$$

where

k_1 deposition constant

P_d deposition probability factor related to velocity and "Stickiness" or adhesion characteristics of the deposit,

n   exponent

Ω water characterization factor,

$(-E_a/R_gT_s)$ the Arrhenius reaction rate function,

E_a the activation energy,

R_g the universal gas constant,

T_s the absolute surface temperature

In this model, the removal rate was postulated to be a function of shear stress, deposit thickness and bonding strength of the deposit. The removal function was given as:

$$\Phi_r = k_2\left(\frac{\tau}{\psi}\right)x_f \tag{11}$$

where;

k_2 removal constant

τ the fluid shear stress exerted on the deposit surface

ψ the strength or toughness of the deposit layer

Substituting for the deposition rate (Eqn.10) and removal rate (Eqn.11) into material balance Eqn. (1) and taking into account Eqn. (3), the resulting equation yields to;

$$R_f = \frac{x_f}{\lambda_f} = \frac{k_1 P_d \Omega^n e^{-E_a/R_gT_s}(1-e^{-k_2\lambda_f t\tau/\psi})}{\dfrac{k_2\tau\lambda_f}{\psi}} \tag{12}$$

and

$$R_f^* = \frac{k_1 P_d \Omega^n e^{-E_a/R_gT_s}}{\dfrac{k_2\tau\lambda_f}{\psi}}, \qquad \beta = \frac{k_2\lambda_f\tau}{\psi} = \frac{1}{t_c} \tag{13}$$

Knudsen Analysis [17]:

As it is known, the fouling process is complicated and dynamic. The fouling resistance is not usually measured directly, but must be determined from the degradation of the overall heat transfer coefficient. The fouling factor, R_f, could be expressed as;

$$R_f = \frac{1}{U_f} - \frac{1}{U_c} \tag{14}$$

Experimental fouling data have been analyzed on the basis of the change in overall heat transfer coefficient of the fouling test section as in equation (16). It is assumed that the thermal hydraulic condition in the test section remains reasonably constant for the duration of the fouling test. The model of Taborek et al. is used and the two parameters R_f^* and t_c can be determined for each fouling situation, where;

R_f^* is the asymptotic fouling resistance contains all the factors that influence fouling.

t_c is the time constant of the fouling resistance exponential curve i.e. the time required for the fouling resistance to reach 63% of its asymptotic value (i.e. $t_c \approx 0.63t^*$, see Fig. 3), it depends on the shear stress, the deposit strength factor   and the deposit thermal conductivity as;

$$t_c = \psi \ / \ \tau \, k_2 \, \lambda_f \tag{15}$$

From the deposition – removal model, which was first presented by Kern and Seaton [13] (Eqn. 6) and from Eqn. (14), the overall heat transfer coefficient of the fouled surface. U_f, may be given as;

$$U_f = \frac{U_c}{1 + U_c R_f} \tag{16}$$

then

$$U_f = \frac{1}{\dfrac{1}{U_c} + R_f^*(1 - e^{-t/tc})} \tag{17}$$

In equation (17), if the two coefficients R_f^* and t_c can be obtained accurately either empirically or analytically, they will be useful for predicting the fouling factor which can be used in practical heat exchanger design.

11. Fouling and heat exchanger design

The heat exchanger designer must consider the effect of fouling upon the exchanger performance during the desired operational lifetime and make provision in his design for sufficient extra capacity to insure that the exchanger will meet process specifications up to shutdown for cleaning. The designer must also consider what suitable arrangements are necessary to permit easy cleaning.

In choosing the fouling resistances to be used in a given heat exchanger, the designer has three main sources:

1.   Past experience of heat exchanger performance in the same or similar environments.
2.   Results from portable test rigs.
3.   TEMA values, which are overall values for a very limited number of environments (table 4).

As it is known, the overall thermal resistance for a heat exchanger involves a series of thermal resistances from the hot fluid to the cold fluid, including thermal resistances due to fouling on both fluid sides, as shown in Fig. (7). Based on the inside heat transfer surface area A_i, the overall heat transfer coefficient is expressed as:

$$\frac{1}{U_i} = (\frac{1}{h_i} + R_{f,i}) + \frac{\delta_w}{\lambda_w} \frac{A_i}{A_w} + (\frac{1}{h_o} + R_{f,o}) \frac{A_i}{A_o} \tag{18}$$

In Eqn. (18), it is assumed that the wall thermal resistance is for a flat plate wall. This equation can be rearranged and simplified as

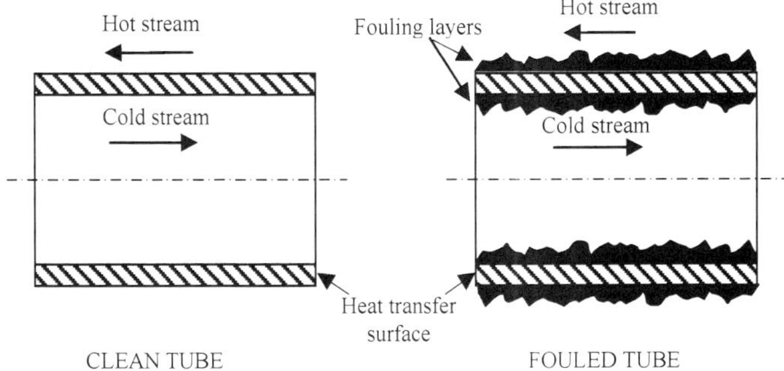

Fig. 7. Thermal resistances for clean and fouled tubes

$$\frac{1}{U_i} = \frac{1}{h_i} + R_{f,i} + R_{f,o}\frac{Ai}{A_o} + \frac{\delta_w}{\lambda_w}\frac{A_i}{A_w} + \frac{1}{h_o}\frac{A_i}{A_o} = \frac{1}{h_i} + R_f + R_w\frac{A_i}{A_w} + \frac{1}{h_o}\frac{A_i}{A_o}$$ (19)

Note that $R_f=R_{f,i}+R_{f,o}(A_i/A_o)$ represents the total fouling resistance, a sum of fouling resistances on both sides of the heat transfer surface, as shown. It should again be reiterated that the aforementioned reduction in the overall heat transfer coefficient due to fouling does not take into consideration the transient nature of the fouling process.

The current practice is to assume a value for the fouling resistance on one or both fluid sides as appropriate and to design a heat exchanger accordingly by providing extra surface area for fouling, together with a cleaning strategy. The complexity in controlling a large number of internal and external factors of a given process makes it very difficult to predict the fouling growth as a function of time using deterministic (well-known) kinetic models.

A note of caution is warranted at this point. There is an ongoing discussion among scholars and engineers from industry as to whether either fouling resistance or fouling rate concepts should be used as the most appropriate tool in resolving design problems incurred by fouling. One suggestion in resolving this dilemma would be that the design fouling-resistance values used for sizing heat exchangers be based on fouling-rate data and estimated cleaning-time intervals.

In current practice, based on application and need, the influence of fouling on exchanger heat transfer performance can be evaluated in terms of either (1) required increased surface area for the same q and ΔT_m, (2) required increased mean temperature difference for the same q and A, or (3) reduced heat transfer rate for the same A and ΔT_m. For these approaches, the expressions; A_f/A_c, $\Delta T_{m,f}/\Delta T_{m,c}$ and q_f/q_c may be determined. In the first two cases, the heat transfer rate in a heat exchanger under clean and fouled conditions are the same. Hence,

$$q = U_c A_c \Delta T_m = U_f A_f \Delta T_m \quad \text{(for constant } \Delta T_m)$$ (20)

Therefore,

$$\frac{A_f}{A_c} = \frac{U_c}{U_f}$$ (21)

Where, the subscript c denotes a clean surface and f the fouled surface.

It must be noted that, the first case of the above mentioned approaches is the design of an exchanger where an allowance for fouling can be made at the design stage by increasing surface area, while the other two cases are for an already designed exchanger in operation, and the purpose is to determine the impact of fouling on exchanger performance.

According to Eqn. (19), the relationships between overall heat transfer coefficients (based on tube outside surface area) and thermal resistances for clean and fouled conditions are defined as follows. For a clean heat transfer surface,

$$\frac{1}{U_c} = \frac{1}{h_{o,f}} + R_w \frac{A_o}{A_w} + \frac{1}{h_{i,f}} \frac{A_o}{A_i} \tag{22}$$

For a fouled heat transfer surface,

$$\frac{1}{U_f} = \frac{1}{h_{o,f}} + R_f + R_w \frac{A_o}{A_w} + \frac{1}{h_{i,f}} \frac{A_o}{A_i} = \frac{1}{h_{o,f}} + R_f + R_w \frac{A_o}{A_w} + \frac{1}{h_{i,c}} \frac{A_o}{A_i} \tag{23}$$

For the ideal conditions that, $h_{o,f} = h_{o,c}$, $h_{i,f} = h_{i,c}$, $A_{i,f} = A_{i,c} = A_i$ and $A_{o,f} = A_{o,c} = A_o$, the difference between Eqns. (22) and (23) yields to Eqn. (14) which is

$$R_f = \frac{1}{U_f} - \frac{1}{U_c} \tag{14}$$

Combining Eqns. (14) and (21), it gets

$$\frac{A_f}{A_c} = U_c R_f + 1 \tag{24}$$

Similarly, when q and A are the same and ΔT_m is different for clean and fouled exchangers, it has

$$q = U_c A_c \Delta T_{m,c} = U_f A_c \Delta T_{m,f} \quad \text{(for constant A)} \tag{25}$$

Hence,

$$\frac{\Delta T_{m,f}}{\Delta T_{m,c}} = \frac{U_c}{U_f} \tag{26}$$

Combining Eqns. (14) and (26), it gets

$$\frac{\Delta T_{m,f}}{\Delta T_{m,c}} = U_c R_f + 1 \tag{27}$$

Finally, if one assumes that heat transfer area and mean temperature differences are fixed, heat transfer rates for the same heat exchanger under fouled and clean conditions are given by $q_f = U_f A \Delta T_m$ and $q_c = U_c A \Delta T_m$, respectively. Combining these two relationships with Eqn. (14), it gets

$$\frac{q_f}{q_c} = \frac{1}{U_c R_f + 1}$$ (28)

Alternatively, Eqn. (28) can be expressed as

$$\frac{q_c}{q_f} = U_c R_f + 1$$ (29)

It is important to be noted that, the right-hand sides of Eqns. (24), (27) and (29) are the same. From this set of equations, it can be concluded that, the percentage increment in A and ΔT_m and the percentage reduction in q due to the presence of fouling are increased by increasing U_c and/or R_f. For this reason and to mitigate and attenuate the effects of fouling, the heat exchanger must be operated with low U_c. That is completely contrary to the well postulated conceptions in the field of heat exchangers design that mostly recommend using high values of U_c. As an example for this fact, in Eqn. (24), if R_f is of order $4{\times}10^{-4}$ $m^2 \cdot K/W$, and U_c of order 1000 $W/m^2 \cdot K$, then the excess surface area will be 40%, where this excess ratio will be reduced to only 20% if the U_c was 500 $W/m^2 \cdot K$ with the same R_f. Therefore, the low overall heat transfer coefficients have been used in some processes in which the fouling resistances are severe such as petrochemical industries to avoid the fouling impact on the exchanger performance.

Another important factor which related to the fouling resistance is the cleanliness factor, CF and is given as

$$CF = \frac{U_f}{U_c} = \frac{1}{U_c R_f + 1}$$ (30)

TEMA fouling resistance values [12] for water and other fluids are given in Table (4).

Fluid	Fouling Resistance (10^4 m^2.K/W)	Fluid	Fouling Resistance (10^4 m^2.K/W)
LIQUID WATER STREAMS		CHEMICAL PROCESS STREAMS	
Artificial spray pond water	1.75–3.5	Acid gas	3.5–5.3
Boiler blowdown water	3.5–5.3	Natural gas	1.75–3.5
Brackish water	3.5–5.3	Solvent vapor	1.75
Closed-cycle condensate	0.9–1.75	Stable overhead products	1.75
Closed-loop treated water	1.75	CRUDE OIL REFINERY	
Distilled water	0.9–1.75	STREAMS	
Engine jacket water	1.75	Temperature ≈ 120°C	3.5–7
River water	3.5–5.3	Temperature ≈ 120–180°C	5.25–7
Seawater	1.75–3.5	Temperature ≈ 180–230°C	7–9
Treated boiler feedwater	0.9	Temperature > 230°C	9–10.5
Treated cooling tower water	1.75–3.5		
INDUSTRIAL LIQUID		PETROLEUM STREAMS	
STREAMS		Lean oil	3.5
Ammonia (oil bearing)	5.25	Liquefied petroleum gases	1.75–3

Engine lube oil	1.75	Natural gasolene	1.75–3.5
Ethanol	3.5	Rich oil	1.75–3.5
Ethylene glycol	3.5	PROCESS LIQUID STREAMS	
Hydraulic fluid	1.75	Bottom products	1.75–3.5
Industrial organic fluids	1.75–3.5	Caustic solutions	3.5
Methanol	3.5	DEA solutions	3.5
Refrigerants	1.75	DEG solutions	3.5
Transformer oil	1.75	MEA solutions	3.5
No. 2 fuel oil	3.5	TEG solutions	3.5
No. 6 fuel oil	0.9	CRUDE AND VACUUM LIQUIDS	12.3
CRACKING AND COKING UNIT STREAMS		Atmospheric tower bottoms	3.5
Bottom slurry oils	5.3	Gasolene	5.3–12.3
Heavy coker gas oil	7–9	Heavy fuel oil	5.3–9
Heavy cycle oil	5.3–7	Heavy gas oil	3.5–5.3
Light coker gas oil	5.3–7	Kerosene	3.5–5.3
Light cycle oil	3.5–5.3	Light distillates and gas oil	3.5–5.3
Light liquid products	3.5	Naphtha	17.6
Overhead vapors	3.5	Vacuum tower bottoms	
LIGHT-END PROCESSING STREAMS		INDUSTRIAL GAS OR VAPOR STREAMS	1.75
Absorption oils	3.5–5.3	Ammonia	3.5
Alkylation trace acid streams	3.5	Carbon dioxide	17.5
Overhead gas	1.75	Coal flue gas	1.75
Overhead liquid products	1.75	Compressed air	2.6–3.5
Overhead vapors	1.75	Exhaust steam (oil bearing)	9
Reboiler streams	3–5.5	Natural gas flue gas	3.5
		Refrigerant (oil bearing)	9
		Steam (non-oil bearing)	

Table 4. TEMA fouling resistance values for water and other fluids [12]

12. Heat exchanger cleaning

In most applications, fouling is known to occur in spite of good design, effective operation, and maintenance. Hence, heat exchangers and associated equipment must be cleaned to restore the heat exchanger to efficient operation. The time between cleaning operations will depend upon the severity of the fouling problem. In some instances, cleaning can be carried out during periodical maintenance programs (say, twice yearly or annually) but in other cases frequent cleaning will be required, perhaps as frequently as monthly or quarterly. For example, locomotive radiators are air blown during their fortnightly schedules.

Cleaning Techniques [18-20]

In general, the techniques used to remove the foulants from the heat exchanger surfaces can be broadly classified into two categories: *mechanical cleaning* and *chemical cleaning*. The cleaning process may be employed while the plant is still in operation, that is named, *on-line* cleaning, but in most situations it will be necessary to shutdown the plant to clean the heat exchangers, known as *off-line* cleaning. In some instances combinations of these cleaning methods may be necessary. Each method of cleaning has advantages and disadvantages with specific equipment types and materials of construction.

Deposit Analysis

Information about the composition of fouling deposits through deposit analysis is extremely helpful to identify the source of the major foulants, to develop proper treatment, and as an aid in developing a cleaning method for a fouling control program. The sample should represent the most critical fouling area. For heat exchangers and boilers, this is the highest heat transfer area. Many analytical techniques are used to characterize deposit analysis. Typical methods include x-ray diffraction analysis, x-ray spectrometry, and optical emission spectroscopy.

Selection of Appropriate Cleaning Method

Before attempting to clean a heat exchanger, the need should be carefully examined. Consider the following factors for selecting a cleaning method:
- Degree of fouling.
- Nature of the foulant, known through deposit analysis.
- The compatibility of the heat exchanger material and system components in contact with the cleaning chemicals (in the case of chemical cleaning which associated with pumping hot corrosives through temporary connections).
- Regulations against environmental discharges.
- Accessability of the surfaces for cleaning.
- Cost factors.
- Precautions to be taken while undertaking a cleaning operation.

These precautions are listed in TEMA [12] as:
1. Individual tubes should not be steam blown because this heats the tube and may result in severe thermal strain and deformation of the tube, or loosening of the tube to tube sheet joint.
2. When mechanically cleaning a tube bundle, care should be exercised to avoid damaging the tubes. Tubes should not be hammered with a metallic tool.

Off-Line Mechanical Cleaning

Techniques using mechanical means for the removal of deposits are common throughout the industry. The various off-line mechanical cleaning methods are

1. Manual cleaning	5. Soot blowing
2. Jet cleaning	6. Thermal cleaning
3. Drilling and Roding of tubes	7. Turbining
4. Blasting	

1. Manual Cleaning

Where there is good access, as with a plate or spiral heat exchanger, or a removable tube bundle, and the deposit is soft, hand scrubbing and washing may be employed, although the labor costs are high.

2. Jet Cleaning

Jet cleaning or hydraulic cleaning with high pressure water jets can be used mostly on external surfaces where there is an easy accessability for passing the high pressure jet. Jet washing can be used to clean foulants such as: (1) airborne contaminants of air-cooled exchangers at a pressure of 2-4 bar, (2) soft deposits, mud, loose rust, and biological growths in shell and tube exchangers at a pressure of 40-120 bar, (3) heavy organic deposits, polymers, tars in condensers and other heat exchangers at a pressure of 300-400 bar, and (4) scales on the tube side and fire side of boilers, pre-heaters, and economizers at a pressure of 300-700 bar. This method consists of directing powerful water jets at fouled surfaces through special guns or lances. A variety of nozzles and tips is used to make most effective use of the hydraulic force. The effectiveness of this cleaning procedure depends on accessibility, and care is needed in application to prevent damage to the tubes and injury to the personnel. Similar to water jet cleaning, pneumatic descaling is employed on the fire side of coal-fired boiler tubes.

3.   Drilling and Roding of Tubes

Drilling is employed for tightly plugged tubes and roding for lightly plugged tubes. Drilling of tightly plugged tubes is known as bulleting. For removing deposits, good access is required, and care is again required to prevent damage to the equipment. A typical example is roding of radiator tubes plugged by solder bloom corrosion products.

4.   Blast Cleaning

Blast cleaning involves propelling suitable abrasive material at high velocity by a blast of air or water (hydroblasting) to impinge on the fouled surface. Hydroblasting is seldom used to clean tube bundles because the tubes are very thin. However, the technique is suitable to descale and clean tube-sheet faces, shells, channel covers, bonnets, and return covers inside and outside.

5.   Soot Blowing

Soot blowing is a technique employed for boiler plants, and the combustion or flue gas heat exchangers of fired equipment. The removal of particles is achieved by the use of air or steam blasts directed on the fin side. Water washing may also be used to remove carbonaceous deposits from boiler plants. **A** similar cleaning procedure is followed for air blowing of radiators on the fin side during periodical schedule attention.

6.   Thermal Cleaning

Thermal cleaning involves steam cleaning, with or without chemicals. This method is also known as hydrosteaming. It can be used to clean waxes and greases in condensers and other heat exchangers.

7.   Turbining

Turbining is a tube-side cleaning method that uses air, steam, or water to send motor-driven cutters, brushes, or knockers in order to remove deposits.

Merits and Demerits of Mechanical Cleaning

The merits of mechanical cleaning methods include simplicity and ease of operation, and capability to clean even completely blocked tubes. However, the demerits of this method may be due to the damage of the equipment, particularly tubes, it does not produce a chemically clean surface and the use of high pressure water jet or air jet may cause injury and/or accidents to personnel engaged in the cleaning operation hence the personnel are to be well protected against injuries.

Chemical Cleaning

The usual practice is to resort to chemical cleaning of heat exchangers only when other methods are not satisfactory. Chemical cleaning involves the use of chemicals to dissolve or loosen deposits. The chemical cleaning methods are mostly off-line. Chemical cleaning methods must take into account a number of factors such as:

1. Compatibility of the system components with the chemical cleaning solutions. If required, inhibitors are added to the cleaning solutions.
2. Information relating to the deposit must be known beforehand.
3. Chemical cleaning solvents must be assessed by a corrosion test before beginning cleaning operation.
4. Adequate protection of personnel employed in the cleaning of the equipment must be provided.
5. Chemical cleaning poses the real possibility of equipment damage from corrosion. Precautions may be taken to reduce the corrosion rate to acceptable levels. On-line corrosion monitoring during cleaning is necessary. Postcleaning inspection is extremely important to check for corrosion damage due to cleaning solvents and to gauge the cleaning effectiveness.
6. Disposal of the spent solution.

Chemical Cleaning Solutions

Chemical cleaning solutions include mineral acids, organic acids, alkaline bases, complexing agents, oxidizing agents, reducing agents, and organic solvents. Inhibitors and surfactant are added to reduce corrosion and to improve cleaning efficiency. Common foulants and cleaning solvents are given in Table (5) and common solvents and the compatible base materials are given in Table (6).

Foulant	Cleaning solvent
Iron oxides	Inhibited hydrofluoric acid, hydrochloric acid, monoammoniated citric acid or sulfamic acid, EDTA
Calcium and magnesium scale	Inhibited hydrochloric acid, citric acid, EDTA
Oils or light greases	Sodium hydroxide, trisodium phosphate with or without detergents, water-oil emulsion
Heavy organic deposits such as tars, asphalts, polymers	Chlorinated or aromatic solvents followed by a thorough rinsing
Coke/carbonaceous deposits	Alkaline solutions of potassium permanganate or steam air decoking

Table 5. Foulants and Common Solvents [2]

General Procedure for Chemical Cleaning

The majority of chemical cleaning procedures follow these steps:
1. Flush to remove loose debris.
2. Heating and circulation of water.
3. Injection of cleaning chemical and inhibitor if necessary in the circulating water.
4. After sufficient time, discharge cleaning solution and flush the system thoroughly.
5. Passivate the metal surfaces.
6. Flush to remove all traces of cleaning chemicals.

It is suggested that one employ qualified personnel or a qualified organization for cleaning services.

Solvent	Base metal/foulant
Hydrochloric acid	Water-side deposits on steels. Inhibited acid can be used for cleaning carbon steels, cast iron, brasses, bronzes, copper-nickels, and Monel 400. This acid is not recommended for austenitic stainless steels, Inconel600, Incoloy 800, and luminum.
Hydrofluoric acid	To remove mill scale.
Inhibited sulfuric acid	Carbon steel, austenitic stainless steels, copper-nickels, admirality brass, aluminum bronze, and Monel 400. It should not be used on aluminum.
Nitric acid	Stainless steel, titanium, and zirconium.
Sulfamic acid	To remove calcium and other carbonate scales and iron oxides. Inhibited acid can be used on carbon steel, copper, admiralty brass, cast iron, and Monel400.
Formic acid with citric acid or HCl.	To remove iron oxide deposit. Can be used on stainless steels.
Acetic acid	To remove calcium carbonate scale.
Citric acid	To clean iron oxide deposit from aluminum or titanium.
Chromic acid	To remove iron pyrite and certain carbonaceous deposits that are insoluble in HC1 on carbon steel and stainless steels. It should not be used on copper, brass, bronze, aluminum, and cast iron.

Table 6. Solvent and Compatible Base Metals [2]

Off-Line Chemical Cleaning

Major off-line chemical cleaning methods are

1. Circulation   4. Vapor-phase organic cleaning
2. Acid cleaning   5. Steam injection cleaning
3. Fill and soak cleaning

1) Circulation; This method involves the filling of the equipment with cleaning solution and circulating it by a pump. While cleaning is in progress, the concentration and temperature of the solution are monitored.

2) Acid Cleaning; Scales due to cooling water are removed by circulating a dilute hydrochloric acid solution. This is discussed in detail with the discussion of cooling-water fouling.

3) Fill and Soak Cleaning; In this method, the equipment is filled with a chemical cleaning solution and drained after a period of time. This may be repeated several times until satisfactory results are achieved. However, this method is limited to small units only.

4) Vapor-Phase Organic Cleaning; This method is used to remove deposits that are organic in nature.

5) Steam Injection Cleaning; This method involves an injection of a concentrated mix of cleaning solution and steam into a fast-moving stream. The steam atomizes the chemicals, increasing their effectiveness and ensuring good contact with metal surfaces.

Merits of Chemical Cleaning

Chemical cleaning offers the following advantages over the mechanical cleaning:

1.   Uniform cleaning and sometimes complete cleaning.
2.   Sometimes chemical cleaning is the only possible method.

3.   No need to dismantle the unit, but it must be isolated from the system.
4.   Capable of cleaning inaccessible areas.
5.   Moderate cleaning cost and longer intervals between cleaning.

Demerits of Chemical Cleaning

Chemicals used for cleaning are often hazardous to use and require elaborate disposal procedures. Noxious gases can be emitted from the cleaning solution from unexpected reactions. Chemical cleaning corrodes the base metal and the possibility of excess corrosion cannot be ruled out. Complete washing of the equipment is a must to eliminate corrosion due to residual chemicals.

On-Line Cleaning Methods

There is an obvious need for an industrial on-line cleaning procedure that can remove fouling deposits without interfering with a plant's normal operation. On-line cleaning methods can be either mechanical or chemical. On-line chemical cleaning is normally achieved by dosing with chemical additives.

On-Line Mechanical Cleaning Methods

Various on-line mechanical cleaning methods to control fouling in practice are:

1. Upstream filtration
2. Flow excursion
3. Air bumping of heat exchangers
4. Reversing flow in heat exchangers
5. Passing brushes through exchanger tubes
6. Sponge rubber balls cleaning system
7. Brush and cage system
8. Grit cleaning
9. Use of inserts
10. Self-cleaning fluidized-bed exchangers
11. Magnetic devices
12. Use of sonic technology

1) Upstream filtration (Debris Filter); Cooling water fouling can be controlled, and in some cases eliminated, by adequately filtering the intake water. Power station condensers are more vulnerable to the intake of debris and biological organisms. One solution to prevent the blockage of condenser tubes is the installation of an upstream filtration system. All particles in the cooling water larger than 10 mm are kept in the filter and rinsed away through the debris discharge.

2) Flow excursion; in this method the instantaneous flow is increased to remove the fouling deposits. This method is particularly applicable to a heat exchanger fouled badly due to the effects of low velocity either on the shell side or the tube side.

3) Air bumping; this technique involves the creation of slugs of air, thereby creating localized turbulence as slugs pass through the equipment. The technique has been applied to the liquid system on the shell side of heat exchangers. Care has to be taken to avoid the possibility of producing explosive mixtures of gases if the process fluid is volatile and flammable.

4) Reversing flow in heat exchangers; this is followed on the water side of the cooling water system by intermittent reversal of flow and intermittent air injection.

5) Passing brushes through exchanger tubes; the ram brush has been developed to clean fouled heat exchanger tubes. The unit consists of a 2.5-in long plastic dowel wrapped with 1.5-in nylon bristles. The brushes are propelled through the tube by an air and water gun and shot right through the tube. Water pressure of 0.3-4.0 bar and air pressure of 6.0-7.0 bar are sufficient to send the ram brush through standard heat exchanger tubes.

6) Sponge rubber ball cleaning system; A large number of sponge rubber balls, slightly larger in diameter than the inner diameter of the tubes and having about the same specific gravity as sea water, are passed continuously into the inlet water box (for instance, the Amertap or Taprogge system). The cooling water flow forces the balls through the tubes and the deposits on the tube walls are wiped out. The method may not be effective on longer runs once hard deposits are formed, or pitted. The balls used for normal operation should have the right surface roughness to gently clean the tubes, without scoring the tube surface. To remove heavy deposits, special abrasive balls that have a coating of carborundum are available.

7) Brush and cage system; The brush and cage or the Mesroc automatic on-load tube cleaning system used for cleaning heat exchangers consists of cleaning sets each containing two baskets and a brush, all within the tubes of the heat exchanger. The baskets are fixed to the ends of the heat exchanger tubes and every tube has its own brush. By reversing the flow direction, every brush is being moved from one end of the tube to the other, where it is retained by the basket. The brushes moving to and fro keep the inner walls clean. An actuator and a control system initiate the cleaning cycles. A major advantage of the system is that it does not require a recirculation system as for rubber ball system and an important disadvantage is the interruption of the flow in the heat exchanger and consequent disturbance of steady state conditions. Unlike the rubber ball circulation system, the brush and cage system has not been used to any extent in power plant condensers. However, it has been applied to single exchangers in process industries.

8) Grit cleaning; in this method, abrasive material, such as sand, glass, or metal spheres, is passed through the tubes. The scouring action removes the deposits from the inside of the tubes. The method has found application in cooling water systems, but it could be used in conjunction with any fouling fluids. A special grit blasting nozzle accelerates the grit and causes it to follow a sinusoidal path through the tube, dislodging the deposits. Velocities more than 3 m/s are probably required for the technique to be effective.

9) Use of inserts; Inserts, as a means of heat transfer augmentation device, located on the inside of tubes, often reduce the incidence of fouling. Wire-wound inserts and twisted tape turbulence promoters have been found to reduce fouling by vibrating axially and radially under the influence of the fluid motion. However, in some instances with the use of wire-wound inserts, the fouling rate has increased with excessive pressure drop due to accumulation of deposits on wire inserts.

10) Self-cleaning fluidized-bed exchangers; A variation of the abrasive cleaning method is to use a fluidized bed of particles to control fouling on the outside or inside of tubular exchangers, known as self-cleaning exchangers. A fluidized-bed exchanger consists of a large number of parallel vertical tubes, in which small solid particles are kept in a fluidized condition by the liquid velocities. The particles have a slightly abrasive effect on the tube walls, so that they remove the deposits. A typical example for fin-side foulant control is the fluidized-bed waste heat recovery (FBWHR) heat exchanger used to preheat combustion air for industrial furnaces. Fluidized-bed heat exchangers consists of horizontal finned heat exchanger tubes with a shallow bed of fine inert particles, which move upward with gas flow and give up heat to the finned tubes. Similarly on the tube side, the fine particles scrub the tube surfaces to minimize fouling deposit.

11) Magnetic devices; the use of magnetic fields to reduce or eliminate scale formation in pipes has been attempted for many years. It could be supposed that slightly soluble compounds such as $CaCO_3$ existing in solution as charged ions, would be affected by the application of an electric field, and that this could form the basis of a technique to alleviate fouling.

12) Use of sonic technology; the principle underlying the application of sound is that the vibration created by the energy associated with the transmission of sound will disturb and dislodge deposits on surfaces, i.e. "to shake" the deposit free. Cavitation produced by the propagation of sonic waves in the continuous phase near the deposit surface, can also assist the removal process. In general the technique has been used for problems in gas systems, particularly flue gases from combustors, i.e. very similar to the application of soot blowers.

Merits of On-Line Cleaning

The merits of on-line cleaning are that it
1. Is more convenient.
2. Does not require any plant shutdown.
3. Can save time and labor.

However, the initial cost may be very high in certain cases.

13. Fouling MITIGATION AND control

Specifying the fouling resistances or oversizing result in added heat transfer surface, the excess surface area can result in problems during startup and bring about conditions that can, in fact, encourage excess fouling due to low velocity. There are a number of techniques that can overcome or mitigate the effects of fouling in heat exchangers, and they include, [21, 22]:
1. Designing the plant or process in such a way that the condition leading to the fouling is limited or reduced.
2. Instituting an on-line mechanical cleaning system, or cleaning the equipment when the effects of the fouling can no longer be tolerated to restore its effectiveness by various offline cleaning techniques.
3. The use of chemical additives or antifoulants in the fouling stream.

Measures to Be Taken During the Design Stages

No hard and fast rules can be applied for heat exchanger design in relation to fouling, but the following points should be kept in mind during the conception and design of a heat exchanger:
1. Make the design simple.
2. Select the heat exchanger type with point 1 in mind. Heat exchangers other than shell and tube units may be better suited to fouling applications. Gasketed plate exchangers and spiral plate exchangers offer better resistance to fouling because of increased turbulence, higher shear, or other factors. Before commissioning a heat exchanger, carry out design checks and ensure that all constructional details and clearances conform to specification.
3. Prevent the possibility of corrosion and fouling during and subsequent to hydrostatic testing.
4. Startup conditions should avoid temperatures higher or velocities lower than the design values.
5. Maximize the flow velocities of process fluids to enhance the removal of the fouling deposits, provided that the fluid velocity is not high enough to cause excessive pressure

drop or flow induced vibration on the shell side. Ensure that velocities in tubes are in general above 2 m/s and about 1 m/s on the shell side. Avoid stagnant areas where the flow velocities are less than those in the bulk of the core.

6. Assume nominal fouling resistance either from past experience or from published standards and design the heat exchanger with nominal oversizing. The oversizing may be of the order of 20-40%. It is generally prudent to avoid large fouling factors, which result in larger equipment. Larger equipment generally results in lower velocities and hence may accelerate fouling.

7. To minimize fouling of finned tube or plate fin heat exchangers, use optimum fin density. Otherwise the initial benefit of increased heat transfer will be offset by fouling in the long run. This is most appropriate for industrial air coolers, radiators of automobiles, and diesel locomotives. Compact heat exchangers functioning in outdoor unit are most prone to fouling due to airborne dirt, flying objects, leaves, and fibrous objects. Other considerations are in-line layouts to provide cleaning lanes for soot blowers, and wide pitches for dirty flue gases.

8. Fouling fluid on the tube side: When the fouling fluid is on the tube side, there are some recommended measures such as (1) using larger diameter tubes (a minimum of 25 mm OD), (2) maintaining high velocity (for cooling water, a minimum velocity of 1.5 m/s for mild steel, 1.2 m/s for nonferrous tubes, and as high as 5 m/s for titanium tubes is recommended), (3) leaving sufficient margin in pressure drop (for high fouling services, leave a margin of 30-40% between the allowable and calculated pressure drop), (4) using a spare tube bundle or spare exchanger, (5) using two shells in parallel (each with 60-70% of total capacity), (6) using wire-fin tube inserts, and (7) using on-line cleaning methods.

9. Fouling fluid on the shell side: When the fouling fluid is on the shell side, use a square or rotated square tube layout, minimize dead spaces by optimum baffle design, and maintain high velocity.

10. If severe fouling is inevitable, it is frequently better to install spare units. Installed spares will permit cleaning while the other unit is in service.

11. Proper selection of cooling medium can frequently avoid problems associated with fouling. For example, air cooling in place of cooling water solves many of the corrosion and fouling problems such as scaling, biological growth, and many of the aqueous corrosion. The cleaning of bare tube or finned tube surfaces fouled by air is easier than surfaces fouled by water.

12. Particulate fouling, scaling, and trace-metal-catalyzed hydrocarbon reaction fouling can often be prevented by pretreatment of the feed streams to a heat exchanger by filtration, softening, and desalting, respectively.

13. Once the unit is onstream, operate at the design conditions of velocity and temperature.

14. Fouling in heat exchanger networks

As mentioned in the previous sections, the presence of fouling in a unique heat exchanger has dramatically serious effects on its performance. In the case of heat exchanger networks (HEN), these effects become more dramatic in which they transferred from one exchanger to another growing up and multiplying through out the network. As an example for the effect of fouling on heat exchanger networks is as the following:

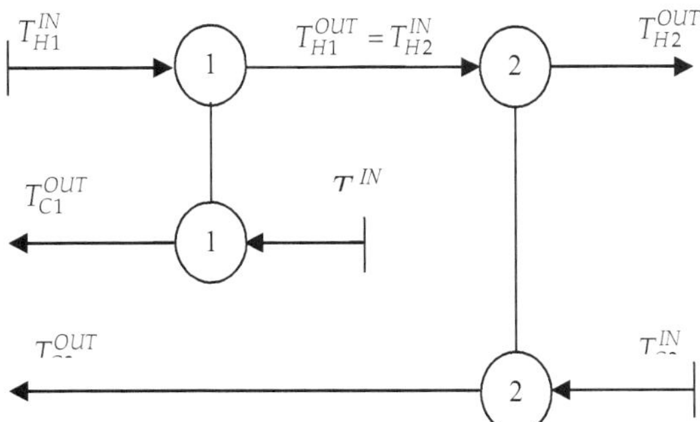

Fig. 8. section of a HEN shows the change in $(T_{H2})^{OUT}$ due to the presence of fouling in exchangers (1) and (2).

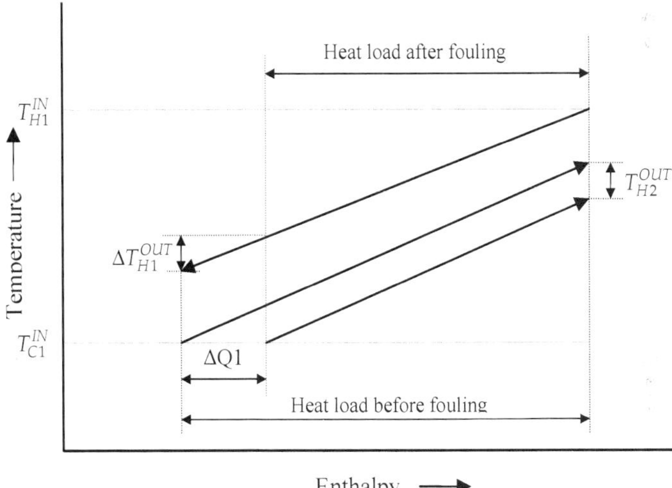

Fig. 9. Fouling effects on the outlet temperatures of cold and hot streams in a single heat exchanger (exchanger 1 in Fig. 8).

Figure (8) represents a part of a big HEN with two heat exchangers and four streams (two cold and two hot). The presence of fouling in exchanger (1) will decrease the outlet temperature of cold stream (1) by $(\Delta T_{C1})^{OUT}$, as shown in Fig. (9), and will increase the outlet temperature of hot stream (1) by $(\Delta T_{H1})^{OUT}$. As $(T_{H1})^{OUT} = (T_{H2})^{IN}$, then the inlet temperature of hot stream (2), $(T_{H2})^{IN}$, will be changed by the value of $(\Delta T_{H1})^{OUT}$ and subsequently the outlet temperature of hot stream (2) will also be changed due the fouling presents in exchanger(1). Beside to this effect, the presence of fouling in exchanger (2) will change the outlet temperature of hot stream (2). Therefore, the outlet temperature of hot stream (2) is affected by the presence of fouling in both heat exchangers, (For details, see Ref. 25).

15. Recently researches in the fouling area

Much research is in underway for fouling. This research will some day enable us to understand the parameters responsible for fouling and hence to devise means to control or eliminate fouling. Today, however, we largely rely upon experience.

Important cooperative research is being done by groups such as Heat Transfer Research, Inc. (USA), and the Heat Transfer and Fluid Flow Service (UK). May be in the near future, it is likely that a number of usable models will be available that are somewhat better than the TEMA fouling resistances.

By using the modern technologies and facilities, the recent researches in the fouling area are aimed to well understanding this phenomenon and subsequently, the fouling problem may be defined as "*unsolved problem*" not as "*unresolved problem*" as mentioned in section (1).

16. Nomenclature

Symbol	Definition	Symbol	Definition
A	heat transfer surface area, m^2	T_s	surface temperature, ^oC
A_c	heat transfer surface area for clean condition, m^2	t	time, s
		t_c	time constant $(1/\beta)$, s
A_f	heat transfer surface area for fouled condition, m^2	t_d	delay time, s
		t^*	required time to achieve asymptote, s
ΔA	percentage change A due to fouling, %	U	overall heat transfer coefficient. $W/m^2 \cdot K$
c	concentration, kg/m^3		
c_b	concentration at fluid bulk, kg/m^3	ΔU	percentage reduction in U due to fouling, %
c_s	concentration at surface, kg/m^3		
c_p	specific heat, $kJ/kg \cdot K$	U_c	overall heat transfer coefficient in clean condition. $W/m^2 \cdot K$
d_i	tube inner diameter, m		
d_o	tube outer diameter, m	U_f	overall heat transfer coefficient in fouled condition. $W/m^2 \cdot K$
E_a	activation energy, $kJ/kmol$		
f	friction factor	v	fluid flow velocity, m/s
h	heat transfer coefficient of flowing fluid, $W/m^2 \cdot K$	x_f	fouling layer thickness, m
		Greek letters:	
h_i	heat transfer coefficient on the tube inside, $W/m^2 \cdot K$	β	f actor = $1/t_c$, s^{-1}
		λ	thermal conductivity, $W/m \cdot K$
h_o	heat transfer coefficient on the tube outside, $W/m^2 \cdot K$	λ_f	thermal conductivity of fouling layer, $W/m \cdot K$
k	transfer coefficient or constant	λ_w	thermal conductivity of wall material, $W/m \cdot K$
k_1	deposition coefficient (equation 10)		
k_2	removal coefficient (equation 11)	μ	fluid viscosity, $N \cdot s/m^2$
k_m	mass transfer coefficient, m/s	ρ	density, kg/m^3
m_f	accumulated mass, kg/m^2	ρ_f	density of fouling layer, kg/m^3
$\dot{m}_f$	rate of mass accumulation, $kg/m^2 \cdot s$	τ	fluid shear stress, N/m^2
$\dot{m}_d$	deposition rate (mass unit), $kg/m^2 \cdot s$	Φ_d	deposition rate (thermal resistance unit), $m^2 \cdot K/W$
$\dot{m}_r$	removal rate (mass unit), $kg/m^2 \cdot s$		
m^*_f	asymptotic value of m_f, kg/m^2	Φ_r	removal rate (thermal resistance unit), $m^2 \cdot K/W$
Nu	Nusselt number		
	sticking probability	Ψ	deposit strength factor, N/m^2

P	pressure drop	Ω	water characterization factor
Δp	Prandtl number	**Subscripts :**	
Pr	heat transfer rate, W	b	bulk
Q	heat flux, W/m^2	c	clean condition
q	percentage reduction in q due to	d	deposition
Δq	fouling, %	f	fouled condition
	Reynolds number	i	inner
Re	fouling resistance, $m^2 \cdot K/W$	o	outer
R_f	asymptotic fouling resistance,	p	particle
R^*_f	$m^2 \cdot K/W$	r	removal
	universal gas constant, $kJ/kmol \cdot K$	s	surface
R_g	total thermal resistance to heat	w	wall
R_t	transfer		
	temperature, ^oC		
T	fluid bulk temperature, ^oC		
T_b	logarithmic mean temperature		
ΔTm	difference, ^oC		

17. References

[1] Bott, T. R., "Fouling of Heat Exchangers", Elsevier Science & Technology Books, 1995.

[2] Kuppan, T., "Heat Exchanger Design Handbook", Marcel Dekker, Inc., New York, 2000.

[3] Shah, R. K. and D. P. Sekulić, "Fundamentals of Heat Exchanger Design", John Wiley & Sons, Inc., 2003.

[4] Pritchard, A. M, ."The Economics of Fouling", in Fouling Science and Technology, (L. F. Melo et al, eds.), Kluwer Academic Publishers, Netherlands, pp 31-45, 1988.

[5] Tay S. N. and C. Yang, "Assessment of The Hydro-Ball Condenser Tube Cleaning System", Hydro-Ball Technics Sea Pte. Ltd, Singapore, 2006.

[6] Puckorius, P. R., "Contolling Deposits in Cooling Water Systems", *Mater. Protect. Perform.*, November, pp19-22, 1972.

[7] Chenoweth, J. M., "Final Report of the HTRI/TEMA Joint Committee to Review the Fouling Section of the TEMA Standards", Heat Transfer Research, Inc., Alhambra, Calif., 1988.

[8] Mukherjee, R., "Conquer heat exchanger fouling", *Hydrocarbon Processing*, January, pp 121-127, 1996.

[9] Bott, T. R., "Fouling Notebook", Institution of Chemical Engineers, Rugby, UK, 1990.

[10] Knudsen, J. G., "Apparatus and Techniques for Measurement of Fouling of Heat Transfer Surface", in *Fouling of Heat Transfer Equipment* (E. F. C. Somerscales and J. G. Knudson, eds.), Proceedings of the International Conference on the Fouling of Heat Transfer Equipment, Hemisphere, Washington, D.C., pp. 57-81, 1981.

[11] Zelver, N., Roe, F. L., and Characklis, W. G., "Monitoring of Fouling Deposits in Heat Transfer Tubes: Case Studies", in *Industrial Heat Exchangers Conference Proceedings* (A. J. Hayes, W. W. Liang, S. L. Richlen, and E. S. Tabb, eds.), American Society for Metals, Metals Park, Ohio, pp 201-208, 1985.

[12] TEMA, "Standard of the Tubular Exchanger Manufacturers Association", 8th ed., Tubular Exchanger Manufacturers Asscociation, New York, 1999.

[13] Zubair, S. M., and R. K. Shah, "Fouling in Plate-and-Frame Heat Exchangers and Cleaning Strategies", in Compact Heat Exchangers and Enhancement Technology for the Process Industries, (R. K. Shah, A. Deakin, H. Honda and T. M. Rudy, eds.), Begell House, New York, pp. 553–565, 2001.

[14] Kern, D.Q. and Seaton, R.E., "A Theoretical Analysis of Thermal Surface Fouling", Brit. Chem. Eng. 14, No. 5, 258, 1959.

[15] Watkinson, A.p., "Process Heat Transfer: Some Practical Problems", Can. J. Chem. Eng., 58, pp 553-559, 1980.

[16] Taborek, J., Aoki, T., Ritter, R.B., Palen, J.W. and Knudsen, J.G., "Predictive Methods for Fouling Behavior". Chem. Eng. Prog. 68, No. 7, 69 – 78, 1972.

[17] Knudsen, J.G. "Fouling of Heat Transfer Surface: An Overview", in Power condenser – Heat transfer technology, (P.J. Marto and R.H . Nunn, eds.), pp 375-424, Hemisphere, London, 1981.

[18] Stegelman, A. F., and Renfftlen, R., "On-Line Mechanical Cleaning of Heat Exchangers", *Hydrocarbon Processing*, 95-97, 1983.

[19] Crittenden, B.D., Kolaczkowski, S.T. and Takemoto, T., "Use of in Tube Inserts to Reduce Fouling from Crude Oils". AIChE Symp. Series, 295, 89, 300- 307, Atlanta, 1993.

[20] Hughes, D.T., Bott, T.R. and Pratt, D.C.F., "Plastic Tube Heat Transfer Surfaces in Falling Film Evaporators". Proc. Second UK Nat. Conf. on Heat Transfer, Vol. 2, 1101 – 1113, 1988.

[21] Epstein, N., "On Minimizing Fouling of Heat Transfer Surfaces", in *Low Reynolds Number Flow Heat Exchangers* (S. Kakac, R. K. Shah, and A. E. Bergles, eds.), Hemisphere, Washington, D.C., pp. 973-979, 1982.

[22] Kim, W.T., Y. I. Cho, and C. Bai, "Effect of Electronic Anti-Fouling Treatment on Fouling Mitigation with Circulating Cooling Tower Water", Int. Comm. Heat and Mass Transfer, vol. 28, No.5, pp. 671-680, 2001.

[23] Knudsen, J. G., "Fouling in Heat Exchangers", in Heat Exchanger Design Handbook, G. F.Hewitt, ed., Begell House, New York, Sec. 3.17,1998.

[24] Epstien, N., "Particulate Fouling of Heat Transfer Surfaces: Mechanisms and Models", in Fouling Science and Technology, (L.F. Melo et al, eds.), 143-164, Kluwer Academic publishers, Netherlands, 1988.

[25] Fryer, P. J., "Modelling Heat Exchanger Fouling", Ph.D Thesis, University of Cambridge, UK., 1986.

[26] Awad, M. M., "Effect of Flow Oscillation on Surface Fouling", Mansoura Engineering Journal (MEJ), Vol.28, No2, pp M47-M56, June 2003.

[27] Awad, M. M., I. Fathy and H. E. Gad "Effect of Surface Temperature on the Fouling of Heat Transfer Surfaces", 11th International Water Technology Conference, Sharm El Sheikh, Egypt, March, 2007.

[28] Awad, M. M., S. M. Abd El-Samad, H.E. Gad and F.I. Asfour "Effect of Flow Velocity on the Surface Fouling", Mansoura Engineering Journal (MEJ), Vol.32, No1, pp M27-M37, March, 2007.

Part 4

Heat Transfer Calculations

21

Spatio-Temporal Measurement of Convective Heat Transfer Using Infrared Thermography

Hajime Nakamura
National Defense Academy
Japan

1. Introduction

Convective heat transfer is, by nature, generally nonuniform and unsteady, a fact reflected by three-dimensional flow near a wall. However, most experimental studies concerning convective heat transfer have been performed in a time-averaged manner or using one-point measurements. This frequently results in poor understanding of the heat transfer mechanisms. Measurement techniques for the temporal and spatial characteristics of heat transfer have been developed using liquid crystals (Iritani, et al., 1983; among others) or using infrared thermography (Hetsroni & Rozenblit, 1994; among others), by employing a thin test surface having a low heat capacity. However, the major problem with these measurements is attenuation and phase delay of the temperature fluctuation due to thermal inertia of the test surface. This becomes serious for higher fluctuating frequencies, for which the fluctuation amplitude weakens and ultimately becomes indistinguishable from noise. In addition, lateral conduction through the test surface attenuates the amplitude of the spatial temperature distribution. This becomes serious for smaller wavelength (higher wavenumber). These attenuations are considerably large, especially for the heat transfer to gaseous fluid such as air for which the heat transfer coefficient is low.

The recent improvement of infrared thermograph with respect to temporal, spatial and temperature resolutions enable us to investigate detailed behavior of the heat transfer caused by flow turbulence. Analytical study indicated that the spatio-temporal distribution of the heat transfer to air caused by flow turbulence can be observed by employing modern infrared thermograph (for example, NETD less than 0.025 K and frame rate more than several hundred Hz) that records temperature fluctuations on a heated thin foil of sufficiently low heat capacity. In the former part of this chapter (sections 2 – 4), an analytical investigation was described on the frequency response and the spatial resolution of a thin foil for the heat transfer measurements. In order to derive general relationships, non-dimensional variables of fluctuating frequency and spatial wavenumber were introduced to formulate the amplitude of temperature fluctuation and/or distribution on the test surface. Based on these relationships, the upper limits on the detectable fluctuating frequency f_{max} and spatial wavenumber k_{max} were formulated using governing parameters of the measurement system, i.e., thermophysical properties of the thin foil and noise-equivalent temperature difference (NETD) of infrared thermograph for a blackbody.

In the latter part of this chapter, this technique was applied to measure the spatio-temporal distribution of turbulent heat transfer to air by employing a high-speed infrared thermograph and a heated thin-foil. At first, as a well-investigated case, the heat transfer on the wall of a turbulent boundary layer was measured in order to verify the applicability of this technique (section 5). Also, this technique was applied to explore the spatio-temporal characteristics of the heat transfer behind a backward-facing step, which represents the separated and reattaching flow (section 6).

Nomenclature

b, b_c	:	spatial wavelength, cut-off wavelength [m]
b_{min}	:	lower limit of spatial wavelength detectable [m]
c	:	specific heat [J/kg K]
f, f_c	:	fluctuating frequency, cut-off frequency [Hz]
f_{max}	:	upper limit of fluctuating frequency detectable [Hz]
H	:	step height [m]
h	:	heat transfer coefficient [W/m²K]
h_t	:	total heat transfer coefficient including conduction and radiation [W/m²K]
k	:	wavenumber $= 2\pi/b$ [m⁻¹]
l_τ	:	wall-friction length $= \nu/u_\tau$
l_z	:	mean spanwise wavelength [m]
Nu	:	Nusselt number; $Nu_H = \bar{h}H / \lambda$
$\dot{q}$	:	heat flux [W/m²]
Re	:	Reynolds number; $Re_H = u_0H/\nu$, $Re_\theta = u_0\delta_\theta/\nu$
St	:	Strouhal number $= fH/u_0$
T	:	temperature [K]
T_0, T_w	:	freestream temperature, wall temperature [K]
ΔT_{IR}	:	noise-equivalent temperature difference of infrared thermography for a non-blackbody [K]
ΔT_{IR0}	:	noise-equivalent temperature difference of infrared thermography for a blackbody [K]
t	:	time [s]
u_0, u_τ	:	freestream velocity, wall-friction velocity [m/s]
x, y, z	:	tangential, normal and spanwise coordinates
α	:	thermal diffusivity $= \lambda/c\rho$ [m²/s]
β	:	space resolution [m]
δ, δ_a	:	thickness, air-layer thickness [m]
δ_θ	:	momentum thickness [m]
ε_t	:	total emissivity
ε_{IR}	:	spectral emissivity for infrared thermograph
κ	:	$= \sqrt{\omega/2\alpha}$ [m⁻¹]
λ	:	thermal conductivity [W/m·K]
ν	:	kinematic viscosity [m²/s]
ρ	:	density [kg/m³]
τ	:	time constant [s]
ω	:	angular frequency $= 2\pi f$ [rad/s]

Subscripts

a, c, i	:	air layer, high-conductivity plate, insulating layer

cd, cv	:	conduction, convection
rd, rdi	:	radiation to outside, radiation to inside
f, s	:	frequency response, space resolution

Other Symbols

$(\bar{\ })$	:	mean value
$\Delta(\ ), (\ )_{rms}$	:	spatial or temporal amplitude, root-mean-square value

2. Analytical solutions without heat losses

2.1 Governing equations

Figure 1 shows a schematic model for heat transfer measurement. The test surface, which is exposed to air flow, is fabricated from a thin metallic foil (thickness δ, specific heat c, density ρ, thermal conductivity λ, and total emissivity ε_t). An instantaneous temperature distribution and its fluctuation on the test surface can be measured using infrared thermography through the air-stream, which is transparent for infrared radiation. Inside the foil, there is a high-conductivity plate (total emissivity ε_{tc}) to impose a thermal boundary condition of a steady and uniform temperature. Between the foil and the high-conductivity plate is some material of low conductivity and low heat capacity, such as still air, forming an insulating layer (thickness δ_i, specific heat c_i, density ρ_i, thermal conductivity λ_i).

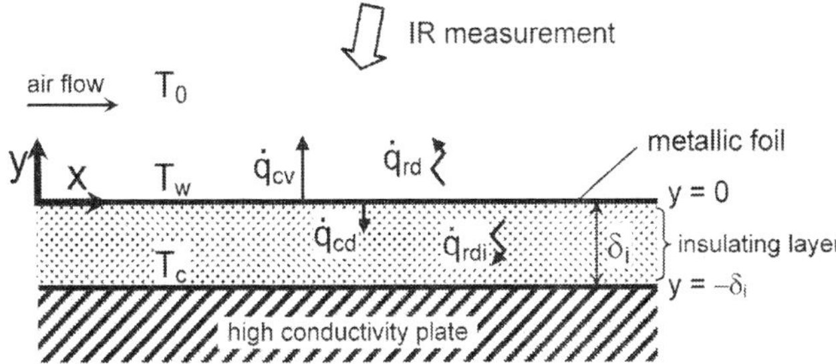

Fig. 1. Schematic model for measurement of heat transfer to air

The tangential and normal directions with respect to the test surface correspond to x and y coordinates, respectively. Assuming that the temperature is uniform along its thickness, the heat balance on the thin foil can be expressed as:

$$c\rho\delta\frac{\partial T_w}{\partial t} = \lambda\delta\left(\frac{\partial^2 T_w}{\partial x^2} + \frac{\partial^2 T_w}{\partial z^2}\right) + \dot{q} , \ (y = 0). \tag{1}$$

Here, T_w is local and instantaneous temperature of the thin foil. The heat flux, $\dot{q}$, is given by:

$$\dot{q} = \dot{q}_{in} - \dot{q}_{cv} - \dot{q}_{cd} - \dot{q}_{rd} - \dot{q}_{rdi} \tag{2}$$

where $\dot{q}_{in}$ is the input heat-flux to the thin foil due to Joule heating, $\dot{q}_{cv}$ and $\dot{q}_{cd}$ are heat fluxes from the thin foil due to convection and conduction, respectively, and $\dot{q}_{rd}$ is radiation heat flux to outside the foil. If the insulating layer is transparent for infrared radiation, as is air, radiation heat flux $\dot{q}_{rdi}$ occurs to the inside. The above heat fluxes are expressed as follows:

$$\dot{q}_{cv} = h\,(T_w - T_0) \tag{3}$$

$$\dot{q}_{cd} = \lambda_i \left(\frac{\partial T}{\partial y}\right)_{y=0-} \tag{4}$$

$$\dot{q}_{rd} = \varepsilon_t\,\sigma(T_w^{\,4} - T_0^{\,4}) \tag{5}$$

$$\dot{q}_{rdi} = \frac{\sigma(T_w^{\,4} - T_c^{\,4})}{1/\varepsilon_t + 1/\varepsilon_{tc} - 1} \tag{6}$$

Here, h is the heat transfer coefficient due to convection to the stream outside, T_0 is the freestream temperature, T_c is surface temperature of the high-conductivity plate, and σ is the Stefan-Boltzmann constant.

Heat conduction in the insulating layer is expressed as:

$$c_i \rho_i \frac{\partial T}{\partial t} = \lambda_i \left(\frac{\partial^2 T}{\partial x^2} + \frac{\partial^2 T}{\partial y^2} + \frac{\partial^2 T}{\partial z^2}\right),\ (-\delta_i < y < 0). \tag{7}$$

The frequency response and the spatial resolution of T_w for the heat transfer to the free stream can be calculated using Eq. (1) – (7) for arbitrary changes in the heat transfer coefficient in time and space.

2.2 Time constant

Assuming that the temperature is uniform in the x–z plane, and that heat conduction $\dot{q}_{cd}$ and radiation $\dot{q}_{rd}$ and $\dot{q}_{rdi}$ are sufficiently small, Eq. (1) – (3) yield the following differential equation:

$$c\rho\delta \frac{\partial T_w}{\partial t} = \dot{q}_{in} - h(T_w - T_0)\ . \tag{8}$$

Solving Eq. (8) yields the time constant τ, which is expressed as:

$$\tau = \frac{c\rho\delta}{h}\ . \tag{9}$$

2.3 Spatial resolution

Assuming that the temperature on the test surface is steady and has a sinusoidal distribution in the x direction (uniform in the z direction), then:

$$T_w = \overline{T_w} + \Delta T_w \sin\left(\frac{2\pi}{b}x\right)\ . \tag{10}$$

where $\overline{T_w}$ and ΔT_w are the mean and spatial amplitude of the temperature of the thin foil, respectively, and b is the wavelength. If $\dot{q}_{cd}$, $\dot{q}_{rd}$ and $\dot{q}_{rdi}$ are sufficiently small, Eq. (1) – (3) yield the following equation:

$$h\,(T_w - T_0) = \dot{q}_{cv} = \dot{q}_{in} + \lambda\delta\,\frac{d^2 T_w}{dx^2} \tag{11}$$

If the thin foil is thermally insulated, Eq. (11) reduces to:

$$h\,(T_{w0} - T_0) = \dot{q}_{cv} = \dot{q}_{in}\,.$$

(12)

Then, the temperature of the insulated surface T_{w0} is calculated from Eq. (10) – (12) as

$$T_{w0} = \overline{T_w} + \left\{ \frac{\lambda\delta}{h}\left(\frac{2\pi}{b}\right)^2 + 1 \right\} \Delta T_w \sin\left(\frac{2\pi}{b}x\right).$$

(13)

A comparison between Eq. (10) and (13) yields the attenuation rate of the spatial amplitude due to lateral conduction through the thin foil:

$$\xi = \frac{1}{\dfrac{\lambda\delta}{h}\left(\dfrac{2\pi}{b}\right)^2 + 1}\,.$$

(14)

A spatial resolution β can be defined as the wavelength b at which the attenuation rate is $1/2$:

$$\beta = b_{(\xi=1/2)} = 2\pi\sqrt{\frac{\lambda\delta}{h}}\,.$$

(15)

Incidentally, if the test surface has a two-dimensional temperature distribution such as:

$$T_w = \overline{T_w} + \Delta T_w \sin\left(\frac{2\pi}{b}x\right)\sin\left(\frac{2\pi}{b}z\right).$$

(16)

then the spatial resolution can be calculated as:

$$\beta_{2D} = 2\pi\sqrt{\frac{2\lambda\delta}{h}}\,.$$

(17)

This indicates that the spatial resolution for the 2D temperature distribution deteriorates by a factor of $\sqrt{2}$.

3. General relations considering heat losses

In this section, general relationship was derived concerning the temporal and spatial attenuations of temperature on the thin foil considering the heat losses. Since the full derivation is rather complicated (Nakamura, 2009), a brief description was made below.

3.1 Temporal attenuation

Assuming that the temperature on the thin foil is uniform and fluctuates sinusoidally in time:

$$T_w = \overline{T_w} + \Delta T_w \sin(\omega t)\,.$$

(18)

Figure 2 shows the analytical solutions of the instantaneous temperature distribution in the insulating layer ($0 \leq |y| \leq \delta_i$) at $\omega t = \pi/2$, at which the temperature of the thin foil ($y = 0$) is maximum. The shape of the distribution depends only on $\kappa_i \delta_i$, where $\kappa_i = \sqrt{\omega/(2\alpha_i)}$, α_i is thermal diffusivity of the insulating layer. For lower frequencies ($\kappa_i \delta_i < 1$), the distribution can be assumed linear, while for higher frequencies ($\kappa_i \delta_i \gg 1$), the temperature fluctuates only in the vicinity of the foil ($|y|/\delta_i \leq 1/\kappa_i \delta_i$).

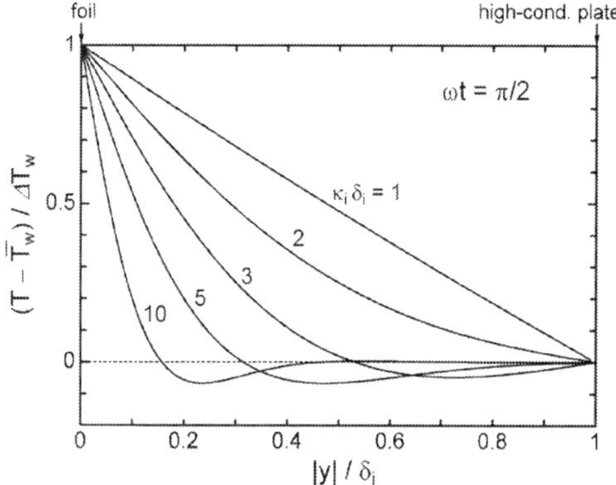

Fig. 2. Instantaneous temperature distribution in the insulating layer at $\omega t = \pi/2$ and $\overline{T}_w = \overline{T}_c$

Introduce the effective thickness of the insulating layer, $(\delta_i^*)_f$, the temperature of which fluctuates with the thin foil:

$$(\delta_i^*)_f \approx 0.5\,\delta_i \ , \ (\kappa_i \delta_i < 1) \tag{19}$$

$$(\delta_i^*)_f \approx 0.5/\kappa_i \ , \ (\kappa_i \delta_i \gg 1). \tag{20}$$

The heat capacity of this region works as an additional heat capacity that deteriorates the frequency response. Thus, the effective time constant considering the heat losses can be defined as:

$$\tau^* \approx \frac{c\rho\delta + c_i\rho_i(\delta_i^*)_f}{h_t} \ , \ h_t = \frac{\dot{q}_{in}}{T_w - T_0}. \tag{21}$$

Here, h_t is total heat transfer coefficient from the thin foil, including the effects of conduction and radiation. Then, the cut-off frequency is defined as follows:

$$f_c^* = \frac{1}{2\pi\tau^*}. \tag{22}$$

We introduce the following non-dimensional frequency and non-dimensional amplitude of the temperature fluctuation:

$$\tilde{f} = f/\overline{f_c^*} \tag{23}$$

$$(\Delta\tilde{T}_w)_f = \frac{(\Delta T_w)_f}{\overline{T_w - T_0}}\frac{\overline{h_t}}{\Delta h}. \tag{24}$$

Here, $(\Delta \tilde{T}_w)_f$ includes the factor $\overline{h}_t / \Delta h$ to extend the value of $(\Delta \tilde{T}_w)_f$ to unity at the lower frequency in the absence of conductive or radiative heat losses (see Fig. 3).

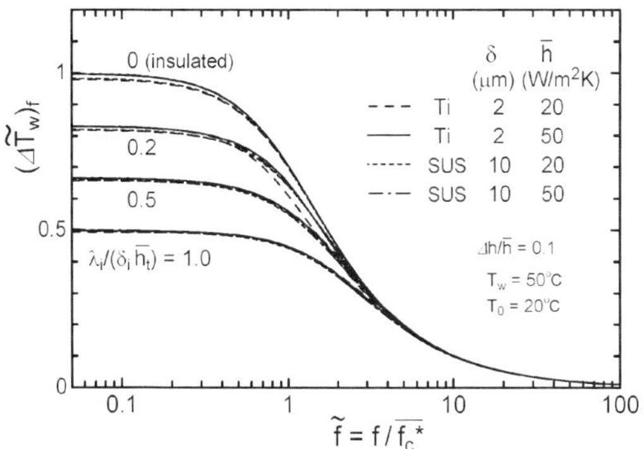

Fig. 3. Relation between non-dimensional frequency $\tilde{f}$ and non-dimensional fluctuating amplitude $(\Delta \tilde{T}_w)_f$

Next, we attempt to obtain the relation between $\tilde{f}$ and $\left(\Delta \tilde{T}_w\right)_f$. The fluctuating amplitude of the surface temperature, $(\Delta T_w)_f$, can be determined by solving the heat conduction equations of Eq. (1) and (7) by the finite difference method assuming a uniform temperature in the x–z plane. Figure 3 plots the relation of $(\Delta \tilde{T}_w)_f$ versus $\tilde{f}$ for practical conditions (see sections 5 and 6). The thin foil is a titanium foil 2 μm thick ($c\rho\delta$ = 4.7 J/m^{2}K, $\lambda\delta$ = 32 μW/K, ε_{IR} = 0.2) or a stainless-steel foil 10 μm thick ($c\rho\delta$ = 40 J/m^{2}K, $\lambda\delta$ = 160 μW/K, ε_{IR} = 0.15), the insulating layer is a still air layer without convection, and the mean heat transfer coefficient is $\overline{h}$ = 20 – 50 W/m^{2}K. A parameter of $\lambda_i/(\delta_i \overline{h}_t)$, which represents the the heat conduction loss from the foil to the high-conductivity plate through the insulating layer, is varied from 0 to 1.

For the lower frequency of $\tilde{f} < 0.1$, $(\Delta \tilde{T}_w)_f$ approaches a constant value:

$$(\Delta \tilde{T}_w)_f \approx \frac{1}{1 + \lambda_i / (\delta_i \overline{h}_t)} \ , \ (\tilde{f} < 0.1). \tag{25}$$

In this case, the fluctuating amplitude decreases with increasing $\lambda_i / (\delta_i \overline{h}_t)$. With increasing $\tilde{f}$, the value of $(\Delta \tilde{T}_w)_f$ decreases due to the thermal inertia. For higher frequency values of $\tilde{f} > 4$, $(\Delta \tilde{T}_w)_f$ depends only on $\tilde{f}$. Consequently, it simplifies to a single relation:

$$(\Delta \tilde{T}_w)_f \approx \frac{1}{\tilde{f}} \ , \ (\tilde{f} > 4). \tag{26}$$

3.2 Spatial attenuation

Assuming that the temperature on the foil is steady and has a sinusoidal temperature distribution in the x direction (1D distribution):

$$T_w = \overline{T_w} + \Delta T_w \sin(kx) \ , \ k = 2\pi/b. \tag{27}$$

Here, k is wavenumber of the spatial distribution.

Figure 4 shows the analytical solutions of the vertical temperature distribution in the insulating layer ($0 \leq |y| \leq \delta_i$) at $kx = \pi/2$, at which the temperature of the thin foil ($y = 0$) is maximum. The shape of the distribution depends only on $k\delta_i$. For the lower wavenumber ($k\delta_i < 1$), the distribution can be assumed linear, while for the higher wavenumber ($k\delta_i >> 1$), the distribution approaches an exponential function.

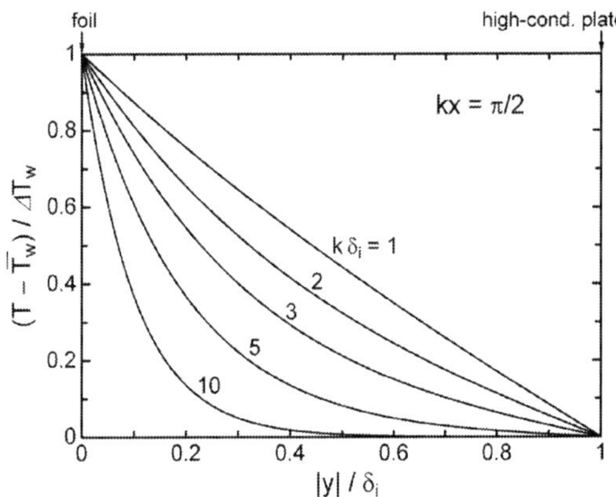

Fig. 4. Temperature distribution in the insulating layer at $kx = \pi/2$ and $\overline{T}_w = \overline{T}_c$

Now, we introduce an effective thickness of the insulating layer, $(\delta_i^*)_s$, the temperature of which is affected by the temperature distribution on the foil:

$$(\delta_i^*)_s \approx \delta_i \, , \, (k\delta_i < 1) \tag{28}$$

$$(\delta_i^*)_s \approx 1 / k \, , \, (k\delta_i >> 1) \tag{29}$$

The heat conduction of this region functions as an additional heat spreading parameter that reduces the spatial resolution. Thus, the effective spatial resolution can be defined as:

$$\beta^* \approx 2\pi \sqrt{\frac{\lambda\delta + \lambda_i(\delta_i^*)_s}{h_t}} \, . \tag{30}$$

Introduce a non-dimensional wavenumber and non-dimensional amplitude of the spatial temperature distribution:

$$\tilde{k} = \frac{k}{(2\pi / \overline{\beta^*})} = \frac{k\overline{\beta^*}}{2\pi} \tag{31}$$

$$(\Delta\tilde{T}_w)_s = \frac{(\Delta T_w)_s}{\overline{T}_w - T_0} \frac{\overline{h_t}}{\Delta h} \, . \tag{32}$$

Here, $2\pi / \overline{\beta^*}$ corresponds to the cut-off wavenumber.

Next, we attempt to obtain a relation between $\tilde{k}$ and $(\Delta\tilde{T}_w)_s$. The spatial amplitude of the surface temperature, $(\Delta T_w)_s$, can be determined by solving a steady-state solution of the heat conduction equations of Eq. (1) and (7) by the finite difference method. Figure 5 plots the relation of $(\Delta\tilde{T}_w)_s$ versus $\tilde{k}$ for practical conditions. For the lower wavenumber of $\tilde{k} < 0.1$, $(\Delta\tilde{T}_w)_s$ approaches a constant value of

$$(\Delta\tilde{T}_w)_s \approx \frac{1}{1 + \lambda_i / (\delta_i \overline{h_t})} \ , \ (\tilde{k} < 0.1). \tag{33}$$

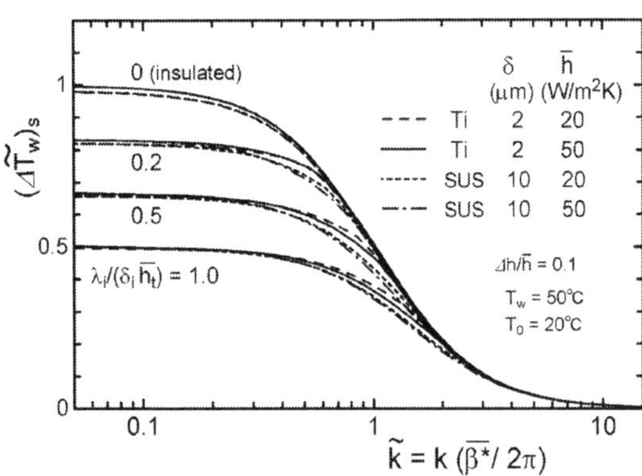

Fig. 5. Relation between non-dimensional wavenumber $\tilde{k}$ and non-dimensional spatial amplitude $(\Delta\tilde{T}_w)_f$

In this case, the spatial amplitude decreases with increasing $\lambda_i / (\delta_i \overline{h_t})$, which represents the vertical conduction. With increasing $\tilde{k}$, the value of $(\Delta\tilde{T}_w)_s$ decreases due to the lateral conduction. For the higher wavenumber of $\tilde{k} > 4$, $(\Delta\tilde{T}_w)_s$ depends only on $\tilde{k}$. It, therefore, corresponds to a single relation.

$$(\Delta\tilde{T}_w)_s \approx \frac{1}{(\tilde{k})^2} \ , \ (\tilde{k} > 4). \tag{34}$$

4. Detectable limits for infrared thermography

4.1 Temperature resolution

The present measurement is feasible if the amplitude of the temperature fluctuation, $(\Delta T_w)_f$, and the amplitude of the spatial temperature distribution, $(\Delta T_w)_s$, is greater than the temperature resolution of infrared measurement, ΔT_{IR}. In general, the temperature resolution of a product is specified as a value of noise-equivalent temperature difference (NETD) for a blackbody, ΔT_{IR0}.

The spectral emissive power detected by infrared thermograph, E_{IR}, can be assumed as follows:

$$E_{IR}(T) = \varepsilon_{IR} C T^n . \tag{35}$$

where ε_{IR} is spectral emissivity for infrared thermograph, and C and n are constants which depend on wavelength of infrared radiation and so forth. For a blackbody, the noise amplitude of the emissive power can be expressed as follows:

$$\Delta E_{IR0}(T) = C(T + \Delta T_{IR0})^n - CT^n .\tag{36}$$

Similarly, for a non-blackbody, the noise amplitude can be expressed as follows:

$$\Delta E_{IR}(T) = \varepsilon_{IR} C(T + \Delta T_{IR})^n - \varepsilon_{IR} CT^n .\tag{37}$$

Since the noise intensity is independent of spectral emissivity ε_{IR}, the values of $\Delta E_{IR0}(T)$ and $\Delta E_{IR}(T)$ are identical. This yields the following relation using the binomial theorem with the assumption of $T >> \Delta T_{IR0}$ and $T >> \Delta T_{IR}$.

$$\Delta T_{IR} = \Delta T_{IR0} / \varepsilon_{IR} .\tag{38}$$

Namely, the temperature resolution (NETD) for a non-blackbody is inversely proportional to ε_{IR}.

4.2 Upper limit of fluctuating frequency

Using Eq. (20) – (24) and (26), the fluctuating amplitude, $(\Delta T_w)_f$, is generally expressed as follows for higher fluctuating frequency:

$$(\Delta T_w)_f \approx \frac{(\overline{T_w} - T_0)\Delta h}{2\pi c\rho\delta f + \sqrt{\pi c_i \rho_i \lambda_i} f^{0.5}} , \quad (\tilde{f} > 4 \text{ and } k_i\delta_i >> 1)\tag{39}$$

The fluctuation is detectable using infrared thermography for $(\Delta T_w)_f > \Delta T_{IR}$. This yields the following equation from Eq. (38) and (39).

$$f < \left(\frac{-B + \sqrt{B^2 - 4AC}}{2A}\right)^2 , \quad A = 2\pi c\rho\delta , \quad B = \sqrt{\pi c_i \rho_i \lambda_i} , \quad C = -\varepsilon_{IR}\Delta h(\overline{T_w} - T_0) / \Delta T_{IR0}\tag{40}$$

The maximum frequency of Eq. (40) at $(\Delta T_w)_f = \Delta T_{IR}$ corresponds to the upper limit of the detectable fluctuating frequency, f_{max}. The value of f_{max} is uniquely determined as a function of $\Delta h(\overline{T_w} - T_0) / \Delta T_{IR0}$ if the thermophysical properties of the thin foil and the insulating layer are specified.

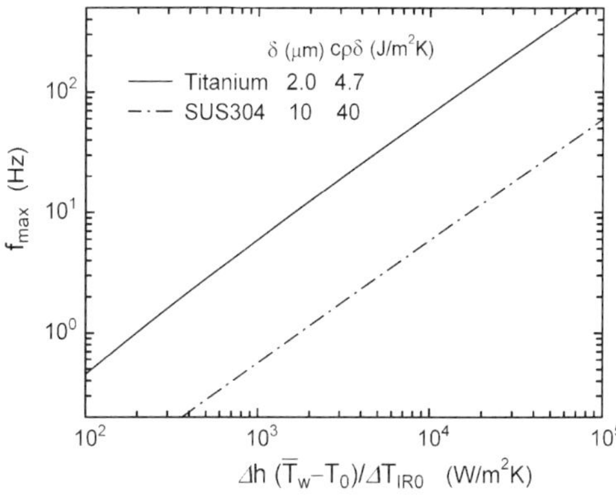

Fig. 6. Upper limit of the fluctuating frequency detectable using infrared measurements

Figure 6 shows the relation of f_{max} for practical metallic foils for heat transfer measurement to air, namely, a titanium foil of 2 µm thick ($c\rho\delta$ = 4.7 J/m²K, ε_{IR} = 0,2) and a stainless-steel foil of 10 µm thick ($c\rho\delta$ = 40 J/m²K, ε_{IR} = 0.15). The insulating layer is assumed to be a still air layer (c_i = 1007 J/kg·K, ρ_i = 1.18 kg/m³, λ_i = 0.0265 W/m·K), which has low heat capacity and thermal conductivity.

For example, a practical condition likely to appear in flow of low-velocity turbulent air (section 6; $\Delta h(\overline{T_w} - T_0) / \Delta T_{IR0}$ = 22000 W/m²K; Δh = 20 W/m²K, $\overline{T_w} - T_0$ = 20 K, and ΔT_{IR0} = 0.018 K), gives the values f_{max} = 150 Hz for the 2 µm thick titanium foil. Therefore, the unsteady heat transfer caused by flow turbulence can be detected using this measurement technique, if the flow velocity is relatively low (see section 6).

The value of f_{max} increases with decreasing $c\rho\delta$ and ΔT_{IR0}, and with increasing ε_{IR}, Δh, and $\overline{T_w} - T_0$. The improvements of both the infrared thermograph (decreasing ΔT_{IR0} with increasing frame rate) and the thin foil (decreasing $c\rho\delta$ and/or increasing ε_{IR}) will improve the measurement.

4.3 Upper limit of spatial wavenumber

Using Eq. (29) – (32) and (34), the spatial amplitude, $(\Delta T_w)_s$, is generally expressed as follows for higher wavenumber:

$$(\Delta T_w)_s \approx \frac{(\overline{T_w} - T_0)\Delta h}{\lambda\delta k^2 + \lambda_i k} \ , \ (\tilde{k} > 4 \text{ and } k\delta_i \gg 1). \tag{41}$$

The spatial distribution is detectable using infrared thermography for $(\Delta T_w)_s > \Delta T_{IR}$. This yields the following equation using Eq. (38) and (41).

$$k < \frac{-\lambda_i + \sqrt{\lambda_i^2 + 4\lambda\delta\,\varepsilon_{IR}\{\Delta h(\overline{T_w} - T_0)/\Delta T_{IR0}\}}}{2\lambda\delta} \tag{42}$$

The maximum wavenumber of Eq. (42) at $(\Delta T_w)_s = \Delta T_{IR}$ corresponds to the upper limit of the detectable spatial wavenumber, k_{max}. If thermophysical properties of the thin foil and the

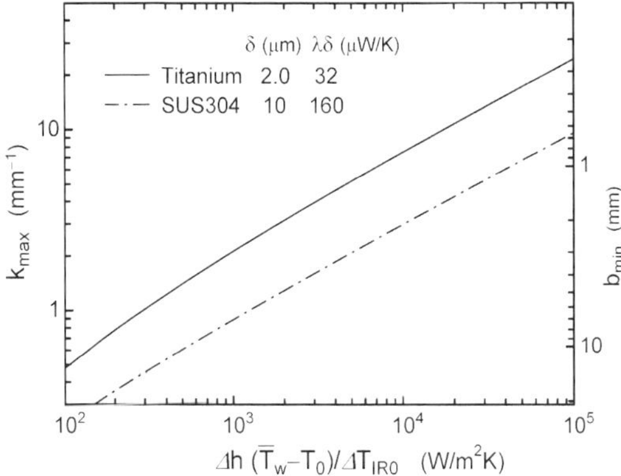

Fig. 7. Upper limit of the spatial wavenumber detectable using infrared measurements

insulating layer are specified, the value of k_{max} is uniquely determined as a function of $\Delta h(\overline{T_w} - T_0) / \Delta T_{IR0}$, as well as f_{max}.

Figure 7 shows the relation for k_{max} for the titanium foil of 2 μm thickness ($\lambda\delta$ = 32 μW/K, ε_{IR} = 0.2), and the stainless-steel foil of 10 μm thickness ($\lambda\delta$ = 160 μW/K, ε_{IR} = 0.15). The insulating layer is assumed to be a still-air layer (λ_i = 0.0265 W/m·K). For example, at a practical condition appeared in section 6, $\Delta h(\overline{T_w} - T_0) / \Delta T_{IR0}$ = 22000 W/m²K, the value of k_{max} (b_{min}) is 11 mm⁻¹ (0.6 mm) for the 2 μm thick titanium foil. Therefore, the spatial structure of the heat transfer coefficient caused by flow turbulence can be detected using this measurement technique. (In general, the space resolution is dominated by rather a pixel resolution of infrared thermograph than k_{max} (b_{min}), see Nakamura, 2007b).

The value of k_{max} increases with decreasing $\lambda\delta$ and ΔT_{IR0}, and with increasing ε_{IR}, Δh, and $\overline{T_w} - T_0$. The improvements of both the infrared thermograph (decreasing ΔT_{IR0} with increasing pixel resolution) and the thin foil (decreasing $\lambda\delta$ and/or increasing ε_{IR}) will improve the measurement.

5. Experimental demonstration (turbulent boundary layer)

In this section, the applicability of this technique was verified by measuring the spatio-temporal distribution of the heat transfer on the wall of a turbulent boundary layer, as a well-investigated case.

5.1 Experimental setup

The measurements were performed using a wind tunnel of 400 mm (H) × 150 mm (W) × 1070 mm (L), as shown in Fig. 8. A turbulent boundary layer was formed on the both-side faces of a flat plate set at the mid-height of the wind tunnel. The freestream velocity u_0 ranged from 2 to 6 m/s, resulting in the Reynolds number based on the momentum thickness was Re_θ = 280 – 930.

The test plate fabricated from acrylic resin (6 mm thick, see Fig. 8 (c)) had a removed section, which was covered with a titanium foil of 2 μm thick on both the lower and upper faces. Both ends of the foil was closely adhered to electrodes with high-conductivity bond to suppress a contact resistance. A copper plate of 4 mm thick was placed at the mid-height of the removed section (see Fig. 8 (b)), to impose a thermal boundary condition of a steady and uniform temperature. On the surface of the copper plate, a gold leaf (0.1 μm thick) was glued to suppress the thermal radiation. The titanium foil was heated by applying a direct current under conditions of constant heat flux so that the temperature difference between the foil and the freestream to be about 30ᵒC. Since both the upper and lower faces of the test plate were heated, the heat conduction loss to inside the plate was much reduced. Under these conditions, air enclosed by both the titanium foil and the copper plate does not convect because the Rayleigh number is below the critical value.

To suppress a deformation of the heated thin-foil due to the thermal expansion of air inside the plate, thin relief holes were connected from the ail-layer to the atmosphere. Also, the titanium foil was stretched by heating it since the thermal expansion coefficient of the titanium is smaller than that of the acrylic resin. This suppressed mechanical vibration of the foil against the fluctuating flow. [The amplitude of the vibration measured using a laser displacement meter was an order of 1 μm at the maximum freestream velocity of u_0 = 6 m/s. This amplitude was one or two orders smaller than the wall-friction length of the turbulent boundary layer].

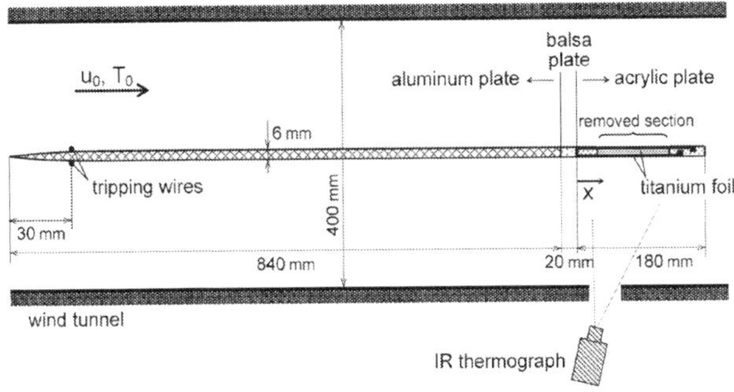

(a) Cross sectional view of the wind tunnel

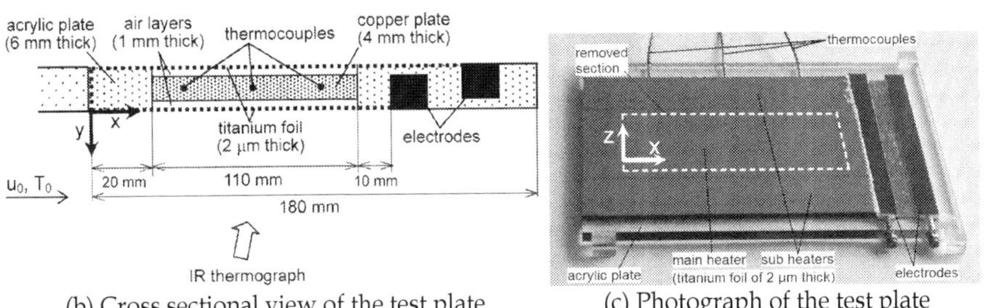

(b) Cross sectional view of the test plate (c) Photograph of the test plate

Fig. 8. Experimental setup (turbulent boundary layer)

The infrared thermograph was positioned below the plate and it measured the fluctuation of the temperature distribution on the lower-side face of the plate. The infrared thermograph used in this section (TVS-8502, Avio) can capture images of the instantaneous temperature distribution at 120 frames per second, and a total of 1024 frames with a full resolution of 256×236 pixels. The value of NETD of the infrared thermograph for a blackbody was $\Delta T_{IR0} = 0.025$ K.

The temperature on the titanium foil T_w was calculated using the following equation:

$$E_{IR} = \varepsilon_{IR} f(T_w) + (1 - \varepsilon_{IR}) f(T_a) \tag{43}$$

Here, E_{IR} is the spectral emissive power detected by infrared thermograph, $f(T)$ is the calibration function of the infrared thermograph for a blackbody, ε_{IR} is spectral emissivity for the infrared thermograph, and T_a is the ambient wall temperature. The first and second terms of the right side of Eq. (43) represent the emissive power from the test surface and surroundings, respectively. In order to suppress the diffuse reflection, the inner surface of the wind tunnel (the surrounding surface of the test surface) was coated with black paint. Also, in order to keep the second term to be a constant value, careful attention was paid to keep the surrounding wall temperature to be uniform. The thermograph was set with an inclination angle of 20º against the test surface in order to avoid the reflection of infrared radiation from the thermograph itself.

The spectral emissivity of the foil, ε_{IR}, was estimated using the titanium foil, which was adhered closely to a heated copper plate. The value of ε_{IR} can be estimated from Eq. (43) by substituting E_{IR} detected by the infrared thermograph, the temperature of the copper plate ($\approx T_w$) measured using such as thermocouples, and the ambient wall temperature T_a.

The accuracy of this measurement was verified to measure the distribution of mean heat transfer coefficient of a laminar boundary layer. The result was compared to a 2D heat conduction analysis assuming the velocity distribution to be a theoretical value. The agreement was very well (within 3 %), indicating that the present measurement is reliable to evaluate the heat transfer coefficient at least for a steady flow condition (Nakamura, 2007a and 2007b).

Also, a dynamic response of this measurement was investigated against a stepwise change of the heat input to the foil in conditions of a steady flow for a laminar boundary layer. The response curve of the measured temperature agreed well to that of the numerical analysis of the heat conduction equation. This indicates that the delay due to the heat capacity of the foil, $c\rho\delta(\partial T_w / \partial t)$ in Eq. (44), and the heat conduction loss to the air-layer, $\dot{q}_{cd} = \lambda_a (\partial T/\partial y)_{y=0-}$ in Eq.(44), can be evaluated with a sufficient accuracy (Nakamura, 2007b).

5.2 Spatio-temporal distribution of temperature

Figure 9 (a) and (b) shows the results of the temperature distribution of laminar and turbulent boundary layers, respectively, measured using infrared thermography. The freestream velocity was $u_0 = 3$ m/s for both cases. Bad pixels existed in the thermo-images were removed by applying a 3×3 median filter (here, intermediate three values were averaged). Also, a low-pass filter (sharp cut-off) was applied in order to remove a high frequency noise more than $f_c = 30$ Hz (corresponds to less than 4 frames) and the small-scale spatial noise less than $b_c = 3.4$ mm (corresponds to less than 6 pixels).

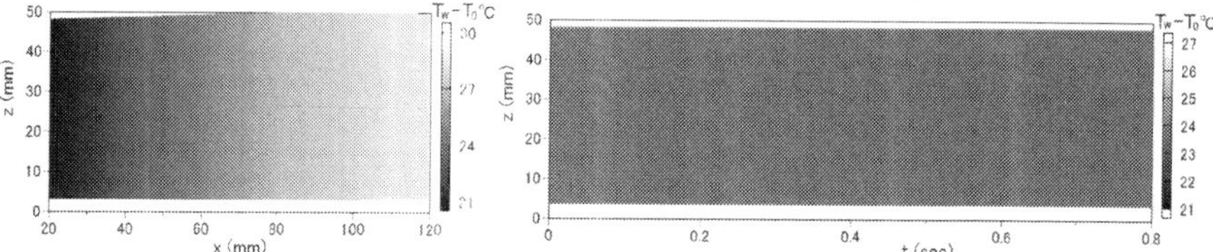

(a) Laminar boundary layer at $u_0 = 3$ m/s; Right – spanwise time trace at $x = 37$ mm

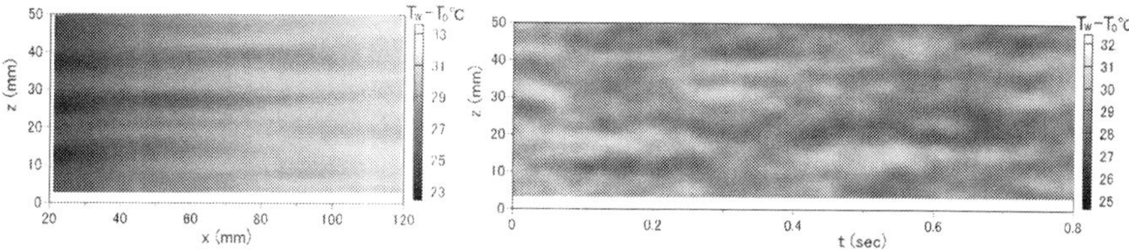

(b) Turbulent boundary layer at $u_0 = 3$ m/s; $Re_\theta = 530$; Right – spanwise time trace at $x = 69$mm

Fig. 9. Temperature distribution T_w-T_0 measured using infrared thermography

As depicted in Figure 9 (b), the temperature for the turbulent boundary layer has large nonuniformity and fluctuation according to the flow turbulence. The thermal streaks appear in the instantaneous distribution, which extend to the streamwise direction. Figure 10 shows the power spectrum of the temperature fluctuation. The S/N ratio of the measurement estimated based on the power spectrum for the laminar boundary layer (noise) was 500 – 1000 (27 – 30 dB) in the lower frequency range of 0.4 – 6 Hz and about 10 (10 dB) at the maximum frequency of $f_c = 30$ Hz after applying the filters.

5.3 Restoration of heat transfer coefficient

The local and instantaneous heat transfer coefficient was calculated using the following equation derived from the heat conduction equation in a thin foil (Eq. (1) – (3)).

$$h = \frac{\dot{q}_{in} - \dot{q}_{cd} - \dot{q}_{rd} - \dot{q}_{rdi} + \lambda\delta \left(\frac{\partial^2 T_w}{\partial x^2} + \frac{\partial^2 T_w}{\partial z^2} \right) - c\rho\delta \frac{\partial T_w}{\partial t}}{T_w - T_0} \tag{44}$$

This equation contains both terms of lateral conduction through the foil, $\lambda\delta(\partial^2 T_w/\partial x^2 + \partial^2 T_w/\partial z^2)$, and the thermal inertia of the foil, $c\rho\delta(\partial T_w/\partial t)$. Heat conduction to the air layer inside the foil, $\dot{q}_{cd} = \lambda_a (\partial T/\partial y)_{y=0-}$, was calculated using the temperature distribution in the air layer, which can be determined by solving the heat conduction equation as follows (the coordinate system is shown in Fig. 8):

$$c_a \rho_a \frac{\partial T}{\partial t} = \lambda_a \left(\frac{\partial^2 T}{\partial x^2} + \frac{\partial^2 T}{\partial y^2} + \frac{\partial^2 T}{\partial z^2} \right), \; (-\delta_a < y < 0) \tag{45}$$

Here, c_a, ρ_a and λ_a are specific heat, density and thermal conductivity of air (This Equation is similar to Eq. (7) only the subscript i is replaced to a). Since the temperature of the copper plate inside the test plate is assumed to be steady and uniform, the boundary condition of Eq. (45) on the copper plate side ($y = -\delta_a$) can be assumed as a mean temperature of the copper plate measured using thermocouples.

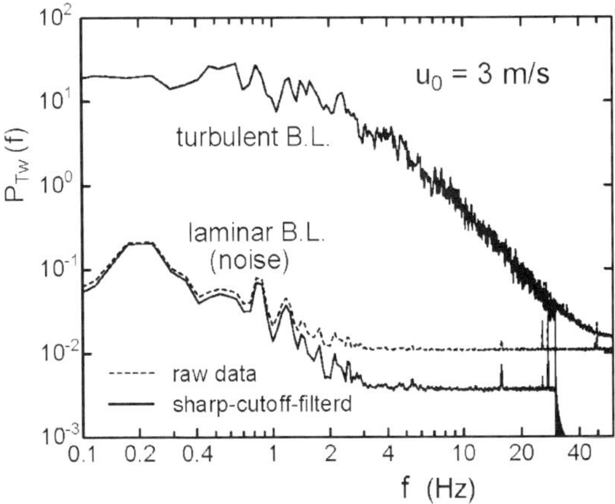

Fig. 10. Power spectrum of temperature fluctuation appeared in Fig. 9

The finite difference method was applied to calculate the heat transfer coefficient h from Eq. (44) and (45). Time differential Δt corresponded to the frame interval of the thermo-images (in this case, $\Delta t = 1/120$ s $= 8.3$ ms). Space differentials Δx and Δz corresponded to the pixel pitch of the thermo-image (in this case, $\Delta x \approx \Delta z \approx 0.56$ mm). The thickness of the air layer ($\delta_a = 1$ mm) was divided into two regions ($\Delta y = 0.5$ mm). [In this case, normal temperature distribution in the air-layer can be assumed to linear within an interval of $\Delta y = 0.5$ mm up to the maximum frequency of $f_c = 30$ Hz, since it satisfies $\kappa_a \Delta y < 1$; see section 3.1] Eq. (45) was solved using ADI (alternative direction implicit) method (Peaceman and Rachford, 1955) with respect to x and z directions.

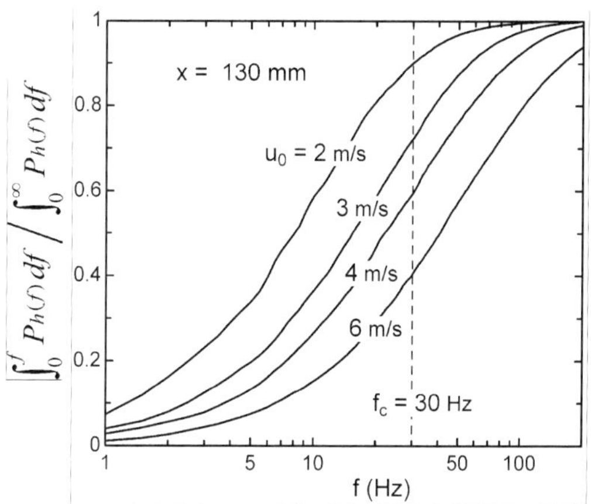

Fig. 11. Cumulative power spectrum of fluctuating heat transfer coefficient

The above procedure (the finite different method including the median and the sharp cut-off filters) restored the heat transfer coefficient up to $f_c = 30$ Hz in time with the attenuation rate of below 20 % and up to $b_c = 3.4$ mm in space with the attenuation rate of below 30 % (Nakamura, 2007b). The wavelength of $b_c = 3.4$ mm corresponded to $20 - 48\ l_\tau$ (for $u_0 = 2 - 6$ m/s), which was smaller than the mean space between the thermal streaks ($\approx 100\ l_\tau$, see Fig. 14).

Figure 11 shows cumulative power spectrum of the fluctuation of the heat transfer coefficient measured using a heat flux sensor (HFM-7E/L, Vatell; time constant faster than 3 kHz) under a condition of steady wall temperature. For the freestream velocity $u_0 = 2$ m/s, the fluctuation energy below $f_c = 30$ Hz accounts for 90 % of the total energy, indicating that the fluctuation can be restored almost completely by the above procedure. However, with an increase in the freestream velocity, the ratio of the fluctuation energy below $f_c = 30$ Hz decreases, resulting in an insufficient restoration.

5.4 Spatio-temporal distribution of heat transfer

The spatio-temporal distribution of the heat transfer coefficient restored using the above procedure is shown in Fig. 12. The features of the thermal streaks are clearly revealed, which extend to the streamwise direction with small spanwise inclinations. The heat transfer coefficient fluctuates vigorously showing a quasi-periodic characteristic in both time and spanwise direction, which is reflected by the unique behavior of the thermal streaks. Although the restoration for $u_0 = 3$ m/s (Fig. 12 (b)) is not sufficient, as shown in Fig. 11, the characteristic scale of the fluctuation seems to be smaller both in time and spanwise direction than that for $u_0 = 2$ m/s, indicating that the structure of the thermal streaks becomes finer with increasing the freestream velocity.

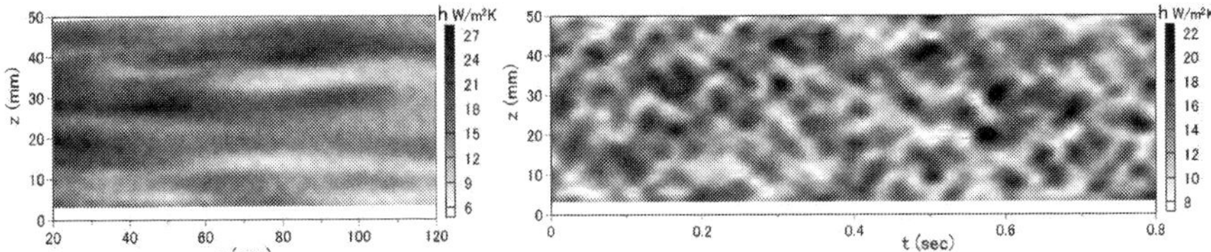

(a) $u_0 = 2$ m/s, $Re_\theta = 280$, $l_\tau = 0.174$ mm; Right – spanwise time trace at $x = 69$ mm

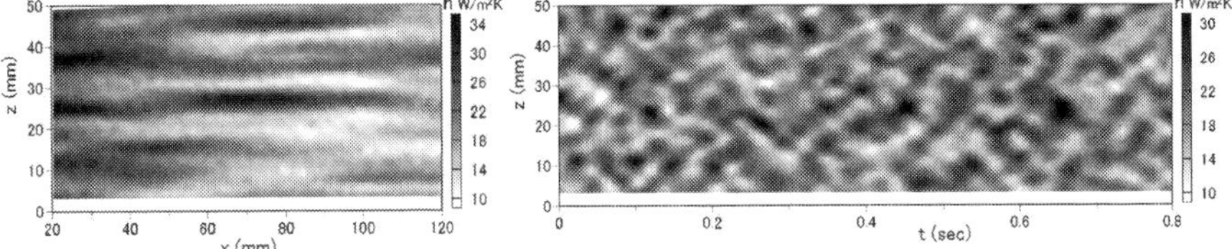

(b) $u_0 = 3$ m/s, $Re_\theta = 530$, $l_\tau = 0.126$ mm; Right – spanwise time trace at $x = 69$ mm

Fig. 12. Time-spatial distribution of heat transfer coefficient (turbulent boundary layer)

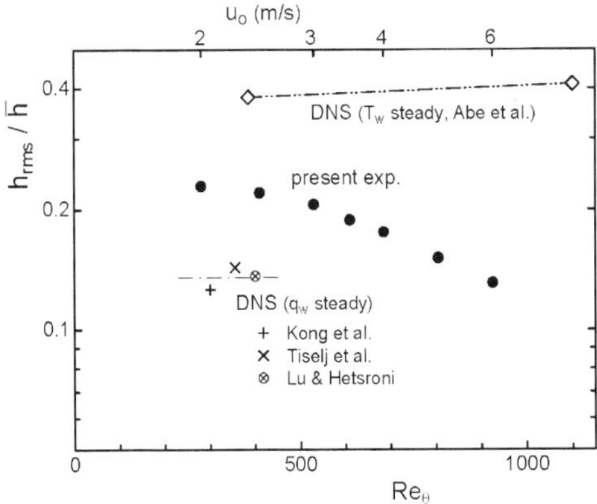

Fig. 13. Rms value of the fluctuating heat transfer coefficient at $x = 69$ mm

Figure 13 plots the rms value of the fluctuation $h_{rms}/\overline{h}$ at $x = 69$ mm. The value at $u_0 = 2$ m/s ($Re_\theta = 280$) was $h_{rms}/\overline{h} = 0.23$, at which the restoration is almost complete. However, it decreases with increasing the freestream velocity due to the insufficient restoration. For $u_0 = 2$ m/s, the value of f_{max} is 37 Hz (see section 4.2), while the frequency restored is $f_c = 30$ Hz. This indicates that the restoration up to $f_c \approx f_{max}$ is possible without exaggerating the noise.

The results of direct numerical simulation (Lu and Hetsroni, 1995, Kong et al, 2000, Tiselj et al, 2001, and Abe et al, 2004) are also plotted in Fig. 13. As shown in this Figure, the value of $h_{rms}/\overline{h}$ greatly depends on the difference in the thermal boundary condition, that is, $h_{rms}/\overline{h}$ ≈ 0.4 for steady temperature condition (corresponds to infinite heat capacity wall), whereas $h_{rms}/\overline{h}$ = 0.13 – 0.14 for steady heat flux condition (corresponds to zero heat capacity wall). Since the present experiment was performed between two extreme conditions, for which the temperature on the wall fluctuates with a considerable attenuation, the value $h_{rms}/\overline{h}$ = 0.23 seems to be reasonable.

Figure 14 plots the mean spanwise wavelength of the thermal streak, $l_z^+ = l_z/l_\tau$, which is determined by an auto-correlation of the spanwise distribution. For the lower velocity of u_0 = 2 – 3 m/s (Re_θ = 280 – 530), the mean wavelength is l_z^+ = 77 – 87, which agrees well to that for the previous experimental data obtained using water as a working fluid (Iritani et al, 1983 and 1985, and Hetsroni & Rozenblit, 1994; l_z^+ = 74 – 89). This wavelength is smaller than that for DNS (Kong et al, 2000, Tiselj et al, 2001, and Abe et al, 2004; l_z^+ = 100 – 150), probably due to the additional flow turbulence in the experiments, such as freestream turbulence. The value of l_z^+ for the present experiment increases with increasing the Reynolds number, the reason of which is not clear at present.

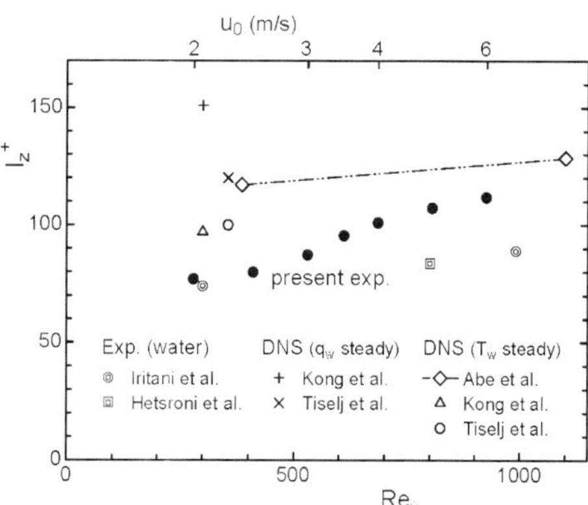

Fig. 14. Mean spanwise wavelength of thermal streaks

In this section, the time-spatial heat transfer coefficient was restored up to 30 Hz in time and 3.4 mm in space at a low heat transfer coefficient of $\overline{h}$ = 10 – 20 W/m²K, by employing a 2 μm thick titanium foil and an infrared thermograph of 120 Hz with NETD of 0.025K. This restoration was, however, not exactly sufficient, particularly for the higher freestream velocity of $u_0 > 2$ m/s. Yet, the higher frequency fluctuation will be restored by employing the higher-performance thermograph (higher frame rate with lower NETD, see section 6), if a condition of $f_c < f_{max}$ is satisfied.

6. Experimental demonstration (separated and reattaching flow)

The recent improvement of infrared thermograph with respect to temporal, spatial and temperature resolutions enable us to investigate more detailed behavior of the heat transfer caused by flow turbulence. In this section, the heat transfer behind a backward-facing step which represents the separated and reattaching flow was explored by employing a higher-performance thermograph. Special attention was devoted to investigate the spatio-temporal characteristics of the heat transfer in the flow reattaching region.

6.1 Experimental setup

Figure 15 shows the test plate used here. The wind tunnel and the flat plate (aluminum plate) is the same as that used in section 5 (see Fig. 8). A turbulent boundary layer was formed on the lower-side face of the flat plate (aluminum plate) followed by a step. The step height was H = 5, 10 and 15.6 mm, thus the aspect ratio was AR = 30, 15, and 9.6 and the expansion ratio was ER = 1.025, 1.05 and 1.08, respectively. The freestream velocity ranged from 2 to 6 m/s, resulting in the Reynolds number based on the step height was Re_H = 570 – 5400.

The test plate fabricated from acrylic resin (6 mm thick) had two removed sections (see Fig. 15 (b)), which were covered with two sheets of titanium foil of 2 μm thick on both the lower and upper faces. A copper plate of 4 mm thick was placed at the mid-height of each removed section. The titanium foil was heated by applying a direct current so that the temperature difference between the foil and the freestream was around 20–30°C. The amplitude of the mechanical vibration of the foil in the flow reattaching region measured using a laser displacement meter was an order of 1 μm at the maximum freestream velocity of u_0 = 6 m/s.

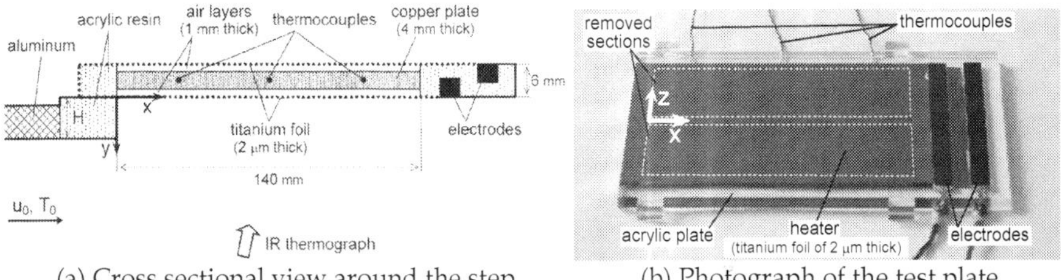

(a) Cross sectional view around the step   (b) Photograph of the test plate

Fig. 15. Experimental setup (backward-facing step)

In this study, a high-speed infrared thermograph of SC4000, FLIR (420 frames per second with a resolution of 320×256 pixels, or 800 frames per second with a resolution of 192×192 pixels, NETD of 0.018 K) was employed in addition to TVS-8502, AVIO (see section 5).

6.2 Time-averaged distribution

Figure 16 shows streamwise distribution of Nusselt number, $Nu_H = \overline{h}H / \lambda$, where $\overline{h}$ is time and spanwise-averaged heat transfer coefficient calculated from the time-spatial distribution of the heat transfer coefficient (shown later in Fig. 20). The x axis is originated from the step. The Nusselt number was normalized by $Re_H^{2/3}$, because the local Nusselt number of the separated and reattaching flows usually proportional to $Re^{2/3}$ (Richardson, 1963; Igarashi, 1986). For the present experiment, the distribution of $Nu_H/Re^{2/3}$ almost corresponded for Re_H > 2000, as shown in Fig. 16.

The Nusselt number distribution has a similar trend as that investigated previously (Vogel and Eaton, 1985; among others); it increases sharply toward the flow reattachment zone ($x/H \approx 5$ for the present experiment), and then it decreases gradually with a development to a turbulent boundary layer. The difference in the peak location of the distribution can be explained by the fact that it moves downstream with an increase in the expansion ratio (ER), as indicated by Durst and Tropea, 1981. Also, it moves upstream with an increase in the turbulent boundary layer thickness upstream of the step (Eaton and Johnston, 1981).

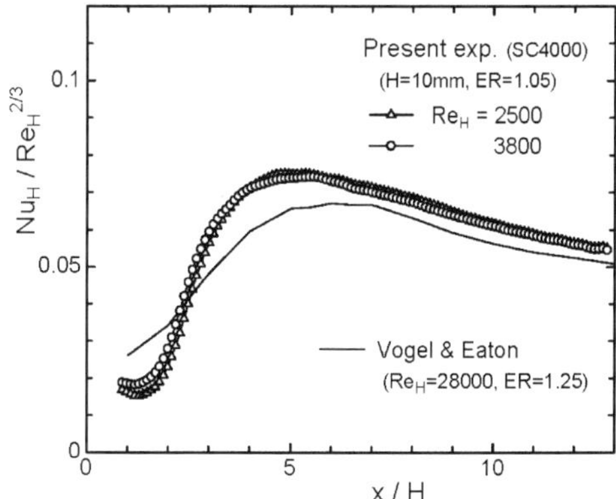

Fig. 16. Streamwise distribution of Nusselt number for the backward-facing step

6.3 Spatio-temporal distribution

Figure 17 shows examples of an instantaneous distribution of temperature on the titanium foil as measured using infrared thermograph (SC4000). The step height was H = 10 mm and the freestream velocity was u_0 = 6 m/s, resulting in the Reynolds number of Re_H = 3800. Bad pixels in the thermo-images were removed by applying a 3×3 median filter (here, intermediate three values were averaged). Also, a low-pass filter (sharp cut-off) was applied in order to remove a high frequency noise (more than f_c = 53 Hz for the wide measurement of Fig. 17 (a) and more than f_c = 133 Hz for the close-up measurement of Fig. 17 (b)) and the small-scale spatial noise (less than b_c = 4.9 mm for the wide measurement and less than b_c = 2.2 mm for the close-up measurement).

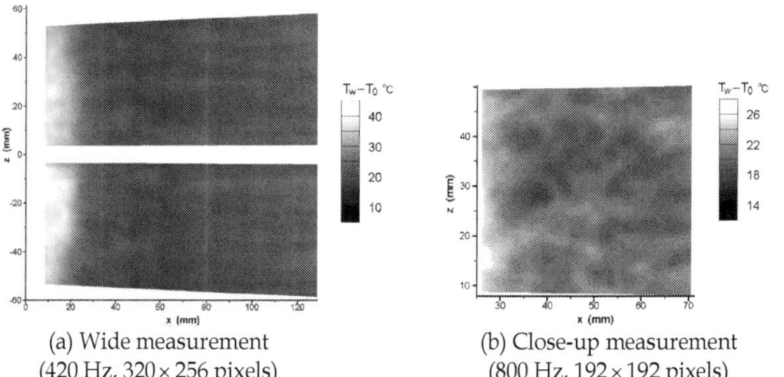

(a) Wide measurement   (b) Close-up measurement
(420 Hz, 320 × 256 pixels)   (800 Hz, 192 × 192 pixels)

Fig. 17. Temperature distribution T_w-T_0 behind the backward-facing step (H = 10 mm, u_0 = 6 m/s, Re_H = 3800; step at x = 0)

Incidentally, the upper limit of the detectable fluctuating frequency (f_{max}, see section 4.2) and the lower limit of the detectable spatial wavelength (b_{min}, see section 4.3) in the reattachment region at u_0 = 6 m/s are f_{max} = 150 Hz and b_{min} = 0.6 mm (Δh = 20 W/m²K, $\overline{T_w} - T_0$ = 20°C, ΔT_{IR0} = 0.018 K, for a 2 μm thick titanium foil). Therefore, both the cutoff frequency of f_c = 133 Hz and the cutoff wavelength of b_c = 2.2 mm are within the detectable range.

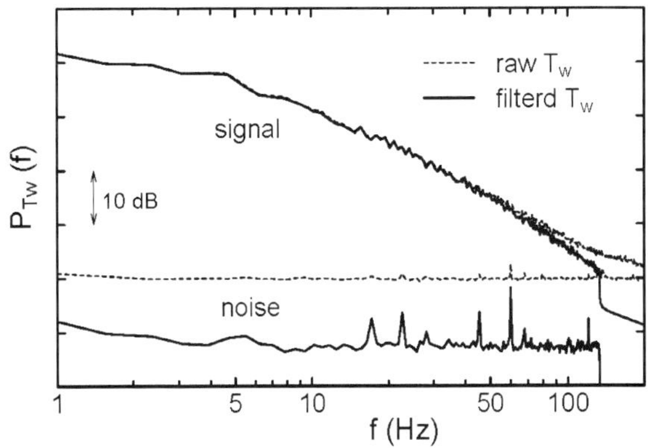

Fig. 18. Power spectrum of the temperature fluctuation: signal – temperature at x = 50 mm; noise – temperature on a steady temperature plate

Figure 18 shows power spectrums for both signal and noise of the temperature detected by the infrared thermograph (SC4000) for the close-up measurement. The noise was estimated by measuring the temperature on the titanium foil glued on a copper plate. The noise was much reduced by about 10 dB by applying the median and the low-pass filters, resulting that the S/N ratio of the measurement was greater than 1000 for f < 30 Hz and 10-20 at the maximum frequency of f_c = 133 Hz.

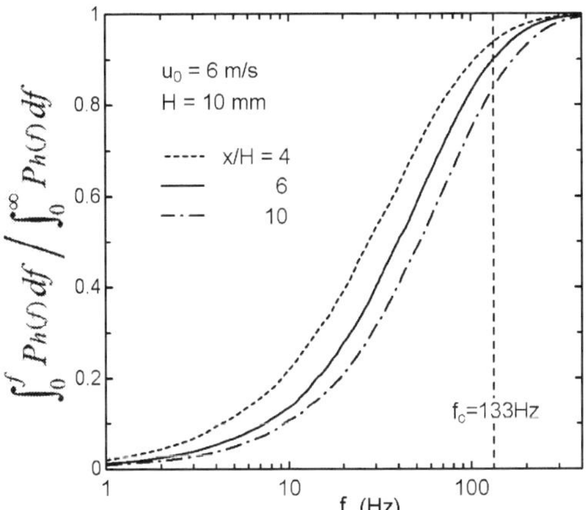

Fig. 19. Cumulative power spectrum of fluctuating heat transfer coefficient

Figure 19 shows a cumulative power spectrum of the fluctuation of the heat transfer coefficient in the flow reattaching region measured using a heat flux sensor (HFM-7E/L, Vatell; time-constant faster than 3 kHz) under a condition of steady wall temperature (its power spectrum is shown later in Fig. 22 (b)). As indicated in Fig. 19, the most part of the fluctuating energy of the heat transfer coefficient (about 90 %) can be restored at the cutoff frequency of f_c = 133 Hz for the maximum velocity of u_0 = 6 m/s.

The spatio-temporal distribution of the heat transfer coefficient corresponding to Fig. 17 (a) and (b) is shown in Figs. 20 and 21, respectively, which were calculated by the similar procedure to that described in section 5.3. These figures reveal some unique characteristics of time-spatial behavior of the heat transfer for the separated and reattaching flow, which has hardly been clarified in the previous experiments. The most impressive feature is that the heat transfer enhancement in the reattachment zone (x = 30 – 70 mm) has a spot-like characteristic, as shown in the instantaneous distribution (Fig. 20 (a) and 21 (a)). The high heat transfer spots appear and disappear almost randomly but have some periodicity in time and spanwise direction, as indicated in the time traces (Fig. 20 (b), (c) and Fig. 21 (b), (c)). Each spot spreads with time, which forms a track of " ∧ " shape in the streamwise time trace (Fig. 21 (b)) corresponding to the streamwise spreading, and forms a track of " ⟨ " shape in the spanwise time trace (Fig. 21 (c)) corresponding to the spanwise spreading. The basic behavior of the spot spreading overlaps with others to form a complex feature in the spatio-temporal characteristics of the heat transfer.

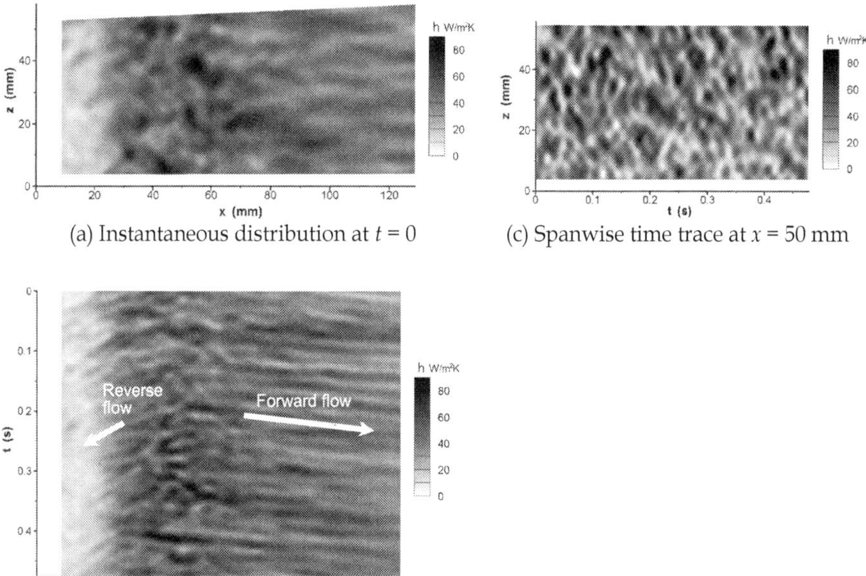

(a) Instantaneous distribution at t = 0

(c) Spanwise time trace at x = 50 mm

(b) Streamwise time trace at z = -10 mm

Fig. 20. Time-spatial distribution of heat transfer coefficient behind the backward-facing step (H = 10 mm, u_0 = 6 m/s, Re_H = 3800; f_c = 53 Hz, b_c = 4.9 mm)

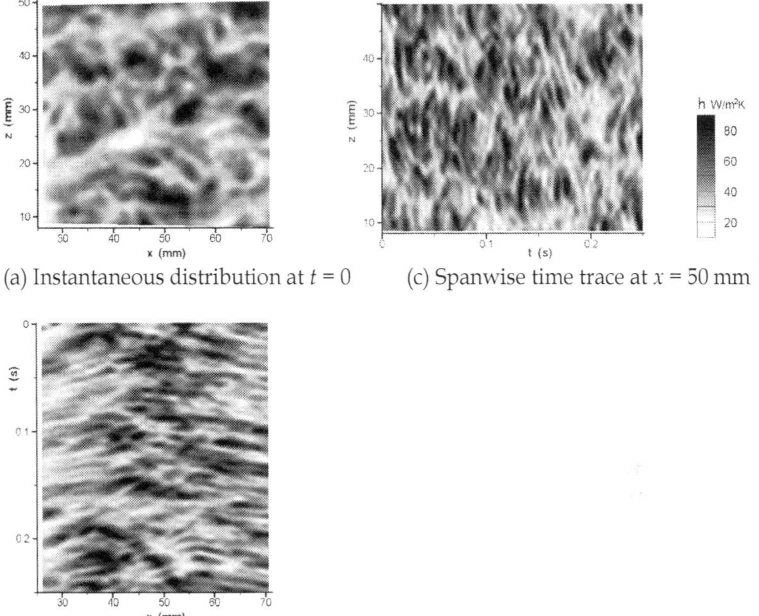

(a) Instantaneous distribution at $t = 0$ (c) Spanwise time trace at $x = 50$ mm

(b) Streamwise time trace at $z = 29$ mm

Fig. 21. Time-spatial distribution of heat transfer coefficient around the reattaching region ($H = 10$ mm, $u_0 = 6$ m/s, $Re_H = 3800$; $f_c = 133$ Hz, $b_c = 2.2$ mm)

The heat transfer coefficient is considerably low beneath the separation region, which is formed between the step and the flow reattachment zone ($x < 30$ mm, see Fig. 20 (a)). The reverse flow occurs from the reattachment zone to this region ($x = 30 - 10$ mm), which is depicted by tracks of high heat transfer regions as shown in the streamwise time trace (Fig. 20 (b)). The velocity of the reverse flow, which was determined by the slope of the tracks, was very slow, approximately $0.05 - 0.1$ of the freestream velocity.

Behind the flow reattachment zone ($x > 70$ mm), the flow gradually develops into a turbulent boundary layer flow. The spot-like structure in the reattachment zone gradually change it form to streaky-structure, as can be seen in the instantaneous distribution of Fig. 20 (a). The characteristic velocity of this structure, which was determined by the slope of the tracks of the streamwise time trace (Fig. 20 (b)), was roughly $0.5u_0$, which varies widely as can be seen in the fluctuation of the tracks. This velocity was similar to the convection speed of vortical structure near the reattachment zone ($0.5u_0$ for Kiya & Sasaki, 1983 and $0.6u_0$ for Lee & Sung, 2002). Kawamura et al., 1994 also indicated that the convection speed of the heat transfer structure is approximately $0.5u_0$ for the constant-wall-temperature condition.

6.4 Temporal characteristics

Figure 22 (a) shows time traces of the fluctuating heat transfer coefficient in the reattaching region measured using the heat flux sensor (HFS) and the infrared thermograph (IR).

Although the time trace of IR does not have sharp peaks as that of HFS probably due to the low-pass filter of f_c = 133Hz, the basic characteristics of the fluctuation seems to be similar. Figure 22 (b) shows power spectrum of the fluctuation corresponding to Fig. 22 (a). The attenuation with frequency for IR is similar to that for HFS up to the sharp-cutoff frequency of f_c = 133Hz, while the thermal boundary condition is different.

The previous studies have indicated that the flow in the reattaching region behind a backward-facing step was dominated by low-frequency unsteadiness. Eaton & Johnston, 1980 measured the energy spectra of the streamwise velocity fluctuations at several locations and reported that the spectral peak occurred at the Strouhal number St = 0.066 – 0.08. The direct numerical simulation performed by Le et al., 1997 also showed the dominant frequency of the velocity was roughly St = 0.06. The origin of this unsteadiness is not completely understood, but it may be caused by the pairing of the shear layer vortices (Schäfer et al., 2007).

In order to explore the effect of the low-frequency unsteadiness on the heat transfer, autocorrelation function of the time trace of Fig. 22 (a) was calculated. The result is shown in Fig. 22 (c), which has some bumps in both HFS and IR measurements. The characteristic

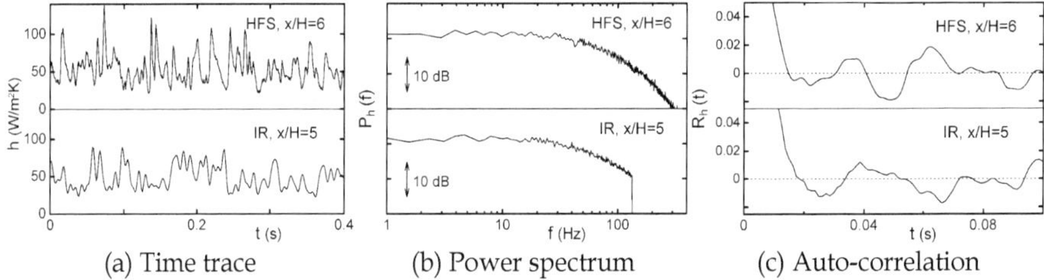

(a) Time trace   (b) Power spectrum   (c) Auto-correlation

Fig. 22. Fluctuation of heat transfer coefficient around the flow reattaching region (H = 10mm, u_0 = 6 m/s, Re_H = 3800)

period of the bumps is roughly 0.02s – 0.04s, corresponding to the fluctuation of St = 0.04 – 0.08. This fluctuation seems to be related to the low-frequency unsteadiness reported in the previous literature although the power spectrum for the present experiment had no dominant peak.

As shown in Fig. 21 (c), the period of 0.02s – 0.04s contains several detailed spots of high heat transfer. This suggests that the low-frequency unsteadiness is originated from a combination of several smaller vortical structures such as the shear layer vortices caused by Kelvin-Helmholtz instability, the Strouhal number of which is 0.2 – 0.4 (Bhattacharjee, 1986).

6.5 Spatial characteristics

As depicted in the spanwise time trace (Fig. 21 (c)), there seems to exist some spanwise periodicity in the heat transfer. Figure 23 shows an example of autocorrelation function of the instantaneous spanwise distribution at the reattachment zone, which is averaged in time. As shown in this figure, there is a clear minimum, at Δz = 6.3 mm in this case. This minimum, which exists for all conditions examined here, is defined to a half spanwise wavelength of $l_z/2$. The typical wavelength estimated here, l_z/H, is plotted in Fig. 24 against Reynolds number Re_H . It is remarkable that all plots are almost concentrated into a single curve regardless of the variation of the step height H. In particular, the wavelength l_z/H has almost a constant value of about 1.2 for $2000 \le Re_H \le 5500$. The measurement performed by

Kawamura et al., 1994 using heat flux sensors also indicated that there is a spanwise periodicity of about 1.2H around the reattaching region behind a backward-facing step at Re_H = 19600. This periodicity corresponds well to that of streamwise vortices formed around the reattaching region behind a backward-facing step, observed by Nakamaru et al., 1980 in their flow visualization, in which the most frequent spanwise wavelength was about (1.2 – 1.5)H. This indicates that the spanwise periodicity appeared in the heat transfer is caused by the formation of the large-scale streamwise vortices, and it is reasonable to consider that the time-spatial distribution of the heat transfer in the reattaching region is dominated by the spatio-temporal behavior of the streamwise vortices.

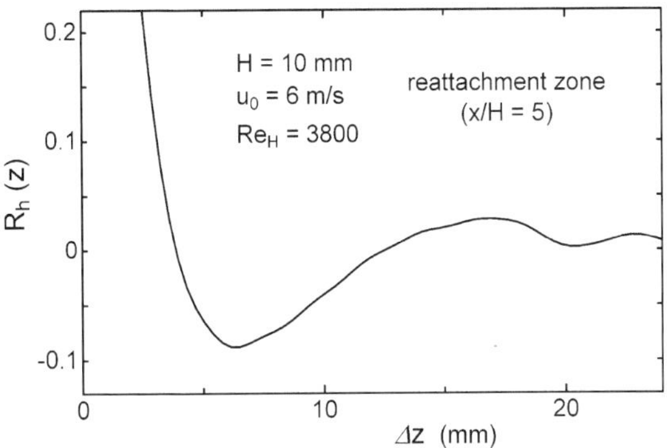

Fig. 23. Auto-correlation of instantaneous spanwise distribution of heat transfer coefficient

As shown in Fig. 24, the spanwise wavelength is closely related to the step height, not to the spacing of streaks of the turbulent boundary layer upstream of the step. This indicates that the origin of the streamwise vortices in the reattaching region is not the spanwise periodicity upstream of the step, but due to some instability behind the step, which may be accompanied by the flow separation and reattachment.

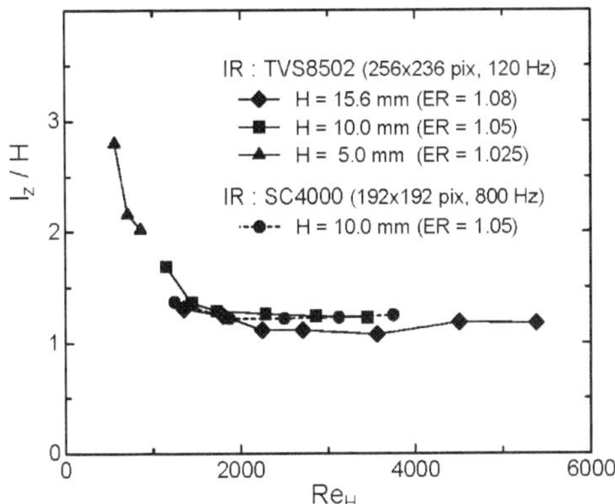

Fig. 24. Mean spanwise wavelength at the reattaching region

7. Summary and future works

In this chapter, a measurement technique was reported to explore the spatio-temporal distribution of turbulent heat transfer. This measurement can be realized using a high-speed infrared thermograph which records the temperature fluctuation on a heated thin-foil with sufficiently low heat capacity. In the existing circumstances at present, the spatial resolution of about 2 mm and the temporal resolution of about 100 Hz were possible to measure the heat transfer to air, by employ a titanium foil of 2 μm thick and a high-performance thermograph (frame rate of more than several hundred Hz and NETD of about 0.02 K). This enables us to investigate the time and spatial characteristics of turbulent heat transfer which has hardly been clarified experimentally so far.

This technique has great merits as listed below:

a.   Non-intrusive measurement which does not disturb the flow and temperature fields.

b.   Permits real-time observation of the spatio-temporal characteristics.

c.   High spatial-resolution corresponding to the pixel pitch of the thermograph.

The spatio and temporal resolution is likely to be improved in the future with an improvement of a performance of the thermograph and a development of quality of the thin-foil.

On the other hand, there are some troublesome aspects as listed below:

d.   Needs special care to treat an extremely thin foil in the fabrication and experimentation.

e.   Susceptible to diffuse reflection from surroundings by using a low emissivity thin-foil.

The above terms (d. and e.) are possible to overcome as demonstrated experimentally (see sections 5 and 6). However, it is desirable to develop a thin-foil, which has a higher emissivity and an enough rigidity and elasticity.

Only a few examples were reported here concerning the forced convection heat transfer to air, yet, this technique is also available to measure the heat transfer for natural convection or mixed convection. Moreover, this technique will be extended to a liquid flow or a multiphase flow, which will be possible by measuring the temperature from the rear of the foil (from the air side), and by using a thin-foil of several tens micro-meter thick to suppress a deformation against the fluid pressure (refer to Hetsroni & Rozenblit, 1994, and Oyakawa, et al., 2000). [Ideally, the frequency-response and spatial-resolution does not deteriorate by using a foil of ten times thick if the heat transfer coefficient becomes ten times higher; see Eq. (9) and (15)].

In the future, it is highly expected that this technique clarifies the heat transfer mechanisms for a complex flow, which has been very difficult to investigate using conventional methods. It is hoped that the knowledge acquired using this technique will be contribute to develop technology in heat transfer control, and also to improve reliability in thermal design of various equipment and machinery.

8. References

Abe, H., Kawamura, H. and Matsuo, Y. (2004). Surface Heat-Flux Fluctuations in a Turbulent Channel Flow up to $Re_\tau = 1020$ with $Pr = 0.025$ and 0.71, *International Journal of Heat and Fluid Flow*, Vol. 25, 404-419.

Bhattacharjee, S., Scheelke, B., and Troutt, T.R. 1986. Modification of Vortex Interactions in a Reattaching Separated Flow, *AIAA Journal*, Vol.24, 623-629.

Durst, F., and Tropea, C. (1981). Turbulent, Backward-Facing Step Flows in Two-Dimensional Ducts and Channels, *Proc. 3rd Int. Symp. on Turbulent Shear Flows*, pp.18.1-18.6, Davis, CA, 1981.

Eaton, J.K. and Johnston, J.P. (1980). Turbulent Flow Reattachment: An Experimental Study of the Flow and Structure behind a Backward-Facing Step. *Rep.*, MD-39, Thermosciences Division, Dept. of Mech. Engng, Stanford University.

Eaton, J.K. and Johnston, J.P. (1981). A Review of on Subsonic Turbulent Flow Reattachment. *AIAA Journal*, Vol.19, No.9, 1093-1100.

Hetsroni, G. and Rozenblit, R. (1994). Heat Transfer to a Liquid-Solid Mixture in a Flume, *International Journal of Multiphase Flow*, Vol.20-4, 671-689, ISSN 03019322

Igarashi, T. (1986). Local Heat Transfer from a Square Prism to an Air Stream, *Int. J. Heat and Mass Transfer*, Vol.29, No.5, 777-784.

Iritani, Y., Kasagi, N. and Hirata, M. (1983). Heat Transfer Mechanism and Associated Turbulent Structure in the Near-Wall Region of a Turbulent Boundary Layer, *4th Symposium on Turbulent Shear Flows*, pp. 17.31-17.36, Karlsruhe, Germany, 1983.9

Iritani, Y., Kasagi, N. and Hirata, M. (1985). Streaky Structure in a Two-Dimensional Turbulent Channel Flow (in Japanese), *Trans. Jpn. Soc. Mech. Eng.*, Vol. 51, No. 470, B, 3092-3101.

Kawamura, T., Tanaka, S., Kumada, M. and Mabuchi, I. (1988). Time and Spatial Unsteady Characteristics of Heat Transfer at the Reattachment Region of a Backward-Facing Step (in Japanese), *Transactions of Japan Society of Mechanical Engineers B*, Vol. 54, No. 504, 1224-1232.

Kawamura, T., Yamamori, M., Mimatsu, J. and Kumada, M. (1994). Three-Dimensional Unsteady Characteristics of Heat Transfer around Reattachment Region of Backward-Facing Step Flow (in Japanese), *Transactions of Japan Society of Mechanical Engineers B*, Vol. 60, No. 576, 2833-2839.

Kiya, M. and Sasaki, K. (1983). Structure of a Turbulent Separation Bubble, *J. Fluid Mech.*, Vol. 137, 83-113.

Kong, H., Choi, H. and Lee, J.S. (2000). Direct Numerical Simulation of Turbulent Thermal Boundary Layers, *Physics of Fluids*, Vol. 12, No. 10, 2555-2568.

Le, H. Moin, P. and Kim, J. (1997). Direct Numerical Simulation of Turbulent Flow Over a Backward-Facing Step. *J. Fluid Mech.*, Vol. 330, 349-374.

Lee, I. and Sung, H.J. (2002). Multiple-Arrayed Pressure Measurement for Investigation of the Unsteady Flow Structure of a Reattaching Shear Layer, *J. Fluid Mech.*, Vol. 463, 377-402.

Lu, D.M. and Hetsroni, G. (1995). Direct Numerical Simulation of a Turbulent Open Channel Flow with Passive Heat Transfer, *Int. J. Heat and Mass Transfer*, Vol. 38, No. 17, 3241-3251.

Nakamaru M, Tsuji M, Kasagi N and Hirata M. (1980). A Study on the Transport Mechanism of Separated Flow behind a Step, 2nd Report (in Japanese), *17th National Heat Transfer Symp. of Japan*, pp.7-9.

Nakamura, H. (2007a). Measurements of time-space distribution of convective heat transfer to air using a thin conductive-film, *Proceedings of 5th International Symposium on Turbulence and Shear Flow Phenomena*, pp. 773-778, München, Germany, 2007.8

Nakamura, H. (2007b). Measurements of Time-Space Distribution of Convective Heat Transfer to Air Using a Thin Conductive Film (in Japanese), *Transactions of Japan Society of Mechanical Engineers B*, Vol. 73, No. 733, 1906-1914.

Nakamura, H. (2009). Frequency response and spatial resolution of a thin foil for heat transfer measurements using infrared thermography. *International Journal of Heat and Mass Transfer,* Vol.52, 5040-5045, ISSN 00179310

Nakamura, H. (2010). Spatio-temporal measurement of convective heat transfer for the separated and reattaching flow, *Proceedings of 14th International Heat Transfer Conference,* IHTC14-22753, Washington, DC, USA, 2010.8.

Oyakawa, K., Miyagi, T., Oshiro, S., Senaha, I., Yaga, M., and Hiwada, M. (2000). Study on Time-Spatial Characteristics of Heat Transfer by Visualization of Infrared Images and Dye Flow, *Proceedings of 9th International Symposium on Flow Visualization,* Pap. no. 233, Edinburgh, Scotland, UK, 2000.

Peaceman, D.W. and Rachford, H.H. (1955). The numerical solution of parabolic and elliptic differential equations, *J. Soc. Ind. Appl. Math.,* Vol.3, 28-41.

Richardson, P.D. (1963). Heat and Mass Transfer in Turbulent Separated Flows, *Chemical Engineering Science,* Vol.18, 149-155.

Schäfer, F., Breuer, M. and Durst, F. (2007). The Dynamics of the Transitional Flow Over a Backward-Facing Step, *J. Fluid Mech.,* Vol. 623, 85-119.

Tiselj, I., Pogrebnyak, E., Li, C., Mosyak, A., and Hetsroni, G. (2001). Effect of Wall Boundary Condition on Scalar Transfer in a Fully Developed Turbulent Flume, *Physics Fluids,* Vol. 13, No. 4, 1028-1039.

Vogel, J.C. and Eaton, J.K. (1985). Combined Heat Transfer and Fluid Dynamic Measurements Downstream of a Backward-Facing Step, *Trans. ASME J. Heat Transfer,* Vol.107, 922-929.

Thermophysical Properties at Critical and Supercritical Conditions

Igor Pioro and Sarah Mokry
University of Ontario Institute of Technology
Canada

1. Introduction

Prior to a general discussion on specifics of thermophysical properties at critical and supercritical pressures it is important to define special terms and expressions used at these conditions. For better understanding of these terms and expressions Fig. 1 is shown below.

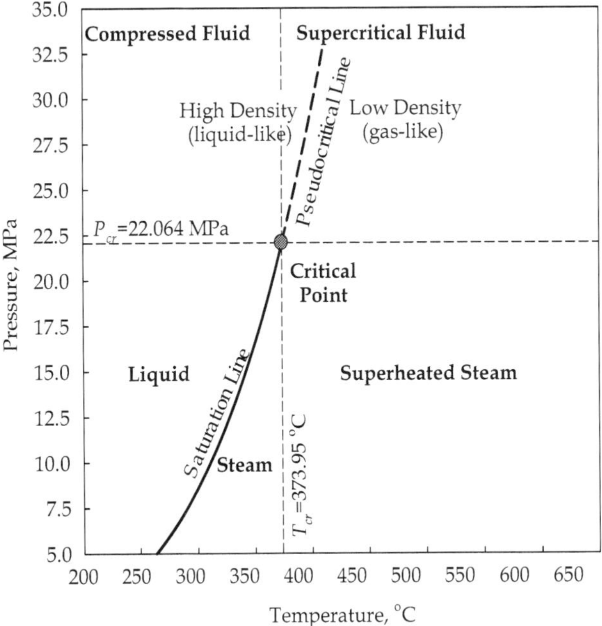

Fig. 1. Pressure-Temperature diagram for water.

Definitions of selected terms and expressions related to critical and supercritical regions
Compressed fluid is a fluid at a pressure above the critical pressure, but at a temperature below the critical temperature.

Critical point (also called a *critical state*) is a point in which the distinction between the liquid and gas (or vapour) phases disappears, i.e., both phases have the same temperature, pressure and volume or density. The *critical point* is characterized by the phase-state parameters T_{cr}, P_{cr} and V_{cr} (or ρ_{cr}), which have unique values for each pure substance.

Near-critical point is actually a narrow region around the critical point, where all thermophysical properties of a pure fluid exhibit rapid variations.

Pseudocritical line is a line, which consists of pseudocritical points.

Pseudocritical point (characterized with P_{pc} and T_{pc}) is a point at a pressure above the critical pressure and at a temperature ($T_{pc} > T_{cr}$) corresponding to the maximum value of the specific heat at this particular pressure.

Supercritical fluid is a fluid at pressures and temperatures that are higher than the critical pressure and critical temperature. However, in the present chapter, a term *supercritical fluid* includes both terms – a *supercritical fluid* and *compressed fluid*.

Supercritical "steam" is actually supercritical water, because at supercritical pressures fluid is considered as a single-phase substance. However, this term is widely (and incorrectly) used in the literature in relation to supercritical "steam" generators and turbines.

Superheated steam is a steam at pressures below the critical pressure, but at temperatures above the critical temperature.

2. Historical note on using supercritical-pressure fluids

The use of supercritical fluids in different processes is not new and has not been invented by humans. Mother Nature has been processing minerals in aqueous solutions at near or above the critical point of water for billions of years (Levelt Sengers, 2000). Only in the late 1800s, scientists started to use this natural process, called hydrothermal processing in their labs for creating various crystals. During the last 50 – 60 years, this process (operating parameters - water pressures from 20 to 200 MPa and temperatures from 300 to 500°C) has been widely used in the industrial production of high-quality single crystals (mainly gem stones) such as quartz, sapphire, titanium oxide, tourmaline, zircon and others.

First works devoted to the problem of heat transfer at supercritical pressures started as early as the 1930s. Schmidt and his associates investigated free-convection heat transfer of fluids at the near-critical point with the application to a new effective cooling system for turbine blades in jet engines. They found that the free convection heat transfer coefficient at the near-critical state was quite high, and decided to use this advantage in single-phase thermosyphons with an intermediate working fluid at the near-critical point (Pioro and Pioro, 1997).

In the 1950s, the idea of using supercritical water appeared to be rather attractive for thermal power industry. The objective was increasing the total thermal efficiency of coal-fired power plants. At supercritical pressures there is no liquid-vapour phase transition; therefore, there is no such phenomenon as Critical Heat Flux (CHF) or dryout. Only within a certain range of parameters a deteriorated heat transfer may occur. Work in this area was mainly performed in the former USSR and in the USA in the 1950s – 1980s (International Encyclopedia of Heat & Mass Transfer, 1998).

In general, the total thermal efficiency of modern thermal power plants with subcritical-parameters steam generators is about 36 – 38%, but reaches 45 – 50% with supercritical parameters, i.e., with a "steam" pressure of 23.5 – 26 MPa and inlet turbine temperature of 535 – 585°C thermal efficiency is about 45% and even higher at ultra-supercritical parameters (25 – 35 MPa and 600 – 700°C) (see Fig. 2).

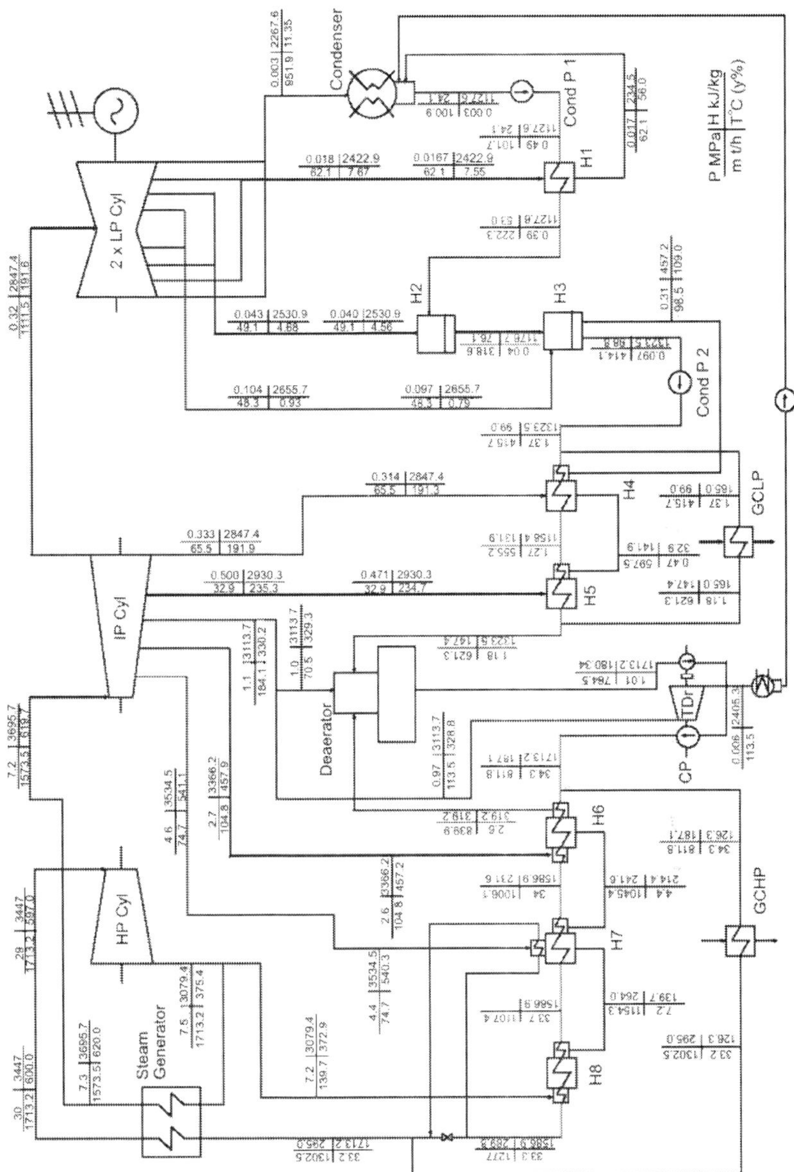

Cyl – Cylinder; H – Heat exchanger (feedwater heater); CP – Circulation Pump; TDr – Turbine Drive; Cond P – Condensate Pump; GCHP – Gas Cooler of High Pressure; and GCLP – Gas Cooler of Low Pressure.

Fig. 2. Single-reheat-regenerative cycle 600-MW$_{el}$ Tom'-Usinsk thermal power plant (Russia) layout (Kruglikov et al., 2009).

At the end of the 1950s and the beginning of the 1960s, early studies were conducted to investigate a possibility of using supercritical water in nuclear reactors (Pioro and Duffey, 2007). Several designs of nuclear reactors using supercritical water were developed in Great Britain, France, USA, and the former USSR. However, this idea was abandoned for almost 30 years with the emergence of Light Water Reactors (LWRs) and regained interest in the 1990s following LWRs' maturation.

Currently, SuperCritical Water-Cooled nuclear Reactor (SCWR) concepts are one of six options included into the next generation or Generation IV nuclear systems. The SCWR concepts therefore follow two main types, the use of either: (a) a large reactor pressure vessel (Fig. 3) with a wall thickness of about 0.5 m to contain the reactor core (fuelled) heat source, analogous to conventional PWRs and BWRs, or (b) distributed pressure tubes or channels analogous to conventional CANDU®1 nuclear reactors (Fig. 4). In general, mainly thermal-spectrum SCWRs are currently under development worldwide. However, several concepts of fast SCWRs are also considered (Oka et al., 2010; Pioro and Duffey, 2007).

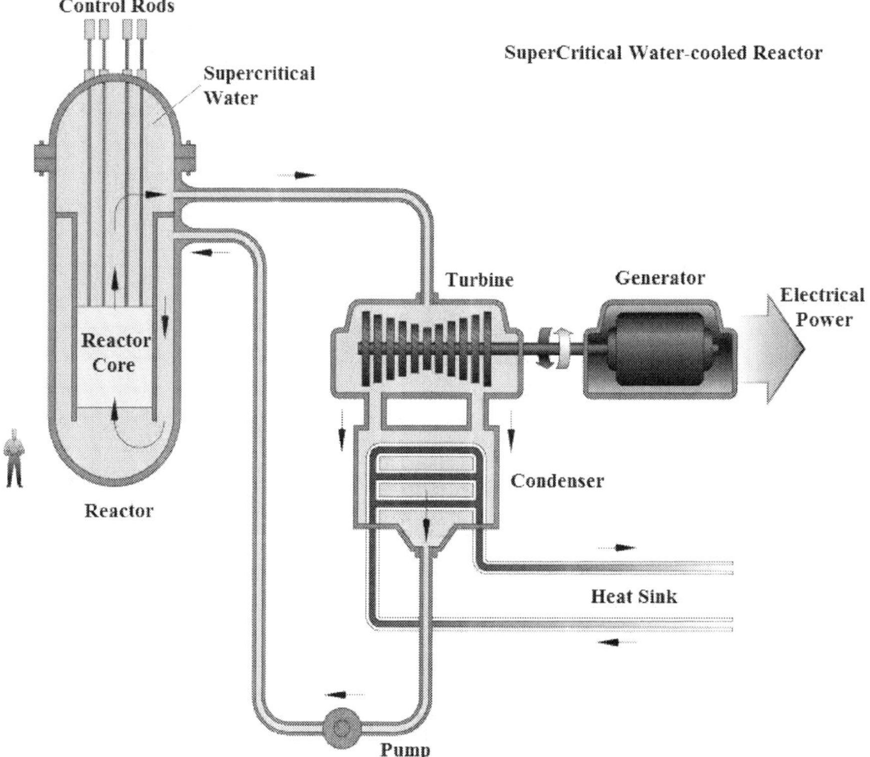

Fig. 3. Pressure-vessel SCWR schematic (courtesy of DOE USA).

---

1 CANDU® (CANada Deuterium Uranium) is a registered trademark of Atomic Energy of Canada Limited (AECL).

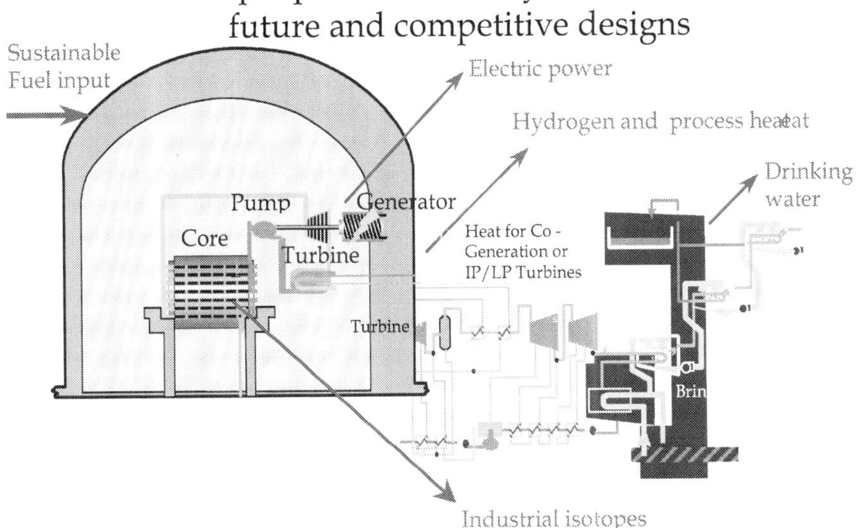

Fig. 4. General scheme of pressure-channel SCW CANDU reactor: IP – intermediate-pressure turbine and LP – low-pressure turbine (courtesy of Dr. Duffey, AECL).

Use of supercritical water in power-plant steam generators is the largest application of a fluid at supercritical pressures in industry. However, other areas exist in which supercritical fluids are used or will be implemented in the near future (Pioro and Duffey, 2007):

- using supercritical carbon-dioxide Brayton cycle for Generation IV Sodium Fast Reactors (SFRs), Lead-cooled Fast Reactors (LFRs) (Fig. 5) and High Temperature helium-cooled thermal Reactors (HTRs);
- using supercritical carbon dioxide for cooling printed circuits;
- using near-critical helium to cool coils of superconducting electromagnets, superconducting electronics and power-transmission equipment;
- using supercritical hydrogen as a fuel for chemical and nuclear rockets;
- using supercritical methane as a coolant and fuel for supersonic transport;
- using liquid hydrocarbon coolants and fuels at supercritical pressures in cooling jackets of liquid rocket engines and in fuel channels of air-breathing engines;
- using supercritical carbon dioxide as a refrigerant in air-conditioning and refrigerating systems;
- using a supercritical cycle in the secondary loop for transformation of geothermal energy into electricity;
- using SuperCritical Water Oxidation (SCWO) technology for treatment of industrial and military wastes;
- using carbon dioxide in the Supercritical Fluid Leaching (SFL) method for removal uranium from radioactive solid wastes and in decontamination of surfaces; and
- using supercritical fluids in chemical and pharmaceutical industries in such processes as supercritical fluid extraction, supercritical fluid chromatography, polymer processing and others.

The most widely used supercritical fluids are water and after that carbon dioxide, helium and refrigerants (Pioro and Duffey, 2007). Usually, carbon dioxide and refrigerants are considered as modeling fluids instead of water due to significantly lower critical pressures and temperatures, which decrease complexity and costs of thermalhydraulic experiments. Therefore, knowledge of thermophysical-properties specifics at critical and supercritical pressures is very important for safe and efficient use of fluids in various industries.

Lead-Cooled Fast Reactor

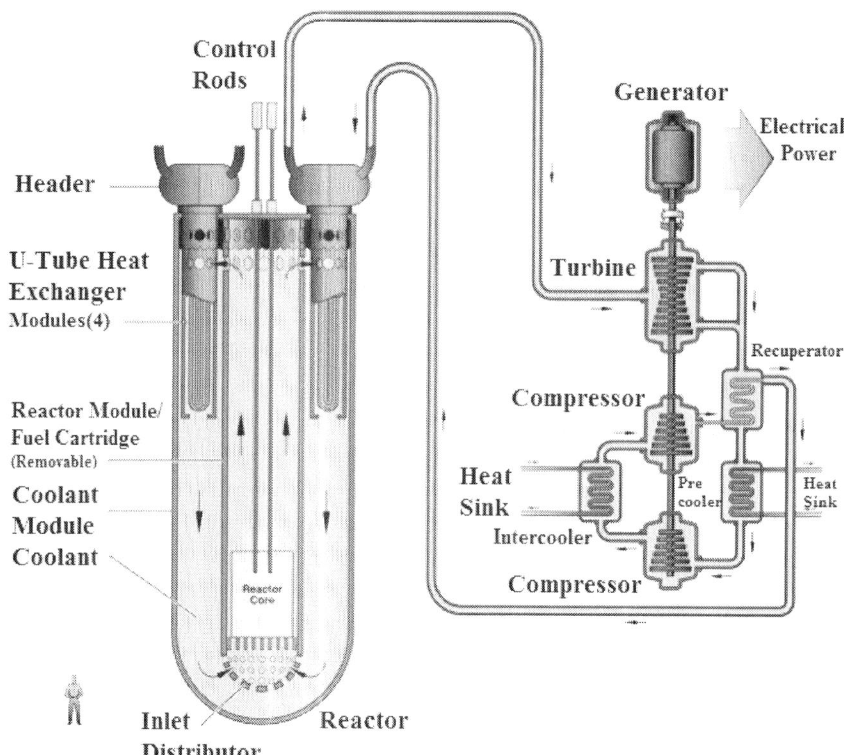

Fig. 5. Lead-cooled Fast Reactor with supercritical carbon dioxide Brayton cycle (courtesy of DOE USA).

3. Thermophysical properties at critical and supercritical pressures

General trends of various properties near the critical and pseudocritical points (Pioro, 2008; Pioro and Duffey, 2007) can be illustrated on a basis of those of water (Figs. 6-9). Figure 6 shows variations in basic thermophysical properties of water at the critical (P_{cr} = 22.064 MPa) and three supercritical pressures (P = 25.0, 30.0, and 35.0 MPa) (also, in addition see Fig. 7). Thermophysical properties of water and other 83 fluids and gases at different pressures and temperatures, including critical and supercritical regions, can be calculated using the NIST REFPROP software (2007). Critical parameters of selected fluids are listed in Table 1.

At the critical and supercritical pressures a fluid is considered as a single-phase substance in spite of the fact that all thermophysical properties undergo significant changes within the critical and pseudocritical regions. Near the critical point, these changes are dramatic (see Fig. 6). In the vicinity of pseudocritical points, with an increase in pressure, these changes become less pronounced (see Figs. 6 and 9).

Fluid	P_{cr}, MPa	T_{cr}, °C	ρ_{cr}, kg/m^3
Carbon dioxide (CO_2)	7.3773	30.978	467.6
Freon-12 (Di-chloro-di-fluoro-methane, CCl_2F_2)	4.1361	111.97	565.0
Freon-134a (1,1,1,2-tetrafluoroethane, CH_2FCF_3)	4.0593	101.06	511.9
Water (H_2O)	22.064	373.95	322.39

Table 1. Critical parameters of selected fluids (Pioro and Duffey, 2007).

Also, it can be seen that properties such as density and dynamic viscosity undergo a significant drop (near the critical point this drop is almost vertical) within a very narrow temperature range (see Figs. 6a,b and 7), while the kinematic viscosity and specific enthalpy undergo a sharp increase (see Figs. 6d,g and 7). The volume expansivity, specific heat, thermal conductivity and Prandtl number have peaks near the critical and pseudocritical points (see Figs. 6c,e,f,h, 7 and 8). Magnitudes of these peaks decrease very quickly with an increase in pressure (see Fig. 9). Also, "peaks" transform into "humps" profiles at pressures beyond the critical pressure. It should be noted that the dynamic viscosity, kinematic viscosity and thermal conductivity undergo through the minimum right after the critical and pseudocritical points (see Fig. 6b,d,f).

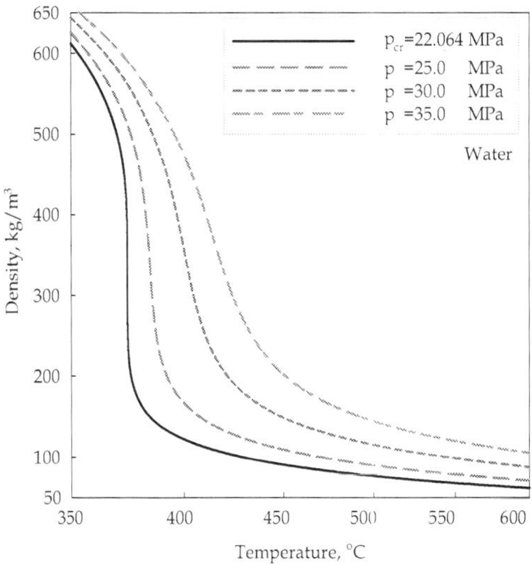

Fig. 6a. Density vs. Temperature: Water.

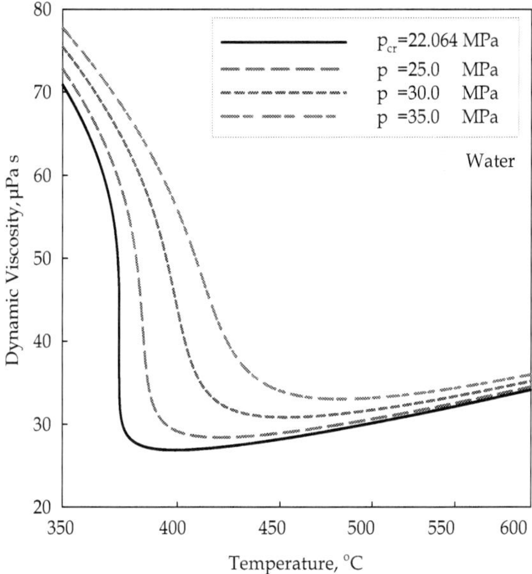

Fig. 6b. Dynamic viscosity vs. Temperature: Water.

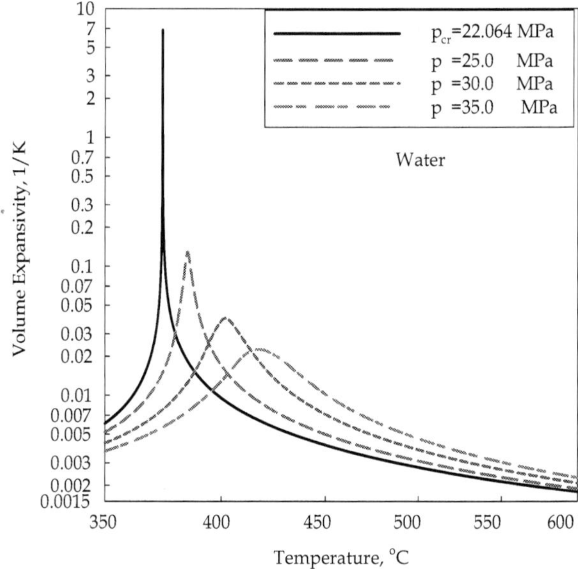

Fig. 6c. Volume expansivity vs. Temperature: Water.

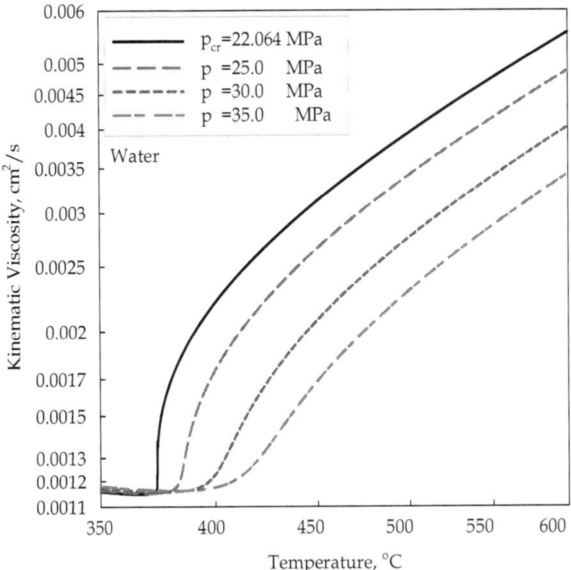

Fig. 6d. Kinematic viscosity vs. Temperature: Water.

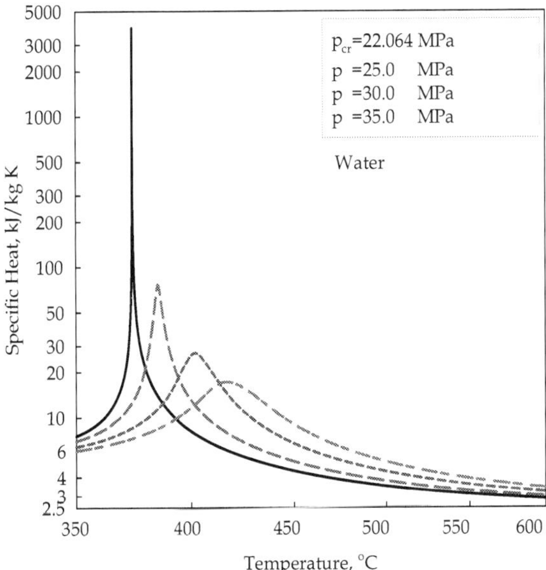

Fig. 6e. Specific heat vs. Temperature: Water.

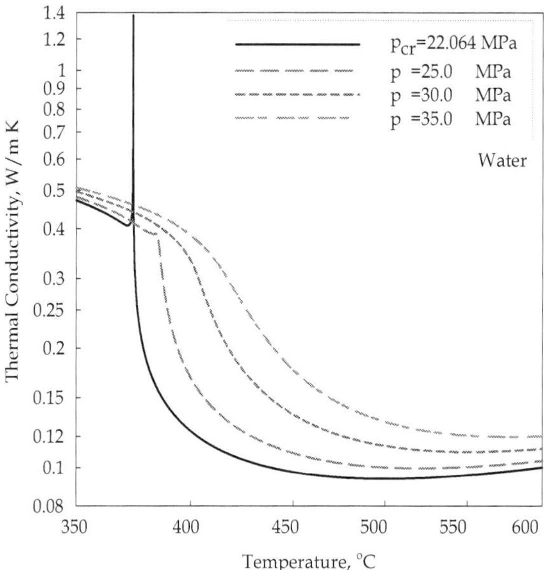

Fig. 6f. Thermal conductivity vs. Temperature: Water.

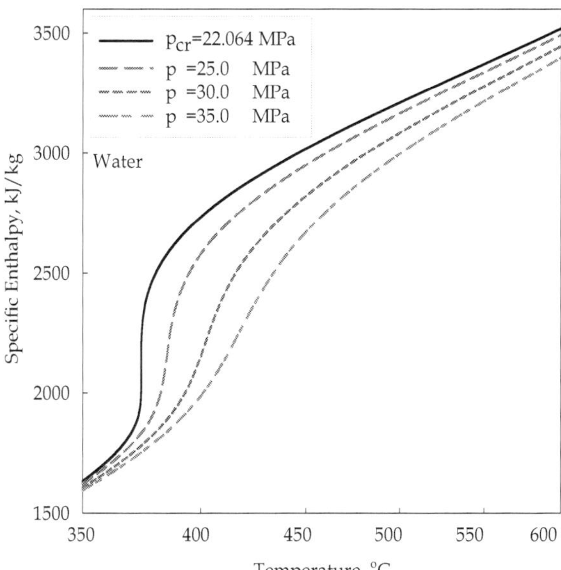

Fig. 6g. Specific enthalpy vs. Temperature: Water.

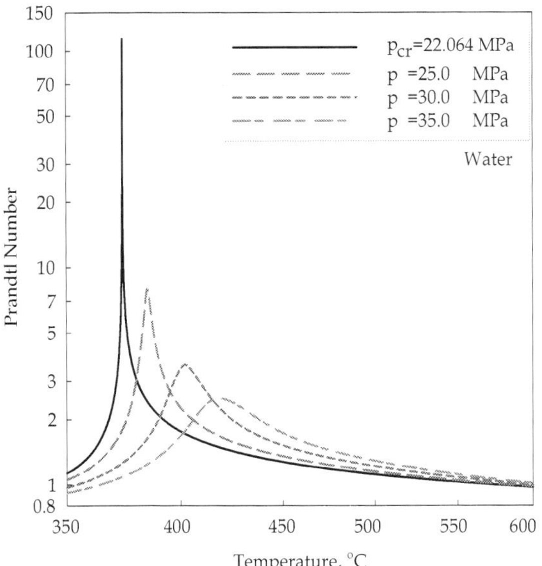

Fig. 6h. Prandtl number vs. Temperature: Water.

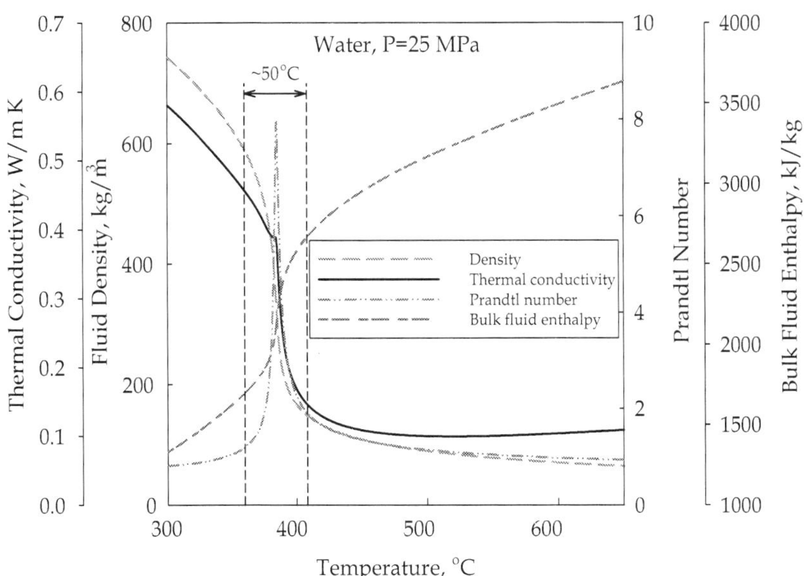

Fig. 7. Variations of selected thermophysical properties of water near pseudocritical point: Pseudocritical region at 25 MPa is about ±25°C around pseudocritical point.

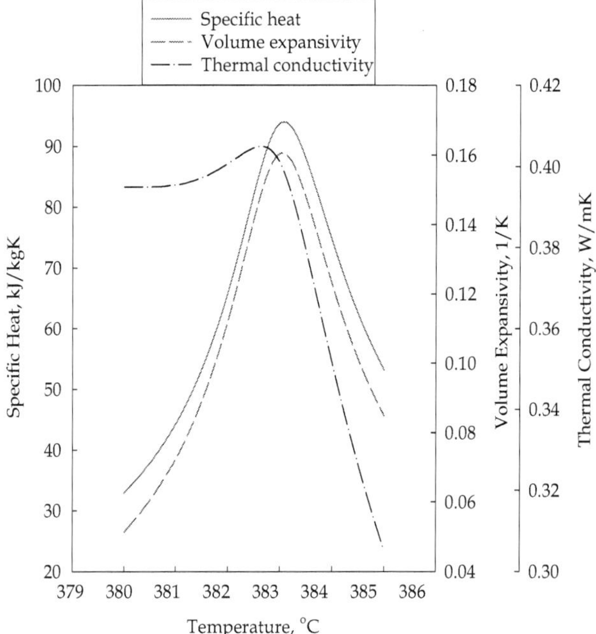

Fig. 8. Specific heat, volume expansivity and thermal conductivity vs. temperature: Water, P = 24.5 MPa.

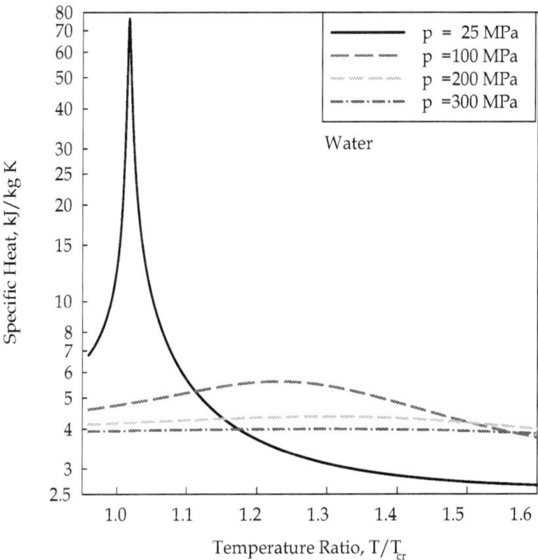

Fig. 9. Specific heat variations at various pressures: Water.

Pressure, MPa	Pseudocritical temperature, °C	Peak value of specific heat, kJ/kg·K
23	377.5	284.3
24	381.2	121.9
25	384.9	76.4
26	388.5	55.7
27	392.0	43.9
28	395.4	36.3
29	398.7	30.9
30	401.9	27.0
31	405.0	24.1
32	408.1	21.7
33	411.0	19.9
34	413.9	18.4
35	416.7	17.2

Table 2. Values of pseudocritical temperature and corresponding peak values of specific heat within wide range of pressures.

Pressure, MPa	Pseudocritical temperature, °C	Temperature, °C	Specific heat, kJ/kg·K	Volume expansivity, 1/K	Thermal conductivity, W/m·K
p_{cr}=22.064	t_{cr}=374.1	–	∞	∞	∞
22.5	375.6	–	690.6	1.252	0.711
23.0	–	377.4	–	–	0.538
	377.5	–	284.3	0.508	–
23.5	–	379.2	–	–	0.468
	–	379.3	–	0.304	–
	379.4	–	171.9	–	–
24.0	–	381.0	–	–	0.429
	381.2	–	121.9	0.212	–
24.5	–	382.6	–	–	0.405
	–	383.0	–	0.161	–
	383.1	–	93.98	–	–
25.0	–	384.0	–	–	0.389
	384.9	–	76.44	–	–
	–	385.0	–	0.128	–
25.5	386.7	–	64.44	0.107	no peak
26.0	388.5	–	55.73	0.090	0.355
27.0	392.0	–	43.93	0.069	0.340
28.0	395.4	–	36.29	0.056	0.329
29.0	398.7	–	30.95	0.046	0.321
30.0	401.9	–	27.03	0.039	0.316

Table 3. Peak values of specific heat, volume expansivity and thermal conductivity in critical and near pseudocritical points.

The specific heat of water (as well as of other fluids) has the maximum value at the critical point (see Fig. 6e). The exact temperature that corresponds to the specific-heat peak above the critical pressure is known as the pseudocritical temperature (see Fig. 1 and Table 2). At pressures approximately above 300 MPa (see Fig. 9) a peak (here it is better to say "a hump") in specific heat almost disappears, therefore, such term as a *pseudocritical point* does not exist anymore. The same applies to the *pseudocritical line*. It should be noted that peaks in the thermal conductivity and volume expansivity may not correspond to the pseudocritical temperature (see Table 3 and Figure 8).

In early studies, i.e., approximately before 1990, a peak in thermal conductivity was not taken into account. Later, this peak was well established (see Fig. 6f) and included into thermophysical data and software. The peak in thermal conductivity diminishes at about 25.5 MPa for water (see Fig. 6f and Table 3).

In general, crossing the pseudocritical line from left to right (see Fig. 1) is quite similar as crossing the saturation line from liquid into vapour. The major difference in crossing these two lines is that all changes (even drastic variations) in thermophysical properties at supercritical pressures are gradual and continuous, which take place within a certain temperature range (see Fig. 6). On contrary, at subcritical pressures we have properties discontinuation on the saturation line: one value for liquid and another for vapour (see Fig. 10). Therefore, supercritical fluids behave as single-phase substances. Also, dealing with supercritical fluids we apply usually a term *"pseudo"* in front of a *critical point, boiling, film boiling*, etc.

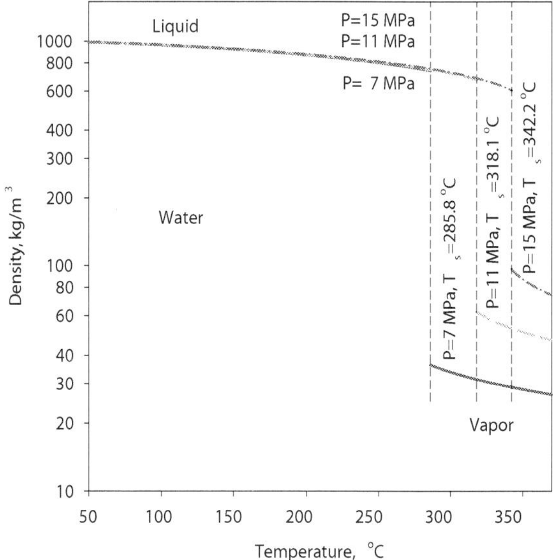

Fig. 10. Density variations at various subcritical pressures for water: Liquid and vapour.

In addition to supercritical-water properties, supercritical properties of R-12 (Richards et al., 2010) are shown in Fig. 11 for reference purposes. Properties of supercritical carbon dioxide, helium and R-134a are shown in Pioro and Duffey (2007).

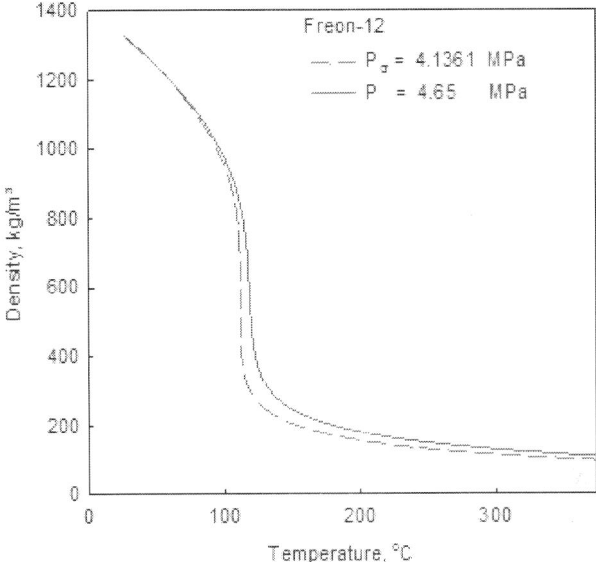

Fig. 11a. Density vs. Temperature: R-12

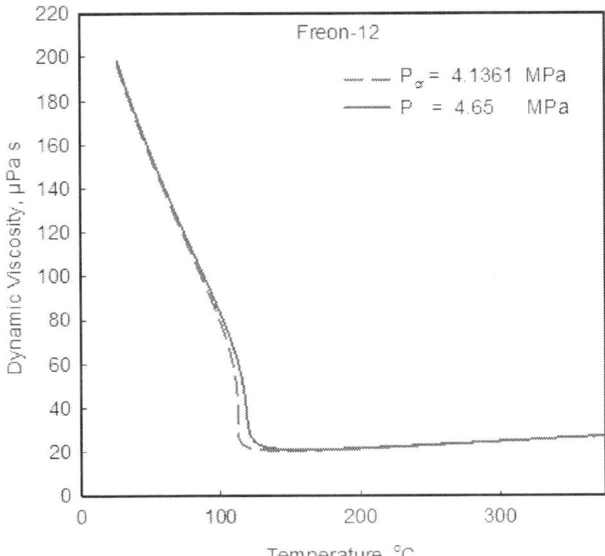

Fig. 11b. Dynamic viscosity vs. Temperature: R-12.

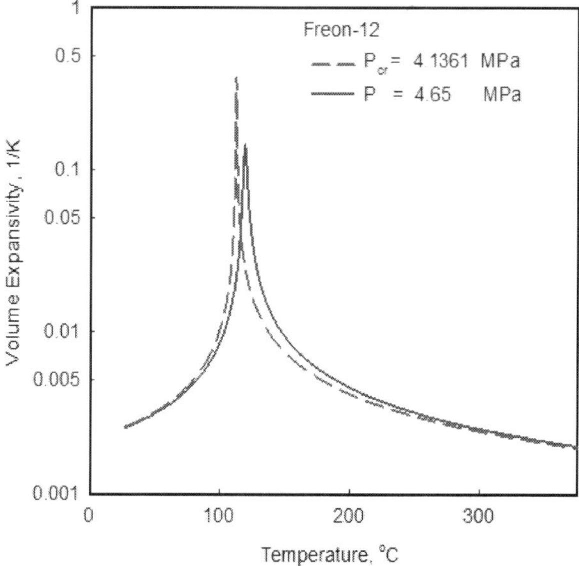

Fig. 11c. Volume expansivity vs. Temperature: R-12.

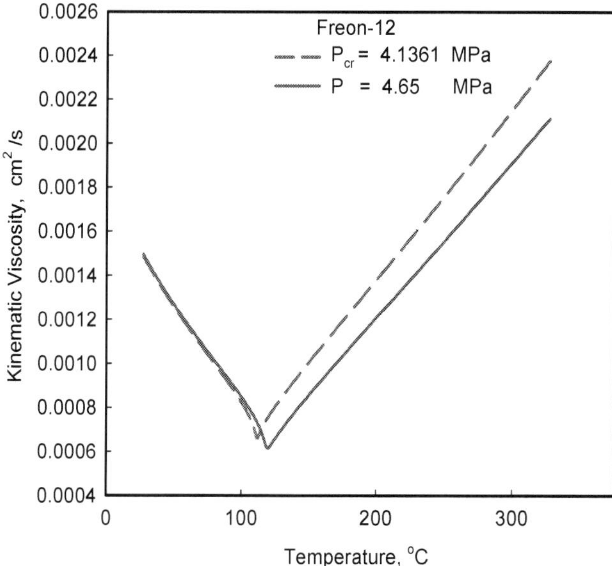

Fig. 11d. Kinematic viscosity vs. Temperature: R-12.

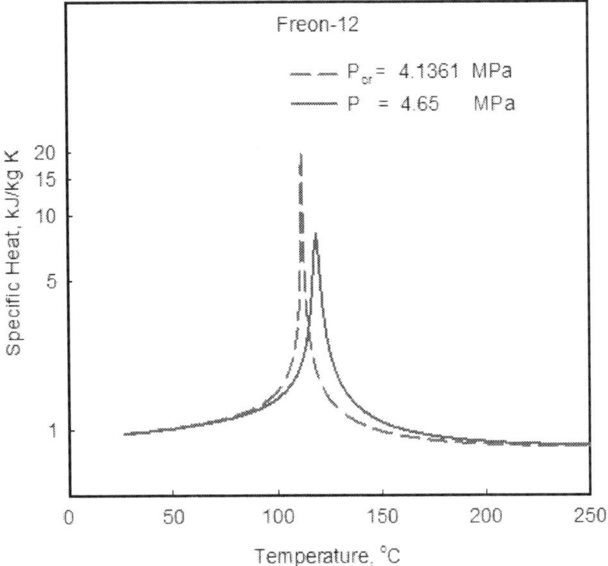

Fig. 11e. Specific heat vs. Temperature: R-12.

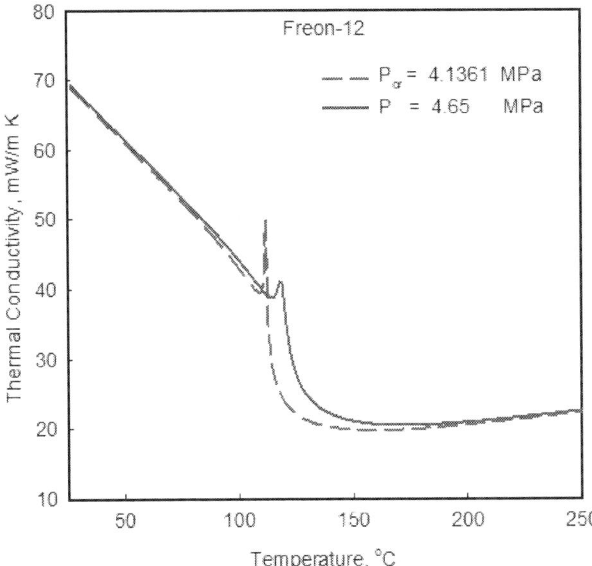

Fig. 11f. Thermal conductivity vs. Temperature: R-12.

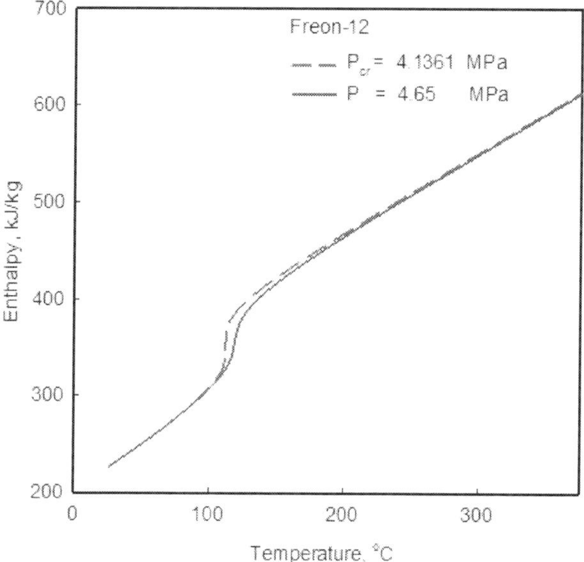

Fig. 11g. Specific enthalpy vs. Temperature.

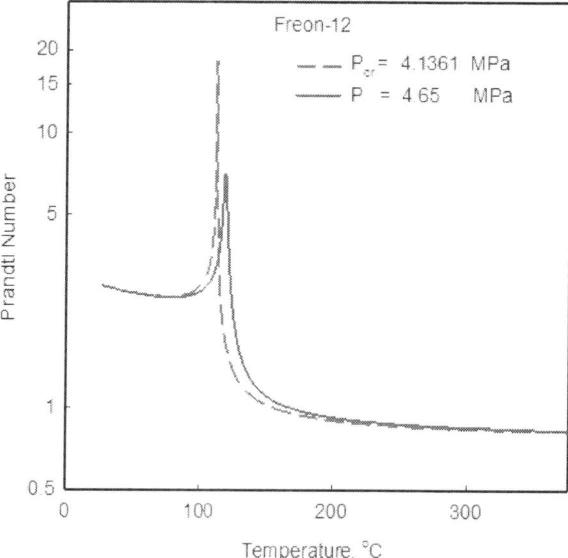

Fig. 11h. Prandlt number vs. Temperature: R-12.

4. Acknowledgements

Financial supports from the NSERC Discovery Grant and NSERC/NRCan/AECL Generation IV Energy Technologies Program are gratefully acknowledged.

5. Nomenclature

P , p pressure, Pa
T , t temperature, °C
V specific volume, m³/kg

Greek letters
ρ density, kg/m³

Subscripts
cr   critical
pc   pseudocritical

Abbreviations:
BWR   Boiling Water Reactor
CHF   Critical Heat Flux
HTR   High Temperature Reactor
LFR   Lead-cooled Fast Reactor
LWR   Light-Water Reactor
NIST   National Institute of Standards and Technology (USA)
PWR   Pressurized Water Reactor
SCWO   SuperCritical Water Oxidation
SFL   Supercritical Fluid Leaching
SFR   Sodium Fast Reactor
USA   United States of America
USSR   Union of Soviet Socialist Republics

6. Reference

International Encyclopedia of Heat & Mass Transfer, 1998. Edited by G.F. Hewitt, G.L. Shires and Y.V. Polezhaev, CRC Press, Boca Raton, FL, USA, pp. 1112–1117 (Title "Supercritical heat transfer").

Kruglikov, P.A., Smolkin, Yu.V. and Sokolov, K.V., 2009. Development of Engineering Solutions for Thermal Scheme of Power Unit of Thermal Power Plant with Supercritical Parameters of Steam, (In Russian), Proc. of Int. Workshop "Supercritical Water and Steam in Nuclear Power Engineering: Problems and Solutions", Moscow, Russia, October 22–23, 6 pages.

Levelt Sengers, J.M.H.L., 2000. Supercritical Fluids: Their Properties and Applications, Chapter 1, in book: Supercritical Fluids, editors: E. Kiran et al., NATO Advanced Study Institute on Supercritical Fluids – Fundamentals and Application, NATO Science Series, Series E, Applied Sciences, Kluwer Academic Publishers, Netherlands, Vol. 366, pp. 1–29.

National Institute of Standards and Technology, 2007. NIST Reference Fluid Thermodynamic and Transport Properties-REFPROP. NIST Standard Reference Database 23, Ver. 8.0. Boulder, CO, U.S.: Department of Commerce.

Oka, Y, Koshizuka, S., Ishiwatari, Y., and Yamaji, A., 2010. *Super Light Water Reactors and Super Fast Reactors*, Springer, 416 pages.

Pioro, I.L., 2008. Thermophysical Properties at Critical and Supercritical Pressures, Section 5.5.16 in Heat Exchanger Design Handbook, Begell House, New York, NY, USA, 14 pages.

Pioro, I.L. and Duffey, R.B., 2007. *Heat Transfer and Hydraulic Resistance at Supercritical Pressures in Power Engineering Applications*, ASME Press, New York, NY, USA, 328 pages.

Pioro, L.S. and Pioro, I.L., *Industrial Two-Phase Thermosyphons*, 1997, Begell House, Inc., New York, NY, USA, 288 pages.

Richards, G., Milner, A., Pascoe, C., Patel, H., Peiman, W., Pometko, R.S., Opanasenko, A.N., Shelegov, A.S., Kirillov, P.L. and Pioro, I.L., 2010. Heat Transfer in a Vertical 7-Element Bundle Cooled with Supercritical Freon-12, Proceedings of the 2nd Canada-China Joint Workshop on Supercritical Water-Cooled Reactors (CCSC-2010), Toronto, Ontario, Canada, April 25-28, 10 pages.

23

Gas-Solid Heat and Mass Transfer Intensification in Rotating Fluidized Beds in a Static Geometry

Juray De Wilde

Université catholique de Louvain, Dept. Materials and Process Engineering (IMAP), Place Sainte Barbe 2, Réaumur building, 1348 Louvain-la-Neuve, Tel.: +32 10 47 2323, Fax: +32 10 47 4028, e-mail: Juray.DeWilde@UCLouvain.be
Belgium

1. Introduction

In different types of reactors, gas and solid particles are brought into contact and gas-solid mass and heat transfer is to be optimized. This is for example the case with heterogeneous catalytic reactions, the porous solid particle providing the catalytic sites and the reactants having to transfer from the bulk flow to the solid surface from where they can diffuse into the pores of the catalyst [Froment et al., 2010]. Gas-solid heat transfer can, for example, be required to provide the heat for endothermic reactions taking place inside the solid catalyst. Intra-particle mass transfer limitations can be encountered as well, but this chapter will focus on interfacial mass and heat transfer. The overall rate of reaction is on the one hand determined by the intrinsic reaction rate, which depends on the catalyst used, and on the other hand by the rates of mass and heat transfer, which depends on the reactor configuration and operating conditions used. Hence, the optimal use of a catalyst requires a reactor in which conditions can be generated allowing sufficiently fast mass and heat transfer. This is not always possible and usually an optimization is carried out accounting for pressure drop and stability limitations. This chapter focuses on fluidized bed type reactors and the limitations of conventional fluidized beds will be explained in more detail in the next section.

To gain some insight in where gas-solid mass and heat transfer limitations come from, consider the flux expression for one-dimensional diffusion of a component A over a film around the solid particle in which the resistance for fluid-to-particle interfacial mass and heat transfer is localized:

$$N_A = -C_t D_{Am} \frac{dy_A}{dz} + y_A \left(N_A + N_B + N_R + N_S + ... \right) \tag{1}$$

In (1), a mean binary diffusivity for species A through the mixture of other species is introduced. For the calculation of the mean binary diffusivity, see Froment et al. [2010]. When a chemical reaction

$$aA + bB + ... \Leftrightarrow rR + sS + ... \tag{2}$$

takes place, the fluxes of the different components are related through the reaction stoichiometry, so that (1) becomes:

$$N_A = -C_t D_{Am} \frac{dy_A}{dz} + y_A N_A \left(1 + \frac{b}{a} - \frac{r}{a} - \frac{s}{a} - ...\right) \tag{3}$$

Solving (3) for N_A:

$$N_A = \frac{-C_t D_{Am}}{1 + \delta_A y_A} \frac{dy_A}{dz} \tag{4}$$

with

$$\delta_A = \frac{r + s + ... - a - b - ...}{a} \tag{5}$$

Integrating (4) over the (unknown) film thickness L for steady state diffusion and using an average constant value for the mean binary diffusivity results in:

$$N_A = \left(\frac{C_t D_{Am}}{L}\right) \frac{y_{A0} - y_A(L)}{y_{fA}} \tag{6}$$

where the film factor, y_{fA}, accounting for non-equimolar counter-diffusion, has been introduced:

$$y_{fA} = \frac{\left(1 + \delta_A y_A\right) - \left(1 + \delta_A y_{As}^s\right)}{\ln \dfrac{1 + \delta_A y_A}{1 + \delta_A y_{As}^s}} \tag{7}$$

Expression (6) shows the importance of the film thickness, L, which depends on the reactor design and the operating conditions. This implies a difficulty for the practical use of (6). In practice, gas-solid mass and heat transfer are modeled in terms of a mass, respectively heat transfer coefficient, noted k_g and h_f. For interfacial mass transfer:

$$N_A = k_g \left(y_A - y_{As}^s\right) = \frac{k_g^0}{y_{fA}} \left(y_A - y_{As}^s\right) \tag{8}$$

where the film factor was factored out, introducing the interfacial mass transfer coefficient for equimolar counter-diffusion, k_g^0. For interfacial heat transfer, the heat flux is written as:

$$Q_A = h_f \left(T - T_s^s\right) \tag{9}$$

For the calculation of k_g^0, correlations in terms of the j_D factor are typically used:

$$k_g^0 = \frac{j_D G}{M_m} Sc^{-2/3} \tag{10}$$

where Sc is the Schmidt number defined as

$$Sc = \frac{\mu}{\rho_f D_{Am}}$$ (11)

and G is the superficial fluid mass flux through the particle bed:

$$G = \varepsilon_g \rho_g (u - v)$$ (12)

Comparing (8) and (10) to (6), it is seen that the mean binary diffusivity is enclosed in Sc. The film thickness, L, is accounted for via the j_D factor which is correlated in terms of the Reynolds number:

$$j_D = f(\mathrm{Re}_p)$$ (13)

with Re_p the particle based Reynolds number:

$$\mathrm{Re}_p = \frac{d_p G}{\mu}$$ (14)

In a similar way, the heat transfer coefficient h_f is usually modeled in terms of the j_H factor and the Prandtl number which contains the fluid conductivity:

$$h_f = j_H c_p G \mathrm{Pr}^{-2/3}$$ (15)

with

$$j_H = f(\mathrm{Re}_p)$$ (16)

and

$$\mathrm{Pr} = \mu c_p / \lambda$$ (17)

Correlations (13) and (16) depend on the reactor type and design - see Froment et al. [2010] and Schlünder [1978] for a comprehensive discussion. For conventional, gravitational fluidized beds, Perry and Chilton [1984] proposed, for example,

$$\mathrm{Re}_p = 2.05 \mathrm{Re}_p^{-0.468}$$ (18)

Balakrishnan and Pei [1975] proposed:

$$j_H = 0.043 \left[\frac{d_p g}{(\varepsilon_g u)^2} \frac{(\rho_s - \rho_g)\varepsilon_s^2}{\rho_g} \right]^{0.25}$$ (19)

Expressions (10)-(19) show in particular the importance of earth gravity, the particle bed density, and the gas-solid slip velocity for the value of the gas-solid heat and mass transfer coefficients.

2. The limitations of conventional fluidized beds

In conventional gravitational fluidized beds, particles are fluidized against gravity, a constant on earth. This limits the window of operating conditions at which gravitational fluidized beds can be operated. The fluidization behavior depends on the type of particles that are fluidized. The typical fluidization behavior of fine particles is illustrated in Figure 1.

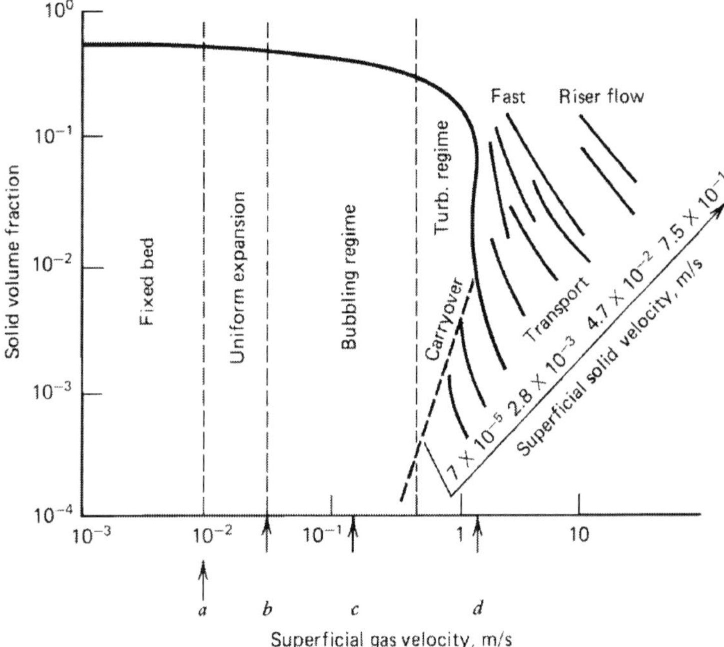

Fig. 1. Fluidization regimes with fine particles. (a) Minimum fluidization velocity; (b) Minimum bubbling; (c) Terminal velocity; (d) Blowout velocity. From Froment et al. [2010] after Squires et al. [1985].

The particle bed is fluidized when the gas-solid slip velocity exceeds the minimum fluidization velocity of the particles. When increasing the gas velocity, the uniformly fluidized state becomes unstable and bubbles appear (Figure 2). These meso-scale non-uniformities are detrimental for the gas-solid contact, but their dynamic behavior improves mixing in the particle bed. Further improvement of the gas-solid contact and the particle bed mixing can be achieved by further increasing the gas velocity and entering into the so-called turbulent regime. The gas-solid slip velocity can not be increased beyond the terminal velocity of the particles, which naturally depends on the gravity field in which the particles are suspended. Particles are then entrained by the gas and a transport regime is reached, meaning the particles are transported with the gas through the reactor. Meso-scale non-uniformities here appear under the form of clusters (Figure 2). The limitation of the gas-solid slip velocity implies limitations on the gas-solid mass and heat transfer.

Fig. 2. Non-uniformity in the particle distribution. Appearance of bubbles and clusters.
From Agrawal et al. [2001].

Macro- to reactor-scale non-uniformities have to be avoided, as they imply complete bypassing of the solids by the gas. In gravitational fluidized beds operated in a non-transport regime, this requires a certain weight of particles above the gas distributor and limits the particle bed width-to-height ratio. This on its turn introduces a constraint on the fluidization gas flow rate that can be handled per unit volume particle bed. In the transport regime, particles have to be returned from the top of the reactor to the bottom. The driving force is the weight of particles in a stand pipe and, hence, the latter should be sufficiently tall. The reactor length has to be adapted accordingly, resulting in very tall reactors. The riser reactors used in FCC, for example, are 30 to 40 m tall. The resulting gas phase and catalyst residence times in some applications limit the catalyst activity.

An important characteristic of the fluidized bed state is the particle bed density. It is directly related to the process intensity that can be reached in the reactor. The process intensity for a given reactor can be defined as how much reactant is converted per unit time and per unit reactor volume. Typically, as the gas velocity is increased, the particle bed expands and the particle bed density decreases. In a transport regime, the average particle bed density decreases significantly. In the riser regime, for example, the reactor is typically operated at 5 vol% solids or less. The process intensity is correspondingly low.

A final important limitation of gravitational fluidized beds comes from the type of particles that can be fluidized. Nano- and micro-scale particles can not be properly fluidized, the Van der Waals forces becoming too important compared to the other forces determining the fluidized bed state, i.e. the weight of the particles and the gas-solid drag force.

Most of the above mentioned limitations of gravitational fluidized beds can be removed by replacing earth gravity with a stronger force - so-called high-G operation. This has led to the development of the cylindrically shaped so-called rotating fluidized beds. A first technology of this type is based on a fluidization chamber which rotates fast around its axis of symmetry by means of a motor [Fan et al., 1985; Chen, 1987]. The moving geometry complicates sealing and continuous feeding and removal of solids and introduces additional challenges related to vibrations. Nevertheless, rotating fluidized beds have been shown successful in removing the limitations of gravitational fluidized beds and, for example, allow the fluidization of micro- and nano-particles [Qian et al., 2001; Watano et al., 2003; Quevedo et al., 2005].

In this chapter, a novel technology is focused on that allows taking advantage of high-G operation in a static geometry. The gas-solid heat and mass transfer properties and the particle bed temperature uniformity are numerically and experimentally studied. The process intensification is illustrated for the drying of biomass and for FCC.

3. The rotating fluidized beds in a static geometry

3.1 Technology description

Figure 3 shows a schematic representation of a rotating fluidized bed in a static geometry (RFB-SG) [de Broqueville, 2004; De Wilde and de Broqueville, 2007, 2008], a vortex chamber [Kochetov et al., 1969; Anderson et al., 1971; Folsom, 1974] based technology. The unique characteristic of the technology is the way the rotational motion of the particle bed is driven, i.e. by the tangential introduction of the fluidization gas in the fluidization chamber through multiple inlet slots in its outer cylindrical wall. As a result, the particle bed is, or better can be, fluidized in two directions. The fluidization gas is forced to leave the fluidization chamber via a centrally positioned chimney. The radial fluidization of the particle bed is then controlled by the radial gas-solid drag force and the solid particles inertia. In a coordinate system rotating with the particle bed, the latter appears as the centrifugal and Coriolis forces. Radial fluidization of the particle bed is, however, not essential to take fully advantage of high-G operation. What is essential for intensifying gas-solid mass and heat transfer is the increased gas-solid slip velocity at which the bed can be operated while maintaining a high particle bed density and being fluidized.

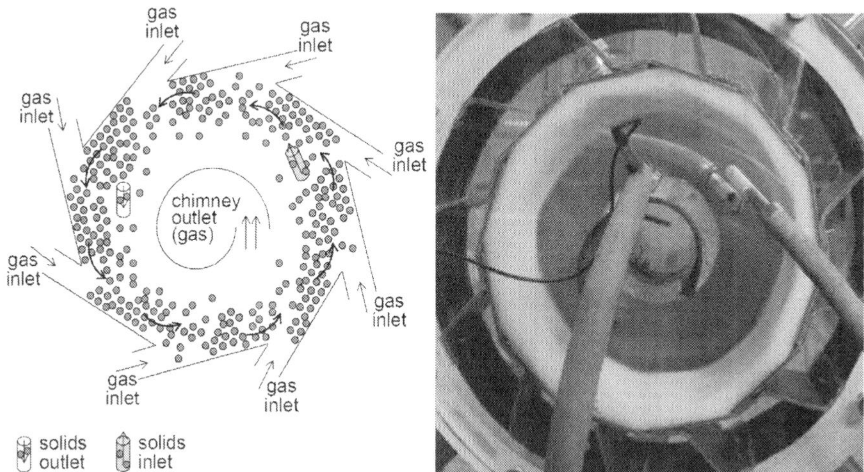

Fig. 3. The rotating fluidized bed in a static geometry. Picture from De Wilde and de Broqueville [2007].

Whether the particle bed will be radially fluidized depends mainly on the type of particles and on the fluidization chamber design, including the fluidization chamber and chimney diameters and the number and size of the gas inlets slots. At fluidization gas flow rates sufficiently high to operate high-G, the influence of the fluidization gas flow rate on the radial fluidization of the particle bed is marginal. This flexibility in the fluidization gas flow rate is an important and unique feature of rotating fluidized beds in a static geometry [De Wilde and de Broqueville, 2007, 2008, 2008b]. The explanation for this comes from the similar influence of the fluidization gas flow rate on the radial gas-solid drag force and the counteracting solid phase inertial forces resulting from the particle bed rotational motion. Experimental observations confirm the absence of radial bed expansion when increasing the fluidization gas flow rate. In some cases, even a radial bed contraction was observed.

Another important characteristic of rotating fluidized beds in a static geometry is the excellent particle bed mixing, resulting from the particle bed rotational motion and the fluctuations in the velocity field of the particles. The particle bed mixing properties were studied by De Wilde [2009] by means of a step response technique with colored particles and a close to well-mixed behavior was demonstrated at sufficiently high fluidization gas flow rates. It should be remarked that the gas phase is hardly mixed and follows a plug flow type pattern.

3.2 Theoretical evaluation of the intensification of gas-solid heat and mass transfer

As mentioned in Section 2 of this chapter, in fluidized beds, the gas-solid slip velocity, essential for the value of the gas-solid heat and mass transfer coefficient - see (10), (12), and (15), can not increase beyond the terminal velocity of the particles. An expression for the latter has been derived for gravitational fluidized beds [Froment et al., 2010] and can be extended to fluidization in a high-G field as:

$$u^{term} = \sqrt{\frac{4c \cdot d_p(\rho_s - \rho_g)}{3\rho_g C_D}} \qquad (20)$$

with c the high-G acceleration. The value of the drag coefficient, C_D, depends on the particle Reynolds number, Re_p. For Re_p below 1000:

$$C_D = \frac{24}{Re_p}\left(1 + 0.15 Re_p^{0.687}\right) \qquad (21)$$

with, for spherical particles:

$$Re_p = \frac{|\overline{u} - \overline{v}| \varepsilon_g \rho_g d_p}{\mu} \qquad (22)$$

, similar to (14). For higher Re_p:

$$C_D = 0.44 \qquad (23)$$

A unique characteristic of rotating fluidized beds in a static geometry is that the high-G acceleration appearing in (20) depends on the fluidization gas flow rate and, hence, on the gas velocity u. An estimation of the high-G acceleration can be calculated from an expression derived by de Broqueville and De Wilde [2009]. Assuming a solid body type motion of the particle bed and neglecting the contribution of the Coriolis effect:

$$c = \omega^2 r = \left(\frac{2 \cdot F_g \cdot \langle n \rangle \cdot \left(\langle v^{\tan g} \rangle \middle/ \langle u^{\tan g} \rangle \right)}{\langle \varepsilon_g \rangle \cdot \left(R^2 - R_f^2 \right) \cdot L} \right)^2 r \tag{24}$$

where r is the radial position in the particle bed, F_g is the fluidization gas flow rate, $\langle n \rangle$ is the average number of rotations made by the fluidization gas in the particle bed, $\langle \varepsilon_g \rangle$ is the average particle bed void fraction, R is the outer fluidization chamber radius, R_f is the particle bed freeboard radius, and L is the fluidization chamber length. In case the fluidization gas injected via a given gas inlet slot leaves the particle bed when approaching the next gas inlet slot, as experimentally observed by De Wilde [2009]:

$$\langle n \rangle \sim [\text{number of gas inlet slots}]^{-1} \tag{25}$$

The average tangential gas-solid slip factor, <v^{tang}>/<u^{tang}>, is determined by the shear resulting from particle-particle and particle-wall collisions and by the tangential gas-solid drag force. In the immediate vicinity of the gas inlet slots, strong variations in its value occur.
Substituting (24) in (20),

$$u^{term} = \left(\frac{2 \cdot F_g \cdot \langle n \rangle \cdot \left(\langle v^{\tan g} \rangle \middle/ \langle u^{\tan g} \rangle \right) \sqrt{r}}{\langle \varepsilon_g \rangle \cdot \left(R^2 - R_f^2 \right) \cdot L} \right) \cdot \sqrt{\frac{4 d_p \cdot (\rho_s - \rho_g)}{3 \rho_g \cdot C_D}} \tag{26}$$

At sufficiently high Re_p, (23) can be applied, and the terminal velocity of the particles is seen to be proportional to the fluidization gas flow rate and to the square root of the radial distance from the fluidization chamber central axis. The proportionality factor depends on the gas and solid phase properties and on the fluidization chamber design.
A similar analysis can be derived for the minimum fluidization velocity. Extending the expression of Wen and Yu [1966] to high-G operation:

$$u_{mf} = \frac{Re_{mf}\, \mu}{d_p \rho_g} \tag{27}$$

with:

$$Re_{mf} = \sqrt{C_1^2 + C_2 Ar} - C_1 \tag{28}$$

and:

$$Ar = \frac{d_p^3 \rho_g \left(\rho_s - \rho_g \right) c}{\mu^2} \tag{29}$$

Wen and Yu [1966] found $C_1 = 33.7$ and $C_2 = 0.0408$. Again using relation (24) between the high-G acceleration and the fluidization gas flow rate results in:

$$Ar = \frac{d_p^3 \rho_g \left(\rho_s - \rho_g \right)}{\mu^2} \left(\frac{2 \cdot F_g \cdot \langle n \rangle \cdot \left(\langle v^{\tan g} \rangle \middle/ \langle u^{\tan g} \rangle \right)}{\langle \varepsilon_g \rangle \cdot \left(R^2 - R_f^2 \right) \cdot L} \right)^2 r \tag{30}$$

Equation (30) shows that, like the terminal velocity, the minimum fluidization velocity of the particles increases with the fluidization gas flow rate and with the radial distance from the central axis of the fluidization chamber. Figure 4 illustrates the theoretical variation with the fluidization gas flow rate of the minimum fluidization and terminal velocities of the particles in rotating fluidized beds in a static geometry [de Broqueville and De Wilde, 2009]. In the case studied, the radial gas(-solid slip) velocity in the particle bed remained well between the minimum fluidization velocity and the terminal velocity of the particles over the entire fluidization gas flow rate range. Figure 5 shows the corresponding gas-solid heat transfer coefficients that can be obtained. Compared to conventional fluidized beds, rotating fluidized beds in a static geometry easily allow a one order of magnitude intensification of gas-solid heat and mass transfer.

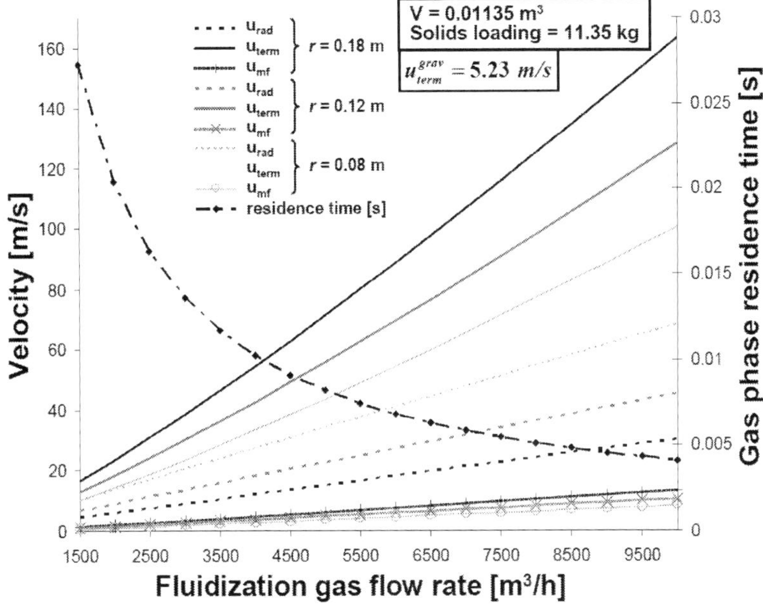

Fig. 4. Theoretical variation with the fluidization gas flow rate of the minimum fluidization and terminal velocities of the particles in rotating fluidized beds in a static geometry. Radial gas(-solid slip) velocity also shown. Conditions: ρ_g = 1.0 kg/m³, ρ_s = 2500 kg/m³, d_p = 700 μm, R = 0.18 m, R_c = 0.065 m, L = 0.135 m, $<\varepsilon_s>$ = 0.4, $<n>$ = 0.042 = 1/24, $<v^{tang}>/<u^{tang}>$ = 0.7. From de Broqueville and De Wilde [2009].

The minimum fluidization and terminal velocities being nearly proportional to the fluidization gas flow rate in rotating fluidized beds in a static geometry results from the counteracting forces - radial gas-solid drag force and solid phase inertial forces - being affected by the fluidization gas flow rate in a similar way. It should be stressed that, as illustrated in Figure 4, this implies a unique flexibility in the fluidization gas flow rate and in the gas-solid slip velocities and related gas-solid heat and mass transfer coefficients at which rotating fluidized beds in a static geometry can be operated.

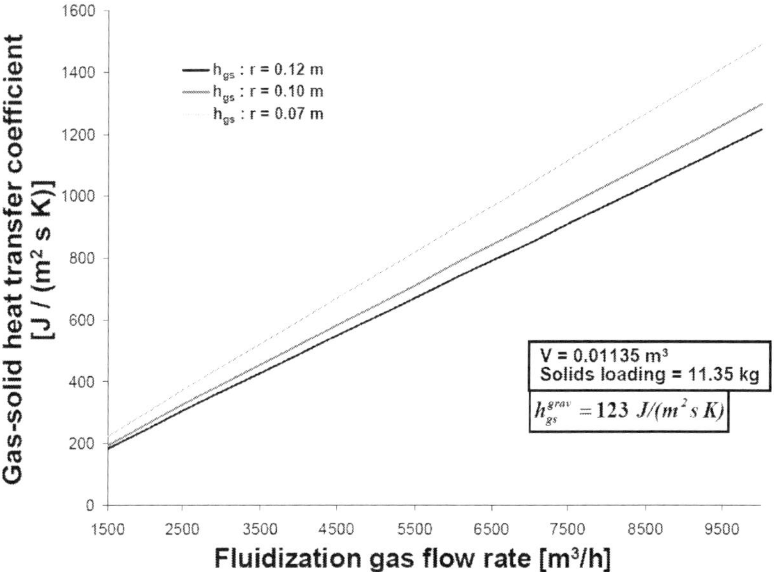

$$h_{gs}^{grav} = 123 \ J/(m^2 s \ K)$$

Fig. 5. Theoretical variation with the fluidization gas flow rate of the gas-solid heat transfer coefficient in rotating fluidized beds in a static geometry. Conditions: see Figure 4. From de Broqueville and De Wilde [2009].

4. A computational fluid dynamics evaluation of the intensification of gas-solid heat transfer in rotating fluidized beds in a static geometry

Recent advances in computational power allow detailed three-dimensional simulations of the dynamic flow pattern in fluidized bed reactors. The most popular model is based on an Eulerian approach for both phases, i.e. the gas and the solid phase [Froment et al., 2010]. The solid phase continuity equations, shown in Table 1, are similar to those of the gas phase and can be derived from the Kinetic Theory of Granular Flow (KTGF) [Gidaspow, 1994]. The solid phase physico-chemical properties, like the solid phase viscosity, depend on the so-called granular temperature, a measure for the fluctuations in the solid phase velocity field at the single particle level. Such fluctuations essentially result in collisions between particles. Expressions for the solid phase physico-chemical properties are also obtained from the KTGF - see Gidaspow [1994] for an overview and more references in this field.

Fluctuations in the flow field also occur at the micro-scale. These are related to turbulence. Their calculation requires Direct Numerical Simulations (DNS) or Large-Eddy Simulations (LES) which are extremely time consuming. Therefore, continuity equations which are averaged over the micro-scales, so-called Reynolds-averaged continuity equations, are derived and solved using a Computational Fluid Dynamics (CFD) routine. Additional terms appear in these equations, expressing the effect of the micro-scale phenomena on the larger-scale behavior. These terms have to be modeled. A popular turbulence model is the k-ε model, also shown in Table 1, which has also been extended in different ways to multi-phase flows.

Gas phase mass balance

$$\frac{\partial}{\partial t}\left(\varepsilon_g \rho_g\right)+\frac{\partial}{\partial r}\cdot\left(\varepsilon_g \rho_g \overline{u}\right)=0$$

Solid phase mass balance

$$\frac{\partial}{\partial t}\left(\varepsilon_s \rho_s\right)+\frac{\partial}{\partial r}\cdot\left(\varepsilon_s \rho_s \overline{v}\right)=0$$

Gas phase momentum balance

$$\frac{\partial}{\partial t}\left(\varepsilon_g \rho_g \overline{u}\right)+\frac{\partial}{\partial r}\cdot\left(\varepsilon_g \rho_g \overline{uu}\right)=-\frac{\partial}{\partial r}\left(P+\frac{2}{3}\rho_g k\right)-\frac{\partial}{\partial r}\cdot\left(\varepsilon_g \overline{\overline{s_g}}\right)-\beta\left(\overline{u}-\overline{v}\right)+\varepsilon_g \rho_g \overline{g}$$

with: $\overline{\overline{s_g}}=-\left[\left(\xi_g-\frac{2}{3}\mu_g\right)\left(\frac{\partial}{\partial r}\cdot\overline{u}\right)\overline{\overline{I}}+\left(\mu_g+\mu_g'\right)\left(\left(\frac{\partial}{\partial r}\overline{u}\right)+\left(\frac{\partial}{\partial r}\overline{u}\right)^T\right)\right]$

Solid phase momentum balance

$$\frac{\partial}{\partial t}\left(\varepsilon_s \rho_s \overline{v}\right)+\frac{\partial}{\partial r}\cdot\left(\varepsilon_s \rho_s \overline{vv}\right)=-\frac{\partial}{\partial r}P_s-\frac{\partial}{\partial r}\cdot\left(\varepsilon_s \overline{\overline{s_s}}\right)+\beta\left(\overline{u}-\overline{v}\right)+\varepsilon_s \rho_s \overline{g}$$

with: $\overline{\overline{s_s}}=-\left[\left(\xi_s-\frac{2}{3}\mu_s\right)\left(\frac{\partial}{\partial r}\cdot\overline{v}\right)\overline{\overline{I}}+\left(\mu_s\right)\left(\left(\frac{\partial}{\partial r}\overline{v}\right)+\left(\frac{\partial}{\partial r}\overline{v}\right)^T\right)\right]$

Gas phase energy balance

$$\frac{\partial}{\partial t}\left(\varepsilon_g \rho_g\left(e_g+k+q_g\right)\right)+\frac{\partial}{\partial r}\cdot\left(\varepsilon_g \rho_g \overline{u}\left(e_g+k+q_g\right)\right)-\frac{\partial}{\partial r}\cdot\left(\varepsilon_g\left(\lambda+\lambda^t\right)\frac{\partial T_g}{\partial r}\right)=$$

$$-\frac{\partial}{\partial r}\cdot\left(\left(P+\frac{2}{3}\rho_g k\right)\overline{u}\right)-\frac{\partial}{\partial r}\cdot\left(\varepsilon_g \overline{\overline{s_g}}\cdot\overline{u}\right)-\frac{\beta}{2}\left(\left(\overline{u}\cdot\overline{u}\right)-\left(\overline{v}\cdot\overline{v}\right)\right)+\varepsilon_g \rho_g \overline{g}\cdot\overline{u}-\varepsilon_s \rho_s a_s h_{gs}\left(T_g-T_s\right)$$

Solid phase energy balance

$$\frac{\partial}{\partial t}\left(\varepsilon_s \rho_s\left(e_s+q_s\right)\right)+\frac{\partial}{\partial r}\cdot\left(\varepsilon_s \rho_s \overline{v}\left(e_s+q_s\right)\right)-\frac{\partial}{\partial r}\cdot\left(\varepsilon_s \lambda_s \frac{\partial T_s}{\partial r}\right)=$$

$$-\frac{\partial}{\partial r}\cdot\left(P_s \overline{v}\right)-\frac{\partial}{\partial r}\cdot\left(\varepsilon_s \overline{\overline{s_s}}\cdot\overline{v}\right)+\frac{\beta}{2}\left(\left(\overline{u}\cdot\overline{u}\right)-\left(\overline{v}\cdot\overline{v}\right)\right)+\varepsilon_s \rho_s \overline{g}\cdot\overline{v}+\varepsilon_s \rho_s a_s h_{gs}\left(T_g-T_s\right)$$

Gas phase k-equation

$$\frac{\partial}{\partial t}\left(\varepsilon_g \rho_g k\right)+\frac{\partial}{\partial r}\cdot\left(\varepsilon_g \rho_g \overline{u}k\right)=\frac{\partial}{\partial r}\cdot\left(\varepsilon_g \frac{\mu_g+\mu_g'}{\sigma_k}\frac{\partial k}{\partial r}\right)+\left[\varepsilon_g \mu_g'\left[\left(\frac{\partial}{\partial r}\overline{u}\right)+\left(\frac{\partial}{\partial r}\overline{u}\right)^T\right]\right]:\left(\frac{\partial}{\partial r}\overline{u}\right)-\varepsilon_g \rho_g \varepsilon$$

Gas phase ε-equation

$$\frac{\partial}{\partial t}\left(\varepsilon_g \rho_g \varepsilon\right)+\frac{\partial}{\partial r}\cdot\left(\varepsilon_g \rho_g \overline{u}\varepsilon\right)=\frac{\partial}{\partial r}\cdot\left(\varepsilon_g \frac{\mu_g+\mu_g'}{\sigma_\varepsilon}\frac{\partial \varepsilon}{\partial r}\right)$$

$$+C_{1\varepsilon}\frac{\varepsilon}{k}\left[\varepsilon_g \mu_g'\left[\left(\frac{\partial}{\partial r}\overline{u}\right)+\left(\frac{\partial}{\partial r}\overline{u}\right)^T\right]:\left(\frac{\partial}{\partial r}\overline{u}\right)\right]-C_{2\varepsilon}\varepsilon_g \rho_g \frac{\varepsilon^2}{k}$$

Granular temperature transport equation

$$\frac{3}{2}\left[\frac{\partial}{\partial t}\left(\varepsilon_s \rho_s \Theta\right)+\frac{\partial}{\partial r}\cdot\left(\varepsilon_s \rho_s \overline{v}\Theta\right)\right]=\frac{\partial}{\partial r}\cdot\left(\varepsilon_s \kappa \frac{\partial}{\partial r}\Theta\right)-\left(P_s \overline{\overline{I}}+\varepsilon_s \overline{\overline{s_s}}\right):\left(\frac{\partial}{\partial r}\overline{v}\right)-\gamma+\beta\left(q_{12}-3\Theta\right)$$

Table 1. Eulerian-Eulerian approach. Continuity equations for each phase.

Such extensions are, however, purely empirical. In gas-solid flows, additional meso-scale structures related to a non-uniform distribution of the particles develop [Agrawal et al., 2001]. Depending on the operating conditions, clusters of particles and gas bubbles are typically observed. Meso-scale structures cover the range from the micro- to the macro-scale, so that there is no separation of scales. The calculation of the dynamics of meso-scale structures is time consuming. Averaging the continuity equations over the meso-scales is theoretically possible, see Agrawal et al. [2001], Zhang and VanderHeyden [2002] and De Wilde [2005, 2007], but modeling the additional terms that appear is challenging. No reliable closure relations exist at this time. Therefore, dynamic simulations using a sufficiently fine spatial and temporal mesh are to be carried out.

Boundary conditions have to be imposed at solid walls. For gas-solid flows, they are usually based on a no-slip behavior for the gas phase and a partial slip behavior for the solid phase. Johnson and Jackson [1987] proposed a model introducing a specularity coefficient and a particle-wall restitution coefficient for which values of 0.2 and 0.9 were used by Trujillo and De Wilde [2010].

	Conventional fluidized bed	Rotating fluidized bed in a static geometry
Gas distribution chamber:	/	Outer diameter [m]: $54 \cdot 10^{-2}$ Number of gas inlets: 12 Gas inlet width [m]: $3.5 \cdot 10^{-2}$
Fluidization chamber:	Width [m]: $15 \cdot 10^{-2}$ Height [m]: $50 \cdot 10^{-2}$ Number of gas inlets: 7 Gas inlet width [m]: $2 \cdot 10^{-3}$	Outer diameter [m]: $36 \cdot 10^{-2}$ Number of tangential gas inlet slots: 24 Gas inlet slot width [m]: $2.3 \cdot 10^{-3}$ Number of solids inlets: 24 Solids inlet slot width [m]: $5 \cdot 10^{-3}$
Chimney:	/	Diameter [m]: $12\text{-}13 \cdot 10^{-2}$ Number of outlet slots: 1 Outlet slot width [m]: $8 \cdot 10^{-2}$
Solid particles:	Diameter [m]: $700 \cdot 10^{-6}$ Density [kg/m^3]: 2500 Restitution coefficient for particle-particle collisions (e): 0.9 Specularity coefficient for particle-wall collisions (ϕ): 0.2 Initial temperature [K]: 300	
	Mass in the fluidization chamber [kg/m$_{\text{length fluid. chamber}}$]: 33.75	Mass fed* to the fluidization chamber [kg/m$_{\text{length fluid. chamber}}$]: 33.75 * fed during first 5.63 s of the simulation
Fluidization gas:	Flow rate [m^3/(h·m$_{\text{length fluid. chamber}}$)]: (a) 195 (b) 540 (c) 1080	Flow rate [m^3/(h·m$_{\text{length fluid. chamber}}$)]: (a) 29800 (b) 59600
	Temperature [K]: $300 \rightarrow 400$ at time t_0	

Table 2. Simulation conditions for the CFD step response study of gas-solid heat transfer in gravitational fluidized beds and rotating fluidized beds in a static geometry by de Broqueville and De Wilde [2009].

By means of CFD, de Broqueville and De Wilde [2009] studied the response of the particle bed temperature to a step change in the fluidization gas temperature. Over a range of fluidization gas flow rates, a comparison between gravitational fluidized beds and rotating fluidized beds in a static geometry was made. The simulation conditions are summarized in Table 2. It should be remarked that the fluidization gas flow rate was close to the maximum possible value, i.e. for avoiding particle entrainment by the gas, for the gravitational fluidized bed, but not for the rotating fluidized bed in a static geometry, due to its unique flexibility explained above.

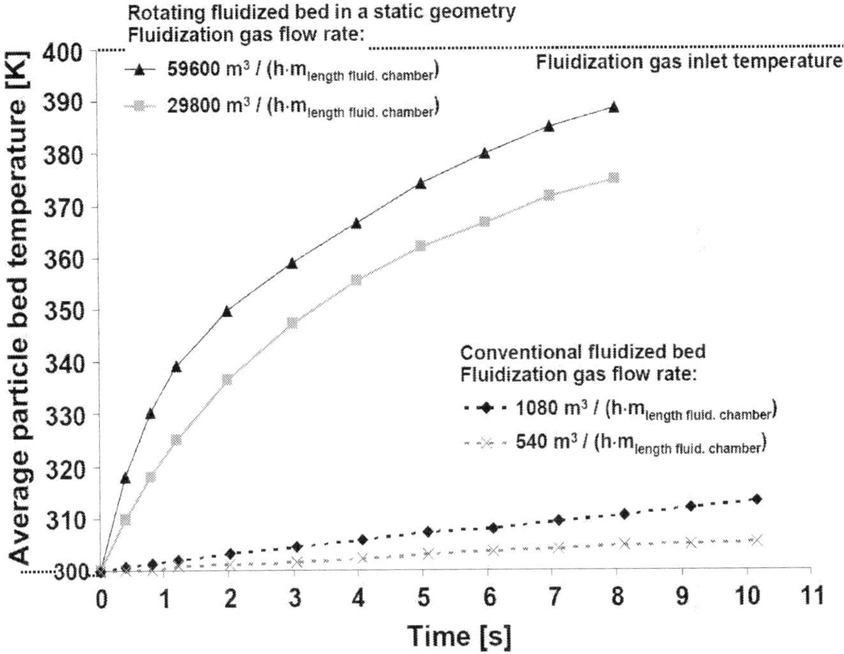

Fig. 6. Response of the average particle bed temperature to a step change in the fluidization gas temperature. Comparison of gravitational fluidized beds and rotating fluidized beds in a static geometry with equal solids loading at different fluidization gas flow rates. Conditions: see Table 2. From de Broqueville and De Wilde [2009].

Figure 6 shows that the particle bed temperature can respond much faster to changes in the fluidization gas temperature in rotating fluidized beds in a static geometry than in gravitational fluidized beds. This is due to a combination of effects. High-G operation allows higher gas-solid slip velocities and, as such, higher gas-solid heat transfer coefficients. The unique flexibility in the fluidization gas flow rate and radial gas-solid slip velocity was demonstrated in Figures 4 and 5. Also, the particle bed is cylindrically shaped in rotating fluidized beds in a static geometry, resulting in a higher particle bed width-to-height ratio than in gravitational fluidized beds. This, combined with the higher allowable radial gas-solid slip velocities, allows much higher fluidization gas flow rates per unit volume particle bed in rotating fluidized beds in a static geometry than in gravitational fluidized beds. The gas-solid contact is also intensified in rotating fluidized beds in a static geometry as a result of the higher particle bed density and the improved particle bed uniformity, i.e. the absence of bubbles. This is demonstrated in Figure 7, showing the calculated solids volume fraction profiles in both the gravitational fluidized bed and the rotating fluidized bed in a static geometry at the highest fluidization gas flow rates studied.

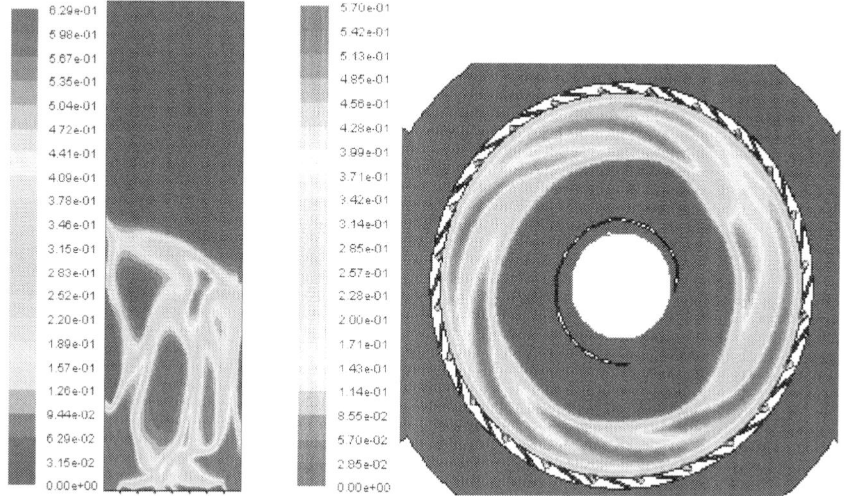

Fig. 7. Calculated solids volume fraction profiles in a gravitational fluidized bed at a fluidization gas flow rate of 1080 m³/(h m_length fluid. chamber) (top) and in a rotating fluidized bed in a static geometry at a fluidization gas flow rate of 59600 m³/(h m_length fluid. chamber) (bottom). Scales shown are different. Conditions: see Table 2. From de Broqueville and De Wilde [2009].

An important feature of rotating fluidized beds in a static geometry is the excellent particle bed mixing. This is reflected in an improved particle bed temperature uniformity, as shown in Figure 8. Such a uniformity may be of particular importance in chemical reactors, where the heat of reaction has to be provided to or removed from the particle bed. This is illustrated in the next section for the Fluid Catalytic Cracking (FCC) process.

5. A computational fluid dynamics study of fluid catalytic cracking of gas oil in a rotating fluidized bed in a static geometry

Fluid Catalytic Cracking (FCC) is a process used in refining to convert heavy Gas Oil (GO) or Vacuum Gas Oil (VGO) into lighter Gasoline (G) and Light Gases (LG). Coke (C) is an inevitable by-product in FCC and is deposited on the catalyst, deactivating it. To restore the catalyst activity, the coke has to be burned off. In view of the reaction and catalyst deactivation time scales, continuous operation requires the catalyst particles to be in a fluidized state, so that they can be easily transported between the cracking reactor and the catalyst regenerator. From the mid-40's on, fluidized bed reactor technology has been

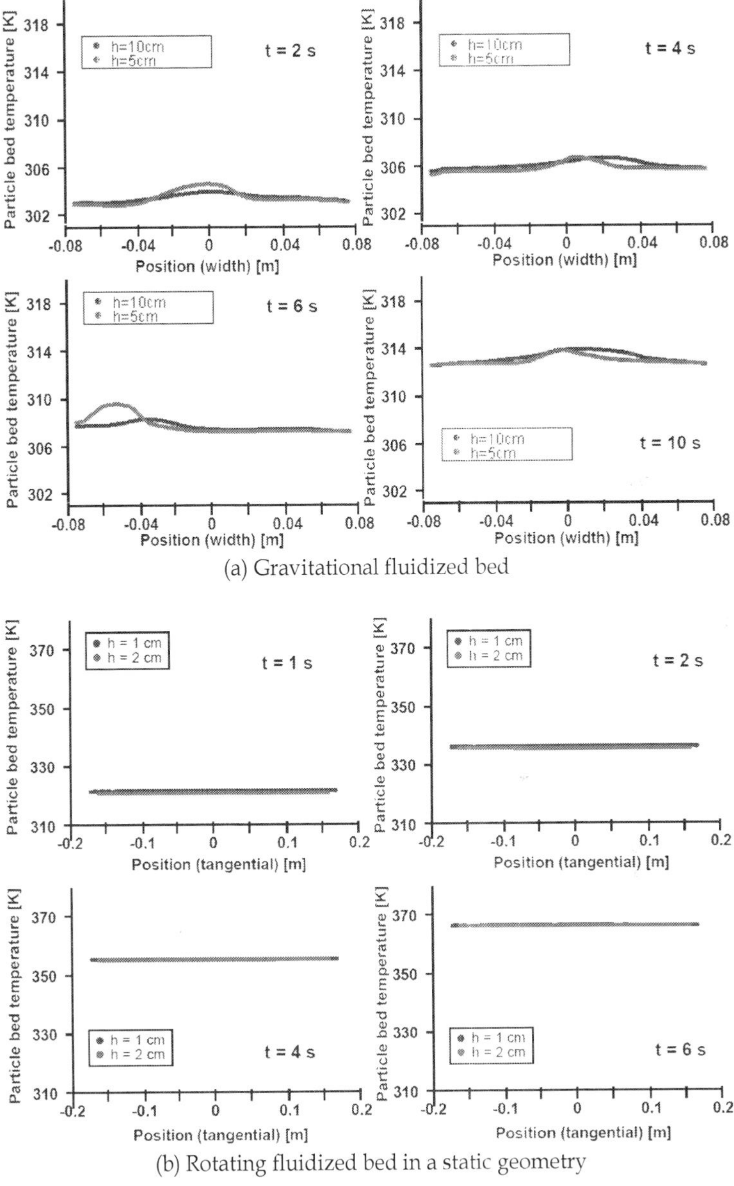

(a) Gravitational fluidized bed

(b) Rotating fluidized bed in a static geometry

Fig. 8. Calculated temperature profiles (a) in a gravitational fluidized bed at a fluidization gas flow rate of 1080 $m^3/(h\ m_{length\ fluid.\ chamber})$ and (b) in a rotating fluidized bed in a static geometry at a fluidization gas flow rate of 29800 $m^3/(h\ m_{length\ fluid.\ chamber})$. Conditions: see Table 2. From de Broqueville and De Wilde [2009].

developed in this context. The process is designed in such a way that the heat of combustion generated by burning off the coke is used for the endothermic cracking reactions, the circulating catalyst being the heat carrier. The original technology made use of a cracking reactor operated in the bubbling or turbulent regime (Figure 1). As catalysts became more active, riser reactors were introduced.

FCC riser reactors are 30 to 40 m tall with a diameter of typically 0.85 m. Operating in the riser regime, the particle bed density is low, with typically a solids volume fraction below 5%. The catalyst leaving the riser reactor is separated from the gaseous product in cyclones and goes via a stripper section to the regenerator. Partial or complete combustion of the coke is possible, depending on the feed cracked and the resulting coke formation and the energy requirements for the cracking reactions. The regenerated catalyst is returned to the riser via standpipes by the action of gravity. A sufficient standpipe height is required and this on its turn imposes a certain riser height. The cracking temperature in the riser reactor is limited to avoid over-cracking and temperature non-uniformities.

Significant intensification of the FCC process is possible. The particle bed density can be drastically increased, that is, by a factor 10. To avoid over-cracking under such conditions, the gas phase residence time has to be sharply decreased. This can be done by increasing the gas phase velocity, reducing the particle bed height, or a combination of these. Also the gas-solid heat transfer has to be intensified. This is possible by increasing the gas-solid slip velocity. Finally, to avoid temperature non-uniformities, the particle bed mixing can be improved. From the previous Sections 3 and 4, the potential of rotating fluidized beds for intensifying the FCC process is clear. Important challenges do, however, remain. Some of these were addressed by Trujillo and De Wilde [2010], who in particular demonstrated that sufficiently high conversions can be achieved in rotating fluidized beds in a static geometry. Furthermore, a one order of magnitude process intensification was predicted.

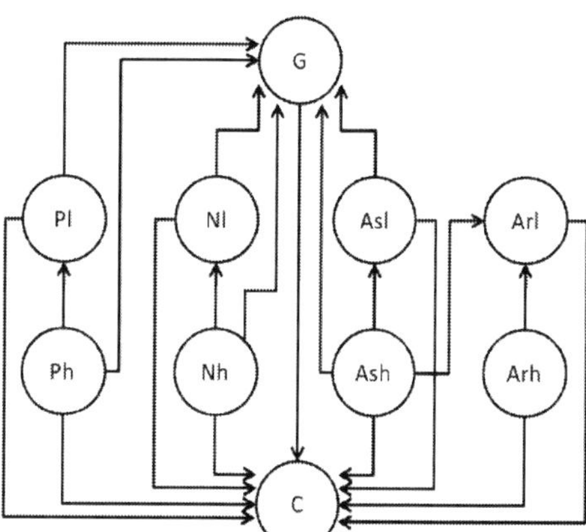

Fig. 9. Ten-lump model for the catalytic cracking of gas oil [Jacob et al., 1976].

The CFD simulations by Trujillo and De Wilde [2010] made use of the Eulerian-Eulerian approach already described in Section 4. The basic set of continuity equations, shown in Table 1, has to be extended to account for the reactions between different species. The catalytic cracking of gas oil was described by a 10-lump model, shown in Figure 9 [Jacob et al., 1976]. The C-lump is a mixture of coke and light gases (C_1 - C_4). The effects of the adsorption of heavy aromatics and of catalyst deactivation by coke on the reaction rates were accounted for. Two-dimensional periodic domain simulations of a 12-slot, 1.2 m diameter, polygonal body RFB-SG type reactor were carried out and a comparison with a 30 m tall riser was made. Details on the simulation model and on the reactor geometries simulated can be found in Trujillo and De Wilde [2010].

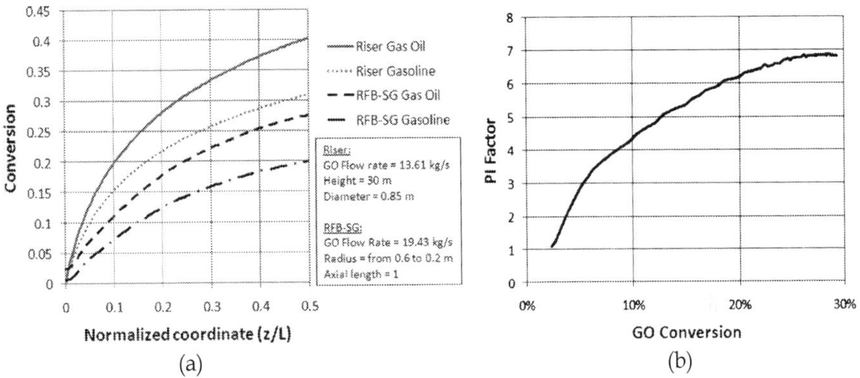

Fig. 10. FCC process intensification using rotating fluidized beds in a static geometry. Comparison with riser technology. (a) Gas Oil conversion as a function of the normalized coordinate; (b) Intrinsic process intensification factor as a function of the Gas Oil conversion. Conventional cracking catalyst and cracking temperature of 775 K. From Trujillo and De Wilde [2010].

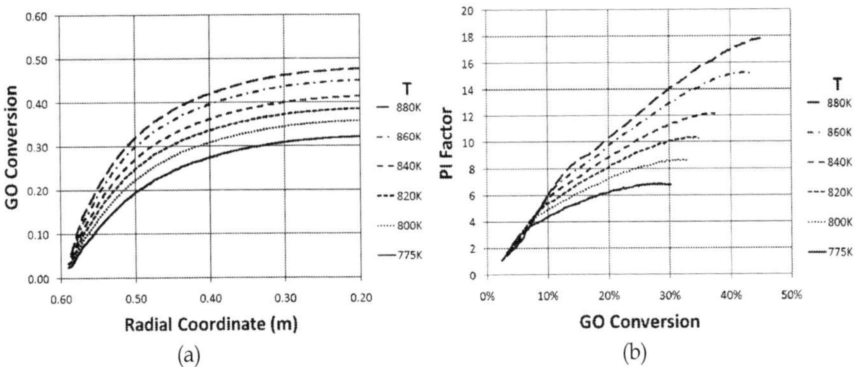

Fig. 11. FCC process intensification using rotating fluidized beds in a static geometry at increased cracking temperatures. Comparison with conventional riser technology. (a) Gas Oil conversion in the RFB-SG as a function of the radial coordinate; (b) Process intensification factor as a function of the Gas Oil conversion for different cracking temperatures. Conventional cracking catalyst and cracking temperature of 775 K. From Trujillo and De Wilde [2010].

The simulations first focused on the intrinsic process intensification potential of rotating fluidized beds in a static geometry, that is, operating at cracking temperatures and with the same catalyst used in riser reactors. Figure 10 shows the gas oil conversion as a function of the normalized coordinate and the intrinsic process intensification factor as a function of the gas oil conversion for a non-optimized RFB-SG reactor design.

With an optimized geometry, intrinsic process intensification by one order of magnitude can be easily achieved. Furthermore, due to the improved temperature uniformity in the particle bed, operation at higher cracking temperatures or working with a ore active catalyst is possible. Simulations at higher cracking temperatures, for example, shown in Figure 11, showed that process intensification by a factor 20 is within reach. This nicely illustrates the advantage that can be taken from the increased gas-solid mass and heat transfer coefficients and the improved particle bed density and uniformity in rotating fluidized beds in a static geometry.

6. Experimental evaluation of the intensification of drying of granular material in rotating fluidized beds in a static geometry

The intensification of gas-solid mass and heat transfer in rotating fluidized beds in a static geometry should allow intensifying the drying of granular material. Eliaers and De Wilde [2010] compared drying of biomass particles in a conventional fluidized bed and in a rotating fluidized bed in a static geometry. Batch and continuous flow experiments were carried out. The fluidization chamber dimensions and the operating conditions for the continuous flow experiments are summarized in Table 3.

	Conventional fluidized bed	Rotating fluidized bed in a static geometry
Dimensions	D = 0.10 m H = 2.00 m	D = 0.43 m D(chimney) = 0.10 m L = 0.24 m
Gas distribution	Cone and perforated plate	72, 30° inclined gas inlet slots
Solids feeding	Via side wall at h = m	Via end plate
Particle characteristics	Pelletized wood, cylindrically shaped, d_p = 4 mm, h_p = 4 mm	
Operating conditions		
T [K]	318	
P_{out} [Pa]	101300	
Gas mass flow rate [Nm³/h]	110	700
Solids mass flow rate [g wet solids / s]	1, 2, 3, 6, 9	3, 6, 9, 12, 15, 18
Solids inlet humidity [g water / kg dry solids]	850	

Table 3. Comparison of the drying of biomass particles in a conventional fluidized bed and in a rotating fluidized bed in a static geometry. Fluidization chamber dimensions and operating conditions.

The pelletized wood particles have mainly macro-pores, so that intra-particle diffusion limitations are not expected to be dominant in the range of drying conditions studied.

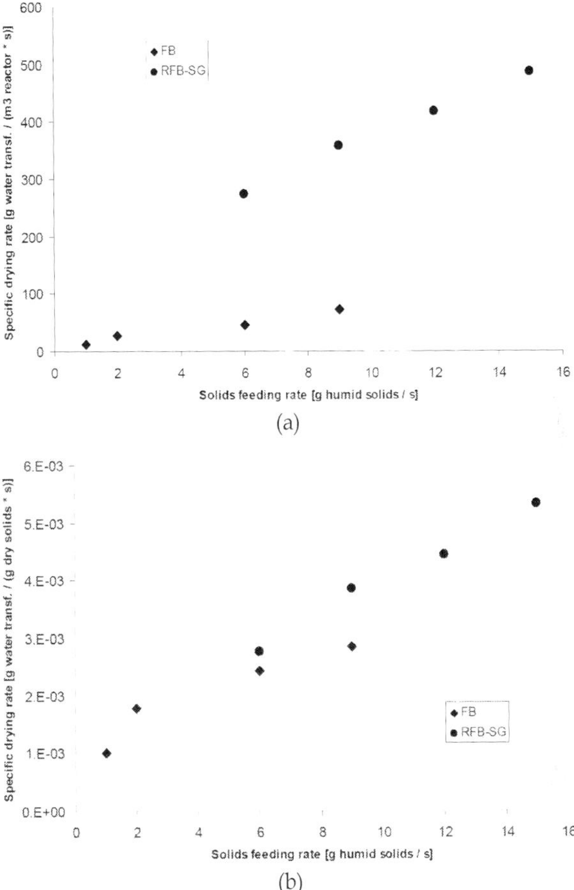

Fig. 12. Comparison of drying of biomass particles in a gravitational fluidized bed and in a rotating fluidized bed in a static geometry. (a) Specific drying rate in g water transferred per m3 reactor and per second as a function of the solids feeding rate; (b) Specific drying rate in g water transferred per g dry solids and per second as a function of the solids feeding rate. Fluidization chamber dimensions and operating conditions: see Table 3. From Eliaers and De Wilde [2010].

Intensification of the drying process can be due to an increased particle bed density and improved particle bed uniformity, on the one hand, and increased gas-solid mass and heat transfer coefficients, on the other hand. For the conditions studied, the value of the gas-solid mass and heat transfer coefficients are comparable in the gravitational fluidized bed and in the RFB-SG. The gas-solid slip velocity in both the gravitational and the rotating fluidized bed in a static geometry is around 4.7 m/s. Hence the intensification comes exclusively from the increased particle bed density and improved particle bed uniformity. Accounting for mean particle bed solids volume fractions of about 4 % in the gravitational fluidized bed and about 25 % in the RFB-SG, there is a theoretical process intensification potential of easily a factor 6. When operating the RFB-SG at higher fluidization gas flow rates, further intensification of the drying process resulting from increased gas-solid mass and heat transfer coefficients may be achieved.

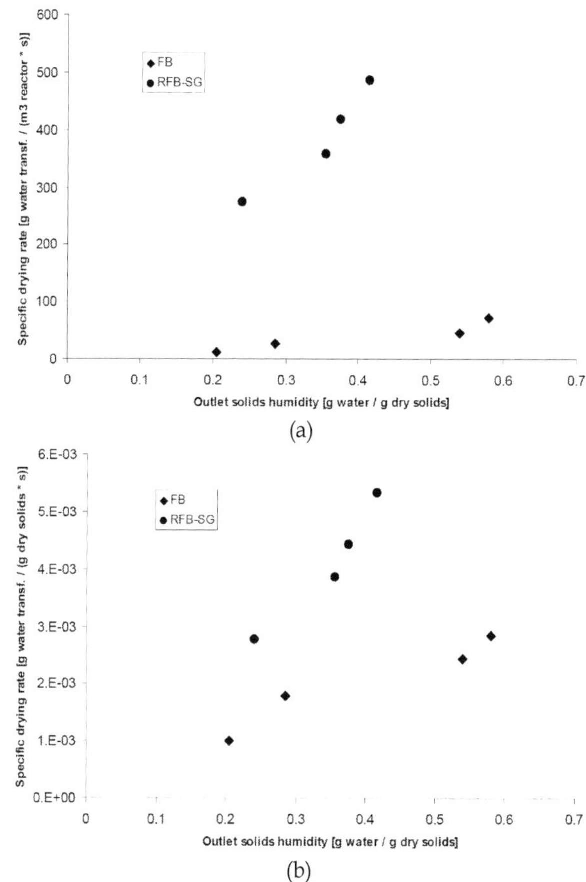

Fig. 13. Comparison of drying of biomass particles in a gravitational fluidized bed and in a rotating fluidized bed in a static geometry. Evaluation of the process intensification factor. (a) Specific drying rate in g water transferred per m3 reactor and per second as a function of the outlet solids humidity; (b) Specific drying rate in g water transferred per g dry solids and per second as a function of the outlet solids humidity. Fluidization chamber dimensions and operating conditions: see Table 3. From Eliaers and De Wilde [2010].

Figures 12 and 13 compare the specific drying rates in the gravitational fluidized bed and the RFB-SG, expressed in g water transferred per second and either per m3 reactor (a) or per g dry solids (b), respectively as a function of the solids feeding rate or as function of the outlet solids humidity. The specific drying rate expressed per g dry solids allows evaluating the process intensification due to increased gas-solid slip velocities and the improved particle bed uniformity. The specific drying rate expressed per m3 reactor then allows evaluating the additional process intensification due to the increased particle bed density. Evaluation of the process intensification is best done at equal outlet solids humidity as shown in Figure 13. The biomass drying process intensification in RFB-SGs is logically the most pronounced at higher outlet solids humidity. With decreasing outlet solids humidity, intra-particle diffusion grows in importance. Figure 13(b) shows that in the range of conditions studied, the improved particle bed uniformity in the RFB-SG results in process intensification with between a factor 2 and 4. As seen from Figure 13(a), the increased particle bed density in the RFB-SG results in additional process intensification and a global process intensification factor of between 10 and 16. As mentioned previously, the process intensification resulting from the use of a RFB-SG can be further increased by operating at higher fluidization gas flow rates.

7. Chapter summary

A theoretical analysis of gas-solid mass and heat transfer shows a significant potential for fluidized bed process intensification by replacing earth gravity with a stronger acceleration. In rotating fluidized beds, the inertia of the solid particles is used to achieve high-G operation. A new type of rotating fluidized bed, i.e., in a static geometry, is studied in this chapter. The rotating motion of the particle bed is generated by the tangential injection of the fluidization gas in the fluidization chamber, via multiple gas inlet slots. A unique flexibility in the fluidization gas flow rate results from the counteracting radial gas-solid drag force and solid phase inertial forces being affected by the fluidization gas flow rate in a similar way. A significant process intensification potential is theoretically expected, resulting from, on the one hand, an increased particle bed density and improved particle bed uniformity and, on the other hand, increased gas-solid mass and heat transfer coefficients. Furthermore, the intense particle bed mixing results in significantly improved particle bed temperature uniformity. Computational Fluid Dynamics (CFD) simulations confirm these promising characteristics of rotating fluidized beds in a static geometry. Application to fluid catalytic cracking (FCC) and to biomass drying is discussed. In FCC, the use of a rotating fluidized bed in a static geometry allows intrinsic process intensification by one order of magnitude. Additional process intensification can be achieved by operating at higher cracking temperatures, which becomes possible due to the improved particle bed temperature uniformity. The use of a more active catalyst can also be considered. Experimental data on drying of biomass particles confirm the theoretically expected process intensification potential. Again, process intensification by one order of magnitude can be easily achieved.

8. Acknowledgments

Waldo Rosales Trujillo is acknowledged for his help with the CFD calculations on the fluid catalytic cracking process. Philippe Eliaers is acknowledged for his help with the drying experiments. Axel de Broqueville is acknowledged for his collaboration.

Notations

a_s	external particle surface area per unit mass particle	[m²/kg solid]
C_t	total molar concentration of active sites	[kmol / kg cat.]
c	acceleration of high-gravity field	[m / s^2]
c_p	specific heat of fluid at constant pressure	[kJ/kg K]
D_{Am}	molecular diffusivity for A in a multicomponent mixture	[m$_f^3$ / m$_f$ s]
d_p	particle diameter	[m]
e_g	gas phase internal energy	[kJ kg^{-1}]
e_s	solid phase internal energy	[kJ kg^{-1}]
F_g	fluidization gas flow rate;	[m³ s^{-1}]
	in 2D, per unit length fluidization chamber	[m$_g^3$/(h m$_{length\ fluid.\ chamber}$)]
G	superficial mass flow velocity;	[kg / m$_r^2$ s]
g	acceleration of gravity	[m / s^2]
h_f , h_{gs}	heat transfer coefficient for film surrounding a particle	[kJ / m$_P^2$ s K]
j_D	j-factor for mass transfer	
j_H	j-factor for heat transfer	
k	gas phase turbulent kinetic energy	[kJ kg^{-1}]
k_g^0	mass transfer coefficient in case of equimolar counterdiffusion	[m$_f^3$ / m$_i$s]
L	length of the fluidization chamber	[m]
L	film thickness	[m]
M_m	mean molecular mass	[kg/kmol]
N_A	molar flux of A with respect to fixed coordinates	[kmol/m² s]
<n>	average number of rotations of the fluidization gas in the particle bed	
P	gas phase pressure	[Pa]
P_s	solid phase pressure	[Pa]
Pr	Prandtl number, $c_p \mu / \lambda$	
q_g	gas phase kinetic energy	[kJ kg^{-1}]
q_s	solid phase kinetic energy	[kJ kg^{-1}]
Q, Q_{gs}	gas-solid heat transfer	[kJ/(s m³$_{reactor}$)]
r	radial distance from the center of the fluidization chamber	[m]
$\bar{r}$	position vector	
R	Outer fluidization chamber radius	[m]
R_f	Particle bed freeboard radius	[m]
R_c	Chimney radius	[m]
$\underline{Sc}$	Schmidt number, $\mu/\rho D$	
$\underline{s}$	viscous stress tensor	[kg m^{-1} s^{-2}]
T	temperature	[K]
T_s^s	temperature inside solid, resp. at solid surface	[K]
t	time	[s]
u	gas phase velocity	[m/s]
V	volume of the fluidization chamber	[m³]
v	solid phase velocity	[m/s]
y_A, y_B	mole fraction of species A, B …	
β	drag coefficient	[kg m^{-3} s^{-1}]
γ	dissipation of kinetic fluctuation energy by inelastic particle-particle collisions	
		[kg m$_r^{-1}$ s^{-3}]

ϵ	dissipation of turbulent kinetic energy of the gas phase	$[m_r^2 \, s^{-3}]$
ϵ_g	gas phase volume fraction	$[m_g^3 / m_r^3]$
ε_s	solid phase volume fraction	$[m_s^3 / m_r^3]$
Θ	granular temperature	$[kJ/kg]$
κ	granular temperature conductivity	$[kg \, m^{-1} \, s^{-1}]$
λ	thermal conductivity	$[kJ/m \, s \, K]$
μ	dynamic viscosity	$[kg/m \, s]$
ω	angular velocity	$[rad \, s^{-1}]$
ρ_f	fluid density	$[kg / m_f^3]$
ρ_g	gas density	$[kg / m_G^3]$
ρ_s	density of catalyst	$[kg \, cat. / m_p^3]$

Subscript / superscript

g	gas phase
p	particle
s	solid phase
r	radial
rad	radial
t, tang	tangential
term	terminal

9. References

Agrawal, K., Loezos, P.N., Syamlal, M., and Sundaresan, S., J. Fluid Mech., 445, 151 (2001).

Anderson, L.A., Hasinger, S., and Turman, B.N., A.I.A.A. paper, 71, 637 (1971).

Balakrishnan, A.R., and Pei, D.C.T., Can. J. Chem. Eng., 53, 231 (1975).

Chen, Y.-M., A.I.Ch.E. J., 33 (5), 722 (1987).

de Broqueville, A., 2004. Belgian Patent 2004/0186, Internat. Classif. : B01J C08F B01F; publication number: 1015976A3.

de Broqueville, A., De Wilde, J., Chem. Eng. Sci., 64 (6), 1232 (2009).

De Wilde, J., Physics of Fluids, 17:(11), Art. No. 113304 (2005).

De Wilde, J., Physics of Fluids, 19:(5), 058103 (2007).

De Wilde, J., de Broqueville, A., A.I.Ch.E. J., 53 (4), 793 (2007).

De Wilde, J., de Broqueville, A., Powder Technol., 183 (3), 426 (2008).

De Wilde, J., de Broqueville, A., A.I.Ch.E. J., 54 (8), 2029 (2008b).

De Wilde, J., proc., OA70, North American Catalysis Society, 21st National (North American) Annual Meeting (21st NAM), San Francisco, CA, USA, June 7-12 (2009).

Eliaers, P., De Wilde, J., UCL report 270810 (2010).

Fan, L.T., Chang, C.C., Yu, Y.S., Takahashi, T., Tanaka, Z., A.I.Ch.E. J., 31 (6), 999 (1985).

Folsom, B.A., PhD thesis, California Institute of Technology (1974).

Froment, G.F., Bischoff, K.B., and De Wilde, J., Chemical Reactor Analysis and Design, Third edition, Wiley (2010).

Gidaspow, D., Multiphase Flow and Fluidization: Continuum and Kinetic Theory Descriptions, Academic Press (1994).

Jacob, S.M., Gross, B., Voltz, S.E., Jr. V.W. W., A.I.Ch.E. J., 22, 701 (1976).

Johnson, P.C., Jackson, R., J. Fluid Mech., Digital Archive, 176, 67 (1987).

Kochetov, L.M., Sazhin, B.S., Karlik, E.A., Khimicheskoe i Neftyanoe Mashinostroenie, 2, 10 (1969).

Perry, R.H., Chilton, C.H., Chemical Engineers Handbook, 6th ed., McGrawHill, NY (1984).

Qian, G.-H., Bagyi, I., Burdick, I.W., Pfeffer, R., Shaw, H., Stevens, J.G., A.I.Ch.E. J., 47 (5), 1022 (2001).

Quevedo, J.A., Nakamura, H., Shen, Y., Dave, R.N., Pfeffer, R., Proceedings of AIChE Annual meeting 2005, Cincinnati, OH, USA (2005).

Schlünder, E.U., in Chemical Reaction Engineering Reviews-Houston, ed. by D. Luss and V.W. Weekman, A.C.S. Symp. Series, 72, Washington, D.C. (1978).

Squires, A. M., Kwauk, M., and Avidan, A. A., Science, 230 (4732), 1329 (1985).

Watano, S., Imada, Y., Hamada, K., Wakamatsu, Y., Tanabe, Y., Dave, R.N., Pfeffer, R., Powder Technol. 131 (2-3), 250 (2003).

Wen, Y.C., Yu, Y.H., Chem. Eng. Progr. Symp. Ser., 62 (62), 100 (1966).

Zhang, D.Z., VanderHeyden, W.B., Int. J. Multiphase Flow, 28 (5), 805 (2002).

24

The Rate of Heat Flow through Non-Isothermal Vertical Flat Plate

T. Kranjc[1] and J. Peternelj[2]

[1]*Department of Physics and Technology, Faculty of Education, University of Ljubljana*
[2]*Faculty of Civil and Geodetic Engineering, University of Ljubljana*
Slovenia

1. Introduction

Any problem of convection consists, basically, in determining the local and/or average heat transfer coefficients connecting the local flux and/or the total transfer rate due to the relevant temperature differences. A wide variety of practical problems may be described by two-dimensional steady flow of a viscous, incompressible fluid for which a compact set of differential equations that govern the velocity and temperature fields in the fluid can be obtained. These can be solved, to a certain degree of approximation, either analytically or numerically.

Consider a flat vertical wall surrounded by air on both sides (Fig. 1). The temperature of the air far from the wall is constant on both sides and denoted by T_{0L} and T_{0R}, respectively. The problem, common in practical engineering situations, is to calculate the rate of heat flow in a stationary situation. Usually, the heat transfer between the wall and the surrounding air is characterized by the heat transfer coefficient h, defined by eqs. (1a) and (1c),

$$\dot{Q}/A = h_L(T_{0L} - T_1)\,, \tag{1a}$$

$$\dot{Q}/A = U(T_1 - T_2)\,, \tag{1b}$$

$$\dot{Q}/A = h_R(T_2 - T_{0R})\,, \tag{1c}$$

where it was assumed that $T_{0L} > T_{0R}$.

As a consequence of continuity $\dot{Q} \equiv dQ/dt$ is, of course, also the heat flow through the wall with thermal transmittance U and surface area A. The temperatures of the left and right surface, T_1 and T_2, respectively, are assumed constant in the phenomenological approach based on heat transfer coefficients. It is then straightforward to show, using the above equations, that the average heat flux density through the wall is

$$\frac{\dot{Q}}{A} = \frac{T_{0L} - T_{0R}}{\dfrac{1}{h_L} + \dfrac{1}{U} + \dfrac{1}{h_R}}\,, \tag{2}$$

In what follows we will employ the laminar boundary-layer theory and free convection equations (Grimson, 1971; Landau & Lifshitz, 1987) in order to determine the surface temperatures T_1 and T_2 and the heat transfer coefficients in such a way that eqs. (1) and (2) describe correctly the total heat flow $\dot{Q}$ across the wall.

Free convection along a vertical flat plate has been studied extensively in the past, however, it has been commonly restricted to one surface of the wall only (Pohlhausen, 1921; Ostrach, 1953; Miyamoto et al., 1980; Pozzi & Lupo, 1988; Vynnycky & Kimura, 1996; Pop & Ingham, 2001). The thermal conditions at the other surface have been prescribed by either constant temperature or constant heat flux. In the situation discussed in this chapter only the temperature of the fluid far away from the wall is prescribed.

2. Free convection equations for a flat vertical wall

To analyze free convection on the right surface where the convective flow is upward, we choose the origin of the coordinate system at the lower edge of the right surface of the wall (Fig. 1).

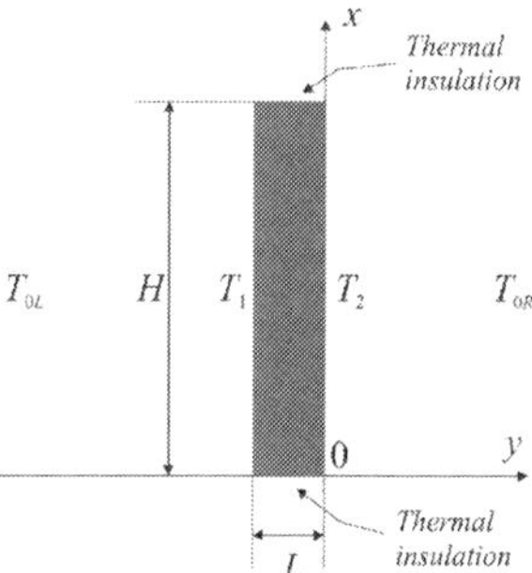

Fig. 1. Free convection at a flat vertical surface.

The x-axis is vertical and the y-axis perpendicular to the wall of height H and width L. Within the framework of the boundary-layer theory the equations of free convection valid for $y > 0$ are (Landau & Lifshitz, 1987):

$$u(\partial u/\partial x) + v(\partial u/\partial y) = \nu(\partial^2 u/\partial y^2) + \beta g(T - T_{0R}), \qquad (3a)$$

$$u(\partial T/\partial x) + v(\partial T/\partial y) = a(\partial^2 T/\partial y^2), \qquad (3b)$$

$$\partial u/\partial x + \partial v/\partial y = 0, \qquad (3c)$$

subject to the boundary conditions

$$u(x, 0) = v(x, 0) = 0, \qquad (4a)$$

$$u(x, \infty) = 0, \; T(x, \infty) = T_{0R}. \qquad (4b)$$

u and v are the x- and y-component of the velocity field, g is the acceleration of gravity, β is the thermal-expansion coefficient of the air, $a = k\rho/c_p$ its thermal diffusivity, and $\nu = \eta/\rho$ the kinematic viscosity.

To satisfy eq. (3c) we introduce the stream function $\psi(x, y)$ such that $u = \partial\psi/\partial y$ and $v = -\partial\psi/\partial x$. Similarity arguments (Landau & Lifshitz, 1987) for free convection suggest that we write the stream function as $\psi(x, y) = \nu\,\psi^*(x/H, y/H, G, P)$, where $G = \beta g(T_s - T_{0R})H^3/\nu^2$ and $P = \nu/a$ are the Grashof and Prandtl numbers, respectively, and T_s is a temperature characteristic of the surface. In addition, since we anticipate the surface temperatures to be close to uniform, except in the vicinity of $x = 0$ (Miyamoto et al., 1980; Pozzi & Lupo, 1988; Vynnycky & Kimura, 1996), we further specify the stream function to have the form ($y > 0$)

$$\psi_R(x, y) = \nu_R G_R^{1/4}(4x^*)^{3/4}\,[\Phi_R(\xi) + \phi_R(x^*, \xi)], \tag{5}$$

where $x^* = x/H$, $y^* = y/H$, $\xi = G_R^{1/4}y^*/(4x^*)^{1/4}$ and $G_R = \beta_R g(T_2 - T_{0R})H^3/\nu_R^2$. $\Phi_R(\xi)$ represents the Pohlhausen solution (Grimson, 1971; Landau & Lifshitz, 1987; Pohlhausen, 1921) describing free convection on the flat vertical wall with uniform temperature $T_2 > T_{0R}$. The temperature distribution in the air to the right of the wall is written correspondingly as

$$T_R(x, y) = T_{0R} + (T_2 - T_{0R})\,[\Theta_R(\xi) + \theta_R(x^*, \xi)], \tag{6}$$

where $\Theta_R(\xi)$ is the Pohlhausen temperature function associated with $\Phi_R(\xi)$. These two functions are obtained as solutions of (see Landau & Lifshitz, 1987)

$$\Phi_R''' + 3\Phi_R\,\Phi_R'' - 2\Phi_R'^2 + \Theta_R = 0, \tag{7a}$$

$$\Theta_R'' + 3P_R\,\Phi_R\,\Theta_R' = 0, \tag{7b}$$

satisfying the boundary conditions

$$\Phi_R(0) = \Phi_R'(0) = \Phi_R'(\infty) = 0, \quad \Theta_R(0) = 1, \Theta_R(\infty) = 0, \tag{7c}$$

and the primes denote differentiation with respect to ξ. $\phi_R(x^*, \xi)$ and $\theta_R(x^*, \xi)$ are, presumably, small corrections due to the fact that the surface temperatures are not exactly uniform.

As pointed out in Ostrach (1953), the choice of the variable ξ defined above essentially implies that the conditions imposed on the velocities and temperature at $y = \infty$ (or $\xi = \infty$) should also be satisfied, for $y \neq 0$, at $x = 0$. This seems reasonable physically if it is understood as a statement that on the right-hand side the convective flow starts at the bottom edge of the wall. Considering the boundary conditions (4b) and (7c) referring to temperature, it follows from (6) that $T_R(x \neq 0, y = 0) = T_2 + (T_2 - T_{0R})\,\theta_R(x \neq 0, \xi = 0)$ and $T_R(x = 0, y \neq 0) = T_{0R}$. In the vicinity of $x = y = 0$, the boundary layer approximation breaks down and the calculation based on this approximation cannot yield reliable results as pointed out already by Miyamoto et al. (1980). However, we believe that this conclusion is not so crucial since the contribution of the surface area near the leading edge of the convective flow to the total heat rate is proportional to x_0/H, where x_0 is small compared to H (Pozzi & Lupo, 1988).

To describe free convection on the left surface of the wall we must reverse the coordinate axes since there the local buoyancy force is directed vertically downward. Thus we choose the origin of the coordinate system at the upper edge of the left surface, the x-axis is oriented vertically downward and the y-axis is perpendicular to the wall and oriented to the left (Fig. 2, left).

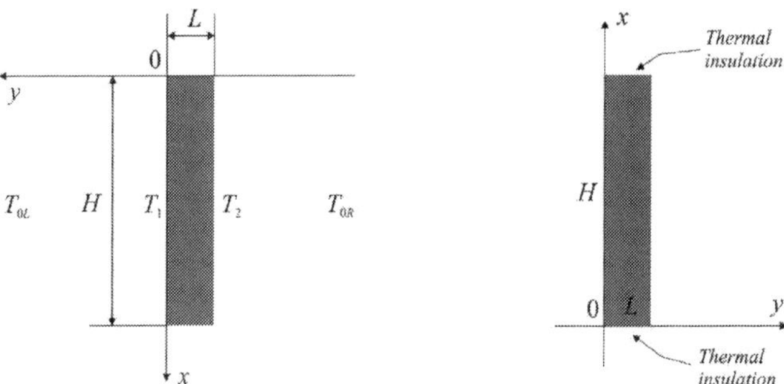

Fig. 2. Left: Coordinate system for convection to the left of the wall. Right: Coordinate system for temperature distribution within the wall.

We write as above,

$$\psi_L(x, y) = \nu_L G_L^{1/4}(4x^*)^{3/4}\,[\Phi_L(\xi) + \phi_L(x^*, \xi)], \tag{8}$$

$$T_L(x, y) = T_{0L} + (T_1 - T_{0L})\,[\Theta_L(\xi) + \theta_L(x^*, \xi)], \tag{9}$$

where $G_L = \beta_L(-g)(T_1 - T_{0L})H^3/\nu_L^2$, $\xi = G_L^{1/4}y^*/(4x^*)^{1/4}$ and $\Phi_L(\xi)$ and $\Theta_L(\xi)$ satisfy identical equations as Φ_R and Θ_R, provided that we replace $g \to -g$ in eq. (3a) and use the parameters characteristic for the air to the left of the wall.

3. Temperature distribution inside the wall

The steady state temperature distribution within the wall satisfies Laplace's equation,

$$\frac{\partial^2 T}{\partial x^2} + \frac{\partial^2 T}{\partial y^2} = 0 . \tag{10}$$

At the bottom and upper side the wall is assumed to be insulated and consequently the boundary conditions, $\partial T/\partial x\,|_{x=0} = \partial T/\partial x\,|_{x=H} = 0$, are imposed (Vynnycky & Kimura, 1996). We choose the origin of the coordinate system, used for the wall only, at the lower edge of the left surface of the wall (Fig. 2, right). With the x-axis vertical and the y-axis perpendicular to the wall, we can write the temperature distribution within the wall as

$$T(x,y) = T_1 + \frac{y}{L}(T_2 - T_1) + \sum_{n=1}^{\infty}\left[(T_1 - T_{0L})A_n \sinh(\tfrac{n\pi(L-y)}{H}) + (T_2 - T_{0R})B_n \sinh(\tfrac{n\pi y}{H})\right]\cos(\tfrac{n\pi x}{H}). \tag{11}$$

The average wall surface temperatures $T_1 = \dfrac{1}{H}\displaystyle\int_0^H T(x,0)dx$, $T_2 = \dfrac{1}{H}\displaystyle\int_0^H T(x,L)dx$, ($T_1 > T_2$) and

the coefficients A_n and B_n are determined by requiring that the temperatures of the two media must be equal at the respective boundaries and, moreover, the heat flux out of one medium must equal the heat flux into the other medium at each of the two boundaries.

Taking into account the relative displacement and orientation of the various coordinate systems, this yields

$$\theta_L(x^*,0) = \sum_{n=1}^{\infty} A_n \sinh\frac{n\pi L}{H}\cos n\pi(1 - x^*) = \sum_{n=1}^{\infty}(-1)^n A_n \sinh\frac{n\pi L}{H}\cos n\pi x^*, \qquad (12a)$$

$$\theta_R(x^*,0) = \sum_{n=1}^{\infty} B_n \sinh\frac{n\pi L}{H}\cos n\pi x^*, \qquad (12b)$$

and

$$-k_L\left(\frac{T_1 - T_{0L}}{H}\right)\left[\left(\frac{G_L}{4x^*}\right)^{1/4}\left(\Theta_L'(0) + \theta_L'(x^*,0)\right)\right] =$$
$$U(T_2 - T_1) + \sum_{n=1}^{\infty}(-1)^n\left\{U(T_{0L} - T_1)A_n\frac{n\pi L}{H}\cosh\frac{n\pi L}{H} + U(T_2 - T_{0R})B_n\frac{n\pi L}{H}\right\}\cos n\pi x^*, \qquad (13a)$$

$$k_R\left(\frac{T_2 - T_{0R}}{H}\right)\left[\left(\frac{G_R}{(4x^*)}\right)^{1/4}\left(\Theta_R'(0) + \theta_R'(x^*,0)\right)\right] =$$
$$U(T_2 - T_1) + \sum_{n=1}^{\infty}\left\{U(T_{0L} - T_1)A_n\frac{n\pi L}{H} + U(T_2 - T_{0R})B_n\frac{n\pi L}{H}\cosh\frac{n\pi L}{H}\right\}\cos n\pi x^*, \qquad (13b)$$

where a uniform composition of the wall was assumed, $U = k_w/L$, and $k_{L,R}$ are thermal conductivities of the air at the left and right surface of the wall, respectively.

As already stressed, we are interested in the total rate of heat flow through the wall. Integrating eqs. (13) with respect to x^* we obtain

$$-k_L\left(\frac{T_{0L} - T_1}{H}\right)\left[\left(\frac{G_L}{4}\right)^{1/4}\left(\frac{4}{3}\Theta_L'(0) + \int_0^1 (x^*)^{\frac{3}{4}-1}\theta_L'(x^*,0)dx^*\right)\right] = U(T_1 - T_2), \qquad (14a)$$

$$-k_R\left(\frac{T_2 - T_{0R}}{H}\right)\left[\left(\frac{G_R}{4}\right)^{1/4}\left(\frac{4}{3}\Theta_R'(0) + \int_0^1 (x^*)^{\frac{3}{4}-1}\theta_R'(x^*,0)dx^*\right)\right] = U(T_1 - T_2). \qquad (14b)$$

Using the definition of the Grashof number and the relations $\beta_{L,R} \approx 1/T_{0L,R}$, we can rewrite eqs. (14) in the form

$$T_{0L} - T_1 = (T_2 - T_{0R})\left(\gamma^4\frac{T_{0L}}{T_{0R}}\right)^{1/5}, \qquad (15a)$$

$$-\frac{\kappa_R}{\sqrt{2}}\left(\frac{gH^3}{v_R^2}\right)^{\frac{1}{4}}\left(\frac{4}{3}\Theta_R'(0) + J_R\right)\left(\frac{T_2 - T_{0R}}{T_{0R}}\right)^{\frac{5}{4}} + \left(1 + \left(\gamma^4\frac{T_{0L}}{T_{0R}}\right)^{\frac{1}{5}}\right)\left(\frac{T_2 - T_{0R}}{T_{0R}}\right) - \left(\frac{T_{0L} - T_{0R}}{T_{0R}}\right) = 0, \qquad (15b)$$

where

$$\gamma = \left(\frac{v_L}{v_R}\right)^{1/2}\left(\frac{k_R}{k_L}\right)\frac{\frac{4}{3}\Theta_R'(0) + J_R}{\frac{4}{3}\Theta_L'(0) + J_L}, \qquad (15c)$$

$$J_{L,R} = \int_0^1 (x^*)^{-\frac{1}{4}} \theta'_{L,R}(x^*,0)dx^* , \tag{15d}$$

and

$$\kappa_{L,R} = \frac{k_{L,R}L}{k_w H} . \tag{15e}$$

The total rate of heat flow per unit area of the wall $q \equiv \dot{Q} / A$ is, referring to eqs. (14), equal to $U(T_1 - T_2)$. To proceed, we use eqs. (1a, c), (2) and (15a) to determine the heat transfer coefficients. From (1a, c) and (15a) it follows

$$\frac{h_R}{h_L} = \left(\gamma^4 \frac{T_{0L}}{T_{0R}} \right)^{1/5} \tag{16}$$

and, equating (2) and (1c), we obtain

$$h_R = \bar{Nu}^{(R)} \frac{k_R}{H} . \tag{17a}$$

Here,

$$\bar{Nu}^{(R)} = \frac{1}{\kappa_R} \left\{ \frac{T_{0L} - T_{0R}}{T_2 - T_{0R}} - \left[1 + \left(\gamma^4 \frac{T_{0L}}{T_{0R}} \right)^{1/5} \right] \right\} \tag{17b}$$

is the average Nusselt number, associated with the right-hand surface of the wall. (Of course, h_R as written in (17a) is also the average heat transfer coefficient, but we shall omit the bar above the symbol.) Writing similarly $h_L = \bar{Nu}^{(L)} \frac{k_L}{H}$ and using eqs. (16, 17a), we obtain the corresponding expression for $\bar{Nu}^{(L)}$,

$$\bar{Nu}^{(L)} = \frac{k_R}{k_L} \left(\gamma^4 \frac{T_{0L}}{T_{0R}} \right)^{-1/5} \bar{Nu}^{(R)} = \frac{1}{\kappa_L} \left\{ \frac{T_{0L} - T_{0R}}{T_{0L} - T_1} - \left[1 + \left(\gamma^4 \frac{T_{0L}}{T_{0R}} \right)^{-1/5} \right] \right\} . \tag{17c}$$

Furthermore, the solution of eq. (15b) can be calculated by Newton's method of successive approximations with the initial approximate solution chosen as

$$\left(\frac{T_2 - T_{0R}}{T_{0R}} \right)_0 = \frac{\left(\dfrac{T_{0L} - T_{0R}}{T_{0R}} \right)}{1 + \gamma^{4/5} \left(\dfrac{T_{0L}}{T_{0R}} \right)^{1/5} - \dfrac{1}{\sqrt{2}} \kappa_R \left(\dfrac{gH^3}{v_R^2} \right)^{1/4} \left(\tfrac{4}{3}\Theta'_R(0) + J_R \right) \left(\dfrac{T_{0L} - T_{0R}}{T_{0R}} \right)^{1/4}} , \tag{18}$$

obtained from (15b) by writing

$$\left(\frac{T_2 - T_{0R}}{T_{0R}}\right)^{5/4} \cong \left(\frac{T_2 - T_{0R}}{T_{0R}}\right)\left(\frac{T_{0L} - T_{0R}}{T_{0R}}\right)^{1/4}.$$

In the lowest approximation, (17b) and (17c) then become

$$\bar{N}u^{(R)} = -\frac{1}{\sqrt{2}}\left(\frac{gH^3}{v_R^{\,2}}\right)^{1/4}\left(\frac{4}{3}\Theta'_R(0) + J_R\right)\left(\frac{T_{0L} - T_{0R}}{T_{0R}}\right)^{1/4}, \tag{19a}$$

$$\bar{N}u^{(L)} = \left(\gamma^4 \frac{T_{0L}}{T_{0R}}\right)^{1/20}\left\{-\frac{1}{\sqrt{2}}\left(\frac{gH^3}{v_L^2}\right)^{1/4}\left(\frac{4}{3}\Theta'_L(0) + J_L\right)\left(\frac{T_{0L} - T_{0R}}{T_{0L}}\right)^{1/4}\right\}. \tag{19b}$$

The average heat flux density $\dot{Q}/A$ can be now calculated easily by using eq. (2), for example, and the expressions for the heat transfer coefficients as determined above.

Using the result obtained by Kao et al. (1977) and quoted by Miyamoto et al. (1980) (eq. (13)), we can calculate $\frac{4}{3}\Theta'_R(0) + J_R$. The heat flux density at the right surface of the wall (the analysis for the left surface is identical)

$$q(x^*) = -k_R\left(\frac{T_2 - T_{0R}}{H}\right)\left(\frac{G_R}{4x^*}\right)^{1/4}\left(\Theta'_R(0) + \theta'_R(x^*,0)\right) \tag{20a}$$

may be, according to Miyamoto et al. (1980), for $P = 0.7$, approximated closely by

$$q(x^*) = k_R\left(\frac{T_2 - T_{0R}}{H}\right)\left(\frac{G_R}{4}\right)^{1/4}\left(c_1 \frac{F_R^{3/2}}{\varsigma_R^{1/4}} + c_2 \frac{\varsigma_R^{3/4}}{F_R^{1/2}}\frac{dF_R}{dx^*}\right), \tag{20b}$$

where $c_1 = 0.4995$, $c_2 = 0.2710$ and

$$F_R(x^*) = \frac{T_R(x,0) - T_{0R}}{T_2 - T_{0R}} = 1 + \theta_R(x^*,0),$$

$$\varsigma_R(x^*) = \int_0^{x^*} F_R dx^*. \tag{20c}$$

Equating the right-hand sides of (20a, b) and integrating the resulting equation with respect to x^* from 0 to 1, we obtain

$$\frac{4}{3}\Theta'_{L,R}(0) + J_{L,R} = -2c_2 F_{L,R}^{1/2}(1) - \left(c_1 - \frac{3}{2}c_2\right)\int_0^1 dx^* \frac{F_{L,R}^{3/2}}{\varsigma_{L,R}^{1/4}}. \tag{21}$$

4. Equations determining $F_{L,R}(x^*)$

In order to calculate (21), we rewrite the boundary conditions (13), using eqs. (12), (14), (15), (20) and (21), as follows:

$$-\kappa_L\left(\frac{G_L}{4}\right)^{1/4}\left[c_1\frac{F_L^{3/2}}{\varsigma_L^{1/4}}+c_2\frac{\varsigma_L^{3/4}}{F_L^{1/2}}\frac{dF_L}{dx*}\right]=-\kappa_L\left(\frac{G_L}{4}\right)^{1/4}\left[2c_2F_L^{1/2}(1)+(c_1-\tfrac{3}{2}c_2)\int\limits_0^1 dx*\frac{F_L^{3/2}}{\varsigma_L^{1/4}}\right]+$$

$$2\sum_{n=1}^{\infty}\left\{\frac{n\pi L}{H}\coth\frac{n\pi L}{H}\int\limits_0^1 dx*\,F_L\cos n\pi x*+(-1)^n\left(\gamma^4\frac{T_{0L}}{T_{0R}}\right)^{-1/5}\frac{\frac{n\pi L}{H}}{\sinh\frac{n\pi L}{H}}\int\limits_0^1 dx*\,F_R\cos n\pi x*\right\}\cos n\pi x*,$$

$$\text{(22a)}$$

$$-\kappa_R\left(\frac{G_R}{4}\right)^{1/4}\left[c_1\frac{F_R^{3/2}}{\varsigma_R^{1/4}}+c_2\frac{\varsigma_R^{3/4}}{F_R^{1/4}}\frac{dF_R}{dx*}\right]=-\kappa_R\left(\frac{G_R}{4}\right)^{1/4}\left[2c_2F_R^{1/2}(1)+(c_1-\tfrac{3}{2}c_2)\int\limits_0^1 dx*\frac{F_R^{3/2}}{\varsigma_R^{1/4}}\right]+$$

$$2\sum_{n=1}^{\infty}\left\{(-1)^n\left(\gamma^4\frac{T_{0L}}{T_{0R}}\right)^{1/5}\frac{\frac{n\pi L}{H}}{\sinh\frac{n\pi L}{H}}\int\limits_0^1 dx*\,F_L\cos n\pi x*+\frac{n\pi L}{H}\coth\frac{n\pi L}{H}\int\limits_0^1 dx*\,F_R\cos n\pi x*\right\}\cos n\pi x*.$$

$$\text{(22b)}$$

If we multiply both sides of eqs. (22) by $\cos(n\pi x^*)$, integrate with respect to x^* from 0 to 1 and rearrange the terms, we obtain

$$\int\limits_0^1 F_L\cos n\pi x*\,dx*=-\kappa_L\left(\frac{G_L}{4}\right)^{1/4}\left[\int\limits_0^1 dx*\left(c_1\frac{F_L^{3/2}}{\varsigma_L^{1/4}}+c_2\frac{\varsigma_L^{3/4}}{F_L^{1/2}}\frac{dF_L}{dx*}\right)\cos n\pi x*\right]\frac{\coth(n\pi L/H)}{(n\pi L/H)}$$

$$+(-1)^n\kappa_R\left(\gamma^4\frac{T_{0L}}{T_{0R}}\right)^{-1/5}\left(\frac{G_R}{4}\right)^{1/4}\left[\int\limits_0^1 dx*\left(c_1\frac{F_R^{3/2}}{\varsigma_R^{1/4}}+c_2\frac{\varsigma_R^{3/4}}{F_R^{1/2}}\frac{dF_R}{dx*}\right)\cos n\pi x*\right]\frac{1}{\frac{n\pi L}{H}\sinh\frac{n\pi L}{H}},$$

$$\text{(23a)}$$

$$\int\limits_0^1 F_R\cos n\pi x*\,dx*=(-1)^n\kappa_L\left(\gamma^4\frac{T_{0L}}{T_{0R}}\right)^{1/5}\left(\frac{G_L}{4}\right)^{1/4}\left[\int\limits_0^1 dx*\left(c_1\frac{F_L^{3/2}}{\varsigma_L^{1/4}}+c_2\frac{\varsigma_L^{3/4}}{F_L^{1/2}}\frac{dF_L}{dx*}\right)\cos n\pi x*\right]\times$$

$$\frac{1}{\frac{n\pi L}{H}\sinh\frac{n\pi L}{H}}-\kappa_R\left(\frac{G_R}{4}\right)^{1/4}\left[\int\limits_0^1 dx*\left(c_1\frac{F_R^{3/2}}{\varsigma_R^{1/4}}+c_2\frac{\varsigma_R^{3/4}}{F_R^{1/2}}\frac{dF_R}{dx*}\right)\cos n\pi x*\right]\frac{\coth(n\pi L/H)}{n\pi L/H)}.$$

$$\text{(23b)}$$

5. Numerical results

We will attempt to solve eqs. (23) by approximating $F_{L,\,R}(x^*)-1=\theta_{L,\,R}(x^*,0)\equiv\theta_{L,\,R}(x^*)$ by a polynomial,

$$\theta_{L,R}(x^*)=-a_{L,R}^{(0)}+a_{L,R}^{(1)}x*+a_{L,R}^{(2)}x*^2+a_{L,R}^{(3)}x*^3+a_{L,R}^{(4)}x*^4+\cdots,\tag{24}$$

where, as a consequence of eqs. (12),

$$0=-a_{L,R}^{(0)}+\left(\tfrac{1}{2}a_{L,R}^{(1)}+\tfrac{1}{3}a_{L,R}^{(2)}+\tfrac{1}{4}a_{L,R}^{(3)}+\tfrac{1}{5}a_{L,R}^{(4)}+\cdots\right).\tag{25}$$

The natural approach to solving for coefficients $a_{L,R}^{(i)}$, $i=1,2,3,\dots$ in eq. (24) is using the Newton method. However, we also employed the iteration method as proposed by Miyamoto et al. (1980). It turned out that the applicability of the simpler iteration procedure is limited to a restricted range of parameters appearing in Eqs. (23) ($G_{L,\,R}$, $\kappa_{L,\,R}$, $T_{0R,\,L}$, γ and aspect ratio L/H). It works well for plates with small aspect ratio and high conductivity (k_w), like stainless steel or aluminum. But it breaks down for walls with large aspect ratio and with the conductivity coefficient 10-100 times lower (such as the thermal conductivity of brick, for example), and a better calculational method has to be applied. This is described in the Mathematical note, Section 7.

The knowledge of $F_{L,R}(x^*)$ makes it possible to calculate the temperatures T_1 and T_2 at the wall surfaces as well as the heat flux (or the heat flux density) through the wall, $\dot{Q}$ (or $\dot{Q}/A$). Using eqs. (17b, c), we can express the temperatures T_1 and T_2 at the wall surfaces in terms of the Nusselt numbers,

$$T_1 = T_{0L} - \frac{T_{0L} - T_{0R}}{\kappa_L \bar{Nu}^{(L)} + 1 + [\gamma^4 (T_{0L}/T_{0R})]^{-1/5}} , \tag{26a}$$

$$T_2 = T_{0R} + \frac{T_{0L} - T_{0R}}{\kappa_R \bar{Nu}^{(R)} + 1 + [\gamma^4 (T_{0L}/T_{0R})]^{1/5}} . \tag{26b}$$

Subtracting the above two equations, we obtain the temperature difference across the wall,

$$T_1 - T_2 =$$

$$(T_{0L} - T_{0R})\left[1 - \left(\frac{1}{\kappa_R \bar{Nu}^{(R)} + 1 + [\gamma^4 (T_{0L}/T_{0R})]^{1/5}} + \frac{1}{\kappa_L \bar{Nu}^{(L)} + 1 + [\gamma^4 (T_{0L}/T_{0R})]^{-1/5}}\right)\right]. \tag{27}$$

The Nusselt numbers $\bar{Nu}^{(R)}$ and $\bar{Nu}^{(L)}$ are given in the lowest approximation by eqs. (19a, b). From (27) and (1b), the average flux density through the wall (equal to the average heat flux density through the air layers at the wall) can be written:

$$\frac{\dot{Q}}{A} = U(T_{0L} - T_{0R})\left[1 - \left(\frac{1}{\kappa_R \bar{Nu}^{(R)} + 1 + [\gamma^4 (T_{0L}/T_{0R})]^{\frac{1}{5}}} + \frac{1}{\kappa_L \bar{Nu}^{(L)} + 1 + [\gamma^4 (T_{0L}/T_{0R})]^{-\frac{1}{5}}}\right)\right]. \tag{28}$$

Finally, through the eq. (17a), the heat transfer (convection) coefficients can be expressed in terms of the Nusselt numbers as

$$h_{R,L} = \left(\frac{k_{R,L}}{H}\right)\bar{Nu}^{(R,L)}. \tag{29}$$

We performed numerical calculations for stainless steel and aluminum plates (compare Miyamoto et al. (1980)), as well as for walls of various dimensions and thermal conductivities comparable to brick or concrete, surrounded by air. We present some of the results for the air temperatures $T_{0L} = 30\ ^\circ C = 303\ K$ and $T_{0R} = 20\ ^\circ C = 293\ K$.

5.1 Stainless steel plate
Thermal conductivities for air and steel are $k_{L,R} = k_a = 2.63 \times 10^{-2}\ W/mK$ (air at ~300 K) and $k_w = 16\ W/m \cdot K$, respectively. For a plate 1 cm thick and 40 cm high (aspect ratio $L/H = 0.025$), $\kappa_L = \kappa_R = \kappa = (k_a/k_w)(L/H) = 4.1 \times 10^{-5}$, and for the temperature difference $T_{0L} - T_1 \cong$

$T_2 - T_{0R} \cong 5.0$ K, the Grashof number takes the value $G_L \approx G_R \approx G = 4.2 \times 10^7$. The coefficient γ (Eq. (15c)) is very close to 1. The function $\theta_L(x^*)$ is shown in Fig. 3.

With the results for $F_{L,R}(x^*)$ and using $\Theta'_{L,R}(0) = -0.4995$ (for the Prandtl number $P = 0.70$), we get $J_L \cong J_R = -0.0041$ and, consequently, using equation (19),

$$\bar{Nu}^{(R)} = 0.474 \left(\frac{gH^3}{v_R^2} \right)^{1/4} \left(\frac{T_{0L} - T_{0R}}{T_{0R}} \right)^{1/4} \tag{30a}$$

and

$$\bar{Nu}^{(L)} = \left(\frac{T_{0R}}{T_{0L}} \right)^{1/5} \bar{Nu}^{(R)} = 0.474 \left(\frac{T_{0L}}{T_{0R}} \right)^{1/20} \left(\frac{gH^3}{v_L^2} \right)^{1/4} \left(\frac{T_{0L} - T_{0R}}{T_{0L}} \right)^{1/4}. \tag{30b}$$

In the present case, the value of the Nusselt number is $\bar{Nu}^{(R)} = \bar{Nu}^{(L)} = 45$. The temperature drop across the plate is negligible ($T_1 - T_2 \sim 10^{-2}$ K), being only a tiny fraction ($\sim 10^{-3}$) of the air temperature difference on the two sides of the plate. Essentially the whole temperature drop takes place within the boundary layers at the plate surfaces.

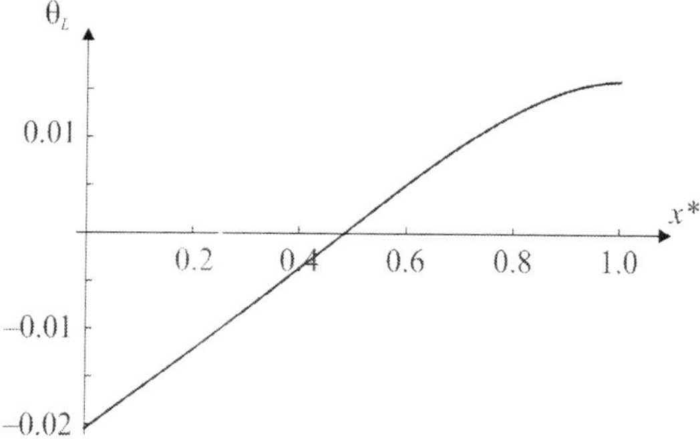

Fig. 3. Stainless steel. Correction $\theta_l(x^*)$ to the Pohlhausen solution for a steel plate ($L/H = 0.05$, $G_L \approx G_R = 4.2 \times 10^7$, $\kappa_L = \kappa_R = 4.1 \times 10^{-5}$, and $T_{0L} = 30\ ^{\circ}C$, $T_{0R} = 20\ ^{\circ}C$). The correction $\theta_R(x^*)$ differs only insignificantly from $\theta_L(x^*)$.

The heat flux density is equal to ~ 15 W/m², and the heat transfer coefficients are ~ 3 W/(m² · K).

If the aspect ratio increases at constant height, the Grashof number increases, while $\theta_{L,R}(x^*)$ decrease. If the height increases at constant aspect ratio, the Grashof number as well as $\theta_{L,R}(x^*)$ increase.

5.2 Aluminum plate

The thermal conductivity of aluminum is $k_w = 203$ W/m · K. If the plate has the same dimensions as the steel plate (1 cm thick, 40 cm high, aspect ratio $L/H = 0.025$), $\kappa_L = \kappa_R = \kappa = (k_a/k_w)(L/H) = 3.24 \times 10^{-6}$, and for the temperature difference of 5.0 K, the Grashof number takes the value $G_L \approx G_R \approx G = 4.2 \times 10^7$. Again, $\gamma \cong 1$. The function $\theta_L(x^*)$ is shown in Fig. 4. In this case, we obtain ($P = 0.70$) $J_L \cong J_R = -0.0035$ and the Nusselt number is

$$\bar{Nu}^{(R)} = 0.473 \left(\frac{gH^3}{v_R^2} \right)^{1/4} \left(\frac{T_{0L} - T_{0R}}{T_{0R}} \right)^{1/4} . \tag{31}$$

In the case of aluminum, the value of the Nusselt number is again $\bar{Nu}^{(R)} = \bar{Nu}^{(L)} = 45$. The temperature drop across the plate is negligible: $T_1 - T_2 \sim 10^{-3}$ K, $(T_1 - T_2)/(T_{0L} - T_{0L}) \sim 10^{-4}$; again almost all of the temperature drop occurs within the boundary layers at the plate surfaces.

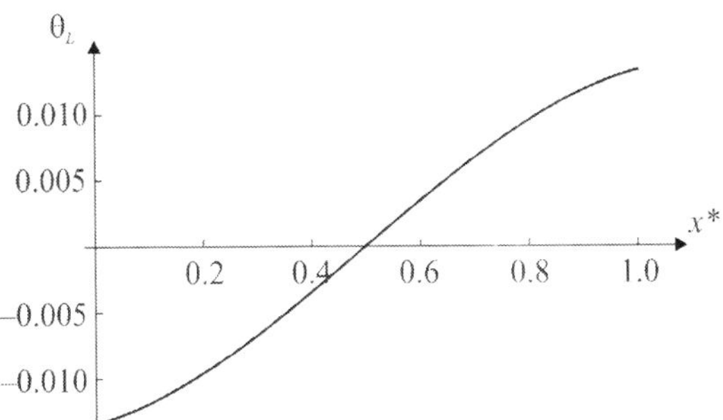

Fig. 4. Aluminum. Correction $\theta_L(x^*)$ to the Pohlhausen solution for an aluminum plate (L/H = 0.05, $G_L \approx G_R = 4.2 \times 10^7$, $\kappa_L = \kappa_R = 3.24 \times 10^{-6}$, and $T_{0L} = 30\ ^\circ$C, $T_{0R} = 20\ ^\circ$C). The correction $\theta_R(x^*)$ differs insignificantly from $\theta_L(x^*)$.

The heat flux density and the heat transfer coefficients are approximately the same as in the previous case, namely 15 W/m² and ~3 W/m² · K, respectively.

5.3 Brick wall

Next we consider a 10 cm thick and 200 cm high wall (aspect ratio L/H = 0.05) with thermal conductivity k_w = 0.72 W/m · K (brick) and surrounded by air; for the temperature difference of 4.3 K, the Grashof numbers take the values $G_L = 4.55 \times 10^9$ and $G_R = 4.52 \times 10^9$, respectively, and $\kappa_L = \kappa_R = \kappa = (k_a/k_w)(L/H) = 0.00183$. The coefficient $\gamma \approx 1$.
While in the previous cases, the iteration method (Miyamoto et al., 1980) was sufficient to solve Eqs. (23), in this example only the Newton method is applicable. The function $\theta_L(x^*)$ ($\theta_R(x^*)$ being essentially identical) is shown in Fig. 5. Here the temperature drop across the wall is $0.13(T_{0L} - T_{0R})$; J_R = -0.0570 and J_L = -0.0565 resulting in

$$\bar{Nu}^{(R)} = 0.511 \left(\frac{gH^3}{v_R^2} \right)^{1/4} \left(\frac{T_{0L} - T_{0R}}{T_{0R}} \right)^{1/4} , \tag{32}$$

and $\bar{Nu}^{(L)}$ is given by a similar expression.
In this case, the value of the Nusselt number is $\bar{Nu}^{(R)} \cong \bar{Nu}^{(L)} \cong 163$. The temperature drop across the plate is $T_1 - T_2$ = 1.3 K. The heat flux density is approximately 9 W/m², and the heat transfer coefficients ~2 W/(m² · K).
If the aspect ratio increases at constant height, the Grashof number increases, while $\theta_{L,R}(x^*)$ decrease. If the height increases at constant aspect ratio, the Grashof number as well as $\theta_{L,R}(x^*)$ increase.

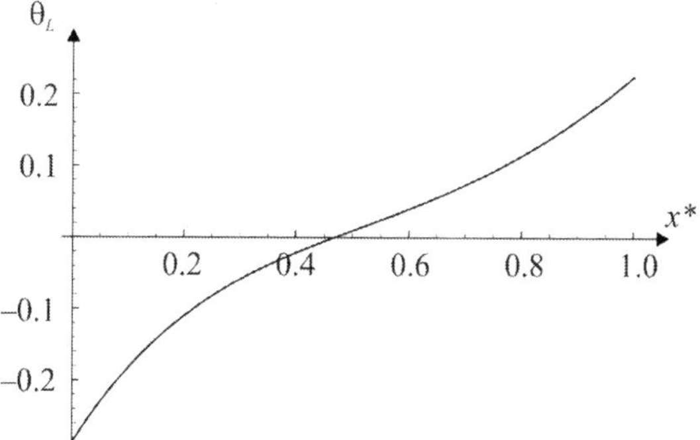

Fig. 5. Brick. The correction $\theta_L(x^*)$ to the Pohlhausen solution for a wall with $k_w = 0.72$ W/m $\cdot$ K ($L/H = 0.05$, $G_L \approx G_R \approx 4.5 \times 10^9$, $\kappa_L = \kappa_R = 0.00183$, and $T_{0L} = 30\,^\circ$C, $T_{0R} = 20\,^\circ$C).

5.4 Concrete wall

Finally, we consider a concrete (stone mix) wall, again 10 cm thick and 2 m high (aspect ratio $L/H = 0.05$) with thermal conductivity $k_w = 1.4$ W/m $\cdot$ K and surrounded by air. For the temperature difference of 4.7 K, the Grashof numbers are $G_L = 4.84 \times 10^9$ and $G_R = 4.97 \times 10^9$, respectively, and $\kappa_L = \kappa_R = \kappa = (k_a/k_w)(L/H) = 0.00094$. The coefficient $\gamma = 1.0005 \approx 1$.

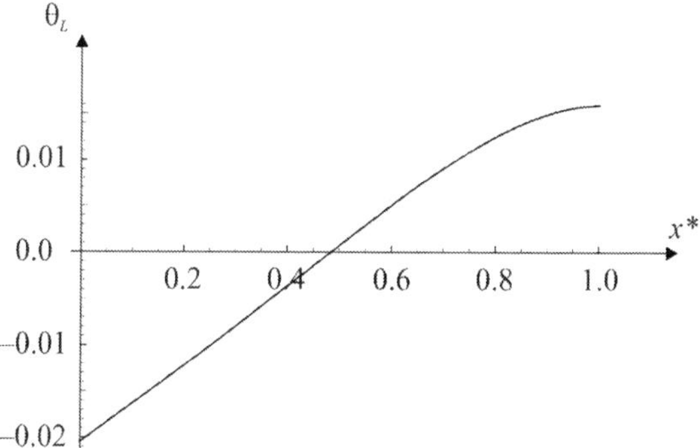

Fig. 6. Concrete (mix stone). The correction $\theta_L(x^*)$ to the Pohlhausen solution for a wall with $k_w = 1.4$ W/m $\cdot$ K ($L/H = 0.05$, $G_L = 4.84 \times 10^9$, $G_R = 4.97 \times 10^9$, $\kappa_L = \kappa_R = 0.00094$, and $T_{0L} = 30\,^\circ$C, $T_{0R} = 20\,^\circ$C).

The function $\theta_L(x^*)$ (or $\theta_R(x^*)$) is shown in Fig. 6. Here the temperature drop across the wall is $0.07(T_{0L} - T_{0R})$; $J_R = -0.0558$ and $J_L = -0.0555$ resulting in

$$\overline{Nu}^{(R)} = 0.507 \left(\frac{gH^3}{v_R^2} \right)^{1/4} \left(\frac{T_{0L} - T_{0R}}{T_{0R}} \right)^{1/4}. \tag{33}$$

In this case, the Nusselt numbers are $\overline{Nu}^{(R)} \cong \overline{Nu}^{(L)} \cong 163$. The temperature drop across the plate is $T_1 - T_2 = 0.7$ K.

The heat flux density is approximately 10 W/m², and the heat transfer coefficients ~2 W/m² · K.

6. Conclusion

Free convection along both sides of a vertical flat wall was considered within the framework of the laminar boundary-layer theory and for the case where only the temperatures of the fluid far away from the wall are known. It has been shown how to determine the average surface temperatures T_1 and T_2 together with the corresponding heat transfer coefficients in order for the equations (1) and (2) to yield the correct value for the total heat flow across the wall. In particular, if the small surface temperature variations $\theta_{L,R}(x^*)$ are neglected, the heat transfer from the wall to the fluid or vice versa is determined by the Pohlhausen solutions $\Theta_{L,R}(\varsigma)$ only. The corresponding Nusselt number $\bar{Nu}^{(R)}$, for example, is obtained from (19a) by neglecting the J_R term. This yields

$$\bar{Nu}^{(R)} = 0.471 \left(\frac{gH^3}{v_R^2} \right)^{1/4} \left(\frac{T_{0L} - T_{0R}}{T_{0R}} \right)^{1/4}. \tag{34}$$

It differs less that one percent as compared to the values given by eqs. (30a) and (31) which are valid for good thermal conductors. Consequently, the Pohlhausen solution can be therefore safely used in this case. For poor thermal conductors, like brick or concrete walls, the corrections may be more substantial. In particular, for a brick wall, the correction, obtained by comparing (32) and (34), is roughly 10 percent and it should be taken into account.

In numerical calculations, the Newton method turned out to be sufficient in solving the equations for the temperature corrections to the Pohlhausen solution. The simple iteration procedure, however, was found to have a rather restricted range of validity (large thermal conductivity, large aspect ratio of the plate).

7. Mathematical note

The system of equations (23a) and (23b) is defined for n = 1, 2,… only. For n = 0, the equations of both parts of the system simplify to a normalization conditions,

$$\int_0^1 F_L(x^*)dx^* = 1, \quad \int_0^1 F_R(x^*)dx^* = 1. \tag{MN.1}$$

It is natural to look for the functions F_L and F_R as elements of some linear subspace $S_m \subset C^r([0, 1])$ where S_m should become dense in $C^r([0, 1])$ as $m \to \infty$. Let us denote the basis of S_m as

$$s_i, \quad i = 0,1,\ldots, m.$$

Then the unknown functions F_L and F_R could be written as

$$F_L = \sum_{i=1}^{m} c_{L,i} s_i, \quad F_R = \sum_{i=1}^{m} c_{R,i} s_i.$$

We expect on physical grounds F_L and F_R to be rather smooth so $r \in N$ could be assumed to be at least 2. This indicates that the Fourier coefficients of the unknown functions

$$\int_0^1 F_L(x^*)\cos(n\pi x^*)dx^*, \quad \int_0^1 F_R(x^*)\cos(n\pi x^*)dx^*,$$

as well as of the other dependent quantities involved should decay at least as $O(1/n^2)$ or faster. As a consequence, only a small part of the infinite system is expected to be significant for F_L and F_R. Thus, for a particular choice of $m \in N$ only the equations $n = 0, 1, ..., m$ are taken into account. This gives a system of $2(m + 1)$ nonlinear equations for the unknown coefficients

$$\mathbf{c}_L := (c_{L,i})_{i=0}^{m}, \quad \mathbf{c}_R := (c_{R,i})_{i=0}^{m},$$

written in short as,

$$\begin{aligned} f(\mathbf{c}_L) &= g_1(\mathbf{c}_L, \mathbf{c}_R), \\ f(\mathbf{c}_R) &= g_2(\mathbf{c}_L, \mathbf{c}_R). \end{aligned} \tag{MN.2}$$

The structure of the system (MN.2) follows from (23) but with (MN.1) added to each equations block. There are two important steps to be considered. The first is the choice of the subspace S_m. We decided to try first perhaps the simplest approach, by choosing the subspace S_m as the space of polynomials $S_m = P_m$ of degree $\leq m$. The numerical results turned out satisfactory. Alternatively, we could always switch to a proper spline space. The second step regards the efficient numerical solution of the system (MN.2). Inspection of the equations (23) reveals that the function f depends linearly on the unknowns. Since the functions g_i are much more complicated, the direct iteration seems to be a cheap shortcut. So, with the starting choice incorporating the conditions (MN.1),

$$c_{L,i}^{(0)} = c_{R,i}^{(0)} = \frac{1}{2(m+1)}, i = 1, 2, ..., m, \tag{MN.3a}$$

$$c_{L,0}^{(0)} = c_{R,0}^{(0)} = 1 - \frac{1}{2(m+1)}\sum_{i=1}^{m}\frac{1}{i}, \tag{MN.3b}$$

the direct iteration reads,

$$f(\mathbf{c}_L^{(k+1)}) = g_1(\mathbf{c}_L^{(k)}, \mathbf{c}_R^{(k)}),$$

$$f(\mathbf{c}_R^{(k+1)}) = g_2(\mathbf{c}_L^{(k)}, \mathbf{c}_R^{(k)}), \quad k = 0, 1, ...$$

This approach was quite satisfactory for some parameter values, but failed to converge for others. Clearly, the map involved in this case ceases to be a contraction. However, the Newton method turned out to be the proper way to solve the system (MN.2). For any consistent data choice and particular m, only several Newton steps were needed. The initial values of the unknowns were again taken as in (MN.3). The Jacobian matrix $J(\mathbf{c}_L, \mathbf{c}_R)$, needed at each Newton step involving the solution of a system of linear equations,

$$J(\mathbf{c}_L^{(k)}, \mathbf{c}_R^{(k)}) = \begin{pmatrix} \Delta\mathbf{c}_L^{(k)} \\ \Delta\mathbf{c}_R^{(k)} \end{pmatrix} = \begin{pmatrix} f(\mathbf{c}_L^{(k)}) - g_1(\mathbf{c}_L^{(k)}, \mathbf{c}_R^{(k)}) \\ f(\mathbf{c}_R^{(k)}) - g_2(\mathbf{c}_L^{(k)}, \mathbf{c}_R^{(k)}) \end{pmatrix},$$

and a correction

$$\begin{pmatrix} \mathbf{c}_L^{(k+1)} \\ \mathbf{c}_R^{(k+1)} \end{pmatrix} = \begin{pmatrix} \mathbf{c}_L^{(k)} \\ \mathbf{c}_R^{(k)} \end{pmatrix} + \begin{pmatrix} \Delta\mathbf{c}_L^{(k)} \\ \Delta\mathbf{c}_R^{(k)} \end{pmatrix}, \quad k = 0, 1, \ldots,$$

admits no close form and has to be computed numerically. It is a simple task to compute all the partial derivatives involved if the three basic terms that depend on the unknown coefficients are determined. A brief outline is as follows. For a given S_m, let $(s_i)_{i=0}^m$ be its basis, and

$$F = \sum_{i=1}^m c_i s_i, \quad \mathbf{c} := (c_i)_{i=0}^m,$$

stands for F_L and F_R, and

$$\zeta(x) = \int_0^x F(u)\,du = \sum_{i=0}^m c_i \int_0^x s_i(u)\,du.$$

Then

$$\frac{\partial}{\partial c_j} F(x) = s_j(x), \quad \frac{\partial}{\partial c_j} \zeta(x) = \int_0^x s_j(u)\,du, \quad j = 0, 1, \ldots, m.$$

Further,

$$\frac{\partial}{\partial c_j} \int_0^1 F(x)\cos(n\pi x)\,dx = \int_0^1 s_j(x)\cos(n\pi x)\,dx, \quad j = 0, 1, \ldots, m,$$

which yields all coefficients in both $n = 0$ equations as well as the parts of the elements in J that contribute by the partial derivatives of the function f. In order to compute $\partial g_i / \partial c_j$, the following two terms have to be determined,

$$\frac{\partial}{\partial c_j}\int_0^1 \frac{F(x)^{3/2}}{\zeta(x)^{1/4}}\cos(n\pi x)dx = \frac{3}{2}\int_0^1 \frac{F(x)^{1/2}}{\zeta(x)^{1/4}}s_j(x)\cos(n\pi x)dx - \frac{1}{4}\int_0^1 \frac{F(x)^{3/2}}{\zeta(x)^{5/4}}\cos(n\pi x)\int_0^x s_j(u)du\,dx$$

and

$$\frac{\partial}{\partial c_j}\int_0^1 \frac{\zeta(x)^{3/4}F'(x)}{F(x)^{1/2}}\cos(n\pi x)dx =$$

$$\int_0^1 \frac{2F(x)s_j{}'(x)-F'(x)s_j(x)}{2F(x)^{3/2}}\zeta(x)^{3/4}\cos(n\pi x)dx - \frac{3}{4}\int_0^1 \frac{F'(x)}{F(x)^{1/2}\zeta(x)^{1/4}}\cos(n\pi x)\int_0^x s_j(u)du\,dx$$

where the prime indicates the ordinary derivative with respect to x. Since n is rather small, it turned out that the use of the Filon's quadrature rules was not necessary.

8. Nomenclature

$a_{L,R}{}^{(i)}$	defined by eq. (24)
A	surface area of the wall
$F_{L,R}(x^*),\ \zeta_{L,R}(x^*)$	defined by eq. (20c)
g	acceleration of gravity
$G = \beta g(T_s - T_{0R})H^3/\nu^2$	Grashof number
$h_{L,R}$	convection transfer coefficients to the left and right of the wall
$J_{L,R}$	defined by eq. (15d)
k	thermal conductivity
$L,\ H$	thickness and height of the wall
$\overline{Nu}^{(L,R)}$	Nusselt numbers associated with the left- and right-hand surface of the wall
$P = \nu/a$	Prandtl number
$\dot{Q},\ q = \dot{Q}/A$	heat flow, heat flow density
$T_{0L,\ 0R}$	air temperature far from the wall to the left and right of the wall
$T_{1,\ 2}$	temperature of the left and right wall surface
T_s	characteristic wall surface temperature
$U = k/L$	thermal transmittance
$u,\ v$	x- and y-component of the velocity field
$x^* = x/H,\ y^* = y/H$	dimensionless coordinates

Greek symbols

a	thermal diffusivity
β	thermal-expansion coefficient of the air

γ	defined by eq. (15c)
$\zeta_{L,R}(x^*)$	defined by eq. (20c)
η	viscosity
$\kappa_{L,R}$	defined in eq. (15e)
$\nu = \eta/\rho$	kinematic viscosity
ξ	$G_R^{1/4}\, y^*/(4x^*)^{1/4}$
ρ	mass density
$\Phi_{L,R}(\xi),$	Pohlhausen solution
$\Theta_{L,R}(\xi)$	temperature function associated with $\Phi_{L,R}(\xi)$
$\phi_R(x^*,\xi),\ \theta_R(x^*,\xi)$	corrections to the Pohlhausen solution introduced in eqs. (5), (6)
ψ	stream function introduced in eq. (5)

Subscripts, superscripts

a	air
L, R	left, right
s	surface
w	wall
$*$	dimensionless coordinate based on H

9. References

Grimson, J. (1971). *Advanced Fluid Dynamics and Heat Transfer*, McGraw-Hill, Maidenhead, pp. 215-219

Kao, T.T.; G.A. Domoto, G.A. & Elrod, H.G. (1977). Free convection along a nonisothermal vertical flat plate, *Transactions of the ASME*, February 1977, 72-78, ISSN: 0021-9223

Landau, L.D.; Lifshitz, E.M. (1987). *Fluid Mechanics*, ISBN 0-08-033933-6, Pergamon Press, Oxford, pp. 219-220

Miyamoto, M.; Sumikawa, J.; Akiyoshi, T. & Nakamura, T. (1980). Effects of axial heat conduction in a vertical flat plate on free convection heat transfer, *International Journal of Heat and Mass Transfer*, 23, 1545-53, ISSN: 0017-9310

Ostrach, S. (1953). An analysis of laminar free convection flow and heat transfer about a flat plate parallel to the direction of the generating body force, *NACA Report* 1111, 63-79

Pohlhausen, H. (1921). Der Wärmeaustausch zwischen festen Körpern und Flüssigkeiten mit Kleiner Wärmeleitung, *ZAMM* 1, 115-21, ISSN

Pop, I. & Ingham, D.B. (2001). *Convective Heat Transfer*, ISBN 0 08 043878 4, Pergamon Press, Oxford, pp. 181-198

Pozzi, A. & Lupo, M. (1988). The coupling of conduction with laminar natural convection along a flat plate, *International Journal of Heat and Mass Transfer*, 31, 1807-14, ISSN: 0017-9310

Vynnycky, M. & Kimura, S. (1996). Conjugate free convection due to a heated vertical plate, *International Journal of Heat and Mass Transfer*, 39, 1067-80, ISSN: 0017-9310

Conjugate Flow and Heat Transfer of Turbine Cascades

Jun Zeng and Xiongjie Qing
China Gas Turbine Establishment
P. R. China

1. Introduction

Heat transfer design of HPT airfoils is a challenging work. HPT usually requires much cooling air to guarantee its life and durability, but that will affect thermal efficiency of turbine and fuel consumption of engine[1]. The amount of blade cooling air depends on the prediction accuracy of temperature field around turbine airfoil surface, which is related to the prediction accuracy of temperature field in laminar-turbulent transition region. The laminar-turbulent transition is very important in modern turbine design. On suction side of turbine airfoil, the flow is relaminarized under the significant negative pressure gradient. In succession, when the relaminarized flow meets enough large positive pressure gradient, laminar-turbulent transition appears. In transition region, the mechanisms of flow and heat transfer are complicated, so it is hard to simulate the region.

Methods of conjugate flow and heat transfer analysis have been discussed extensively. In 1995, Bohn[2] simulated the heat transfer along Mark II[3] cascade using a two-dimensional (2D) conjugate method at the transonic condition (the exit isentropic Mach number is 1.04). In his simulation, the turbulence model used was Baldwin-Lomax model. The result was that the max difference between the predicted temperature along the cascade and the test data was not larger than 15K. The method was also used to simulate C3X[3] turbine cascade at 2D boundary conditions by Bohn[4], and a good agreement between the prediction results and the test data was gotten. Bohn also published a paper[5] in which Mark II turbine cascade with thermal barrier coatings was calculated in 2D cases. The ZrO_2 coatings with a thickness of 0.125mm bonded a 0.06mm MCrA1Y layer were applied. There were two configurations of the coatings. The problems and the influence of coatings on the thermal efficiency were solved by the same solver and evaluated. On the basis, the 3D numerical investigation of conjugate flow and heat transfer about Mark II with thermal barrier coating was done[6]. The uncoated vane was also used to validate the 3D method. The influence of the reduced cooling fluid mass flow on the thermal stresses was discussed in detail. York[7] used 3D conjugate method to simulate C3X turbine cascade. Because no transition model was used, the simulated external HTC (EHTC) at the leading edge stagnation point and laminar region had low precision. Facchini[8] made another 3D conjugate heat transfer simulation of C3X, however there was an obvious difference of HTC between simulation results and experimental data. Sheng[9] researched 3D conjugate flow and heat transfer method of turbine. In some reference, the effect of transition on conjugate flow and heat transfer was not mentioned.

Because of laminar-turbulent transition on suction side of turbine airfoil and the transition consequentially effects conjugate flow and heat transfer, high precision transition models must be researched.

The best method for the simulation of conjugate flow and heat transfer case is Large Eddy Simulation (LES) or Direct Numerical Simulations (DNS). The two methods have high accuracy of predicting flow and heat transfer in transition region, but they are not suitable for engineering application nowadays, because they are too costly. At present, the better method for conjugate simulation is still using two-equation turbulence model with transition model.

In this paper, the numerical method considering transition was used to predict 2D and 3D conjugate flow and heat transfer. T3A flat plate, VKI HPT stator, VKI HPT rotor and MARK II stator were calculated. T3A flat plate was used to validate the accuracy of aerodynamic simulation, and the conjugate flow and heat transfer cases of other blades were calculated to validate the method. In the conjugate simulations, the effect of various turbulence models and inlet turbulence intensities on heat transfer were investigated.

2. Numerical method

2.1 Governing equations

The governing equations were:
Continuity equations:

$$\frac{\partial \rho}{\partial t} + \frac{\partial \rho U_i}{\partial x_i} = 0$$

Momentum equations (N-S equations):

$$\frac{\partial \rho U_i}{\partial t} + \frac{\partial \rho U_i U_j}{\partial x_j} = -\frac{\partial p}{\partial x_i} + \frac{\partial}{\partial x_j}\left(\mu_{eff}\left(\frac{\partial U_i}{\partial x_j} + \frac{\partial U_j}{\partial x_i}\right)\right) - \frac{2}{3}\frac{\partial}{\partial x_i}\left(\mu_{eff}\frac{\partial U_j}{\partial x_j}\right)$$

Energy equations:

$$\frac{\partial \rho h^*}{\partial t} + \frac{\partial \rho U_i h^*}{\partial x_i} = \frac{\partial p}{\partial t} + \frac{\partial}{\partial x_i}\left(\lambda\frac{\partial T}{\partial x_i}\right) + \frac{\partial}{\partial x_j}\left(\mu_{eff}U_i\left(\frac{\partial U_i}{\partial x_j} + \frac{\partial U_j}{\partial x_i}\right)\right) - \frac{2}{3}\frac{\partial}{\partial x_i}\left(\mu_{eff}U_i\frac{\partial U_j}{\partial x_j}\right)$$

For thermal conduction of solid, when there is no thermal source, the governing equations were:

$$\frac{\partial \rho c T}{\partial t} = \frac{\partial}{\partial x_i}\left(\lambda\frac{\partial T}{\partial x_i}\right)$$

On the interface of fluid and solid, heat flux is equivalent.

The governing equations were discretized with finite volume method. By means of solving continuity equations and Momentum equations simultaneously, the uncoupling of pressure and temperature was resolved. Convection term has second-order precision.

2.2 Turbulence model

Advanced turbulence model with high accuracy of describing turbulence nature must be used to get exact flow field, especially for engineering problems. In order to improve accuracy of flow and heat transfer analysis, Menter[10] developed SST turbulence model. The model assimilated the advantages of k-ω model and k-ε model. It used k-ω model near wall and k-ε model far from wall, having high accuracy of predicting flow field near wall and avoiding strong sensitivity to free stream conditions. A number of test cases were predicted by means of the model, proving that the model has high accuracy of conjugate flow and heat transfer problem especially for large adverse pressure gradient[11].

2.3 Transition model

Transition has significant effect on heat transfer. In order to predict conjugate flow and heat transfer of turbine cascades, turbulence model must be coupled with transition model. Experience modified transition models include zero-equation model, one-equation model and two-equation model. Intermittency is given in zero-equation model. In one-equation model, user-defined transition Reynolds number is used to solve intermittency, avoiding solving another equation so that reduce computation time, but the model does not consider effect of turbulence intensity and pressure gradient on transition. Two-equation model connects free stream turbulence intensity with transition momentum thickness Reynolds number at the onset of transition and solves two transport equations. One is used to calculate intermittency and the other is used to calculate momentum thickness Reynolds number. The two equations couple with production terms in SST turbulence model. The two-equation model can solve the transition caused by shock wave or separation. In order to predict complex cascade flow field with high accuracy and improve the solving precision of temperature and external heat transfer coefficient along airfoils, the modified two-equation transition model developed by Menter[12] was used.

The transport equation of intermittency γ in the two-equation transition model is:

$$\frac{\partial(\rho\gamma)}{\partial t} + \frac{\partial(\rho U_j \gamma)}{\partial x_j} = P_{\gamma 1} - E_{\gamma 1} + P_{\gamma 2} - E_{\gamma 2} + \frac{\partial}{\partial x_j}\left((\mu + \mu_t)\frac{\partial\gamma}{\partial x_j}\right)$$

where:

$$P_{\gamma 1} = 2F_{length}\rho S(\gamma F_{onset})^{0.5} \; ; E_{\gamma 1} = P_{\gamma 1}\gamma \; ;$$

$$P_{\gamma 2} = 0.06\rho\Omega\gamma F_{turb} \; ; E_{\gamma 2} = 50 P_{\gamma 2}\gamma \; ;$$

$$F_{onset} = \max\left(F_{onset2} - F_{onset3}, 0\right) \; ; F_{turb} = e^{-(0.25R_T)^4} \; ;$$

$$F_{onset1} = \frac{Re_v}{2.193\,Re_{\theta c}} \; ; F_{onset2} = \min\left(\max\left(F_{onset1}, F_{onset1}^4\right), 2.0\right) ; F_{onset3} = \max\left(1 - \left(\frac{R_T}{2.5}\right)^3, 0\right) ;$$

$$Re_v = \frac{\rho y^2 S}{\mu} \; ; R_T = \frac{\rho k}{\mu\omega} \; ; S = \sqrt{2S_{ij}S_{ij}} \; ; \Omega = \sqrt{2\Omega_{ij}\Omega_{ij}} \; ; S_{ij} = \frac{1}{2}\left(\frac{\partial U_i}{\partial x_j} + \frac{\partial U_j}{\partial x_i}\right) ; \Omega_{ij} = \frac{1}{2}\left(\frac{\partial U_i}{\partial x_j} - \frac{\partial U_j}{\partial x_i}\right)$$

For the transition caused by separation, the correction is:

$$\gamma_{eff} = \max\left(\gamma, \gamma_{sep}\right)$$

where:

$$\gamma_{sep} = \min\left(2 \bullet \max\left(\left(\frac{Re_v}{3.235 Re_{\theta c}}\right) - 1,0\right)F_{reattach}, 2\right)F_{\theta t}; \ F_{reattach} = e^{-\left(\frac{R_T}{20}\right)^4} \ ;$$

The transport equation of transition momentum thickness Reynolds number $\tilde{Re}_{\theta t}$ in the two-equation transition model is:

$$\frac{\partial(\rho \tilde{Re}_{\theta t})}{\partial t} + \frac{\partial(\rho U_j \tilde{Re}_{\theta t})}{\partial x_j} = P_{\theta t} + \frac{\partial}{\partial x_j}\left(2 \bullet (\mu + \mu_t)\frac{\partial \tilde{Re}_{\theta t}}{\partial x_j}\right)$$

where:

$$P_{\theta t} = 0.03\frac{\rho}{t_1}\left(Re_{\theta t} - \tilde{Re}_{\theta t}\right)\left(1 - F_{\theta t}\right); \ t_1 = \frac{500\mu}{\rho U^2} \ ;$$

$$F_{\theta t} = \min\left(\max\left(F_{wake} \bullet e^{-\left(\frac{y}{\delta}\right)^4}, 1 - \left(\frac{\gamma - 0.02}{1 - 0.02}\right)^2\right), 1\right)$$

$$\delta = \frac{50\Omega y}{U} \bullet \delta_{BL}; \ \delta_{BL} = 7.5\theta_{BL}; \ \theta_{BL} = \frac{\tilde{Re}_{\theta t}\mu}{\rho U} \ ;$$

$$F_{wake} = e^{-\left(\frac{Re_\omega}{1\times10^5}\right)^2}; \ Re_\omega = \frac{\rho \omega y^2}{\mu} \ ;$$

$Re_{\theta t}$ is the transition momentum thickness Reynolds number in experiment.

3. Validation

3.1 T3A flat plate

The geometric configuration and the boundary conditions of ERCOFTAC T3A[13] flat plate were shown in figure 1. The case was used to validate SST turbulence model with transition.

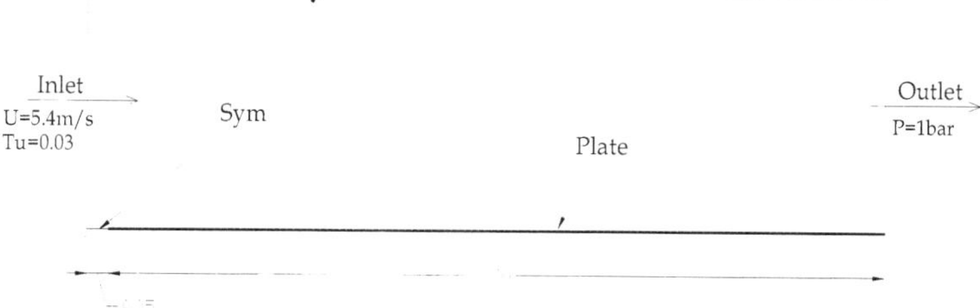

Fig. 1. Geometric configuration and boundary conditions of T3A

The computational mesh was shown in figure 2. Near the leading edge of T3A, the mesh density is the largest. The mesh included both tetrahedral and prismatic mesh. Total number of nodes was 67538 and total number of elements was 115090.

Fig. 2. Tetrahedral mesh of T3A

Figure 3 showed comparison between skin friction coefficient of predicted result and the one of test. It was shown that SST turbulence model with transition model ("transition" in the figure) has high accuracy of predicting the onset of transition in flat plate case. Around the leading edge of the plate, flow is laminar. In this region, SST model without transition model ("fully turbulent" in the figure) over predicted the friction coefficient. In the region of fully turbulent flow, the effect of transition model on friction coefficient is insignificant.

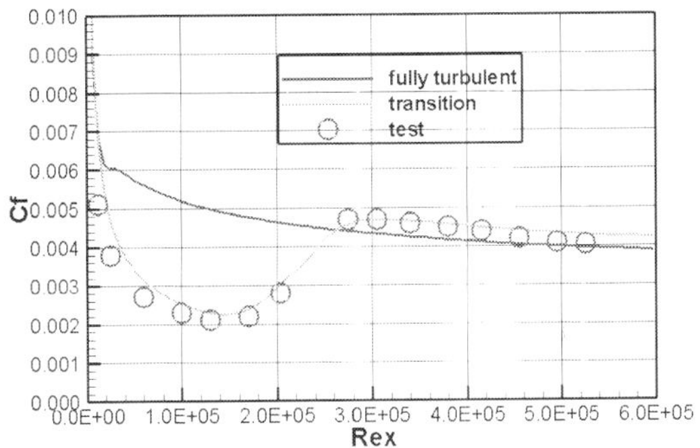

Fig. 3. Skin friction coefficient of T3A

The influence of the mesh shape on the simulation was also researched. A mesh only including hexahedral elements was created, as shown in figure 4. Total number of nodes was 123328 and total number of elements was 91233.

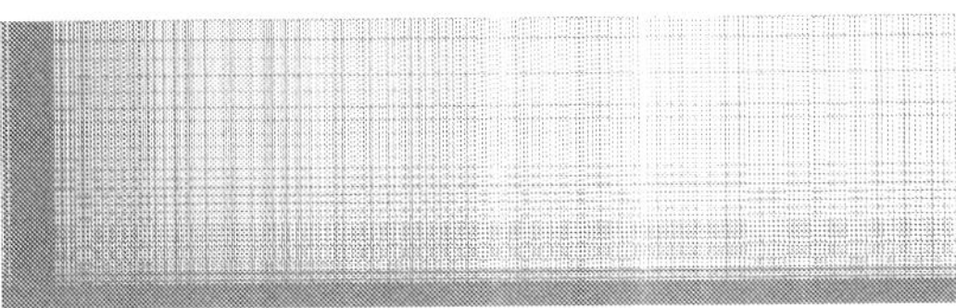

Fig. 4. Hexahedral mesh of T3A

Figure 5 showed the simulation results by means of hexahedral mesh. The result with transition model was also in good agreement with the test data. The trend of the predicted skin friction with transition model was similar to the result of the tetrahedral mesh. Figure 5b imply that the mesh shape had insignificant influence on the results.

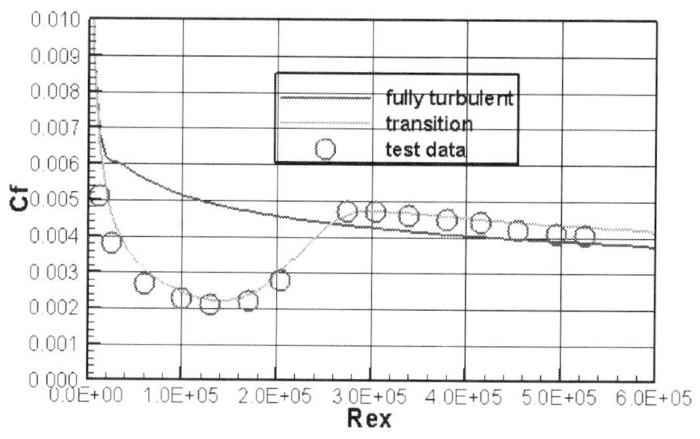

a) Simulation results

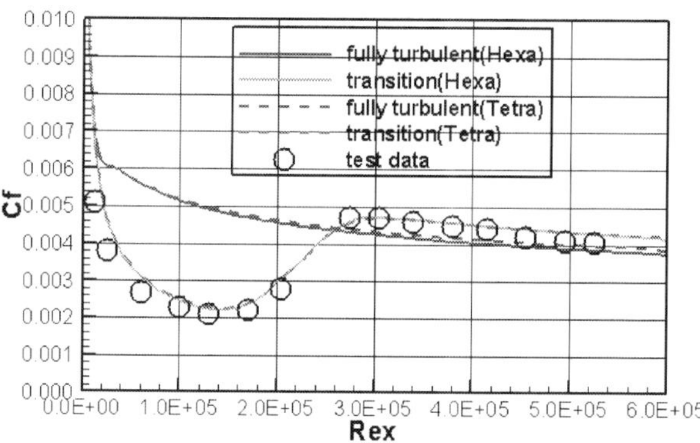

b) Compared to tetrahedral mesh

Fig. 5. Skin friction coefficient of T3A with hexahedral mesh

3.2 VKI HPT stator

VKI HPT stator was shown in figure 6. The geometric data was also given in the figure, and detail geometric data could be found in the reference 14. The computational mesh (in figure 7) was hexahedron. Total number of nodes was 109272. Altogether two predictions were made, and the boundary conditions were shown in table 1.

Condition	Tu	M_{2is}	Re_2	P_1^* /bar	T_1^* /K	P_2 /bar	T_w /K
1	1%	0.92	5×10^5	0.803	420	0.464	300
2	1%	0.92	2×10^6	3.21	420	1.86	300

Table 1. Boundary conditions of the cascade

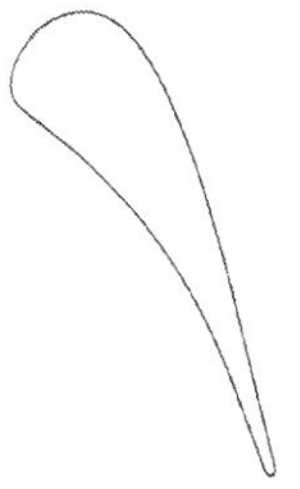

Geometric data	
Chord/mm	67.647
Pitch/mm	57.500
Stagger Angle/°	55.0
Throat/mm	17.568
Leading edge thickness/mm	8.252
Trailing edge thickness/mm	1.420

Fig. 6. Sketch and geometric data of VKI stator cascade

Fig. 7. Computational mesh of VKI stator cascade

Figure 8 showed the results at condition 1. On suction side of the airfoil, a large adverse pressure gradient was found near the trailing edge. The pressure gradient resulted in flow separation, inducing laminar-turbulent transition. The predicted onset of transition was closer to trailing edge as compared to the test data.

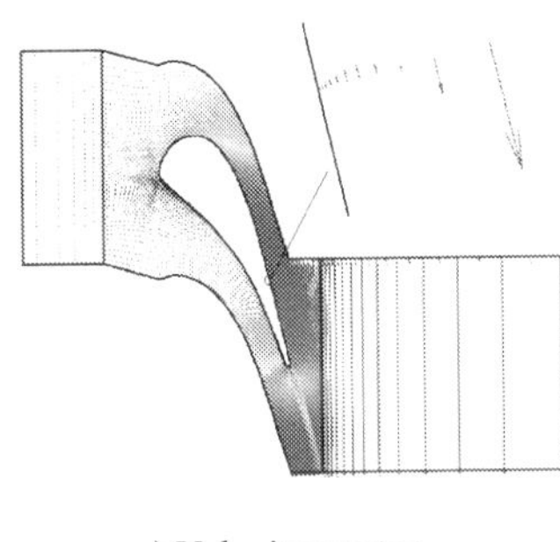

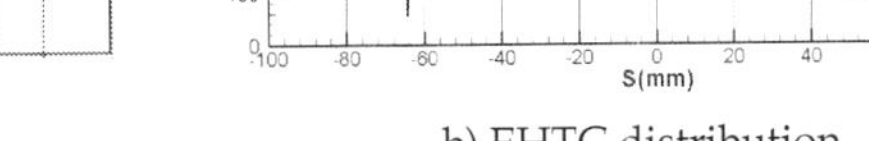

a) Velocity vector   b) EHTC distribution

Fig. 8. Results at condition 1

Figure 9 showed the EHTC distribution of the cascade at condition 2. In the most regions, the predicted EHTC agreed with the test one. The onset of transition was downstream predicted and the transition was predicted more sharply than test. In the low turbulence intensity, the method has high accuracy of predicting EHTC with various Reynolds number.

Comparing figure 8 with figure 9, it can be seen that EHTC increased with Reynolds number increment when other conditions were equivalent.

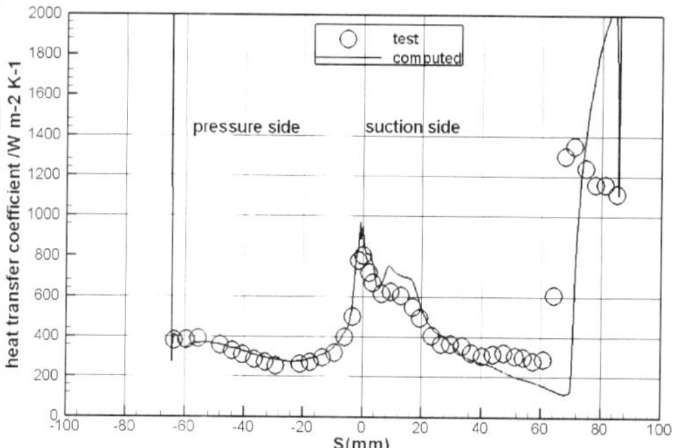

Fig. 9. EHTC distribution at condition 2

3.3 VKI HPT rotor

VKI HPT rotor was shown in figure 10. The geometric data was also given in the figure, and detail geometric data could be found in the reference 15. The hexahedral computational mesh was shown in figure 11. Total number of nodes was 39042. According to various turbulence intensities, two predictions were made.. The boundary conditions were shown in table 2.

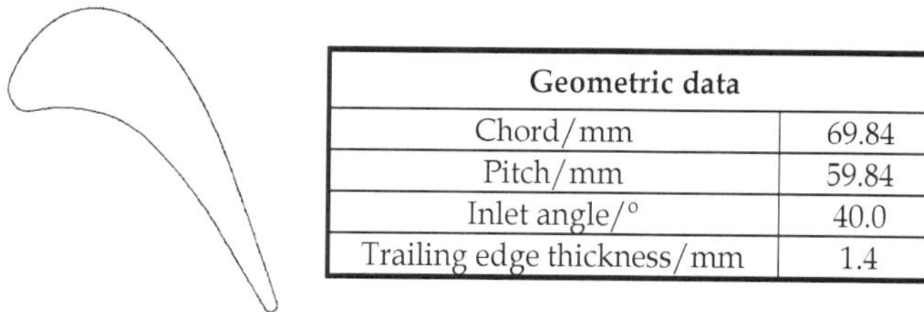

Geometric data	
Chord/mm	69.84
Pitch/mm	59.84
Inlet angle/°	40.0
Trailing edge thickness/mm	1.4

Fig. 10. Sketch and Geometric data of VKI rotor cascade

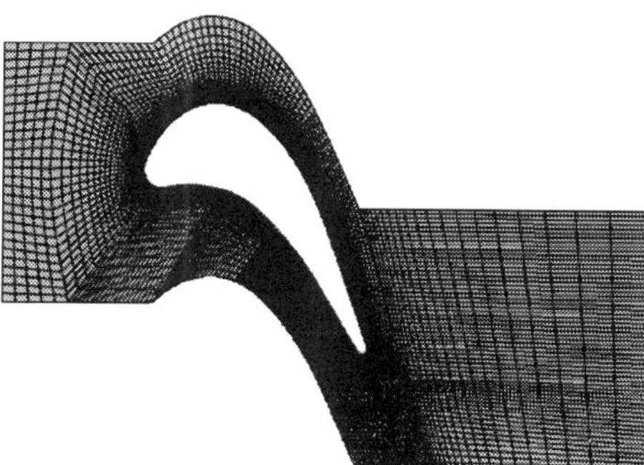

Fig. 11. Computational grid of VKI rotor cascade

condition	Tu	M_{2is}	Re_2	P_1^* /bar	T_1^* /K	P_2 /bar	T_w /K
1	1%	0.92	5×10^5	0.803	420	0.464	300
2	4%	0.92	5×10^5	0.803	420	0.464	300

Table 2. Boundary conditions of the cascade

Figure 12 showed the comparison between the predicted EHTC and the test data at the two conditions. For the onset of transition, the predicted one agreed with the test one at condition 1, and the predicted one was upstream of the test one at condition 2. Under the two turbulence intensities, the predicted EHTC in the laminar region located on suction side agreed with the test one. On pressure side, the predicted EHTC agreed well with the test one when intensity was 1%, and the predicted one was lower than the test one when intensity was 4%. The method has high accuracy of predicting EHTC under low turbulence intensity. That intensity increment resulted in EHTC growth.

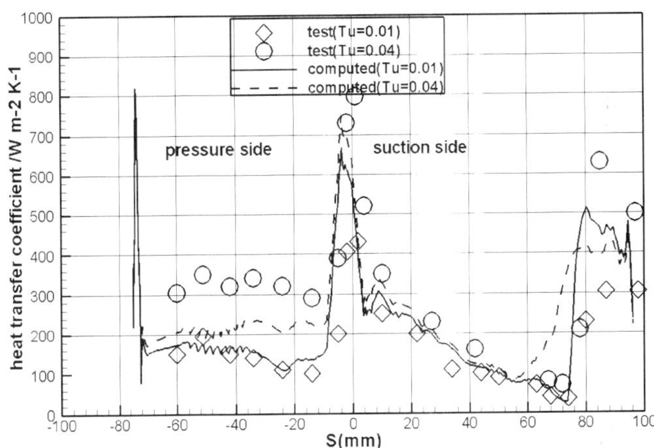

Fig. 12. EHTC distribution (condition 1 and 2)

4. Mark II stator

Mark II HPT stator is convectively cooled by ten cooling channels. The cooling medium is air. The geometric data and cooling channels were shown in figure 13. The detail geometric data could be found in the reference 3. The material of the stator is ASTM 310 stainless steel, the density of which is 7900 kg m-3 and the specific pressure heat capacity is 586.5 J kg-1 K-1. The thermal conductivity varies as temperature changes, the function is:

$$\lambda = 6.811 + 0.020176 \bullet T$$

Hylton L. D[3] researched the airfoil with experiment method, and got detail aerodynamic and thermal test data. Consequently MARK II stator becomes a typical case for conjugate flow and heat transfer simulations, which is applied to validating codes or the accuracy of methods.

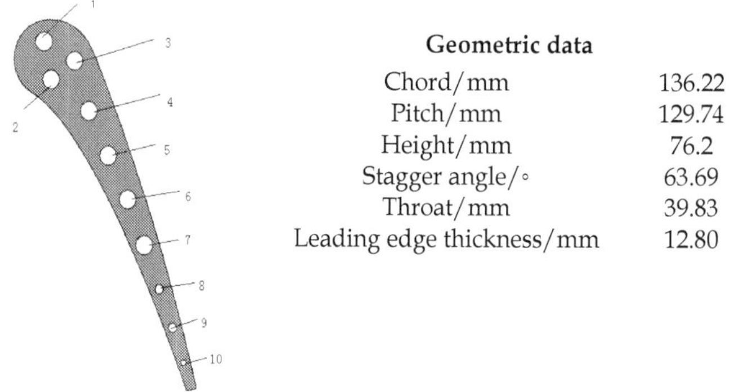

Geometric data

Chord/mm	136.22
Pitch/mm	129.74
Height/mm	76.2
Stagger angle/°	63.69
Throat/mm	39.83
Leading edge thickness/mm	12.80

Fig. 13. Sketch and geometric data of MARK II cascade

4.1 2D simulation

The 2D simulation used unstructured grid (shown in figure 14), increasing the grid density around wall in the fluid domain and the solid domain. Total number of nodes was 35279 and total number of elements was 58618. Total number of nodes in the fluid domain was 23143 and total number of elements in the fluid domain was 38158. Total number of nodes in the solid domain was 12136 and total number of elements in the solid domain was 20460. The boundary conditions were shown in table 3. The boundary conditions of cooling channel were given cooling temperature and heat transfer coefficient. All boundary conditions could be seen in the reference 3.

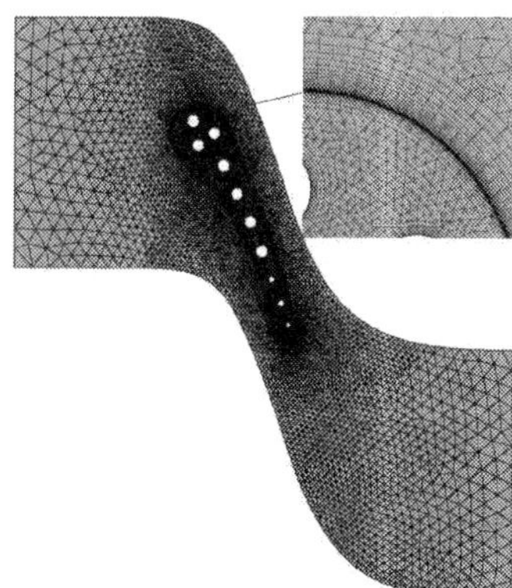

Fig. 14. Computational grid of MARK II cascade

Condition	Tu	M_{2is}	Re_2	P_1^* /bar	T_1^* /K	P_2 /bar
1	6.5%	0.89	1.98×10⁶	3.44	784	2.08
2	6.5%	1.04	2.01×10⁶	3.39	788	1.73

Table 2. Boundary conditions of MARK II cascade

4.1.1 The simulation result of SST with transition model

Figure 15 showed the pressure distributions at subsonic (condition 1) and transonic (condition 2) conditions by using SST with transition model. Along the entire surface of the cascade, the predicted pressure distribution agreed with the test results. In the most regions on pressure side, the distributions of the two conditions were the same, and had no relationship with exit pressure. On suction side, from the location of stagnation point to the location of the first shock wave, a large negative pressure gradient made flow relaminarized. In the next region, shock wave appeared, resulting in laminar-turbulent transition, and the flow finally became turbulent. At the subsonic condition, the predicted onset of the shock wave was slightly on the upstream of that at the transonic condition. However, at the subsonic condition, the intersection of the shock wave and boundary layer resulted in separation. In the laminar regions under the two conditions, the various exit pressures had no effect on surface pressure. In the region downstream of shock wave, the surface pressure distributions at the two conditions differed from each other. At the subsonic conditions, after shock wave, flow accelerated slowly and then decelerated until flow arriving in the trailing edge. At the transonic conditions, after shock wave, flow straightly accelerated to the trailing edge until trailing shock wave was met.

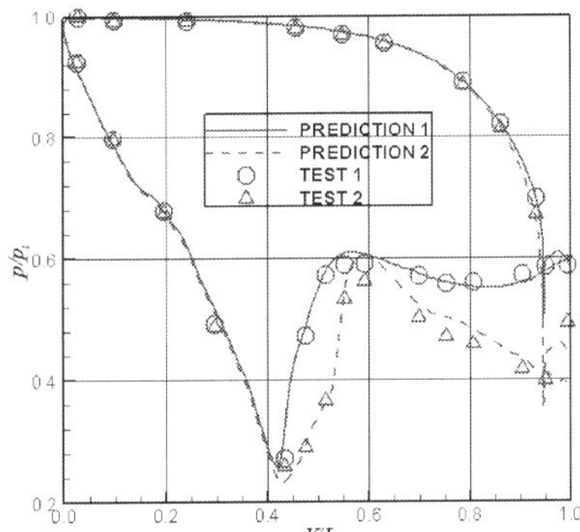

Fig. 15. Pressure distributions (condition 1 and 2)

Figure 16 showed Mach number contour and temperature contour at the subsonic and transonic conditions. At the subsonic condition, separation appeared at the location of 0.43 axial chord on suction side. At the transonic condition, there was no separation in the same location and shock wave appeared at the trailing edge. Upstream of the trailing edge on pressure side, flow accelerated fast resulting in gas temperature reduced rapidly in the solid domain, temperature was kept at a low level because of cooling air in cooling channels, the minimum value of which appeared between cooling channel 2 and cooling channel 3, and the maximum value of which appeared at the trailing edge.

Figure 17 demonstrated surface temperature distributions at the two conditions. The predicted distribution trend agreed with the test data, and the predicted distributions agreed with the test data in most regions. At the stagnation point and the laminar region on suction side, the predicted temperature had high accuracy at the two conditions. On the suction, shock wave appeared resulting in laminar-turbulent transition. Strong heat transfer made temperature curve becoming abrupt in the region of transition, and the maximum

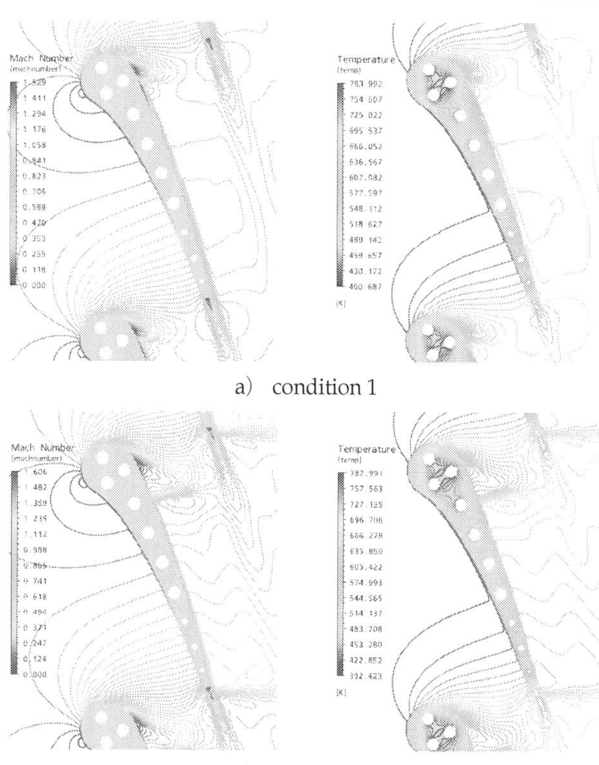

a) condition 1

b) condition 2

Fig. 16. Contours of Mach Number and Temperature (condition 1 and 2)

appeared at the end of the transition region. The numerical method over predicted the maximum temperature, especially at the subsonic condition, and the difference between the maximum temperature and the test data was 30K. In the fully turbulent region, the difference between the predicted temperature and the test data became smaller. On entire pressure side, the predicted temperature agreed with the test results. On pressure side and suction side, temperature distribution fluctuated, and the maximum and minimum values appeared in some locations. The minimum values located around cooling channels and the maximum value located between two channels. The maximum temperature on the entire surface located in the trailing edge. The EHTC was high at this location and the channel 10 was far from the location, so that the air could not effectively cool the trailing edge.

EHTC distribution along the airfoil at two conditions was shown in figure 18. As same as the temperature distribution, in the laminar region of suction side and the one of pressure side, the predicted EHTC agreed with the test data. Consequently the method predicted the onset of transition accurately. In transition region, EHTC increased abruptly and encountered the maximum at the end of the transition region where the method over predicted the EHTC. In the turbulent region downstream the transition one, the difference between the predicted EHTC and the test one reduced gradually.

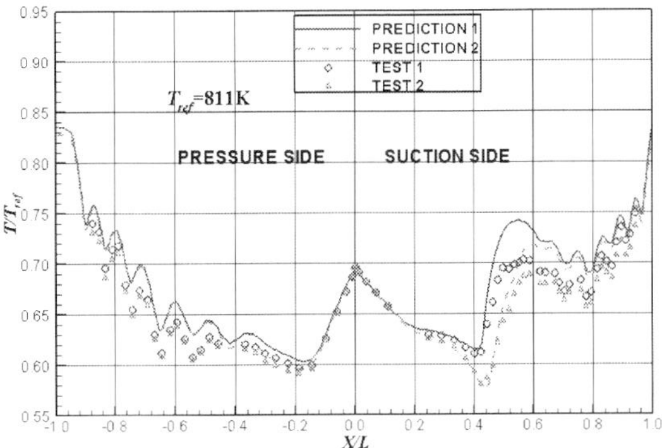

Fig. 17. Temperature distribution (condition 1 and 2)

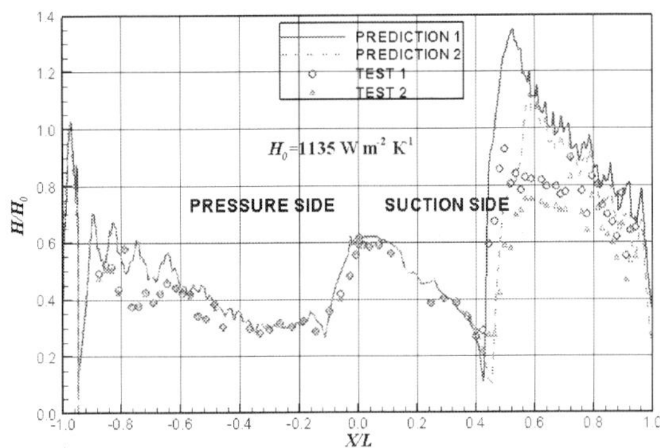

Fig. 18. EHTC distribution (condition 1 and 2)

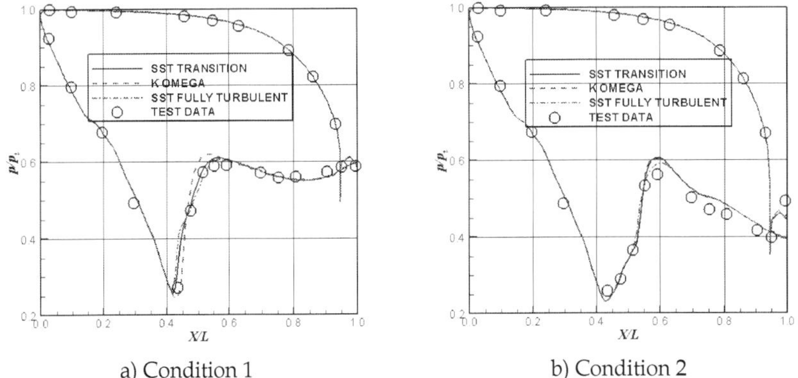

a) Condition 1b) Condition 2

Fig. 19. Pressure distribution of different turbulence models

4.1.2 The effect of turbulence models

Figure 19 showed the surface pressure distributions of simulations by using k-ω turbulence model (model 1), SST model with fully turbulent flow (model 2) and SST turbulent model with transition model (model 3) at the two conditions. Except around shock wave, the pressure distributions using the three models agreed with each other. At the subsonic condition, the predicted pressure distributions around shock wave by means of model 2 and model 3 agreed with the test data, and the predicted intensity of shock wave by means of model 1 was stronger than the test results. At the transonic condition, the predicted pressure distribution using model 2 was closest to the test data. However, as a whole, the three models had high accuracy of predicting surface pressure distributions.

Figure 20 showed EHTC distributions of the three models. Whether subsonic or transonic condition, the predicted EHTC by means of model 1 or model 2 at the stagnation point and in the laminar region had large difference from the test data. In the regions, the predicted results by means of model 3 agreed with the test data. In the fully turbulent region, the results of the three models were close to each other, agreeing with the test data.

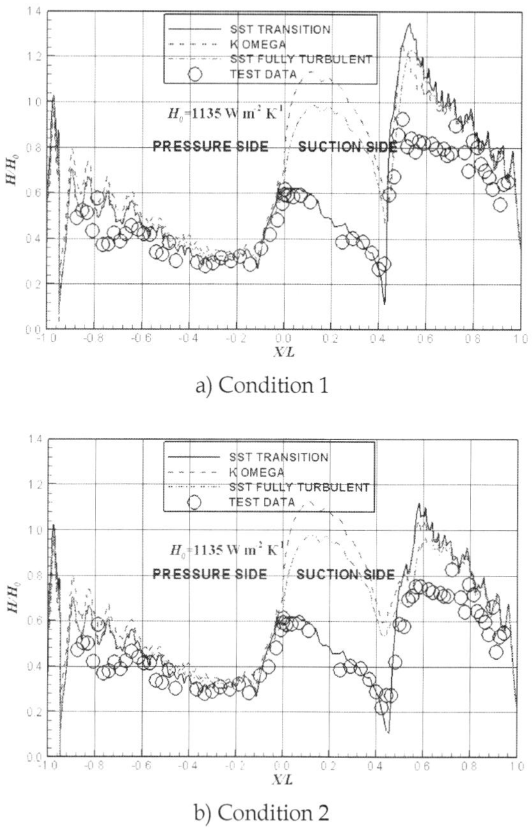

a) Condition 1

b) Condition 2

Fig. 20. EHTC distributions of different turbulence models

4.1.3 The effect of inlet turbulence intensities

Model 3 was used to research the effect of inlet turbulence intensities on predicted surface pressure and temperature. Figure 21 illustrated surface pressure distributions with various inlet turbulence intensities at the two conditions. Whether subsonic or transonic condition, when inlet intensity located in the range of 1% ~ 20%, the surface pressure and the location of shock wave were unaffected by inlet turbulence intensities.

Figure 22 showed surface temperature distributions with various inlet turbulence intensities at the two conditions. For various intensities, the trends of temperature along the surface were consistent, but the temperature values had some difference. With the intensity increasing, surface temperature increased in the region from the stagnation point to 0.43 axial chord on suction side and in the region from the stagnation point to 0.56 axial chord on pressure side. The intensity had insignificant effect on temperature distribution in the latter regions. In the two conditions, the predicted temperature under the low intensity (1%) had a large difference from the one under the medium intensity (6.5%). In the range of intensity 3% ~ 20%, the surface temperature using intensity 6.5% could be predicted well, with the maximum difference 18K. In the range of intensity 6.5% ~ 20%, the surface temperature using intensity 15 % could be predicted well, with the maximum difference 12K. These opinions might only be adequate for the airfoils.

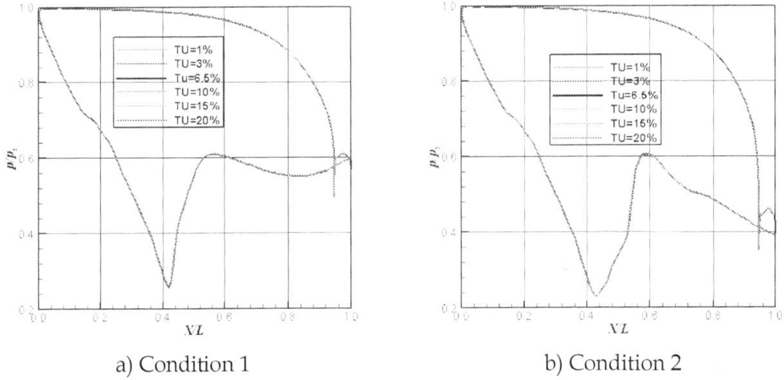

a) Condition 1   b) Condition 2

Fig. 21. Pressure distribution under different inlet turbulence intensities

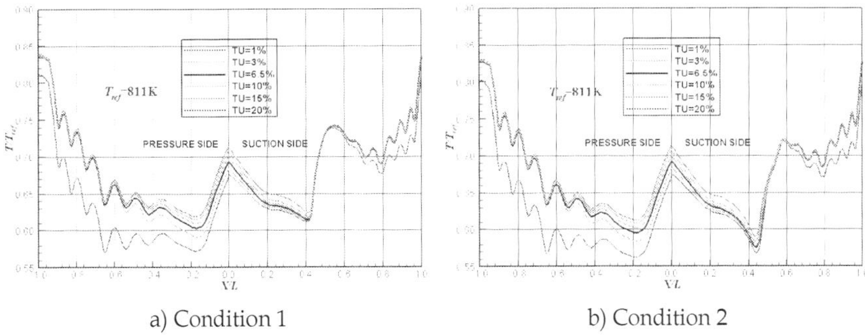

a) Condition 1   b) Condition 2

Fig. 22. Temperature distribution under different inlet turbulence intensities

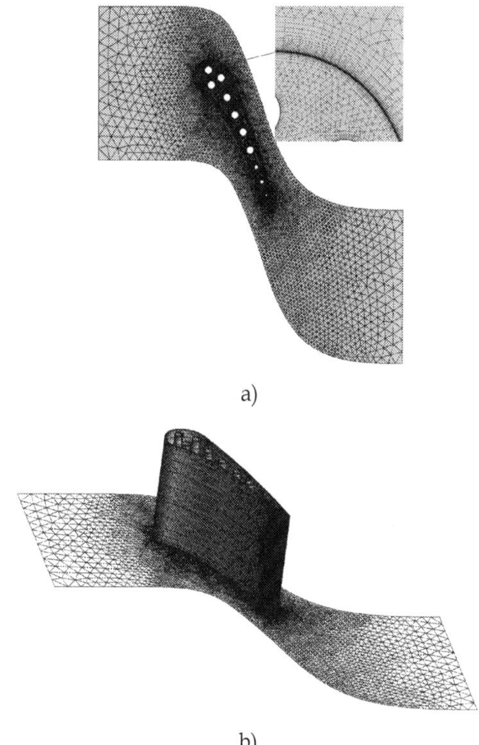

a)

b)

Fig. 23. Computational grid(3D) of MARK II cascade

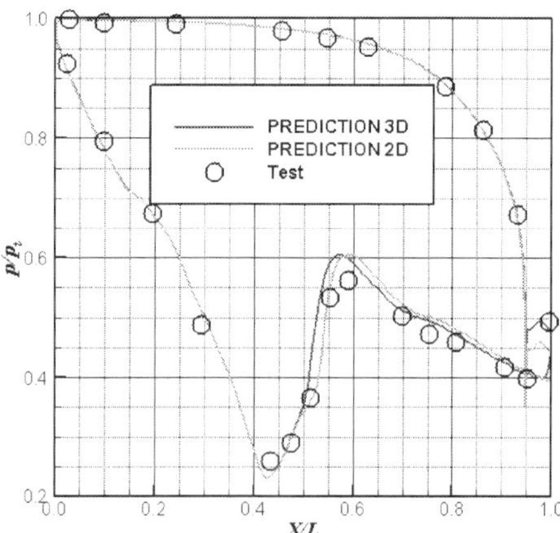

Fig. 24. Pressure distribution (2D and 3D)

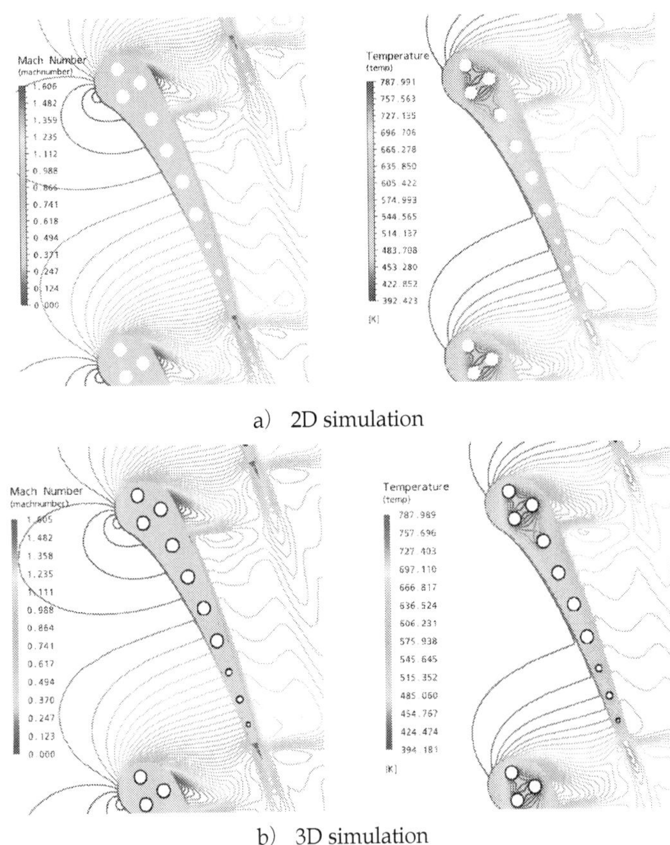

a) 2D simulation

b) 3D simulation

Fig. 25. Contours of Mach Number and Temperature (2D and 3D)

4.2 3D simulation

The 3D simulation used unstructured grid (shown in figure 23), which was got by extruding the 2D grid used in the 2D simulation above. Total number of nodes was 659310 and total number of elements was 788945. Total number of nodes in the fluid domain was 572370 and total number of elements in the fluid domain was 671147. Total number of nodes in the solid domain was 86940 and total number of elements in the solid domain was 117798.

The boundary condition was the same as condition 2 shown in table 3. The boundary conditions of cooling channel were given cooling temperature and heat transfer coefficient, too. The simulation was made using SST turbulence model with transition model. The 3D results on the mean plane were compared to the 2D one.

Figure 24 showed the pressure distribution by using 2D and 3D method. Comparing to 2D result, the location of the end of shock wave was upstream. The reason may be that the endwall had some influence on the flow. The difference between Mach contours were shown in figure 25.

The temperature distribution of 2D and 3D was shown in figure 26 (also can be seen in figure 25). All the results agreed well with the test data, but there were some dissimilarities which may arise from the endwall. The dissimilarities focused on the transition region. For the cascade, the influence of the endwall on the blade surface temperature and HTC distribution could be ignored.

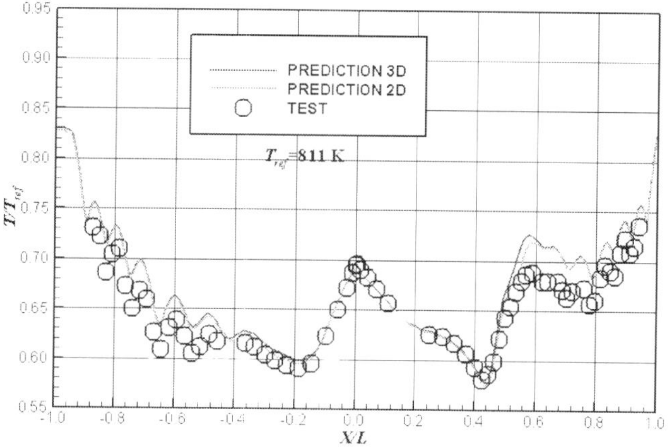

Fig. 26. Temperature distribution (2D and 3D)

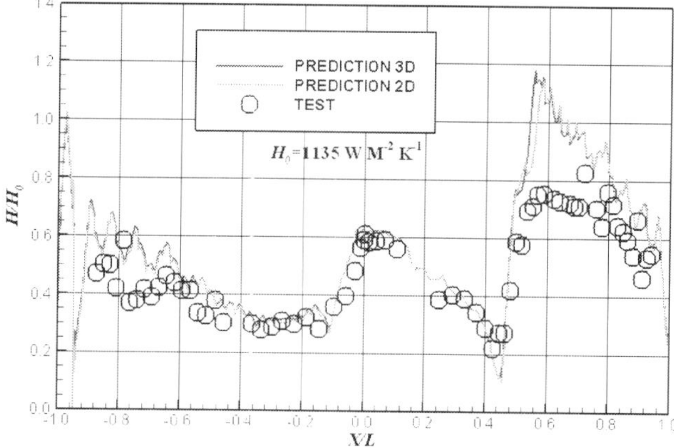

Fig. 27. EHTC distribution (2D and 3D)

5. Conclusion

The 3D method which was used to simulate some cases should be validated. The method calculated URANS equations by employing SST turbulence model with transition model. The cases were T3A flat plate, VKI HPT stator, VKI HPT rotor and Mark II stator. The result of T3A flat plate showed that the method had high accuracy on predicting flow. All the other results indicated that the method had high accuracy on simulating conjugate flow and heat transfer.

For the conjugate flow and heat transfer problems, two steps were applied. The first step was that no internal cooling cascades, VKI HPT stator and rotor, were calculated. With the good results, the second step was that Mark II stator with internal cooling channel was calculated. Using SST turbulence model with transition model, the accuracy of surface pressure distribution and temperature loading was high. The effect of turbulence models and the one of inlet turbulence intensities were researched, too. The results showed that trend of the temperature along the profile under inlet low turbulence intensity (1%) was different from the one under other intensities. However, there was a shortcoming that the temperature or heat transfer coefficient at the end of the transition region was over predicted results.

For the above simulations, 3D CFD method has high accuracy on predicting conjugate flow and heat transfer problems. But the method should be improved in the transition region.

6. Nomenclature

ρ	Density	$kg\ m^{-3}$
U	Velocity	$m\ s^{-1}$
t	time	s
x	Cartesian coordinate	m
p	Pressure	$kg\ m^{-1}\ s^{-2}$
μ_{eff}	Effective viscosity	$kg\ m^{-1}\ s^{-1}$
h	Enthalpy	$m^2\ s^{-2}$
T	Temperature	K
c	Solid specific heat	$m^2\ s^{-2}\ K^{-1}$
λ	Thermal conductivity	$kg\ m\ s^{-3}\ K^{-1}$
γ	Turbulence intermittency	
μ	Dynamic viscosity	$kg\ m^{-1}\ s^{-1}$
μ_t	Turbulence viscosity	$kg\ m^{-1}\ s^{-1}$
k	Turbulence kinetic energy	$m^2\ s^{-2}$
ω	Turbulence eddy frequency	s^{-1}
Re	Reynolds number	
Tu	Inlet turbulence intensity	
M_2	Exit Mach number	
P_1	Inlet Pressure	$kg\ m^{-1}\ s^{-2}$
T_1	Inlet temperature	K
P_2	Exit Pressure	$kg\ m^{-1}\ s^{-2}$
T_w	Wall temperature	K

subscripts

i,j	Cartesian coordinate

superscripts

*	total

7. References

[1] Mahmoud L. Mansour; Khosro Molla Hosseini & Jong S. Liu. "Assessment of the impact of laminar-turbulent transition on the accuracy of heat transfer coefficient prediction in high pressure turbines".ASME GT2006-90273.

[2] Bohn D.; Bonhoff B. & Schonenborn H. "Combined aerodynamic and thermal analysis of a high-pressure turbine nozzle guide vane". 95-YOKOHAMA-IGTC-108

[3] Hylton L.D.; Mihelc M.S.; Turner E.R.; Nealy D.A. & York R.E. "Analytical and experimental evaluation of the heat distribution over the surfaces of turbine vanes". NASA CR168015, 1983.

[4] Bohn D. & Heuer T. "Conjugate flow and heat transfer calculation of a high pressure turbine nozzle guide vane". AIAA 2001-3304

[5] Bohn D.; Heuer T. & Kortmann. "Numerical conjugate flow and heat transfer investigation of a transonic convection-cooled turbine guide vane with stress adapted thicknesses of different thermal barrier coatings". AIAA 2000-1034

[6] Bohn D. & Tümmers C. "Numerical 3-D conjugate flow and heat transfer investigation of a transonic convection-cooled thermal barrier coated turbine guide vane with reduced cooling fluid mass flow". ASME GT2003-38431

[7] William D.York & James H.Leylek. "Three-dimensional conjugate heat transfer simulation of an internally-cooled gas turbine vane".ASME GT2003-38551

[8] Facchini B.; Magi A. & Greco A. S. D. " Conjugate heat transfer simulation of a radially cooled gas turbine vane". ASME GT2004-54213

[9] Chunhua Sheng & Qingluan Xue."Aerothermal analysis of turbine blades using an unstructured flow solver-U2NCLE".AIAA 2005-4683.

[10] Menter F.R.. "Two-equation eddy-viscosity turbulence models for engineering applications". AIAA Journal Vol. 32, No. 8, pp.269-289, 1994.

[11] Menter F.R.; Kuntz M. & Langtry R. "Ten years of industrial experience with SST turbulence model".Turbulence Heat and Mass Transfer 4 Begell House Inc., 2003

[12] Menter F.R.; Langtry R.B.; Likki S.R.; Suzen, Y.B.; Huang, P.G. & Völker, S. "A correlation based transition model using local variables part 1 - model formulation". ASME GT2004-53452.

[13] Roach, P.E. & Briley, D.H. (1990). "The influence of a turbulent free stream on zero pressure gradient transitional boundary layer development. Part 1: testcases T3A and T3B". Cambridge University Press.

[14] Tony Arts, "Aero-Thermal Performance of a Two Dimensional Highly Loaded Transonic Turbine Nozzle Guide Vane", ASME 90-GT-358.

[15] Tony Arts, "External Heat Transfer Study on a HP Turbine Rotor Blade", Heat Transfer and Cooling in Gas Turbines, AGARD-CP-390: 5-1-5-12.